MARK B. TISCHLER

Multivariable Control

ELECTRICAL ENGINEERING AND ELECTRONICS

A Series of Reference Books and Textbooks

1. Rational Fault Analysis, *edited by Richard Saeks and S. R. Liberty*
2. Nonparametric Methods in Communications, *edited by P. Papantoni-Kazakos and Dimitri Kazakos*
3. Interactive Pattern Recognition, *Yi-tzuu Chien*
4. Solid-State Electronics, *Lawrence E. Murr*
5. Electronic, Magnetic, and Thermal Properties of Solid Materials, *Klaus Schröder*
6. Magnetic-Bubble Memory Technology, *Hsu Chang*
7. Transformer and Inductor Design Handbook, *Colonel Wm. T. McLyman*
8. Electromagnetics: Classical and Modern Theory and Applications, *Samuel Seely and Alexander D. Poularikas*
9. One-Dimensional Digital Signal Processing, *Chi-Tsong Chen*
10. Interconnected Dynamical Systems, *Raymond A. DeCarlo and Richard Saeks*
11. Modern Digital Control Systems, *Raymond G. Jacquot*
12. Hybrid Circuit Design and Manufacture, *Roydn D. Jones*
13. Magnetic Core Selection for Transformers and Inductors: A User's Guide to Practice and Specification, *Colonel Wm. T. McLyman*

14. Static and Rotating Electromagnetic Devices, *Richard H. Engelmann*
15. Energy-Efficient Electric Motors: Selection and Application, *John C. Andreas*
16. Electromagnetic Compossibility, *Heinz M. Schlicke*
17. Electronics: Models, Analysis, and Systems, *James G. Gottling*
18. Digital Filter Design Handbook, *Fred J. Taylor*
19. Multivariable Control: An Introduction, *P. K. Sinha*

Other Volumes in Preparation

Multivariable Control

An Introduction

P. K. Sinha
University of Warwick
Coventry, England

MARCEL DEKKER, INC. New York and Basel

Library of Congress Cataloging in Publication Data

Sinha, P. K. (Pradip K.), [date]
Multivariable Control.

(Electrical engineering and electronics ; 19)
Includes bibliographies and index.
1. Control theory. 2. Automatic control. I. Title.
II. Series.
QA402.3.S554 1984 629.8'312 84-3232
ISBN 0-8247-1858-5

MARCEL DEKKER, INC.
270 Madison Avenue, New York, New York 10016

Current printing (last digit):
10 9 8 7 6 5 4 3 2 1

PRINTED IN THE UNITED STATES OF AMERICA

To Alex and Jay

Preface

This book is intended for use as a reference work for control engineers, systems engineers, systems analysts, and applied mathematicians, and as a graduate and senior undergraduate textbook. The material is arranged to form a logical progression from the elementary concepts in system theory to the more advanced synthesis and design techniques. Although some of the structural properties of multivariable systems are now being expressed within the framework of abstract geometric concepts, an algebraic approach has been adopted throughout the book to establish a direct correspondence between the analytical results and their applications. For the student, it presents an evolution of the basic concepts of some of the fundamental aspects of multivariable theory and their relevance in design. For the practicing engineer, the book provides a systematically arranged source of reference. Although many basic aspects of control have been included, the reader is assumed to have an acquaintance with the rudiments of feedback theory.*

P. K. Sinha

*Covered in the companion volume, An Introduction to Linear Control Systems by T. E. Fortmann and K. L. Hitz (Marcel Dekker, 1977), Control and Systems Theory series, Vol. 5.

Acknowledgments

Although a certain amount of original material has been included in this book, a vast majority of material has been drawn from the literature. I have attempted to acknowledge my sources as much as possible. However, in view of the extensive research activities in this area of control theory, the included list must remain incomplete. I am deeply indebted to the authors of many books, published papers, and unpublished reports, and to the originators of many now-classical concepts associated with the analysis and design of multivariable systems. I am particularly grateful to the editors of *Automatica*, the *IEE Proceedings*, the *IEEE Transactions*, and the *International Journal of Control* for permitting me to use material from many papers published in these journals. I have derived a number of classical results from *State Space Analysis* by K. Ogata (Prentice-Hall), *State Variables for Engineers* by P. M. DeRusso, J. R. Roy, and C. M. Close (Wiley), and *Introduction to Linear Systems* by C. T. Chen (Holt, Rinehart and Winston). I would like to thank these publishers, and many others too numerous to mention individually, for their permission to use many mathematical examples.

I am grateful to Professor J. C. West for encouraging me to write this book, and to Professor J. L. Douce, and the late Professor J. A. Shercliff for their support during the preparation of this book. A one-term sabbatical from the University of Warwick to complete the final version of the book is gratefully acknowledged. I express my appreciation to Sandra Callanan, Nada Gvero, Terri Moss, Dinah Staples, and Linda Wooldridge for typing the various sections of the manuscript, and to Christine Allsopp for her skillful drawing of the line diagrams. I thank my wife for freeing me from the domestic chores during the writing of the book.

I owe a significant debt of gratitude to Dr. A. D. G. Hazlerigg for introducing me to the problems of designing interacting control systems and to Professor B. V. Jayawant for providing the opportunity to write this book. I dedicate this book to them for their moral support which often went beyond academic interests.

Contents

Preface v

Acknowledgments vii

Introduction xv

Short Table of Laplace Transform Pairs xxi

PART I: ANALYSIS

1 MATHEMATICAL PRELIMINARIES 3

1.1. Matrices and Vectors 3

1.1.1. Special Types of Matrices 6
1.1.2. Basic Matrix Operations 11
1.1.3. Rules of Matrix Operations 11
1.1.4. Vectors and Linear Spaces 13
1.1.5. Special Matrix Forms and Norms 18

1.2. Inverse, Determinants, and Eigenvalues 20

1.2.1. Determinants 20
1.2.2. Adjoint and Inverse of a Matrix 24
1.2.3. Computation of Inverse Matrices 25
1.2.4. Characteristic Equations and Eigenvector 31

1.3. Transformation and Diagonalization 38

1.3.1. Diagonalization 41
1.3.2. Similarity Transformation 45

1.4. Functions of Constant Matrices 49

1.4.1. Powers of Matrices 49
1.4.2. Minimal Polynomial 51

1.5. Polynomial and Rational Matrices 58

1.5.1. Polynomials 58
1.5.2. Polynomial Matrix 60
1.5.3. Rational Matrices 69

References 75

2 SYSTEM DESCRIPTION 77

2.1. System Classification 78

2.2. State-Space Representation 84

2.2.1. State and Output Equations 85
2.2.2. Matrix Representation 93
2.2.3. Free and Forced Responses 95
2.2.4. Modal Solution 101

2.3. Input/Output Description 111

2.3.1. Transfer-Function Matrix 113
2.3.2. State-Space and Transfer Functions 116
2.3.3. Poles and Zeros of G(s) 119

2.4. Polynomial-Function Description 133

2.4.1. Standard Forms 134
2.4.2. Equivalence and Similarity 136
2.4.3. Transmission Blocking 145
2.4.4. System Zeros 148

References 156

3 CONTROLLABILITY AND OBSERVABILITY 159

3.1. Basic Concepts 159

3.1.1. Pole-Zero Cancellation 169
3.1.2. Effect of Transformation 172

3.2. Canonical Forms 179

3.2.1. Controllable Form 179
3.2.2. Observable Form 185
3.2.3. Canonical Decomposition 187
3.2.4. Irreducible Dynamic Equation 190

3.3. Algebraic Interpretations 192

3.3.1. Input-Decoupling Zeros 194
3.3.2. Output-Decoupling Zeros 203
3.3.3. Input/Output Decoupling Zeros 209
3.3.4. Reduction of Order 219

References 228

4 TRANSFER-FUNCTION REALIZATIONS 231

4.1. Realization Problem 231

4.1.1. Concepts and Definitions 233
4.1.2. Standard Forms 234

4.2. Nonminimal Realizations 240

4.2.1. Block Forms 240
4.2.2. Controllable Companion Form 244
4.2.3. Observable Companion Form 248
4.2.4. Normal Form 252

4.3. Irreducible Realizations 256

4.3.1. Direct Sum 256
4.3.2. Jordan Form 260
4.3.3. Companion Form 269

4.4. Realization from Matrix Fractions 276

4.4.1. Matrix-Fraction Description 277
4.4.2. Observable Canonical Form 287
4.4.3. Controllable Canonical Form 295

References 302

5 STABILITY AND CONTROL 305

5.1. Multivariable Systems 305

5.1.1. Feedback Systems 310
5.1.2. State Feedback 315
5.1.3. Output Feedback 323
5.1.4. Optimal Control 325

5.2. Generalized Stability Criteria 331

5.2.1. Characteristic Loci 334
5.2.2. Nyquist Stability Criterion 341
5.2.3. Inverse Nyquist Criterion 351
5.2.4. Diagonal Dominance 359

5.3. Design Requirements 366

5.3.1. Sensitivity 366
5.3.2. Interaction 375
5.3.3. Integrity 380

References 385

Bibliography for Part I 389

Problems for Part I 397

PART II: SYNTHESIS

6 STATE FEEDBACK CONTROL 407

6.1. Full-Rank Feedback 408

6.2. Unity-Rank State Feedback 416

6.3. Zero Assignment 421
6.4. Linear Optimal Control 426
6.5. Adaptive Control 431
References 439

7 OUTPUT FEEDBACK CONTROL 441
7.1. Extension of State Feedback 441
7.2. Unity-Rank Output Feedback 448
7.3. Proportional-Plus-Error Integral Control 454
7.4. Proportional-Plus-Derivative Control 464
7.5. Model-Following Control 470
7.6. Feedback Compensation 476
References 478

8 DESIGN WITH STATE OBSERVER 481
8.1. Full-Order Observer 482
8.2. Identity Observer 483
8.3. Reduced-Order Observer 490
8.4. Control Through Observer 517
References 522

9 DECOUPLING CONTROL 525
9.1. State Feedback Decoupling 525
9.2. Output Feedback Decoupling 536
9.2.1. Extension of State Feedback Analysis 537
9.2.2. Equivalence Between Decoupling F and H 540
9.3. Precompensation and Decoupling 545
9.3.1. Precompensation for State Feedback Decoupling 545
9.3.2. Precompensation for Output Feedback Decoupling 551
9.4. Decoupling Through Matrix Fraction 556
References 560

10 GENERALIZED DESIGN TECHNIQUES 563

10.1. Inverse-Nyquist-Array (INA) Technique 564

10.2. Characteristic-Loci Method 574

10.3. Sequential-Return-Difference Method 584

10.4. Design Through Dyadic Transformation 592

References 602

Bibliography for Part II 605

Problems for Part II 615

APPENDIXES

A1 STABILITY CONCEPTS 621

A1.1. Stability Definitions 621

A1.2. Stability Through $\phi(\cdot)$ 629

A1.3. Stability Through $H(\cdot)$ 632

A1.4. Concept of Feedback 634

A2 TESTS FOR STABILITY 641

A2.1. Routh-Hurwitz Criterion 641

A2.2. Nyquist Criterion 642

A2.3. Root-Locus Plot 653

A2.4. Bode Diagram 657

A2.5. Lyapunov Stability Analysis 658

A2.5.1. First Method of Lyapunov 658

A2.5.2. Second Method of Lyapunov 660

A2.5.3. Lyapunov Function 662

A3 CONTROLLABILITY, OBSERVABILITY, AND DECOUPLING 671

A3.1. Output Controllability and Decoupling 671

A3.2. Existence of a Decoupling Pair (F, G) 674

A3.3. Computation of G 675

A3.4. Computation of F 677

References for the Appendixes 681

Index 683

Introduction

The development as well as the choice of topics in this volume have been the result of a compromise between the formulation of a sufficiently broad analytical framework and the presentation of a number of specific applicable design procedures. In this rather compact treatment of an extensive literature covering many significant contributions, many points of detail and some relevant topics had to be omitted. The book is divided into two parts: Part I deals with analysis by using the concept of state variables (time domain) and by extending the transform methods (frequency domain) to multi-input/multi-output systems; Part II presents a systematic development of a number of time- and frequency-domain synthesis methods which are conceptually simple and easier to implement rather than to present a state-of-the-art survey of multivariable synthesis methods. Wherever appropriate, illustrative examples have been included to outline the importance or usefulness of the analytical results. An overview of these two parts is given below.

Topics in the first part have been developed to establish a mathematical framework for the analysis of linear time-invariant multivariable systems. In view of the wide-ranging concepts necessary to introduce the many facets of multivariable control, topics with related themes have been grouped into five self-contained chapters. This approach, while creating a certain amount of compartmentalization, has been adopted primarily to present a (large) number of related concepts without too much reference to their implications in the broader (analytical) context. Each chapter, however, has been structured and cross-referenced to enable the reader to appreciate the interrelationships among the various analytical results. The interdependence of the different sections in this part is shown in the accompanying figure.

A brief review of a number of essential results of matrix theory is given in Chapter 1. The contents here have been organized into five sections covering the introductory concepts of matrices and vectors, followed by the definition and computation of matrix inverses and eigenvectors. The notions of transformation and function of matrices are then outlined. The final section contains some salient results associated with polynomial and rational function matrices. Material in this section forms the basis for algebraic control theory covered in subsequent chapters.

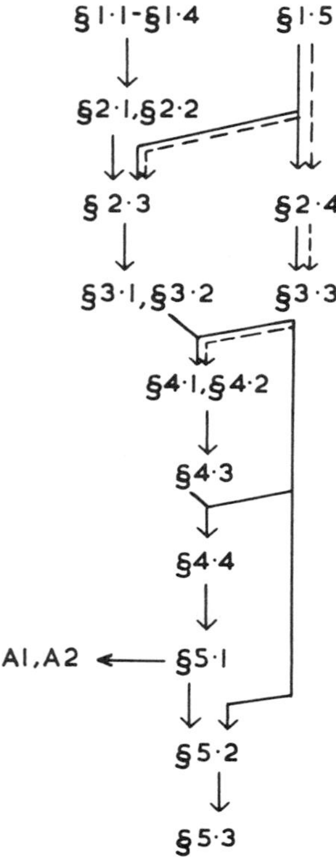

Development of sections in Part I (dashed line shows the initial progression for readers with introductory control background).

The basic theory of system description is developed in Chapter 2. The first three sections cover the fundamentals of system representations in the time and frequency domains, and the final section introduces the polynomial-function descriptions of linear multivariable systems. A significant part of the results in this section has been developed during the past fifteen years and forms the basis of algebraic control theory now being extensively used in the generalization of classical stability analyses.

Chapter 3 begins with an introductory section covering the basic concepts and definitions of controllability and observability and their relationship to pole-zero cancellation. A short account of controllable and observable forms is given in the second section. Since many of these concepts and the related transformations are covered in most introductory books on control, treatment here is brief, aimed primarily at presenting the main results within the framework of the book. The last section presents a fairly extensive algebraic interpretation of controllable and observable modes through decoupling zeros. The material in this section has been structured to establish, as far as possible, a one-to-one correspondence between the time- and frequency-domain results and to relate them to the reduction of system order through the removal of modes (zeros) which do not appear in the input/output description.

In view of the growing use of a combination of state-space and transfer-function representations in synthesis, transformation of one form to another plays an important role in many computer-aided-design techniques. This is likely to remain so as more user-oriented software support of distributed computer networks becomes available. The criterion for the selection of material in Chapter 4 has been not merely to cover the principles, but also to help the reader appreciate the mechanisms of realization and develop some expertise in using "abstract" analytical results. The first two sections present the basic definitions and direct (nonminimal) realization methods, and the last two sections present a selection of irreducible realization methods. The section of realization from matrix-fraction description has been included to complement previous results on matrix polynomials.

The single-input/single-output frequency domain approaches of Nyquist and Bode are used extensively by control engineers in the analysis of a system's ability to achieve the desired operating requirements, due mainly to their conceptual simplicity. Because of this significant advantage, a considerable amount of research has been undertaken over the past fifteen years to extend these "classical" methods to multivariable systems. This has resulted in the establishment of a number of practically useful methods of analyzing the behavior of multi-input/multi-output systems under a wide range of operating conditions. The selection of material in the three sections in Chapter 5 reflects the relevance of algebraic control theory in the development of generalized stability theory. The first section covers the basic stability definitions and some associated properties. This is followed by a brief review of some "more established" generalized stability criteria reflecting the relevance of the classical concepts in the analysis of multivariable systems. (The analytical results presented in this chapter form the basis of the frequency-domain synthesis methods discussed in Chapter 10.) The final section presents an outline description of three principal design criteria usually needed in synthesis.

The second part of the book complements the first through the development of a number of directly applicable methods of designing linear time-invariant multivariable systems. The subdivision of synthesis methods considered in this half of the book is shown schematically in the accompanying figure. Most of the material in Chapters 6 to 10 has been collected from research papers published over the past decade, and structured to present self-contained derivations of the analytical results and their application. The illustrative examples have been worked out to help the reader develop adequate skill to select the most appropriate method for a given design

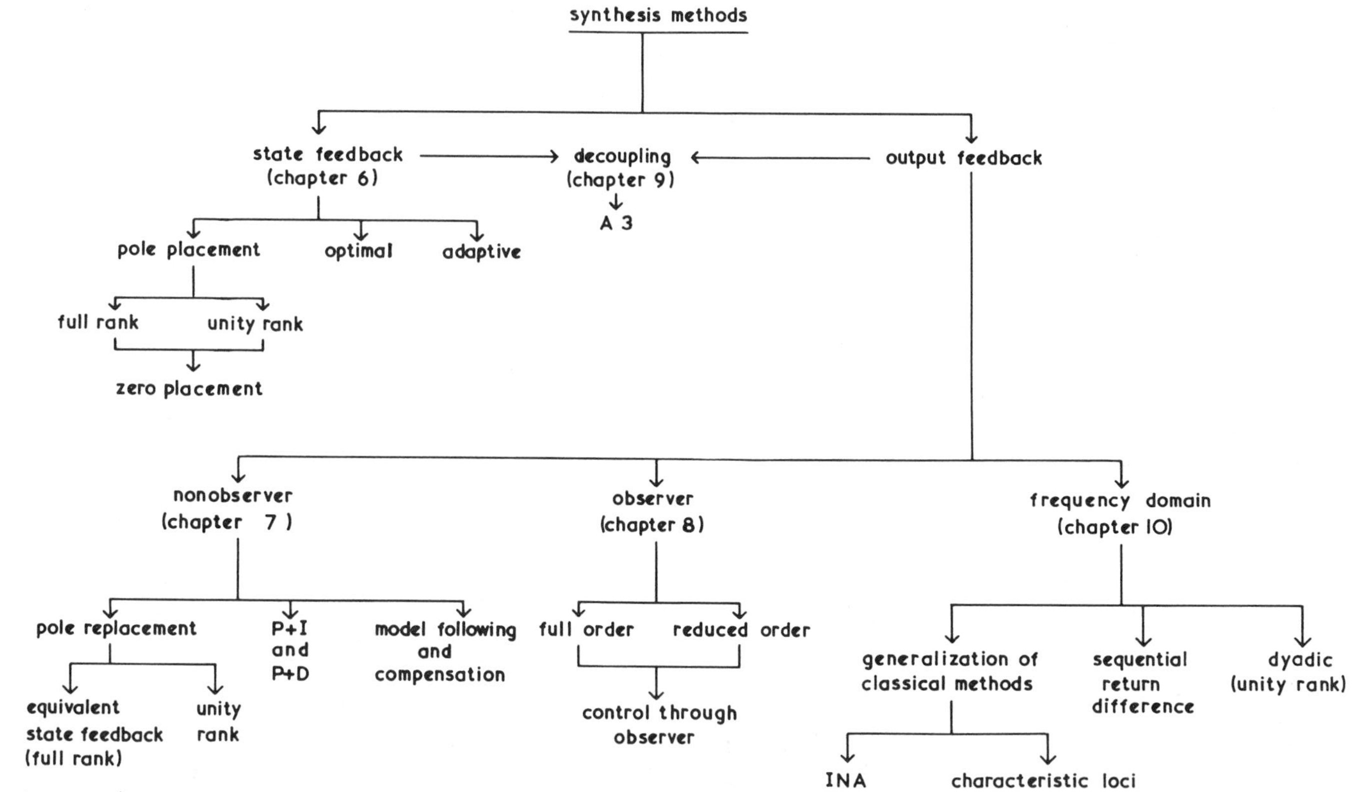

Classification of synthesis methods in Part II.

problem. A number of problems have been included at the end of each part rather than at the end of the individual chapters to provide a collection of numerical exercises which may be used in the application of results derived in more than one chapter.

Many of the analytical results developed here have numerous computational implications, and despite the importance of numerical analysis in modern control theory, these have not been covered in this volume. This is due mainly to the fact that this introductory book is aimed at providing a framework for the appreciation of the mechanisms of using "abstract" results. Literature covering the relevant numerical routines has been referenced wherever appropriate, and a selection of references has been included in the bibliography* for each part. It is hoped that the reader will find the illustrative examples, many worked out in algorithmic form, suitable bases for the development of software for computer-aided analysis/synthesis.

*An introductory treatment of numerical methods may be found in Mathematics of Finite-Dimensional Control Systems by D. L. Russell (New York: Marcel Dekker, 1979). A number of computer programs for basic control system design are contained in Computer Programs for Computational Assistance in the Study of Linear Control Theory by J. L. Melsa and S. K. Jones (New York: McGraw-Hill, 1970).

Short Table of Laplace Transform Pairs

$$\mathcal{L}\{f(t)\} = F(s) = \int_0^{\infty} -e^{-st} f(t)\, dt$$

	Function $f(t)$	Transform $F(s)$	
1	$u_i(t)$	1	Unit impulse
2	$u_s(t)$	$\frac{1}{s}$	Unit step
3	$u_r(t)$	$\frac{1}{s^2}$	Unit ramp
4	e^{-at}	$\frac{1}{s+a}$	Exponential
5	$\sin \omega t$	$\frac{\omega}{s^2+\omega^2}$	Sine
6	$\cos \omega t$	$\frac{s}{s^2+\omega^2}$	Cosine
7	$\frac{1}{(n-1)!} t^{n-1} e^{-at}$	$\frac{1}{(s+a)^n}$; n = positive integer	Repeated roots
8	$\frac{e^{-\zeta\omega_n t}}{\omega_n\sqrt{1-\zeta^2}} \sin \omega_n\sqrt{1-\zeta^2}\, t$	$\frac{1}{s^2+2\zeta\omega_n s+\omega_n^2}$	Damped sine
9	$\frac{df(t)}{dt}$	$sF(s) - f(0^-)$; $F(s) = \mathcal{L}\{f(t)\}$	First derivative
10	$\frac{d^n f(t)}{dt^n}$	$s^n F(s) - s^{n-1} f(0^-) - s^{n-2} \frac{df(0^-)}{dt} \cdots - \frac{d^{n-1} f(0^-)}{dt^{n-1}}$	nth derivative
11	$\int_0^t f(t)\, dt$	$\frac{1}{s} F(s)$	Integration
12	$\int_0^t \cdots \int_0^t f(t)\, dt^n$	$\frac{1}{s^n} F(s)$	n integrations
13	$\int_0^t f_1(\tau) f_2(t-\tau)\, d\tau$	$F_1(s)F_2(s)$; $F_1(s) = \mathcal{L}\{f_1(t)\}$, $F_2(s) = \mathcal{L}\{f_2(t)\}$	Convolution
14	$f(t-\tau)u_s(t-\tau)$	$e^{-\tau s}F(s)$	Delay
15	$f(at)$	$\frac{1}{a} F\left(\frac{s}{a}\right)$	Scale change
16	$e^{-at} f(t)$	$F(s+a)$	Exponential attenuation
17	$t f(t)$	$-\frac{d}{ds}\{F(s)\}$	Time multiplication

Multivariable Control

I
ANALYSIS

Chapter 1: MATHEMATICAL PRELIMINARIES

Introduces the main results of matrix theory which are used in the subsequent chapters.

Chapter 2: SYSTEM DESCRIPTION

Presents an overview of state-space and transfer-function representations of linear time-invariant multivariable systems.

Chapter 3: CONTROLLABILITY AND OBSERVABILITY

Reviews the definitions and salient results associated with canonical forms and irreducible equations.

Chapter 4: TRANSFER-FUNCTION REALIZATIONS

Presents the basic concepts and a selection of realization methods.

Chapter 5: STABILITY AND CONTROL

Presents a unified account of basic stability and control concepts in the time and frequency domains, a brief account of a selection of multivariable stability criteria, and an outline of design criteria.

Bibliography and Problems

1
Mathematical Preliminaries

This chapter presents the terminology, definitions, and basic concepts of matrix algebra and linear vector spaces. The analytical results discussed here usually come under the heading of linear algebra. The reason for using these rather abstract results in the study of control systems is due primarily to the vast amount of information necessary to describe and analyze large-scale systems. This information, which often consists of a set of high-order differential or difference equations, can be conveniently expressed in the notation of matrices, determinants, linear vector spaces, linear transformation and characteristic value problems, to name but a few, and these thus play important roles in system theory. Once the particular notation is adopted, the analysis of a system consists mainly of the analysis of the properties of the system matrices.

The topics included in this introductory chapter do not exhaust the detailed knowledge that exists about matrix algebra, but are chosen due to their special relevance to multivariable control theory. Some of the special concepts not included here are mentioned as and when necessary in the subsequent chapters. The contents of this chapter are presented with a view to developing the mathematical framework necessary for an understanding of the main body of the text. Although some of the fundamentals are assumed to be known, this chapter has been developed to supplement this knowledge and to familiarize the reader with the notation and philosophy. The treatment here is fairly short, and to a certain extent incomplete, in that most results are given without proof, and have been oriented toward use in the remainder of the book.

1.1 MATRICES AND VECTORS

A <u>matrix</u> is a rectangular array of elements that are members of a field of real numbers (real field), imaginary numbers (imaginary field), polynomials, functions, and so on. Matrices, usually denoted by capital or boldface letters, are closely related to linear transformation. Consider the set of m algebraic linear equations in terms of n unknowns x_j ($j \in 1, \ldots, n$):

$$\begin{aligned}
&a_{11}x_1 + a_{12}x_2 + \cdots + a_{1j}x_j + \cdots + a_{1n}x_n = y_1\\
&\vdots\\
&a_{i1}x_1 + a_{i2}x_2 + \cdots + a_{ij}x_j + \cdots + a_{in}x_n = y_i\\
&\vdots\\
&a_{m1}x_1 + a_{m2}x_2 + \cdots + a_{mj}x_j + \cdots + a_{mn}x_n = y_m
\end{aligned} \tag{1.1}$$

where y_i, $i \in 1, \ldots, m$, are the dependent variables and the elements of a_{ij} are the elements of the field F.* This set of equations assigns a unique set of n variables x_j to every set of m variables y_i. This assignment is called a linear transformation of variables x_j into the set of variables y_i, and is characterized by the array of mn coefficients a_{ij}. The array of these numbers is called the matrix of linear transformation. Equations (1.1) may also be conveniently expressed by

$$\sum_{i=1,\, j=1}^{m,\, n} a_{ij}x_j = y_i \tag{1.2}$$

or by the simpler notation

$$Ax = y \tag{1.3a}$$

where A is the $(m \times n)$ rectangular array of the coefficients a_{ij} and represented as

$$A = \begin{bmatrix} a_{11} & a_{12} & \cdots & a_{1j} & \cdots & a_{1n}\\ \vdots & & & & & \\ a_{i1} & a_{i2} & \cdots & a_{ij} & \cdots & a_{in}\\ \vdots & & & & & \\ a_{m1} & a_{m2} & \cdots & a_{mj} & \cdots & a_{mn} \end{bmatrix} \tag{1.3b}$$

Here A is the $(m \times n)$ matrix with elements a_{ij}, $i \in 1, \ldots, m$; $j \in 1, \ldots, n$, where the first subscript denotes the row number and the second the column number, $m \times n$ being the dimension (or size) of A. The matrix may also be defined as

*The fields that are relevant to the topics considered in this book are the real fields (containing all real numbers) and the complex fields (containing all complex numbers).

$$A = \{a_{ij}\} \quad i \in 1, \ldots, m; \; j \in 1, \ldots, n \tag{1.4}$$

where the term in braces indicates the array, a typical element of which is a_{ij}; x and y are (n × 1) and (m × 1) matrices or simply n and m column matrices, respectively, and are defined by

$$x = \begin{bmatrix} x_1 \\ \vdots \\ x_j \\ \vdots \\ x_n \end{bmatrix} \quad \text{and} \quad y = \begin{bmatrix} y_1 \\ \vdots \\ y_i \\ \vdots \\ y_m \end{bmatrix} \tag{1.5}$$

or $x = \{x_j\}$, $j \in 1, \ldots, n$, and $y = \{y_i\}$, $i \in 1, \ldots, m$.

A matrix with m rows and n columns is called an <u>(m × n) matrix</u> or is said to be of <u>order m by n</u>; when m = n, the matrix is called an <u>(n × n) square matrix</u> or a <u>square matrix of order n</u> or an <u>n-square matrix</u>.

(a) Column matrix

An (m × 1) matrix is called a <u>column matrix</u> or a <u>column vector</u>, since it consists of a single column of m elements: for example,

$$x = \begin{bmatrix} x_1 \\ x_2 \\ \vdots \\ x_m \end{bmatrix}; \quad y = \begin{bmatrix} y_1 \\ y_2 \\ \vdots \\ y_m \end{bmatrix}$$

(b) Row matrix

A matrix containing a single row of elements, such as a (1 × n) matrix x,

$$x = [x_1 \; x_2 \; \cdots \; x_n]$$

is called a <u>row matrix</u> or a <u>row vector</u>.

(c) Diagonal matrix

The <u>principal diagonal</u> of a square matrix consists of the elements a_{ii}. A <u>diagonal matrix</u> is a square matrix all of whose elements which do not lie on the principal diagonal are zero. For example, if $A = \{a_{ij}\}$, $i, j \in 1, 2, \ldots, n$, is a diagonal matrix, then

$$A = \begin{bmatrix} a_{11} & & & \\ & a_{22} & & \\ & & \ddots & \\ & & & a_{nn} \end{bmatrix} = \{a_{ij}\}, \text{ where } a_{ij} \begin{cases} = 0, & i \neq j \\ = 1, & i = j \end{cases}$$

(d) Scalar matrix

A diagonal matrix whose elements are equal to one another is called a scalar matrix.

(e) Unit matrix

A unit (or identity) matrix is a diagonal matrix whose principal diagonal elements are equal to unity. The unit matrix is usually denoted by I_n, n indicating its order.

$$I_n = \underbrace{\begin{bmatrix} 1 & 0 & & 0 \\ 0 & 1 & & \\ & & \ddots & \\ 0 & & & 1 \end{bmatrix}}_{n \text{ columns}} \left. \vphantom{\begin{bmatrix} 1 \\ 0 \\ \\ 0 \end{bmatrix}} \right\} n \text{ rows}$$

(f) Null matrix

A matrix that has all of its elements identically equal to zero is called a null (or zero) matrix, and usually denoted by $0_{m \times n}$, the suffix indicating its order.

(g) Transpose matrix

Transposing a matrix A is the operation whereby the rows and columns are interchanged. The transpose of a matrix $A = \{a_{ij}\}$ is defined by $A^T = \{a_{ji}\}$. Thus if A is an $(m \times n)$ matrix, A^T is an $(n \times m)$ matrix.

(h) Triangular matrix

A matrix whose elements below (above) the principal diagonal are all zero is called a lower (upper) triangular matrix.

1.1.2 Special Types of Matrices

(a) Symmetric matrix

A square matrix all of whose elements are real is said to be symmetric if it is equal to its transpose, that is, if

$$A = A^T \quad \text{or} \quad \{a_{ij}\} = \{a_{ji}\}, \quad i, j \in 1, 2, \ldots, n$$

(b) Antisymmetric matrix

A real square matrix is said to be <u>antisymmetric</u> if

$$A = -A^T \quad \text{or} \quad \{a_{ij}\} = \{-a_{ji}\}, \quad i, j \in 1, 2, \ldots, n$$

This implies that the elements on the principal diagonal are identically zero.

(c) Conjugate matrix

If the elements of the matrix A are complex ($a_{ij} = \alpha_{ij} + j\beta_{ij}$), then the <u>conjugate matrix</u> B has elements $b_{ij} = \alpha_{ij} - j\beta_{ij}$. This is written in the form

$$B = A^*$$

(d) Associate matrix

The <u>associate matrix</u> of A is the transposed conjugate of A; that is,

$$\text{associate of } A = (A^*)^T$$

(e) Real matrix

If $A = A^*$, then A is a <u>real matrix</u>.

(f) Imaginary matrix

If $A = -A^*$, then A is pure <u>imaginary</u>.

(g) Hermitian matrix

If a matrix is equal to its associate matrix, it is said to be <u>Hermitian</u>; that is, the matrix A is Hermitian if

$$A = (A^*)^T$$

(h) Skew-Hermitian

If $A = -(A^*)^T$, then A is said to be <u>skew-Hermitian</u>.

In addition to the types listed above, a few matrices that have special structural properties and are common in the analysis of control systems are defined below.

(i) Companion form

The square matrix A with the structure

$$A_c = \underbrace{\begin{bmatrix} 0 & 1 & 0 & \cdots & 0 \\ 0 & 0 & 1 & \cdots & 0 \\ \vdots & & & & \\ 0 & 0 & 0 & \cdots & 1 \\ -a_n & -a_{n-1} & -a_{n-2} & \cdots & -a_1 \end{bmatrix}}_{\text{n columns}} \left.\vphantom{\begin{bmatrix} 0 \\ 0 \\ \vdots \\ 0 \\ 0 \end{bmatrix}}\right\} \text{n rows}$$

is said to be in <u>companion</u> form. The characteristic polynomial* associated with this special matrix is

$$\det(\lambda I - A) = \lambda^n + a_1\lambda^{n-1} + a_2\lambda^{n-2} + \cdots + a_n$$

An alternative form of A, with the same characteristic polynomial, is

$$A_c^1 = \underbrace{\begin{bmatrix} 0 & 0 & 0 & \cdots & 0 & -a_n \\ 1 & 0 & 0 & \cdots & 0 & -a_{n-1} \\ 0 & 1 & 0 & \cdots & 0 & -a_{n-2} \\ \vdots & & & & & \\ 0 & 0 & 0 & \cdots & 1 & -a_1 \end{bmatrix}}_{\text{n columns}} \left.\vphantom{\begin{bmatrix} 0 \\ 0 \\ 0 \\ \vdots \\ 0 \end{bmatrix}}\right\} \text{n nows}$$

(j) Jordan normal form

If the characteristic polynomial of an n-square matrix A is given by

$$\det(\lambda I - A) = (\lambda - \lambda_1)^{k_1}(\lambda - \lambda_2)^{k_2} \cdots (\lambda - \lambda_r)^{k_r}$$

where $\Sigma_{i=1}^{r} k_i = n$, then the <u>Jordan normal form</u>† of A is

*A formal definition is given in Sec. 1.2.4.

†Also called <u>Jordan canonical form</u>.

$$J = \begin{bmatrix} J_1 & & & 0 \\ & J_2 & & \\ & & \ddots & \\ 0 & & & J_r \end{bmatrix}$$

where the $(k_i \times k_i)$ block matrix J_i, $i \in 1, \ldots, r$, has the following structure:

$$J_i = \begin{bmatrix} \lambda_i & 1 & 0 & \cdots & 0 \\ 0 & \lambda_i & 1 & \cdots & 0 \\ \vdots & & & & \\ 0 & 0 & 0 & \cdots & 1 \\ 0 & 0 & 0 & \cdots & \lambda_i \end{bmatrix}$$

(k) Frobenius canonical form

A square matrix A is said to be in Frobenius canonical form if it has the structure

$$A = \begin{bmatrix} F_1 & & & 0 \\ & F_2 & & \\ & & \ddots & \\ 0 & & & F_r \end{bmatrix}$$

where the α_r-square block matrix F_i, $i \in 1, \ldots, r$, has the form

$$F_i = \begin{bmatrix} 0 & 1 & 0 & \cdots & 0 \\ 0 & 0 & 1 & \cdots & 0 \\ \vdots & & & & \\ -f_{\alpha_r - 1} & -f_{\alpha_r - 2} & & \cdots & -f_0 \end{bmatrix}$$

and the characteristic polynomial for each F_r divides that of all preceding matrices (i.e., $F_{r-1}, \ldots, F_1$).

(l) Schwarz form

An m-square matrix A is said to be in Schwarz form if it has the structure

$$A = \begin{bmatrix} 0 & 1 & 0 & 0 & 0 & 0 \\ -a_n & 0 & 1 & 0 & 0 & 0 \\ 0 & -a_{n-1} & 0 & 1 & 0 & 0 \\ \vdots & & \ddots & & & \\ 0 & & & \ddots & 0 & 1 \\ 0 & & & & -a_2 & -a_1 \end{bmatrix}$$

If none of the a_k, $k \in 1, \ldots, n$, are zero, the number of eigenvalues* of A with negative real parts is equal to the number of positive terms in the following sequence:

$$a_1;\ a_1a_2;\ a_1a_2a_3;\ \ldots;\ a_1a_2 \cdots a_n$$

(m) Toeplitz matrix

A matrix $A = \{a_{ij}\}$ is said to be a Toeplitz matrix if the value of any element a_{ij} is dependent only on the value of (i - j); the matrix is thus constant along the diagonal. The elements of a lower (upper) triangular Toeplitz matrix is completely defined by the elements of the first column (row).

(n) Jacobi matrix

A tridiagonal matrix A defined by

$$A = \begin{bmatrix} \beta_1 & \nu_1 & 0 & \cdots & 0 & 0 \\ \alpha_1 & \beta_1 & \nu_1 & \cdots & 0 & 0 \\ 0 & \alpha_2 & \beta_3 & \cdots & 0 & 0 \\ \vdots & & & & & \\ 0 & 0 & 0 & \cdots & \alpha_{n-1} & \beta_n \end{bmatrix}$$

is said to be a Jacobi matrix if $\alpha_i \nu_i > 0$, $\forall\, i \in 1, \ldots, n-1$. If the elements of A are real, one consequence is that the eigenvalues of the Jacobi matrix are all real.

*Defined in Sec. 1.2.4.

1.1.2 Basic Matrix Operations

1. The sum of two $(m \times n)$ matrices $A = \{a_{ij}\}$ and $B = \{b_{ij}\}$ is the $(m \times n)$ matrix $C = \{c_{ij}\}$, where

$$c_{ij} = a_{ij} + b_{ij}$$

for all $i = 1, \ldots, m$, and all $j = 1, \ldots, n$.
2. Given the $(m \times n)$ matrix A and the scalar α, the product of αA is the $(m \times n)$ matrix C, where

$$c_{ij} = \alpha a_{ij}$$

for all $i = 1, \ldots, m$, and all $j = 1, \ldots, n$.
3. Two $(m \times n)$ matrices $A = \{a_{ij}\}$ and $B = \{b_{ij}\}$ are said to be equal if

$$a_{ij} = b_{ij}$$

for all $i = 1, \ldots, m$, and for all $j = 1, \ldots, n$. The definition of matrix equality is valid only for matrices of equal dimensions.
4. The product of the $(n \times m)$ matrix $A = \{a_{ik}\}$ and the $(m \times \ell)$ matrix $B = \{b_{ki}\}$ is the matrix $C = \{c_{ij}\}$, where

$$c_{ij} = \sum_{k=1}^{m} a_{ik} b_{kj} \tag{1.6}$$

for all $i = 1, \ldots, n$, and for all $j = 1, \ldots, \ell$. The two matrices to be multiplied must be conformable; that is, the number of columns of the first must be equal to the number of rows of the second; in general, $AB \neq BA$.
5. Two matrices of the same dimension are said to commute if AB (A postmultiplied by B) equals BA (A premultiplied by B). If AB = BA, the matrices A and B are said to anticommute. If for a matrix A there is a least positive integer k such that $A^{k+1} = A$, then A is said to be periodic and of period k.

1.1.3 Rules of Matrix Operations

(a) Distributive rules

$(A + B)C = AC + BC$, where A and B are $(n \times m)$ and C is $(m \times \ell)$.
$A(B + C) = AB + AC$, where A is $(n \times m)$ and B and C are $(m \times \ell)$.

$(\alpha + \beta)A = \alpha A + \beta A$ and $\alpha(A + B) = \alpha A + \alpha B$, where α and β are scalars and A and B are $(n \times m)$.

(b) Associative rules

$\alpha A = A\alpha$ or $(\alpha A)B = \alpha AB = A(\alpha B)$, where α is a scalar and A is $(n \times m)$ and B is $(m \times \ell)$.
$A(BC) = (AB)C$, where A is $(n \times m)$, B is $(m \times \ell)$, and C is $(\ell \times k)$.
$(A + B) + C = A + (B + C)$, where A, B, and C are $(n \times m)$.

(c) Commutative rules

$A + B = B + A$, where A and B are $(n \times m)$.

(d) Matrix cancellation

Cancellation of matrices is not valid in matrix algebra. If two matrices A and B are such that $AB = 0$, there are three possibilities:

(1) $A = 0$
(2) $B = 0$
(3) $A = B = 0$

It can be proved that if both A and B are nonzero matrices, and $AB = 0$, then both A and B are singular.*

(e) Matrix differentiation and integration

The time derivative of a matrix A(t) is defined by the matrix whose (i, j)th element is the time derivative of the (i, j)th element of the original matrix, provided that all the elements $a_{ij}(t)$ have derivatives with respect to t, that is,

$$\frac{d}{dt} A(t) = \left\{ \frac{d}{dt} a_{ij}(t) \right\}$$

Similarly, the time integral of A(t) is given by

$$\int A(t)\, dt = \int \{a_{ij}(t)\, dt\}$$

provided that $a_{ij}(t)$ are integratable as functions of t.

(f) Matrix rank

An $n \times m$ matrix A is said to have rank r if the maximum number of linearly independent† rows (columns) is r, $r < n$, $r < m$. Hence $r \times r$ is the highest order of the nonsingular submatrix N that can be formed out of the rows (columns) of A. A is said to have full rank (or be nonsingular) if $r = n$.

(g) Ranks of products of matrices

If the orders of two matrices A and B are such that the product AB is conformable, then

*Defined in Sec. 1.2.1(e).
†Defined in Sec. 1.1.4(j).

$$\text{rank}(AB) < \text{rank}(A)$$
$$< \text{rank}(B)$$

(h) Matrix operations by partitioning

A matrix can be partitioned by drawing vertical and/or horizontal line between two rows or columns; the subset of elements formed as individual matrices are called <u>submatrices</u>. As long as the submatrices are compatible, the submatrices can be treated as elements in basic matrix operations as illustrated below.

$$A = \left[\begin{array}{cc|cc} a_{11} & a_{12} & a_{13} & a_{14} \\ a_{21} & a_{22} & a_{23} & a_{24} \\ \hline a_{31} & a_{32} & a_{33} & a_{34} \\ a_{41} & a_{42} & a_{43} & a_{44} \end{array}\right] = \left[\begin{array}{c|c} A_{11} & A_{12} \\ \hline A_{21} & A_{22} \end{array}\right] \tag{1.7}$$

$$B = \left[\begin{array}{cc|cc} b_{11} & b_{12} & b_{13} & b_{14} \\ b_{21} & b_{22} & b_{23} & b_{24} \\ \hline b_{31} & b_{32} & b_{33} & b_{34} \\ b_{41} & b_{42} & b_{43} & b_{44} \end{array}\right] = \left[\begin{array}{c|c} B_{11} & B_{12} \\ \hline B_{21} & B_{22} \end{array}\right] \tag{1.8}$$

$$A + B = \left[\begin{array}{c|c} A_{11} + B_{11} & A_{12} + B_{12} \\ \hline A_{21} + B_{21} & A_{22} + B_{22} \end{array}\right] \tag{1.9}$$

$$AB = \left[\begin{array}{c|c} A_{11}B_{11} + A_{12}B_{21} & A_{11}B_{12} + A_{12}B_{22} \\ \hline A_{21}B_{11} + A_{22}B_{21} & A_{21}B_{12} + A_{22}B_{22} \end{array}\right] \tag{1.10}$$

1.1.4 Vectors and Linear Spaces [1, 2]

The row and column vectors in (1.5) are simply extensions of the more familiar concept of vectors in two- or three-dimensional spaces to an n-dimensional space. When n is greater than 3, the geometric visualization becomes obscured, but the terminology associated with the familiar coordinate system is still useful. For example, the coordinate system having the unit vectors

$$I_1 = \begin{bmatrix} 1 \\ 0 \\ 0 \\ \vdots \\ 0 \end{bmatrix}, \quad I_2 = \begin{bmatrix} 0 \\ 1 \\ 0 \\ \vdots \\ 0 \end{bmatrix}, \quad \ldots, \quad I_n = \begin{bmatrix} 0 \\ \vdots \\ 0 \\ 1 \end{bmatrix}$$

can be thought of as an n-dimensional system with mutually orthogonal coordinate system. Some of the definitions associated with vectors and linear spaces are given below.

(a) Vector

An n-dimensional vector x, called an n vector, over a field (F) is an ordered set of n numbers $(x_1, x_2, \ldots, x_n)$ that belong to F and is denoted by

$$x = \begin{bmatrix} x_1 \\ x_2 \\ \vdots \\ x_i \\ \vdots \\ x_n \end{bmatrix}$$

The n numbers are called components or elements of x and are denoted by $x_i \in x$, where $x \in F$. x is called a real or a complex vector depending on whether F is real or complex.

Two vectors x and y are equal if their corresponding components are equal, that is, $x_i = y_i$, $\forall\, i$.

(b) Scalar

One-dimensional vectors are scalars: these may be real or complex numbers.

(c) Null vector

The vector whose components are all zero is called a null vector, and is denoted by 0.

(d) Linear vector space

A linear vector space (or set) S over a field F is defined as a set of all vectors over F for which the operations of addition and multiplication by scalars are defined and satisfy the following rules:

1. The sum of two vectors in A is also a vector in the space S.
2. Every scalar multiple of a vector in the set is a vector in the set.
3. For each vector x in S and each scalar a in F, the product ax belongs to S. Multiplication of vectors by scalars is associative and distributive.

The product of the identity matrix I and the vector x is x itself; that is, $Ix = x$ for any $x \in S$.

S is called a real (or complex) space if F is the field of real (or complex) numbers. In general, all vectors and scalars are assumed to be elements of a complex vector space unless stated otherwise.

(e) Scalar product

The scalar (or inner) product of two vectors x and y is written as $\langle x, y \rangle$ and is defined as

$$\langle x, y \rangle = \langle x^* \rangle^T y = \sum_{i=1}^{n} x_i^* y_i = y^T x^* \neq x^T y^*$$

For the case where x and y are real,

$$\langle x, y \rangle = x^T y = y^T x = \langle y, x \rangle$$

(f) Dyadic product†

If the $(n \times 1)$ column vector x is denoted by $x\rangle$ and the $(1 \times m)$ row vector $(y^*)^T$ denoted by $\langle y$, then the dyadic (or outer) product $x\rangle\langle y$ is the $(n \times m)$ matrix.

$$x\rangle\langle y = x(y^*)^T = \begin{bmatrix} x_1 y_1^* & \cdots & x_1 y_i^* & \cdots & x_1 y_m^* \\ \vdots \\ x_i y_1^* & \cdots & x_i y_i^* & \cdots & x_i y_m^* \\ \vdots \\ x_n y_1^* & \cdots & x_n y_i^* & \cdots & x_n y_m^* \end{bmatrix}$$

(g) Orthogonal vectors

Two vectors x and y are said to be orthogonal if their scalar product is zero, that is, if

†Also known as tensor (or outer) product of two vectors.

$$\langle x, y \rangle = 0$$

(h) Length of a vector

The length (or norm) of a vector x, denoted by $\| x \|$, is defined as the nonnegative square root of the scalar product of x and x, that is,

$$\| x \| = \langle x, x \rangle = \sum_{i=1}^{n} [x_i^* x]^{1/2}$$

As a consequence of this definition, it can be shown that, for two vectors x and y,

$$\| x + y \| \leq \| x \| + \| y \| \qquad \text{(triangular inequality)}$$

$$\| \langle x, y \rangle \| \leq \| x \| \cdot \| y \| \qquad \text{(Schwarz inequality)}$$

If x is a real vector, the quantity $\| x \|^2$ can be interpreted geometrically as the square of the distance from the origin to the point represented by the vector x.

(i) Unit vector

A vector is said to be a unit vector if its length is unity, so that $\langle \hat{x} \cdot \hat{x} \rangle = 1$. A unit vector can be obtained from the vector x by its length; thus

$$\hat{x} = \frac{x}{\langle x, x \rangle}$$

(j) Linear dependence and independence

A set of m vectors $x_1, x_2, \ldots, x_m$ is said to be linearly dependent if there is a set of constants $k_1, k_2, \ldots, k_m$ (at least one k_i must be nonzero) such that

$$k_1 x_1 + k_2 x_2 + \cdots + k_m x_m = 0$$

If no such scalars k_i exist, the set of m vectors is said to be linearly independent.

(k) Dimension

The dimension of a vector space is the maximum number of linearly independent vectors in the space.

In an n-dimensional space, the set of n linearly independent vectors is said to "span the space." These n vectors can also be used to form a basis for the space. If n is finite, then S, denoted by S^n, is called a finite-dimensional vector space.

(ℓ) Basis

A basis of a space is a set of vectors such that any vector in the space is a linear combination of the basis vector. A basis is essentially a coordinate system.

(m) Subspace

A nonempty set V of vectors S is called a subspace of S if it is closed with respect to vector addition and multiplication of vectors by scalars. A subspace of a vector space is itself a vector space.

(n) Dual space and dual basis

The vector space S' consisting of all linear functionals on S is called the dual vector space of S. If a set of vectors $x_1, x_2, \ldots, x_n$ is the basis of S, there exists a uniquely determined basis $y_1, y_2, \ldots, y_n$ in S' such that

$$(y_i, x_j) = \delta_{ij}, \quad i, j \in 1, 2, \ldots, n; \quad \delta_{ij} = \begin{cases} 0, & i \neq j \\ 1, & i = j \end{cases}$$

The basis $y_1, y_2, \ldots, y_n$ is called the dual (or reciprocal) basis of $x_1, x_2, \ldots, x_n$. The dual (or reciprocal) vector space of an n-dimensional vector space is also n-dimensional.

(o) Sum of vector spaces

Let u be a vector in a space U (i.e., $u \in U$) and v be a vector in a space V (i.e., $v \in V$), where U and V are subspaces of a vector space S. The sum of U and V, denoted by U + V, consists of all sums u + v:

$$U + V = \{u + v: u \in U, v \in V\}$$

Let u + v and u' + v' belong to U + V, where $u, u' \in U$ and $v, v' \in V$; then, from above,

$$(u + v) + (u' + v') = (u + u') + (v + v') \in U + V$$

and for any scalar k,

$$k(u + v) = ku + kv \in U + V$$

The vector space S is said to be the direct sum of its subspaces U and V, denoted by $S = U \oplus V$, if every vector $s \in S$ can be written in only one way as

$$s = u + v, \quad \text{where } u \in U \text{ and } v \in V$$

that is, $S = U + V$ and $U \cap V = \{0\}$.

1.1.5 Special Matrix Forms and Norms

(a) Bilinear form

An expression of the form

$$B = \sum_{i}^{n} \sum_{j}^{n} a_{ij} x_i y_j \tag{1.11}$$

where A is an $(n \times n)$ real matrix and x and y^T are row vectors, is called a real bilinear form in the variables x_i, y_j. (1.11) can also be written as

$$B = x^T Ay = \langle x, Ay \rangle \tag{1.12}$$

The matrix A is called the coefficient matrix of the form, and the rank of A is called the rank of the form.

(b) Quadratic form

If the vector x is equal to the vector y, then (1.12) becomes

$$Q = \sum_{i} \sum_{j} a_{ij} x_i x_j = x^T Ax = \langle x, Ax \rangle, \quad i, j = 1, \ldots, n \tag{1.13}$$

Q is called the real quadratic form in $x_1, x_2, \ldots, x_n$.

In (1.13) the coefficient of the term $x_i x_j$ $(i \neq j)$ is equal to $(a_{ij} + a_{ji})$. This coefficient would be unchanged if both a_{ij} and a_{ji} are set equal to $\frac{1}{2}(a_{ij} + a_{ji})$. Thus without any loss in generality, the matrix A can be said to be a symmetric matrix.

(c) Hermitian form

If the matrix A is a Hermitian matrix, such that $a_{ij}^* = a_{ji}$, the corresponding Hermitian form is defined as

$$H = x^{T*} Ax = \sum_{i} \sum_{j} a_{ij} x_i^* x_j = \langle x, Ax \rangle, \quad i, j = 1, \ldots, n \tag{1.14}$$

(d) Definite and semidefinite forms

A quadratic form $Q = \langle x, Ax \rangle$ (Hermitian form $Q = \langle x^*, Ax \rangle$), where A is a real symmetric matrix (a Hermitian matrix), is said to be

$$\text{positive definite if } Q \begin{cases} > 0 & \text{for } x \neq 0 \\ = 0 & \text{for } x = 0 \end{cases} \tag{1.15}$$

$$\text{positive semidefinite if } Q \begin{cases} \geq 0 & \text{for } x \neq 0 \\ = 0 & \text{for } x = 0 \end{cases} \tag{1.16}$$

Table 1.1 Associated Conditions of the Quadratic Form

	$\|A\|$	Rank (r) and index (p)	Characteristic values[a]	Form	Leading principal minors[a]
Real quadratic form	$\|A\| > 0$	$r = p = n$	$\lambda_i > 0$	$Q = x_1^2 + x_2^2 + \cdots + x_n^2$	$\Delta_1, \Delta_2, \ldots, \Delta_n$ all positive
Hermitian form				$Q = x_1^* x_1 + x_2^* x_2 + \cdots + x_n^* x_n$	
Real quadratic form	$\|A\| = 0$	$r = p < n$	n - r roots equal zero; others positive	$Q = x_1^2 + x_2^2 + \cdots + x_r^2$	$\Delta_1, \Delta_2, \ldots, \Delta_r$ all positive;
Hermitian form				$Q = x_1^* x_1 + x_2^* x_2 + \cdots + x_r^* x_r$	$\Delta_{r+1}, \ldots, \Delta_{n-1}$ nonnegative
Real quadratic form	$(-1)^n \|A\| > 0$	$r = n$, $p = 0$	$\lambda_i < 0$	$Q = -x_1^2 - x_2^2 - \cdots - x_n^2$	$-\Delta_1, \Delta_2, \ldots, (-1)^n \Delta_n$ all positive
Hermitian form				$Q = -x_1^* x_1 - x_2^* x_2 - \cdots - x_n^* x_n$	
Real quadratic form	$\|A\| = 0$	$r < n$, $p = 0$	n - r roots equal zero; others negative	$Q = -x_1^2 - x_2^2 - \cdots - x_r^2$	$-\Delta_1, \Delta_2, \ldots, (-1)^r \Delta_r$ all positive
Hermitian form				$Q = -x_1^* x_2 - x_2^* x_2 - \cdots - x_r^* x_r$	$(-1)^r \Delta_{r+1}, \ldots, (-1)^n \Delta_{n-1}$ nonpositive

[a] Section 1.2.

The consequences of positive definiteness are that A must be nonsingular and that the index (number of positive terms) and rank of the quadratic form must be equal; (1.15) also implies that the eigenvalues of A are all positive. It can be shown that the quadratic form is positive semidefinite for an $n \times n$ matrix A when it is singular and of rank $r < n$, with $(n - r)$ eigenvalues equal to zero.

Analogous to the definitions above, the quadratic form may be negative definite and negative semidefinite. The associated conditions are listed in Table 1.1.

(e) Norms of matrix

The concepts of norms of a vector can be extended to matrices. Two commonly used definitions of norm of an $(n \times n)$ matrix, denoted by $\|A\|$, are

(1) $\|A\| = \sum_{i=1}^{n} \sum_{j=1}^{n} |a_{ij}|$; (2) $\|A\| = \max \{|a_{ij}|\}$

Norms of a matrix A have the following properties:

(3) $\|A\| = \|A^*\|$; (4) $\|A + B\| < \|A\| + \|B\|$

(5) $\|AB\| < \|A\| \cdot \|B\|$; (6) $\|Ax\| < \|A\| \cdot \|x\|$

1.2 INVERSE, DETERMINANTS, AND EIGENVALUES [3-7]

A vector-matrix equation $y = Ax$, where A is a square matrix and y and x are column vectors, may be viewed as a transformation of the vector x into vector y $(x \rightarrow y)$. The question then arises whether there exists an inverse transformation that yields y from x $(y \rightarrow x)$. The existence of this inverse transformation is related to the existence of the inverse of the matrix A and nonzero determinant of A. These are considered here and the concept of eigenvalues and eigenvectors introduced.

1.2.1 Determinants

The determinant of an $(n \times n)$ square matrix A is written as $|A|$ [or det $\{A\}$], and is a scalar whose value is the algebraic sum of all possible products of n elements that contain one and only one element from each row and column, where each product is either positive or negative depending on the position of the elements in the matrix A. The determinant $|A|$ of a square matrix A, of order 3,

$$A = \begin{bmatrix} a_{11} & a_{12} & a_{13} \\ a_{21} & a_{22} & a_{23} \\ a_{31} & a_{32} & a_{33} \end{bmatrix}$$

is given by

$$|A| = \begin{vmatrix} a_{11} & a_{12} & a_{13} \\ a_{21} & a_{22} & a_{23} \\ a_{31} & a_{32} & a_{33} \end{vmatrix} = a_{11}a_{21}a_{33} + a_{12}a_{23}a_{31} + a_{13}a_{21}a_{32} - (a_{11}a_{23}a_{32} + a_{12}a_{21}a_{33} + a_{13}a_{22}a_{31}) \quad (1.17)$$

(a) Minors

If the ith row and jth column of an $(n \times n)$ determinant $|A|$ are deleted, the remaining $(n - 1)$ rows and $(n - 1)$ columns form a determinant $|M_{ij}|$. This determinant is called the minor of element a_{ij}. A minor of $|A|$ whose diagonal elements are also diagonal elements of $|A|$ is called a principal minor of $|A|$.

(b) Cofactors

The cofactor of the element a_{ij} is equal to the minor of a_{ij}, with the sign $(-1)^{i+j}$ affixed to it. Thus the cofactor of a_{ij}, written as C_{ij}, is defined as

$$C_{ij} = (-1)^{i+j}|M_{ij}| \quad (1.18)$$

The determinant of a square matrix A is given by the sum or the products of the elements of any single row or column and their respective cofactors. Thus

$$|A| = \sum_{i=1}^{n} a_{ij}C_{ij}, \qquad j = 1, \text{ or } 2, \text{ or } \ldots, \text{ or } n \text{ (column expansion)} \quad (1.19a)$$

$$|A| = \sum_{j=1}^{n} a_{ij}C_{ij}, \qquad i = 1, \text{ or } 2, \text{ or } \ldots, \text{ or } n \text{ (row expansion)} \quad (1.19b)$$

An alternative form of this expansion is given below.

(c) Laplace expansion

The expansion of a determinant $|A|$ of order n along a row or column is a special case of the Laplace expansion. Instead of selecting one row of $|A|$, let m rows numbered $i_1, i_2, \ldots, i_m$, when arranged in order of magnitude, be selected. From these m rows,

$$\rho = \frac{n(n-1)\cdots(n-m+1)}{1\cdot 2\cdots m} \quad \text{minors} \quad \left| A_{i_1, j_2, \ldots, i_m}^{j_i, j_2, \ldots, j_m} \right|$$

can be found by making all possible selections of m columns from the n columns. Using these minors and their algebraic components, the Laplace expansion is expressed as

$$|A| = \sum_{\rho} (-1)^s \left| A_{i_1, j_2, \ldots, i_m}^{j_1, j_2, \ldots, j_m} \right| \cdot \left| A_{i_{m+1}, i_{m+2}, \ldots, i_m}^{j_{m+1}, j_{m+2}, \ldots, j_m} \right| \tag{1.20}$$

where $s = i_1 + i_2 + \cdots + i_m + j_1 + j_2 + \cdots + j_m$, and the summation extends over the ρ selections of the column indices taken m at a time.

Example 1.1 [6]: Computation of

$$|A| = \begin{vmatrix} 2 & 3 & -2 & 4 \\ 3 & -2 & 1 & 2 \\ 3 & 2 & 3 & 4 \\ -2 & 4 & 0 & 5 \end{vmatrix} \tag{1.21}$$

Cofactor method (1.19a) and (1.19b):

$$A = 2(-1)^{1+1} \begin{vmatrix} -2 & 1 & 2 \\ 2 & 3 & 4 \\ 4 & 0 & 5 \end{vmatrix} + (3)(-1)^{1+2} \begin{vmatrix} 3 & 1 & 2 \\ 3 & 3 & 4 \\ -2 & 0 & 5 \end{vmatrix}$$

$$+ (-2)(-1)^{1+3} \begin{vmatrix} 3 & -2 & 2 \\ 3 & 2 & 4 \\ -2 & 4 & 5 \end{vmatrix} + (4)(-1)^{1+4} \begin{vmatrix} 3 & -2 & 1 \\ 3 & 2 & 3 \\ -2 & 4 & 0 \end{vmatrix}$$

$$\begin{vmatrix} -2 & 1 & 2 \\ 2 & 3 & 4 \\ 4 & 0 & 5 \end{vmatrix} = -2 \begin{vmatrix} 3 & 4 \\ 0 & 5 \end{vmatrix} - 1 \begin{vmatrix} 2 & 4 \\ 4 & 5 \end{vmatrix} + 2 \begin{vmatrix} 2 & 3 \\ 4 & 0 \end{vmatrix} = -2(15) - 1(10 - 16) + 2(-12) = -48$$

$$\begin{vmatrix} 3 & 1 & 2 \\ 3 & 3 & 4 \\ -2 & 0 & 5 \end{vmatrix} = 3 \begin{vmatrix} 3 & 4 \\ 0 & 5 \end{vmatrix} - 1 \begin{vmatrix} 3 & 4 \\ -2 & 5 \end{vmatrix} + 2 \begin{vmatrix} 3 & 3 \\ -2 & 0 \end{vmatrix} = 3(15) - 1(15 + 8) + 2(+6) = 34$$

$$\begin{vmatrix} 3 & -2 & 2 \\ 3 & 2 & 4 \\ -2 & 4 & 5 \end{vmatrix} = 3\begin{vmatrix} 2 & 4 \\ 4 & 5 \end{vmatrix} + 2\begin{vmatrix} 3 & 4 \\ -2 & 5 \end{vmatrix} + 2\begin{vmatrix} 3 & 2 \\ -2 & 4 \end{vmatrix} = 3(10-16)+2(15+8)+2(12+4) = 60$$

$$\begin{vmatrix} 3 & -2 & 1 \\ 3 & 2 & 3 \\ -2 & 4 & 0 \end{vmatrix} = 3\begin{vmatrix} 2 & 3 \\ 4 & 0 \end{vmatrix} + 2\begin{vmatrix} 3 & 3 \\ -2 & 0 \end{vmatrix} + 1\begin{vmatrix} 3 & 2 \\ -2 & 4 \end{vmatrix} = 3(-12) + 2(+6) + 1(12+4) = -8$$

$$|A| = 2(-48) - 3(34) - 2(60) - 4(-8) = -286$$

<u>Laplace expansion (1.20)</u>:

$$\begin{aligned} |A| = {} & (-1)^{1+2+1+2} A^{1,2}_{1,2} \times A^{3,4}_{3,4} + (-1)^{1+2+1+3} A^{1,3}_{1,2} \times A^{2,4}_{3,4} \\ & + (-1)^{1+2+1+4} A^{1,4}_{1,2} \times A^{2,3}_{3,4} + (-1)^{1+2+2+3} A^{2,3}_{1,2} \times A^{1,4}_{3,4} \\ & + (-1)^{1+2+2+4} A^{2,4}_{1,2} \times A^{1,3}_{3,4} + (-1)^{1+2+3+4} A^{3,4}_{1,2} \times A^{1,2}_{3,4} \\ = {} & \begin{vmatrix} 2 & 3 \\ 3 & -2 \end{vmatrix} \times \begin{vmatrix} 3 & 4 \\ 0 & 5 \end{vmatrix} - \begin{vmatrix} 2 & -2 \\ 3 & 1 \end{vmatrix} \times \begin{vmatrix} 2 & 4 \\ 4 & 5 \end{vmatrix} + \begin{vmatrix} 2 & 4 \\ 3 & 2 \end{vmatrix} \times \begin{vmatrix} 2 & 3 \\ 4 & 0 \end{vmatrix} \\ & + \begin{vmatrix} 3 & -2 \\ -2 & 1 \end{vmatrix} \times \begin{vmatrix} 3 & 4 \\ -2 & 5 \end{vmatrix} - \begin{vmatrix} 3 & 4 \\ -2 & 2 \end{vmatrix} \times \begin{vmatrix} 3 & 3 \\ -2 & 0 \end{vmatrix} + \begin{vmatrix} -2 & 4 \\ 1 & 2 \end{vmatrix} \times \begin{vmatrix} 3 & 2 \\ -2 & 4 \end{vmatrix} = -286 \end{aligned}$$

Another method of evaluating $|A|$ is to use the property that the value of a determinant is unchanged by adding to the elements of any row (column) a scalar multiple of the corresponding elements of another row (column). This method is used below to compute $|A|$.

$$|A| = \begin{vmatrix} 2 & 3 & -2 & 4 \\ 3 & -2 & 1 & 2 \\ 3 & 2 & 3 & 4 \\ -2 & 4 & 0 & 5 \end{vmatrix} \quad \begin{vmatrix} 2+2(3) & 3+2(-2) & -2+2(1) & 4+2(2) \\ 3 & -2 & 1 & 2 \\ 3-3(3) & 2-3(-2) & 3-3(1) & 4-3(2) \\ -2 & 4 & 0 & 5 \end{vmatrix}$$

$$= \begin{vmatrix} 8 & -1 & 0 & 8 \\ 3 & -2 & 1 & 2 \\ -6 & 8 & 0 & -2 \\ -2 & 4 & 0 & 5 \end{vmatrix} = (-1)^{2+3} \begin{vmatrix} 8 & -1 & 8 \\ -6 & 8 & -2 \\ -2 & 4 & 5 \end{vmatrix}$$

$$= -\begin{vmatrix} 8+8(-1) & -1 & 8+8(-1) \\ -6+8(8) & 8 & -2+8(8) \\ -2+8(4) & 4 & 5+8(4) \end{vmatrix} = -\begin{vmatrix} 0 & -1 & 0 \\ 58 & 8 & 62 \\ 30 & 4 & 37 \end{vmatrix}$$

$$= -(-1)^{1+2}(-1)\begin{vmatrix} 58 & 62 \\ 30 & 37 \end{vmatrix} = -286$$

(d) Partitioned determinants

Determinants of large-order matrices may be conveniently evaluated through partitioning. Sometimes, even for smaller matrices, partitioning may illuminate the effects of some of the elements of the matrices on their determinants. Some commonly encountered block matrices and their determinants are given below.

1. When A_{11} and A_{22} are square matrices and A_{21} and 0_{12} have appropriate dimensions,

$$\det \left[\begin{array}{c|c} A_{11} & 0_{12} \\ \hline A_{21} & A_{22} \end{array}\right] = \det A_{11} \det A_{22}$$

2. When A_{11} is nonsingular

$$\det \left[\begin{array}{c|c} A_{11} & A_{12} \\ \hline A_{21} & A_{22} \end{array}\right] = \begin{cases} \det(A_{11}) \det(A_{22} - A_{21}A_{11}^{-1}A_{12}) & \text{if } \det(A_{11}) \neq 0 \\ \det(A_{22}) \det(A_{11} - A_{22}A_{22}^{-1}A_{21}) & \text{if } \det(A_{22}) \neq 0 \end{cases}$$

(e) Singular matrix

A square matrix whose determinant is identically equal to zero is called a <u>singular matrix</u>.

1.2.2 Adjoint and Inverse of a Matrix

If A is a square matrix and C_{ij} is the cofactor of a_{ij}, the matrix formed by the cofactor C_{ji} is defined as the <u>adjoint matrix</u> of A, denoted by adj A: that is,

$$\text{adj } A = \{C_{ji}\} \tag{1.22}$$

The adjoint matrix is thus the transpose of the matrix formed by replacing the elements a_{ij} by their cofactors. Combining (1.19), (1.20), and (1.22), we have

$$\{a_{ij}\}\{C_{ji}\}^T = |A| I \tag{1.23}$$

which is a diagonal matrix with all its elements equal to the determinant of the coefficient matrix A. If $|A| \neq 0$, then (1.22) and (1.23) give

$$A \frac{\text{adj } A}{|A|} = I \tag{1.24}$$

or

$$A(A^{-1}) = I, \quad \text{where } \frac{\text{adj } A}{|A|} = A^{-1} \quad (|A| \neq 0) \tag{1.25}$$

The matrix A^{-1} is called the inverse (or reciprocal) of A; it is apparent that A and A^{-1} commute. Only square matrices possess inverses.

The product of a string of inverse matrices satisfies the same rules of transposition as does the product of transpose matrices; that is,

$$\text{if } C = AB, \quad B^{-1}A^{-1} = C^{-1}$$

1.2.3 Computation of Inverse Matrices

(a) Inversion by partitioning

This procedure is illustrated through the partitioning of a square matrix $A = \{a_{ij}\}$ of order n and its inverse $B = \{b_{ij}\}$.

$$\left[\begin{array}{c|c} A_{11} & A_{12} \\ (p \times p) & (p \times q) \\ \hline A_{21} & A_{22} \\ (q \times p) & (q \times q) \end{array}\right]; \quad B = A^{-1} = \left[\begin{array}{c|c} B_{11} & B_{12} \\ (p \times p) & (p \times q) \\ \hline B_{21} & B_{22} \\ (q \times p) & (q \times q) \end{array}\right]; \quad p + q = n \tag{1.26}$$

Provided that A_{11}^{-1} is nonsingular, the elements of B are

$$\begin{aligned} B_{11} &= A_{11}^{-1} + (A_{11}^{-1}A_{12})Z^{-1}(A_{21}A_{11}^{-1}) \\ B_{12} &= -(A_{11}^{-1}A_{12})Z^{-1} \\ B_{21} &= -Z^{-1}(A_{21}A_{11}^{-1}) \\ B_{22} &= Z^{-1}, \quad \text{where } Z = A_{22} - A_{21}(A_{11}^{-1}A_{22}) \end{aligned} \tag{1.27}$$

(b) Inversion through elementary transformation

The following notations are used below:

1. Interchange of the ith and jth rows (columns) is denoted by $H_{ij}(K_{ij})$.

2. Multiplication of every element of the ith row (column) by a nonzero scalar k is denoted by $H_i(k)(K_i(k))$.
3. Addition to the elements of the ith row (column) of k, a scalar, times the corresponding elements of the jth row (column) is denoted by $H_{ij}(k)(K_{ij}(k))$.

The transformations H are called elementary row transformations; the transformations K are called the elementary column transformations.

(c) Inversion of a symmetric matrix

When a matrix $A = \{a_{ij}\}$, $i, j \in 1, \ldots, n$, is symmetric (i.e., $a_{ij} = a_{ji}$), only $\frac{1}{2}n(n+1)$ cofactors need be computed.

If there is to be any gain in computing A^{-1} as the product of elementary matrices, the elementary transformations must be performed in pairs, that is, a row transformation followed by the same column transformation. This would preserve symmetry. If A were symmetric, inversion by partitioning reduces (1.27) to

$$\begin{aligned} B_{11} &= A_{11}^{-1} + (A_{11}^{-1}A_{12})Z^{-1}(A_{11}^{-1}A_{12})^T \\ B_{12} &= -(A_{11}^{-1}A_{12})Z^{-1} \\ B_{21} &= B_{12}^T \\ B_{22} &= Z^{-1} \\ Z &= A_{22} - A_{21}(A_{11}^{-1}A_{12}) \end{aligned} \tag{1.28}$$

(d) Generalized inverse

When A is not symmetric, the procedure described above may be used to find the inverse of A^TA, which is symmetric, and then the inverse of A, obtained through the generalized inverse matrix equation*

$$A^{-1} = (A^TA)^{-1}A^T \tag{1.29}$$

Example 1.2 [6]: Inverse of matrix A computed by four methods

$$A = \begin{bmatrix} 1 & 3 & 3 \\ 1 & 4 & 3 \\ 1 & 3 & 4 \end{bmatrix} \tag{1.30}$$

Since $|A| = 1 \neq 0$, A^{-1} exists.

*The method of computing generalized inverse of a nonsquare matrix is given in Chap. 7.

Inverse from the adjoint:

$$A^{-1} = \frac{\text{adj } A}{|A|} = \frac{1}{A}\begin{bmatrix} c_{11} & c_{12} & c_{13} \\ c_{21} & c_{22} & c_{23} \\ c_{31} & c_{32} & c_{33} \end{bmatrix} = \begin{bmatrix} 7 & -3 & -3 \\ -1 & 1 & 0 \\ -1 & 0 & 1 \end{bmatrix}$$

Inverse by partitioning: Let A be partitioned as

$$A = \left[\begin{array}{cc|c} 1 & 3 & 3 \\ 1 & 4 & 3 \\ \hline 1 & 3 & 4 \end{array}\right] = \left[\begin{array}{c|c} A_{11} & A_{12} \\ \hline A_{21} & A_{22} \end{array}\right]$$

$$A_{11}^{-1} = \begin{bmatrix} 4 & -3 \\ -1 & 1 \end{bmatrix}; \quad A_{11}^{-1}A_{12} = \begin{bmatrix} 4 & -3 \\ -1 & 1 \end{bmatrix}\begin{bmatrix} 3 \\ 2 \end{bmatrix} = \begin{bmatrix} 3 \\ 0 \end{bmatrix},$$

$$A_{21}A_{11}^{-1} = [1 \quad 3]\begin{bmatrix} 4 & -3 \\ -1 & 1 \end{bmatrix} = [1 \quad 0]; \quad Z = A_{22} - A_{21}(A_{11}^{-1}A_{22}) = [4] - [1 \quad 3]\begin{bmatrix} 3 \\ 0 \end{bmatrix} = [1]$$

$$Z^{-1} = [1]$$

From (1.27),

$$B_{11} = A_{11}^{-1} + (A_{11}^{-1}A_{22})Z^{-1}(A_{21}A_{11}^{-1}) = \begin{bmatrix} 4 & -3 \\ -1 & 1 \end{bmatrix} + \begin{bmatrix} 3 \\ 0 \end{bmatrix}[1][1 \quad 0] = \begin{bmatrix} 7 & -3 \\ -1 & 1 \end{bmatrix};$$

$$B_{12} = -(A_{11}^{-1}A_{12})Z^{-1} = \begin{bmatrix} -3 \\ 0 \end{bmatrix}; \quad B_{21} = -Z^{-1}(A_{21}A_{11}^{-1}) = [-1 \quad 0]; \quad B_{22} = Z^{-1} = [1]$$

Thus

$$A^{-1} = \begin{bmatrix} B_{11} & B_{12} \\ B_{21} & B_{22} \end{bmatrix} = \begin{bmatrix} 7 & -3 & -3 \\ -1 & 1 & 0 \\ -1 & 0 & 1 \end{bmatrix}$$

Inverse through row transformation: An alternative method of deriving the inverse of a matrix is considered below by a sequence of row transformation of the augmented matrix $[A \mid I_3]$. Row operations are carried out such that the rows of A are transformed into the rows of an identity matrix of order 3.

$$[A \mid I_3] = \left[\begin{array}{ccc|ccc} 1 & 3 & 3 & 1 & 0 & 0 \\ 1 & 4 & 3 & 0 & 1 & 0 \\ 1 & 3 & 4 & 0 & 0 & 1 \end{array}\right] \rightarrow \left[\begin{array}{ccc|ccc} 1 & 3 & 3 & 1 & 0 & 0 \\ 0 & 1 & 0 & -1 & 1 & 0 \\ 0 & 0 & 1 & -1 & 0 & 1 \end{array}\right]$$

$$\rightarrow \left[\begin{array}{ccc|ccc} 1 & 0 & 3 & 4 & -3 & 0 \\ 0 & 1 & 0 & -1 & 1 & 0 \\ 0 & 0 & 1 & -1 & 0 & 1 \end{array}\right] \rightarrow \left[\begin{array}{ccc|ccc} 1 & 0 & 0 & 7 & -3 & -3 \\ 0 & 1 & 0 & -1 & 1 & 0 \\ 0 & 0 & 1 & -1 & 0 & 1 \end{array}\right]$$

Thus A is reduced to I_3 and I_3 is carried into

$$A^{-1} = \begin{bmatrix} 7 & -3 & -3 \\ -1 & 1 & 0 \\ -1 & 0 & 1 \end{bmatrix}$$

Inversion through elementary transformation: As the first step toward computing A^{-1}, the elementary row and column operations are carried out such that A is transformed into I_3. From (1.30), the elementary transformations $H_{21}(-1) \times H_{31}(-1)$ and $K_{21}(-3) \times K_{31}(-3)$ reduce A to I_3; that is

$$H_{31}(-1) \times H_{21}(-1) \times A \times K_{21}(-3) \times K_{31}(-3) = PAQ = I$$

Therefore,

$$A^{-1} = (P^{-1}Q^{-1})^{-1} = QP$$

$$= \begin{bmatrix} 1 & -3 & 0 \\ 0 & 1 & 0 \\ 0 & 0 & 1 \end{bmatrix} \begin{bmatrix} 1 & 0 & -3 \\ 0 & 1 & 0 \\ 0 & 0 & 1 \end{bmatrix} \begin{bmatrix} 1 & 0 & 0 \\ -1 & 1 & 0 \\ 0 & 0 & 1 \end{bmatrix} \begin{bmatrix} 1 & 0 & 0 \\ 0 & 1 & 0 \\ -1 & 0 & 1 \end{bmatrix} = \begin{bmatrix} 7 & -3 & -3 \\ -1 & 1 & 0 \\ -1 & 0 & 1 \end{bmatrix}$$

<u>Example 1.3 [6]: Inversion through row operations of</u>

$$A = \begin{bmatrix} 2 & 4 & 3 & 2 \\ 3 & 6 & 5 & 2 \\ 2 & 5 & 2 & -3 \\ 4 & 5 & 14 & 14 \end{bmatrix} \tag{1.31}$$

$$[A \mid I_4] = \left[\begin{array}{cccc|cccc} 2 & 4 & 3 & 2 & 1 & 0 & 0 & 0 \\ 3 & 6 & 5 & 2 & 0 & 1 & 0 & 0 \\ 2 & 5 & 2 & -3 & 0 & 0 & 1 & 0 \\ 4 & 5 & 14 & 14 & 0 & 0 & 0 & 1 \end{array}\right] \rightarrow \left[\begin{array}{cccc|cccc} 1 & 2 & \frac{3}{2} & 1 & \frac{1}{2} & 0 & 0 & 0 \\ 3 & 6 & 5 & 2 & 0 & 1 & 0 & 0 \\ 2 & 5 & 2 & -3 & 0 & 0 & 1 & 0 \\ 4 & 5 & 14 & 14 & 0 & 0 & 0 & 1 \end{array}\right]$$

$$\rightarrow \left[\begin{array}{cccc|cccc} 1 & 2 & \frac{3}{2} & 1 & \frac{1}{2} & 0 & 0 & 0 \\ 0 & 0 & \frac{1}{2} & -1 & -\frac{3}{2} & 1 & 0 & 0 \\ 0 & 1 & -1 & -5 & -1 & 0 & 1 & 0 \\ 0 & -3 & 8 & 10 & -2 & 0 & 0 & 1 \end{array}\right] \rightarrow \left[\begin{array}{cccc|cccc} 1 & 2 & \frac{3}{2} & 1 & \frac{1}{2} & 0 & 0 & 0 \\ 0 & 1 & -1 & -5 & -1 & 0 & 1 & 0 \\ 0 & 0 & \frac{1}{2} & -1 & -\frac{3}{2} & 1 & 0 & 0 \\ 0 & -3 & 8 & 10 & -2 & 0 & 0 & 1 \end{array}\right]$$

$$\rightarrow \left[\begin{array}{cccc|cccc} 1 & 0 & \frac{7}{2} & 11 & \frac{5}{2} & 0 & -2 & 0 \\ 0 & 1 & -1 & -5 & -1 & 0 & 1 & 0 \\ 0 & 0 & 1 & -2 & -3 & 2 & 0 & 0 \\ 0 & 0 & 5 & -5 & -5 & 0 & 3 & 1 \end{array}\right] \rightarrow \left[\begin{array}{cccc|cccc} 1 & 0 & 0 & 18 & 13 & -7 & -2 & 0 \\ 0 & 1 & 0 & -7 & -4 & 2 & 1 & 0 \\ 0 & 0 & 1 & -2 & -3 & 2 & 0 & 0 \\ 0 & 0 & 0 & 5 & 10 & -10 & 3 & 1 \end{array}\right]$$

$$\rightarrow \left[\begin{array}{cccc|cccc} 1 & 0 & 0 & 18 & 13 & -7 & -2 & 0 \\ 0 & 1 & 0 & -7 & -4 & 2 & 1 & 0 \\ 0 & 0 & 1 & -2 & -3 & 2 & 0 & 0 \\ 0 & 0 & 0 & 1 & 2 & -2 & -\frac{3}{5} & \frac{1}{5} \end{array}\right] \rightarrow \left[\begin{array}{cccc|cccc} 1 & 0 & 0 & 0 & -23 & 29 & -\frac{64}{5} & -\frac{18}{5} \\ 0 & 1 & 0 & 0 & 10 & -12 & \frac{26}{5} & \frac{7}{5} \\ 0 & 0 & 1 & 0 & 1 & -2 & \frac{6}{5} & \frac{2}{5} \\ 0 & 0 & 0 & 1 & 2 & -2 & \frac{3}{5} & \frac{1}{5} \end{array}\right]$$

$$= [I_4 \mid A^{-1}]$$

$$A^{-1} = \begin{bmatrix} -23 & 29 & -\frac{64}{5} & -\frac{18}{5} \\ 10 & -12 & \frac{26}{5} & \frac{7}{5} \\ 1 & -2 & \frac{6}{5} & \frac{2}{5} \\ 2 & -2 & -\frac{3}{5} & \frac{1}{5} \end{bmatrix}$$

Example 1.4 [6]: Inversion through partitioning of

$$A = \left[\begin{array}{ccc|c} 2 & 1 & -1 & 2 \\ 1 & 3 & 2 & -3 \\ -1 & 2 & 1 & -1 \\ \hline 2 & -3 & -1 & 4 \end{array}\right] \equiv \left[\begin{array}{c|c} D_{11} & D_{12} \\ \hline D_{21} & D_{22} \end{array}\right] \tag{1.32}$$

Stage I: D_{11} is first considered:

$$D_{11} = \left[\begin{array}{cc|c} 2 & 1 & -1 \\ 1 & 3 & 2 \\ \hline -1 & 2 & 1 \end{array}\right] \equiv \left[\begin{array}{c|c} A_{11} & A_{12} \\ \hline A_{21} & A_{22} \end{array}\right]$$

To compute the inversion of D_{11}, the following are obtained from (1.27):

$$B_{11} = \begin{bmatrix} \frac{3}{5} & -\frac{1}{5} \\ -\frac{1}{5} & \frac{2}{5} \end{bmatrix} + \begin{bmatrix} -1 \\ 1 \end{bmatrix} \left[-\frac{1}{2}\right] [-1 \quad 1] = \frac{1}{10} \begin{bmatrix} 1 & 3 \\ 3 & -1 \end{bmatrix};$$

$$B_{12} = \begin{bmatrix} -\frac{1}{2} \\ \frac{1}{2} \end{bmatrix}; \quad B_{21} = \left[-\frac{1}{2} \ \frac{1}{2}\right]; \quad B_{22} = \left[-\frac{1}{2}\right]$$

Therefore,

$$D_{11}^{-1} = \frac{1}{10} \begin{bmatrix} 1 & 3 & -5 \\ 3 & -1 & 5 \\ -5 & 5 & -5 \end{bmatrix}$$

Stage II: The matrix A is now considered to have the following partitioned form:

$$A = \left[\begin{array}{ccc|c} 2 & 1 & -1 & 2 \\ 1 & 3 & 2 & -3 \\ -1 & 2 & 1 & -1 \\ \hline 2 & -3 & -1 & 4 \end{array}\right] = \left[\begin{array}{c|c} A_{11} & A_{12} \\ \hline A_{21} & A_{22} \end{array}\right]$$

For above, from (1.27),

$$A_{11}^{-1} = \frac{1}{10} \begin{bmatrix} 1 & 3 & -5 \\ 3 & -1 & 5 \\ -5 & -5 & -5 \end{bmatrix}; \quad A_{11}^{-1}A_{12} = \frac{1}{5} \begin{bmatrix} -1 \\ 2 \\ -2 \end{bmatrix}; \quad Z = \left[\frac{18}{5}\right]; \quad Z^{-1} = \left[\frac{5}{18}\right]$$

$$B_{11} = \frac{1}{18}\begin{bmatrix} 2 & 5 & -7 \\ 5 & -1 & 5 \\ -7 & 5 & 11 \end{bmatrix}; \quad B_{12} = \frac{1}{18}\begin{bmatrix} 1 \\ -2 \\ 10 \end{bmatrix}; \quad B_{21} = \frac{1}{18}[1 \;\; -2 \;\; 10];$$

$$B_{22} = \left[\frac{5}{18}\right]$$

Combining the above gives us

$$A^{-1} = \frac{1}{18}\begin{bmatrix} 2 & 5 & -7 & 1 \\ 5 & -1 & 5 & -2 \\ -7 & 5 & 11 & 10 \\ 1 & -2 & 10 & 5 \end{bmatrix}$$

(e) Derivative of inverse matrix

For a t-value at which a square matrix A(t) is differentiable and possesses an inverse, the derivative of $A^{-1}(t)$ is given by

$$\frac{d}{dt}[A^{-1}(t)] = -A^{-1}(t)\left\{\frac{d}{dt}[A(t)]\right\}A^{-1}(t)$$

(f) Special inverse matrices

Involuntary matrix: If a square matrix A is equal to its own inverse, that is, if $AA^{-1} = I$, it is said to be involuntary.

Orthogonal matrix: A matrix A is called orthogonal if it is real and satisfies the relationship $AA^T = A^T A = I$.

Unitary matrix: A unitary matrix is a complex matrix in which the inverse is equal to the conjugate of the transpose, that is, $A^{-1} = A^*$ or $A^*A = AA^* = I$.

1.2.4 Characteristic Equations and Eigenvector

The vector-matrix equation

$$y = Ax \tag{1.33}$$

where A is a square (n × n) matrix and x and y are column vectors, may be viewed as a transformation of the vector x into the vector y. The question then arises whether there exists a vector x such that the transformation A produces a vector y, which has the same direction in vector space as x. If such a vector x exists, then y is proportional to x, or

$$y = Ax = \lambda x \tag{1.34}$$

where λ is a scalar of proportionality. The nth-order polynomial $P(\lambda)$ given by the determinant $|A - \lambda I|$

$$P(\lambda) = |A - \lambda I| = (-1)^n(\lambda^n + a_1\lambda^{n-1} + \cdots + a_{n-1}\lambda + a_n) \tag{1.35}$$

is called the characteristic polynomial of A, and the equation

$$P(\lambda) = 0 \tag{1.36}$$

is called the characteristic equation corresponding to the matrix A.

(a) Eigenvalue

The n roots of the characteristic equation in (1.36) and the coefficients of the characteristic polynomial of A are related to the trace* of A by

$$a_k = -\frac{1}{k}(a_{k-1}T_1 + a_{k-2}T_2 + \cdots + a_1T_{k-1} + T_k) \tag{1.37}$$

where $T_k = \text{trace}(A^k)$, $k \in 1, \ldots, n$. This equation is often used for numerical computation of the characteristic polynomials.

The matrix equation in (1.33) and (1.34) can be written in the form of a homogeneous set of linear equations

$$\begin{aligned}
(a_{11} - \lambda)x_1 + a_{12}x_2 + \cdots + a_{1n}x_n &= 0 \\
a_{21}x_1 + (a_{22} - \lambda)x_2 + \cdots + a_{2n}x_n &= 0 \\
&\vdots \\
a_{n1}x_1 + a_{n2}x_2 + \cdots + (a_{nn} - \lambda)x_n &= 0
\end{aligned}$$

or

$$[\lambda I - A]x = 0 \tag{1.38}$$

which has a nontrivial solution only if the determinant of the coefficients vanishes, that is, if

$$|\lambda I - A| = 0 \tag{1.39}$$

*The trace of a square matrix is the sum of its diagonal elements.

The roots of λ in (1.39) are called the eigenvalues or characteristic roots of A. The eigenvalues of A are said to be distinct, repeated, or complex (occurring in conjugate pairs) depending on whether the roots of (1.39) are distinct, repeated, or complex.

From the definition above,it follows that for a diagonal or triangular matrix, the n diagonal elements are the n eigenvalues of the matrix.

(b) Eigenvector

Any nonzero vector x_i such that $Ax_i = \lambda_i x_i$ is called an eigenvector (or characteristic vector) associated with the eigenvalue λ_i, $i \in 1, 2, \ldots, n$, of an $n \times n$ matrix A. Since the components of x_i are determined from n linear homogeneous algebraic equations within a constant factor, if x_i is an eigenvector, then for any scalar $c \neq 0$, cx_i is also an eigenvector.

The eigenvector is called a normalized eigenvector if the length or absolute value of the eigenvector is unity.

(c) Modal matrix

For each of the n eigenvalues λ_i, (1.34) gives a column vectors x_i. The matrix M formed by these n column vectors (or their scalar multiples) is called the modal matrix. Thus $M = \{x_i\}$ or $\{k_i x_i\}$, $i \in 1, \ldots, n$.

If A has distinct eigenvalues, the columns of the modal matrix can be taken to be equal, or proportional, to any nonzero column of adj $[\lambda_i I - A]$. Thus

$$x_i = \text{adj}\,[\lambda_i I - A] \tag{1.40}$$

gives an alternative to (1.38) to compute x_i if A has distinct eigenvalues, and if A has the form

$$A = \begin{bmatrix} a_{11} & a_{12} & a_{13} & a_{14} \\ a_{21} & a_{22} & a_{23} & a_{24} \\ a_{31} & a_{32} & a_{33} & a_{34} \\ a_{41} & a_{42} & a_{43} & a_{44} \end{bmatrix} \quad \text{and} \quad \det(\lambda I - A) = \prod_{i=1}^{4} (\lambda - \lambda_i)$$

then, from (1.40), the ith column of the modal matrix is given by

$$
M_i = \begin{bmatrix}
a_{22} - \lambda_i & a_{23} & a_{24} \\
a_{32} & a_{33} - \lambda_i & a_{34} \\
a_{42} & a_{43} & a_{44} - \lambda_i \\
\hline
a_{21} & a_{23} & a_{24} \\
a_{31} & a_{33} - \lambda_i & a_{34} \\
a_{41} & a_{43} & a_{44} - \lambda_i \\
\hline
a_{21} & a_{22} - \lambda_i & a_{24} \\
a_{31} & a_{32} & a_{34} \\
a_{41} & a_{42} & a_{44} - \lambda_i \\
\hline
a_{21} & a_{22} - \lambda_i & a_{23} \\
a_{31} & a_{32} & a_{33} - \lambda_i \\
a_{41} & a_{42} & a_{43}
\end{bmatrix}
$$

If A has repeated eigenvalues, the method described above may be extended to obtain the eigenvectors by using differentiated adjoints. This method, although not developed here, is illustrated through numerical example (Example 1.7).

(d) Special form of A

If

$$
A = \begin{bmatrix}
0 & 1 & 0 & \cdots & 0 \\
0 & 0 & 1 & \cdots & 0 \\
\vdots & & & & \\
0 & 0 & 0 & \cdots & 1 \\
-a_n & -a_{n-1} & -a_{n-2} & \cdots & -a_1
\end{bmatrix} \tag{1.41a}
$$

has n distinct eigenvalues λ_i, $i \in 1, \ldots, n$, the modal matrix of A is given by

$$
M = \begin{bmatrix}
1 & 1 & \cdots & 1 & 1 \\
\lambda_1 & \lambda_2 & \cdots & \lambda_{n-1} & \lambda_n \\
\vdots & & & & \\
\lambda_1^{n-1} & \lambda_2^{n-1} & \cdots & \lambda_{n-1}^{n-1} & \lambda_n^{n-1}
\end{bmatrix} \tag{1.41b}
$$

Methods of evaluating M through similarity transformations are presented in the following section.

(e) Eigenvalues and symmetric matrices

A fundamental property of real symmetric matrices is that their eigenvalues must be real. A second property of real symmetric matrices is that the eigenvectors form an orthogonal set. These two can be easily proved from the basic property of a symmetric matrix $A = A^T$.

It can also be shown that the eigenvalues of a Hermitian matrix are real and that the eigenvectors are orthogonal.

Example 1.5 [5]: Computation of eigenvalues and eigenvectors of

$$A = \begin{bmatrix} 2 & -2 & 3 \\ 1 & 1 & 1 \\ 1 & 3 & -1 \end{bmatrix} \tag{1.42}$$

Computation of eigenvalues (1.35) to (1.37): Trace (A) = $T_1 = 2+1-1$; $a_1 = -T_1 = -2$.

$$A^2 = AA = \begin{bmatrix} 5 & 3 & 1 \\ 4 & 2 & 3 \\ 4 & -2 & 7 \end{bmatrix}; \quad \text{trace}\,(A^2) = T_2 = 5 + 2 + 7 = 14; \quad a_2 = -\frac{1}{2}(a_1T_1 + T_2) = -5$$

$$A^3 = A^2A = \begin{bmatrix} 14 & -4 & 17 \\ 13 & 3 & 11 \\ 13 & 11 & 3 \end{bmatrix}; \quad \text{trace}\,(A^3) = 14 + 3 + 3 = 20; \quad a_3 = -\frac{1}{3}(a_2T_1 + a_1T_2 + T_3) = 6$$

The characteristic polynomial of (1.42) is, therefore,

$$P(\lambda) = \lambda^3 - 2\lambda - 5\lambda + 6 = (\lambda - 1)(\lambda + 2)(\lambda - 3)$$

Hence $\lambda_1 = 1$, $\lambda_2 = -2$, and $\lambda_3 = 3$.

Computation of eigenvectors (1.42):

$$\text{adj}\,[\lambda I - A] = \begin{bmatrix} \lambda^2 - 4 & -2\lambda + 7 & 3\lambda - 5 \\ \lambda + 2 & \lambda^2 - \lambda - 5 & \lambda + 1 \\ \lambda + 2 & 3\lambda - 8 & \lambda^3 - 3\lambda + 4 \end{bmatrix}$$

$$\text{adj}\,[\lambda_1 I - A] = \begin{bmatrix} -3 & 5 & -2 \\ 3 & -5 & 2 \\ 3 & -5 & 2 \end{bmatrix}; \quad \text{adj}\,[\lambda_2 I - A] = \begin{bmatrix} 0 & 11 & -11 \\ 0 & 1 & -1 \\ 0 & -14 & 14 \end{bmatrix};$$

$$\text{adj}\,[\lambda_3 I - A] = \begin{bmatrix} 5 & 1 & 4 \\ 5 & 1 & 4 \\ 5 & 1 & 4 \end{bmatrix}$$

Eigenvectors are uniquely determined only in direction. These vectors can therefore be multiplied by any scalar and still satisfy (1.38). Consequently, a modal matrix is

$$M = \begin{bmatrix} -1 & 11 & 1 \\ 1 & 1 & 1 \\ 1 & 14 & 1 \end{bmatrix}$$

Each column of the modal matrix is an eigenvector that spans a one-dimensional vector space. The three columns of the modal matrix form a basis in the corresponding three-dimensional space.

Example 1.6 [4]: Computation of modal matrix of

$$A = \begin{bmatrix} 1 & 1 + j\sqrt{2} & 0 \\ 1 - j\sqrt{2} & 2 & j\sqrt{3} \\ 0 & -j\sqrt{3} & 1 \end{bmatrix} \tag{1.43}$$

$$P(\lambda) = (\lambda - 1)(\lambda + 1)(\lambda - 4); \quad \lambda_1 = 1, \ \lambda_2 = -1, \ \lambda_3 = 4$$

$$[\lambda I - A] = \begin{bmatrix} \lambda - 1 & -(1 + j\sqrt{2}) & 0 \\ -(1 - j\sqrt{2}) & \lambda - 2 & -j\sqrt{3} \\ 0 & j\sqrt{3} & \lambda - 1 \end{bmatrix}$$

The eigenvectors associated with the three eigenvalues are obtained by solving (1.34):

$$\lambda = \lambda_1 = +1: \begin{bmatrix} 0 & -(1 + j\sqrt{2}) & 0 \\ -(1 - j\sqrt{2}) & -1 & -j\sqrt{3} \\ 0 & j\sqrt{3} & 0 \end{bmatrix} \begin{bmatrix} x_{11} \\ x_{12} \\ x_{13} \end{bmatrix} = 0 \rightarrow \begin{bmatrix} x_{11} \\ x_{12} \\ x_{13} \end{bmatrix} = \begin{bmatrix} -j\sqrt{3}\,a \\ 0 \\ (1 - j\sqrt{2})a \end{bmatrix}$$

$$\lambda = \lambda_2 = -1: \begin{bmatrix} -2 & -(1+j\sqrt{2}) & 0 \\ -(1-j\sqrt{2}) & 0 & -j\sqrt{3} \\ 0 & j\sqrt{3} & -2 \end{bmatrix} \begin{bmatrix} x_{21} \\ x_{22} \\ x_{23} \end{bmatrix} = 0 \rightarrow \begin{bmatrix} x_{21} \\ x_{22} \\ x_{23} \end{bmatrix} = \begin{bmatrix} (1+j\sqrt{2})b \\ -2b \\ -j\sqrt{3}\,b \end{bmatrix}$$

$$\lambda = \lambda_3 = 4: \begin{bmatrix} 3 & -(1+j\sqrt{2}) & 0 \\ -(1-j\sqrt{2}) & 2 & -j\sqrt{3} \\ 0 & j\sqrt{3} & 3 \end{bmatrix} \begin{bmatrix} x_{31} \\ x_{32} \\ x_{33} \end{bmatrix} = 0 \rightarrow \begin{bmatrix} x_{31} \\ x_{32} \\ x_{33} \end{bmatrix} = \begin{bmatrix} (1+j\sqrt{2})c \\ 3c \\ -j\sqrt{3}\,c \end{bmatrix}$$

where a, b, and c are nonzero scalars. Thus the modal matrix of A after normalizing each eigenvector is given by

$$M = \{x_i\} = \begin{bmatrix} \dfrac{-j\sqrt{3}}{\sqrt{6}} & \dfrac{1+j\sqrt{2}}{\sqrt{10}} & \dfrac{1+j\sqrt{2}}{\sqrt{15}} \\ 0 & \dfrac{-2}{\sqrt{10}} & \dfrac{3}{\sqrt{15}} \\ \dfrac{1-j\sqrt{2}}{\sqrt{6}} & \dfrac{-j\sqrt{3}}{\sqrt{10}} & \dfrac{-j\sqrt{3}}{\sqrt{15}} \end{bmatrix}$$

Example 1.7 [5]: Repeated eigenvalues—computation of eigenvalues of

$$A = \begin{bmatrix} 2 & 1 & 1 \\ 1 & 2 & 1 \\ 0 & 0 & 1 \end{bmatrix} \tag{1.44}$$

$$P(\lambda) = (\lambda - 1)^2(\lambda - 3): \quad \lambda_1 = 1, \quad \lambda_2 = 1, \quad \lambda_3 = 3$$

$$\text{adj}\,[\lambda I - A] = \begin{bmatrix} (\lambda-2)(\lambda-1) & \lambda - 1 & \lambda - 1 \\ \lambda - 1 & (\lambda-2)(\lambda-1) & \lambda - 1 \\ 0 & 0 & (\lambda-1)(\lambda-3) \end{bmatrix}$$

The matrix A has two repeated eigenvalues and the matrix $[\lambda I - A]$ (sometimes called the characteristic matrix) has only two linearly independent columns.* However, it is possible to obtain a linearly independent vector solution for each of the repeated roots. This is derived through the first derivative of the adjoint matrix.†

*The characteristic matrix in this case is said to have "full degeneracy."
†In general, if the degeneracy of $[\lambda I - A]$ is q, the eigenvector x_i (for $\lambda = \lambda_i$) is $(d^{q-1}/d\lambda^{q-1})\{\text{adj}\,[\lambda I - A]\}$ at $\lambda = \lambda_i$, where $(d^{q-2}/d\lambda^{q-2})\{\cdot\} = 0$; see Ref. 5, pp. 235-239.

$$\frac{d}{dt}\{\text{adj }[\lambda I - A]\} = \begin{bmatrix} 2\lambda - 3 & 1 & 1 \\ 1 & 2\lambda - 3 & 1 \\ 0 & 0 & 2\lambda - 4 \end{bmatrix}$$

For $\lambda_{1,2} = 1$,

$$\frac{d}{dt}\{\text{adj }[\lambda I - A]\} = \begin{bmatrix} -1 & 1 & 1 \\ 1 & -1 & 1 \\ 0 & 0 & -2 \end{bmatrix} = \begin{bmatrix} -1 & 1 \\ 1 & 1 \\ 0 & -2 \end{bmatrix} \begin{bmatrix} 1 & -1 & 0 \\ 0 & 0 & 1 \end{bmatrix}$$

The corresponding eigenvectors are

$$x_1 = \begin{bmatrix} -1 \\ 1 \\ 0 \end{bmatrix}; \quad x_2 = \begin{bmatrix} 1 \\ 1 \\ -2 \end{bmatrix}$$

The eigenvector for $\lambda_3 = 3$ is

$$x_3 = \text{adj }[\lambda_3 I - A] = \begin{bmatrix} 1 \\ 1 \\ 0 \end{bmatrix}$$

Therefore,

$$M = \begin{bmatrix} -1 & 1 & 1 \\ 1 & 1 & 1 \\ 0 & -2 & 0 \end{bmatrix}$$

1.3 TRANSFORMATION AND DIAGONALIZATION [4-7]

Suppose that an $(n \times m)$ matrix A is a linear transformation from a vector space S to a vector space S'. The effect of a change of basis in S is to replace A by PA, where P is a nonsingular n-square matrix. The effect of a change of basis in S' is to replace A by AQ, where Q is a nonsingular m-square matrix. By choosing suitable bases for S and S', the matrix A can be given the form

$$PAQ = \begin{bmatrix} I_r & 0 \\ 0 & 0 \end{bmatrix} \tag{1.45}$$

where I_r is an $r \times r$ identity matrix ($r < n$, $r < m$). The matrix A can be given this simple form by performing a suitable sequence of elementary transformations of the types:

1. Interchange of any two basis vectors x_i and x_j in S (or y_i and y_j in S')
2. Replacement of a basis vector x_i by cx_i (or y_i by cy_i), where c is a nonzero scalar
3. Replacement of a basis vector x_i by $x_i + cx_j$ (or y_i by cy_i), where c is a nonzero scalar

The effect of elementary transformations is to carry out following analogous operations called elementary row (column) operations:

1. Interchange of any two rows (or columns)
2. Multiplication of a row (column) by a nonzero constant
3. Addition of one row (column) multiplied by a constant to another row (column)

(a) Equivalent matrices

Two matrices A and B are said to be equivalent if one can be obtained from the other by a series of elementary operations. This, together with the definitions of elementary operations, suggests that A and B are equivalent if there exists a pair of nonsingular matrices P and Q such that

$$B = PAQ \tag{1.46}$$

Two matrices that are equivalent have the same rank.

(b) Normal form

A matrix A of rank $r > 0$ can be reduced to an equivalent matrix of the form

$$I_r, \quad \left[\begin{array}{c|c} I_r & 0 \\ \hline 0 & 0 \end{array}\right], \quad [I_r \quad 0], \quad \text{or} \quad \left[\begin{array}{c} I_r \\ \hline 0 \end{array}\right] \tag{1.47}$$

These forms are called normal (or canonical) forms.

The transformation in (1.46) is the most general kind of transformation using the elementary operations described. Other transformations defined in terms of P and Q are given below:

1. Similarity: $B = Q^{-1}AQ$ or $P = Q^{-1}$
2. Orthogonal: $B = Q^{T}AQ = Q^{-1}AQ$ or $P = Q^{T} = Q^{-1}$
3. Congruent: $B = Q^{T}AQ$ or $P = Q^{T}$

If A is a Hermitian matrix, the following transformations are defined:

4. Conjunctive: $B = Q^{*T}AQ$ or $P = Q^{*T}$

5. Unitary: $B = Q^{*T}AQ = Q^{-1}AQ$ or $P = Q^{*T} = Q^{-1}$

(c) Similar matrices

Two $(n \times n)$ matrices A and B are called similar if there exists a nonsingular matrix P such that

$$P^{-1}AP = B \tag{1.48}$$

The matrix B is obtained from A by a similarity transformation in which P is the transformation matrix. Two similar matrices have the same eigenvalues.

(d) Diagonable matrix

If a matrix A is similar to a diagonal matrix, it is called diagonable.

Example 1.8 [5]: Elementary transformation. Here we reduce the singular matrix

$$A = \begin{bmatrix} 1 & 2 & -3 \\ -1 & 2 & -1 \\ -1 & -3 & 4 \end{bmatrix} \tag{1.49}$$

to the normal form in (1.47) and (1.48).

$$A = \begin{bmatrix} 1 & 2 & -3 \\ -1 & 2 & -1 \\ -1 & -3 & 4 \end{bmatrix} \rightarrow \underbrace{\begin{bmatrix} 1 & 0 & 0 \\ 0 & 1 & 0 \\ 1 & 0 & 1 \end{bmatrix}}_{P_1} \begin{bmatrix} 1 & 2 & -3 \\ -1 & 2 & -1 \\ -1 & -3 & 4 \end{bmatrix} = \begin{bmatrix} 1 & 2 & -3 \\ -1 & 2 & -1 \\ 0 & -1 & 1 \end{bmatrix}$$

$$\rightarrow \underbrace{\begin{bmatrix} 1 & 0 & 0 \\ 1 & 1 & 0 \\ 0 & 0 & 1 \end{bmatrix}}_{P_2} \begin{bmatrix} 1 & 2 & -3 \\ -1 & 2 & -1 \\ 0 & -1 & 1 \end{bmatrix} = \begin{bmatrix} 1 & 2 & -3 \\ 0 & 4 & -4 \\ 0 & -1 & 1 \end{bmatrix}$$

$$\rightarrow \begin{bmatrix} 1 & 2 & -3 \\ 0 & 4 & -4 \\ 0 & -1 & 1 \end{bmatrix} \underbrace{\begin{bmatrix} 1 & 0 & 0 \\ 0 & 1 & 1 \\ 0 & 0 & 1 \end{bmatrix}}_{Q_1} = \begin{bmatrix} 1 & 2 & -1 \\ 0 & 4 & 0 \\ 0 & -1 & 0 \end{bmatrix}$$

$$\rightarrow \begin{bmatrix} 1 & 2 & -1 \\ 0 & 4 & 0 \\ 0 & -1 & 0 \end{bmatrix} \underbrace{\begin{bmatrix} 1 & 0 & 1 \\ 0 & 1 & 0 \\ 0 & 0 & 1 \end{bmatrix}}_{Q_2} = \begin{bmatrix} 1 & 2 & 0 \\ 0 & 4 & 0 \\ 0 & -1 & 0 \end{bmatrix}$$

$$\rightarrow \begin{bmatrix} 1 & 2 & 0 \\ 0 & 4 & 0 \\ 0 & -1 & 0 \end{bmatrix} \underbrace{\begin{bmatrix} 1 & -2 & 0 \\ 0 & 1 & 0 \\ 0 & 0 & 1 \end{bmatrix}}_{Q_3} = \begin{bmatrix} 1 & 0 & 0 \\ 0 & 4 & 0 \\ 0 & -1 & 0 \end{bmatrix}$$

$$= \underbrace{\begin{bmatrix} 1 & 0 & 0 \\ 0 & \frac{1}{4} & 0 \\ 0 & 0 & 1 \end{bmatrix}}_{P_3} \begin{bmatrix} 1 & 0 & 0 \\ 0 & 4 & 0 \\ 0 & -1 & 0 \end{bmatrix} = \begin{bmatrix} 1 & 0 & 0 \\ 0 & 1 & 0 \\ 0 & -1 & 0 \end{bmatrix}$$

$$= \underbrace{\begin{bmatrix} 1 & 0 & 0 \\ 0 & 1 & 0 \\ 0 & 1 & 1 \end{bmatrix}}_{P_4} \begin{bmatrix} 1 & 0 & 0 \\ 0 & 1 & 0 \\ 0 & -1 & 0 \end{bmatrix} = \left[\begin{array}{cc|c} 1 & 0 & 0 \\ 0 & 1 & 0 \\ \hline 0 & 0 & 0 \end{array}\right]$$

The elementary matrices are

$$P = P_4 P_3 P_2 P_1 = \frac{1}{4} \begin{bmatrix} 4 & 0 & 0 \\ 1 & 1 & 0 \\ 5 & 1 & 4 \end{bmatrix}; \quad Q = Q_1 Q_2 Q_3 = \begin{bmatrix} 1 & -2 & 1 \\ 0 & 1 & 1 \\ 0 & 0 & 1 \end{bmatrix}$$

The transformation PAQ yields the desired result.

1.3.1 Diagonalization

Quite frequently it is useful to perform coordinate transformation such that a linear transformation in the new coordinate system is dependent on a diagonal matrix. Congruent transformation is considered here to outline the conditions under which the matrix $B = PAQ$, in (1.46), can be reduced to a diagonal form.

(a) Real symmetric matrices

The real symmetric matrix A of rank r ($< n$) can be reduced to a congruent diagonal having the canonical form

$$Q^T A Q = \begin{bmatrix} I_p & 0 & 0 \\ 0 & -I_{r-p} & 0 \\ 0 & 0 & 0 \end{bmatrix} \tag{1.50}$$

The integer p is called the <u>index</u>, and the integer $s = p - (r - p) = 2p - r$ is called the <u>signature</u> of the matrix A. Two $(n \times n)$ real symmetric matrices are congruent if they have the same rank and the same signature or index.

(b) Antisymmetric matrices

An antisymmetric matrix A of rank r can be reduced to the canonical form below by a congruent transformation.

$$B = -Q^T A Q = \begin{bmatrix} D_1 & & & & & & \\ & \ddots & & & & & \\ & & D_i & & & & \\ & & & \ddots & & & \\ & & & & D_m & & \\ & & & & & \ddots & \\ & & & & & & 0 \\ & & & & & & & 0 \end{bmatrix} \tag{1.51}$$

where

$$D_i = \begin{bmatrix} 0 & 1 \\ -1 & 1 \end{bmatrix} \quad \text{and} \quad m = \tfrac{1}{2} \text{ rank } A \tag{1.52}$$

Two antisymmetric matrices are congruent only if they have the same rank.

(c) Complex symmetric matrices

An $(n \times n)$ complex symmetric matrix of rank r can be reduced by a congruency transformation to the canonical form

$$Q^T A Q = \left[\begin{array}{c|c} I_r & 0 \\ \hline 0 & 0 \end{array}\right] \tag{1.53}$$

Two $(n \times n)$ complex symmetric matrices are congruent only if they have the same rank.

(d) Hermitian matrices

An (n × n) Hermitian matrix of rank r can be reduced to the canonical matrix below by a conjunctive transformation.

$$B = Q^{*T}AQ = \begin{bmatrix} I_p & 0 & 0 \\ 0 & -I_{r-p} & 0 \\ 0 & 0 & 0 \end{bmatrix} \tag{1.54}$$

The index p and signature s are the same as defined for a real symmetric matrix.

(e) Skew-Hermitian matrix

An (n × n) skew-Hermitian matrix of rank r can be reduced to the canonical matrix below by a conjunctive transformation.

$$B = Q^{*T}AQ = \begin{bmatrix} jI_p & 0 & 0 \\ 0 & -jI_{r-p} & 0 \\ 0 & 0 & 0 \end{bmatrix} \tag{1.55}$$

Example 1.9 [6]. Here we transform

$$A = \begin{bmatrix} 0 & 0 & 2 & 4 \\ 0 & 0 & 1 & -3 \\ -2 & -1 & 0 & -2 \\ -4 & 3 & 2 & 0 \end{bmatrix} \tag{1.56}$$

into the canonical form (1.51)

$$[A \mid I] = \left[\begin{array}{cccc|cccc} 0 & 0 & 2 & 4 & 1 & 0 & 0 & 0 \\ 0 & 0 & 1 & -3 & 0 & 1 & 0 & 0 \\ -2 & -1 & 0 & -2 & 0 & 0 & 1 & 0 \\ -4 & 3 & 2 & 0 & 0 & 0 & 0 & 1 \end{array}\right] \rightarrow \left[\begin{array}{cccc|cccc} 0 & 2 & 0 & 4 & 1 & 0 & 0 & 0 \\ -2 & 0 & -1 & -2 & 0 & 0 & 1 & 0 \\ 0 & 1 & 0 & -3 & 0 & 1 & 0 & 0 \\ -4 & 2 & 3 & 0 & 0 & 0 & 0 & 1 \end{array}\right]$$

$$\rightarrow \left[\begin{array}{cccc|cccc} 0 & 1 & 0 & 2 & \frac{1}{2} & 0 & 0 & 0 \\ -1 & 0 & -1 & -2 & 0 & 0 & 1 & 0 \\ 0 & 1 & 0 & -3 & 0 & 1 & 0 & 0 \\ -2 & 2 & 3 & 0 & 0 & 0 & 0 & 1 \end{array}\right] \rightarrow \left[\begin{array}{cccc|cccc} 0 & 1 & 0 & 0 & \frac{1}{2} & 0 & 0 & 0 \\ -1 & 0 & 0 & 0 & 0 & 0 & 1 & 0 \\ 0 & 0 & 0 & -5 & -\frac{1}{2} & 1 & 0 & 0 \\ 0 & 0 & 5 & 0 & -1 & 0 & -2 & 1 \end{array}\right]$$

$$\to \left[\begin{array}{cc:cc:cccc} 0 & 1 & 0 & 0 & \frac{1}{2} & 0 & 0 & 0 \\ -1 & 0 & 0 & 0 & 0 & 0 & 1 & 0 \\ \hdashline 0 & 0 & 0 & 1 & \frac{1}{10} & -\frac{1}{5} & 0 & 0 \\ 0 & 0 & -1 & 0 & -1 & 0 & -2 & 1 \end{array}\right] = \left[\begin{array}{c:c} D_1 & 0 \\ \hdashline 0 & D_2 \end{array}\right] Q^T$$

Hence for

$$Q = \begin{bmatrix} \frac{1}{2} & 0 & \frac{1}{10} & -1 \\ 0 & 0 & -\frac{1}{5} & 0 \\ 0 & 1 & 0 & -2 \\ 0 & 0 & 0 & 1 \end{bmatrix}$$

(1.56) becomes $Q^T AQ = \text{diag}\,[D_1, D_2]$.

<u>Example 1.10</u> [6]. Now let us transform

$$A = \begin{bmatrix} 1 & 1 - j & -3 + 2j \\ 1 + j & 2 & -j \\ -3 - 2j & j & 0 \end{bmatrix} \tag{1.57}$$

into the canonical form (1.54)

$$[A \mid I] = \left[\begin{array}{ccc:ccc} 1 & 1 - j & -3 + 2j & 1 & 0 & 0 \\ 1 + j & 2 & -j & 0 & 1 & 0 \\ -3 - 2j & j & 0 & 0 & 0 & 1 \end{array}\right] \to \left[\begin{array}{ccc:ccc} 1 & 0 & 0 & 1 & 0 & 0 \\ 0 & 0 & 5 & -1 - j & 1 & 0 \\ 0 & 5 & -13 & 3 + 2j & 0 & 1 \end{array}\right]$$

$$\to \left[\begin{array}{ccc:ccc} 1 & 0 & 0 & 1 & 0 & 0 \\ 0 & \frac{25}{13} & 0 & \frac{2 - 3j}{13} & 1 & \frac{5}{13} \\ 0 & 0 & -13 & 3 + 2j & 0 & 1 \end{array}\right] \to \left[\begin{array}{ccc:ccc} 1 & 0 & 0 & 1 & 0 & 0 \\ 0 & 1 & 0 & \frac{2 - 3j}{5\sqrt{13}} & \frac{13}{5\sqrt{13}} & \frac{1}{\sqrt{13}} \\ 0 & 0 & -1 & \frac{3 + 2j}{\sqrt{13}} & 0 & \frac{1}{\sqrt{13}} \end{array}\right]$$

$(p = 3,\ r = 3,\ s = 3)$

$$Q = \begin{bmatrix} 1 & 0 & 0 \\ \frac{2-3j}{5\sqrt{13}} & \frac{13}{5\sqrt{13}} & \frac{1}{\sqrt{13}} \\ \frac{3+2j}{\sqrt{13}} & 0 & \frac{1}{\sqrt{13}} \end{bmatrix} \rightarrow Q^{*T} = \begin{bmatrix} 1 & \frac{2+3j}{5\sqrt{13}} & \frac{3-2j}{\sqrt{13}} \\ 0 & \frac{13}{5\sqrt{13}} & 0 \\ 0 & \frac{1}{\sqrt{13}} & \frac{1}{\sqrt{13}} \end{bmatrix}$$

1.3.2 Similarity Transformation

(a) Matrices with distinct eigenvalue

In the case where the modal matrix (M) is nonsingular, solution of (1.34) can be combined to form the single equation

$$\{\lambda_j x_{ij}\} = \{a_{ij}\}\{x_{ij}\}$$

or $M\Lambda = AM$ where

$$A = \begin{bmatrix} \lambda_1 & & & & 0 \\ & \ddots & & & \\ & & \lambda_i & & \\ & & & \ddots & \\ 0 & & & & \lambda_n \end{bmatrix} \tag{1.58}$$

is a diagonal matrix composed of the characteristic values $\lambda_1, \ldots, \lambda_n$. Since M^{-1} exists, from (1.58),

$$\Lambda = M^{-1}AM \tag{1.59}$$

Higher powers of A can be diagonalized in the same manner; for example,

$$\Lambda^2 = M^{-1}A^2M \tag{1.60}$$

It is thus apparent that the square matrix A can be diagonalized through similarity transformation with the choice

$$Q = M \tag{1.61}$$

The significance of this transformation is briefly discussed here. If the transformation

$$x = Mq \tag{1.62}$$

is made, then $y = Ax$ can be written as

$$y = AMq$$

or

$$M^{-1}y = M^{-1}AMq = \Lambda q \quad \text{or} \quad z = \Lambda q \quad (y = M^{-1}z) \tag{1.63}$$

In terms of the new coordinate system q_i, the set of equations described by (1.63) is uncoupled, these q_i, $i \in 1, \ldots, n$, coordinates being in the direction of eigenvectors. These coordinates are called normal coordinates. This special form of similarity transformation to isolate the eigenvalues (or modes) of the system is called normalization or modal transformation, with the matrix M being the modal matrix (Sec. 2.2).

The columns of M form a basis, and the rows of M^{-1} a reciprocal basis, in the original system.

(b) Matrices with repeated eigenvalues

For the case where the characteristic polynomials of A have repeated roots, normalization is still possible provided that the matrix $[\lambda I - A]$ has full degeneracy. This similarity transformation is always possible if the matrix A is real and symmetric (or Hermitian), since there always exists a real orthogonal matrix such that

$$Q^{-1}AQ = Q^{T}AQ = \begin{bmatrix} \lambda_1 & & & \\ & \lambda_2 & & \\ & & \ddots & \\ & & & \lambda_n \end{bmatrix} \tag{1.64}$$

However, most matrices found in system theory are nonsymmetric, and hence (1.64) is not always applicable.

When an $n \times n$ nonsymmetric matrix has repeated eigenvalues, there may be less than n linearly independent characteristic vectors; thus normalization may be impossible. In such a case, however, the square matrix A can be transformed through a similarity transformation to a Jordan canonical matrix having the following properties:

1. All diagonal elements of the matrix are eigenvalues of A.
2. All the elements below the principal diagonal are zero.
3. A certain number of unit elements are contained in the superdiagonal when the adjacent elements in the principal diagonal are equal.

A typical Jordan (canonical) form is

$$J = \begin{bmatrix} \lambda_1 & 1 & 0 & & & & \\ 0 & \lambda_1 & 1 & & & & \\ 0 & 0 & \lambda_1 & & & & \\ & & & \lambda_2 & 1 & & \\ & & & 0 & \lambda_2 & & \\ & & & & & \lambda_3 & \\ & & & & & & \lambda_4 \end{bmatrix} \tag{1.65}$$

The 1's appear in blocks, known as Jordan blocks. A diagonal matrix, it is to be noted, is a special case of Jordan canonical form.

The number of Jordan blocks associated with a given characteristic value λ_i in the Jordan form resulting from a similarity transformation of A is equal to the number of eigenvectors associated with the eigenvalues (i.e., q, the degeneracy of $[\lambda_i I - A]$). Unfortunately, however, evaluation of the orders of the Jordan blocks is a complicated problem and is not easily accomplished. The result is that it is not clear whether the Jordan form given above or the form

$$J = \begin{bmatrix} \lambda_1 & 1 & & & & \\ 0 & \lambda_1 & & & & \\ & & \lambda_1 & 1 & & \\ & & 0 & \lambda_1 & & \\ & & & & \lambda_2 & 1 \\ & & & & 0 & \lambda_2 \end{bmatrix} \tag{1.66}$$

would be result of the transformation $J = M^{-1}AM$. The number of 1's associated with a given λ_i is the order of λ_i in the characteristic equation minus the degeneracy of $[\lambda I - A]$.

It is useful to note that in the case of full degeneracy no 1's are present, and in the case of simple degeneracy (q = 1) all superdiagonal elements are unity.

Computation of M such that AM = MJ is illustrated below. Let the columns of M be denoted by $x_1, x_2, \ldots, x_n$. Then there is a Jordan block of order m associated with λ_i only if the m linearly independent vectors x_i satisfy

$$(\lambda_i I - A)x_j = -x_{j-1}, \quad j = 1, \ldots, m \tag{1.67}$$

Example 1.11 [5]: Derivation of Jordan form. Here we transform matrix

$$A = \begin{bmatrix} 0 & 1 & 0 \\ 0 & 0 & 1 \\ 1 & -3 & 3 \end{bmatrix} \tag{1.68}$$

into the Jordan form (1.66). The eigenvalues of the matrix above and $\lambda_{1,2,3} = 1$.

$$\text{adj}\,[\lambda_1 I - A] = \begin{bmatrix} 1 & -2 & 1 \\ 1 & -2 & 1 \\ 1 & -2 & 1 \end{bmatrix} \equiv x_1 = \begin{bmatrix} x_{11} \\ x_{21} \\ x_{31} \end{bmatrix}$$

Hence the eigenvector x_1 associated with λ_1 is $x = [1 \;\; 1 \;\; 1]^T$. From (1.67), for $i = 1$ and $j = 2$, $Ax_2 = \lambda_1 x_2 + x_1$ or

$$\begin{bmatrix} 0 & 1 & 0 \\ 0 & 0 & 1 \\ 1 & -3 & 3 \end{bmatrix} \begin{bmatrix} x_{12} \\ x_{22} \\ x_{32} \end{bmatrix} = \begin{bmatrix} x_{12} \\ x_{22} \\ x_{32} \end{bmatrix} + \begin{bmatrix} 1 \\ 1 \\ 1 \end{bmatrix}$$

from which

$$x_2 = \begin{bmatrix} x_{12} \\ x_{22} \\ x_{33} \end{bmatrix} = \begin{bmatrix} 1 \\ 2 \\ 3 \end{bmatrix}$$

Similarly, for $i = 2$ and $j = 3$, from (1.67), $Ax_3 = \lambda_2 x_3 + x_2$ and

$$\begin{bmatrix} 0 & 1 & 0 \\ 0 & 0 & 1 \\ 1 & -3 & 3 \end{bmatrix} \begin{bmatrix} x_{13} \\ x_{23} \\ x_{33} \end{bmatrix} = \begin{bmatrix} x_{13} \\ x_{23} \\ x_{33} \end{bmatrix} + \begin{bmatrix} 1 \\ 2 \\ 3 \end{bmatrix}$$

from which the solution is

$$x_3 = \begin{bmatrix} x_{13} \\ x_{23} \\ x_{33} \end{bmatrix} = \begin{bmatrix} -1 \\ 0 \\ 2 \end{bmatrix}$$

Hence the modal matrix of (1.68) is

$$M = (x_1 \mid x_2 \mid x) = \begin{bmatrix} x_{11} & x_{12} & x_{13} \\ x_{21} & x_{22} & x_{23} \\ x_{31} & x_{32} & x_{33} \end{bmatrix} = \begin{bmatrix} 1 & 1 & -1 \\ 1 & 2 & 0 \\ 1 & 3 & 2 \end{bmatrix}$$

and

$$M^{-1}AM = J = \begin{bmatrix} 1 & 1 & 0 \\ 0 & 1 & 0 \\ 0 & 0 & 1 \end{bmatrix}$$

1.4 FUNCTIONS OF CONSTANT MATRICES [4-9]

Some of the basic notions regarding matrix polynomials and infinite series are presented here. With the introduction of a few modifications the concepts of matrix polynomials and infinite series are shown to be analogous to those of scalar variables.

1.4.1 Powers of Matrices

The matrix product $A \cdot A \cdot A \cdots A$, where A is a square matrix of order n, can be written as A^k, where k is the number of factors involved in the product. The multiplication of powers of a matrix follows the usual rules for scalar algebra, with

$$A^k A^m = A^{k+m}; \quad (A^k)^m = A^{km}; \quad (A^{-1})^m = A^{-m}; \quad A^0 = I \tag{1.69}$$

A set of similar rules applies in the case where a fraction power of a matrix is to be completed. Thus if $A^m = B$, where A is a square matrix, then B is an mth root of A. The number of mth roots of matrix depends on the nature of the matrix, there being no general rules as to how many mth roots the matrix A possesses.

(a) Matrix polynomial

A polynomial of order n, where the argument of the polynomial is a scalar variable x, may be expressed as

$$N(x) = p_n x^n + p_{n-1} x^{n-1} + \cdots + p_1 x + p_0 \tag{1.70}$$

If the scalar variable is replaced by a square matrix of order n, the corresponding matrix polynomial is defined by

$$N(A) = p_n A^n + p_{n-1} A^{n-1} + \cdots + p_1 A + p_0 I_n \tag{1.71}$$

with the coefficient of p_0 being $A^0 = I_n$.

The polynomial N(x) may be written in factored form,

$$N(x) = p_n (x - \lambda_1)(x - \lambda_2) \cdots (x - \lambda_n) \tag{1.72}$$

where λ_1, λ_2, and λ_n are the n roots of the polynomial $N(x) = 0$, and assumed to be distinct. Similarly, the factored form of a matrix polynomial is*

$$N(A) = p_n (A - \lambda_1 I)(A - \lambda_2 I) \cdots (A - \lambda_n I) \tag{1.73}$$

(b) Infinite series of matrices

If the argument x of the infinite series S(x),

$$S(x) = a_0 + a_1 x + \cdots = \sum_{k=0}^{\infty} a_k x^k \tag{1.74}$$

is replaced by a square matrix of order n, the infinite series can be written as

$$S(A) = a_0 I_n + a_1 A + \cdots = \sum_{k=0}^{\infty} a_k A^k \tag{1.75}$$

It may be shown that this series converges as k approaches infinity if all the corresponding scalar series $S(\lambda_i)$, $i = 1, 2, \ldots, n$, converge, where the λ_i's are the eigenvalues of A.

Two of the important infinite series of square matrices are

Geometric series: $$G(A) = \sum_{k=0}^{\infty} a^k A^k$$

Exponential functions: $$e^A = \sum_{k=0}^{\infty} \frac{A^k}{k!}; \qquad e^{-A} = \sum_{k=0}^{\infty} (-1)^k \frac{A^k}{k!}$$

Although in scalar multiplication $e^x e^y = e^y e^x$, this equality does not apply to the corresponding matrix product $e^A e^B$ unless A and B commute; thus

*Considered in detail under "Sylvester's Theorem" in Sec. 1.4.2.

$$e^A e^B = e^B e^A = e^{A+B} \quad \text{if } AB = BA$$

Hence if $B = -A$, $e^A e^{-A} = I$ (i.e., $e^A = e^{-A}$).

(c) Cayley-Hamilton theorem

This theorem plays an important role in simplifying polynomials in an $n \times n$ matrix A.

Every square matrix satisfies its own characteristic equation. Then if A is a square matrix of order n, with characteristic polynomial

$$p(\lambda) = |A - \lambda I| = (-1)^n(\lambda^n + c_1\lambda^{n-1} + \cdots + c_{n-1}\lambda + c_n)$$

then

$$A^n + c_1A^{n-1} + \cdots + c_{n-1}A + c_n = 0 \tag{1.76}$$

1.4.2 Minimal Polynomial

Although every square matrix A satisfies its own characteristic equation, the characteristic equation is not, however, necessarily the scalar equation of least degree that the matrix A satisfies. The least-degree polynomial having the matrix A as a root is called the minimal polynomial; that is, the minimal polynomial of a square matrix A of order n is defined as the polynomial $\phi(\lambda)$ of least degree

$$\phi(\lambda) = \lambda^m + a_1\lambda^{m-1} + \cdots + a_m, \quad m < n$$

such that

$$\phi(A) = A^m + a_1A^{m-1} + \cdots + a_m = 0 \tag{1.77}$$

An n-square matrix A whose characteristic polynomial and minimal polynomial are identical is called nonderogatory; otherwise, derogatory.

By using the minimal polynomial $\phi(A)$ in (1.77), the inverse of a nonsingular square matrix A can be expressed as a polynomial in A with scalar coefficients:

$$\phi(A) = A^m + a_1A^{m-1} + \cdots + a_mI = 0$$

For a nonsingular max A, $a_m \neq 0$, that is,

$$I = -\frac{1}{a_m}(A^m + a_1 A^{m-1} + \cdots + a_{m-1}A)$$

or

$$A^{-1} = -\frac{1}{a_m}(A^{m-1} + a_1 A^{m-2} + \cdots + a_{m-1}I) \tag{1.78}$$

(a) Sylvester's theorem

This theorem is a useful method for obtaining a function of a matrix if the function can be expressed as a matrix polynomial.

Distinct eigenvalue case: If N(A) is a matrix polynomial in A, and if the square matrix A possesses n distinct eigenvalues, the polynomial N(A) can be written as

$$N(A) = \sum_{i=1}^{n} N(\lambda_i)Z_0(\lambda_i), \quad \text{where } Z_0(\lambda_i) = \frac{\prod_{\substack{j=1 \\ i \neq j}}^{n} (A - \lambda_i I)}{\prod_{\substack{j=1 \\ i \neq j}}^{n} (\lambda_i - \lambda_j)} \tag{1.79}$$

Repeated eigenvalue case: When the matrix A contains repeated eigenvalues, (1.79) is modified to a form called the confluent form of Sylvester's theorem. Assuming that the matrix A has repeated eigenvalue of order s, the contribution to N(A) from the ith root λ_i can be shown to be

$$\frac{1}{(s-1)!}\left[\frac{d^{s-1}}{d\lambda^{s-1}} \frac{N(\lambda)\,\mathrm{adj}\,(\lambda I - A)}{\prod_{\substack{j=1 \\ j \neq i}} (\lambda - \lambda_j)^s}\right]_{\lambda=\lambda_i} \tag{1.80}$$

The sum of the contributions of all the distinct eigenvalues is then N(A). Thus

$$N(A) = \sum_i \frac{1}{(s-1)!}\left[\frac{d^{s-1}}{d\lambda^{s-1}} \frac{N(\lambda)\,\mathrm{adj}\,(\lambda I - A)}{\prod_{\substack{j=1 \\ j \neq i}} (\lambda - \lambda_j)^s}\right]_{\lambda=\lambda_i} \tag{1.81}$$

where the summation is taken over all the roots, with repeated roots taken only once.

(b) Cayley-Hamilton technique

An alternative, and often simpler procedure for evaluating a function of a matrix is obtained by making use of the Cayley-Hamilton theorem. In the case where N(A) is a matrix polynomial that is of higher degree than the order of A, if $N(\lambda)$ is divided by the characteristic polynomial $[P(\lambda)]$ of A, then

$$\frac{N(\lambda)}{P(\lambda)} = Q(\lambda) + \frac{R(\lambda)}{P(\lambda)} \tag{1.82}$$

where $R(\lambda)$ is the remainder. If (1.82) is multiplied by $P(\lambda)$,

$$N(\lambda) = Q(\lambda)P(\lambda) + R(\lambda)$$

Correspondingly, since P(A) = 0 by the Cayley-Hamilton theorem,

$$N(A) = R(A) \tag{1.83}$$

The methods of deriving functions of a matrix described above are illustrated through the following examples.

Example 1.12 [6]: Derivation of characteristic and minimal polynomials of

$$A = \begin{bmatrix} 2 & 2 & 1 \\ 1 & 3 & 1 \\ 1 & 2 & 2 \end{bmatrix} \tag{1.84}$$

By direct expansion, $P(\lambda) = \det(\lambda I - A) = \lambda^2 - 7\lambda + 11\lambda - 5$.

Cayley-Hamilton theorem:

$$A^2 = \begin{bmatrix} 7 & 12 & 6 \\ 6 & 13 & 6 \\ 6 & 12 & 7 \end{bmatrix}; \quad A^3 = \begin{bmatrix} 32 & 62 & 31 \\ 31 & 63 & 31 \\ 31 & 62 & 32 \end{bmatrix}$$

and

$$\begin{bmatrix} 32 & 62 & 31 \\ 31 & 63 & 31 \\ 31 & 62 & 32 \end{bmatrix} - 7\begin{bmatrix} 7 & 12 & 6 \\ 6 & 13 & 6 \\ 6 & 12 & 7 \end{bmatrix} + 11\begin{bmatrix} 2 & 2 & 1 \\ 1 & 3 & 1 \\ 1 & 2 & 2 \end{bmatrix} - 5\begin{bmatrix} 1 & 0 & 0 \\ 0 & 1 & 0 \\ 0 & 0 & 1 \end{bmatrix} = 0$$

Derivation of minial polynomial—Method 1: The most elementary method of finding the minimal polynomial $\phi(\lambda)$ of $A \not\equiv 0$ involves the following stages:

1. If $A = a_0 I$, then $\phi(\lambda) = \lambda - a_0$ (for some a_0).
2. If $A \neq aI$ for all a but $A^2 = a_1 I + a_0 I$, then $\phi(\lambda) = \lambda^2 - a_1\lambda - a_0$.
3. If $A^2 \neq aA + bI$ for all a and b, but $A^3 = a_2 A^2 + a_1 A + a_0 I$, then $\phi(\lambda) = \lambda^3 - a_2\lambda^2 - a_1\lambda - a_0$, and so on.

For the matrix in (1.84), $A - a_0 I = 0$ is impossible, so the following identity is set up:

$$A^2 = \begin{bmatrix} 7 & 12 & 6 \\ 6 & 13 & 6 \\ 6 & 12 & 7 \end{bmatrix} = a_1 \begin{bmatrix} 2 & 2 & 1 \\ 1 & 3 & 1 \\ 1 & 2 & 2 \end{bmatrix} + a_0 \begin{bmatrix} 1 & 0 & 0 \\ 0 & 1 & 0 \\ 0 & 0 & 1 \end{bmatrix}$$

Using the first two elements of the first row of each matrix, we have

$$\left.\begin{aligned} 7 &= 2a_1 + a_0 \\ 12 &= 2a_1 \end{aligned}\right\} \rightarrow a_1 = 6, \quad a_0 = -5$$

After (and not before) checking for every element of A^2, it is seen that $A^2 = 6A - 5I$, and the required minimal polynomial is $\phi(\lambda) = \lambda^2 - 6\lambda + 5$.

Method 2: An alternative method of deriving $\phi(\lambda)$ is based on the following lemma.

<u>Lemma</u>. The minimal polynomial $\phi(\lambda)$ of an n-square matrix A is that similarity invariant polynomial $f_n(\lambda)$ of A which has the highest degree.

Proof of the lemma is not considered here. Only the basic expression for the computation of $\phi(\lambda)$ is given below, and illustrated through an example. From the definition of inverse of a matrix,

$$(\lambda I - A) \operatorname{adj} (\lambda I - A) = \det (\lambda I - A) I$$

Let $d_{n-1}(\lambda)$ denote the greatest common diviser of the (n - 1) square minors of $(\lambda I - A)$; then

$$\det (\lambda I - A) = P(\lambda) = d_{n-1}(\lambda) f_n(\lambda)$$

and

$$\operatorname{adj} (\lambda I - A) = d_{n-1}(\lambda) B(\lambda)$$

where the greatest common divisor of the elements of $B(\lambda)$ is 1. The minimal polynomial of A is then given by

$$\phi(\lambda) = \frac{P(\lambda)}{d_{n-1}(\lambda)} \tag{1.85}$$

From (1.84),

$$P(\lambda) = \det(\lambda I - A) = \begin{bmatrix} \lambda - 2 & -2 & -1 \\ -1 & \lambda - 3 & -1 \\ -1 & -2 & \lambda - 2 \end{bmatrix} = \lambda^3 - 7\lambda^2 + 11\lambda - 5$$

$$= (\lambda - 1)(\lambda^2 - 6\lambda + 5) = (\lambda - 1)f_n(\lambda)$$

$$\operatorname{adj}(\lambda I - A) = \begin{bmatrix} (\lambda - 4)(\lambda - 1) & 2(\lambda - 1) & \lambda - 1 \\ -(\lambda - 1) & (\lambda - 1)(\lambda - 3) & \lambda - 1 \\ (\lambda - 1) & -2(\lambda - 1) & (\lambda - 4)(\lambda - 1) \end{bmatrix} = (\lambda - 1)B(\lambda)$$

Therefore, the minimal polynomial of A is $\phi(\lambda) = f_n(\lambda) = \lambda^2 - 6\lambda + 5$.

Example 1.13 [5]: Function of a matrix. We now compute e^A and A^k by using Sylvester's theorem, where

$$A = \begin{bmatrix} 0 & 1 \\ -2 & -3 \end{bmatrix} \tag{1.86}$$

For the matrix above, $P(\lambda) = (\lambda + 1)(\lambda + 2)$. From (1.79),

(1) $$e^A = N(A) = \sum_{i=1}^{2} e^{\lambda_i} Z_0(\lambda_i) = \sum_{i=1}^{2} e^{\lambda_i} \frac{\prod_{\substack{j=1 \\ i \neq j}}^{2} (A - \lambda_j I)}{\prod_{\substack{j=1 \\ i \neq j}}^{2} (\lambda_i - \lambda_j)}$$

$$= e^{\lambda_1} \frac{A - \lambda_2 I}{\lambda_1 - \lambda_2} + e^{\lambda_2} \frac{A - \lambda_1 I}{\lambda_2 - \lambda_1} = e^{-1} \begin{bmatrix} 2 & 1 \\ -2 & -1 \end{bmatrix} + e^{-2} \begin{bmatrix} -1 & -1 \\ 2 & 2 \end{bmatrix}$$

$$= \begin{bmatrix} 2e^{-1} - e^{-2} & e^{-1} - e^{-2} \\ 2e^{-2} - 2e^{-1} & 2e^{-2} - e^{-1} \end{bmatrix}$$

$$(2)\qquad A^k = \sum_{i=1}^{2} (\lambda_i)^k Z_0(\lambda_i) = (-1)^k \begin{bmatrix} 2 & 1 \\ -2 & -1 \end{bmatrix} + (-2)^k \begin{bmatrix} -1 & -1 \\ 2 & 2 \end{bmatrix}$$

[$Z_0(\lambda_i)$ is the same as above.]

$$= \begin{bmatrix} 2(-1)^k - (-2)^k & (-1)^k - (-2)^k \\ 2[(-2)^k - (-1)^k] & 2(-2)^k - (-1)^k \end{bmatrix}$$

Example 1.14. Derivation of the general form of a matrix function N(A) for

$$A = \begin{bmatrix} 2 & 1 & 1 \\ 1 & 2 & 1 \\ 0 & 0 & 1 \end{bmatrix} \tag{1.87}$$

$$P(\lambda) = \det(\lambda I - A) = (\lambda - 1)^2(\lambda - 3); \quad \lambda_1, \lambda_2 = 1, \quad \lambda_3 = 3$$

The contribution to N(A) from the distinct eigenvalue at $\lambda = \lambda_3$ is given by [from (1.79)]

$$N(A)|_{\lambda=3} = \frac{N(3)\ \mathrm{adj}\,[\lambda I - A]|_{\lambda=3}}{(3-1)^2}$$

$$\mathrm{adj}\,[\lambda I - A] = \begin{bmatrix} (\lambda-2)(\lambda-1) & \lambda - 1 & \lambda - 1 \\ \lambda - 1 & (\lambda-2)(\lambda-1) & \lambda - 1 \\ 0 & 0 & (\lambda-1)(\lambda-3) \end{bmatrix}$$

$$\mathrm{adj}\,[\lambda I - A]|_{\lambda=3} = \begin{bmatrix} 2 & 2 & 2 \\ 2 & 2 & 2 \\ 0 & 0 & 0 \end{bmatrix}$$

$$N(A)|_{\lambda=3} = \frac{N(3)}{2} \begin{bmatrix} 1 & 1 & 1 \\ 1 & 1 & 1 \\ 0 & 0 & 0 \end{bmatrix}$$

Combining (1.79) and (1.80), the contribution to N(A) from the repeated eigenvalue at $\lambda = 1$ is given by

$$N(\lambda)Z_1(\lambda)|_{\lambda=1} + \frac{d}{d\lambda} N(\lambda)|_{\lambda=1} Z_0(\lambda)$$

where

$$Z_1(\lambda)|_{\lambda=1} = \frac{d}{d\lambda} \left. \frac{\text{adj}\,[\lambda I - A]}{\lambda - 3} \right|_{\lambda=1} = [-\text{adj}\,[\lambda I - A] - \frac{d}{d\lambda}\,\text{adj}\,[\lambda I - A]|_{\lambda=1}$$

$$= \frac{1}{2}\begin{bmatrix} 1 & -1 & -1 \\ -1 & 1 & -1 \\ 0 & 0 & 2 \end{bmatrix}$$

$$Z_0(\lambda)|_{\lambda=1} = -\text{adj}\,[\lambda I - A]|_{\lambda=1} = \begin{bmatrix} 0 & 0 & 0 \\ 0 & 0 & 0 \\ 0 & 0 & 0 \end{bmatrix}$$

The sum of these contributions is, therefore,

$$N(A) = \frac{N(3)}{2}\begin{bmatrix} 1 & 1 & 1 \\ 1 & 1 & 1 \\ 0 & 0 & 0 \end{bmatrix} + \frac{N(1)}{2}\begin{bmatrix} 1 & -1 & -1 \\ -1 & 1 & -1 \\ 0 & 0 & 2 \end{bmatrix} + \frac{dN(\lambda)}{d\lambda}\begin{bmatrix} 0 & 0 & 0 \\ 0 & 0 & 0 \\ 0 & 0 & 0 \end{bmatrix}$$

Let $N(A) = e^{At}$ or $N(\lambda) = e^{\lambda t}$; then

$$N(3) = e^{3\lambda}, \qquad N(1) = e^{t}, \qquad \left. \frac{dN(\lambda)}{d\lambda} \right|_{\lambda=1} = te^{t}$$

$$N(A) = e^{At} = \frac{1}{2}\begin{bmatrix} e^{3t} + e^{t} & e^{3t} - e^{t} & e^{3t} - e^{t} \\ e^{3t} - e^{t} & e^{3t} + e^{t} & e^{3t} - e^{t} \\ 0 & 0 & 2e^{t} \end{bmatrix}$$

<u>Example 1.15 [5]: Reduction of polynomials</u>. Here we derive $N(A) = A^4 + A^3 + A^2 + A + I$, where

$$A = \begin{bmatrix} 0 & 1 \\ -2 & -3 \end{bmatrix} \tag{1.88}$$

Cayley-Hamilton theorem: The characteristic equation of A is $\lambda^2 + 3\lambda + 2 = 0$. Therefore, $3A^2 + 3A + 2I = [0]$, or $A^2 = -3A - 2I$. Therefore,

$$A^3 = A[-3A - 2I] = -3A^2 - 2A = -3(-3A - 2I) - 2A = 7A + 6I$$

Similarly, $A^4 = (A^2)^2 = -15A - 14I$. Combining the above gives us

$$N(A) = -10A - 9I = \begin{bmatrix} -9 & -10 \\ 20 & 21 \end{bmatrix}$$

Cayley-Hamilton technique: The characteristic polynomial is $P(\lambda) = \lambda^2 + 3\lambda + 2$. The required polynomial is

$$N(\lambda) = \lambda^4 + \lambda^3 + \lambda^2 + \lambda + I$$

From (1.82),

$$\frac{N(\lambda)}{P(\lambda)} = \frac{\lambda^4 + \lambda^3 + \lambda^2 + \lambda + I}{\lambda^2 + 3\lambda + 2} = (\lambda^2 - 2\lambda + 5) + \frac{-10\lambda - 9}{\lambda^2 + 3\lambda + 2} \equiv Q(\lambda)P(\lambda) + R(\lambda)$$

The remainder, $R(\lambda) = -10\lambda - 9$. From (1.83), the required polynomial in A is

$$N(A) = R(A) = -10A - 9I = \begin{bmatrix} -9 & -10 \\ 20 & 21 \end{bmatrix}$$

1.5 POLYNOMIAL AND RATIONAL MATRICES [10-15]

Matrices considered so far contain elements that are real or imaginary constant numbers. Although such matrices are common in the analysis of linear systems, there are other forms that are useful in algebraic control theory and analysis. Two forms now being extensively used in multivariable system theory are polynomial and rational matrices; some basic definitions and properties associated with these matrices are given here.

1.5.1 Polynomials

The notion of a polynomial was introduced earlier through the definition of characteristic polynomial or characteristic equation. In the context of this section, polynomials may be viewed either in a polynomial form or as a polynomial function.

(a) Polynomial form

This is an expression given by

$$p(s) = p_n s^n + p_{n-1} s^{n-1} + \cdots + p_1 s + p_0 \tag{1.89}$$

in which p_k, $k \in 0, 1, \ldots, n$, belong to a real or complex field F, and s is indeterminate* such that $p_k s^k$ is defined whenever $p \in F$ and $ps = sp$.

If $p_n \neq 0$, p(s) is said to be a polynomial of degree n, and this is expressed as $\delta\{p(s)\} = n$; if $p_n = 1$, the polynomial is said to be monic. On the other hand, if every coefficient p_k is zero, p(s) is called a zero polynomial. Two polynomials p(s) and q(s) of the same degree, given by

$$p(s) = \sum_{k=0}^{n} p_k s^k \quad \text{and} \quad q(s) = \sum_{k=0}^{n} q_k s^k \tag{1.90}$$

are said to be equal if $p_k = q_k$, $\forall\, k \in 0, 1, \ldots, n$. For the two nonzero polynomials p(s) and q(s),

$$\delta\{p(s) + q(s)\} \leq \delta\{p(s)\} + \delta\{q(s)\} \tag{1.91}$$

$$\delta\{p(s) \cdot q(s)\} = \delta\{p(s)\} \cdot \delta\{q(s)\} \tag{1.92}$$

(b) Polynomial function

This function is an expression of the form

$$p(s) = p_n s^n + p_{n-1} s^{n-1} + \cdots + p_1 s + p_0$$

in which p_k, $k \in 0, 1, \ldots, n$, and s all belong to F; any member s_0 of F associates p(s) with another member $p(s_0)$ of F. The values of s for which a polynomial function p(s) takes the value of $0 \in F$ [i.e., $p(s) \equiv 0$] are called the roots or zeros of p(s). When a real or complex field F is infinite, no distinction is necessary between polynomial forms and polynomial functions. The ratio of two polynomial functions, with the degree of numerator not greater than the degree of denominator, is called a rational function. A rational function is said to be proper if the degree of the numerator polynomial is equal to the degree of the denominator polynomial.

If $p_1(s)$ and $p_2(s)$ are polynomials over a field F with $p_2(s) \neq 0$, then there exists two polynomials $p_3(s)$ and r(s), where either $\delta\{r(s)\} = 0$ or $< \delta\{p_1(s)\}$; and $\delta\{p_2(s)\} < \delta\{p_1(s)\}$ such that

$$p_3(s) = \frac{p_1(s)}{p_2(s)} + r(s) \tag{1.93}$$

If $r(s) \equiv 0$, then $p_2(s)$ is said to divide $p_1(s)$, sometimes denoted $p_1(s) | p_2(s)$.

*In subsequent sections, s is assumed to be a complex number ($s = \sigma + j\omega$) denoting the Laplace operator.

(c) Greatest common divisor

If p(s) and q(s) are two polynomials of arbitrary degree and d(s) is a third polynomial of highest degree that divides both p(s) and q(s), then d(s) is called the greatest common divisor (gcd) of p(s) and q(s). A consequence of this definition is that if there is another polynomial $d_1(s)$ that divides both p(s) and q(s), then $d_1(s)$ divides d(s).

If the gcd of p(s) and q(s) has degree zero [$\delta\{d(s)\} = 0$, i.e., d(s) = 1], the polynomials {p(s), q(s)} are said to be relatively prime (or co-prime).

1.5.2 Polynomial Matrix

A nonnull (m × n) matrix

$$P(s) = \{p_{ij}(s)\}, \quad i \in 1, \ldots, m; \; j \in 1, \ldots, n \tag{1.94}$$

is called a polynomial matrix if the elements $p_{ij}(s)$ are polynomials in s. Such a matrix P(s) can always be expressed as a matrix polynomial of degree α in s, where α is the maximum degree in s of the polynomial $p_{ij}(s)$ for all (i, j); that is,

$$P(s) = P_\alpha s^\alpha + P_{\alpha-1} s^{\alpha-1} + \cdots + P_1 s + P_0 \tag{1.95}$$

where P_k, $k \in 0, 1, \ldots, \alpha$, are constant matrices of order (m × n). The degree of P(s), $\delta\{P(s)\}$, is the degree of the polynomial or polynomials of highest degree comprising P(s). Many of the notions of (scalar) polynomials in the preceding section can be extended to polynomial matrices; some of these are considered here.

If P(s) is an n-space polynomial matrix, it is said to be

1. Nonsingular, or proper, if $\det\{P(s)\} \neq 0$
2. Singular, or improper, if $\det\{P(s)\} \equiv 0, \; \forall s$
3. Unimodular if $\det\{P(s)\}$ is independent of s (i.e., constant or scalar)

The normal rank of a polynomial matrix P(s) is equal to the dimension of the largest minor of P(s) whose determinant is a nonzero polynomial.* The inverse of P(s) is denoted by $[P(s)]^{-1}$ or $P^{-1}(s)$, and is given by

$$[P(s)]^{-1} = \frac{\text{adj}\,(s)}{\det\{P(s)\}} \tag{1.96}$$

*For a finite number of specific values of s, the rank of P(s) may be less than the normal rank; such an exceptional value is called the local rank of P(s) for those specific values of s.

Therefore, the polynomial matrices whose inverses are polynomial matrices are unimodular matrices.

(a) Elementary operations

These are defined along the lines of Sec. 1.3 except that the indeterminate s may be treated as a constant in elementary operations of polynomial matrices. For $P(s) = \{p_{ij}(s)\}$:

1. Interchange of the ith and jth rows (columns)
2. Multiplication of any ith row (column) by a nonzero scalar
3. Replacement of any ith row (column) by itself plus or minus any polynomial q(s) multiplied by any other jth row (column)

With these elementary row or column operations and the definition of unimodular matrix above, it is clear that a unimodular matrix U(s) may be obtained from an identity matrix by a finite number of elementary operations. Furthermore, any sequence of elementary row operations in P(s) is equivalent to pre (left)-multiplication by a unimodular matrix $U_\ell(s)$, and any sequence of elementary column operations in P(s) is equivalent to post (right)-multiplication by a unimodular matrix $U_r(s)$.

(b) Remainder theorem

If P(s) and Q(s) are two n-square polynomial matrices, there exist two pairs of unique matrix polynomials $\{U(s),\ R_1(s)\}$ and $\{V(s),\ R_2(s)\}$ such that

$$\begin{aligned} P(s) &= U(s)Q(s) + R_1(s) \\ P(s) &= Q(s)V(s) + R_2(s) \end{aligned} \tag{1.97}$$

where U(s) and V(s) are unimodular matrices and $R_1(s)$ and $R_2(s)$ are either null or of degree less than that of Q(s); the following definitions are given:

(1) If $R_1(s) \equiv 0$ $\begin{cases} Q(s) \text{ is a right divisor of } P(s) \\ P(s) \text{ is a left multiplier of } Q(s) \end{cases}$

(2) If $R_2(s) \equiv 0$ $\begin{cases} Q(s) \text{ is a left divisor of } P(s) \\ P(s) \text{ is a right multiplier of } Q(s) \end{cases}$

The notations left divisor and right divisor in matrix polynomials stem from the fact that post- and premultiplication of matrices do not give the same results. The significance of the definitions above is that Q(s) is a right (left) divisor of P(s) if post (pre)-multiplication of P(s) by $Q^{-1}(s)$ yields a unimodular matrix (multiplication by an inverse is division, and post ≡ right, pre ≡ left in this context).

(c) Row and column equivalents and primeness

Two polynomial matrices P(s) and Q(s) are said to be

1. Row equivalent If P(s) may be obtained from Q(s) through a series of row operations, that is, if

$$P(s) = U_\ell(s)Q(s)$$

2. Column equivalent If P(s) may be obtained from Q(s) through a series of column operations, that is, if

$$P(s) = Q(s)U_r(s)$$

3. Row and column equivalent Equivalent if

$$P(s) = U_\ell(s)Q(s)U_r(s)$$

that is, if P(s) may be obtained from Q(s) through a series of elementary row and column operations

where $U_\ell(s)$ and $U_r(s)$ are unimodular matrices of appropriate dimensions. The two polynomial matrices P(s) and Q(s) are said to be right co-prime or relatively right prime if there exists two polynomial matrices $B_1(s)$ and $B_2(s)$ such that

$$B_1(s)P(s) + B_2(s)Q(s) = I_{n,n} \tag{1.98}$$

where the dimensions of various matrices are P(s), $n \times n$; Q(s), $m \times n$; $B_1(s)$, $n \times n$; $B_2(s)$, $n \times m$.

Similarly, $B_1(s)$ and $B_2(s)$ are said to be left co-prime or relatively left-prime if there exist two polynomials P(s) and Q(s) for which (1.98) holds.

(d) Greatest common divisors

A polynomial matrix $Q_r(s)$ is said to be the greatest common right divisor (gcrd) of two polynomial matrices $\{P_1(s), P_2(s)\}$ with the same number of columns if there exist two polynomial matrices $\{R_1(s), R_2(s)\}$ such that

$$P_1(s) = R_1(s)Q_r(s) \quad \text{and} \quad P_2(s) = R_2(s)Q_r(s) \tag{1.99}$$

Gcrds are not unique, but if $Q_r^1(s)$ is another right divisor of $\{P_1(s), P_2(s)\}$, then $Q_r^1(s)$ is a right divisor of $Q_r(s)$; that is, there exists a polynomial matrix $R_3(s)$ such that

$$Q_r(s) = R_3(s)Q_r^1(s)$$

In a similar way, $Q_\ell(s)$ is said to be the greatest common left divisor (gcld) of $\{P_1(s), P_2(s)\}$ if there are two polynomial matrices $\{L_1(s), L_2(s)\}$ for which

$$P_1(s) = Q_\ell(s)L_1(s) \quad \text{and} \quad P_2(s) = Q_\ell(s)L_2(s) \tag{1.100}$$

Gclds are not unique, but if $Q_\ell^1(s)$ is another left divisor of $\{P_1(s), P_2(s)\}$, then there is a polynomial matrix $L_3(s)$ such that

$$Q_\ell(s) = Q_\ell^1(s)L_3(s)$$

Thus if one gcrd (gcld) is nonsingular, so are all other gcrd's (gcld's), and they can only differ by a unimodular left (right) factor. Furthermore, if one gcrd (gcld) is unimodular, all other gcrd's (gcld's) are unimodular. These lead to the following property: Every pair of polynomial matrices $\{P_1(s), P_2(s)\}$ with the same number of columns (rows) has a gcrd (gcld) $Q_r(s)$ $[Q_\ell(s)]$ such that

$$\begin{aligned} \text{gcrd:} \quad Q_r(s) &= R_1(s)P_1(s) + R_2(s)P_2(s) \\ \text{gcld:} \quad Q_\ell(s) &= P_1(s)L_1(s) + P_2(s)L_2(s) \end{aligned} \tag{1.101}$$

Comparison of (1.98) and (1.101) leads to the alternative definition of primeness.* Two polynomial matrices with the same number of columns (rows) are said to be relatively right (left) prime if all their gcrd's (gcld's) are unimodular.

Thus, from (1.101), $\{P_1(s), P_2(s)\}$ are relatively right (left) prime if there exist two polynomial matrices $Q_1(s)$ and $Q_2(s)$ such that

$$\begin{aligned} \text{Right:} \quad & Q_1(s)P_1(s) + Q_2(s)P_2(s) = I \\ \text{Left:} \quad & P_1(s)Q_1(s) + P_2(s)Q_2(s) = I \end{aligned}$$

Derivation of the gcrd of two given polynomial matrices $\{P_1(s), P_2(s)\}$ is based on the property that if $P_1(s)$ is of order $n \times n$ and $P_2(s)$ of order $m \times n$, there exists a unimodular matrix $U_\ell(s)$ of order $(m + n) \times (m + n)$ (equivalent to a finite number of row operations) for which

$$U_\ell(s) \underbrace{\begin{bmatrix} P_1(s) \\ \hline P_2(s) \end{bmatrix}}_{n} \begin{matrix} \}\, n \\ \}\, m \end{matrix} = \underbrace{\begin{bmatrix} Q_r(s) \\ \hline 0 \end{bmatrix}}_{n} \begin{matrix} \}\, n \\ \}\, m \end{matrix} \tag{1.102}$$

*Also known as <u>relatively prime</u> or <u>co-prime</u>.

In a similar manner, for a gcld, there exists a unimodular matrix $U_r(s)$ of order $n \times n$ (equivalent to a finite number of column operations) such that

$$\begin{matrix} n\{ \\ m\{ \end{matrix} \underbrace{\begin{bmatrix} P_1(s) \\ ---- \\ P_2(s) \end{bmatrix}}_{n} U_r(s) = \underbrace{\begin{bmatrix} Q_\ell(s) \\ ---- \\ 0 \end{bmatrix}}_{n} \begin{matrix} \}n \\ \}m \end{matrix} \tag{1.103}$$

A consequence of the equations above is that the gcrd (gcld) of two polynomial matrices has the same number of rows (columns).

(e) Row and column proper

Let d_{cj}, $j \in 1, \ldots, n$, be the degree of the jth column of the n-square nonsingular polynomial matrix $P(s) = \{p_{ij}(s)\}$. Then P(s) may be expressed as

$$P(s) = \begin{bmatrix} p_{11}s^{d_{c1}} + \cdots & \cdots & p_{1j}s^{d_{cj}} + \cdots & \cdots & p_{1n}s^{d_{cn}} + \cdots \\ p_{i1}s^{d_{c1}} + \cdots & \cdots & p_{ij}s^{d_{cj}} + \cdots & \cdots & p_{in}s^{d_{cn}} + \cdots \\ \\ p_{n1}s^{d_{c1}} + \cdots & \cdots & \underbrace{p_{nj}s^{d_{cj}} + \cdots}_{\text{ith column}} & \cdots & p_{nn}s^{d_{cn}} + \cdots \end{bmatrix} \begin{matrix} \\ \} \text{ ith row} \\ \\ \\ \end{matrix} \tag{1.104}$$

$$\equiv C_{hc}\Lambda_c(s) + L_c(s) \tag{1.105}$$

where $\Lambda_c(s) = \text{diag}\,\{s^{d_{c1}}, \ldots, s^{d_{cj}}, \ldots, s^{d_{cn}}\}$

C_{hc} = highest-column-degree-coefficient matrix or the leading column coefficient matrix of P(s)

$L_c(s)$ = polynomial matrix containing elements whose degrees are less than those of P(s)

Example 1.16 [12]. Following is an example of (1.105):

$$\underbrace{\begin{bmatrix} s^3+s & s+2 \\ s^2+s+1 & 1 \end{bmatrix}}_{P(s)} = \underbrace{\begin{bmatrix} 1 & 1 \\ 0 & 0 \end{bmatrix}}_{C_{hc}} \underbrace{\begin{bmatrix} s^3 & 0 \\ 0 & s \end{bmatrix}}_{\Lambda(s)} + \underbrace{\begin{bmatrix} s & 2 \\ s^2+s+1 & 1 \end{bmatrix}}_{L_c(s)} \qquad (1.106)$$

With the notation of (1.105) the nonsingular matrix P(s) is said to be <u>column proper</u> or <u>column reduced</u> if C_{hc} is nonsingular. Since det $\{C_{hc}\}$ in the example above is zero, P(s) given by (1.106) is not column proper or column reduced.

In a similar manner, by defining d_{ri} as the degree of the ith row of the n-square nonsingular matrix P(s), it may be expressed as

$$P(s) = \Lambda_r(s)C_{hr} + L_r(s) \qquad (1.107)$$

where C_{hr} is the highest-row-degree coefficient matrix or the leading row coefficient matrix of P(s), $\Lambda_r(s) = \text{diag}\{s^{d_{r1}}, \ldots, s^{d_{ri}}, \ldots, s^{d_{rn}}\}$ and $L_r(s)$ denotes the remaining terms. P(s) is then said to be <u>row proper</u> or <u>row reduced</u> if C_{hr} is nonsingular, where, from (1.105) and (1.107), $C_{hc} = [C_{hr}]^T$. It can easily be verified that

$$\det[P(s)] = \det[C_{hc}] \cdot s^d + \text{lower-order terms in } s \qquad (1.108)$$

where

$$\begin{aligned} d &= d_{c1} + d_{c2} + \cdots + d_{cn} \\ &= d_{r1} + d_{r2} + \cdots + d_{rn} \end{aligned}$$

Therefore, in polynomial notation,

$$\delta\{\det[P(s)]\} = \begin{cases} \delta\{\det[\Lambda_c(s)]\} = \sum_{j=1}^{n} d_{cj} \begin{cases} = d & \text{if P(s) is column proper} \\ < d & \text{otherwise} \end{cases} \\ \delta\{\det[\Lambda_r(s)]\} = \sum_{i=1}^{n} d_{ci} \begin{cases} = d & \text{if P(s) is row proper} \\ < d & \text{otherwise} \end{cases} \end{cases}$$

For the example in (1.106), $d_{c1} = 3$, $d_{c2} = 1$, and $\delta\{\det[P(s)]\} = 5 < d_{c1} + d_{c2}$, since C_{hc} is singular.

The preceding definitions and relationships lead to the property that any square nonsingular polynomial matrix may be reduced to a column (row)-proper matrix, or a column (row)-reduced matrix, by a sequence of elementary column (row) operations. The derivation of this property is not considered here, but is illustrated through the following example.

Example 1.17 [12]: Column reduction of P(s)

$$P(s) = \begin{bmatrix} (s+1)^2(s+2)^2 & -(s+1)^2(s+2) \\ 0 & s+2 \end{bmatrix} \tag{1.109a}$$

$$P(s) = \underbrace{\begin{bmatrix} 1 & 1 \\ 0 & 0 \end{bmatrix}}_{C_{hc}} \underbrace{\begin{bmatrix} s^4 & 0 \\ 0 & s^3 \end{bmatrix}}_{\Lambda(s)} + \underbrace{\begin{bmatrix} 6s^3 + 13s^2 + 12s + 4 & -(4s^2 + 5s + 2) \\ 0 & s+2 \end{bmatrix}}_{L_c(s)}$$

$$P(s)\begin{bmatrix} 1 & 0 \\ s & 1 \end{bmatrix} = \begin{bmatrix} 1 & 1 \\ 0 & 0 \end{bmatrix}\begin{bmatrix} 0 & -s^3 \\ 0 & 0 \end{bmatrix} + \begin{bmatrix} 2s^3 + 8s^2 + 10s + 4 & -(4s^2 + 5s + 2) \\ s(s+2) & s+2 \end{bmatrix}$$

$$P(s)\begin{bmatrix} 1 & 0 \\ s & 1 \end{bmatrix}\begin{bmatrix} 1 & 0 \\ 2 & -1 \end{bmatrix} = \begin{bmatrix} 1 & 1 \\ 0 & 0 \end{bmatrix}\begin{bmatrix} -2s^3 & s^3 \\ 0 & 0 \end{bmatrix} + \begin{bmatrix} 2s^3 & 4s^2 + 5s + 2 \\ s^2 + 4s + 4 & -(s+2) \end{bmatrix}$$

$$= \begin{bmatrix} 0 & s^3 + 4s^2 + 5s + 2 \\ s^2 + 4s + 4 & -(s+2) \end{bmatrix}$$

$$\equiv \underbrace{\begin{bmatrix} 0 & 1 \\ 1 & 0 \end{bmatrix}}_{C'_{hc}} \underbrace{\begin{bmatrix} s^2 & 0 \\ 0 & s^3 \end{bmatrix}}_{\Lambda'_c(s)} + \underbrace{\begin{bmatrix} 0 & 4s^2 + 5s + 2 \\ 4s + 4 & -(s+2) \end{bmatrix}}_{L'_c(s)}$$

$$= P_c(s)$$

$$P_c(s) = P(s) \underbrace{\begin{bmatrix} 1 & 0 \\ s+2 & -1 \end{bmatrix}}_{U_r(s)} \rightarrow \text{unimodular} \tag{1.109b}$$

which is column proper. Another unimodular matrix $U_\ell(s)$ may be shown to exist such that $P_r(s) = U_\ell(s)P(s)$ is row proper.

(f) Smith normal form

The transformation to column proper or row proper matrices considered above do not apply to nonsquare matrices. When P(s) is an $m \times \ell$ polynomial matrix of rank r, it may be transformed through a series of elementary row and column operations using only polynomials into a matrix S(s) of the form

$$S(s) = \left[\begin{array}{c|c} S^*(s) & 0 \\ \hline \underbrace{0}_{r} & \underbrace{0}_{\ell - r} \end{array}\right]\begin{array}{l} \}\, r \\ \}\, m - r \end{array} \quad \text{where} \quad S^*(s) = \begin{bmatrix} p_1(s) & & 0 \\ & \ddots & \\ 0 & & p_r(s) \end{bmatrix} \tag{1.110}$$

where $p_i(s)$, $i \in 1, \ldots, r$, are unique monic polynomials in s having the property that $p_i(s)$ divides $p_{i+1}(s)$ without remainder. S(s) is known as the <u>Smith normal form</u> or <u>Smith form</u> of P(s).

The sequence of row and column operations involved in this case may, as before, be represented by post (right)- and pre (left)-multiplication of P(s) by appropriate unimodular matrices so that

$$S(s) = U_\ell(s)P(s)U_r(s) \tag{1.111}$$

If $g_i(s)$ is defined as the greatest common divisor (gcd) of all $i \times i$ minors of P(s), it can be shown from (1.111) that

$$p_i(s) = \frac{g_i(s)}{g_{i-1}(s)} \qquad \text{with } g_0(s) = 1,\ i \in 1, \ldots, r$$

$\{g_i(s)\}$ are called the <u>determinantal divisors</u> of P(s) and the $p_i(s)$ are called the <u>invariant polynomials</u> or <u>invariant factors</u> of P(s), $i \in 1, 2, \ldots, r$. The derivation of the Smith form of P(s) is illustrated through the following example.

<u>Example 1.18 [11]: Computation of Smith form of</u>

$$P(s) = \begin{bmatrix} s + 2 & s + 1 & s + 3 \\ s(s + 1)^2 & s(s^2 + s + 1) & s(s + 1)(2s + 1) \\ (s + 1)(s + 2) & (s + 1)^2 & 3(s + 1)^2 \end{bmatrix} \tag{1.112}$$

The transformation is based on the property that

$$p_i(s) = \frac{g_i(s)}{g_{i-1}(s)} \quad \text{and hence} \quad g_i(s) = \prod_{k=1}^{i} p_k(s), \quad i \in 1, 2, \ldots, r$$

1. The gcd of the first-order minors (element) of P(s) is $g_1(s) = 1$; therefore, $p_1(s) = g_1(s)/g_0(s) = 1$. The next step is to bring this $p_1(s)$ in the leading element of P(s) through appropriate row and column operations. This is achieved by subtracting column 2 from column 1, giving

$$P_1(s) = \begin{bmatrix} 1 & s+1 \leftarrow & s+3 \leftarrow \\ s^2 \leftarrow & s^3+s^2+s & 2s^3+3s^2+s \\ s+1 \leftarrow & s^2+2s+1 & 3s^2+6s+3 \end{bmatrix}$$

$P_1(s)$ is now to be transformed into a form where the elements marked with $\leftarrow$ are zeros. This is obtained by the following sequence of elementary operations:

Rows	Columns
-	C1 = C1 - C2
R2 = R2 - s^2R1	-
R3 = R3 - (s + 1)R1	-
-	C3 = C3 - (s + 1)C1

The resulting equivalent form is

$$P_2(s) = \begin{bmatrix} 1 & 0 & 0 \\ 0 & s & s^3+s \leftarrow \\ 0 & 0 & 2s^2+2s \end{bmatrix}$$

2. The gcd of the second-order minors of the matrix above is $g_2(s) = s$, and hence $p_2(s) = g_2(s)/g_1(s) = s$; column operations to obtain zero in the position marked $\leftarrow$ gives the following transformed (equivalent) matrix:

$$P_3(s) = \begin{bmatrix} 1 & 0 & 0 \\ 0 & s & 0 \\ 0 & 0 & 2s(s+1) \end{bmatrix}$$

3. The gcd of the third-order minors, by inspection, is $2s^2(s + 1)$, but as a gcd is monic by definition, this is to be divided by 2, giving $g_3(s) = s^2(s + 1)$, and hence $p_3(s) = g_3(s)/g_2(s) = s(s + 1)$.

The Smith form of P(s) is thus

$$S(s) = \begin{bmatrix} p_1(s) & & 0 \\ & p_2(s) & \\ 0 & & p_3(s) \end{bmatrix} = \begin{bmatrix} 1 & 0 & 0 \\ 0 & s & 0 \\ 0 & 0 & s(s+1) \end{bmatrix}$$

$$\equiv U_\ell(s)P(s)U_r(s)$$

1.5.3 Rational Matrices

A matrix each of whose elements is a rational function (i.e., ratio of two polynomial functions) is called a rational matrix. An $(m \times \ell)$ rational transfer function R(s) may thus be expressed as

$$R(s) = \{r_{ij}(s)\} = \left\{ \frac{n_{ij}(s)}{d_{ij}(s)} \right\}, \quad i \in 1, \ldots, m; \; j \in 1, \ldots, \ell$$

that is,

$$\{n_{ij}(s)\} = \{r_{ij}(s)\}\{d_{ij}(s)\} \tag{1.113}$$

when $n_{ij}(s)$ and $d_{ij}(s)$ are the numerator and denominator polynomials of the (i, j)th element of R(s). Any element $r_{ij}(s)$ is said to be proper (strictly proper) if all elements $n_{ij}(s)$ have degree less than or equal to (less than) that of the highest-degree polynomial in the jth column of $\{d_{ij}(s)\}$. Since (1.113) may be expressed in matrix form as*

$$N(s) = R(s)D(s) \tag{1.114}$$

a consequence of $r_{ij}(s)$ being proper (strictly proper) is that every column of N(s) has degree less than or equal to (strictly less than) that of the corresponding column of D(s). The physical significance of these definitions is that a rational transfer function matrix R(s) is

1. Proper if $\lim_{s\to\infty} R(s) < \infty$ (finite)
2. Strictly proper if $\lim_{s\to\infty} R(s) \to 0$
3. Nonproper if neither (1) nor (2) applies

Every rational matrix R(s) may thus be expressed as

$$R(s) = N(s)[D(s)]^{-1} = R_{sp}(s) + P(s)$$

*Another form of (1.114) is $R(s) = [D_1(s)]^{-1}N(s)$ as in (1.121). R(s) need not be square, but $D_1(s)$ or D(s) must be.

where $R_{sp}(s)$ is a strictly proper rational matrix and P(s) is a polynomial matrix where the degree of any jth column of P(s) is less than the degree of the jth column of D(s).

(a) Smith-McMillan form

If R(s) is an $(m \times \ell)$ rational function matrix of rank r, and d(s) the monic least common denominator of all its elements, then from (1.113), R(s) may be expressed as

$$R(s) = \frac{\{n_{ij}(s)\}}{d(s)} \equiv \frac{N(s)}{d(s)} \tag{1.115}$$

where N(s) is an $(m \times \ell)$ polynomial matrix which may be reduced to Smith form by elementary operations in (1.111):

$$S(s) = U_\ell(s)N(s)U_r(s) \tag{1.116}$$

where $U_\ell(s)$ and $U_r(s)$ are appropriate unimodular matrices. A rational matrix S(s)/d(s) may now be formed and any common factors between the numerators and denominators in the elements of the leading diagonal canceled. The resultant matrix is called the Smith-McMillan form of R(s), denoted by M(s):

$$\begin{aligned} M(s) &= \frac{S(s)}{d(s)} \equiv \frac{1}{d(s)} U_\ell(s)N(s)U_r(s) \\ &= U_\ell(s)\frac{N(s)}{d(s)}U_r(s) \\ &= U_\ell(s)R(s)U_r(s) \end{aligned} \tag{1.117}$$

and has the form

$$M(s) = \left[\begin{array}{c|c} M^*(s) & 0 \\ \hline \underbrace{0}_{r} & \underbrace{0}_{\ell - r} \end{array}\right]\begin{array}{l} \}\,r \\ \}\,m-r \end{array}, \quad \text{where } M^*(s) = \begin{bmatrix} \dfrac{\psi_1(s)}{\phi_1(s)} & 0 \\ 0 & \dfrac{\psi_r(s)}{\phi_r(s)} \end{bmatrix}$$

with $\psi_i(s)$ dividing $\psi_{i+1}(s)$ and $\phi_j(s)$ dividing $\phi_{j-1}(s)$ without remainder,* $\forall\, i,\ j \in 1, \ldots, r$. M(s) is said to be proper if $\psi_i(s)/\phi_i(s) \to 0$ as $s \to \infty$,

*As a memonic rule the numerators divide up and the denominators divide down.

$\forall$ i. Derivation of the Smith-McMillan form of R(s) follows from the argument that since d(s) is the monic least common multiplier (lcm) of the denominators of the elements of R(s), N(s) = d(s)R(s) is a polynomial matrix; thus, by (1.111), appropriate elementary row and column operations will yield

$$U_\ell(s)R(s)U_r(s) = \frac{S(s)}{d(s)} = \mathrm{diag}\left\{\frac{p_i(s)}{d(s)}\right\} \tag{1.118}$$

If after cancellation of common factors any ith element reduces to

$$\frac{p_i(s)}{d(s)} \equiv \frac{\psi_i(s)\,\alpha_i(s)}{\phi_i(s)\,\alpha_i(s)} = \frac{\psi_i(s)}{\phi_i(s)}, \qquad i \in 1, \ldots, r$$

where $\psi_i(s)$ and $\phi_i(s)$ are relatively prime, R(s) may be rewritten as

$$R(s) = U_\ell^{-1}(s)M(s)U_r^{-1}(s) \tag{1.119}$$

where

$$M(s) = \left[\begin{array}{c|c} \mathrm{diag}\left\{\dfrac{\psi_i(s)}{\phi_i(s)}\right\} & 0 \\ \hline \underbrace{\quad 0 \quad}_{r} & \underbrace{0}_{\ell - r} \end{array}\right] \begin{array}{l} \left.\vphantom{\dfrac{\psi_i(s)}{\phi_i(s)}}\right\} r \\ \} m - r \end{array}$$

Since d(s) and all the elements of N(s) are assumed to have no common factors, it can easily be shown that $\phi_1(s) = d(s)$. The method of deriving the Smith-McMillan form is illustrated below.

Example 1.19 [12]: Computation of the Smith-McMillan form of

$$R(s) = \begin{bmatrix} \dfrac{s}{(s+1)^2(s+2)^2} & \dfrac{s}{(s+2)^2} \\ -\dfrac{s}{(s+2)^2} & -\dfrac{s}{(s+2)^2} \end{bmatrix}$$

$$\equiv \frac{1}{(s+1)^2(s+2)^2}\begin{bmatrix} s & s(s+1)^2 \\ -s(s+1)^2 & -s(s+1)^2 \end{bmatrix} \equiv \frac{N(s)}{d(s)} \tag{1.120}$$

The transformation of N(s) into Smith form follows the same procedure as Example 1.18, and is

$$N(s) = \underbrace{\begin{bmatrix} 1 & 0 \\ -(s+1)^2 & 1 \end{bmatrix}}_{U_\ell(s)} \underbrace{\begin{bmatrix} s & 0 \\ 0 & s^2(s+1)^2(s+2) \end{bmatrix}}_{S(s)} \underbrace{\begin{bmatrix} 1 & (s+1)^2 \\ 0 & 1 \end{bmatrix}}_{U_r(s)}$$

Thus

$$M(s) = \frac{S(s)}{d(s)} = \begin{bmatrix} \dfrac{s}{(s+1)^2(s+2)^2} & 0 \\ 0 & \dfrac{s^2}{s+2} \end{bmatrix}$$

It is to be noted that even though R(s) is strictly proper, its Smith-McMillan form may not be so; this is due to the structure of unimodular matrices $U_\ell(s)$ and $U_r(s)$, which do not appear in M(s).

(b) Matrix-fraction description

If R(s) is an $(m \times \ell)$ rational matrix and d(s) is the lcm of all its elements, then from (1.113) R(s) may be expressed in two forms:

$$R(s) = \begin{cases} N(s)\, D_r^{-1}(s) \\ D_\ell^{-1}(s) N(s) \end{cases} \tag{1.121}$$

$N(s)D_r^{-1}(s)$ is called the right matrix-fraction description of R(s) and $D_\ell^{-1}(s)N(s)$ is the left matrix-fraction description of R(s), $D_r(s)$ and $D_\ell(s)$ being the right and left denominator matrices. Since in the descriptions above, a large number of combinations of the elements of $D_r(s)$ and $D_\ell(s)$ exist, there may be many different right and left matrix-fraction descriptions of any rational polynomial matrix with varying degrees of det $\{D_r(s)\}$ and det $\{D_\ell(s)\}$. This is illustrated through the following example.

Example 1.20 [12]: Matrix-fraction description of R(s) given by (1.120)

Right matrix-fraction description: $R(s) = N_r(s)D_r^{-1}(s)$.

(1) $R(s) = N_{r1}(s)D_{r1}^{-1}(s)$, where

$$N_{r1}(s) = \begin{bmatrix} s & s(s+1)^2 \\ -s(s+1)^2 & -s(s+1)^2 \end{bmatrix}; \quad D_{r1}(s) = \begin{bmatrix} (s+1)^2(s+2)^2 & 0 \\ 0 & (s+1)^2(s+2)^2 \end{bmatrix}$$

$$\delta\{\det[D_{r1}(s)]\} = 8$$

(2) $R(s) = N_{r2}(s)D_{r2}^{-1}(s)$, where

$$N_{r2}(s) = \begin{bmatrix} s & 0 \\ -s(s+1)^2 & s^2 \end{bmatrix}; \quad D_{r2}(s) = \begin{bmatrix} (s+1)^2(s+2)^2 & (s+1)^2(s+2) \\ 0 & s+2 \end{bmatrix}$$

$$\delta\{\det[D_{r2}(s)]\} = 5$$

Left matrix-fraction description: $R(s) = D_\ell^{-1}(s)N_\ell(s)$.

(3) $R(s) = D_{\ell 1}^{-1}(s)N_{\ell 1}(s)$, where

$$N_{\ell 1}(s) = \begin{bmatrix} s & s(s+1)^2 \\ -s & -s \end{bmatrix}; \quad D_{\ell 1}(s) = \begin{bmatrix} (s+1)^2(s+2)^2 & 0 \\ 0 & s+2 \end{bmatrix}$$

$$\delta\{\det[D_{\ell 1}(s)]\} = 6$$

The degree of the matrix-fraction description of R(s) is equal to the minimal degree of the determinant of its denominator matrix. As the example above shows, the degree of R(s) may not always be obtained through a direct derivation using (1.121); this leads to the concept of irreducible matrix-fraction description.

Let T(s) be a nonsingular matrix such that

$$\begin{aligned} R(s) &= N_r(s)D_r^{-1}(s) = [N_1(s)T(s)][D_1(s)T(s)]^{-1} \\ &= N_1(s)D_1^{-1}(s) \end{aligned} \tag{1.122}$$

where $N(s) = N_1(s)T(s)$ and $D_1(s) = D_1(s)T(s)$, and by definition in (1.97) T(s) is the right divisor of N(s) and D(s). Consequently,

$$\delta\{\det[D(s)]\} = \delta\{\det[D_1(s)]\} + \delta\{\det[T(s)]\} \geq \delta\{\det[D_1(s)]\} \tag{1.123}$$

and therefore the degree of the denominator matrix D(s) may be reduced by eliminating the right divisor of the numerator and denominator polynomials of R(s). The minimum degree of the matrix-fraction description of R(s) may thus be obtained by removing the gcrd of $\{N_r(s), D_r(s)\}$ [or gcld of $\{N_\ell(s), D_\ell(s)\}$], yielding

$$\delta\{\det[D(s)]\} = \delta\{\det[D_1(s)]\} \tag{1.124}$$

for all nonsingular right divisors T(s) of $\{N_r(s), D_r(s)\}$. From (1.123) and the definitions in Sec. 1.5.2(b), (1.124) holds good if T(s) is unimodular. The consequences of the derivations above are:

1. If $N_r(s)$ and $D_r(s)$ are relatively right prime, the matrix-fraction description $R(s) = N_r(s)D_r^{-1}(s)$ is irreducible.
2. There is no unique matrix fraction description of R(s) since if $N_r(s)D_r^{-1}(s)$, is a matrix-fraction description of R(s), then for any unimodular T(s), so is $N_r(s)T(s)[D_r(s)T(s)]^{-1}$. (This is due to the fact that gcrd's are not unique.)

The method of reducing the order of R(s) through the elimination of gcrd of N(s) and D(s) is illustrated below.

Example 1.21: Irreducible realization of

$$R(s) = \underbrace{\begin{bmatrix} (s+2)^2 & (s+2)(s+1) \\ (s+1)(s+2) & (s+1)^2(s+3) \end{bmatrix}}_{N(s)} \underbrace{\begin{bmatrix} (s+1)^2(s+2) & 0 \\ 0 & (s+1)(s+2)^2 \end{bmatrix}}_{D(s)}^{-1} \tag{1.125}$$

In the given matrix-fraction description, $\delta\{D(s)\} = 6$.

$$N(s) \equiv \underbrace{\begin{bmatrix} s+2 & s+2 \\ s+1 & (s+1)(s+3) \end{bmatrix}}_{N_1(s)} \underbrace{\begin{bmatrix} s+2 & 0 \\ 0 & s+1 \end{bmatrix}}_{T(s)}$$

$$D(s) = \underbrace{\begin{bmatrix} (s+1)^2 & 0 \\ 0 & (s+2)^2 \end{bmatrix}}_{D_1(s)} \underbrace{\begin{bmatrix} s+2 & 0 \\ 0 & s+1 \end{bmatrix}}_{T(s)} \qquad \begin{aligned} \delta\{D_1(s)\} &= 4 \\ \delta\{T(s)\} &= 2 \end{aligned}$$

$$N_1(s)[D_1(s)]^{-1} = \begin{bmatrix} \dfrac{s+2}{s+1} & \dfrac{1}{s+2} \\ \dfrac{1}{s+1} & \dfrac{(s+1)(s+3)}{(s+2)^2} \end{bmatrix} \equiv N(s)[D(s)]^{-1} = R(s)$$

Thus $\delta\{D(s)\} = \delta\{D_1(s)\} + \delta\{T(s)\}$.

(c) Inversion through Smith-McMillan form

It is interesting to note that once the two unimodular matrices in (1.117) are computed, the inverse of M(s) may be derived by the following identity:

$$[M(s)]^{-1} = [U_r(s)]^{-1}[R(s)]^{-1}[U_\ell(s)]^{-1} \tag{1.126}$$

Alternatively, if the representation $R(s) = \bar{U}_\ell(s)M(s)U_r(s)$ is known, then

$$[R(s)]^{-1} = [T\bar{U}_r(s)]^{-1}T[M(s)]^{-1}T[\bar{U}_\ell(s)T]^{-1} \tag{1.127}$$

where T is a constant symmetric matrix given by $T = \{T_j\} = \{e^n_{\alpha n}\}$, $j \in 1, \ldots, n$, $e^n \rightarrow$ an $(1 \times n)$ column vector, with a 1 in the position $\alpha = (n + 1 - j)$ for any $j \in 1, \ldots, n$, that is,

$$T = \begin{bmatrix} 0 & 0 & \cdots & 0 & 1 \\ 0 & 0 & \cdots & 1 & 0 \\ & & \iddots & & \\ 0 & 1 & \cdots & 0 & 0 \\ 1 & 0 & \cdots & 0 & 0 \end{bmatrix} = T^{-1} \tag{1.128}$$

Thus

$$\bar{M}(s) = T[M(s)]^{-1}T \equiv \text{diag}\left\{\frac{\phi_i(s)}{\psi_i(s)}\right\} \equiv [M(s)]^{-1}$$

which is the inverse of the Smith-McMillan form of R(s), which proves the consistency of (1.127). Thus once M(s) is established, $[R(s)]^{-1}$ may be computed directly from (1.126) or (1.127).

REFERENCES

1. Bellman, R., Introduction to Matrix Analysis, McGraw-Hill, New York, 1960.
2. Halmos, P. R., Finite Dimensional Vector Spaces, D. Van Nostrand, Princeton, N.J., 1956.
3. Gantmatchan, F. R., Applications of the Theory of Matrices, Chelsea, New York, 1957.
4. Ogata, K., State Space Analysis of Control Systems, Prentice-Hall, Englewood Cliffs, N.J., 1967.
5. DeRusso, P. M., Roy, J. R., and Close, C. M., State Variables for Engineers, Wiley, New York, 1965.
6. Ayres, F., Matrices, Schaum's Outline Series, McGraw-Hill, New York, 1974.
7. Barnett, S., Matrices in Control Theory, Van Nostrand, London, 1971.
8. Faddeev, D. K., and Faddeeva, V. N., Computational Methods of Linear Algebra, W. H. Freeman, San Francisco, 1963.
9. Smiley, M. F., Algebra of Matrices, Allyn and Bacon, Boston, 1965.
10. Rosenbrock, H. H., State Space and Multivariable Theory, Thomas Nelson, London, 1970.
11. Bell, D. J., "Linear algebra," in Modern Approaches to Control System Design, N. Munro (Ed.), Peter Peregrinus, London, 1979.
12. Kailath, T., Linear Systems, Prentice-Hall, Englewood Cliffs, N.J., 1980.

13. McDuffee, C. C., Theory of Matrices, Chelsea, New York, 1956.
14. Coppel, W. A., "Matrices of rational functions," Bull. Austral. Math. Soc., 11: 89-113 (1974).
15. Wolovich, W. A., Linear Multivariable Systems, Springer-Verlag, New York, 1974.

2
System Description

In practical engineering control problems, analysis starts with the formulation of a mathematical model (or model, for short) of the physical system under investigation. The exact form of the model, often a collection of a set of analytical relations or equations, would depend on some observed properties of the system, the judgment and experience of the design engineer, the characteristics of the system to be studied, or a combination of these. The solution of these equations is called the response of the model, which for a "good" model is essentially identical to that of the physical system. The design process thus generates a model whose response has useful engineering properties. Broadly speaking, there are two methods available for mathematical representation of physical systems: the input/output and state-variable methods. In the former, the system is identified through the terminal or external variables—namely, inputs (causes) and outputs (effects)—using Laplace transformation.* The latter method still deals with the interrelationship of the external inputs and outputs, but involves an additional set of internal parameters called the state variables. This chapter develops these equations for linear systems from some basic concepts and studies their solution. The relation between the input/output description and the state-variable method is established and the concepts of poles and zeros introduced.

Before proceeding, however, it is appropriate to define some terms commonly used in system analysis and design [1].

Physical: An adjective that refers to any quantity that is associated with the basic quantities of physics: length, mass, time, charge.
Mapping: A rule, formula, or set of formulas for going from one set of objects to another.
Function: A mapping from one set (called the domain) to another set (called the range) that assigns only one member of the range to each

*For an introductory treatment of Laplace transform theory, see any book on basic control theory (e.g., Ref. 10).

member of the domain. The word function is also used to mean a member of the range of the function.
Signal: A physical function of time: for example, voltage, current, force, velocity, temperature, or pressure.
System: A signal processor (a mapping from a set of input signals to a set of output signals). By tying systems and signals to the physical world the generality is somewhat limited, although from a mathematical viewpoint, the word system can mean almost anything.
Element or component: A system that is part of another system.

2.1 SYSTEM CLASSIFICATION [1-5]

Before the various methods of analyzing systems are considered, it is essential to define some of the terms used in describing and specifying systems. For this, it is only necessary to consider a general representation of a system as shown in Fig. 2.1, where the operator $\mathscr{L}$ is a function that characterizes the system with an input control u(t) (excitation) and an output variable y(t) (response). Assuming that the system does not contain any independent sources and is at rest with no internal energy before the input signal is applied,* the input/output relation may be symbolically expressed by

$$y(t) = \mathscr{L}u(t) \tag{2.1}$$

This is an analytical means of representing the cause-and-effect relationship between u(t) and y(t).

(a) Causal

A system is causal (or physical or nonanticipative) if the present output does not depend on future values of the input. A noncausal (or nonanticipative) system would, on the other hand, violate the normal cause-and-effect relationship.

(b) Realizable

A system is realizable if it is nonanticipative and if y(t) is a real function of time for all real u(t). This definition, of course, does not imply that there is necessarily a known procedure for combining components to yield a given realizable system.

(c) Deterministic

A system is deterministic if for each input u(t) there is a unique output y(t). In a nondeterministic (or probabilistic or stochastic) system there may be

*Such a system is said to be initially relaxed.

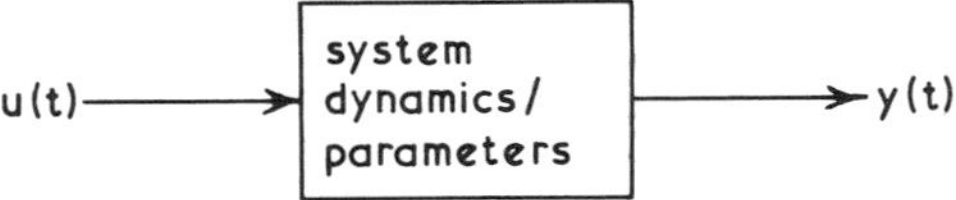

Fig. 2.1 "Black-box" representation of a system with input u(t) and output y(t): $y(t) = \mathscr{L}u(t)$.

several possible outputs, each with a certain probability of occurrence, for a given input.

(d) Linear and nonlinear

A system is said to be linear if the response to $u(t) = c_1u_1(t) + c_2u_2(t)$ is $y(t) = c_1y_1(t) + c_2y_2(t)$, where c_1 and c_2 are two constants, and $y_1(t)$ and $y_2(t)$ are responses to two different inputs, $u_1(t)$ and $u_2(t)$. This definition is expressed symbolically using (2.1) by

$$\mathscr{L}[c_1u_1(t) + c_2u_2(t)] = c_1[\mathscr{L}u_1(t)] + c_2[\mathscr{L}u_2(t)] \tag{2.2}$$

which is known as the superposition principle. If this relation holds within a certain range, the system is linear only within that range. A system that does not satisfy the principle of superposition is said to be nonlinear.

(e) Time-invariant and time-varying systems

A system is time invariant (or fixed) if the relationship between the input and output is independent of time. If the response to u(t) is y(t), the response to $u(t - \tau) = y(t - \tau)$, where τ is a finite time; symbolically,

$$y(t - \tau) = \mathscr{L}u(t - \tau) \tag{2.3}$$

In such a system, the size and shape of the output are independent of the time at which input is applied. Systems that do not satisfy (2.3) are time varying.

(f) Stationary and dynamic systems

A stationary (or instantaneous or memoryless) system is one in which the response at any time t_1 is dependent only on the excitation at that time t_1 and not on any future or past values of excitation. If the response depends on past values of input signal, the system is said to be dynamic. (With this definition, an electric circuit consisting of pure resistors only is a stationary system, whereas capacitive/inductive circuits are dynamic systems.)

(g) Lumped-parameter and distributed-parameter systems

A system that can be described by a finite number of scalar variables is known as a lumped-parameter (or finite-order) system. The term is used to express the concept that each point in a system embodies the properties of the region immediately surrounding it. Systems whose states are described by functions are generally called distributed-parameter systems. The characteristics of a lumped-parameter system may be described through ordinary differential equations, whereas partial differential equations are to be used to describe distributed-parameter systems.

(h) Continuous-time and discrete-time systems

A continuous-time system is one in which the inputs and outputs are defined as functions of the continuous independent variable time (t). The system must be uniquely defined at all values of t within a given range. Continuous systems are described by differential equations. Systems with inputs and outputs defined at a sequence of discrete values of the independent variable t are discrete-time systems. In many cases of practical interest, the instants at which the input (output) variables are defined are equally spaced and can be defined by $t = t_0 + kT$, where T is the time between instants and k is an independent variable that assumes only integer values. Discrete-time systems are defined by difference equations.

(i) Block diagrams and signal flow graphs

Block diagrams are widely used by control engineers to provide a pictorial representation of the cause-and-effect relationship between the input control and the output response. Block diagrams consist of a combination of the four basic types of elements shown in Fig. 2.2; the arrows imply the unidirectional property of the signals. The primary advantage of the block diagram representation is that it provides a picture of the complex interconnection of various (self-contained) elements in a feedback system, so that the effects of the individual elements on the overall performance of the system may be evaluated.

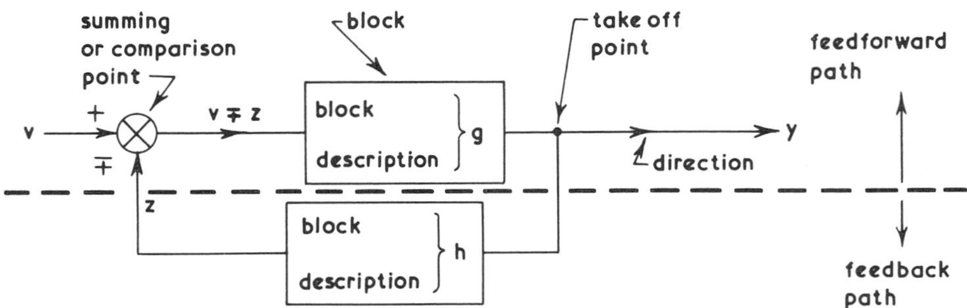

Fig. 2.2 Four basic elements of a block diagram.

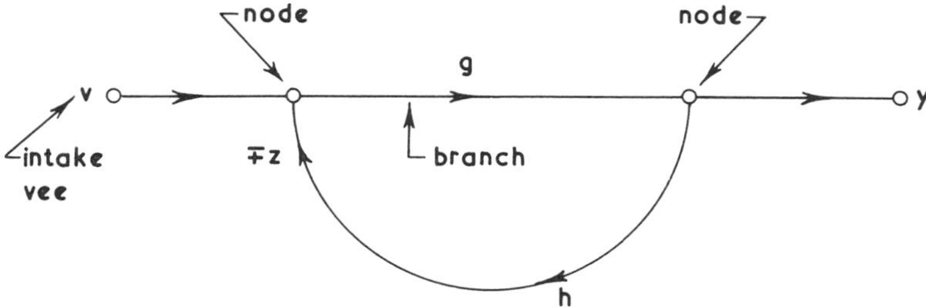

Fig. 2.3 Elementary signal flow graph.

Signal flow graphs,* on the other hand, provide a pictorial representation of a set of simultaneous algebraic equations describing a system. As for block diagrams, signal flow graphs display the propagation of (control) signals through the system. The construction of signal flow graphs is essentially a matter of following the cause-and-effect relationship relating each variable in terms of itself and other variables in the system, using the basic elements shown in Fig. 2.3 and the following analogy with block diagrams:

Block diagram		Signal flow graph
Line carrying a signal	→	Node
Block g	→	Gain g
Summer/comparator	→	Addition/subtraction at the node

The advantage of signal flow graphs is that they not only illustrate the passage of signals through the systems, but also give a precise indication of the interdependence of the various signals within a system. Although similar representations are possible using block diagrams, which essentially depict the relationship between various noninteracting elements, the resulting representation would be more complex. This comparison is shown in Fig. 2.4, which represents the general linear algebraic differential equations

$$\frac{dx_1(t)}{dt} = a_{11}x_1(t) + a_{12}x_2(t) + \alpha_{11}u_1(t) + \alpha_{12}u_2(t)$$

$$\frac{dx_2(t)}{dt} = a_{21}x_1(t) + a_{22}x_2(t) + \alpha_{21}u_1(t) + \alpha_{22}u_2(t)$$

and

*S. J. Mason, "Feedback theory—some properties of signal flow graphs," Proc. IRE, 41(9): 1144-1156 (1953).

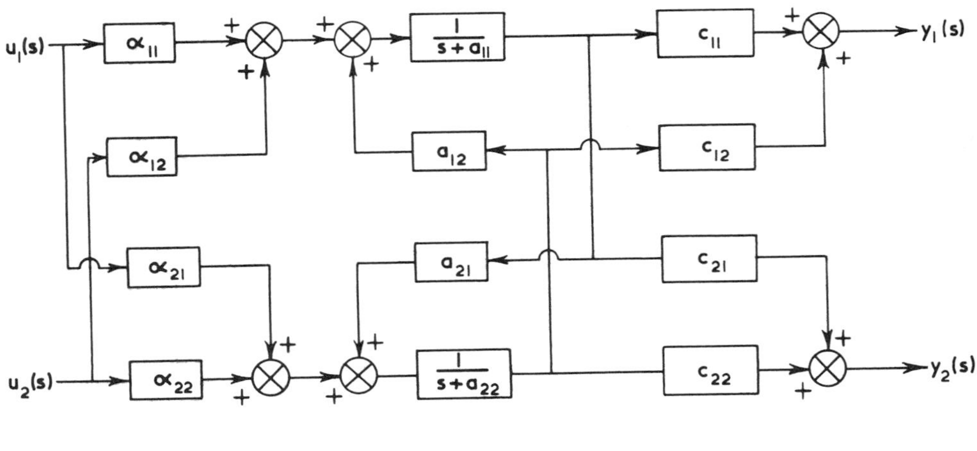

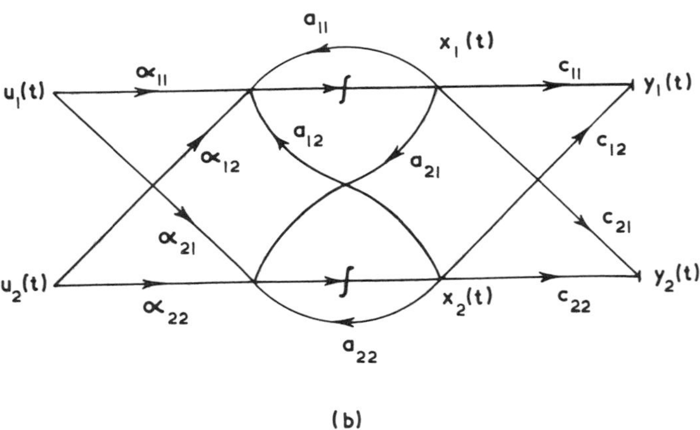

Fig. 2.4 Representation of (2.4): (a) block diagram; (b) signal flow graph.

$$
\begin{aligned}
y_1(t) &= c_{11}x_1(t) + c_{12}x_2(t) \\
y_2(t) &= c_{21}x_1(t) + c_{22}x_2(t)
\end{aligned}
\tag{2.4}
$$

or

$$\frac{dx_i(t)}{dt} = \sum_{j=1}^{2} a_{ij}x_j(t) + \sum_{j=1}^{2} \alpha_{ij}u_j(t)$$

and

$$y_i(t) = \sum_{j=1}^{2} c_{ij} x_j(t), \quad i \in 1, 2$$

Although both representations may be used to convey the same representations, signal flow graphs are normally used in the subsequent chapters due mainly to their graphical simplicity, except where block diagrams are more appropriate.

(j) Single-input/single-output and multivariable systems

A system with only one input control and only one output variable is called a single-input/single-output (scalar, or rarely, single-variable) system. Thus a scalar function is required to describe a single-input/single-output system. A system with more than one input control and output variable is known as a multivariable (or multi-input/multi-output) system. The inputs and outputs of a multivariable system may be arranged to form an input

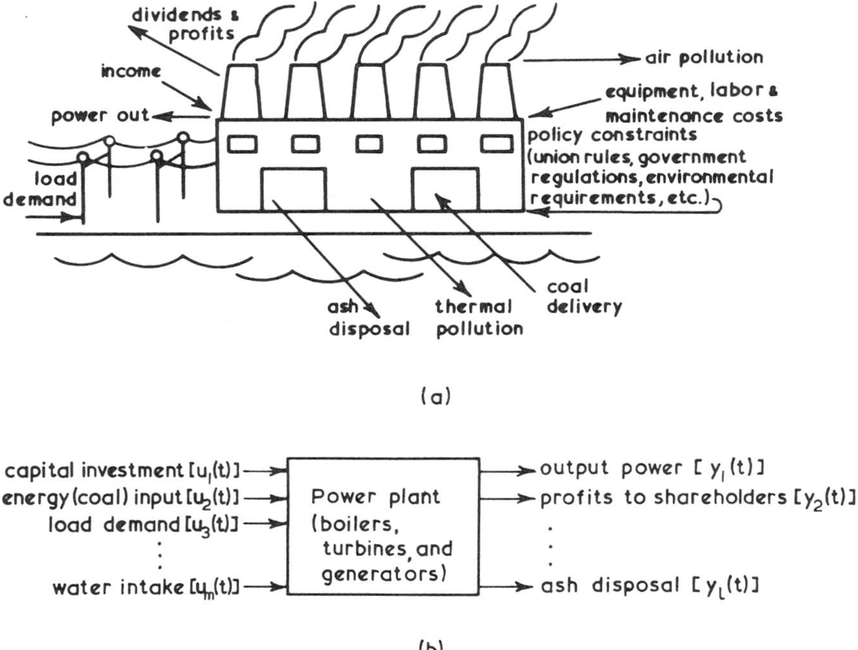

Fig. 2.5 Representation of a power plant: (a) input/output variables; (b) simplified block diagram.

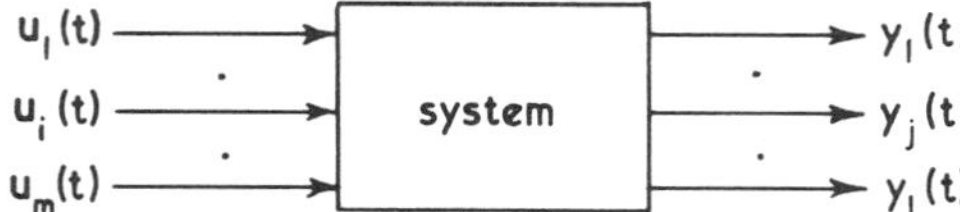

Fig. 2.6 Block diagram of an m-input/ℓ-output system.

vector and an output vector, respectively; characteristics of such systems are therefore to be represented by a system of equations relating the various inputs to the various outputs.

In reality all systems are multivariable in that there is more than one input control (some of which may not be under the control of the designer, e.g., environment effects) and more than one output response. Only through suitable assumptions, often formulated for analytical (and design) simplicity, are these reduced to single-input/single-output systems. Such reductions are thus not unique and are very dependent on the properties of the system to be investigated. An example of this reduction is given here with reference to a power plant (Fig. 2.5a), a complex linear time-varying (and discrete) dynamic system whose input/output relationship is governed by technical, economic, and social conditions.

A block diagram of the system as a multivariable system is shown in Fig. 2.5b. If, however, one is interested in the analysis with a view to investigating the effects of changing input thermal energy (coal input) to output electrical power while all other inputs and outputs are kept constant, a much simpler input/output relationship can be established. It is apparent that if the effects of input coal energy content to ash disposal were to be studied, a completely different input/output characterization would be required. A general configuration of an m-input/ℓ-output system is shown in Fig. 2.6.

2.2 STATE-SPACE REPRESENTATION [1-9]

A <u>state</u> of a dynamic system is the smallest collection of parameters which when defined at any initial time $t = t_0$ enable the unique prediction of behavior (input/output relation) for any future time $t_1 > t_0$, the parameters being called the <u>state variables</u>. The state is a compact representation of the past activity of the system, compact enough to predict, on the basis of the inputs, the exact behavior of future outputs, and to update itself.

An example is an RLC circuit (Fig. 2.7) in which the energy storage in the capacitor and the inductor at an initial time (t_0) is sufficient to determine the total response at any future time ($t_1 > t_0$) if the excitation is known between t_0 and t_1. A suitable set of parameters to describe the state of the system is thus the voltage across the capacitor $v_C(t)$ and the current through

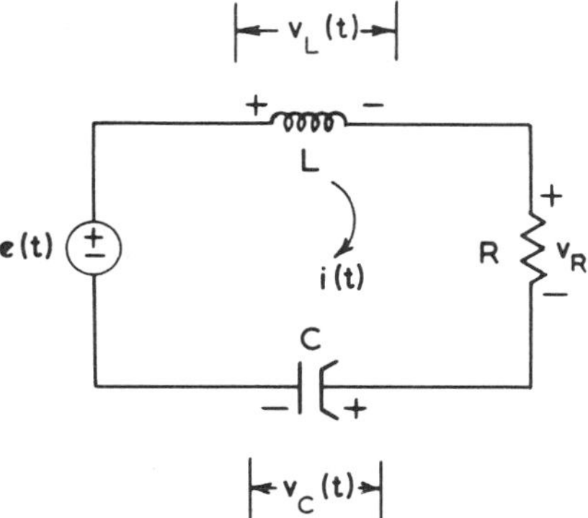

Fig. 2.7 Second-order RLC network.

the inductor i(t); these are the state variables of the system in Fig. 2.8. It should, however, be stressed here that, in general, representation of a state is not unique; there are many different collections of state variables that would convey the same information about the system.

The idea of state is a fundamental concept, and therefore can be illustrated or defined only by means of the properties of a system whose behavior involves the notion of state. In state-space representation, the order of a system is the degree of the highest derivative in the differential equation that describes the system. In physical terms, this is equal to the number of energy storage components or elements in a system. In mathematical terms, the order represents the number of initial conditions required to determine the response from specified input signals.

2.2.1 State and Output Equations

Although the concept of the state of a physical system is of prime importance in systems analysis, there are no guidelines for defining "the state of a

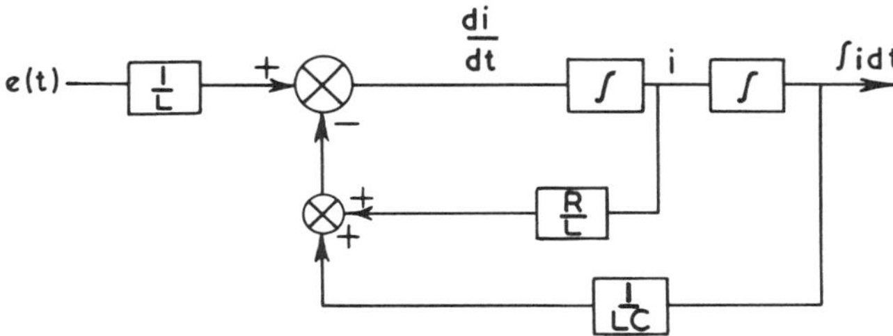

Fig. 2.8 Block diagram of the network in Fig. 2.7.

system." In most practical cases state variables are chosen by intuition (voltage across a capacitor, current through an inductor, output of an integrator, etc.), while in mathematical terms it is convenient to associate the state variables with physical parameters. By using very abstract generalization, let the m-input/ℓ-output nth-order system in Fig. 2.6 be described by the following set of differential equations:

$$\begin{aligned} h_1[y_1, \overset{(1)}{y_1}, \ldots, \overset{(n)}{y_1}, y_2, \overset{(1)}{y_2}, \ldots, \overset{(n)}{y_2}, \ldots, \overset{(n)}{y_\ell}] &= 0 \\ h_2[y_1, \overset{(1)}{y_1}, \ldots, \overset{(n)}{y_1}, y_2, \overset{(1)}{y_2}, \ldots, \overset{(n)}{y_2}, \ldots, \overset{(n)}{y_\ell}] &= 0 \\ \cdots\cdots\cdots\cdots\cdots\cdots\cdots \\ h_m[y_1, \overset{(1)}{y_1}, \ldots, \overset{(n)}{y_1}, y_2, \overset{(1)}{y_2}, \ldots, \overset{(n)}{y_2}, \ldots, \overset{(n)}{y_\ell}] &= 0 \end{aligned} \tag{2.5}$$

where $\overset{(i)}{y_j} = d^i y_j(t)/dt^i$, $i \in 1, \ldots, n$, $j \in 1, \ldots, \ell$, y being the output of the system and t in dependent variable time.*

A particular set of initial conditions characterize the state of the physical system if that set of initial data can, at least in principle, construct a unique solution for the system.

A conceptual (and notational) framework may be developed by reducing all differential equations in (2.5) involving second- and higher-order derivative to first-order vector equations. Assuming that at least n variables, $x_1(t)$, $x_2(t)$, $\ldots$, $x_n(t)$ are necessary to describe the system, (2.5) may be reduced to the following set of n first-order simultaneous equations:

$$\begin{aligned} \frac{dx_1(t)}{dt} &= \dot{x}_1(t) = f_1[x_1(t), x_2(t), \ldots, x_n(t), t, u_1(t), u_2(t), \ldots, u_m(t)] \\ \frac{dx_2(t)}{dt} &= \dot{x}_2(t) = f_2[x_1(t), x_2(t), \ldots, x_n(t), t, u_1(t), u_2(t), \ldots, u_m(t)] \\ &\vdots \\ \frac{dx_n(t)}{dt} &= \dot{x}_n(t) = f_n[x_1(t), x_2(t), \ldots, x_n(t), t, u_1(t), u_2(t), \ldots, u_m(t)] \end{aligned} \tag{2.6}$$

For convenience, the simultaneous equations above are represented as a single vector differential equation

*The "dot notation" is commonly used in state-space representation: $\dot{x}(t) = dx(t)/dt$.

$$\frac{dx(t)}{dt} = \dot{x}(t) = f[x(t), t, u(t)] \tag{2.7}$$

where $x = [x_1, x_2, \ldots, x_n]^T$ is a column vector, $f(\cdot)$ is a single-valued matrix function, and $u(t)$ is the input control vector with m elements, $u_1(t), \ldots, u_m(t)$.

The $(n \times 1)$ vector x is called the state vector ($\equiv$ state) of the system and the elements $x_i \in x$, $i \in 1, \ldots, n$) the state variables.* The differential equations represented by (2.7) are known as the state equations of the system described by the set of differential equations in (2.6).

A state space is defined as the n-dimensional space whose coordinates are the components x_i, $i \in 1, \ldots, n$, of the state vector x. In notations of linear algebra, a state space is the vector space spanned by the state variables of a system. Thus the vectors $x = \{x_i\}$, $i \in 1, \ldots, n$, form the basis for the n-dimensional state space R^n.† The motion of the tip of the state vector in state space is called the trajectory of the state vector.

By the very definition of state variables, it is always possible to define (or choose) a set of parameters that contain information about the past history of a system, and hence obtain (2.7). In many practical cases these variables are selected to simplify formulation of the state equations, thus making it necessary to use abstract parameters that resemble the physical variables associated with the system being analyzed. Since the choice of state variables is not unique, the exact nature of the state equations is likely to be dependent on the characteristic features under investigation. However, since the aim of system analysis is to evaluate appropriate control laws to achieve the desired input/output behavior, it is necessary to express these state variables in terms of measurable information about the system. The equation that relates the state variables to physical output variables of a system is called the output-state equation, and may be described by

$$\begin{aligned} y_1(t) &= g_1[x_1(t), x_2(t), \ldots, x_n(t), t, u_1(t), \ldots, u_m(t)] \\ y_2(t) &= g_2[x_1(t), x_2(t), \ldots, x_n(t), t, u_1(t), u_2(t), \ldots, u_m(t)] \\ &\vdots \\ y_\ell(t) &= g_\ell[x_1(t), x_2(t), \ldots, x_n(t), t, u_1(t), u_2(t), \ldots, u_m(t)] \end{aligned} \tag{2.8}$$

In vector notation (2.8) can be expressed as

$$y(t) = g[x(t), t, u(t)] \tag{2.9}$$

*The notation (t) to represent time dependency is dropped for convenience.
†The symbol R will be used subsequently to denote the field of real numbers ($\equiv$ F in Sec. 1.1).

where $y(t) = [y_1(t), y_2(t), \ldots, y_\ell(t)]^T$ is the $(\ell \times 1)$ output vector and g[·] is a single-valued matrix function.

The set of equations that describes the unique relationships between the input controls and the output variables is called the dynamic (or system) equations. It can be shown that given an initial vector x_0 at time $t = t_0$ there exists a unique solution of (2.7) that passes through x_0 at t_0. Hence

$$\begin{aligned} \dot{x}(t) &= f\{x(t), t, u(t)\} \quad \text{(state equation)} \\ \dot{y}(t) &= g\{x(t), t, u(t)\} \quad \text{(output equation)} \end{aligned} \tag{2.10}$$

qualify as dynamic equations for an m-input/ℓ-output system with n state variables. It is to be noted that there is no loss of generality in writing the output equation in the form (2.9), since by definition of the state, knowledge of x(t) and u(t) is sufficient to determine y(t). The state space of (2.10) is an n-dimensional vector space; hence the equation is called an n-dimensional dynamic equation. The system represented by (2.10) is said to be linear or nonlinear depending on whether f[·] and g[·] are linear or nonlinear functions of the state vector and the input control. In the subsequent development, f[·] and g[·] are assumed to be linear functions [i.e., they are independent of x(t) and u(t)] unless otherwise stated.

Example 2.1: State-space representation—single-input/single-output

The physical significance of state variables and the derivation of state and output equations are illustrated below. Let it be required to represent the "dynamics" of the second-order RLC network in Fig. 2.7 through state equations. Kirchhoff's voltage loop equation for this "system" is

$$e(t) - v_L(t) - v_R(t) - v_C(t) = 0 \tag{2.11}$$

Substituting the values for v_L, v_R, and v_C, Kirchhoff's current law gives

$$L \frac{d^2 i(t)}{dt^2} + R \frac{di(t)}{dt} + \frac{1}{C} i(t) = \frac{de(t)}{dt} \tag{2.12}$$

In order to arrive at expressions with the desired state form, some "parameters" will have to be chosen as the state variables of the system. To help in this procedure, it is appropriate to identify the physical quantities that are most likely to be specified as initial conditions for the system (network) in Fig. 2.7. In this case, as the initial current through the inductor $[i(t_0)]$ and the initial voltage across the capacitor $[v_C(t_0)]$ are most likely to be given,* a "possible" choice for the state variables would appear to be the capacitance voltage $[x_1(t) = v_C(t)]$ and the inductor current $[x_2(t) = i(t)]$. With this choice of state variables, the pair of first-order differential equations describing the network becomes

*An alternative choice of initial conditions: $i(t_0)$ and $di(t_0)/dt$.

$$\dot{x}_1(t) = -\frac{1}{C}x_2(t)$$
$$\dot{x}_2(t) = -\frac{1}{L}x_1(t) - \frac{R}{L}x_2(t) + \frac{1}{L}e(t) \tag{2.13}$$

which is of the same form as (2.6). In matrix notation this becomes*

$$\begin{bmatrix} \dot{x}_1(t) \\ \dot{x}_2(t) \end{bmatrix} = \begin{bmatrix} -\frac{1}{C} & 0 \\ -\frac{1}{L} & -\frac{R}{L} \end{bmatrix} \begin{bmatrix} x_1(t) \\ x_2(t) \end{bmatrix} + \begin{bmatrix} 0 \\ \frac{1}{L} \end{bmatrix} e(t) \tag{2.14}$$

which has the form (2.7). A block diagram of the network is given in Fig. 2.8 and a signal flow graph of the state equations in Fig. 2.9. The output equation for the system is

$$y(t) = x_2(t) \tag{2.15}$$

In a similar manner, the state equations of the spring-mass system in Fig. 2.10 may be derived as a

$$\dot{x}_1(t) = x_2(t)$$
$$\dot{x}_2(t) = -\frac{K}{m}x_1(t) - \frac{D}{m}x_2(t) + \frac{1}{m}F \tag{2.16}$$

where the state variables are chosen as $x_1(t) = y(t)$ and $x_2(t) = \dot{x}_1(t)$. It is interesting to note the similarity between the state equations in (2.13) describing the electrical system in Fig. 2.9 and the state equations in (2.16) describing the mechanical system in Fig. 2.10. The systems may be seen to be described by equivalent state equations if the following table of correspondence is set up [3]:

Electrical system	Mechanical system
Voltage, V	Force, F
Current, i	Velocity, v
Charge, q	Displacement, y
Inductance, L	Mass, M
Resistance, R	Damping coefficient, D
Capacitance, C	Compliance, K

*An alternative set of initial conditions [i.e., $i(t_0)$ and $di(t_0)/dt$] would be required to solve (2.12). Such a choice for state variables, however, would lead to a set of state equations [different from the structure of (2.13)].

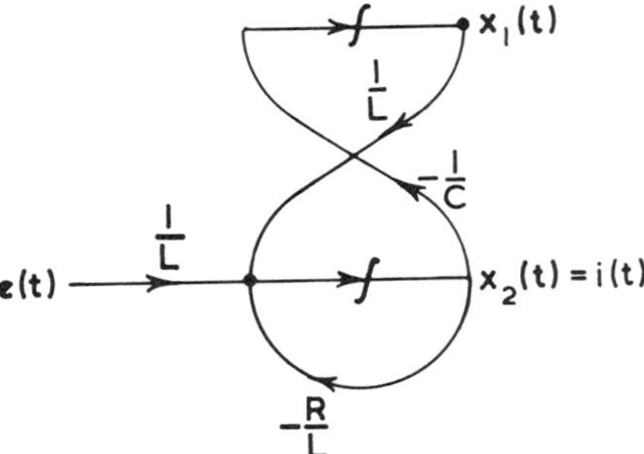

Fig. 2.9 Signal flow graph of (2.14).

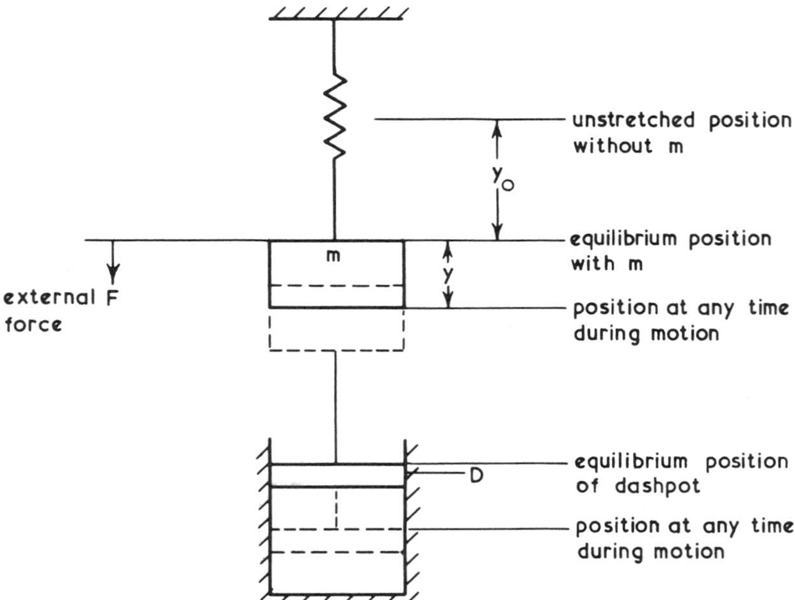

Fig. 2.10 Spring-mass-dashpot system.

This correspondence, known as the force-voltage analogy, allows the derivation of the solution of an electrical problem from the solution of an analogous mechanical problem, or vice versa.

The output equation for the system in Fig. 2.10 is

$$y(t) = x_1(t) \tag{2.17}$$

Example 2.2: State-space representation—multi-input/multi-output
The system in Fig. 2.11a is considered here. For this RC system, the two mode equations are

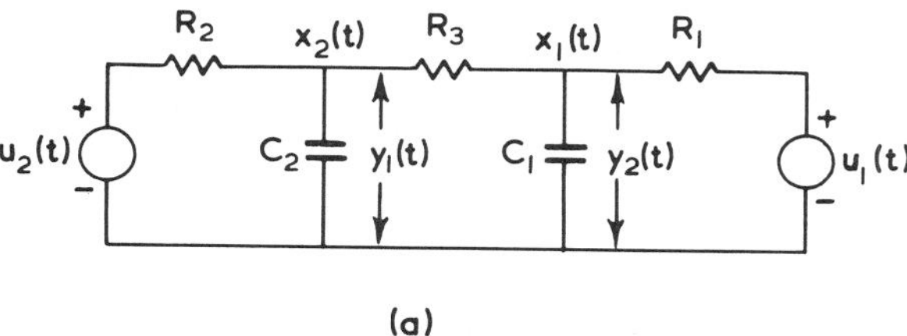

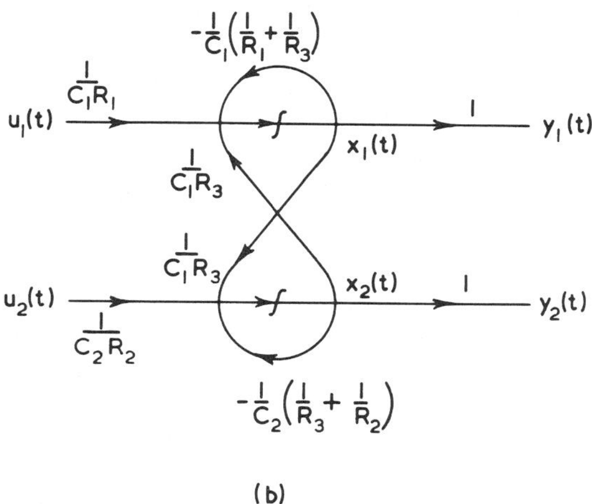

Fig. 2.11 Two-input/two-output system: (a) circuit diagram; (b) signal flow graph.

$$\frac{x_1 - x_2}{R_3} + \frac{x_1 - u_1}{R_1} + C_1 \frac{dx_1}{dt} = 0: \; x_1 \to x_1(t),\; x_2 \to x_2(t)$$
$$\frac{x_2 - x_1}{R_3} + \frac{x_2 - u_2}{R_2} + C_2 \frac{dx_2}{dt} = 0: \; u_1 \to u_1(t),\; u_2 \to u_2(t) \tag{2.18a}$$

In matrix notation, (2.18a) may be written as

$$\begin{bmatrix} \dot{x}_1(t) \\ \dot{x}_2(t) \end{bmatrix} = \begin{bmatrix} -\frac{1}{C_1}\left(\frac{1}{R_1} + \frac{1}{R_3}\right) & \frac{1}{C_1R_3} \\ \frac{1}{C_2R_3} & -\frac{1}{C_2}\left(\frac{1}{R_3} + \frac{1}{R_2}\right) \end{bmatrix} \begin{bmatrix} x_1(t) \\ x_2(t) \end{bmatrix} + \begin{bmatrix} \frac{1}{C_1R_1} & 0 \\ 0 & \frac{1}{C_2R_2} \end{bmatrix} \begin{bmatrix} u_1(t) \\ u_2(t) \end{bmatrix} \tag{2.18b}$$

The signal flow graph of the system is given in Fig. 2.11b and the output equation for this system is

$$\begin{bmatrix} y_1(t) \\ y_2(t) \end{bmatrix} = \begin{bmatrix} x_1(t) \\ x_2(t) \end{bmatrix} = \begin{bmatrix} 1 & 0 \\ 0 & 1 \end{bmatrix} \begin{bmatrix} x_1(t) \\ x_2(t) \end{bmatrix} \tag{2.19}$$

Example 2.3: Differential equations and state equations. To illustrate the relationship between system differential equations (representing the input/output dynamics) and state equations, the following example is considered. The simultaneous equations relating the two input controls, $u_1(t)$

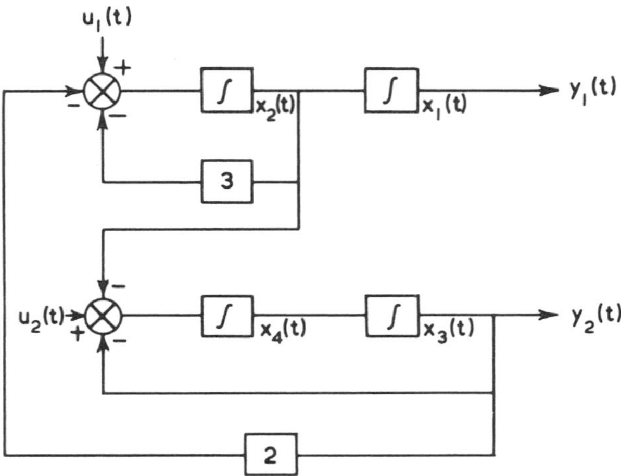

Fig. 2.12 Block diagram of the "coupled" differential equations in (2.20).

and $u_2(t)$, and the two output variables, $y_1(t)$ and $y_2(t)$, are given by

$$\frac{d^2y_1(t)}{dt^2} + 3\frac{dy_1(t)}{dt} + 2y_2(t) = u_1(t)$$
$$\frac{d^2y_2(t)}{dt^2} + \frac{dy_1(t)}{dt} + y_2(t) = u_2(t) \tag{2.20}$$

A block diagram of the system is shown in Fig. 2.12; with the choice of state variables shown there, the state equations may be formulated as

$$\begin{aligned}\dot{x}_1(t) &= x_2(t)\\ \dot{x}_2(t) &= -3x_2(t) - 2x_3(t) = u_1(t)\\ \dot{x}_3(t) &= x_4(t)\\ \dot{x}_4(t) &= -x_2(t) - x_3(t) + u_2(t)\end{aligned} \tag{2.21}$$

The output equations are

$$\begin{aligned}y_1(t) &= x_1(t)\\ y_2(t) &= x_3(t)\end{aligned} \tag{2.22}$$

In matrix notation, the equations above may be expressed as

$$\begin{bmatrix}\dot{x}_1(t)\\ \dot{x}_2(t)\\ \dot{x}_3(t)\\ \dot{x}_4(t)\end{bmatrix} = \begin{bmatrix}0 & 1 & 0 & 0\\ 0 & -3 & -2 & 0\\ 0 & 0 & 0 & 1\\ 0 & -1 & -1 & 0\end{bmatrix}\begin{bmatrix}x_1(t)\\ x_2(t)\\ x_3(t)\\ x_4(t)\end{bmatrix} + \begin{bmatrix}0 & 0\\ 1 & 0\\ 0 & 0\\ 0 & 1\end{bmatrix}\begin{bmatrix}u_1(t)\\ u_2(t)\end{bmatrix} \tag{2.23a}$$

$$\begin{bmatrix}y_1(t)\\ y_2(t)\end{bmatrix} = \begin{bmatrix}1 & 0 & 0 & 0\\ 0 & 0 & 1 & 0\end{bmatrix}\begin{bmatrix}x_1(t)\\ x_2(t)\\ x_3(t)\\ x_4(t)\end{bmatrix} + \begin{bmatrix}0 & 0\\ 0 & 0\end{bmatrix}\begin{bmatrix}u_1(t)\\ u_2(t)\end{bmatrix} \tag{2.23b}$$

2.2.2 Matrix Representation

As shown in (2.23), the linear dynamical equation (2.10) can be written, in general, in matrix form as

$$S(A, B, C, D, t): \quad \begin{aligned}\dot{x}(t) &= A(t)x(t) + B(t)u(t) \quad \text{(state equation)}\\ y(t) &= C(t)x(t) + D(t)u(t) \quad \text{(output equation)}\end{aligned} \tag{2.24}$$

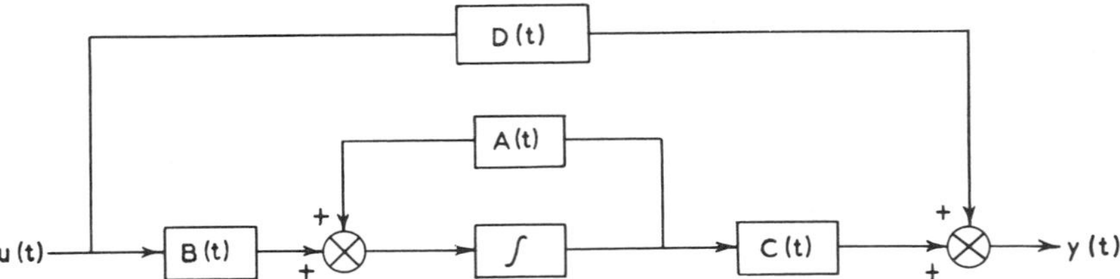

Fig. 2.13 General block diagram of a system S(A, B, C, D, t).

The n-square matrix A(t) describes the dynamics of the system and controls the trajectory of the state vector x(t) in R^n; A(t) is sometimes called the dynamics matrix of the system. The elements of the $(n \times m)$ matrix B(t) show how each control input affects the state variables of systems, and is called the input matrix. The $(\ell \times n)$ output matrix C(t) transforms the state vector x(t) into the output vector y(t), and the $(\ell \times m)$ transmission matrix D(t) indicates the direct (feedforward) effect of control inputs to output variables. These four matrices are usually known as system matrices, and the system defined by these is denoted by S(A, B, C, D, t).

A sufficient condition for (2.24) to have a unique solution is that every element of A(t) is a continuous function of t for $-\infty < t < \infty$. For convenience the entries B(t), C(t), and D(t) are usually assumed to be continuous in $(-\infty, \infty)$. Since the values of the system matrices A(t), B(t), C(t), and D(t) are dependent on time, a system S(A, B, C, D, t) described by (2.24) is known as a linear time-varying system; a general block diagram of such a system is shown in Fig. 2.13.

If the system matrices are independent of time, the system described by (2.24) is called a linear time-invariant (or fixed) system with m inputs, ℓ outputs, and n state variables, and is represented by

$$S(A,B,C,D): \quad \begin{aligned} \dot{x}(t) &= Ax(t) + Bu(t) \quad \text{(state equation)} \\ y(t) &= Cx(t) + Du(t) \quad \text{(output equation)} \end{aligned} \tag{2.25}$$

In the time-invariant case the parameters of the system S(A, B, C, D) do not change with time; thus there is no loss of generality in choosing the initial time to be 0. The time interval of interest for analysis then becomes $(0, \infty)$.

As mentioned before, the state space of S(A, B, C, D, t) in (2.24) or S(A, B, C, D) in (2.25) is defined in the n-dimensional finite space R^n. Hence the n-square matrix A may be looked upon as a linear operator that maps x(t) onto R^n. The vector representations in (2.10), (2.24), and (2.25) are called state-space representations of dynamical systems. Large-scale complex systems can be easily analyzed through numerical techniques by the systematic format of this vector-matrix equation.

The dynamical system is said to be free (or unforced) if the state equation does not contain an input control (forcing function), that is, if u(t) in (2.24) and (2.25) is identically equal to zero. The solution of the dynamic equation of a free system is known as zero-input (or natural or unforced) response. If $u(t) \not\equiv 0$, the system is said to be forced. The solution of a forced system with zero initial state $[x(t_0) \equiv 0]$ is known as the zero-state response. For linear systems, the total response is the addition of these responses for any given forcing function.

2.2.3 Free and Forced Responses

The state-space representation of linear dynamic systems, expressed in vector differential equations in (2.24) and (2.25) are useful in the analysis of the structural properties of the system. However, it is also necessary in system analysis to study the time response of the output for a specified input and some initial state $x(t_0)$; this can be obtained only by solving the state and output equations describing the system dynamics.

(a) Free (unforced) response

Since the system is linear, it is decomposible; that is, the total solution can be written as a sum of the zero-input response and zero-state response. The zero-input response for the state vector can be obtained from (2.24) by letting u(t) = 0; the resulting homogeneous state equation is

$$\dot{x}(t) = A(t)x(t) \tag{2.26}$$

which can be solved by using the Peano-Baker method of integration [1]. Let $x(t_0)$ be the initial state vector at $t = t_0$; then integration of (2.26) gives

$$x(t) = x(t_0) + \int_{t_0}^{t} A(\tau)x(\tau)\, d\tau \tag{2.27}$$

This equation is known as a vector Volterra integral equation, and can be solved by repeated substitution of the right-hand side of the integral equation into the integral for x(t). If $\phi(t, t_0)$ is a solution of (2.26), that is,

$$x(t) = \phi(t, t_0)x(t_0) \tag{2.28}$$

and A is constant (independent of time), then from (2.27),

$$\phi(t, t_0) = e^{A(t-t_0)} = I + (t - t_0)A + \frac{(t - t_0)^2}{2!} A^2 + \cdots \tag{2.29}$$

or

$$x(t) = e^{A(t-t_0)}x(t_0) = \phi(t,t_0)x(t_0) \tag{2.30}$$

which is the solution of the homogeneous input-state equation of (2.26) for time-invariant systems; for time-varying systems, the solution is given by (2.28), where $\phi(t,t_0)$ is the solution of the matrix differential equation

$$\frac{d\phi(t,t_0)}{dt} = A(t)\phi(t,t_0)x(t_0), \quad \forall\, t > t_0 \tag{2.31}$$

The matrix $\phi(t,t_0)$ is called the state transition matrix (or the fundamental matrix, in matrix theory) of the system in (2.26), and describes the trajectory of the state vector x(t) from its initial position (at $t = t_0$) in state space. In other words, the state transition matrix $\phi(t,t_0)$ takes the state of a system at time t_0 and yields the state to which the system will move under zero input by time t, represented in the vector notation

$$\phi(t,t_0) : x(t_0) \to x(t)$$

$\phi(t,t_0)$ is thus a linear transformation that maps $x(t_0)$ at t_0 into x(t) at t in R^n. Since $x(t) = \{x_i(t)\}$, $i \in 1, 2, \ldots, n$, $\phi(t,t_0)$ contains a great deal of information relevant to analyzing the behavior of the system in the time domain, some of the commonly used properties of $\phi(t,t_0)$ are as follows (these apply for all t_0, t_1, t_2, or t, as appropriate):

1. By definition: $\phi(t,t_0)$ is nonsingular, and $\phi(t_0,t_0) = I$.
2. Group property: $\phi(t_2,t_0) = \phi(t_2,t_1)\phi(t_1,t_0)$.
3. Inverse property: $\phi(t_1,t_2) = [\phi(t_2,t_1)]^{-1}$.
4. For time-invariant systems: $\phi(t+t_0) = \phi(t)\phi(t_0)$ and $\phi(-t) = [\phi(t)]^{-1}$.
5. For time-varying systems: $d\phi(t,t_0)/dt = A(t)\phi(t,t_0)$.

(b) Forced system response

By direct substitution into the input-state differential equation, the response of the state vector of the forced system may be obtained as

$$x(t) = \phi(t,t_0)x(t_0) + \int_{t_0}^{t} \phi(t,p)B(p)u(p)\,dp \tag{2.32}$$

The output response is obtained by combining (2.32) and the state-output equation of (2.24), and is given by

$$y(t) = C(t)\phi(t,t_0)x(t_0) + \int_{t_0}^{t} C(t)\phi(t,p)B(p)u(p)\,dp + D(t)u(t) \tag{2.33}$$

The integral is superposition integral. In the time-invariant case, it becomes a convolution integral:

$$x(t) = \phi(t, t_0)x(t_0) + \int_{t_0}^{t} \phi(t - p)Bu(p)\, dp$$

$$y(t) = C\phi(t, t_0)x(t_0) + C\int_{t_0}^{t} \phi(t - p)Bu(p)\, dp + Du(t) \tag{2.34}$$

and the response depends on $(t - t_0)$ or the time difference between the application of cause and observation of effect; the state transition matrix in this case has only one independent variable, t.

For the time-varying case, however, the solution depends on both the independent variable (t) and the initial time (t_0) and thus the cause-and-effect relationship varies with time. The exact instances of applying the input and observation of the output response are significant. Since the state transition matrix for a time-varying system is dependent on two independent variables t_0 and t (the time of application of input signal and time of observing the output response, respectively), the solutions in (2.32) and (2.33) depend on superposition integral.

If the initial state is zero [i.e., $x(t_0) = 0$], the output response is given by

$$\begin{aligned} y(t) &= \int_{t_0}^{t} C(t)\phi(t, p)B(p)u(p)\, dp + D(t)u(t) \\ &\equiv \int_{t_0}^{t} [C(t)\phi(t, p)B(p) + u_0(t - p)D(t)]u(p)\, dp \\ &\equiv \int_{t_0}^{t} H(t, p)u(p)\, dp, \qquad t > t_0 \end{aligned} \tag{2.35}$$

where, for a time-varying system

$$H(t, p) = \begin{cases} C(t)\phi(t, p)B(p) + u_0(t - p)D(t), & t > t_0 \\ 0, & t < t_0 \end{cases} \tag{2.36}$$

and for a time-invariant system,

$$H(t, p) = \begin{cases} C\phi(t)B + u_0(t)D, & t > t_0 \\ 0, & t < t_0 \end{cases} \tag{2.37}$$

The matrix H(t, p) is called the impulse response matrix of the system [$u_0(t)$ being an impulse input function], since the (i, j)th element of this matrix is the response of $y_i(t)$ to an impulse applied at $u_j(t)$ at $t > t_0$, while all other elements of the input control vector are zero. The step response matrix S(t, p) is defined as

$$S(t,p) = \int_p^t H(t,\alpha)\, d\alpha, \qquad t < p \tag{2.38}$$

The (i, j)th element of the step response matrix is the response $y_i(t)$ for a step input in $u_j(t)$ at time $t > t_0$, while all other elements of the input control vector are zero, and $x(t_0) = 0$.

The following theorems may now be stated:

Theorem 2.1. Two equivalent linear time-invariant systems are zero-state equivalent and zero-input equivalent.

Proof: Let T be an (n × n) nonsingular matrix with coefficients in the field of complex numbers and let $\bar{x}(t) = Tx(t)$. Then, by definition, the system

$$\bar{S}(\bar{A},\bar{B},\bar{C},\bar{D}): \quad \begin{aligned} \dot{\bar{x}}(t) &= \bar{A}x(t) + \bar{B}u(t) \\ \bar{y}(t) &= \bar{C}x(t) + \bar{D}u(t) \end{aligned} \tag{2.39}$$

where $\bar{A} = TAT^{-1}$, $\bar{B} = TD$, $\bar{C} = CT^{-1}$, and $D = \bar{D}$, is said to be equivalent to S(A, B, C, D) in (2.25) and T is an equivalent transformation. Theorem 2.1 then can be proved by using the impulse response matrix, for the zero-input response of $\bar{S}$ is

$$\bar{y}(t) = \bar{C}e^{\bar{A}(t-t_0)}\bar{x}(t_0) = Ce^{A(t-t_0)}T^{-1}\bar{x}(t_0) \tag{2.40}$$

Hence for any $x(t_0)$, if $\bar{x}(t_0) = Tx(t_0)$ is chosen, S and $\bar{S}$ have the same zero-input response. It should be noted, however, that although equivalence implies zero-state equivalence and zero-input equivalence, the converse is not true. Furthermore, two dynamical systems can be zero-state equivalent without being zero-input equivalent.

Example 2.4: Free system—time-domain solution. Solution of state equations through the state transition matrix is illustrated below for the free system described by

$$\begin{bmatrix} \dot{x}_1(t) \\ \dot{x}_2(t) \end{bmatrix} = \begin{bmatrix} 0 & -2 \\ 1 & -3 \end{bmatrix} \begin{bmatrix} x_1(t) \\ x_2(t) \end{bmatrix} \tag{2.41}$$

Infinite series method: From the definition of e^{At} in (2.30), with $t_0 = 0$, the state transition matrix $\phi(t)$ is given by the infinite series

$$\phi(t) = \sum_{k=0}^{\infty} \frac{A^k t^k}{k!} \tag{2.42}$$

Unless A^k disappears for some small value of k, this is the most laborious method. Once the summation is performed, the closed form of each series for each element of $\phi(t)$ will have to be found as shown below. For A in (2.41),

$$\phi(t) = \underbrace{\begin{bmatrix} 1 & 0 \\ 0 & 1 \end{bmatrix}}_{I} + \underbrace{\begin{bmatrix} 0 & -2 \\ 1 & -3 \end{bmatrix}}_{A} t + \underbrace{\begin{bmatrix} -2 & 6 \\ -3 & 7 \end{bmatrix}}_{A^2} \frac{t^2}{2!} + \underbrace{\begin{bmatrix} 6 & 14 \\ 7 & -15 \end{bmatrix}}_{A^3} \frac{t^3}{3!} + \cdots$$

$$= \left[\begin{array}{c|c} 1 - \frac{2t^2}{2!} + \frac{6t}{3!} + \cdots & -2t + \frac{6t^2}{2!} + \frac{14t^2}{3!} + \cdots \\ \hline t - \frac{3t^2}{2!} + \frac{7t^3}{3!} + \cdots & 1 - 3t + \frac{7t^2}{2!} - \frac{15t^3}{3!} + \cdots \end{array}\right] \tag{2.43}$$

$$\equiv \begin{bmatrix} 2e^{-t} - e^{-2t} & 2(e^{-2t} - e^{-t}) \\ e^{-t} - e^{-2t} & 2e^{-2t} - e^{-t} \end{bmatrix} \tag{2.44}$$

The transition from (2.43) to (2.44) is the most difficult part if $\phi(t)$ were to be computed without the aid of any numerical algorithm.

Using the Cayley-Hamilton Theorem: By using the Cayley-Hamilton theorem (Sec. 1.4), the state transition matrix may be expressed as

$$\phi(t) = \sum_{k=0}^{n-1} \alpha_k(t) A^k \tag{2.45}$$

where the coefficients $\alpha_k(t)$ are to be determined by substituting the eigenvalues of A into the scalar equivalent of (2.45) and obtaining a set of simultaneous equations. For the system in (2.41), the characteristic equation is

$$|\lambda I - A| = \begin{bmatrix} \lambda & 2 \\ -1 & \lambda + 3 \end{bmatrix} = 0 \quad \text{or} \quad \lambda^2 + 3\lambda + 2 = 0$$

from which the two eigenvalues of A are $\lambda_1 = -1$ and $\lambda_2 = -2$. The state transition matrix can then be expressed as

$$\phi(t) = \alpha_0(t) I + \alpha_1(t) A = e^{At} \tag{2.46}$$

and the scalar form of this is

$$\alpha_0(t) + \alpha_1(t)\lambda = e^{\lambda t}$$

Substituting $\lambda = \lambda_1 = -1$ and $\lambda = \lambda_2 = -2$ gives

$$\alpha_0(t) - \alpha_1(t) = e^{-t} \quad \text{and} \quad \alpha_0(t) - 2\alpha_1(t) = e^{-2t} \tag{2.47}$$

Solving the equations above for α_0 and α_1, we obtain

$$\alpha_0(t) = 2e^{-t} - e^{-2t} \quad \text{and} \quad \alpha_1(t) = e^{-t} - e^{-2t}$$

Inserting these values into (2.46), we have

$$\phi(t) = (2e^{-t} - e^{-2t}) \begin{bmatrix} 1 & 0 \\ 0 & 1 \end{bmatrix} + (e^{-t} - e^{-2t}) \begin{bmatrix} 0 & -2 \\ 1 & -3 \end{bmatrix}$$

$$= \begin{bmatrix} 2e^{-t} - e^{-2t} & 2(e^{-2t} - e^{-t}) \\ e^{-t} - e^{-2t} & 2e^{-2t} - e^{-t} \end{bmatrix} \tag{2.48}$$

Solution Through Transformation: The derivation of $\phi(t)$ here is based on a similarity transformation as Theorem 2.1. The transformation matrix T is chosen such that TAT^{-1} is a diagonal (or block diagonal) matrix.* Such a transformation matrix for A in (2.41) is

$$T = \begin{bmatrix} 1 & -1 \\ -1 & 2 \end{bmatrix}$$

The transformed form of (2.41) is, therefore,

*This is known as normalization. The resulting solution for $\bar{x}(t)$ is called the modal solution. T is derived by using (2.51). This is considered in detail in Sec. 2.2.4.

$$\dot{\bar{x}}(t) = TAT^{-1}\bar{x}(t) = \Lambda\bar{x}(t) = \begin{bmatrix} -1 & 0 \\ 0 & -2 \end{bmatrix}\bar{x}(t) \tag{2.49}$$

Assuming a zero initial condition $[x_0(t_0) = 0 = \bar{x}(t_0)]$, the state transition matrix of (2.49) is

$$\bar{\phi}(t) = e^{\Lambda t} = \begin{bmatrix} e^{-t} & 0 \\ 0 & e^{-2t} \end{bmatrix} = Te^{At}T^{-1} = T\phi(t)T^{-1}$$

Therefore,

$$\phi(t) = T\bar{\phi}(t)T^{-1} = \begin{bmatrix} 1 & -1 \\ -1 & 2 \end{bmatrix}\begin{bmatrix} e^{-t} & 0 \\ 0 & e^{-2t} \end{bmatrix}\begin{bmatrix} 2 & 1 \\ 2 & 1 \end{bmatrix}$$

$$= \begin{bmatrix} 2e^{-t} - e^{-2t} & 2(e^{-2t} - e^{-t}) \\ e^{-t} - e^{-2t} & 2e^{-2t} - e^{-t} \end{bmatrix}$$

When the order of A is high, Leverrier's algorithm may be used to develop a computer program [8] for the derivation of $\phi(t)$.

2.2.4 Modal Solution

The derivation of the free and forced solutions of the state equations in (2.25) may be made simpler if the following linear transformation (i.e., a change of basis for the state vector) is chosen:

$$x(t) = Mz(t) \tag{2.50}$$

where M is the modal matrix of A and z(t) is the new state vector after this coordinate transformation. This is a special case of the similarity transformation in Sec. 1.3 and is called the normalization of S(A, B, C, D), with the new representation

$$\begin{aligned} \dot{z}(t) &= M^{-1}AMz(t) + M^{-1}Bu(t) = \Lambda z(t) + B_n u(t) \\ y(t) &= CMz(t) + Du(t) = C_n z(t) + D_n(t) \end{aligned} \tag{2.51}$$

known as the <u>normal form</u> of (2.25), and B_n, C_n, and D_n are the normalized input, normalized output, and transmission matrices.

If A has distinct eigenvalues, then

$$\Lambda = \text{diag}\,\{\lambda_i\}, \quad i \in 1, \ldots, n$$

and the resulting state equation may therefore be expressed as a set of modally uncoupled set of equations:

$$\begin{aligned} \dot{z}_i(t) &= \lambda_i z_i(t) + B_n u(t) \\ &= \lambda_i z_i(t) + f_i(t) \end{aligned} \tag{2.52}$$

where $f_i(t) = \Sigma_{j=1}^{m} b_{nij} u_j(t)$ [b_{nij} = (i, j) element of B_n] in the forcing function applied to the ith normalized state variable; the new coordinates $z_i(t)$ are state vectors of the normalized system. The columns of the modal matrix M form a basis, and the rows of M^{-1} form a reciprocal basis. If the columns of the modal matrix (i.e., the eigenvectors of A) are $v_1, v_2, \ldots, v_n$ (where v_i is the eigenvector associated with the eigenvalue λ_i, $i \in 1, \ldots, n$), and the reciprocal basis is denoted by $r_1, r_2, \ldots, r_n$, the solution of (2.25) may be expressed as

$$x(t) = \left[\sum_{i=1}^{n} \langle r_i, x(t_0) \rangle e^{\lambda_i t} v_i \right] + \left[\int_0^t \sum_{i=1}^{n} \langle r_i, Bu(\tau) \rangle e^{\lambda_i (t-\tau)} v_i \, \delta\tau \right] \tag{2.53}$$

where $x(t_0)$ is the initial state.

In the case of repeated eigenvalues of A, Λ is to be replaced by the block diagonal Jordan matrix J (Sec. 1.3):

$$\begin{aligned} \dot{z}(t) &= Jz(t) + B_n u(t) \\ y(t) &= C_n z(t) + D_n u(t) \end{aligned} \tag{2.54}$$

where $J = M^{-1}AM$, $B_n = M^{-1}B$, $C_n = CM$, and $D_n = D$. These matrices play important roles in the study of the internal behavior of the system, considered in Chap. 3. Examples of the derivation of normal form and modal solutions of the state equation are given below.

Example 2.5: Derivation of normal form—distinct eigenvalue case

$$\begin{aligned} \begin{bmatrix} \dot{x}_1(t) \\ \dot{x}_2(t) \\ \dot{x}_3(t) \end{bmatrix} &= \begin{bmatrix} 2 & -2 & 3 \\ 1 & 1 & 1 \\ 1 & 3 & -1 \end{bmatrix} \begin{bmatrix} x_1(t) \\ x_2(t) \\ x_3(t) \end{bmatrix} + \begin{bmatrix} 1 & 1 \\ 0 & 1 \\ 2 & 1 \end{bmatrix} \begin{bmatrix} u_1(t) \\ u_2(t) \end{bmatrix} \\ \begin{bmatrix} y_1(t) \\ y_2(t) \end{bmatrix} &= \begin{bmatrix} 1 & 2 & 3 \\ 1 & 0 & 2 \end{bmatrix} \begin{bmatrix} x_1(t) \\ x_2(t) \\ x_3(t) \end{bmatrix} + \begin{bmatrix} 1 & 2 \\ 3 & 1 \end{bmatrix} \begin{bmatrix} u_1(t) \\ u_2(t) \end{bmatrix} \end{aligned} \tag{2.55}$$

The characteristic equation (see Example 1.5) is

$$\det\{\lambda I - A\} = \begin{bmatrix} \lambda - 2 & 2 & -3 \\ -1 & \lambda - 1 & -1 \\ -1 & -3 & \lambda + 1 \end{bmatrix} = \lambda^3 - 2\lambda^2 - 5\lambda + 6 = (\lambda - 1)(\lambda + 2)(\lambda - 3)$$

The eigenvalues are $\lambda_1 = 1$, $\lambda_2 = -2$, and $\lambda_3 = 3$.

Method 1: Let

$$\begin{bmatrix} \alpha_{1i} \\ \alpha_{2i} \\ \alpha_{3i} \end{bmatrix}$$

be the eigenvector associated with λ_i, $i \in 1, 2, 3$. From Sec. 1.2 we have

$$[\lambda_i I - A] \begin{bmatrix} \alpha_{1i} \\ \alpha_{2i} \\ \alpha_{3i} \end{bmatrix} = 0 \quad \text{or} \quad \begin{bmatrix} \lambda_i - 2 & 2 & -3 \\ -1 & \lambda_i - 1 & -1 \\ -1 & -3 & \lambda_i + 1 \end{bmatrix} \begin{bmatrix} \alpha_{1i} \\ \alpha_{2i} \\ \alpha_{3i} \end{bmatrix} = 0$$

The eigenvectors are, for $\lambda_1 = 1$, $\lambda_2 = -2$, and $\lambda_3 = 3$, respectively,

$$\begin{bmatrix} -1 \\ 1 \\ 1 \end{bmatrix} \quad \begin{bmatrix} 11 \\ 1 \\ -14 \end{bmatrix} \quad \begin{bmatrix} 1 \\ 1 \\ 1 \end{bmatrix}$$

Consequently, a modal matrix is

$$M = \begin{bmatrix} -1 & 11 & 1 \\ 1 & 1 & 1 \\ 1 & -14 & 1 \end{bmatrix}; \quad M^{-1} = \frac{1}{30} \begin{bmatrix} -15 & 25 & -10 \\ 0 & 2 & -2 \\ 15 & 3 & 12 \end{bmatrix} \tag{2.56}$$

The linear transformation for normalization (2.50) is, therefore,

$$\begin{bmatrix} x_1(t) \\ x_2(t) \\ x_3(t) \end{bmatrix} = \begin{bmatrix} -1 & 11 & 1 \\ 1 & 1 & 1 \\ 1 & -14 & 1 \end{bmatrix} \begin{bmatrix} z_1(t) \\ z_2(t) \\ z_3(t) \end{bmatrix} \tag{2.57}$$

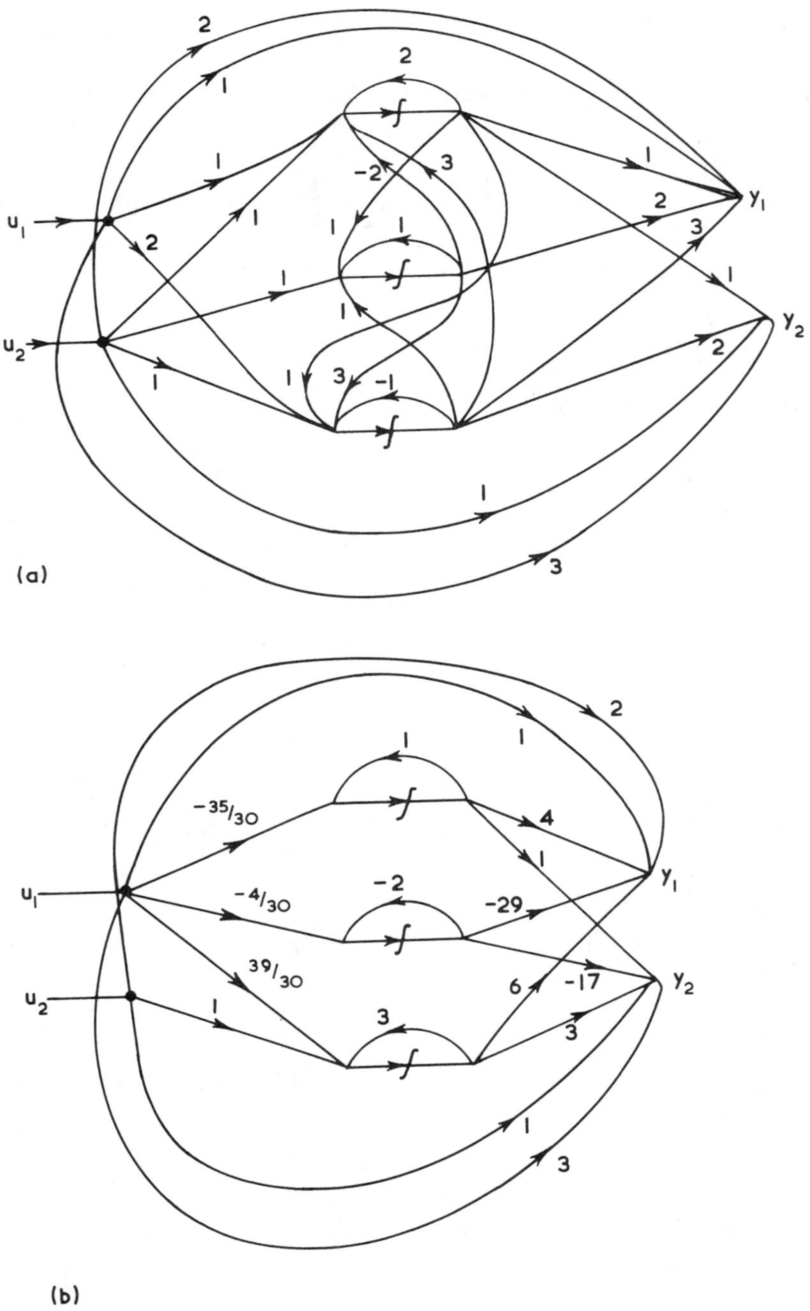

Fig. 2.14 Signal flow graphs of Example 2.5: (a) signal flow graph of the state and output equations in standard form (2.55); (b) signal flow graph of the state and output equations in normal form (2.58).

The normalized state and output equations are

$$\dot{z}(t) = M^{-1}AMz(t) + M^{-1}Bu(t)$$

$$= \begin{bmatrix} 1 & 0 & 0 \\ 0 & -2 & 0 \\ 0 & 0 & 3 \end{bmatrix} z(t) + \frac{1}{30}\begin{bmatrix} -35 & 0 \\ -4 & 0 \\ 39 & 30 \end{bmatrix} u(t)$$

$$y(t) = CMz(t) + Du(t) \tag{2.58}$$

$$= \begin{bmatrix} 4 & -29 & 6 \\ 1 & -17 & 3 \end{bmatrix} z(t) + \begin{bmatrix} 1 & 2 \\ 3 & 1 \end{bmatrix} u(t)$$

The signal flow graphs of (2.55) and (2.58) are shown in Fig. 2.14. The signal flow graph of the normalized equations in (2.58) consists of three "modally uncoupled" first-order subsystems in parallel, each driven by the input controls $u_1(t)$ and $u_2(t)$. Since this normalization (also called diagonalization) is accomplished by the change of variable $x(t) = Mz(t)$, where the columns of M are the eigenvalues of A, the new state variables $z_1(t)$, $z_2(t)$, and $z_3(t)$ are just the coordinates of the original state vector x(t) with respect to the eigenvector basis

$$x(t) = z_1(t)\begin{bmatrix} \alpha_{11} \\ \alpha_{12} \\ \alpha_{13} \end{bmatrix} + z_2(t)\begin{bmatrix} \alpha_{21} \\ \alpha_{22} \\ \alpha_{23} \end{bmatrix} + z_3(t)\begin{bmatrix} \alpha_{31} \\ \alpha_{32} \\ \alpha_{33} \end{bmatrix}$$

Thus the outputs of the integrators in the "uncoupled" representation decompose the state trajectory into its components along the eigenvectors of A—hence the name fundamental modes for these fundamental motions along the eigenvectors of A.

Method 2: The adjoint matrix adj $[\lambda I - A]$ is

$$\begin{bmatrix} \lambda^2 - 4 & -2\lambda + 7 & 3\lambda - 5 \\ \lambda + 2 & \lambda^2 - \lambda - 5 & \lambda + 1 \\ \lambda + 2 & 3\lambda - 8 & \lambda^2 - 3\lambda + 4 \end{bmatrix}$$

For $\lambda = \lambda_i$, $i \in 1, 2, 3$, the adjoint matrices are

$$\lambda_1 = 1 \qquad\qquad \lambda_2 = -2 \qquad\qquad \lambda_3 = 3$$

$$\begin{bmatrix} -3 & 5 & -2 \\ 3 & -5 & 2 \\ 3 & 5 & 2 \end{bmatrix}; \quad \begin{bmatrix} 0 & 11 & -11 \\ 0 & 1 & -1 \\ 0 & -14 & 14 \end{bmatrix}; \quad \begin{bmatrix} 5 & 1 & 4 \\ 5 & 1 & 4 \\ 5 & 1 & 4 \end{bmatrix}$$

The eigenvectors are uniquely determined only in direction; these vectors can be multiplied by any scalar and still remain the eigenvectors of A. By definition,

$$[\lambda_i - A]\begin{bmatrix}\alpha_{1i}\\ \alpha_{2i}\\ \alpha_{3i}\end{bmatrix} = 0$$

Consequently, a modal matrix is $M = \begin{bmatrix}-1 & 11 & 1\\ 1 & 1 & 1\\ 1 & -14 & 1\end{bmatrix}$

Example 2.6: Repeated eigenvalues

$$\begin{bmatrix}\dot{x}_1(t)\\ \dot{x}_2(t)\\ \dot{x}_3(t)\end{bmatrix} = \begin{bmatrix}2 & 1 & 1\\ 1 & 2 & 1\\ 0 & 0 & 1\end{bmatrix}\begin{bmatrix}x_1(t)\\ x_2(t)\\ x_3(t)\end{bmatrix} + \begin{bmatrix}1 & 1\\ 0 & 1\\ 2 & 1\end{bmatrix}\begin{bmatrix}u_1(t)\\ u_2(t)\end{bmatrix} \tag{2.59a}$$

$$\begin{bmatrix}y_1(t)\\ y_2(t)\end{bmatrix} = \begin{bmatrix}1 & 2 & 3\\ 1 & 0 & 2\end{bmatrix}\begin{bmatrix}x_1(t)\\ x_2(t)\\ x_3(t)\end{bmatrix} + \begin{bmatrix}1 & 2\\ 2 & 1\end{bmatrix}\begin{bmatrix}u_1(t)\\ u_2(t)\end{bmatrix} \tag{2.59b}$$

The roots of the characteristic equation det $\{\lambda I - A\} = 0$ are $\lambda_{1,2} = 1$ and $\lambda_3 = 3$. The degeneracy of $[\lambda_i - A]$ for $\lambda_i = 1$ is equal to 2. Since the characteristic matrix has "full degeneracy," it is possible to obtain a linearly independent vector solution for each of the repeated roots. The adjoint matrix (Example 1.7) is

$$\text{adj}\,[\lambda I - A] = \begin{bmatrix}(\lambda - 2)(\lambda - 1) & \lambda - 1 & \lambda - 1\\ \lambda - 1 & (\lambda - 2)(\lambda - 1) & \lambda - 1\\ 0 & 0 & (\lambda - 1)(\lambda - 3)\end{bmatrix}$$

Substituting $\lambda_i = 1$ in any column of the adjoint matrix yields a null column. The first derivative of the adjoint matrix is

$$\frac{d}{d\lambda}\{\text{adj}\,[\lambda I - A]\} = \begin{bmatrix}2\lambda - 3 & 1 & 1\\ 1 & 2\lambda - 3 & 1\\ 0 & 0 & 2\lambda - 3\end{bmatrix} = \begin{bmatrix}-1 & 1 & 1\\ 1 & -1 & 1\\ 0 & 0 & -1\end{bmatrix} \quad \text{for } \lambda = 1$$

$$= \begin{bmatrix} -1 & 1 \\ 1 & 1 \\ 0 & -2 \end{bmatrix} \begin{bmatrix} 1 & -1 & 0 \\ 0 & 0 & 1 \end{bmatrix}$$

Hence the two eigenvectors corresponding to the repeated root $\lambda = 1$ are given by

$$\begin{bmatrix} \alpha_{11} \\ \alpha_{21} \\ \alpha_{31} \end{bmatrix} = \begin{bmatrix} 1 \\ 1 \\ 0 \end{bmatrix} \quad \text{and} \quad \begin{bmatrix} \alpha_{21} \\ \alpha_{22} \\ \alpha_{23} \end{bmatrix} = \begin{bmatrix} 1 \\ 1 \\ -2 \end{bmatrix}$$

The eigenvector corresponding to $\lambda = 3$ can be chosen to be proportional to any nonzero column of adj $[\lambda I - A]$ and is

$$\begin{bmatrix} \alpha_{31} \\ \alpha_{32} \\ \alpha_{33} \end{bmatrix} = \begin{bmatrix} 1 \\ 1 \\ 0 \end{bmatrix}$$

Hence the modal matrix of (2.59) is

$$M = \begin{bmatrix} -1 & 1 & 1 \\ 1 & 1 & 1 \\ 0 & -2 & 0 \end{bmatrix} ; \quad M^{-1} = \frac{1}{2} \begin{bmatrix} -1 & 1 & 0 \\ 0 & 0 & -1 \\ 1 & 1 & 1 \end{bmatrix}$$

The normalized state and output equations of (2.59) are

$$\dot{z}(t) = \begin{bmatrix} 1 & 0 & 0 \\ 0 & 1 & 0 \\ 0 & 0 & 3 \end{bmatrix} z(t) + \frac{1}{2} \begin{bmatrix} -1 & 0 \\ -2 & -1 \\ 3 & 3 \end{bmatrix} u(t)$$

$$y(t) = \begin{bmatrix} 3 & -3 & 3 \\ -1 & -3 & 1 \end{bmatrix} z(t) + \begin{bmatrix} 1 & 2 \\ 2 & 1 \end{bmatrix} u(t) \tag{2.60}$$

The signal flow graphs of (2.59) and (2.60) are shown in Fig. 2.15.

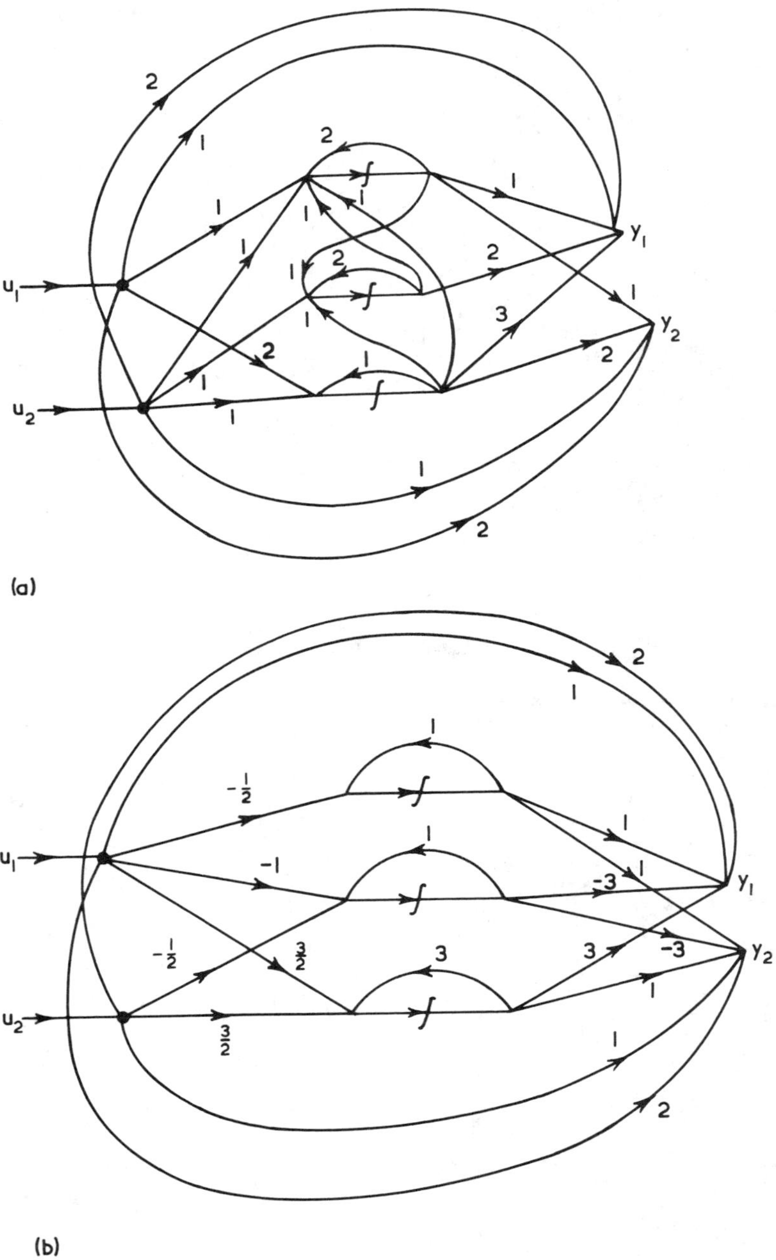

Fig. 2.15 Signal flow graphs of Example 2.6: (a) signal flow graph of the standard state and output equations (2.59); (b) signal flow graph of the normalized state equations (2.60).

Example 2.7: Modal solution. The following two-input/two-output third-order system is considered:

$$\dot{x}(t) = \begin{bmatrix} -2 & 1 & 0 \\ -1 & -1 & 0 \\ 0 & 2 & -1 \end{bmatrix} x(t) + \begin{bmatrix} 1 & 0 \\ 0 & 0 \\ 0 & 2 \end{bmatrix} u(t)\,; \quad y(t) = \begin{bmatrix} 2 & 0 & 0 \\ 0 & 0 & 1 \end{bmatrix} x(t) \tag{2.61}$$

The state transition matrix for this system is obtained here by using Theorem 2.1 and the transformation

$$T = \begin{bmatrix} 0 & -1 & -1 \\ 0 & \frac{1}{2} + j\frac{3}{2} & \frac{1}{2} - j\frac{3}{2} \\ 1 & 0 & 0 \end{bmatrix}; \quad \lambda_1 = -1, \quad \lambda_{2,3} = -\frac{3}{2}(1 \pm j)$$

$$\phi(\tau) = T\bar{\phi}(\tau)T^{-1} = \left[\begin{array}{c|c|c} \frac{1}{2}(1 + j/\sqrt{3})e^{\lambda_2 \tau} + \frac{1}{2}(1 - j/\sqrt{3})e^{\lambda_3 \tau} & j(e^{\lambda_2 \tau} - e^{\lambda_3 \tau})/\sqrt{3} & 0 \\ \hline j(e^{\lambda_2 \tau} - e^{\lambda_3 \tau})/2\sqrt{3} & \frac{1}{2}(1 - j/\sqrt{3})e^{\lambda_2 \tau} + \frac{1}{2}(1 + j/\sqrt{3})e^{\lambda_3 \tau} & 0 \\ \hline 0 & 0 & e^{\lambda_1 \tau} \end{array}\right],$$

where $\tau = t - t_0$. From (2.34),

$$y(t) = C\phi(t - t_0)x(t_0) + C\int_{t_0}^{t} \phi(t - p)Bu(p)dp \quad (D \equiv 0)$$

$$= \left[\begin{array}{c|c|c} (1 + j/\sqrt{3})e^{\lambda_2 \tau} + (1 - j/\sqrt{3})e^{\lambda_3 \tau} & j2(e^{\lambda_2 \tau} - e^{\lambda_3 \tau})/3 & 0 \\ \hline 0 & 0 & e^{\lambda_1 \tau} \end{array}\right] \begin{bmatrix} x_1(0) \\ x_2(0) \\ x_3(0) \end{bmatrix}$$

$$+\int_{t_0}^{t}\left[\begin{array}{c|c}(1+j/\sqrt{3})e^{\lambda_2\tau} & \\ +(1-j/\sqrt{3})e^{\lambda_3\tau} & 0 \\ \hline 0 & 1\end{array}\right]\begin{bmatrix}u_1(t)\\ u_2(t)\end{bmatrix} \tag{2.62}$$

Example 2.8: Modal solution of $\dot{x}(t) = Ax(t)$ using (2.53)

$$A = \begin{bmatrix} 0 & 1 \\ -2 & -3 \end{bmatrix} \tag{2.63}$$

The eigenvalues are $\lambda_1 = -1$ and $\lambda_2 = -2$. The eigenvectors are given by $Av_i = \lambda_i v_i$, $i = 1, 2$. For

$$\lambda = \lambda_1:\ v_1 = \begin{bmatrix} \frac{1}{\sqrt{2}} \\ -\frac{1}{\sqrt{2}} \end{bmatrix}; \quad \lambda = \lambda_2:\ v_2 = \begin{bmatrix} \frac{1}{\sqrt{5}} \\ -\frac{2}{\sqrt{5}} \end{bmatrix}$$

The reciprocal bases are given by $\langle r_i, v_j \rangle = \delta_{ij}$ and are

$$\lambda = \lambda_1:\ r_1 = \begin{bmatrix} 2\sqrt{2} \\ \sqrt{2} \end{bmatrix}; \quad \lambda = \lambda_2:\ r_2 = \begin{bmatrix} -\sqrt{5} \\ -\sqrt{5} \end{bmatrix}$$

The initial response of the system is thus

$$x(t) = \sum_{i=1}^{2} \langle r_i, x(t_0) \rangle e^{\lambda_i t} v_i$$

$$= [2\sqrt{2}\,y(t_0) + \sqrt{2}\,\dot{y}(t_0)]e^{-t}v_1 - [\sqrt{5}\,y(t_0) + \sqrt{5}\,\dot{y}(t_0)]\,e^{-2t}v_2$$

Example 2.9: Time-varying system. The free system

$$\begin{bmatrix} \dot{x}_1(t) \\ \dot{x}_2(t) \end{bmatrix} = \begin{bmatrix} a & e^{-t} \\ -e^{-t} & a \end{bmatrix} \begin{bmatrix} x_1(t) \\ x_2(t) \end{bmatrix} \tag{2.64}$$

is considered here to illustrate the validity of the method above for the computation of $\phi(t, \tau)$.

The eigenvalues are $\lambda_{1,2} = a \pm je^{-t}$, and the matrix

$$T = \begin{bmatrix} j & j \\ -1 & -1 \end{bmatrix}$$

transforms the system above to (Theorem 2.1)

$$\dot{\bar{x}}(t) = \begin{bmatrix} a - je^{-t} & 0 \\ 0 & a + je^{-t} \end{bmatrix} \bar{x}(t) = \Lambda(t)\bar{x}(t)$$

For the system above,

$$\bar{\phi}(t,\tau) = T[e^{\int_\tau^t \Lambda(p)dp}]T^{-1}$$

$$= e^{a(t-\tau)} \begin{bmatrix} \cos(e^{-\tau} - e^{t}) & \sin(e^{-\tau} - e^{-t}) \\ -\sin(e^{-\tau} - e^{-t}) & \cos(e^{-\tau} - e^{-t}) \end{bmatrix}$$

2.3 INPUT/OUTPUT DESCRIPTION [1-10]

In developing this description, knowledge of the internal structure or parameters of the system is assumed to be unavailable, the only access to the system being by means of the input/output terminals, shown in Fig. 2.16.

The time intervals in which the inputs and outputs will be defined are from $-\infty$ to $+\infty$. In special cases where the input control is defined over a finite interval (t_0, t_1), the notation $u_{(t_0, t_1)}$ will be used, with similar notation being used for output variable y. Thus if an input $u_{(t_1, \infty)}$ is applied to a dynamic system, by definitions in Sec. 2.1, unless the input applied before t_1 is known, the output $y_{(t_1, \infty)}$ is generally not uniquely determinable. Hence in developing the input/output relation, before an input is applied the system

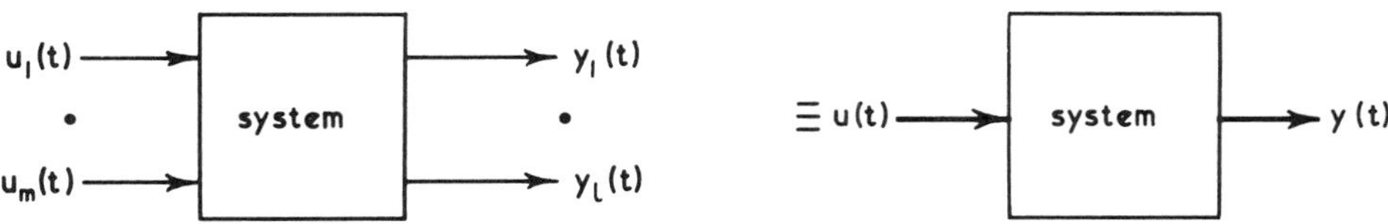

Fig. 2.16 Input/output representation of an m-input/ℓ-output system.

must be assumed to be relaxed or at rest, and the output excited solely and and uniquely by the input applied thereafter. If the concept of energy is applicable to a system, the system is said to be relaxed at time t_1 if no energy is stored in the system at that instant. Under the relaxedness assumption it is thus possible to write

$$y(t) = Gu(t) \tag{2.65}$$

where G is a matrix operator or function that uniquely specifies the output y(t) in terms of the input u(t) of the system. Systems satisfying (2.65) are said to be initially relaxed at $-\infty$, or simply, relaxed. Thus for a linear, causal, and time invariant system, the output response may be expressed as

$$y(t) = \int_{t_0}^{\infty} H(t - \tau)u(\tau)\, d\tau \tag{2.66}$$

where H is the impulse response matrix (2.36).

In the time-invariant case, the initial time t_0 may always be chosen to be 0, without any loss of generality; from (2.66), then, the output is given by the convolution integral

$$y(t) \triangleq \int_{0}^{\infty} H(t - \tau)u(\tau)\, d\tau = \int_{0}^{\infty} H(\tau)u(t - \tau)\, d\tau \tag{2.67}$$

In a study of the class of systems that are describable by convolution integrals, it is a great advantage to use the Laplace transform, since this transform changes a convolution integral in the time domain into an algebraic equation in the frequency domain. Thus

$$\begin{aligned} y(s) &= \mathscr{L}y(t) = \int_{-\infty}^{\infty} y(t)e^{-st}\, dt = \int_{-\infty}^{\infty} \left[\int_{-\infty}^{\infty} H(t - \tau)u(\tau)\, d\tau \right] e^{-st}\, dt \\ &= \int_{-\infty}^{\infty} \left[\int_{-\infty}^{\infty} H(t - \tau)e^{-s(t-\tau)}\, d\tau \right] u(\tau)e^{-s\tau}\, d\tau \\ &= \int_{-\infty}^{\infty} H(p)e^{-sp}\, dp \int_{-\infty}^{\infty} u(\tau)e^{-s\tau}\, d\tau \\ &= G(s)u(s) \qquad [H(t) = 0 \text{ for } t < 0] \end{aligned} \tag{2.68}$$

where $G(s) = \int_{0}^{\infty} H(t)e^{-st}\, dt$. G(s), the Laplace transform of the impulse-response matrix, is called the <u>transfer-function matrix</u> of the system. For single-input/single-output systems, G(s) reduces to a scalar and is given by

$$g(s) = \left.\frac{\mathscr{L}y(t)}{\mathscr{L}u(t)}\right|_{\substack{\text{system is}\\ \text{relaxed at}\\ t=0}} = \left.\frac{y(s)}{u(s)}\right|_{\substack{\text{system is}\\ \text{relaxed at}\\ t=0}}$$

2.3.1 Transfer-Function Matrix

In the study of the class of systems that are describable by convolution integrals, it is of great advantage to use the Laplace transform, because, as mentioned earlier, it will change a convolution integral in the time domain into an algebraic equation in the frequency domain. If y(s) is the Laplace transform of y(t),* then using (2.25), we have

$$y(s) \triangleq \mathscr{L}[y(t)] = \int_0^\infty y(t)e^{-st}\,dt = \int_0^\infty [Cx(t) + Du(t)]e^{-st}\,dt$$

$$= C\int_0^\infty x(t)e^{-st}\,dt + D\int_0^\infty u(t)e^{-st}\,dt$$

$$= Cx(s) + Du(s) \tag{2.69}$$

where x(s) and u(s) are the Laplace transform of x(t) and u(t), respectively.

Also, from (2.25), taking the Laplace transform of the state equation with x(0) as the initial value of the state vector, we obtain

$$(sI_n - A)x(s) = x(0) + Bu(s)$$

$$x(s) = (sI_n - A)^{-1}x(0) + (sI_n - A)^{-1}Bu(s) \tag{2.70}$$

Combining (2.69) and (2.70) gives us

$$y(s) = C(sI_n - A)^{-1}x(0) + C(sI_n - A)^{-1}Bu(s) + Du(s) \tag{2.71}$$

For any arbitrary initial state x(0) and an input control u(s), the state and output responses in the frequency domain can be computed from (2.70) and

*If y contains delta functions at t = 0, the lower limit of the integration should start from 0- to include the delta functions in the transform.

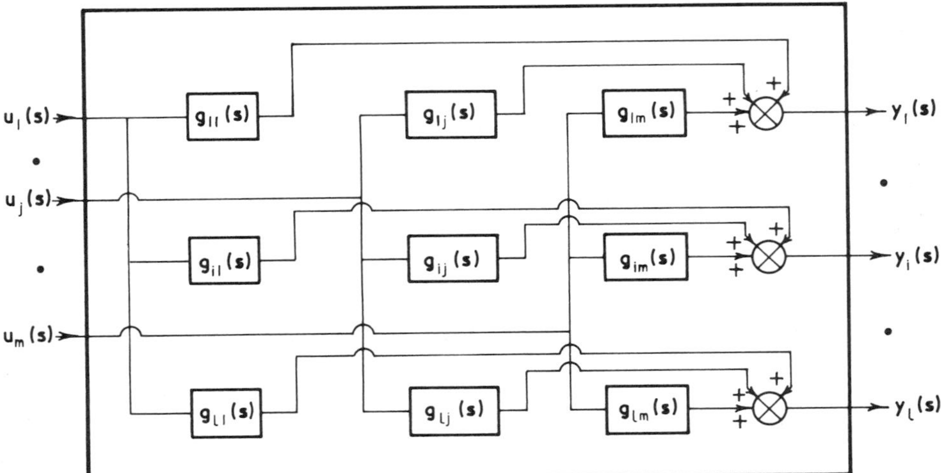

Fig. 2.17 Transfer-function-matrix representation of an m-input/ℓ-output system

(2.71) provided that $[sI - A]^{-1}$, called the resolvent matrix,* exists. If the initial state x(0) is 0, and if the system is relaxed at t = 0, then (2.71) reduces to

$$y(s) = [C(sI_n - A)^{-1}B + D]u(s) = G(s)u(s) \tag{2.72}$$

where

$$G(s) = [C(sI_n - A)^{-1}B + D], \quad \text{an } (\ell \times m) \text{ matrix} \tag{2.73}$$

G(s) is known as the transfer-function matrix of the m-input/ℓ-output linear time-invariant system S(A, B, C, D) described by (2.25). The general form of G(s) is

$$G(s) = \begin{bmatrix} g_{11}(s) & g_{12}(s) & \cdots & g_{1m}(s) \\ g_{21}(s) & g_{22}(s) & \cdots & g_{2m}(s) \\ \vdots & & & \\ g_{\ell 1}(s) & g_{\ell 2}(s) & \cdots & g_{\ell m}(s) \end{bmatrix} \tag{2.74}$$

*$\phi(s) = (sI - A)^{-1}$, where $\phi(s)$ is the state transition matrix as defined in Sec. 2.2.3.

$$y(s) = \{y_i(s)\}^T, \quad i \in 1, \ldots, \ell$$

$$u(s) = \{u_j(s)\}^T, \quad j \in 1, \ldots, m$$

where $g_{ij}(s) = c_i(sI - A)^{-1}b_j + d_{ij}(s)$, c_i being the ith row of C, b_j the jth column of B, and $D = \{d_{ij}\}$. The transfer from multi-input control to multi-output response in the frequency domain is thus accomplished by the matrix in (2.74), in contrast to a scalar transfer function necessary for similar description for single-input/single-output systems; a schematic representation of (2.73) is shown in Fig. 2.17, which, in contrast with Fig. 2.1, reflects the effect of any one input control on more than one (generally all) output variables.

By using the properties associated with the functions of a matrix from Sec. 1.4, for $|s| > 0$,

$$(sI - A)^{-1} = s^{-1}\left(I - \frac{A}{s}\right)^{-1} = \frac{1}{s}\sum_{k=0}^{\infty}\frac{A^k}{s^k} \tag{2.75}$$

Hence from (2.73),*

$$\lim_{s \to \infty} G(s) = D \tag{2.76}$$

which implies that the transmission matrix D indicates the high-frequency gain of the system. If D(s) is finite, the system is said to be proper. In the special case when the matrix D(s) is null [lim $s \to \infty$ G(s) $\to$ 0] such a system is said to be strictly proper. If all entries in the system matrix G(s) are rational functions in s, it is said to be in rational form. In cases where $D \not\equiv 0$, the system is said to be improper or nonproper.

The use of transfer-function matrices for system analysis has two advantages. One is that transfer functions allow the use of algebraic equations rather than differential equations. The second is due to the smaller size of the transfer-function matrix. For example, if A has n state vari- m(< n) inputs, and ℓ(< n) outputs, the size of A is $(n \times n)$, whereas G(s) is of the order $(\ell \times m) < (n \times n)$. However, there are limitations in using transfer-function matrices. One is that the transfer function of a composite system (a collection of a number of subsystems) may mask one or more mode(s) of the subsystem(s) through pole-zero cancellations. The other limitation is that although the elements of the transfer-function matrix may indicate all modes of the subsystem, the matrix may fail to indicate the modes of the composite system. Methods of deriving state equations from transfer functions, and vice versa, are presented in Chap. 4.

*The notation D(s) is used subsequently to indicate that the transmission matrix may, in some representation, be dependent on s.

2.3.2 State-Space and Transfer Functions

For a one-input/one-output system of order n, it can be shown from (2.11) to (2.17) that the input/output transfer function has the form

$$g(s) = \frac{n(s)}{d(s)} = \frac{\beta_m s^m + \beta_{m-1} s^{m-1} + \cdots + \beta_1 s + \beta_0}{\alpha s^n + \alpha_{n-1} s^{n-1} + \cdots + \alpha_1 s + \alpha_0} = \frac{\prod_{i=1}^{m} (s - z_i)}{\prod_{i=1}^{n} (s - p_i)}, \qquad m < n \tag{2.77}$$

where z_i and p_i are the zeros and poles of the system, and n(s) and d(s) are coprime. The difference between $\delta(n)$ and $\delta(d)$ is known as the relative order of g(s).

In the multivariable case, poles and zeros can be evaluated in a similar, but computationally rather involved manner. The transfer-function matrix between $\{u_j(s)\}$, $j \in 1, \ldots, m$, and $\{y_i(s)\}$, $i \in 1, \ldots, \ell$, from (2.74) is

$$\begin{aligned} G(s) = \{g_{ij}(s)\} &= \frac{C \text{ adj } (sI - A)B + \Delta(s)D}{\Delta(s)} \\ &= \frac{\{c_i^r\} \text{ adj } (sI - A)\{b_j^c\} + \Delta(s)\{d_{ij}\}}{\Delta(s)} \end{aligned} \tag{2.78}$$

That is,

$$g_{ij}(s) = \frac{c_i^r \text{ adj } (sI - A)b_j^c + \Delta(s)d_{ij}}{\Delta(s)} \tag{2.79}$$

represents the transfer function between any jth input $u_j(s)$ and ith output $y_i(s)$, c_i^r and b_j^c being the ith row of C and the jth column of B, respectively.

Hence the poles of the (i, j)th transfer function are given by

$$\Delta(s) = |sI - A| = 0, \qquad \forall (i, j) \tag{2.80}$$

and its zeros by

$$n_{ij}(s) = c_i^r \text{ adj } (sI - A)b_j^c + \Delta(s)d_{ij} \tag{2.81}$$

where $n_{ij}(s)$ is an analytic polynomial in s whose degree is at most n - 1.

Since the characteristic equation (2.80) is the same for each of the ($\ell \times m$) subsystems of (2.74), the matrix G(s) is said to have n poles given by its characteristic equation (2.80). This result and (1.36-1.37) suggest that the poles of the transfer-function matrix G(s) in (2.79) and the eigenvalues of A are identical. A similar statement, however, does not apply to the zeros of a transfer-function matrix given by (2.81). It should be noted here that if the number of zeros of $g_{ij}(s)(z_{ij})$ is the same as the number of poles (n), then $d_{ij} \neq 0$. If $z_{ij} < n$, then $d_{ij} = 0$, since $\Delta(s)$ is a polynomial of degree n.

Example 2.10: Impulse-response matrix and transfer-function matrix [1]. The system in Fig. 2.18 is used to illustrate the relationship between the impulse-response matrix and the transfer-function matrix. By using the selection of state and output variables shown, the following equations may be formulated:

$$
\begin{aligned}
\dot{x}_1(t) &= -\frac{2}{(R_1 + R_2)C}x_1(t) + \frac{1}{(R_1 + R_2)C}u_1(t) + \frac{1}{(R_1 + R_2)C}u_2(t) \\
\dot{x}_2(t) &= -\frac{2R_1R_2}{(R_1 + R_2)L}x_2(t) + \frac{R_2}{(R_1 + R_2)L}u_1(t) - \frac{R_2}{(R_1 + R_2)L}u_2(t) \\
y_1(t) &= -\frac{1}{R_1 + R_2}x_1(t) - \frac{R_2}{R_1 + R_2}x_2(t) + \frac{1}{R_1 + R_2}u_1(t) \\
y_2(t) &= -\frac{1}{R_1 + R_2}x_1(t) + \frac{R_2}{R_1 + R_2}x_2(t) + \frac{1}{R_1 + R_2}u_2(t)
\end{aligned}
\tag{2.82}
$$

The impulse-response matrix, from (2.37), is

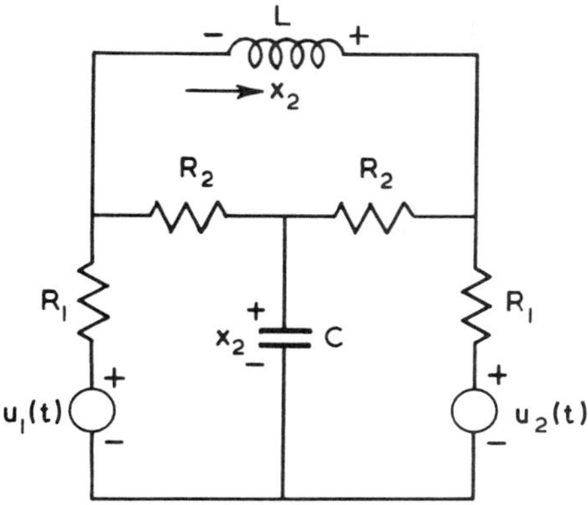

Fig. 2.18 Electrical network with two input voltages and two output variables ($R_1 = R_2 = 1\ \Omega$, $L = 1$ H, $C = 1$ F).

$$H(t) = C\phi(t)B + u_0(t)D$$

$$= \begin{bmatrix} -\frac{1}{R} & -\frac{R_2}{R} \\ -\frac{1}{R} & \frac{R_2}{R} \end{bmatrix} \begin{bmatrix} e^{-\frac{2}{RC}t} & 0 \\ 0 & e^{-\frac{2R_1R_2}{RL}t} \end{bmatrix} \begin{bmatrix} \frac{1}{RC} & \frac{1}{RC} \\ \frac{R_2}{RL} & -\frac{R_2}{RL} \end{bmatrix} + u_0(t) \begin{bmatrix} \frac{1}{R} & 0 \\ 0 & \frac{1}{R} \end{bmatrix}$$

$$= \begin{bmatrix} -k_1e^{-\alpha t} - k_2e^{-\beta t} & -k_1e^{-\alpha t} + k_2e^{-\beta t} \\ -k_1e^{-\alpha t} + k_2e^{-\beta t} & -k_1e^{-\alpha t} - k_2e^{-\beta t} \end{bmatrix} + u_0(t) \begin{bmatrix} \frac{1}{R} & 0 \\ 0 & \frac{1}{R} \end{bmatrix}$$

where $k_1 = 1/R^2C$, $k_2 = R_2^2/R^2L$, $\alpha = (2/RC)t$, $\beta = 2R_1R_2/RL$, and $R = R_1 + R_2$. Then the transfer-function matrix, from (2.68), is

$$G(s) = \int_{-\infty}^{\infty} H(t)e^{-st}\, dt$$

$$= \begin{bmatrix} -\frac{k_1}{s+\alpha} - \frac{k_2}{s+\beta} & -\frac{k_1}{s+\alpha} + \frac{k_2}{s+\beta} \\ -\frac{k_1}{s+\alpha} + \frac{k_2}{s+\beta} & -\frac{k_1}{s+\alpha} - \frac{k_2}{s+\beta} \end{bmatrix} + \begin{bmatrix} \frac{1}{R} & 0 \\ 0 & \frac{1}{R} \end{bmatrix} \tag{2.83}$$

Example 2.11: Derivation of transfer-function matrix from S(A,B,C,D)

$$\begin{bmatrix} \dot{x}_1(t) \\ \dot{x}_2(t) \\ \dot{x}_3(t) \end{bmatrix} = \begin{bmatrix} 0 & 1 & 0 \\ 0 & 0 & 1 \\ -6 & -11 & -6 \end{bmatrix} \begin{bmatrix} x_1(t) \\ x_2(t) \\ x_3(t) \end{bmatrix} + \begin{bmatrix} 0 & 1 \\ 0 & 0 \\ 1 & 1 \end{bmatrix} \begin{bmatrix} u_1(t) \\ u_2(t) \end{bmatrix}$$

$$y(t) = \begin{bmatrix} 1 & 1 & 0 \\ 0 & 0 & 1 \end{bmatrix} \begin{bmatrix} x_1(t) \\ x_2(t) \\ x_3(t) \end{bmatrix} + \begin{bmatrix} 1 & 0 \\ 0 & 1 \end{bmatrix} \begin{bmatrix} u_1(t) \\ u_2(t) \end{bmatrix} \tag{2.84}$$

The initial condition is $x(t_0) = 0$. For this system,

$$sI - A = \begin{bmatrix} s & -1 & 0 \\ 0 & s & -1 \\ 6 & 11 & s+6 \end{bmatrix} \quad \text{and} \quad \det\{sI - A\} = \Delta = (s+1)(s+2)(s+3)$$

$$[sI - A]^{-1} = \frac{\text{adj}\,(sI - A)}{\Delta} = \frac{1}{\Delta}\begin{bmatrix} s^2 + 6s + 11 & s + 6 & 1 \\ -6 & s^2 + 6s & s \\ -6s & -11s - 6 & s^2 \end{bmatrix} \tag{2.85}$$

$$\begin{aligned}
G(s) &= C(sI - A)^{-1}B + D \\
&= \begin{bmatrix} 1 & 1 & 0 \\ 0 & 0 & 1 \end{bmatrix}\begin{bmatrix} s^2 + 6s + 11 & s + 6 & 1 \\ -6 & s^2 + 6s & s \\ -6s & -11s - 6 & s^2 \end{bmatrix}\begin{bmatrix} 0 & 1 \\ 0 & 0 \\ 1 & 1 \end{bmatrix} + \begin{bmatrix} 1 & 0 \\ 0 & 1 \end{bmatrix} \\
&= \frac{1}{\Delta}\begin{bmatrix} s^2 + 6s - 5 & s^2 + 7s - 5 \\ -6s & s^2 \end{bmatrix} + \begin{bmatrix} 1 & 0 \\ 0 & 1 \end{bmatrix} \\
&= \begin{bmatrix} \dfrac{s^2 + 6s - 5 + (s + 1)(s + 2)(s + 3)}{(s + 1)(s + 2)(s + 3)} & \dfrac{s^2 + 7s - 5}{(s + 1)(s + 2)(s + 3)} \\ -\dfrac{6s}{(s + 1)(s + 2)(s + 3)} & \dfrac{s^2 + (s + 1)(s + 2)(s + 3)}{(s + 1)(s + 2)(s + 3)} \end{bmatrix} \\
&= \begin{bmatrix} \dfrac{s \ + 7s^2 + 17s + 1}{(s + 1)(s + 2)(s + 3)} & \dfrac{s^2 + 7s - 5}{(s + 1)(s + 2)(s + 3)} \\ -\dfrac{6s}{(s + 1)(s + 2)(s + 3)} & \dfrac{s^3 + 7s^2 + 11s + 6}{(s + 1)(s + 2)(s + 3)} \end{bmatrix}
\end{aligned} \tag{2.86}$$

2.3.3 Poles and Zeros of G(s) [11-14]

A single-input/single-output system with transfer function

$$g(s) = \frac{n(s)}{d(s)}, \quad \text{where } \delta\{n(s)\} < \delta\{d(s)\} \tag{2.87}$$

is said to have a zero at z if it vanishes at s = z [i.e., if $g(z) \equiv 0$]. In the multivariable case, however, a zero of a transfer-function matrix G(s) is not a value of s for which G(s) becomes a null matrix, but a value that reduces the rank of G(s) below its normal rank.* Computation of the zeros of a multivariable system may thus not always be possible using (2.81).

*It is to be noted in this context that the rank of a rational function matrix G(s) is evaluated by first substituting a specific real or complex value of s (= z) and then by computing the rank of G(z). Any value of $\rho[G(z)]$ less than the normal rank of G(s) is called a <u>local rank</u> of G(s) for that specific value of z.

By following the same argument, a real or complex number p is said to be a pole of the scalar transfer function g(s) in (2.87) if $g(p) = \infty$. If the eigenvalues of A are distinct, for any eigenvalue λ_i, the transfer-function matrix may be written as*

$$G(s) = \sum_{i=1}^{n} \frac{c_{ni}^{c} b_{ni}^{r}}{s - \lambda_i} = \sum_{i=1}^{n} \frac{G_i}{s - \lambda_i} \tag{2.88}$$

where $G_i = \lim s \to \lambda_i \; (s - \lambda_i)G(s)$ is called the residue matrix for G(s) at λ_i. From the equation above for finite values of λ_i,

$$\lim_{s \to \lambda_i} \|G(s)\| = \lim_{s \to \infty} \left\| \sum_{i=1}^{n} \frac{G_i}{s - \lambda_i} \right\| \to \infty, \qquad i \in 1, \ldots, n$$

Thus a natural extension of the single-input/single-output system provides the following definition of the pole of a transfer-function matrix: A complex or real number p is said to be a pole of order n of a transfer-function matrix G(s) if some element of G(s) has a pole of order n at $s = p$ and no element has a pole of larger order than n at p.

With this notion the poles of an m-input/ℓ-output transfer-function matrix† is assumed to include the set of all the poles of its ℓm scalar transfer-function matrices

$$g_{ij}(s) = \frac{c_i^r \operatorname{adj}(sI - A) b_j^c}{\det(sI - A)} \tag{2.89}$$

where c_i^r and b_j^c are the ith row and the jth column of C and B, respectively. Since there may be cancellations between the numerator and denominator polynomials in the expression for $g_{ij}(s)$ above, the set of poles of G(s) given by

$$G(s) = \frac{C \operatorname{adj}(sI - A)B}{\det(sI - A)} \tag{2.90}$$

is generally a subset of the set of eigenvalues of A; that is, some of the

*This follows directly through the modal form of S(A, B, C), where $\Lambda = \operatorname{diag}\{\lambda_i\}$, $i \in 1, \ldots, n$, are distinct eigenvalues, $B_n = \{b_{ni}^r\}$, $c_n = \{c_{ni}^c\}$, and $y(s) = C_n(sI - \Lambda)^{-1}B_n$.

†The transmission matrix D is not included here for the sake of clarity in concepts and definitions.

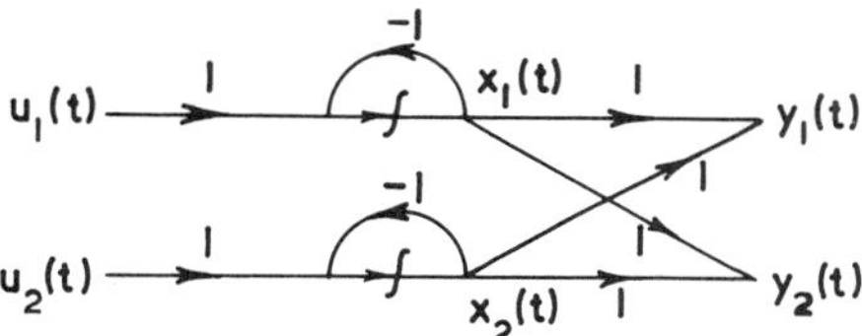

Fig. 2.19 Signal flow graph of the system in (2.91).

eigenvalues of A, obtained by solving det (sI - A) = 0, may not appear among the set of poles of G(s). This is apparent from (2.88), where if any of the normalized column vectors c^c_{ni} is null, there will be no coupling between the output and the ith mode, and consequently λ_i will not appear as a pole of G(s). Similarly, if any of the row vectors b^r_{nj} is null, there will be no coupling between the jth mode and the input, and λ_i will not appear* in G(s). Thus in a rigorous analysis it is necessary to distinguish between the set of poles of G(s) and the eigenvalues of A. This distinction is sometimes made by calling the eigenvalues of A characteristic frequencies [12]. Characteristic frequencies are therefore the roots of the characteristic polynomial of A. In a similar analogy, the poles of G(s) are said to be the roots of the pole polynomial, where the pole polynomial† is defined as the least common denominator of all the nonidentically zero minors of all orders of G(s). To see the difference between these two polynomials, consider the following example.

Example 2.12: Characteristic and pole polynomials of the system defined by the signal flow graph in Fig. 2.19. The state-space representation of the system is

$$A = \begin{bmatrix} -1 & 0 \\ 0 & -1 \end{bmatrix}; \quad B = \begin{bmatrix} 1 & 0 \\ 0 & 1 \end{bmatrix}; \quad C = \begin{bmatrix} 1 & 1 \\ 1 & 1 \end{bmatrix}$$

and the transfer-function matrix is

$$G(s) = \begin{bmatrix} \dfrac{1}{s+1} & \dfrac{1}{s+1} \\ \dfrac{1}{s+1} & \dfrac{1}{s+1} \end{bmatrix} \tag{2.91}$$

*These are related to the controllability and observability properties discussed in Sec. 3.1.

†In the literature this is also called the characteristic polynomial [p(s)] of G(s). For a system that is controllable and observable, $\Delta(s) \equiv p(s)$ (see Sec. 3.2).

The minors of order 1 of G(s) are 1/(s + 1), 1/(s + 1), 1/(s + 1), and 1/(s + 1). The minor of order 2 of G(s) is 0. Hence the pole polynomial of G(s) is p(s) = s + 1.

The characteristic polynomial of A is det (sI - A) = $\Delta(s) = (s + 1)^2$. The discrepancy between the degrees of p(s) and $\Delta(s)$ is due to the fact that modes $x_1(t)$ and $x_2(t)$ are not connected to controls $u_2(t)$ and $u_1(t)$, respectively. These reasons are considered in detail in the following chapter (Sec. 3.2), but it is already clear that poles and zeros of transfer-function matrices need much more careful consideration than what may be suggested by (2.78).

In conceptual forms, poles are characteristic of the ways in which dynamical systems behave, and may be thought of as being associated with system resonances coupled to the input and output. In a similar manner, zeros are characteristic of ways in which a dynamical system is linked to the external world and associated with certain specific values of the complex frequency s at which transmission through the system is zero (blocked). Zeros may thus be associated with antiresonances at which there is no signal propagation through the system [12]. The location of these poles and zeros therefore has a significant effect on the dynamic behavior of multivariable systems. Some of the conceptually easier methods of evaluating the poles and zeros of a transfer-function matrix are presented below.

(a) Smith-McMillan form of G(s)

Let G(s) be an ($\ell \times m$) proper transfer-function matrix of normal rank r. Then from the derivations in Sec. 1.5 it may be expressed in Smith-McMillan form as

$$G(s) = L(s)M(s)R(s) \tag{2.92}$$

where L(s) and R(s) are two unimodular matrices of dimensions $\ell \times \ell$ and $m \times m$, respectively.

With appropriate partitioning of L(s), M(s), and R(s), (2.92) may be expressed as

$$\begin{aligned} G(s) &= [\underbrace{L_1(s)}_{r} \mid \underbrace{L_2(s)}_{\ell - r}] \left[\begin{array}{c|c} M^*(s)_{r,r} & 0 \\ \hline 0 & 0_{\ell-r,m-r} \end{array}\right] \left[\begin{array}{c} R_1(s) \\ \hline R_2(s) \end{array}\right] \begin{array}{l} \}r \\ \}m - r \end{array} \\ &= L_1(s)M^*(s)R_1(s) \end{aligned} \tag{2.93}$$

where

$$M^*(s) = \text{diag}\left\{\frac{\psi_i(s)}{\phi_i(s)}\right\}; \quad \psi_i(s) \mid \psi_{i+1}; \ \phi_i(s) \mid \phi_{i-1}(s), \quad i \in 1, 2, \ldots, r, \quad r < \min(\ell, m)$$

Let the submatrices $L_1(s)$ and $R_1(s)$ be defined by the following column and row vectors:

$$L_1(s) = \{\ell_{ci}(s)\}; \quad i \in 1, \ldots, r \rightarrow r \text{ columns of } L_1(s)$$

$$R_1(s) = \{r_{ri}(s)\}^T; \quad i \in 1, \ldots, r \rightarrow r \text{ rows of } R_1(s)$$

Since by definition $\rho\{L(s)\} = r$ and $\rho\{R(s)\} = r$ for all s, the set of r column vectors $\{\ell_{ci}(s)\}$ are linearly independent for all values of s, and the set of r row vectors $\{r_{ri}(s)\}$ are linearly independent for all values of s. In terms of these column and row vectors, the transfer function matrix may, from (2.93), be expressed as

$$G(s) = \sum_{i=1}^{r} \ell_{ci}(s) \frac{\psi_i(s)}{\phi_i(s)} [r_{ri}(s)]^T, \qquad r \leq \min(\ell, m)$$

For any input control vector $u(s) = \{u_i(s)\}^T$, the output response $y(s) = \{y_i(s)\}^T$, $i \in 1, \ldots, r$, is therefore given by

$$y(s) = \left\{ \sum_{i=1}^{r} \ell_{ci}(s) \frac{\psi_i(s)}{\phi_i(s)} [r_{ri}(s)]^T \right\} u(s) \tag{2.94}$$

By following the basic concepts of poles and zeros described earlier, a simple way to characterize the poles and zeros of the transfer function matrix given by (2.93) is to identify those values of s for which

(1) $\|y(s)\| \rightarrow \infty$ for $\|u(s)\| < \infty$ (poles)

(2) $\|y(s)\| \rightarrow 0$ for $\|u(s)\| \neq 0$ (zeros)

Thus if the input control vector u(s) is chosen such that

$$r_{ri}^T(s)u(s) = \delta_{ij}f(s) = \begin{cases} f(s), & i = j \\ 0, & i \neq j \end{cases}$$

where f(s) is a nonzero scalar function, the ideas outlined above, when applied to (2.94), lead to the following observations:

1. $y(s) = \{y_i(s)\} \rightarrow \infty$, for finite s if and only if some $\phi_i(s) \rightarrow 0$. [By definition of $M^*(s)$, all cancellations between $\psi_i(s)$ and $\phi_i(s)$ have been carried out.]
2. $y = \{y_i(s)\} \rightarrow 0$ if and only if some $\psi_i(s) \rightarrow 0$.

The derivations above lead to the following general definition of poles and zeros for a proper rational transfer-function matrix G(s).

1. Poles of G(s) are the roots of the denominator polynomials $\{\phi_i(s)\}$, $i \in 1, \ldots, r$, in the Smith-McMillan form of G(s), where r is the normal rank of G(s). Thus if $\{p_1, p_2, \cdots, p_p\}$ are the poles of G(s), then the pole polynomial,* p(s), is given by

$$p(s) = \prod_{i=1}^{p} \phi_i(s) = (s - p_1)(s - p_2) \cdots (s - p_p) \tag{2.95}$$

2. Zeros of G(s) are the roots of the numerator polynomials $\{\psi_i(s)\}$, $i \in 1, \ldots, r$, in the Smith-McMillan form of G(s), where r is the normal rank of G(s). If $\{z_1, z_2, \ldots, z_z\}$ are the zeros of G(s), the zero polynomial† of G(s) is given by

$$z(s) = \prod_{i=1}^{z} \psi_i(s) = (s - z_1)(s - z_2) \cdots (s - z_z) \tag{2.96}$$

If d(s) is the monic least common denominator of all the elements of G(s) = $\{g_{ij}(s)\}$, $i \in 1, \ldots, \ell$, $j \in 1, \ldots, m$, then, from Sec. 1.5, G(s) may be expressed as

$$G(s) = \frac{N(s)}{d(s)} \tag{2.97}$$

where N(s) is a polynomial matrix which, by using two unimodular matrices L'(s) and R'(s), may be expressed as

$$N(s) = L'(s)S(s)R'(s) \tag{2.98}$$

where S(s) is the Smith form of N(s). Thus combining (2.97) and (2.98), we have

*$\Sigma \deg \phi_i(s)$ is called the McMillan degree of G(s) [denoted by $\delta_M(\cdot)$], and corresponds to the minimum number of energy-storing devices required to represent a system.

†Another definition of the zero polynomial z(s) (as opposed to the pole polynomial defined earlier) is that it is the greatest common divisor of the numerators of all the minors of G(s) of order r when these minors have all been adjusted to have the pole polynomial p(s) as their common denominator. This is considered later in this section.

$$\det\{G(s)\} = \frac{1}{d(s)} \det\{N(s)\} = \frac{1}{d(s)} \alpha_1 \det\{S^*(s)\} \alpha_2$$

$$= \alpha \frac{\eta(s)}{d(s)} \tag{2.99}$$

where $\alpha = \alpha_1\alpha_2$ is a scalar independent of s and $S(s) = \text{diag}\{\eta_i(s)\}$, $i \in 1, \ldots, r$ and $\eta(s) = \Pi_{i=1}^{r} \eta_i(s)$. In a similar manner, from (2.95) and (2.96),

$$\det\{G(s)\} = \beta \frac{z(s)}{p(s)} \tag{2.100}$$

where β is a scalar independent of s and z(s) and p(s) are not necessarily relatively prime.

Thus for any rational transfer function G(s) of normal rank r, from (2.99) and (2.100),

$$\alpha \frac{\eta(s)}{d(s)} = \beta \frac{z(s)}{p(s)} \tag{2.101}$$

Consequently, if $d(s) \not\equiv p(s)$, cancellations of common factors between z(s) and p(s) are possible in evaluating M(s) and the determinant in (2.100). This makes the Smith-McMillan form unsuitable for general use in evaluating or defining the poles and zeros of G(s). The method and its limitation are illustrated by the following example [11, 12].

Example 2.13: Pole-zero computation of G(s)

$$G(s) = \begin{bmatrix} \frac{1}{s+1} & 0 & \frac{s-1}{(s+1)(s+2)} \\ -\frac{1}{s-1} & \frac{1}{s+2} & \frac{1}{s+2} \end{bmatrix} \tag{2.102}$$

The Smith-McMillan form of G(s) may be shown to be

$$M^*(s) = \begin{bmatrix} \frac{1}{(s+1)(s+2)(s-1)} & 0 \\ 0 & \frac{s-1}{s+2} \end{bmatrix} \equiv \begin{bmatrix} \frac{\psi_1(s)}{\phi_1(s)} & 0 \\ 0 & \frac{\psi_2(s)}{\phi_2(s)} \end{bmatrix}$$

$$\det\{M^*(s)\} = \frac{\psi_1(s)\psi_2(s)}{\phi_1(s)\phi_2(s)} = \frac{z(s)}{p(s)} = \frac{1}{(s+1)(s+2)^2}$$

G(s) may also be expressed as

$$G(s) = \frac{N(s)}{d(s)} = \frac{1}{(s^2 - 1)(s + 2)} \begin{bmatrix} (s - 1)(s + 2) & 0 & (s - 1)^2 \\ -(s + 1)(s + 2) & s^2 - 1 & s^2 - 1 \end{bmatrix} \tag{2.103}$$

The Smith-McMillan form of N(s) is given by

$$S^*(s) = \begin{bmatrix} 1 & 0 \\ 0 & s - 1 \end{bmatrix} \equiv \begin{bmatrix} \eta_1(s) & 0 \\ 0 & \eta_2(s) \end{bmatrix}$$

$$\frac{1}{d(s)} \det \{S(s)\} = \frac{\eta_1(s)\eta_2(s)}{d(s)} = \frac{s - 1}{(s^2 - 1)(s + 2)}$$

which shows that the pole and zero polynomials do not contain information about all poles and zeros of the system.

(b) Laplace expansion of G(s) [11, 14]

An alternative method of computing the poles and zeros of G(s) may be derived by extending the Laplace expansion outlined in Sec. 1.2, and is described here. Let $\gamma_i(s)$ be the monic greatest common divisor of all minors of N(s) in (2.97) with $\gamma_0(s) = 1$. Then from the definition of the Smith form

$$\gamma_1(s) = \eta_1(s) \tag{2.104}$$

where $S(s) = \text{diag}\{\eta_i(s)\}$, $i \in 1, \ldots, r$, and $\eta_i(s) \mid \eta_{i+1}(s)$. Furthermore, combining the Smith-McMillan form [using (2.99)], we obtain

$$\frac{\eta_1(s)}{d(s)} = \frac{\gamma_1(s)}{d(s)} \equiv \frac{\psi_1(s)}{\phi_1(s)} \tag{2.105}$$

that is,

$$\gamma_1(s) = \frac{\psi_1(s)}{\phi_1(s)} d(s) \tag{2.106}$$

Since $\gamma_1(s)$ is defined to be a polynomial and all cancellations between the common factors of $\psi_1(s)$ and $\phi_1(s)$ have been carried out during the derivation of $M^*(s)$, (2.106) is valid only if

$$\phi_1(s) = d(s)$$

Thus $\phi_1(s)$ is the least common denominator of all 1×1 minors of G(s). Furthermore, by definition of $\gamma_i(s)$,

$$\frac{\psi_1(s)}{\phi_1(s)}\frac{\psi_2(s)}{\phi_2(s)} = \frac{\eta_1(s)}{d(s)}\frac{\eta_2(s)}{d(s)} = \frac{\gamma_2(s)}{[d(s)]^2} \tag{2.107}$$

where $\gamma_2(s) = \eta_1(s)\eta_2(s)$.

The derivation above leads to the following observations:

1. Since $\gamma_2(s)$ is, by definition, the greatest common divisor of all 2×2 minors of N(s) in (2.97), the denominator term $\phi_1(s)\phi_2(s)$ must be at least common denominator of all minors of G(s).
2. Since $\psi_1(s)$ divides $\psi_2(s)$ and $\phi_2(s)$ divides $\phi_1(s)$, and $\psi_i(s)$ and $\phi_i(s)$ are relatively prime, cancellations on the left side of (2.107) can occur only between $\psi_2(s)$ and $\phi_1(s)$.
3. Since $\psi_2(s) \mid \phi_2(s)$, the significance of the observations above is that all factors of $\phi_2(s)$ are "added" to the least common denominator of all 2×2 minors of G(s). Therefore, the least common denominator of all 1×1 and 2×2 minors of G(s) is

$$\gamma_2(s) = \phi_1(s)\phi_2(s) \tag{2.108}$$

4. If this process is continued, it follows that the least common denominator of all nonzero minors of all orders (1×1, 2×2, ..., $r \times r$) of G(s) is given by

$$\begin{aligned} \gamma_r(s) &= \phi_1(s)\phi_2(s) \cdots \phi_r(s) \\ &= \prod_{i=1}^{r} \phi_i(s) \\ &\triangleq p(s) \quad \text{from (2.95)} \end{aligned}$$

Thus the pole polynomial p(s) of a rational transfer-function matrix G(s) may be defined as the least common denominator of all the nonidentically zero minors of all orders of G(s). To derive a similar conclusion for the zero polynomial, let $\delta(s)$ denote the greatest common divisor of all minors of order r of the polynomial matrix N(s), $r \le \min(m, \ell)$. For any $q < r$, the greatest common divisor of all q-square minors of the N(s) is also the greatest common divisor of its Smith form S(s). If $q = r$, then

$$\begin{aligned} \delta(s) &= \det\{S^*(s)\} \\ &= \eta_1(s)\eta_2(s) \cdots \eta_r(s) \\ &= \frac{\psi_1(s)}{\phi_1(s)} d(s) \frac{\psi_2(s)}{\phi_2(s)} d(s) \cdots \frac{\psi_r(s)}{\phi_r(s)} d(s) \end{aligned}$$

$$= \frac{\prod_{i=1}^{r} \psi_i(s)}{\prod_{i=1}^{r} \phi_i(s)} [d(s)]^r$$

$$\triangleq \frac{z(s)}{p(s)}$$

that is,

$$z(s) = \frac{p(s)}{[d(s)]^r} \delta(s) \tag{2.109}$$

Furthermore, for any $q < r$, the greatest common divisor (gcd) of all numerators of the q-square minor of G(s) is equal to the gcd of all the numerators of the q-square minors of its Smith-McMillan form M(s), provided that all these minors [of M(s) and G(s)] have been "adjusted" to have a common denominator. If this common denominator is chosen to be the pole polynomial [p(s)] of G(s), for $q = r$, the Smith-McMillan form of G(s) will have only one nonzero minor, given by

$$\det \{M^*(s)\} = \frac{z(s)}{p(s)} \tag{2.110}$$

This leads to the following alternative method of deriving the zero polynomial of G(s).

The zero polynomial of a rational matrix G(s) of normal rank r is given by the gcd of the numerator of all 1×1, 2×2, $\ldots$, $r \times r$ minors of G(s), provided that these minors have all been adjusted in such a way as to have the pole polynomial p(s) as their common denominator.

The development above is illustrated next using Example 2.13.

<u>Example 2.14: Pole-zero computation</u>

$$G(s) = \begin{bmatrix} \dfrac{(s-1)(s+2)}{\Delta(s)} & 0 & \dfrac{(s-1)^2}{\Delta(s)} \\ -\dfrac{(s+1)(s+2)}{\Delta(s)} & \dfrac{(s-1)(s+1)}{\Delta(s)} & \dfrac{(s-1)(s+1)}{\Delta(s)} \end{bmatrix} \tag{2.111}$$

where $\Delta(s) = (s+1)(s+2)(s-1)$. Minors of order 2 are defined as

$$G^{\text{row } 1,2}_{\text{col. } 1,2} \triangleq \begin{cases} \text{determinant of the submatrix formed} \\ \text{by the rows (superscripts) 1 and 2,} \\ \text{and columns (subscripts) 1 and 2} \end{cases}$$

Thus

$$G_{1,2}^{1,2} = \frac{(s-1)(s+2)(s-1)(s+1)}{(s+1)^2(s-1)^2(s+2)^2} = \frac{1}{(s+1)(s+2)}$$

$$G_{1,3}^{1,2} = \frac{(s-1)(s+2)(s-1)(s+1) + (s-1)^2(s+1)(s+2)}{[\Delta(s)]^2} = \frac{2}{(s+1)(s+2)}$$

$$G_{2,3}^{1,2} = -\frac{(s-1)(s+1)(s-1)^2}{[\Delta(s)]^2} = -\frac{s-1}{(s+1)(s+2)^2}$$

The least common denominator of the minors above is

$$p(s) = (s+1)(s+2)^2(s-1) \rightarrow \text{pole polynomial}$$

On adjusting the denominators of the foregoing minors of order 2 to be p(s), we obtain

$$\bar{G}_{1,2}^{1,2} = \frac{(s-1)(s+2)}{p(s)}, \quad \bar{G}_{1,3}^{1,2} = \frac{2(s-1)(s+2)}{p(s)}, \quad \bar{G}_{2,3}^{1,2} = -\frac{s(s-1)^2}{p(s)}$$

The greatest common numerator of the unions above is $z(s) = (s - 1) \rightarrow$ zero polynomial. Thus the pole and zero polynomials contain information about all the poles and zeros of G(s).

(c) Matrix fraction [17-21]

In order to develop a close analogy to single-input/single-output systems, a transfer-function matrix may be factorized into the product of a polynomial matrix and the inverse of another polynomial matrix, as shown earlier. The fact that any proper G(s) can be factorized this way follows from the derivation below.

It has been established that for any rational proper G(s) of dimension $\ell \times m$, there exists a Smith-McMillan form M(s) such that

$$M(s) = L(s)G(s)R(s) \tag{2.112}$$

where

$$M(s) = \left[\begin{array}{c|c} \text{diag}\left\{\dfrac{\psi_i(s)}{\phi_i(s)}\right\} & 0 \\ \hline 0 & 0 \end{array}\right] \begin{array}{l} \}\, r \\ \}\, \ell - r \end{array}$$

with the column blocks of widths r and $m - r$.

Using the matrix-fraction description of Sec. 1.5, M(s) may be expressed in either of the forms

$$
\begin{aligned}
M(s) &= M_{\psi}(s)[M_{\phi R}(s)]^{-1} \\
&= [M_{\phi L}(s)]^{-1} M_{\psi}(s)
\end{aligned}
\tag{2.113}
$$

where

$$
M_{\psi}(s) = \left[\begin{array}{c|c} \text{diag}\,\{\psi_i(s)\} & 0 \\ \hline 0 & 0 \end{array}\right]
\begin{array}{l} \}\, r \\ \}\, \ell - r \end{array}
\qquad \text{(column blocks: } r,\ m - r\text{)}
$$

$$
M_{\phi R}(s) = \left[\begin{array}{c|c} \text{diag}\,\{\phi_i(s)\} & 0 \\ \hline 0 & I \end{array}\right]
\begin{array}{l} \}\, r \\ \}\, m - r \end{array}
\ (\text{columns: } r,\ m-r), \qquad
M_{\phi L}(s) = \left[\begin{array}{c|c} \text{diag}\,\{\phi_i(s)\} & 0 \\ \hline 0 & I \end{array}\right]
\begin{array}{l} \}\, r \\ \}\, \ell - r \end{array}
\ (\text{columns: } r,\ \ell-r)
$$

Since $\psi_i(s)$ and $\phi_i(s)$ are assumed to be relatively prime, so are the matrices $\{M_{\psi}(s), M_{\phi R}(s)\}$ or $\{M_{\phi L}(s), M_{\psi}(s)\}$. Thus if $U_{L1}(s)$ and $U_{R1}(s)$ are two unimodular matrices of appropriate dimensions, the matrices $\{N_1(s), D_1^{-1}(s)\}$ are relatively prime, where

$$
N_1(s) = U_{L1}(s) M_{\psi}(s) \quad \text{and} \quad D_1^{-1}(s) = M_{\phi R}^{-1}(s) U_{R1}(s)
$$

Therefore,

$$
\begin{aligned}
N_1(s) D_1^{-1}(s) &= U_{L1}(s) M_{\psi}(s) M_{\phi R}^{-1}(s) U_{R1}(s) \\
&= U_{L1}(s) M(s) U_{R1}(s) \\
&\equiv G(s)
\end{aligned}
\tag{2.114}
$$

Since for any G(s), there is a unique Smith-McMillan form, comparison of (2.112), (2.113), and (2.114) yields

$$
\begin{aligned}
N_1(s) &= L^{-1}(s) M_{\psi}(s) \\
D_1(s) &= R(s) M_{\phi R}(s)
\end{aligned}
\tag{2.115}
$$

These derivations lead to the following observations:

1. Poles of G(s): The roots of the polynomial det $\{D_1(s)\} = 0$, where $D_1(s)$ is the denominator polynomial matrix of G(s). This is a direct consequence of (2.115) since

$$\det\{D_1(s)\} = \det\{R(s)\} \det\{M_{\phi R}(s)\} = \alpha \prod_{i=1}^{r} \phi_i(s) = \alpha p(s) \tag{2.116}$$

where α is a scalar independent of s, p(s) being the pole polynomial of G(s) (Sec. 2.3.3).

2. Zeros of G(s): Since G(s) may not be square, observation here has to be made for two cases, where $m = \ell$: G(s) is square and $m \neq \ell$: G(s) rectangular.

Square G(s): If the transfer-function matrix G(s) is square and nonsingular, its zeros are the roots of the polynomial det $\{N_1(s)\} = 0$, where $N_1(s)$ is the numerator polynomial matrix of G(s). This follows from

$$\det\{N_1(s)\} = \det\{L^{-1}(s)\} \det\{M_\psi(s)\} = \beta \prod_{i=1}^{r} \psi_i(s) = \beta z(s) \tag{2.117}$$

where β is a scalar independent of s and z(s) is the zero polynomial.

Rectangular G(s): Zeros are those values of s for which the local rank of $N_1(s)$ is less than its normal rank. That is, the zeros of G(s) are the roots of the nonzero elements of the numerator polynomial of $N_1(s)$. The derivations above, when combined with (1.121) to (1.124)* lead to the observation that if N(s) and $D^{-1}(s)$ in

$$G(s) = N(s)D^{-1}(s)$$

are not relatively prime, they have a common divisor† matrix T(s) which is related to $N_1(s)$ and $D_1(s)$ as

$$N(s) = N_1(s)T(s) \quad \text{and} \quad D(s) = D_1(s)T(s)$$

If T(s) is unimodular, the definitions of poles and zeros of G(s) may be expressed in terms of normal ranks of N(s) and D(s) as follows:

*Section 1.5.

†A similar comment applies for a common left divisor if G(s) is expressed as $[D'(s)]^{-1}N'(s)$.

1. Pole: A complex number s = p is a pole of G(s) if the rank of D(p) is less than the normal rank of D(s).
2. Zero: A complex number s = z is a zero of G(s) if the rank of N(z) is less than the normal rank of N(s).

This leads to the following theorem:

Theorem 2.2 If the polynomials $\{N(s), D(s)\}$ are relatively left or right prime, then all the roots of N(s) and D(s), respectively, are taken to represent the zeros and poles of G(s).

The derivations above are used below for pole-zero computations.

Example 2.15: Poles and zeros of relatively prime $\{N(s), D(s)\}$ with

$$G(s) = \begin{bmatrix} (s+1)(s+2) & \dfrac{(s+1)^2}{s+2} \\ -\dfrac{s+2}{(s+1)^2} & -\dfrac{1}{s+2} \end{bmatrix} \equiv \underbrace{\begin{bmatrix} 0 & (s+1)^2(s+2) \\ -(s+2)^2 & -(s+2) \end{bmatrix}}_{N(s)} \underbrace{\begin{bmatrix} -(s+1)^2 & 0 \\ (s+1)(s+2)^2 & (s+2)^2 \end{bmatrix}^{-1}}_{D^{-1}(s)} \quad (2.118)$$

Through appropriate row/column operations, N(s) and D(s) may be expressed as

$$N(s) = \underbrace{\begin{bmatrix} -(s+1)^2(s+2)^2 & (s+1)^2(s+2) \\ 0 & -(s+2) \end{bmatrix}}_{N_1(s)} \underbrace{\begin{bmatrix} -1 & 0 \\ s+2 & 1 \end{bmatrix}}_{T(s)}$$

$$D(s) = \underbrace{\begin{bmatrix} -1 & 0 \\ s+2 & 1 \end{bmatrix}}_{T(s)} \underbrace{\begin{bmatrix} (s+1)^2 & 0 \\ 0 & (s+2)^2 \end{bmatrix}}_{D_1(s)}$$

which shows that the divisors of N(s) and D(s) are unimodular, and hence N(s) and D(s) are relatively prime. Consequently, pole-zero computations made by $\{N(s), D(s)\}$ and $\{N_1(s), D_1(s)\}$ give the same results. It is clear that any other representation of $G(s) = N_2(s)D_2^{-1}(s)$, with $\{N_2(s), D_2(s)\}$ not relatively prime, would yield different poles and zeros; this is apparent from Example 1.21.

2.4 POLYNOMIAL-FUNCTION DESCRIPTION [14, 22-32, 35]

In state-space representation original equations of the physical system written as (2.10)

$$\begin{aligned} \dot{x}(t) &= f\{x(t), t, u(t)\} \\ y(t) &= g\{x(t), t, u(t)\} \end{aligned} \tag{2.119}$$

were transformed into a set of first-order linear differential equations to obtain the matrix representation of the state-space equations. As seen in Sec. 2.2, for mechanical and electrical systems, this transformation may be derived by following certain "standard" rules. Many physical systems, however, are nonlinear and linearized models may only be derived by considering small perturbations around steady state. In these cases, the transformation from (2.119) to matrix state space from S(A, B, C, D) may not be based on standard notions. Thus for a more general description it is convenient to express (2.119) in the form of a set of linear algebraic and differential equations. The transfer-function matrix was such a description for linear time-invariant systems, but was rather special in that it connected the input controls directly to output variables without any explicit relationship between the input and the state vector or the state vector and the output. Although the notion of the state variables may be described as "artificial," it contained some significant amount of information regarding the actual way the input energy was converted into the output variable. Although the equations in Sec. 2.2 may be used to bring in the state variables in input/output description, such representations, as discussed in Sec. 2.2, may not always be satisfactory. The representation that may be used to combine the advantage of having an internal variable (e.g., state variable or the state vector) with the use of Laplace transformation in input/output description may be obtained directly (after linearization, if appropriate) from (2.119) when expressed as

$$S\{P(s), Q(s), R(s), W(s)\}: \quad \begin{aligned} P(s)\xi(s) &= Q(s)u(s) \\ v(s) &= R(s)\xi + W(s)u(s) \end{aligned} \tag{2.120}$$

If (2.120) describes an m-input/ℓ-output system of order n [in the sense of $\xi(s) \in R^r(s)$ in "state-space" representation*], then P(s), Q(s), R(s), and W(s) are all polynomial matrices of dimension $r \times r$, $r \times m$, $\ell \times r$, and $\ell \times m$, respectively. The only constraint in this polynomial-function description is that the determinant of the polynomial matrix P(s) is not identically equal to zero, as this would make (2.120) indeterminant. If n is defined to be $\delta\{P(s)\}$, P(s) may be replaced by (sI - A) to obtain the state-space representation

*R(s) is an r-dimensional space of polynomials in s.

and then $\xi(s)$ becomes equivalent to the n-state vector. The notation ξ is used throughout this section to denote the state vector in polynomial-function description $S\{P(s), Q(s), R(s), W(s)\}$, and x the n-state vector in state-space representation $S(A, B, C, D)$.

Descriptions of the type (2.120) appear in many physical systems, and are derived directly from their high-order differential description. For example, the loop equations of an LCR network, written in terms of voltages and charges, are of this form, with entries in P(s) of second degree. There are many different forms in which these polynomial matrices may be used to describe system behavior. Some of these forms and their associated properties are considered in this section.

Example 2.16: Derivation of polynomial-matrix description of the system in Example 2.3 (2.20). Laplace transformation of the equations, assuming zero initial conditions, gives

$$\begin{aligned} s^2 y_1(s) + 3sy_1(s) + 2y_2(s) &= u_1(s) \\ s^2 y_2(s) + sy_1(s) + y_2(s) &= u_2(s) \end{aligned} \tag{2.121}$$

or in matrix notation

$$\underbrace{\begin{bmatrix} s^2 + 3s & 2 \\ s & s^2 + 1 \end{bmatrix}}_{P(s)} \begin{bmatrix} y_1(s) \\ y_2(s) \end{bmatrix} = \underbrace{\begin{bmatrix} 1 & 0 \\ 0 & 1 \end{bmatrix}}_{Q(s)} u(s)$$

which is a polynomial-matrix description where $\xi_1(s) \equiv y_1(s)$ and $\xi_2(s) = y_2(s)$.

As can be seen, the advantage of the representation above is that the "state variable" $\xi_i(s)$ could be chosen to represent the variables that appear in the basis system equations. The polynomial state vector $\xi(s)$ may therefore have a direct relationship to the physical parameters of the system, unlike the state vector, which may not have any physical significance.

2.4.1 Standard Forms [14, 22-24]

Since P(s) in (2.120) is nonsingular, a direct relationship between the three forms of multivariable system representations (2.25), (2.73), and (2.120) may be directly obtained as, with zero initial conditions,

$$\begin{aligned} y(s) &= G(s)u(s) \\ &= [R(s)P^{-1}(s)Q(s) + W(s)]u(s) && (2.122a) \\ &= N_r(s)D_r^{-1}(s)u(s) \equiv D_\ell^{-1}(s)N_\ell(s)u(s) && (2.122b) \\ &= [C(sI - A)^{-1}B + D]u(s) && (2.122c) \end{aligned}$$

It is to be noted that (2.122a) does not necessarily imply a proper rational matrix, whereas (2.122c) is always proper.

The relationship between the input u(s), output y(s), and state $\xi(s)$ for a zero initial condition may therefore be described by*

$$\underbrace{\begin{bmatrix} P(s) & -Q(s) \\ R(s) & W(s) \end{bmatrix}}_{T_1(s)} \begin{bmatrix} \xi(s) \\ u(s) \end{bmatrix} = \begin{bmatrix} 0 \\ y(s) \end{bmatrix} = \begin{bmatrix} sI - A & -B \\ C & D(s) \end{bmatrix} \begin{bmatrix} x(s) \\ u(s) \end{bmatrix} \qquad (2.123)$$

The $(r + \ell) \times (r + m)$ matrix $T_1(s)$, defined by

$$T_1(s) = \begin{bmatrix} P(s) & -Q(s) \\ R(s) & W(s) \end{bmatrix} \equiv \begin{bmatrix} sI - A & -B \\ C & D(s) \end{bmatrix} \qquad (2.124)$$

where A, B, and C are matrices of real numbers and D(s) is a polynomial matrix, is called the <u>system matrix in the first form</u> of the dynamical system described by $S\{P(s), Q(s), R(s)W(s)\}$ or S(A, B, C, D) if $r > n \triangleq \delta\{P(s)\}$.

If $r < n$, the $T_1(s)$ is not itself a system matrix, but another matrix $T_2(s)$, obtained by appropriate row/column operations,

$$T_1(s) = \left[\begin{array}{c|c} I_\alpha & 0 \\ \hline 0 & T_2(s) \end{array}\right] = \left[\begin{array}{c|cc} I_\alpha & 0 & \\ \hline 0 & P_2(s) & -Q_2(s) \\ & R_2(s) & W_2(s) \end{array}\right] \qquad (2.125)$$

The order n of $T_2(s)$ is defined to be the degree of det $\{P_2(s)\}$, and for $T_2(s)$ to be accepted as a system matrix it is required that det $\{P_2(s)\} \neq 0$ and $\alpha = n - r$. $T_2(s)$ is called the <u>second form</u> of a system matrix $T_1(s)$. The corresponding polynomial function representation of the system is

$$\begin{aligned} P_2(s)\xi_2(s) &= Q_2(s)u(s) \\ y(s) &= R_2(s)\xi_2(s) + W_2(s)u(s) \end{aligned} \qquad (2.126)$$

The <u>third form</u> includes the second form as a special case, as given by

*The notation D(s) is used for conformity.

$$T(s) = \left[\begin{array}{c|c} I_\beta & 0 \\ \hline 0 & T_3(s) \end{array}\right] = \left[\begin{array}{c|cc} I_\beta & 0 & 0 \\ \hline 0 & P_3(s) & -Q_3(s) \\ 0 & R_3(s) & W_3(s) \end{array}\right] \qquad (2.127)$$

where $\beta > \alpha$ and $P_3(s)$, $Q_3(s)$, $R_3(s)$, and $W_3(s)$ are $r \times r$, $r \times m$, $\ell \times r$, and $\ell \times m$ polynomial matrices, with $P_3(s)$ being nonsingular. The corresponding equations are

$$\begin{aligned} P_3(s)\xi_3(s) &= Q_3(s)u(s) \\ y(s) &= R_3(s)\xi_3(s) + W_3(s)u(s) \end{aligned} \qquad (2.128)$$

A consequence of this representation is that $T_1(s)$ in (2.124) may be expressed as*

$$\begin{bmatrix} M(s) & 0 \\ X(s) & I_\ell \end{bmatrix} \begin{bmatrix} P(s) & -Q(s) \\ R(s) & W(s) \end{bmatrix} \begin{bmatrix} N(s) & Y(s) \\ 0 & I_m \end{bmatrix} = \begin{bmatrix} I_\beta & 0 & 0 \\ 0 & sI_n - A & -B \\ 0 & C & D(s) \end{bmatrix}$$

A comparison of this equation with (2.127) gives

$$\begin{bmatrix} 0_{r-n,1} \\ \hline x(s) \end{bmatrix} = N^{-1}(s)[\xi(s) + Y(s)u(s)]$$

which is the relationship between the polynomial representation state vector $\xi(s)$ and the state-space state vector $x(s)$. Different choice of $N(s)$ and $Y(s)$ will produce different state vectors.

2.4.2 Equivalence and Similarity

Since the third and second forms are derived from the standard form through elementary operations using polynomials in s, it is reasonable to assume that all three forms represent the same transfer function (or state-space equivalence) of any given system. This and some properties associated with such equivalence are considered here.

*$X(s)$ and $Y(s)$ are different from $x(s)$ and $y(s)$, respectively. $X(s)$ and $Y(s)$ are polynomial matrices; $M(s)$ and $N(s)$ are $r \times r$ unimodular polynomial matrices.

Definition 2.1. Two system matrices T(s) and $T_1(s)$ given by

$$T(s) = \begin{bmatrix} P(s) & -Q(s) \\ R(s) & W(s) \end{bmatrix} \quad \text{and} \quad T_1(s) = \begin{bmatrix} P_1(s) & -Q_1(s) \\ R_1(s) & W(s) \end{bmatrix}$$

where T(s) is an $(r + \ell) \times (r + m)$ polynomial system matrix, are said to be strictly system equivalent (sse) if there exists a set of polynomial matrices M(s), N(s), X(s), and Y(s) of appropriate dimensions, of which M(s) and N(s) are r-square unimodular matrices, such that

$$\begin{bmatrix} P_1(s) & -Q_1(s) \\ R_1(s) & W_1(s) \end{bmatrix} = \underbrace{\begin{bmatrix} M(s) & 0 \\ X(s) & I_\ell \end{bmatrix}}_{A_\ell(s)} \begin{bmatrix} P(s) & -Q(s) \\ R(s) & W(s) \end{bmatrix} \underbrace{\begin{bmatrix} N(s) & Y(s) \\ 0 & I_m \end{bmatrix}}_{A_r(s)} \tag{2.129}$$

Theorem 2.3. If two sse system matrices produce the same transfer function, they have the same order.

Proof: (1) Transfer function. From (2.129), multiplication of the left side gives

$$\begin{bmatrix} P_1(s) & -Q_1(s) \\ R_1(s) & W_1(s) \end{bmatrix} = \begin{bmatrix} M(s)P(s) & -M(s)Q(s) \\ X(s)P(s) + R(s) & -X(s)Q(s) + W(s) \end{bmatrix} \begin{bmatrix} N(s) & Y(s) \\ 0 & I \end{bmatrix}$$

$$= \begin{bmatrix} M(s)P(s)N(s) & -M(s)[Q(s) - P(s)Y(s)] \\ [X(s)P(s) + R(s)]N(s) & [X(s)P(s) + R(s)]Y(s) - X(s)Q(s) + W(s) \end{bmatrix}$$

Thus by using (2.122), the transfer function of the representation above is

$$\begin{aligned} G_1(s) &= R_1(s)P_1^{-1}(s)Q_1(s) + W_1(s) \\ &= [X(s)P(s) + R(s)]N(s)[M(s)P(s)N(s)]^{-1}M(s)[Q(s) - P(s)Y(s)] \\ &\quad + [X(s)P(s) + R(s)]Y(s) - X(s)Q(s) + W(s) \\ &= [X(s)P(s) + R(s)]P^{-1}(s)[Q(s) - P(s)Y(s)] \\ &\quad + X(s)[P(s)Y(s) - Q(s)] + R(s)Y(s) + W(s) \\ &= R(s)P^{-1}(s)Q(s) + W(s) \end{aligned} \tag{2.130}$$

(2) Order. Since M(s) and N(s) are r-square unimodular matrices, from (2.129),

$$\delta\{\det P_1(s)\} = \delta\{\det P(s)\} = n \tag{2.131}$$

Since M(s) and N(s) are unimodular matrices, $A_\ell(s)$ and $A_r(s)$ in (2.129) can be represented by a series of row and column operations. This leads to the following theorem [14, 25]:

Theorem 2.4. Any polynomial system matrix T(s) may be transformed by sse to a system matrix in state-space form.

Proof: In the representation $S\{P(s), Q(s), R(s), W(s)\}$, let P(s) be an r-square matrix and $\delta\{P(s)\} = n$. Then it is necessary to show that

$$\underbrace{\begin{bmatrix} P(s) & -Q(s) \\ R(s) & W(s) \end{bmatrix}}_{T(s)} \xrightarrow{\text{sse}} \begin{bmatrix} I_{r-n} & 0 & 0 \\ 0 & sI - A_n & -B \\ 0 & C & D(s) \end{bmatrix} \tag{2.132}$$

where → denotes equivalent to (obtained from the matrix through row/column operations), B and C are constant matrices (independent of s), and D(s) is a polynomial matrix. To prove (2.132) it is sufficient to show that there are unimodular matrices such that a series of post- and premultiplication according to (2.129) yields (2.132).

Stage 1: Since P(s) is a polynomial matrix, there are two unimodular matrices $U_{\ell 1}(s)$ and $U_{r1}(s)$ which transform P(s) into its Smith form; that is,

$$S_1(s) = U_{\ell 1}(s)P(s)U_{r1}(s) \equiv U_{\ell 1}(s)(sI - A)U_{r1}(s) \tag{2.133}$$

Thus by using (2.129), the following equivalence may be obtained:

$$T(s) \xrightarrow{\text{sse}} T_1(s)$$

where

$$T_1(s) = \begin{bmatrix} U_{\ell 1}(s) & 0 \\ 0 & I \end{bmatrix} \begin{bmatrix} P(s) & -Q(s) \\ R(s) & W(s) \end{bmatrix} \begin{bmatrix} U_{r1}(s) & 0 \\ 0 & I \end{bmatrix}$$

$$= \begin{bmatrix} U_{\ell 1}(s)P(s)U_{r1}(s) & -U_{\ell 1}(s)Q(s) \\ R(s)U_{r1}(s) & W(s) \end{bmatrix}$$

Since $n = \delta\{\deg P(s)\} < r$, there will be at least $(r - n)$ unit entries in the diagonal matrix $S_1(s)$ which may be rearranged to assume the following form:

$$S_1(s) = \text{diag}\{\underbrace{1,1,\ldots,1}_{r-n},\underbrace{p_{r-n+1}(s),\ldots,p_k(s),\ldots,p_r(s)}_{n}\}$$

where $\Sigma\, d_k = n < p$, $d_k = \delta\{p_k(s)\}$, $k \in r-n+1, \ldots, r$. $T_1(s)$ may therefore be written as

$$T_1(s) = \begin{bmatrix} I_{r-n} & 0 & -\bar{Q}_1(s) \\ 0 & \bar{S}(s) & -\bar{Q}_1(s) \\ \bar{R}_1(s) & \bar{R}_2(s) & W(s) \end{bmatrix} \tag{2.134}$$

where $\bar{S}(s) = \text{diag}\{p_k(s)\}$, $k \in r-n+1, \ldots, r$, and $Q(s)$ and $R(s)$ are appropriately partitioned with

$$U_{\ell 1}(s)Q(s) = \begin{bmatrix} \bar{Q}_1(s) \\ \hline \bar{Q}_2(s) \end{bmatrix}\begin{matrix} \}\, r-n \\ \\ \}\, n \end{matrix} \quad ; \quad R(s)U_{r1}(s) = [\underbrace{\bar{R}_1(s)}_{r-n} \mid \underbrace{\bar{R}_2(s)}_{n}] \tag{2.135}$$

Stage 2: Two further unimodular matrices $U_{\ell 2}(s)$ and U_{r2} are chosen as

$$U_{\ell 2}(s) = \begin{bmatrix} I & 0 & 0 \\ 0 & I & 0 \\ -\bar{R}_1(s) & 0 & I \end{bmatrix}; \quad U_{r2}(s) = \begin{bmatrix} I & 0 & \bar{Q}_1(s) \\ 0 & I & 0 \\ 0 & 0 & I \end{bmatrix}$$

Then by using (2.129), the following equivalence may be obtained:

$$T_1(s) \xrightarrow{\text{sse}} T_2(s)$$

where

$$T_2(s) = \begin{bmatrix} I & 0 & 0 \\ 0 & I & 0 \\ -\bar{R}_1(s) & 0 & I \end{bmatrix}\begin{bmatrix} I_{r-n} & 0 & -\bar{Q}_1(s) \\ 0 & \bar{S}(s) & -\bar{Q}_2(s) \\ \bar{R}_1(s) & \bar{R}_2(s) & W(s) \end{bmatrix}\begin{bmatrix} I & 0 & \bar{Q}_1(s) \\ 0 & I & 0 \\ 0 & 0 & I \end{bmatrix}$$

$$= \begin{bmatrix} I & 0 & 0 \\ 0 & \bar{S}(s) & -\bar{Q}_2(s) \\ 0 & \bar{R}_2(s) & W(s) \end{bmatrix} \tag{2.136}$$

Stage 3: Since for any $P(s)$ there is a unique Smith form $S(s)$, from (2.133),

$$U_{\ell 1}(s)P(s)U_{r1}(s) \equiv U_{\ell 1}(s)(sI - A)U_{r1}(s) \equiv S_1(s) = \left[\begin{array}{c|c} I_{r-n} & 0 \\ \hline 0 & \bar{S}(s) \end{array}\right]$$

Hence there exists a transformation such that

$$[\underbrace{U_{\ell 2}(s)}_{r-n} \mid \underbrace{U_{\ell 3}(s)}_{n}] \underset{\begin{array}{cc} \underbrace{\hspace{2.5cm}}_{r-n} & \underbrace{\hspace{2cm}}_{n} \end{array}}{\left[\begin{array}{c|c} sI - A_{r-n} & 0 \\ \hline 0 & sI - A_n \end{array}\right]} \left[\begin{array}{c} U_{r2}(s) \\ \hline U_{r3}(s) \end{array}\right] \begin{array}{l} \}\, r-n \\ \}\, n \end{array} \equiv \underset{\begin{array}{cc} \underbrace{\hspace{0.6cm}}_{r-n} & \underbrace{\hspace{1cm}}_{n} \end{array}}{\left[\begin{array}{c|c} I & 0 \\ \hline 0 & \bar{S}(s) \end{array}\right]} \begin{array}{l} \}\, r-n \\ \}\, r \end{array}$$

that is,

$$U_{\ell 3}(s)(sI - A_n)U_{r3}(s) = \bar{S}(s)$$

where the partitioning in $U_{\ell 1}(s)$ and $U_{r1}(s)$ correspond to that in (2.135).

Thus by suitable choice of two unimodular matrices $U_{\ell 4}(s)$ and $U_{r4}(s)$, the following transformation may be obtained:

$$T_2(s) \xrightarrow{sse} T_3(s)$$

where

$$T_3(s) = \underbrace{\begin{bmatrix} I & 0 & 0 \\ 0 & U_{\ell 3}(s) & 0 \\ 0 & 0 & I \end{bmatrix}}_{U_{\ell 4}(s)} T_2(s) \underbrace{\begin{bmatrix} I & 0 & 0 \\ 0 & U_{r3}(s) & 0 \\ 0 & 0 & I \end{bmatrix}}_{U_{r4}(s)}$$

$$= \begin{bmatrix} I & 0 & 0 \\ 0 & sI - A_n & -U_{\ell 3}(s)\bar{Q}_2(s) \\ 0 & \bar{R}_2(s)U_{r3}(s) & W(s) \end{bmatrix} \qquad (2.137)$$

Stage 4: Since det $\{\bar{S}(s)\} \neq 0$, it can be seen from the property of Smith form, det $(sI - A_n) \neq 0$; hence unimodular matrices of the following form exist:

$$U_{\ell 5}(s) = \begin{bmatrix} I & 0 & 0 \\ 0 & I & 0 \\ 0 & [C - \bar{R}_2(s)U_{r3}(s)][sI - A_n]^{-1} & I \end{bmatrix}$$

$$U_{r5}(s) = \begin{bmatrix} I & 0 & 0 \\ 0 & I & [sI - A_n]^{-1}[U_{\ell 3}(s)\bar{Q}_2(s) - B] \\ 0 & 0 & I \end{bmatrix}$$

where B and C are constant matrices. Thus the following equivalent transformation exists:

$$T_3(s) \xrightarrow{\text{sse}} T_4(s)$$

where

$$T_4 = U_{\ell 5}(s)T_3(s)U_{r5}(s) = \begin{bmatrix} I & 0 & 0 \\ 0 & sI - A_n & -B \\ 0 & C & W(s) \end{bmatrix} \tag{2.138}$$

Thus the following equivalences have been established:

$$T(s) \xrightarrow[U_{\ell 1}(s),\, U_{r1}(s)]{\text{sse}} T_1(s) \xrightarrow[U_{\ell 2}(s),\, U_{r2}(s)]{\text{sse}} T_2(s) \xrightarrow[U_{\ell 4}(s),\, U_{r4}(s)]{\text{sse}} T_3(s)$$

$$\xrightarrow[U_{\ell 5}(s),\, U_{r5}(s)]{\text{sse}} T_4(s)$$

that is,

$$\underbrace{\begin{bmatrix} P(s) & -Q(s) \\ R(s) & W(s) \end{bmatrix}}_{T(s)} \xrightarrow[U_\ell(s),\, U_r(s)]{\text{sse}} \underbrace{\begin{bmatrix} I_{r-n} & 0 & 0 \\ 0 & sI - A_n & -B \\ 0 & C & D(s) \end{bmatrix}}_{T_4(s)} \tag{2.139}$$

This transformation of sse to take T(s) into $T_4(s)$ can be written as

$$\begin{bmatrix} M(s) & 0 \\ X(s) & I \end{bmatrix} \begin{bmatrix} P(s) & -Q(s) \\ R(s) & W(s) \end{bmatrix} = \begin{bmatrix} I_{r-n} & 0 & 0 \\ 0 & sI - A_n & -B \\ 0 & C & D(s) \end{bmatrix} \begin{bmatrix} N(s) & Y(s) \\ 0 & I \end{bmatrix} \tag{2.140}$$

where N(s) is unimodular and Y(s) is polynomial. If $M_n(s)$, $N_n(s)$ and $Y_n(s)$ consist of the last n rows of M(s), N(s), and Y(s), respectively, (2.140) reduces to

$$\begin{bmatrix} M_n(s) & 0 \\ X(s) & I_\ell \end{bmatrix} \begin{bmatrix} P(s) & -Q(s) \\ R(s) & W(s) \end{bmatrix} = \begin{bmatrix} sI_n - A & -B \\ C & W(s) \end{bmatrix} \begin{bmatrix} N_n(s) & Y_n(s) \\ 0 & I_m \end{bmatrix} \tag{2.141}$$

The derivation above also establishes a method of evaluating the state-space representation of a system matrix. This method, however, will not visually give the simplest way of obtaining the desired form. It is often more convenient to obtain the transformations above through elementary operations, as illustrated below [14].

<u>Example 2.17: Transformation of T(s) into state-space form</u>

$$T(s) = \left[\begin{array}{cc|c} s+1 & s^3 & 0 \\ 0 & s+1 & 1 \\ \hline -1 & & 0 \end{array}\right] \tag{2.142}$$

$$T(s) \rightarrow R1 = R1 - (s^2 - s + 1)R2 = \left[\begin{array}{cc|c} s+1 & -1 & -(s^2 - s - 1) \\ 0 & s+1 & 1 \\ \hline -1 & 0 & 0 \end{array}\right] = T_2(s)$$

$$T_2(s) \rightarrow C3 = (s - 2)C1 + C3 = \left[\begin{array}{cc|c} s+1 & -1 & -3 \\ 0 & s+1 & 1 \\ \hline -1 & 0 & -(s - 2) \end{array}\right]$$

$n = \delta\{T(s)\} = 2 < r = 3$, hence the partitioning as below:

$$\left[\begin{array}{c|c} sI - A_2 & b_2 \\ \hline c_2 & d(s) \end{array}\right] = \left[\begin{array}{cc|c} s+1 & -1 & -3 \\ 0 & s+1 & 1 \\ \hline -1 & 0 & -(s + 2) \end{array}\right]$$

If the transformation procedure described above is used, the end result may be a description S[A, B, C, D(s)] in terms of a vector x, whereas the system may have a different state vector $\bar{x}$. In such a case a further transformation of the type

$$x = T\bar{x}$$

would be needed to obtain the required form $\bar{S}[\bar{A}, \bar{B}, \bar{C}, \bar{D}(s)]$, given by

$$\begin{bmatrix} sI - \bar{A} & \bar{B} \\ \bar{C} & \bar{D}(s) \end{bmatrix} = \begin{bmatrix} T^{-1} & 0 \\ 0 & I_\ell \end{bmatrix} \begin{bmatrix} sI - A & B \\ C & D(s) \end{bmatrix} \begin{bmatrix} T & 0 \\ 0 & I_m \end{bmatrix} \tag{2.143}$$

This transformation is called system similarity (ss); sse is thus a special case of ss, and the system transformation and system order are both invariant under ss. As discussed earlier, this transformation merely indicates a change of basis for the state vector, and leads to the following definition

Definition 2.2. Two system matrices $T_1(s)$ and $T_2(s)$ in state-space form are ss if they are sse.

Theorem 2.5. The transfer-function matrix G(s) is a standard form for system matrices under system equivalence.

Proof: Follows from the following sse:

$$\begin{bmatrix} P(s) & -Q(s) \\ R(s) & W(s) \end{bmatrix} \to \begin{bmatrix} I & -P^{-1}(s)Q(s) \\ R(s) & W(s) \end{bmatrix} \to \begin{bmatrix} I & -P(s)Q(s) \\ 0 & R(s)P^{-1}(s)Q(s) + W(s) \end{bmatrix}$$

$$= \begin{bmatrix} I & P^{-1}(s)Q(s) \\ 0 & G(s) \end{bmatrix} \to \begin{bmatrix} I & 0 \\ 0 & G(s) \end{bmatrix} \to G(s) \tag{2.144}$$

A direct consequence of this derivation is the following definition:

Definition 2.3. Two system matrices are system equivalent (se) if they yield the same transfer-function matrix. This is illustrated through the following example [14]:

Example 2.18: Equivalence of two system matrices

$$T_1(s) = \begin{bmatrix} 1 & 0 & 0 & 0 \\ 0 & s(s+1) & 0 & s \\ 0 & 0 & s(s+2) & -s \\ 0 & 0 & -1 & 0 \end{bmatrix} \quad \text{and} \quad T_2(s) = \begin{bmatrix} 1 & 0 & 0 & 0 \\ 0 & s^2(s+1) & s(s+2) & -s \\ 0 & 0 & s+2 & -1 \\ 0 & 0 & -1 & 0 \end{bmatrix}$$

Transfer-function matrix of $T_1(s)$:

$$g_1(s) = [0 \;\; 0 \;\; -1]\begin{bmatrix} 1 & 0 & 0 \\ 0 & s(s+1) & 0 \\ 0 & 0 & s(s+2) \end{bmatrix}^{-1}\begin{bmatrix} 0 \\ s \\ -s \end{bmatrix} = \frac{1}{s+2}$$

Transfer-function matrix of $T_2(s)$:

$$g_2(s) = [0 \;\; 0 \;\; -1]\begin{bmatrix} 1 & 0 & 0 \\ 0 & s^2(s+1) & s(s+2) \\ 0 & 0 & s+2 \end{bmatrix}^{-1}\begin{bmatrix} 0 \\ -s \\ -1 \end{bmatrix} = \frac{1}{s+2}$$

Thus by definition $T_1(s)$ and $T_2(s)$ are se. By appropriate row/column operations, $T_1(s)$ and $T_2(s)$ may be expressed as

$$T_1(s) = \begin{bmatrix} 1 & 0 & 0 & 0 \\ 0 & s & -s & s \\ 0 & 0 & s & 0 \\ 0 & 0 & 0 & -1 \end{bmatrix} \underbrace{\begin{bmatrix} 1 & 0 & 0 & 0 \\ 0 & s+1 & 0 & 0 \\ 0 & 0 & s+2 & -1 \\ 0 & 0 & 1 & 0 \end{bmatrix}}_{T(s)} \begin{bmatrix} 1 & 0 & 0 & 0 \\ 0 & 1 & 1 & 0 \\ 0 & 0 & 1 & 0 \\ 0 & 0 & 0 & 1 \end{bmatrix}$$

$$T_2(s) = \begin{bmatrix} 1 & 0 & 0 & 0 \\ 0 & s^2 & s & 0 \\ 0 & 0 & 1 & 0 \\ 0 & 0 & 0 & -1 \end{bmatrix} \underbrace{\begin{bmatrix} 1 & 0 & 0 & 0 \\ 0 & s+1 & 0 & 0 \\ 0 & 0 & s+2 & -1 \\ 0 & 0 & 1 & 0 \end{bmatrix}}_{T(s)}$$

Thus $T_1(s)$ and $T_2(s)$ may be derived through elementary operations from another system matrix T(s) given above, that is,

$$T_1(s) \xrightarrow{se} T(s) \xrightarrow{se} T_2(s)$$

hence,

$$T_1(s) \xrightarrow{se} T_2(s)$$

Theorem 2.6. Any matrix T(s) in third standard form is equivalent to a system matrix in second standard form.

Proof: This follows directly from the proof of Theorem 2.5 in the reverse form by assuming that G(s) can be expressed as

$$G(s) = \frac{N(s)}{d(s)}$$

Thus

$$G(s) \xrightarrow{se} \begin{bmatrix} I_m & 0 \\ 0 & G(s) \end{bmatrix} \xrightarrow{se} \begin{bmatrix} I_m & 0 \\ I_m & G(s) \end{bmatrix} \xrightarrow{se} \begin{bmatrix} I_m & -G(s) \\ I_m & 0 \end{bmatrix}$$

$$\xrightarrow{sse} \begin{bmatrix} -d(s)I_m & N(s) \\ I_m & 0 \end{bmatrix} \xrightarrow{sse} \begin{bmatrix} d(s)I_m & N(s) \\ -I_m & 0 \end{bmatrix}$$

$$\xrightarrow{sse} \begin{bmatrix} I_\alpha & 0 & 0 \\ 0 & d(s)I_m & N(s) \\ 0 & -I_m & 0 \end{bmatrix} \qquad (2.145)$$

which is in second standard form if $\alpha = m[\delta\{d(s)\} - 1]$.

2.4.3 Transmission Blocking [11-14]

Earlier discussions on zeros suggest that these are related to the physical condition where the system has an identically zero output [i.e., $y(s) = \{y_i(s)\} \equiv 0, \forall i \in 1, \ldots, \ell$], but neither the state nor the input control vectors are null vectors. The values of the complex frequency s for which there is no propagation of signal through G(s) have, in Sec. 2.3.3, been called the zeros of G(s). The matrix G(s) may thus be assumed to provide an input/output or external description of the system behavior. In a similar way, another set of definitions of zeros may be arrived at by examining the relationship between the matrices A, B, C, and D(s) in the system matrix T(s). Such an analysis, which may be termed as internal structural analysis, would provide the relationships underlying the motion of the state in relation to zero output for nonzero input control. The phenomenon of no signal propagation between input and output, called transmission blocking, forms the basis for pole-zero analysis of multivariable systems. The conditions for transmission blocking are derived here [11, 12].

As shown earlier, the transmission properties of the zeros of G(s) are quite similar to those of single-input/single-output systems with scalar g(s). This may be broadly defined as follows: Any complex number s_0 is a zero of G(s) if it completely blocks the transmission of any input control proportional to $e^{s_0 t}$. Since the output vector is identically equal to zero, the state vector must follow a specific path to satisfy the nonzero input and zero output constraint. This is examined below for a proper system S(A, B, C).

Taking the Laplace transformation of both sides of (2.25), we have

$$
\begin{aligned}
x(s) &= [sI - A]^{-1}x(t_0) + [sI - A]^{-1}Bu(s) \\
y(s) &= C[sI - A]^{-1}x(t_0) + C[sI - A]^{-1}Bu(s)
\end{aligned}
\tag{2.146}
$$

where $x(t_0)$ is the initial state at $t = t_0$. As the system is assumed to be proper, with $u(t_0) = 0$,

$$
y(t_0) = Cx(t_0) = 0 \tag{2.147}
$$

The following identities are used in the subsequent derivations:

$$
s_0 I - A = (sI - A) - (s - s_0)I \tag{2.148}
$$

that is,

$$
I = \frac{sI - A}{s - s_0} - \frac{s_0 I - A}{s - s_0} \qquad \forall \text{ complex } s_0
$$

or

$$
\begin{aligned}
(sI - A)^{-1} &= \frac{1}{s - s_0} I - (sI - A)^{-1} \frac{s_0 I - A}{s - s_0} \\
&\equiv \frac{1}{s - s_0} I - (sI - A)^{-1}(s_0 I - A) \frac{1}{s - s_0}
\end{aligned}
\tag{2.149}
$$

The derivations are now aimed at evaluating the trajectory of x(t) for an exponential input control

$$
u(t) \propto e^{s_0 t} = \text{unit step}\,(t) e^{s_0 t} g \tag{2.150}
$$

where g is a complex nonzero vector, such that

$$
y(s) = 0, \qquad \forall\, s \tag{2.151}
$$

Combining (2.146), (2.149), and (2.151) gives us

$$C\underbrace{\left[\frac{1}{s - s_0}I - (sI - A)^{-1}(s_0I - A)\frac{1}{s - s_0}\right]}_{(sI - A)^{-1}}x(t_0) + C(sI - A)^{-1}Bu(s) = 0$$

that is,

$$C(sI - A)^{-1}\left[(s_0I - A)\frac{1}{s - s_0}x(t_0) - Bu(s)\right] = 0 \quad \text{since } Cx(t_0) = 0$$

or

$$C(sI - A)^{-1}[(s_0I - A)x(t_0) - Bu(s)(s - s_0)] = 0 \tag{2.152}$$

Since neither C nor (sI - A) are null matrices, the equation above yields

$$(s_0I - A)x(t_0) - Bu(s)(s - s_0) \equiv 0, \quad \forall s \tag{2.153}$$

To make the identity above valid for all values of s, the input control should satisfy the constraint

$$(s_0I - A)x(t_0) = Bu(s)(s - s_0) = Bg \tag{2.154}$$

where g is related to input u(s) by

$$g = u(s)(s - s_0) \tag{2.155}$$

Thus, combining (2.147) and (2.155), the condition for zero output for exponential input control is that the constant complex vector g in (2.155) be chosen such that it is related to the initial state through the matrix equation

$$\begin{bmatrix} s_0I - A & -B \\ C & 0 \end{bmatrix}\begin{bmatrix} x(t_0) \\ g \end{bmatrix} \equiv 0 \tag{2.156}$$

and any complex number s_0 that satisfies (2.156) is a zero of the system. For such value of s, there is complete input/output transmission blocking in the system.

The derivation for a nonproper system S(A, B, C, D) follows along the lines above with the modification for the output equation

$$y(t) = Cx(t) + \left(\sum_{i=0}^{k} D_i \frac{d^i}{dt}\right)u(t), \quad k < n \tag{2.157}$$

with this equation the condition for complete input/output blocking becomes

$$\begin{bmatrix} s_0 I - A & -B \\ C & D(s_0) \end{bmatrix} \begin{bmatrix} x(t_0) \\ g \end{bmatrix} \equiv 0 \qquad (2.158)$$

The value of $s = s_0$ for which (2.158) holds true is a zero of S(A, B, C, D). This derivation leads to the following theorem.

Theorem 2.7. A necessary and sufficient condition for an input u(t) = [unit step (t)$e^{s_0 t}$ g] to produce a rectilinear motion in state space x(t) = [unit step (t)]$e^{s_0 t}$, such that $y(t) \equiv 0$, $\forall\, t > 0$, is that

$$\begin{bmatrix} sI - A & -B \\ C & D(s) \end{bmatrix} \begin{bmatrix} x(t_0) \\ g \end{bmatrix} = 0 \quad \text{for } s = s_0$$

The concept of zeros of multivariable systems in thus closely related to the physical condition where the system has an identically zero output, while the states and inputs are not themselves identically zero. Because of this very important criterion, the zeros of systems have been studied extensively during the past decade, resulting in several definitions of the term "zero" and different structural properties associated with it [11, 34]. Some of the salient features of these results are considered in the next section.

2.4.4 System Zeros

Since the descriptions G(s) in Sec. 2.3.1 and T(s) in Sec. 2.4.1 are concerned essentially with the same physical system, it is relevant to establish their relationship in terms of the signal propagation properties. In view of the transmission blocking criteria developed in the preceding section, it is clear that such relationships can be studied through the zeros of the two descriptions. This section presents a brief account of the definitions and structural properties associated with the internal description T(s) which, in general, cannot be derived directly from G(s).

For the existence of the vector

$$\begin{bmatrix} x(t_0) \\ g \end{bmatrix}$$

which satisfies (2.156), it is necessary for the matrix T(s) to lose column rank at $s = s_0$. If the Smith form of T(s) is $S_T(s)$, where

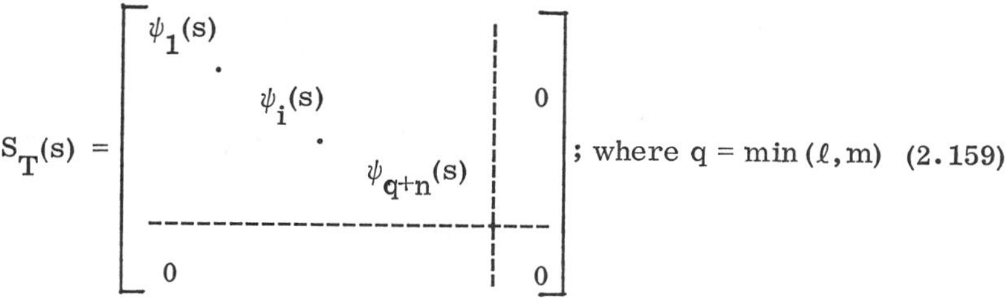

$$S_T(s) = \left[\begin{array}{ccc|c} \psi_1(s) & & & \\ & \psi_i(s) & & 0 \\ & & \psi_{q+n}(s) & \\ \hline 0 & & & 0 \end{array}\right]; \text{ where } q = \min(\ell, m) \quad (2.159)$$

then a necessary and sufficient condition for the loss of column rank of T(s) at $s = s_0$ is that s_0 be a root of one of the invariant polynomials $\psi_i(s)$, $i \in 1, \ldots, q+n$. The rank deficiency ρ_d of T(s) at $s = s_0$ is called the <u>geometric multiplicity</u> (σ_{s0}) of the corresponding root s_0; σ_{s0} is thus associated with the number of elementary divisors of T(s) which are associated with this particular value of $s = s_0$. The degree δ_{s0} of the product of the elementary divisors corresponding to $s = s_0$ is called the <u>algebraic multiplicity</u> of $s = s_0$; in general, $\delta_s > \sigma_s$. T(s) is said to have simple structure if the algebraic and geometric multiplicities are equal for all the roots of $\psi_i(s)$, $i \in 1, \ldots, q+n$; otherwise, T(s) is said to have nonsimple structure, even if $\delta_s \neq \sigma_s$ for only a single value of $s = s_0$. The following definitions may now be given:

(a) Input-decoupling zeros

These are values of s for which the column rank of $T_i(s)$ is less than its normal column rank, where

$$T_i(s) = [(sI - A) \mid B] \quad (2.160)$$

Thus if $s = s_i$ is an input-decoupling zero (idz) of T(s), there exists a row vector (one of the eigenvectors of A) such that

$$\alpha^T(s_iI - A) = 0 \quad \underline{\text{and}} \quad \alpha^T B = 0$$

that is,

$$\alpha^T[(s_iI - A), B] = 0 \quad (2.161)$$

Input-decoupling zeros may therefore be alternatively defined as the roots of the invariant polynomials of the matrix $T_i(s)$. By using polynomials in s and elementary row/column operations, it can be shown that S(A, B, C, D) has no decoupling zeros if the matrix $T_i(s)$ in (2.160) can be transformed into a Smith form $[I_n \mid 0]$. If $T_i(s)$ does not have such a Smith form, there exists a unimodular matrix U_ℓ which is a common left divisor of $T_i(s)$, that is

$$T_i(s) = U_\ell(s)T_i'(s)$$

where $T_i'(s)$ has no idz. The idz of $T_i(s)$ can therefore be eliminated by removing the common left divisor $U_\ell(s)$ through elementary operations.

(b) Output-decoupling zeros

These are values of s for which the row rank of $T_0(s)$ is less than its normal row rank, where

$$T_0(s) = \begin{bmatrix} sI - A \\ C \end{bmatrix} \tag{2.162}$$

Thus if $s = s_0$ is an output-decoupling zero of $T_0(s)$, there is an eigenvector α_0 of A such that

$$(s_0 I - A)\alpha_0 = 0 \quad \text{and} \quad C\alpha_0 = 0$$

that is,

$$\begin{bmatrix} s_0 I - A \\ C \end{bmatrix} \alpha_0 = 0 \tag{2.163}$$

Output-decoupling zeros (odz's) may thus also be defined as the roots of the invariant polynomials of the matrix $T_0(s)$. As with idz's, it can be shown* that a system S(A, B, C, D) will have no output-decoupling zero if it has a Smith form $[I_n \mid 0]$. If $T_0(s)$ does not have such a Smith form, there exists a unimodular matrix $U_r(s)$ which is a common right divisor of $T_0(s)$:

$$T_0(s) = T_0'(s)U_r(s)$$

As with idz's, odz's may be eliminated by removing the common right divisor $U_r(s)$ through elementary operations.

(c) Input/output decoupling zeros

These are values of s for which the ranks of the matrices

$$[(sI - A) \mid B] \quad \text{and} \quad \begin{bmatrix} sI - A \\ \hline C \end{bmatrix}$$

*See the definition of transmission zeros later in this section.

are less than their normal ranks. This is equivalent to the existence of two vectors α_i and α_o such that for $s = s_{io}$,

$$\alpha_i^T[(s_{io}I - A) \mid B] = 0 \quad \text{and} \quad \begin{bmatrix} s_{io}I - A \\ \hline C \end{bmatrix} \alpha_o = 0 \qquad (2.164)$$

The input/output decoupling zeros (iodz's) are therefore the output (input) decoupling zeros of T(s) which disappear when the input (output) decoupling zeros are eliminated. A consequence of this definition is that if T(s) has iodz's, then there are two unimodular matrices $U_\ell(s)$ and $U_r(s)$ which are common left and right divisors of $T_i(s)$ and $T_o(s)$, respectively. The iodz's may therefore be eliminated by removing $U_\ell(s)$ and $U_r(s)$ through elementary row and column operations using polynomials in s.

Example 2.19: Computation of decoupling zeros of T(s)

$$T(s) = \left[\begin{array}{ccc|c} s+3 & 0 & 0 & 1 \\ 0 & s^2(s+1) & s(s+2) & -s \\ 0 & 0 & s+2 & 1 \\ \hline 1 & 0 & -1 & 0 \end{array}\right] \qquad (2.165)$$

The system representation is in matrix polynomial form

$$T(s) = \begin{bmatrix} P(s) & -Q(s) \\ R(s) & W(s) \end{bmatrix}$$

Therefore, the definitions of decoupling zeros are to be extended to $\{P(s), Q(s), R(s)W(s)\}$ using (2.139):

1. Input-decoupling zeros are those values of s for which the rank of $[P(s) \mid -Q(s)]$ is less than its normal rank. That is, if $s = s_i$ is an idz, then

$$\rho[P(s_i) \mid -Q(s_i)] < \rho[P(s) \mid -Q(s)]$$

From (2.165),

$$T_i(s) = [P(s) \mid -Q(s)] = \begin{bmatrix} s+3 & 0 & 0 & 1 \\ 0 & s^2(s+1) & s(s+2) & -s \\ 0 & 0 & s+2 & 1 \end{bmatrix}$$

$$\rightarrow \text{normal rank} = 3$$

It is seen that for $s = 0$, $\rho[T_i(s)] = 2 < 3$; hence, (0) is an input decoupling zeros of the system.

2. Output-decoupling zeros are those values of s for which the rank of

$$\begin{bmatrix} P(s) \\ --- \\ R(s) \end{bmatrix}$$

is less than its normal rank; that is, for $s = s_o$ to be an odz,

$$\rho \begin{bmatrix} P(s_o) \\ ---- \\ R(s_o) \end{bmatrix} < \rho \begin{bmatrix} P(s) \\ \\ R(s) \end{bmatrix}$$

For the given system,

$$T_o(s) = \begin{bmatrix} P(s) \\ --- \\ R(s) \end{bmatrix} = \begin{bmatrix} s+3 & 0 & 0 \\ 0 & s^2(s+1) & s(s+2) \\ 0 & 0 & s+2 \\ 1 & 0 & -1 \end{bmatrix}$$

$\rightarrow$ normal rank = 3

The output-decoupling zeros of the system are (0, 0, -1). [The multiplicity (0, 0) in the odz appears from the fact that the highest-degree term which can divide any column of $T_o(s)$ is s^2 [≡ postmultiplication by $U_r(s)$], whereas the highest-degree term that divides any row [≡ premultiplication by $U_\ell(s)$] of $T_i(s)$ is only s, giving a single idz (0)].

3. Input/output-decoupling zeros are those values of s for which the ranks of $T_i(s)$ and $T_o(s)$ drop below their normal ranks. A simple method of identifying these is to remove the idz form T(s) to obtain $T_1(s)$, then identifying the odz of $T_1(s)$. Then the odz's that appear in (2) above but not in $T_1(s)$ are the iodz's of T(s). When the idz of T(s) is removed, T(s) becomes

$$T_1(s) = \left[\begin{array}{ccc|c} s+3 & 0 & 0 & 0 \\ 0 & s(s+1) & s+2 & -1 \\ 0 & 0 & s+2 & 1 \\ \hline 1 & 0 & -1 & 0 \end{array}\right]$$

which has an odz at $s = 0$. Since $\rho[T_o(s_o)] < 3$ for $s_o = 0, 0$, the system has only one iodz (0).

(d) Invariant zeros

The set of roots of the invariant polynomials $\psi_1(s)$, $\psi_2(s)$, $\cdots$, $\psi_{q+n}(s)$ [$q = \min(\ell, m)$] in (2.159) are called the invariant zeros (invz's) of T(s). Alternatively, invariant zeros of T(s) are the roots, counting multiplicities, of a monic polynomial $\psi_z(s)$ which is the greatest common divisor of all minors of T(s) of maximum order.

(e) Transmission zeros

The zeros of the transfer-function matrix as defined in Sec. 2.3.3 are called the set of transmission zeros (tz's). The relationship between invz's and tz's may be derived as follows. Let s_o be a complex frequency and define

$$\rho(s_o) = \text{normal rank of } T(s) - \text{rank of } T(s_o)$$

Then $\rho(s_o) > 0$ if

$$T(s) = \left[\begin{array}{c|c} sI - A & -B \\ \hline C & D(s) \end{array}\right] \begin{array}{l} \}n \\ \}m \end{array}$$
$$\qquad\qquad \underbrace{\qquad}_{n} \quad \underbrace{\qquad}_{\ell}$$

has elementary divisors $(s - s_o)^k$, $k > 0$ [i.e., if s_o is an invariant zero of T(s)]. In such a case, $\rho(s_o)$ is the number of finite elementary divisors associated with s_o [i.e., $\psi_i(s_o) = 0$ for some values of $s = s_o$, for some i, in (2.159).

To establish a relationship between the normal ranks of T(s) and G(s), the following identity is used:

$$\text{normal rank}\left\{\begin{bmatrix} sI - A & -B \\ C & D(s) \end{bmatrix}\right\}$$

$$= \text{normal rank}\left\{\begin{bmatrix} I & 0 \\ C(sI - A)^{-1} & B \end{bmatrix}\begin{bmatrix} sI - A & -B \\ -C & D(s) \end{bmatrix}\begin{bmatrix} I & (sI - A)^{-1}B \\ 0 & I \end{bmatrix}\right\}$$

$$= \text{normal rank}\left\{\begin{bmatrix} sI - A & 0 \\ 0 & G(s) \end{bmatrix}\right\}$$

$$= \text{normal rank}\{[sI - A]\} + \text{normal rank}\{[G(s)]\} \tag{2.166}$$

that is,

$$\rho[T(s)] = n + \rho[G(s)]$$

or

$$\rho[G(s)] = n - \rho[T(s)] \tag{2.167}$$

where [·] denotes the normal rank of [·], and $\rho[G(s)] = \min(\ell, m)$.

Thus for a system without any input/output decoupling zeros, (2.167) gives the relationship between invz's and tz's.

(f) System zeros

These are the roots of the monic polynomial $\phi(s)$, which is the greatest divisor of all the minors of T(s) formed from the columns 1, 2, ..., r, $r + j_1$, $r + j_2$, ..., $r + j_k$ and rows 1, 2, ..., r, $r + i_1$, $r + i_2$, ..., $r + i_k$, denoted by

$$T^{c1,2,\ldots,r,r+j_1,\ldots,r+j_k}_{r1,2,\ldots,r,r+i_1,\ldots,r+i_k}$$

where k (> 0) has a maximum value which is less than or equal to min (ℓ, m), such that at least one minor of that order is not identically zero. Such system zeros (sz's) are the roots of all nonzero minors of T(s), the set of invariant zeros defined above is a subset of system zeros. Furthermore, if $m = \ell$ and P(s) is nonsingular, these two sets coincide. The relationship between the various sets of zeros may now be given:

$$\begin{aligned}\{\text{system zeros}\} = \{&\text{input-decoupling zeros}\} + \{\text{output-decoupling zeros}\} \\ &+ \{\text{transmission zeros}\} - \{\text{input/output decoupling zeros}\}\end{aligned}$$

or

$$\{sz\} = \{idz\} + \{odz\} + \{tz\} - \{iodz\} \tag{2.168}$$

where {·} denotes "the set of." A similar relationship for system poles may be derived as

$$\{\text{system poles}\} = \{sp\} = \{idz\} + \{odz\} + \{tp\} - \{iodz\} \tag{2.169}$$

where tp denotes the transmittance poles (as opposed to transmission zeros) of G(s) and includes the set of poles of G(s) defined in Sec. 2.3.3.

Example 2.20: Computation of various zeros of T(s) in (2.164). Through elementary row operation T(s) may be expressed as

$$T(s) = \underbrace{\begin{bmatrix} 1 & 0 & 0 & 0 \\ 0 & 1 & s & 0 \\ 0 & 0 & 1 & 0 \\ 0 & 0 & 0 & 1 \end{bmatrix}}_{\text{unimodular}} \underbrace{\left[\begin{array}{cccc} s+3 & 0 & 0 & 1 \\ 0 & s^2(s+1) & 0 & 0 \\ 0 & 0 & s+2 & 1 \\ \hline 1 & 0 & -1 & 0 \end{array}\right]}_{T_1(s)} \xrightarrow{se} T_1(s) \tag{2.170}$$

The decoupling zeros of the system are

$$\{idz\} = \{0\}; \quad \{odz\} = \{0, 0, -1\}; \quad \{iodz\} = \{0\}$$

The set of system zeros is thus given by

$$\{sz\} = \{-3, 0, 0, -1, -2\} = \{-1, -2, -3, 0, 0\}$$

The transfer-function matrix corresponding to $T_1(s)$ is

$$g_1(s) = -\frac{1}{(s+3)(s+2)}$$

Thus

$$\{tz\} = \{-3, -2\}$$

which, of course, is consistent with (2.168), since

$$\begin{aligned} \{tz\} &= \{sz\} - \{idz\} - \{odz\} + \{iodz\} \\ &= \{-3, 0, 0, -1, -2\} - \{0\} - \{0, 0, -1\} + \{0\} \\ &= \{-3, -2\} \end{aligned}$$

From (2.169):

$$\begin{aligned} \text{system poles} &= \{sp\} = \{0\} + \{0, 0, -1\} + \{-3, -2\} - \{0\} \\ &= \{-1, -2, -3, 0, 0\} \\ &\equiv \text{roots of the characteristic equation det } \{sI - A\} \\ &\quad\ \text{of the system} \end{aligned}$$

This illustrates the significance of transmission and system zeros (tz and sz) defined by (2.168) or transmission and system poles (tp and sp) defined by (2.169).

Three different methods of describing linear time-invariant multivariable systems have been outlined in this chapter and their interrelationships developed. Until recently, the time-domain representation has been used extensively because of the numerical advantages associated with the processing of constant finite-order matrices. During the past decade, however, this situation has changed considerably, due mainly to the advent of fast interactive computing and graphics facilities. As a result, the analysis of multi-input/multioutput systems through either transfer-function matrix or polynomial-function representations does not pose any serious problem for the analyst.* This has helped to remove the traditional barrier between the time- and frequency-domain analyses of multivariable systems. This has given the engineer the flexibility of combining the best features of any one representation in his analysis, through the use of suitable transformation algorithms, for specific design specifications. Some of the key aspects of multivariable system analysis are considered in Chaps. 3 to 5, and various design methods developed in Chaps. 6 to 10.

*Although the software support needed for analysis in the frequency domain is likely to be more complicated than that for the time domain, but once the system package has been developed, there need be no significant difference in running either of the packages.

REFERENCES

1. DeRusso, P. M., Roy, R. J., and Close, C. M., State Variables for Engineers, Wiley, New York, 1965.
2. Ogata, K., State Space Analysis of Control Systems, Prentice-Hall, Englewood Cliffs, N.J., 1967.
3. Director, S. W., Introduction to Systems Theory, McGraw-Hill, New York, 1972.
4. Chen, C. T., Introduction to Linear System Theory, Holt, Rinehart and Winston, New York, 1970.
5. McGillem, C. D., and Cooper, G. R., Continuous and Discrete Signal and System Analysis, Holt, Rinehart and Winston, New York, 1974.
6. Brockett, R. W., Finite Dimensional Linear Systems, Wiley, New York, 1970.
7. Pipes, L. A., Matrix Methods for Engineering, Prentice-Hall, Englewood Cliffs, N.J., 1963.
8. Melsa, J. L., and Jones, S. K., Computer Programs for Computational Assistance in the Study of Linear Control Theory, McGraw-Hill, New York, 1970.
9. Wiberg, D. M., State Space and Linear Systems, McGraw-Hill, New York, 1971.
10. Fortman, T. E., and Hitz, K. L., An Introduction to Linear Control Systems, Marcel Dekker, New York, 1977.

11. MacFarlane, A. G. J., "Relationship between recent developments in linear control theory and classical design techniques," Part 1, Meas. Control, 8: 179-187 (1975).
12. MacFarlane, A. G. J., and Karcanias, N., "Poles and zeros of linear multivariable systems: a survey of the algebraic, geometric and complex variable theory," Int. J. Control, 24: 33-74 (1976); see also 26: 157-161 (1977).
13. MacFarlane, A. G. J., "Complex variable design methods," in Modern Approaches to Control System Design, N. Munro (Ed.), Peter Peregrinus, London, 1979.
14. Rosenbrock, H. H., State Space and Multivariable Theory, Nelson, London, 1970.
15. Chen, C. T., "Representation of linear time-invariant composite systems," Trans. IEEE, AC-13: 277-283 (1968).
16. Wolovich, W. A., Linear Multivariable Systems, Springer-Verlag, New York, 1974.
17. Kailath, T., Linear Systems, Prentice-Hall, Englewood Cliffs, N.J., 1980.
18. Wolovich, W. A., "On the numerators and zeros of rational transfer matrices," Trans. IEEE, AC-19: 544-546 (1973).
19. Wolovich, W. A., "On determining the zeros of state-space systems," Trans. IEEE, AC-19: 542-544 (1973).
20. Desoer, C. A., and Schulman, J. D., "Zeros and poles of matrix transfer functions and their dynamical interpretation," Trans. IEEE, CAS-21: 3-8 (1974).
21. Francis, B. A., and Wonham, W. M., "The role of transmission zeros in linear multivariable regulators," Int. J. Control, 22: 657-681 (1975).
22. Rosenbrock, H. H., "On linear system theory," Proc. IEE, 114: 1353-1359 (1967).
23. Rosenbrock, H. H., "Properties of linear constant systems," Proc. IEE, 117: 1717-1720 (1970).
24. Rosenbrock, H. H., "Transformation of linear constant system equations," Proc. IEE, 114: 541-544 (1967).
25. Pernebo, L., "Notes on strict system equivalence," Int. J. Control, 25: 21-38 (1977).
26. Rosenbrock, H. H., "The transformation of strict system equivalence," Int. J. Control, 25: 11-19 (1977).
27. Fuhrmann, P. A., "On strict system equivalence and similarity," Int. J. Control, 25: 5-10 (1977).
28. Rosenbrock, H. H., and Rowe, A., "Allocation of poles and zeros," Proc. IEE, 117: 1879-1886 (1970).
29. Pugh, A. C., "Transmission and system zeros," Int. J. Control, 26: 315-324 (1977).
30. Rosenbrock, H. H., "Structural properties of linear dynamical systems," Int. J. Control, 20: 191-202 (1974).

31. Rosenbrock, H. H., "Order, degree and complexity," Int. J. Control, 19: 323-331 (1974).
32. Suda, N., and Mutsuyoshi, E., "Invariant zeros and input-output structure of linear time-invariant systems," Int. J. Control, 28: 525-535 (1978).
33. Kouvaritakis, B., and MacFarlane, A. G. J., "Geometric approaches to analysis and synthesis of system zeros," Int. J. Control, 23: 149-166, 167-181 (1976).
34. Karkanias, N., and Kouvaritakis, B., "The use of frequency transmission concepts in linear multivariable system analysis," Int. J. Control, 26: 197-240 (1978).
35. Rosenbrock, H. H., "The zeros of a system," Int. J. Control, 18: 297-299 (1973); see also 20: 525-527 (1974).

3
Controllability and Observability

Once an acceptable mathematical description of a system has been established, the designer usually examines certain structural properties of the system (as represented by its model) to ascertain the degree of control that can be exerted on the system. This is accomplished through the controllability and observability properties of the system, which form the basis of many design techniques. The first two sections of this chapter briefly outline these concepts. The second half of the chapter presents some recently developed results in algebraic control theory with a view to establishing the notion of decoupling zeros and formulating a method of deriving the least order of a system from either state-space or polynomial function representation.

3.1 BASIC CONCEPTS [1-7]

In the analysis of multivariable systems, it is essential to ascertain the amount of information necessary to achieve a desired control of a system. Once the state and output equations of a system have been evaluated, some of this information is conveyed by the system matrices in that the nature of "connectivity" between input/state/output are specified by these matrices. For the purposes of system design, however, additional information about the system's behavior is required; these are associated with the following properties of the system.

1. Given a system in its initial state $[x(t_0)]$, can inputs be applied to move (or transform) it to a desired arbitrary state $[x(t)]$ in a finite time, or vice versa?
2. Will the observation of the output $[y(t)]$ for a finite time enable the initial state $[x(t_0)]$ to be identified?

These are the properties of a system associated with the concepts of state controllability (1) and observability (2). These may be interpreted as necessary, and sometimes as sufficient, conditions for the existence of a solution to most control problems. Physical significance of these concepts for time-invariant systems and their relationship with pole-zero cancellations and the effects of coordinate transformations are discussed briefly here.

Definition 3.1: State Controllability A dynamic system is said to be completely state controllable if for any time t_0, it is possible to construct an unconstrained control vector u(t) that will transfer any given initial state $x(t_0)$ to any final state x(t) in a finite time interval* $t_0 < t < T$.

For analytical convenience, the final state is often taken to be the origin of the state space. Derivation of the condition for complete state controllability is based on the solution of the state equation (Sec. 2.2.3) where the convolution integral (for the time-invariant case) may be substituted by summation. Thus for the time-invariant system S(A, B, C, D),

$$\begin{aligned} \dot{x}(t) &= Ax(t) + Bu(t) \\ y(t) &\equiv Cx(t) + Du(t) \end{aligned} \tag{3.1}$$

$x \in X^n$, $y \in Y^\ell$, $u \in U^m$, and A, B, C, and D being $(n \times n)$, $(n \times m)$, $(m \times n)$, and $(\ell \times m)$ constant matrices, the zero state response may be expressed as

$$x(t) = \sum_{k=0}^{n-1} \sum_{i=1}^{m} A^k b_i^c \gamma_{ik}(t), \qquad t \in (t_0, T) \tag{3.2}$$

where

$$\gamma_{ik}(t) = \int_{t_0}^{t} \alpha_k(\tau) u_i(t - \tau)\, d\tau = \text{a scalar}; \qquad u_i(t) = 0, \quad t < t_0$$

$$e^{-At} = \sum_{k=0}^{n-1} \alpha_k(t) A^k, \quad x(t_0) \equiv 0 \quad \text{and} \quad b_i^c = i\text{th column of B}, \quad i \in 1, \ldots, m$$

Since the system is linear, the forcing function [≡ input control u(t)] can be expressed as the sum of its components

$$Bu(t) = \sum_{i=1}^{m} b_i^c u_i(t) \tag{3.3}$$

Combining (3.2) and (3.3), the state vector x(t) may be expressed as

*The initial time t_0 is often assumed to be 0 without any loss of generality.

$$
x(t) \equiv \underbrace{[b_1 \cdots b_i \cdots b_m]}_{\substack{\text{m columns} \\ \text{n rows}}} \underbrace{\begin{bmatrix} \int_{t_0}^{t} \alpha_0(\tau)u_1(t-\tau)\,d\tau \\ \vdots \\ \int_{t_0}^{t} \alpha_0(\tau)u_i(t-\tau)\,d\tau \\ \vdots \\ \int_{t_0}^{t} \alpha_0(\tau)u_m(t-\tau)\,d\tau \end{bmatrix}}_{1} \left.\vphantom{\begin{bmatrix} \int \\ \vdots \\ \int \\ \vdots \\ \int \end{bmatrix}}\right\} m
$$

$$
+ \underbrace{[A^k b_1 \cdots A^k b_i \cdots A^k b_m]}_{\substack{\text{m columns} \\ \text{n rows}}} \underbrace{\begin{bmatrix} \int_{t_0}^{t} \alpha_{n-1}(t)u_1(t-\tau)\,d\tau \\ \vdots \\ \int_{t_0}^{t} \alpha_{n-1}(t)u_i(t-\tau)\,d\tau \\ \vdots \\ \int_{t_0}^{t} \alpha_{n-1}(t)u_m(t-\tau)\,d\tau \end{bmatrix}}_{1} \left.\vphantom{\begin{bmatrix} \int \\ \vdots \\ \int \\ \vdots \\ \int \end{bmatrix}}\right\} m
$$

$$
+ \underbrace{[A^{n-1} b_1 \cdots A^{n-1} b_i \cdots A^{n-1} b_m]}_{\substack{\text{m columns} \\ \text{n rows}}} \underbrace{\begin{bmatrix} \int_{t_0}^{t} \alpha_{n-1}(\tau)u_1(t-\tau)\,d\tau \\ \vdots \\ \int_{t_0}^{t} \alpha_{n-1}(\tau)u_i(t-\tau)\,d\tau \\ \vdots \\ \int_{t_0}^{t} \alpha_{n-1}(\tau)u_m(t-\tau)\,d\tau \end{bmatrix}}_{1} \left.\vphantom{\begin{bmatrix} \int \\ \vdots \\ \int \\ \vdots \\ \int \end{bmatrix}}\right\} m \qquad (3.4)
$$

Therefore, for any jth channel, $j \in 1, \ldots, n$, the state variable $x_j(t)$ for any input u(t) is

$$x_j(t) = b_j \begin{bmatrix} \int_{t_0}^{t} \alpha_0(t)u_1(t-\tau)\,d\tau \\ \vdots \\ \int_{t_0}^{t} \alpha_0(t)u_i(t-\tau)\,d\tau \\ \vdots \\ \int_{t_0}^{t} \alpha_0(t)u_m(t-\tau)\,d\tau \end{bmatrix} + \cdots + A^k b_j \begin{bmatrix} \int_{t_0}^{t} \alpha_k(t)u_1(t-\tau)\,d\tau \\ \vdots \\ \int_{t_0}^{t} \alpha_k(t)u_i(t-\tau)\,d\tau \\ \vdots \\ \int_{t_0}^{t} \alpha_k(t)u_m(t-\tau)\,d\tau \end{bmatrix}$$

$$+ \cdots + A^{n-1} b_j \begin{bmatrix} \int_{t_0}^{t} \alpha_{n-1}(t)u_1(t-\tau)\,d\tau \\ \vdots \\ \int_{t_0}^{t} \alpha_{n-1}(t)u_i(t-\tau)\,d\tau \\ \vdots \\ \int_{t_0}^{t} \alpha_{n-1}(t)u_m(t-\tau)\,d\tau \end{bmatrix}$$

$$\equiv [b_j \cdots A^k b_j \cdots A^{n-1} b_j] \begin{bmatrix} \int_{t_0}^{t} \alpha_0(\tau)u(t-\tau)\,d\tau \\ \vdots \\ \int_{t_0}^{t} \alpha_k(\tau)u(t-\tau)\,d\tau \\ \vdots \\ \int_{t_0}^{t} \alpha_{n-1}(\tau)u(t-\tau)\,d\tau \end{bmatrix}$$

$$\equiv P_j V \tag{3.5}$$

where

$$P_j = [b_j \cdots A^k b_j \cdots A^{n-1} b_j]$$

and

$$V = \begin{bmatrix} \int_{t_0}^{t} \alpha_0(\tau)u(t - \tau)\, d\tau \\ \vdots \\ \int_{t_0}^{t} \alpha_k(\tau)u(t - \tau)\, d\tau \\ \vdots \\ \int_{t_0}^{t} \alpha_{n-1}(\tau)u(t - \tau)\, d\tau \end{bmatrix}$$

The significance of state controllability is now clear. As for complete state controllability, one should be able to choose n variables $x_1(t), \ldots, x_j(t), \ldots, x_n(t)$ arbitrarily, and evaluate the corresponding input control vector u(t). Such a solution for u(t) exists only when (3.5) yields a set of n linearly independent equations, that is, when the rows of the matrix P are linearly independent, where

$$P = [P_1 \cdots P_j \cdots P_m] \tag{3.6}$$

and the $(n \times n)$ block matrices P_j are linearly independent.

To derive a condition for complete state controllability according to Definition 3.1, let the initial state $x(t_0)$ be included in (3.2). The response of the state vector for any arbitrary input control u(t) then becomes

$$x(t) = e^{At}\left[x(t_0) + \int_{t_0}^{T} d^{A(t-\tau)}Bu(\tau)\, d\tau\right] \tag{3.7}$$

If the system above is completely state controllable, its initial state $x(t_0)$ can be transferred to the origin of the state space in a finite time interval (t_0, T), say; then for $t = T$,

$$\begin{aligned} x(t_0) &= -\sum_{k=0}^{n-1} \int_0^t \alpha_k A^k Bu(\tau)\, d\tau \\ &= -\sum_{k=0}^{n-1} \sum_{i=1}^{m} A^k b_i \beta_{ki} \end{aligned} \tag{3.8}$$

where

$$\beta_{ki} = \int_0^T \alpha_k(\tau)u_i(\tau)\, d\tau \quad \text{a scalar}$$

By using the argument as before, from (3.5) it is seen that the initial state $x(t_0)$ can be transferred to the origin if the rank of the matrix P is n, where

$$P = [B \cdots A^k B \cdots A^{n-1} B] \tag{3.9}$$

(meaning that there is a set of n linearly independent vectors which constitute a basis for the n-dimensional state space). The $(n \times mn)$ matrix P is known as the state controllability matrix.

To check state controllability, in many instances, however, the state controllability matrix P need not be computed in full but only a matrix P_k with a smaller number of columns.

$$P_k = [B \;\; AB \cdots A^{k-1} B], \quad k < n \tag{3.10}$$

This is based on the following theorem.

Theorem 3.1 If there is an integer j such that $\rho(P_j) = \rho(P_{j+1})$, then $\rho(P_k) = \rho(P_j)$ for all integers $k > j$ and $j < \min(n - m, p - 1)$, where m is the rank B and p is the degree of minimal polynomial of A.

A consequence of this theorem is that the pair* (A, B) is completely controllable if

$$\rho[P_{n-m}] = \rho[B \;\; AB \cdots A^{n-m} B] = n \tag{3.11}$$

or the $(n \times m)$ matrix $P_{n-m}[P_{n-m}]^T$ is nonsingular. The theorem, whose proof is not considered here, shows that for some systems, there is a smallest integer k in (3.10) called the (state) controllability index of the pair (A, B). [The matrix P_k in (3.10) is sometimes called the partial state controllability matrix.]

Definition 3.2: Observability A system is said to be completely observable on $t_0 < t < T$ if, for every t_0 and some T, every state vector $x(t_0)$ can be determined from the knowledge of the output vector $y(t)$ on $t_0 < t < T$. In physical terms, a system is completely observable if every transition of the system state eventually affects the output.

The notion of observability being associated with the processing of data from observations of the system behavior, the problem here is to evaluate the condition under which the state of a system at any time t can be determined by observing the output response over a finite time interval $t_0 < t < t_0 + T$. From (2.34), the output at time t is given by

*(A, B) forms a part of the complete system S(A, B, C) and is often called the pair (A, B), meaning that the output matrix C has no influence on the consequent results.

$$\tilde{y}(t) = Ce^{At}x(t_0) + C\int_0^t e^{A(t-\tau)}Bu(\tau)\,dt + Du(t) \tag{3.12}$$

Since the matrices A and B are known, and the input control u(t) is also known, the integral term in (3.12) is a known quantity. Therefore, this quantity may be subtracted from the observed value of y(t); hence

$$y(t) = \tilde{y}(t) - C\int_0^t e^{A(t-\tau)}Bu(\tau)\,d\tau - Du(t) = Ce^{At}x(t_0) \tag{3.13}$$

which shows that the properties associated with observability may be studied by considering the purely dynamic unforced part of the system in (3.1):

$$\dot{x}(t) = Ax(t); \quad y(t) = Cx(t) \tag{3.14}$$

for which any ith output response is, from (3.5),

$$y_i(t) = \left\{ C_i x(t_0) + \sum_{k=0}^{n-1} \alpha_k(t)[A^T]^k[C_i]^T \right\}, \quad \text{where } e^{-At} = \sum_{k=0}^{n-1} \alpha_k(t)e^{At} \tag{3.15}$$

$\alpha_k(t)$ being linearly independent.

By following derivations similar to state controllability, it can be seen that the set of ℓn equations in ℓ unknown variables $\tilde{y}$ (t), $\tilde{y}_2(t)$, ..., $\tilde{y}_\ell(t)$ given by (3.13) have a unique solution if the rank of the ($\ell n \times n$) coefficient matrix R is n, where

$$R = [C^T (A^T)C^T \cdots (A^T)^k C^T \cdots (A^T)^{n-1}C^T] = \begin{bmatrix} C \\ \vdots \\ CA^k \\ \vdots \\ CA^{n-1} \end{bmatrix} \tag{3.16}$$

When $\rho(R) = n$, there exists a set of n linearly independent vectors among the ℓn columns of R. These vectors form a basis of the state space of the system. Hence the initial state $x(t_0)$ can be determined by the measurement of y(t) for a finite time interval (t_0, T). The matrix R is called the observability matrix of the pair (C, A).

In the case where $p < n$, by following an argument similar to Theorem 3.1, the pair (C, A) can be shown to be completely observable if the rank of R_{k-1} is n, where

$$R_{k-1} = \begin{bmatrix} C \\ CA \\ \vdots \\ CA^{k-1} \end{bmatrix} \tag{3.17}$$

where the integer $k < \min(n - \ell, p - 1)$, is called the observability index of the system, p being the degree of minimal polynomial of A. It can also be shown by using the above that a system is completely observable if

$$\rho[R_{n-\ell}] = \rho \begin{bmatrix} C \\ AC \\ \vdots \\ A^{n-\ell}C \end{bmatrix} \tag{3.18}$$

or the matrix $[R_{n-\ell}]^T[R_{n-\ell}]$ is nonsingular.

If a system is state controllable and observable, then for any initial state $x(t_0)$ it is possible to drive the output to zero at time t_2 and keep it at zero thereafter. This is achieved in two stages. First, the initial state is estimated [through the observation of y(t) from t_0 to t_1] and then an appropriate input is applied to bring the state vector to zero. As the second step, an input $u_{(t_1, t_2)}$ is applied to transfer $x(t_1)$ to the zero state at time t_2. If the state at time t_2 is the zero state, and if no input in applied for $t < t_2$, the output will be identically zero for all $t > t_2$. This cannot be achieved if the dynamical system is uncontrollable or unobservable.

The conditions in (3.9) and (3.16) show that state controllability is related to the linear independence of the rows of A^kB, whereas observability to the linear independence of the columns of CA^k, $k \in 1, \ldots, n - 1$. This structural similarity between the two concepts leads to the following theorem.

Theorem 3.2: Duality Theorem* The system $S^T(-A^T, C^T, B^T, D^T)$, defined by

$$S^T: \quad \begin{aligned} \dot{z}(t) &= -A^T z(t) + C^T v(t) \\ q(t) &= B^T z(t) + D^T v(t) \end{aligned} \tag{3.19}$$

where $(\cdot)^T$ denotes the complex conjugate of $(\cdot)$, is said to be completely

*Based on Kalman's principle of duality [7].

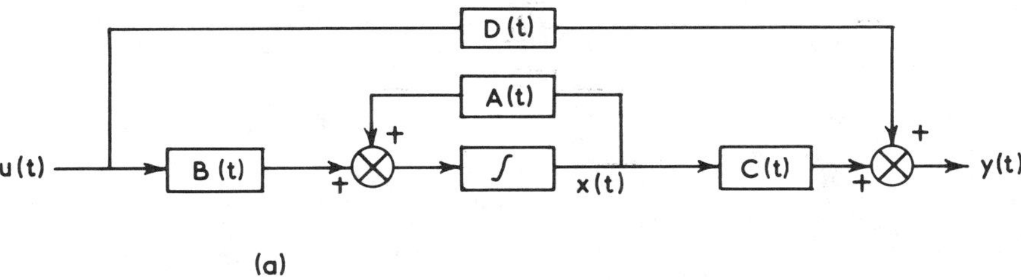

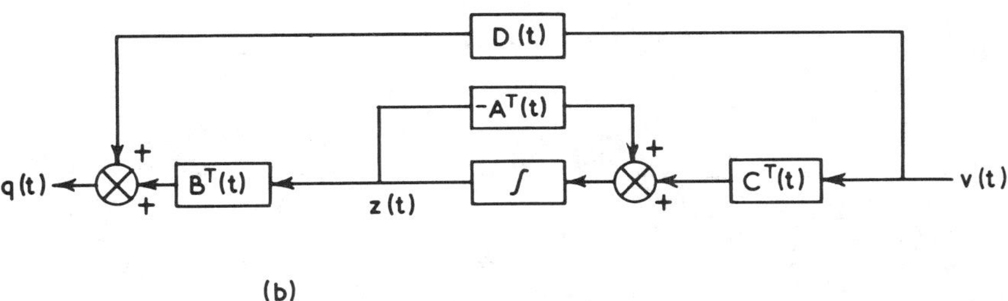

Fig. 3.1 (a) System S(A, B, C, D, t) and (b) its adjoint $S^T(-A^T, B^T, C^T, D^T)$.

(state) controllable (observable) if the system S(A, B, C, D) defined by (3.1) is completely observable (controllable).

Proof of the theorem follows from the preceding development. The system $S^T(-A^T, C^T, B^T, D^T)$ in (3.19) is called the dual or adjoint of S(A, B, C, D) (Fig. 3.1).

Controllability and observability properties of a time-invariant system with distinct eigenvalues may also be examined from its normal form in (2.51). Controllability is a function of the coupling between the input controls and the various modes (state variables) of the system. A particular state variable cannot be controlled if there is no connection between the input and this state variable. This would be the case if there were a null row in B_n, the input control matrix, in (2.51). A particular state variable would not appear in any of the outputs of the system if there were no connection between that state variable and any of the outputs, which would be the case if C_n, the output matrix, in (2.51) contained any null columns. A system in which all state variables are controllable (observable) is said to be completely controllable (observable).

Complete state controllability is neither necessary nor sufficient for the existence of a solution to the problem of controlling the output y(t) of the system (Fig. 3.2). For this reason, it is useful to define separately

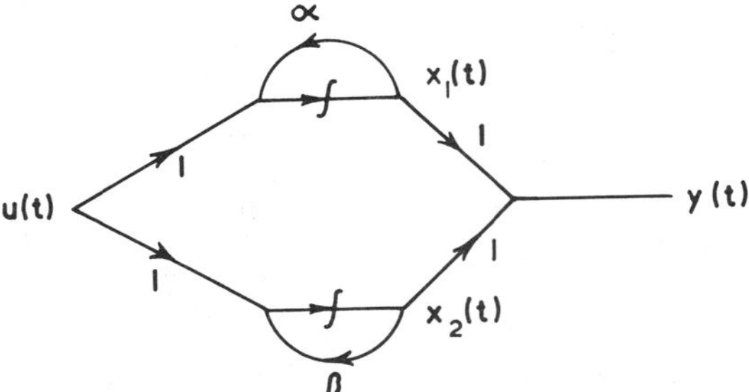

Fig. 3.2 System that is not completely state controllable.

complete output controllability which allows for the consideration of a system with time delays and direct transmission from control input to output.

Definition 3.3: Output Controllability A system is said to be completely output controllable if it is possible to construct an unconstrained input control vector u(t) which will transfer any initial output $y(t_0)$ to any final output y(T) in a finite time interval $t_0 < t < T$.

The definition of output controllability of the system S(A, B, C, D) implies that the output y(t) starting any initial value y(0) can be transferred to the origin of the output space in a finite time interval* $0 < t < T$, that is,

$$y(T) = Cx(T) + Du(T) = 0$$

which from (2.34) and (3.7) gives

$$y(T) = Ce^{AT}x(0) + \int_0^T e^{A\tau}Bu(\tau)\, d\tau + Du(T) = 0 = 0 \tag{3.20}$$

Also, $y(0) = x(0) + Du(0)$.

The definition of complete controllability means that the vector y(0) in (3.9) spans the ℓ-dimensional output space. Since e^{AT} is nonsingular, if y(0) spans the ℓ-dimensional output space, so does $Ce^{AT}x(0)$, and vice versa. From (3.13),

$$Ce^{AT}x(0) = -Ce^{AT}\int_0^T e^{-A\tau}Bu(\tau)\, d\tau$$

$^*t_0 = 0$ for subsequent derivations.

$$= -C \sum_{k=0}^{p-1} \sum_{j=1}^{\ell} \int_0^T \alpha_k(\tau) A^k b_j^c u_j(T - \tau)\, d\tau + \sum_{j=1}^{\ell} d_j^c u_j(T)$$

(3.21)

where, as before, $e^{At} = \Sigma_{k=0}^{p-1} \alpha_k(t) A^k$ and p is the degree of minimal of A; b_j^c and d_j^c are the jth columns of B and D, respectively. From (3.21) it is clear that $Ce^{AT}x(0)$ is a linear combination of $CA^i b_j^c$, d_j^c ($i \in 0, 1, \ldots, p - 1$; $j \in 1, 2, \ldots$). Thus the rank of the output controllability matrix A, given by

$$Q = [CB \;\; CAB \cdots CA^{p-1}B \;\; D]$$

is ℓ; this is obvious if p = n. If p < n, then the $[CA^h b_j^c, d_j^c]$, where p < h < n - 1, are linearly dependent on Cb_j^c, CAb_j^c, ..., $CA^{p-1}b_j^c$, d_j^c. Hence the rank of Q is equal to ℓ.

Conversely, if the system is completely output controllable but $\rho(Q) = h < \ell$, the set of all initial outputs that can be transferred to the origin in the h-dimensional space. Hence the dimension of this set is less than ℓ, which contradicts the assumption that the system is completely output controllable.

If the rank of Q is ℓ, when $Ce^{AT}x(0)$ spans the ℓ-dimensional output space. This implies that if $\rho(Q) = \ell$, y(0) also spans the ℓ-dimensional output space and the system is completely output controllable. If p = n, the condition for complete output controllability is that $\rho(Q) = \ell$, where

$$Q = [CB \;\; CAB \cdots \;\; CA^{n-1}B \;\; D] \tag{3.22}$$

This shows that complete state controllability of S(A, B, C, D) on 0 < t < T implies complete output controllability on 0 < t < T if the ℓ rows of C are linearly independent. If D = 0, the output controllability matrix Q in (3.22) becomes

$$Q = [CB \;\; CAB \cdots CA^{n-1}B] \tag{3.23}$$

which shows that the existence of direct transmission from input to output does not affect state controllability, but aids output controllability.

3.1.1 Pole-Zero Cancellation [1, 3, 22-24]

The concepts of controllability and observability in time domain are associated with the ability to control the state vector in a desired manner, or to evaluate the state vector from the available output variables. Similar notions of these two concepts are also applicable in frequency-domain analysis; these highlight some of the limitations of transfer-function representation of dynamic systems.

Let $g(s) = n(s)/d(s)$ be the transfer function of a stable single-input/single-output nth-order system where the polynomials n(s) and d(s) are of order m and n ($m < n$), respectively. If n(s) and d(s) have a factor (or factors) in common, and cancellation occurs, the order of the system becomes $n - k$ ($0 < k < m$), which is lower than n. Such a system is not completely controllable in that the canceled pole-zero(s) [or mode(s)] do not appear in the input/output description and hence cannot be excited independently through the input control.*

(a) State controllability

Theorem 3.3 If a system is completely state controllable, that is no pole-zero cancellation in the input-state transfer function (matrix).

Proof: For multivariable systems, the transfer-function matrix between the m input controls and n state variables is given by

$$G_s(s) = (sI - A)^{-1}B = \frac{\text{adj }(sI - A)B}{\Delta(s)} = \frac{\text{adj }(sI - A)}{\Delta(s)}[b_1^c \; b_2^c \cdots b_m^c] \qquad (3.24)$$

$$\equiv \frac{1}{\Delta(s)}\{n_{ij}(s)\}, \qquad i \in 1, \ldots, n, \quad j \in 1, \ldots, m \qquad (3.25)$$

If there is to be no pole-zero cancellation in the transfer-function matrix above, there should be no common factor (or factors) between $\Delta(s)$ and the elements $n_{ij}(s)$, $\forall$ (i, j), which from (3.9) suggests that the matrix

$$P = [b_1 \;\; Ab_1 \cdots b_m \cdots A^{n-1}b_m]$$

$$= [B \;\; AB \cdots A^{n-1}B] \qquad (3.26)$$

should have rank n. This completes the proof.

(b) Observability

The significance of complete observability is that there is no cancellation in the transfer function between the state variables and the output. This is briefly illustrated below for a single-input/single-output system.

For the dynamic system in (3.1) the transfer function between the state variables and the output is

*This is also considered using the notion of decoupling zeros in Sec. 3.3.

$$y(s) = c(sI - A)^{-1}$$
$$= \frac{1}{\Delta(s)}[q_1(s) \quad q_2(s) \cdots q_n(s)] \tag{3.27}$$

where $q_i(s)$ are polynomials in s of degree less than n. If y(s) has a cancellation [i.e., if the polynomials $q_1(s)$, $q_2(s)$, ..., $q_n(s)$ and $\Delta(s)$ have a common factor], this canceled mode (state variable) cannot be observed in the output y.

For multivariable systems, the condition for no mode cancellation between the state vector and the output is that the observability matrix

$$R = [C \quad CA \cdots CA^{n-1}]^T \tag{3.28}$$

have rank n.

(c) Output (function) controllability

As concluded earlier, if a system is output controllable, its output can be transferred to any desired value at a certain instant of time. This leads to the problem of evaluating whether it is possible to steer the output following a preassigned curve over any interval of time. A system whose output can be so steered is said to be output-function controllable or functional reproducible.

Definition 3.4: Functional controllability A system with an $(\ell \times m)$ proper rational-function transfer-function matrix G(s) is said to be functional controllable if $\rho[G(s)] = \ell$.

If the system is initially relaxed, then

$$y(s) = G(s)u(s) \tag{3.29}$$

If $\rho[G(s)] = \ell$ [i.e., if all rows of G(s) are linearly independent over the field of rational functions], the $m \times m$ matrix $[G^T(s)G(s)]$ is nonsingular. Thus for any y(s), the choice

$$u(s) = [G^T(s)G(s)]^{-1}G^T(s)y(s) \tag{3.30}$$

satisfies (3.29). Consequently, if $\rho[G(s)] = \ell$, the system is output-function controllable. If $\rho[G(s)] < \ell$, there exists no y(s) in the range of G(s) for which (3.29) has a solution for u(s).

By extending the results of state controllability and observability, it can be shown that a system S(A, B, C, D) with a strictly proper rational transfer function matrix G(s) is output controllable if the rows of G(s) are linearly independent (i.e., if $\rho[G(s)] = \ell$). This combined with the result above suggests that output controllability and output functional controllability are synonymous. The condition for the latter can also be stated in terms of A, B, C, D since

$$G(s) = C(sI - A)^{-1}B + D \tag{3.31}$$

but is rather more involved, and is omitted here.

3.1.2 Effect of Transformation

The properties of state controllability and observability of linear systems are invariant under nonsingular transformation. This is illustrated here through an equivalent transformation

$$x(t) = T\bar{x}(t) \tag{3.32}$$

of the linear time-invariant system S(A, B, C, D)

$$\begin{aligned} \dot{x}(t) &= Ax(t) + Bu(t) \\ y(t) &= Cx(t) + Du(t) \end{aligned} \tag{3.33}$$

the transformed system $\bar{S}(\bar{A}, \bar{B}, \bar{C}, \bar{D})$ being

$$\begin{aligned} \dot{\bar{x}}(t) &= T^{-1}AT\bar{x}(t) + T^{-1}Bu(t) = \bar{A}\bar{x}(t) + \bar{B}u(t) \\ y(t) &= CT\bar{x}(t) + Du(t) = \bar{C}\bar{x}(t) + \bar{D}u(t) \end{aligned} \tag{3.34}$$

From (3.9), the state controllability matrix for the system in (3.34) is

$$\bar{P} = [\bar{B} \;\; \bar{A}\bar{B} \cdots \bar{A}^{n-1}\bar{B}] = T^{-1}[B \;\; AB \cdots A^{n-1}B]T \quad (\text{since } \bar{A}^k = T^{-1}A^kT) \tag{3.35}$$

Since T is nonsingular, the rank of the controllability matrix does not change after equivalent transformation. The observability part can be proved similarly. The assertion above can also be extended to linear time-varying systems.

If (3.33) is transformed into Jordan form, the state controllability and observability properties of the equivalent system can be determined almost by inspection. If the matrix A in (3.33) has q distinct eigenvalues $\lambda_1, \lambda_2, \ldots, \lambda_q$, the system matrices after transformation into Jordan form (from Sec. 1.3) may be expressed as

$$\bar{A}_{(n\times n)} \begin{bmatrix} A_1 & & 0 \\ & A_2 & \\ 0 & & A_q \end{bmatrix}; \quad \bar{B}_{(n\times m)} \begin{bmatrix} B_1 \\ B_2 \\ \vdots \\ B_q \end{bmatrix}; \quad \bar{C}_{(\ell\times n)} = [C_1 \;\; C_2 \cdots C_q] \tag{3.36}$$

where

$$A_i = \underbrace{\begin{bmatrix} \lambda_k & 1 & 0 & \cdots & 0 \\ 0 & \lambda_k & 1 & \cdots & 0 \\ & & \ddots & \ddots & \\ 0 & 0 & \cdots & \lambda_k & 1 \\ 0 & 0 & \cdots & 0 & \lambda_k \end{bmatrix}}_{n_k} \Bigg\} n_k \tag{3.37}$$

$$B_i = \underbrace{\begin{bmatrix} b_{k11} & \cdots & b_{k1m} \\ b_{kj1} & \cdots & b_{kjm} \\ b_{kn_k1} & \cdots & b_{kn_km} \end{bmatrix}}_{m} = \begin{bmatrix} b_k^{r1} \\ b_k^{rj} \\ b_k^{rn_i} \end{bmatrix} \Bigg\} n_k \tag{3.38}$$

$$C_i = \ell \left\{ \begin{bmatrix} c_{k11} & \cdots & c_{k1j} & \cdots & c_{k1n_j} \\ \vdots & & & & \\ c_{k\ell 1} & \cdots & c_{k\ell j} & \cdots & c_{k\ell n_k} \end{bmatrix} \right. = \underbrace{[c_k^{c1} \cdots c_k^{cj} \cdots c_k^{cn_k}]}_{n_k} \tag{3.39}$$

$j \in 1, 2, \ldots, n_k$ and $k \in 1, 2, \ldots, q$. A_k denotes all the Jordan blocks associated with the eigenvalue λ_k, n_k is the order of the Jordan block ($\equiv$ multiplicity of λ_k), and $n = \Sigma_{k=1}^{q} n_k$.

Corresponding to A_k, the matrices B and C are partitioned as shown above. The first and the last rows of B_k are denoted by b_k^{r1} and $b_k^{rn_k}$, respectively. The first and the last columns of C_k are denoted by c_k^{c1} and $c_k^{cn_k}$.

<u>Theorem 3.4</u> The linear time-invariant system S(A, B, C, D) defined by (3.36) is state controllable if for each $k \in 1, 2, \ldots, q$, the rows of the $n_k \times m$ matrix B_k in (3.38) are linearly independent, and observable if for each $k \in 1, 2, \ldots, q$, the columns of the $\ell \times n_k$ matrix C_k in (3.39) are linearly independent.

Proof of this theorem is not given here, but a few comments are relevant. The significance of the row (column) rank condition of $B_k(C_k)$ is that for complete controllability (observability) each of the q subsystems described by (A_k, B_k, C_k) must be controllable (observable). The linear independence of the rows (columns) of B_k (C_k) ensures that the inputs to the n_k-state variables (for any $k \in 1, \ldots, q$) may be chosen independently (controllability) and the corresponding output variables are not linearly related (observability). The usefulness of this transformation is illustrated below.

Example 3.1: Controllable but unobservable system

$$S:\ A = \begin{bmatrix} -1 & & & 0 \\ & -2 & & \\ & & -3 & \\ 0 & & & -4 \end{bmatrix};\quad B = \begin{bmatrix} 1 & 1 & 0 \\ 0 & 1 & 0 \\ 0 & 1 & 0 \\ 0 & 0 & 1 \end{bmatrix};\quad C = \begin{bmatrix} 1 & 0 & 1 & 1 \\ 0 & 1 & 0 & 1 \\ 1 & -1 & 1 & 0 \end{bmatrix} \tag{3.40}$$

The state controllability matrix is

$$P = [B\ \ AB\ \ A^2B\ \ A^3B]$$

$$= \left[\begin{array}{ccc|ccc|ccc|ccc} 1 & 1 & 0 & -1 & -1 & 0 & 1 & 1 & 0 & -1 & -1 & 0 \\ 0 & 1 & 0 & 0 & -2 & 0 & 0 & 4 & 0 & 0 & -8 & 0 \\ 0 & 1 & 0 & 0 & -3 & 0 & 0 & 9 & 0 & 0 & -27 & 0 \\ 0 & 0 & 1 & 0 & 0 & -4 & 0 & 0 & 16 & 0 & 0 & -64 \end{array}\right];\quad \rho(P) = 4 = n \tag{3.41}$$

$$(sI - A)^{-1}B = \begin{bmatrix} \frac{1}{s+1} & \frac{1}{s+1} & 0 \\ 0 & \frac{1}{s+2} & 0 \\ 0 & \frac{1}{s+3} & 0 \\ 0 & 0 & \frac{1}{s+4} \end{bmatrix} \rightarrow \text{no pole-zero cancellation}$$

The system is therefore state controllable. This is also apparent from the signal flow graph of the system in Fig. 3.3, which shows that each state can be controlled independently through suitable combination of the input controls.

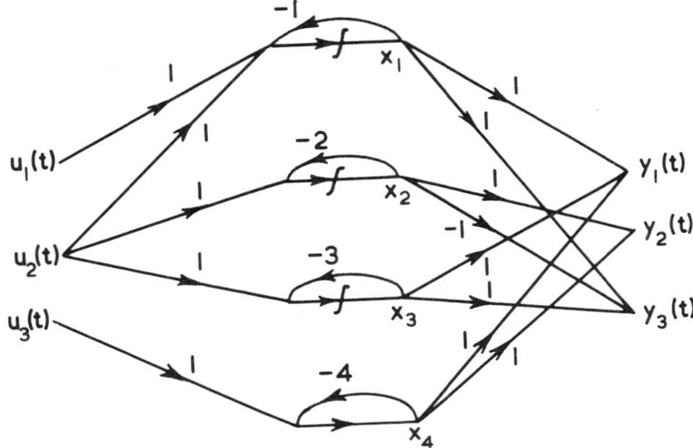

Fig. 3.3 Signal flow graph of (3.40).

The observability matrix is

$$R = \begin{bmatrix} C \\ CA \\ CA^2 \\ CA^3 \end{bmatrix} = \left[\begin{array}{cccc} 1 & 0 & 1 & 0 \\ 0 & 1 & 0 & 1 \\ 1 & -1 & 1 & 0 \\ \hline -1 & 0 & -3 & -4 \\ 0 & -2 & 0 & -4 \\ -1 & 2 & -3 & 0 \\ \hline 1 & 0 & 9 & 16 \\ 0 & 4 & 0 & 16 \\ 1 & -4 & 9 & 0 \\ \hline -1 & 0 & -27 & -64 \\ 0 & -8 & 0 & -64 \\ -1 & 8 & -27 & 0 \end{array}\right]; \quad \rho(R) = 2 < n = 3 \tag{3.42}$$

The system is therefore not completely observable, as is apparent from Fig. 3.3, which shows that $y_1 = y_2 + y_3$. The output controllability matrix is

$$Q = [CB\ \ CAB\ \ CA^2B\ \ CA^3B] = \left[\begin{array}{ccc|ccc|ccc|ccc} 1 & 2 & 1 & -1 & -4 & -4 & 1 & 10 & 16 & -1 & -28 & -64 \\ 0 & 1 & 1 & 0 & -2 & -4 & 0 & 4 & 16 & 0 & -8 & -64 \\ 1 & 1 & 0 & -1 & -2 & 0 & 1 & 6 & 0 & -1 & -20 & 0 \end{array}\right]$$

$$\rho(Q) = 2 < \ell$$

The system is therefore not output controllable, as expected from (3.42). This is also apparent from the transfer-function matrix of the system

$$G(s) = \begin{bmatrix} \frac{1}{s+1} & \frac{2(s+2)}{(s+1)(s+3)} & \frac{1}{s+4} \\ 0 & \frac{1}{s+2} & \frac{1}{s+4} \\ \frac{1}{s+1} & \frac{2(s+2)^2 - (s+1)(s+3)}{(s+1)(s+2)(s+3)} & 0 \end{bmatrix}; \quad \rho[G(s)] = 2 < \ell; \quad y_1(s) = y_2(s) + y_3(s) \tag{3.43}$$

It is worth noting at this stage that the system in (3.40) is in normalized form. Since A has distinct eigenvalues, the conditions for state controllability and observability may be reduced to

$$\rho(B_n) = m \text{ with no null rows in } B_n \text{ and } \rho(C_n) = \ell \text{ with no null column in } C_n$$

where m and ℓ are the number of input controls and the number of output variables. This is illustrated below.

Example 3.2: Controllability and observability through normalization

$$S:\ A = \begin{bmatrix} 8 & -3 & -6 \\ 10 & -5 & -6 \\ 7 & -3 & -5 \end{bmatrix}; \quad B = \begin{bmatrix} 6 & 3 \\ 7 & 4 \\ 7 & 4 \end{bmatrix}; \quad C = \begin{bmatrix} -6 & -2 & -3 \\ -1 & 1 & 0 \end{bmatrix}$$

A modal matrix for this system is

$$M = \begin{bmatrix} 1 & 3 & 3 \\ 1 & 4 & 3 \\ 1 & 3 & 4 \end{bmatrix}$$

The normalized equations are

$$A = \begin{bmatrix} -1 & & 0 \\ & -2 & \\ 0 & & -3 \end{bmatrix}; \quad B_n = \begin{bmatrix} 0 & 0 \\ 1 & 1 \\ 1 & 0 \end{bmatrix}; \quad C_n = \begin{bmatrix} 1 & 1 & 0 \\ 0 & 1 & 0 \end{bmatrix} \tag{3.44}$$

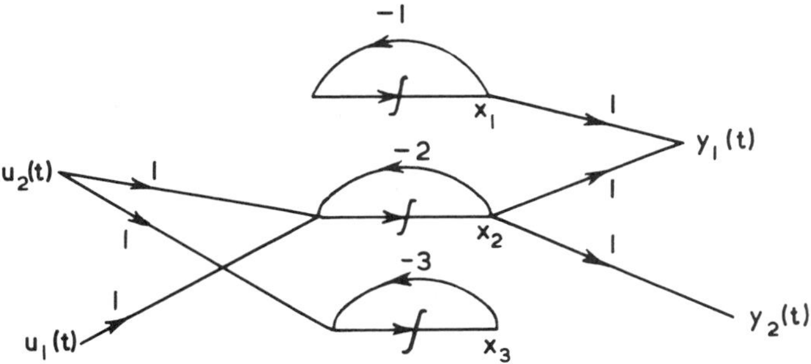

Fig. 3.4 Signal flow graph of (3.44).

The system is neither controllable nor observable, as shown by the signal flow graph in Fig. 3.4.

Example 3.3: Pole-zero cancellation The following single-input/single-output second-order system is considered here to illustrate the relation between pole-zero cancellation and state controllability.

$$\dot{x}(t) = \begin{bmatrix} -3 & 1 \\ -2 & 1.5 \end{bmatrix} x(t) + \begin{bmatrix} 1 \\ 4 \end{bmatrix} u(t); \quad y(t) = x_1(t) \tag{3.45}$$

The state controllability matrix

$$P = [B : Ab] = \begin{bmatrix} 1 & 1 \\ 4 & 4 \end{bmatrix}; \quad \rho(P) = 1 < n = 2$$

The transfer function between y(t) and u(t) is

$$y(s) = c(sI - A)^{-1}b = [1 \quad 0] \begin{bmatrix} \dfrac{s - 1.5}{(s+2.5)(s-1)} & \dfrac{-1}{(s+2.5)(s-1)} \\ \dfrac{2}{(s+2.5)(s-1)} & \dfrac{s+3}{(s+2.5)(s-1)} \end{bmatrix} \begin{bmatrix} 1 \\ 4 \end{bmatrix}$$

$$= \frac{s+2.5}{(s+2.5)(s-1)} \tag{3.46}$$

which shows that control over one mode is lost due to the cancellation of the common factor in the numerator and denominator of the transfer function.

Example 3.4: Pole-zero cancellation The following system is considered to illustrate pole-zero cancellation in a transfer-function matrix.

$$A = \begin{bmatrix} 0 & 1 & 1 & 0 \\ \alpha_1 & \beta_1 & \beta_1 & 0 \\ \alpha_1 & \beta_1 & \beta_1 - 1 & -1 \\ 0 & -1 & 0 & {}_2 - 1 \end{bmatrix}; \quad B = \begin{bmatrix} 0 & 0 \\ 1 & 0 \\ 1 & 0 \\ -1 & 2 \end{bmatrix}; \quad C = \begin{bmatrix} 1 & 0 & 0 & 0 \\ 0 & 0 & 0 & 0 \end{bmatrix} \tag{3.47}$$

The transfer-function matrix of the system is

$$G(s) = C(sI - A)^{-1}B = \begin{bmatrix} \dfrac{2}{s^2 - \beta_1 s - \alpha_1} & 0 \\ 0 & \dfrac{1}{s - \alpha_2} \end{bmatrix}$$

Derivation of the state controllability and observability matrices so relate them to pole-zero cancellation is left to the reader.

Example 3.5: Controllability and observability from Jordan form

$$A = \begin{bmatrix} 5 & 4 & 0 \\ 0 & 1 & 0 \\ -4 & 4 & 1 \end{bmatrix}; \quad B = \begin{bmatrix} -2 & 0 \\ -1 & 1 \\ -4 & 0 \end{bmatrix}; \quad C = \begin{bmatrix} 1 & 0 & 1 \\ 0 & 1 & 1 \end{bmatrix} \tag{3.48}$$

The system has eigenvalues at $\lambda_1 = \lambda_2 = 1$, $\lambda_3 = 5$. The transformation matrix M may be derived by following Sec. 1.3. The resulting Jordan form of the system is [1]

$$J = M^{-1}AM = \frac{1}{32}\begin{bmatrix} -1 & -2 & -1 \\ 0 & -8 & 0 \\ 2 & 2 & 0 \end{bmatrix}\begin{bmatrix} 5 & 4 & 0 \\ 0 & 1 & 0 \\ -4 & 4 & 1 \end{bmatrix}\begin{bmatrix} 0 & 4 & 16 \\ 0 & -4 & 0 \\ -32 & 4 & -16 \end{bmatrix}$$

$$= \left[\begin{array}{cc|c} 1 & 1 & 0 \\ 0 & 1 & 0 \\ \hline 0 & 0 & 5 \end{array}\right] = \left[\begin{array}{c|c} A_1 & 0 \\ \hline 0 & A_2 \end{array}\right]$$

$$\bar{B} = M^{-1}B = \frac{1}{3}\left[\begin{array}{cc} 8 & -2 \\ 8 & -8 \\ \hline -20 & 0 \end{array}\right] \begin{array}{l} \left.\vphantom{\begin{array}{c} 8 \\ 8 \end{array}}\right\} B_1 \\ \} B_2 \end{array} \quad \text{and} \quad \bar{C} = CM = \left[\begin{array}{cc|c} -32 & 8 & 0 \\ -32 & 0 & -16 \end{array}\right]$$

$$\underbrace{}_{C_1} \quad \underbrace{}_{C_2}$$

The system is completely (state) controllable and completely observable.

3.2 CANONICAL FORMS [3, 5, 7-11]

A dynamic system S(A, B, C, D) which is not completely state controllable (observable) can in general be transformed into an equivalent controllable (observable) system with the same transfer function as the original system.

In the following analysis, a superscript c is used to stand for controllable, a $\bar{c}$ for uncontrollable, an o for observable, and an $\bar{o}$ for unobservable.

3.2.1 Controllable Form

If the system S(A, B, C, D) is not completely controllable, its controllability matrix P in (3.9) will have rank less than n; let $\rho(P) = n_1 < n$. There exists an equivalence transformation $x = T\bar{x}$, where T is a constant nonsingular matrix which transforms S(A, B, C, D) into $\bar{S}$, defined as

$$\bar{S}: \quad \begin{aligned} \begin{matrix} n_1\{ \\ n-n_1\{ \end{matrix} \begin{bmatrix} \dot{\bar{x}}^c(t) \\ \dot{\bar{x}}^{\bar{c}}(t) \end{bmatrix} &= \begin{bmatrix} \overbrace{\bar{A}_c}^{n_1} & \overbrace{\bar{A}_{12}}^{n-n_1} \\ 0 & \bar{A}_{\bar{c}} \end{bmatrix} \begin{bmatrix} \bar{x}^c(t) \\ \bar{x}^{\bar{c}}(t) \end{bmatrix} + \begin{bmatrix} \bar{B}_c \\ 0 \end{bmatrix} u(t) \\ y(t) &= [\underbrace{\bar{C}_c}_{n_1} \mid \underbrace{\bar{C}_{\bar{c}}}_{n-n_1}] \begin{bmatrix} \bar{x}^c(t) \\ \bar{x}^{\bar{c}}(t) \end{bmatrix} + Du(t) \end{aligned} \tag{3.49}$$

and the n_1-dimensional subsystem $\bar{S}^c$:

$$\bar{S}^c: \quad \begin{aligned} \dot{\bar{x}}^c(t) &= \bar{A}_c \bar{x}^c(t) + \bar{B}_c u(t) \\ \bar{y}(t) &= \bar{C}_c \bar{x}^c(t) + Du(t) \end{aligned} \tag{3.50}$$

is completely controllable and has the same transfer-function matrix as $\bar{S}$ in (3.49).

The assertion above is based on the following identities:

(1) $\rho[B \;\; AB \;\; A^{n-1}B] = \rho[\bar{B}_c \;\; \bar{A}_c\bar{B}_c \;\; \bar{A}^{n_1-1}\bar{B}_c] = n_1 < n$

and

(2) $$[(\bar{C}_c \;\; \bar{C}_{\bar{c}})] \begin{bmatrix} sI - \bar{A}_c & -\bar{A}_{12} \\ 0 & sI - A_{\bar{c}} \end{bmatrix}^{-1} \begin{bmatrix} \bar{B}_c \\ 0 \end{bmatrix} + D$$

$$= [\bar{C}_c \mid \bar{C}_{\bar{c}}] \left[\begin{array}{c|c} (sI - \bar{A}_c)^{-1} & (sI - \bar{A}_c)^{-1}\bar{A}_{12} \quad (sI - \bar{A}_{\bar{c}})^{-1} \\ \hline 0 & (sI - \bar{A}_{\bar{c}})^{-1} \end{array} \right] \left[\begin{array}{c} \bar{B}_c \\ \hline 0 \end{array} \right] + D$$

$$= \bar{C}_c[sI - \bar{A}_c]^{-1}\bar{B}_c + D \tag{3.51}$$

which is the transfer-function matrix of $\bar{S}_c$ in (3.49).

In the equivalence transformation $x = T\bar{x}$, the state space $\bar{R}^n$ of $\bar{S}$ is divided into two subspaces. One is the n_1-dimensional subspace $\bar{R}_1$ which consists of all the vectors

$$\begin{bmatrix} \bar{x}^c \\ 0 \end{bmatrix}$$

The other is the $(n - n_1)$-dimensional subspace $\bar{R}_2$, which consists of all the vectors

$$\begin{bmatrix} 0 \\ \bar{x}^{\bar{c}} \end{bmatrix}$$

Since $\bar{S}^c$ is controllable, all the vectors $\bar{x}^c$ in $\bar{R}_1$ are controllable, the subspace $\bar{R}_1$ is called the <u>controllable subspace</u>. Thus the state variables in $\bar{x}^{\bar{c}}$ are not affected directly by the input $\underline{u}(t)$ or indirectly through the state vector $\bar{x}^c$: therefore, the state vector $\bar{x}^{\bar{c}}$ is not controllable and is dropped in the reduced equation (3.49).

Thus if a linear time-invariant system is not completely state controllable, by a proper choice of a basis, the state vector can be decomposed into two groups: one controllable and the other uncontrollable. By dropping the uncontrollable state vector, a controllable dynamical system of lesser dimension is obtained. This reduced system, which is zero-state equivalent to the original system, is known as the <u>controllable form</u> of S(A, B, C, D); a relatively simple method of deriving this form is described below [10, 11]. The basic requirement for the applicability of this algorithm is that the arbitrary system S(A, B, C, D) be controllable with a controllability index n_1; the rank of the $(n \times mn)$ state controllability matrix P in (3.9) being n_1, the m columns of B being linearly independent. Since the transmission matrix D does not affect state controllability (Sec. 3.1.1), it is not included in the derivation below.

The algorithm to transform S(A, B, C) into a controllable form $S_c(A_c, B_c, C_c)$ is based on the formation of two rectangular transformation matrices T_c and W_c of dimensions $n \times n_c$ and $n_c \times n$, such that*

*If $n = n_c$, then $T_c^{-1} = W_c$; $n_c \equiv n_1$ in the above derivations.

$$W_c T_c = I_{nc} \tag{3.52}$$

where n_c is the controllability index of S(A, B, C) in (3.1). The first step in this development is the selection of a family of $(n \times n_c)$ matrices T_j from the reduced-order controllability matrix

$$P_{nc} = [B \;\; AB \;\; A^{n_c-1} \;\; B] \tag{3.53}$$

Choice of the n_c columns of T_j matrices, in general, is not unique, but may be accomplished by using the following structure:

$$T_j = [b_j^c \;\; Ab_j^c \cdots A^{p_j-1} \; b_j^c], \qquad b_j^c = \text{jth column of B}, \;\; j \in 1, \ldots, m \tag{3.54}$$

where p_j is the least integer such that $A^{p_j} b_j$ is linearly dependent on b_j, $Ab_j, \ldots, A^{b_j-1} b_j$, that is

$$A^{p_j} b_j = \sum_{k=0}^{p_j} \alpha_k A^{k-1} b_j \tag{3.55}$$

The selection procedure starts with j = 1, choosing b_1^c first and then proceeding to Ab_1^c, $A^2 b_1^c$, and so on, until $A^{n_c-1} b_1^c$ is obtained [in which case the system is controllable from the first input $u_1(t)$ alone], or until a dependency arises in which case $p_1 < n_c$. The procedure is then continued with b_j^c, j = 2, 3, ..., until n_c linearly independent column vectors are obtained. In these subsequent selections, the new vectors of the form $A^k b_j^c$ are retained only if it is linearly independent of all previously selected vectors; otherwise, it is discounted in the selection procedure. The resulting $n \times n_c$ transformation matrix will therefore have the form

$$T_c = [\underbrace{b_1^c Ab_1^c \cdots A^{p_1-1} \; b_1^c}_{T_1} \cdots \underbrace{b_j^c Ab_j^c \cdots A^{p_j-1} \; b_j^c}_{T_j} \cdots \underbrace{b_q^c Ab_q^c \cdots A^{p_q-1} \; b_q^c}_{T_q}] \tag{3.56}$$

where

$$p_1 + \cdots + p_j + \cdots + p_q = n_c$$

and q (< m) is the number of column vectors of B used in the construction of T_c. The physical significance of $q < m$ is that the remaining $(m - q)$ columns of the input control matrix B do not play any special role in the associated canonical form, and will appear in an arbitrary fashion in the final result. It is to be noted that, since the selections above depend on the a priori ordering of the columns of B, there are many other structures for the column vectors of T_c.

Once T_c is formulated, the $(n_c \times n)$ matrix W_c may be derived from (3.52), which may be accomplished by more than one method. One simple procedure is based on setting up arrays

$$T_c \quad I_n$$

By successive row operations, on I_n, the arrays above may be transformed into the following forms:

$$T_c \rightarrow \underbrace{\begin{bmatrix} I_{nc} \\ T_A \end{bmatrix}}_{n_c} \begin{matrix} \}n_c \\ \}n - n_c \end{matrix} \qquad I_n \rightarrow Z \equiv \underbrace{\begin{bmatrix} Z_1 \\ Z_2 \end{bmatrix}}_{n} \begin{matrix} \}n_c \\ \}n - n_c \end{matrix} \tag{3.57}$$

Multiplying the two arrays gives us

$$ZT_c = \begin{bmatrix} Z_1 \\ Z_2 \end{bmatrix} T_c = \begin{bmatrix} Z_1 T_c \\ Z_2 T_c \end{bmatrix} \equiv \begin{bmatrix} I_{nc} \\ T_A \end{bmatrix} \tag{3.58}$$

Therefore, $Z_1 T_c = I_{nc}$, which when compared with (3.52) gives

$$W_c = Z_1 \tag{3.59}$$

The controllable canonical form $S_c(A_c, B_c, C_c)$ is then given by

$$A_c = W_c A T_c = \begin{bmatrix} A_1 & & & & 0 \\ & \ddots & & & \\ & & A_j & & \\ & & & \ddots & \\ 0 & & & & A_q \end{bmatrix} (n_c \times n_c) \tag{3.60}$$

where

$$A_j = \begin{bmatrix} 0 & 0 & \cdots & 0 & \alpha_0 \\ 1 & 0 & \cdots & 0 & \alpha_1 \\ \vdots & & & & \\ 0 & 0 & \cdots & 1 & \alpha_{p_j-1} \end{bmatrix} (n_c \times p_j), \quad j \in 1, \ldots, q$$

$B_c = W_c B$, and $C_c = CT_c$.

The number of arbitrary elements in this transformation is

$$\text{in } A_j\text{: } q \times n_c; \quad \text{in } B_c\text{: } n_c \times (m - q); \quad \text{in } C_c\text{: } \ell \times n_c$$

The mechanism of deriving the controllable form is demonstrated through the following example.

Example 3.6: Derivation of the controllable form of S(A, B, C), where

$$A = \begin{bmatrix} 1 & 1 & 0 \\ 0 & 1 & 0 \\ 0 & 1 & 1 \end{bmatrix}; \quad B = \begin{bmatrix} 0 & 1 \\ 1 & 0 \\ 0 & 1 \end{bmatrix}; \quad C = \begin{bmatrix} 1 & 1 & 2 \\ 1 & 0 & 2 \end{bmatrix} \tag{3.61}$$

$$[b_1 \;\; Ab_1 \;\; A^2b_1] = \begin{bmatrix} \underbrace{\begin{matrix} 0 & 1 \\ 1 & 1 \\ 0 & 1 \end{matrix}}_{T_1} & \begin{matrix} 2 \\ 1 \\ 2 \end{matrix} \end{bmatrix} \rightarrow p_1 = 2$$

$$[b_2 \;\; Ab_2 \;\; Ab_2^2] = \begin{bmatrix} \underbrace{\begin{matrix} 1 \\ 0 \\ 1 \end{matrix}}_{T_2} & \begin{matrix} 1 & 1 \\ 0 & 0 \\ 1 & 1 \end{matrix} \end{bmatrix} \rightarrow p_1 = 0$$

since the column of T_2 is linearly related to the columns of T_1. Thus the system is not completely state controllable; consequently, $q = 1$ and $n_C = p_1 + p_2 = 2 < n$. Thus

$$T_c = \begin{bmatrix} 0 & 1 \\ 1 & 0 \\ 0 & 1 \end{bmatrix} \tag{3.62}$$

Computation of W_c follows from (3.52) by using the following procedure:

$$
T_c = \begin{bmatrix} 0 & 1 \\ 1 & 0 \\ 0 & 1 \end{bmatrix} \rightarrow \begin{bmatrix} 1 & 0 \\ 0 & 1 \\ 1 & 0 \end{bmatrix} \equiv \underbrace{\begin{bmatrix} I_{nc} \\ \\ T_A \end{bmatrix}}_{2 = n_c} \begin{matrix} \}2 = n_c \\ \\ \}1 = (n - n_c) \end{matrix}
$$

$$
I_n = \begin{bmatrix} 1 & 0 & 0 \\ 0 & 1 & 0 \\ 0 & 0 & 1 \end{bmatrix} \rightarrow \begin{bmatrix} 0 & 1 & 0 \\ 1 & 0 & 0 \\ \hline 0 & 0 & 1 \end{bmatrix} = \underbrace{\begin{bmatrix} Z_1 \\ -- \\ Z_2 \end{bmatrix}}_{3 = n} \begin{matrix} \}\ 2 = n_c \\ \\ \}\ 1 = (n - n_c) \end{matrix}
$$

Therefore,

$$
W_c = Z_1 = \begin{bmatrix} 0 & 1 & 0 \\ 1 & 0 & 0 \end{bmatrix} \tag{3.63}
$$

The controllable form of (3.61) is therefore given by

$$
A_c = W_c A T_c = \begin{bmatrix} 1 & 0 \\ 1 & 1 \end{bmatrix}; \quad B_c = W_c B = \begin{bmatrix} 1 & 0 \\ 0 & 1 \end{bmatrix}; \quad C_c = C T_c \tag{3.64}
$$

Since T_2 is linearly dependent on T_1, the second input control $u_2(t)$ has no influence over the controllable form of the system.

In addition to deriving the controllable form, the method above may also be used to check for controllability of a system; this is shown below.

Example 3.7: Controllable form of the system in (3.48) For this system:

$$
[b_1 \;\; Ab_1 \;\; A^2 b_1] = \begin{bmatrix} -2 & -14 & -74 \\ -1 & -1 & -1 \\ -4 & 0 & 52 \end{bmatrix} \rightarrow p_1 = 3
$$

$$
[b_2 \;\; Ab_2 \;\; A^2 b_2] = \begin{bmatrix} 0 & 4 & 24 \\ 1 & 1 & 1 \\ 0 & 4 & -8 \end{bmatrix} \rightarrow p_2 = 3
$$

Thus the system is completely state controllable with $n_c = 3$. (The system is controllable from either of the two inputs.) For the sake of convenience, the transformation matrix T_c is chosen as

$$T_c = \begin{bmatrix} -2 & 0 & 4 \\ -1 & 1 & 1 \\ -4 & 0 & 4 \end{bmatrix} \quad \text{and} \quad T_c^{-1} = \frac{1}{8}\begin{bmatrix} 4 & 0 & -4 \\ 0 & 8 & -2 \\ 8 & 0 & -2 \end{bmatrix} \tag{3.65}$$

The controllable form of the system is given by

$$A_c = T_c^{-1}AT_c; \quad B_c = T_c^{-1}B; \quad C_c = CT_c$$

Detailed computations are left to the reader.

3.2.2 Observable Form

If for the n-dimensional system S(A, B, C, D) the observability matrix R in (3.16) has rank $n_o < n$, there exists an equivalence transformation $x = T\bar{x}$ that transforms S(A, B, C, D) into $\bar{S}$.

$$\bar{S}: \begin{bmatrix} \dot{\bar{x}}^o(t) \\ \hline \dot{\bar{x}}^{\bar{o}}(t) \end{bmatrix} = \begin{bmatrix} \bar{A}_o & 0 \\ \hline \bar{A}_{21} & \bar{A}_{\bar{o}} \end{bmatrix} \begin{bmatrix} \bar{x}^o(t) \\ \hline \bar{x}^{\bar{o}}(t) \end{bmatrix} + \begin{bmatrix} \bar{B}_o \\ \hline \bar{B}_{\bar{o}} \end{bmatrix} u(t) \tag{3.66}$$

and the n_o-dimensional subsystem $\bar{S}^o$:

$$\bar{S}^o: \begin{aligned} \dot{\bar{x}}^o(t) &= \bar{A}_o\bar{x}^o(t) + \bar{B}_o u(t) \\ \bar{y}(t) &= \bar{C}_o\bar{x}^o(t) + Du(t) \end{aligned} \tag{3.67}$$

is observable with the same transfer function as S(A, B, C, D). Proof of this assertion follows along the lines similar to the controllable form in the preceding section. From (3.67) it is apparent that the vector $\bar{x}^{\bar{o}}$ does not appear directly in the output y nor indirectly through $\bar{x}^o$. Hence the vector $\bar{x}^{\bar{o}}$ is not observable and is dropped in the reduced equation. This reduced system is known as the <u>observable form</u> of the original system S(A, B, C, D). A method of deriving an observable form is given below. The derivation here is based on the assumption that the arbitrary system S(A, B, C, D) is observable with an observability index n_o. Then from Sec. 3.1.2, the rank of the reduced observability matrix R_{no} is n_o, where

$$R_{no} = [C^T \;\; A^TC^T \cdots (A^T)^{n_o-1}C^T] \tag{3.68}$$

The transformation matrix V^T is constructed by following a procedure similar to the controllable form through replacing A by A^T and b_j^c by c_j, c_j being the jth row of C. The selection starts by choosing

$$V_j^T = [c_j^T \;\; A^T c_j^T \cdots (A^T)^{r_t - 1} c_j^T](n \times r_j) \tag{3.69}$$

with j from 1, 2, ...; the process stops when no further independent vectors are found (i.e., when $j = n_o$), yielding

$$V_o^T = [V_1^T \cdots V_j^T \cdots V_t^T] \tag{3.70}$$

where

$$r_1 + \cdots + r_j + \cdots + r_t = n_o \quad \text{and} \quad (A^T)^{r_j} c_j = \sum_{k=1}^{r_j} \beta_k (A^T)^{k-1} c_j, \quad \beta = \text{constant}$$

Once V_o^T is formulated, the $n \times n_o$ matrix W_o is computed by using (3.52):

$$W_o V_o = I_{no} \tag{3.71}$$

The observable form $S_o(A_o, B_o, C_o)$ of S(A, B, C) is then given by

$$A_o = W_o A V_o = \begin{bmatrix} A_1 & & & & 0 \\ & \ddots & & & \\ & & A_j & & \\ & & & \ddots & \\ 0 & & & & A_t \end{bmatrix} (n_o \times n_o) \tag{3.72}$$

$$A_j = \begin{bmatrix} 0 & 1 & 0 & \cdots & 0 \\ 0 & 0 & 1 & \cdots & 0 \\ \vdots & & & & \\ \beta_0 & \beta_1 & \beta_2 & \cdots & \beta_{r_j - 1} \end{bmatrix} (n_o \times r_j)$$

$$B_o = W_o B \; (n_o \times m) \quad \text{and} \quad C_o = C V_o \; (\ell \times n_o)$$

The number of arbitrary parameters associated with this transformation is $n_o(m + \ell)$.

Derivation of the observable form is illustrated through the following example.

Example 3.8: Derivation of observable form of S(A, B, C) in (3.61)

$$V_1 = \begin{bmatrix} C_1 \\ C_1A \end{bmatrix} = \begin{bmatrix} 1 & 1 & 2 \\ 1 & 4 & 2 \end{bmatrix} \rightarrow r_1 = 2$$

$$V_2 = [C_2] = [1 \;\; 0 \;\; 1] \rightarrow r_2 = 1$$

$$V_o = \begin{bmatrix} V_1 \\ V_2 \end{bmatrix} = \begin{bmatrix} 1 & 1 & 2 \\ 1 & 4 & 2 \\ 1 & 0 & 2 \end{bmatrix} \tag{3.73}$$

Computation of W_o may follow the procedure as W_c, but since $n_o = r_1 + r_2 = 3$, the system is completely observable and V_o is nonsingular. Thus

$$W_o = V_o^{-1} = \frac{1}{3}\begin{bmatrix} -4 & 1 & 6 \\ -1 & 1 & 0 \\ 4 & -1 & -3 \end{bmatrix} \tag{3.74}$$

The observable form of (3.61) is therefore given by

$$A_o = W_oAV_o = \frac{1}{3}\begin{bmatrix} 5 & 8 & 4 \\ -1 & -1 & -2 \\ 1 & 4 & 5 \end{bmatrix}; \quad C_o = CV_o = \begin{bmatrix} 4 & 5 & 6 \\ 2 & 1 & 3 \end{bmatrix};$$
$$B_o = W_oB \tag{3.75}$$

This method may, as in controllability, also be used to check the observability of a system.

3.2.3 Canonical Decomposition

The preceding derivations show that for any given system there exist two transformation matrices which transform the system into its controllable and observable forms separately. After a system has been "taken" through these transformations, the resulting form will be controllable and observable, and this leads to the canonical decomposition theory below [3, 6, 18-20].

Theorem 3.5 A linear time-invariant system S(A, B, C, D) can be transformed by an appropriate equivalent transformation $x = T\bar{x}$ into the canonical form $\bar{S}$ defined as

$$
\bar{S}: \quad \begin{bmatrix} \dot{\bar{x}}^{c\bar{o}}(t) \\ \dot{\bar{x}}^{co}(t) \\ \dot{\bar{x}}^{\bar{c}o}(t) \\ \dot{\bar{x}}^{\bar{c}\bar{o}}(t) \end{bmatrix} = \begin{bmatrix} \bar{A}_{c\bar{o}} & \bar{A}_{12} & \bar{A}_{13} & \bar{A}_{14} \\ 0 & \bar{A}_{co} & \bar{A}_{23} & 0 \\ 0 & 0 & A_{\bar{c}o} & 0 \\ 0 & 0 & \bar{A}_{43} & A_{\bar{c}\bar{o}} \end{bmatrix} \begin{bmatrix} \bar{x}^{c\bar{o}}(t) \\ \bar{x}^{co}(t) \\ \bar{x}^{\bar{c}o}(t) \\ \bar{x}^{\bar{c}\bar{o}}(t) \end{bmatrix} + \begin{bmatrix} \bar{B}_{c\bar{o}} \\ \bar{B}_{co} \\ 0 \\ 0 \end{bmatrix} u(t)
$$

$$
y(t) = [0 \;\; \bar{C}_{co} \;\; \bar{C}_{\bar{c}o} \;\; 0] \begin{bmatrix} \bar{x}^{c\bar{o}} \\ \bar{x}^{co} \\ \bar{x}^{\bar{c}o} \\ \bar{x}^{\bar{c}\bar{o}} \end{bmatrix} + Du(t) \tag{3.76}
$$

where the state vector $\bar{x}^{c\bar{o}}$ is controllable but not observable, $\bar{x}^{co}$ is controllable and observable, $\bar{x}^{\bar{c}o}$ is not controllable but observable, and $\bar{x}^{\bar{c}\bar{o}}$ is neither controllable nor observable. Furthermore, the transfer function of S(A, B, C, D) is

$$
\bar{C}_{co}(sI - \bar{A}_{co})^{-1}\bar{B}_{co} + D
$$

which depends only on the controllable and observable part of the system.

The decomposition theory suggests that every system admits a canonical decomposition into four subsystems, corresponding to the state variables, which are both controllable and observable, either controllable or observable, and neither controllable nor observable, as shown in Fig. 3.5.

The subsystem $\bar{S}^{co}$, which is both controllable and observable, is zero-state equivalent to S(A, B, C, D) and is represented in the input/output transfer function of the system. In other words, the minimal realization* of the transfer function $C(sI - A)^{-1}B + D$ depends solely on the controllable and observable parts of the system equation in (3.1). This emphasizes the reason why the transfer-function representation of a system could be insufficient to describe the complete input/output properties of a system containing uncontrollable (unobservable) modes. If a system does not contain any uncontrollable and unobservable modes (subsystem $S^{\bar{c}\bar{o}}$ in Fig. 3.5), then (3.76) assumes the following special form:

*Realization is considered in Chap. 4.

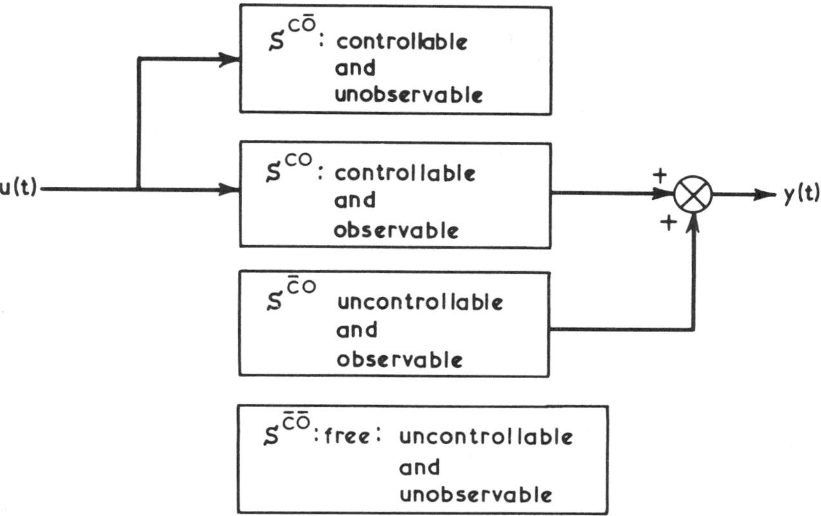

Fig. 3.5 Canonical decomposition of a system S as shown in (3.76).

$$\begin{bmatrix} \dot{\bar{x}}^{c\bar{o}}(t) \\ \dot{\bar{x}}^{co} \\ \dot{\bar{x}}^{\bar{c}o} \end{bmatrix} = \begin{bmatrix} \bar{A}_{c\bar{o}} & \bar{A}_{12} & \bar{A}_{13} \\ 0 & \bar{A}_{co} & \bar{A}_{23} \\ 0 & 0 & \bar{A}_{\bar{c}o} \end{bmatrix} \begin{bmatrix} \bar{x}^{c\bar{o}}(t) \\ \bar{x}^{co}(t) \\ \bar{x}^{\bar{c}o}(t) \end{bmatrix} + \begin{bmatrix} \bar{B}_{c\bar{o}} \\ \bar{B}_{co} \\ 0 \end{bmatrix} u(t)$$

$$y(t) = [0 \;\; \bar{C}_{co} \;\; \bar{C}_{\bar{c}o}] \begin{bmatrix} \bar{x}^{c\bar{o}}(t) \\ \bar{x}^{co}(t) \\ \bar{x}^{\bar{c}o}(t) \end{bmatrix} + Du(t) \tag{3.77}$$

If the system S(A, B, C, D) is not completely controllable, it can be transformed into the controllable form in (3.49). If this controllable part $\bar{S}^c$ is not completely observable, $\bar{S}^c$ can be transformed into the observable form in (3.66), which can also be written as

$$\begin{bmatrix} \dot{\bar{x}}^{c\bar{o}}(t) \\ \dot{\bar{x}}^{co}(t) \end{bmatrix} = \begin{bmatrix} \bar{A}_{c\bar{o}} & \bar{A}_{12} \\ 0 & \bar{A}_{co} \end{bmatrix} \begin{bmatrix} \bar{x}^{c\bar{o}}(t) \\ \bar{x}^{co}(t) \end{bmatrix} + \begin{bmatrix} \bar{B}_{c\bar{o}} \\ \bar{B}_{co} \end{bmatrix} u(t) \tag{3.78}$$

$$y(t) = [0 \;\; \bar{C}_{co}] \begin{bmatrix} \bar{x}^{c\bar{o}}(t) \\ \bar{x}^{co}(t) \end{bmatrix} + Du(t)$$

Combining these two transformations, (3.76) is obtained. Following Sec. 3.2.1, it can be shown that the transfer function of S(A, B, C, D), through the transformation above, is given by

$$\bar{C}_{co}(sI - \bar{A}_{co})^{-1}\bar{B}_{co} + D$$

3.2.4 Irreducible Dynamic Equation

The canonical decomposition theory shows that if a linear time-invariant system is either uncontrollable or unobservable, the order of the dynamic equation can be reduced without any change in the system transfer-function matrix (i.e., the reduced dynamic equation has the same zero-state response). This leads to the following definition and theorem.

Definition 3.5: Irreducible system A linear time-invariant system is said to be reducible if there exists a linear time-invariant dynamic equation of lesser dimension that is zero-state equivalent to the original equation. Otherwise, the system is said to be irreducible.

Theorem 3.6 A linear time-invariant system S(A, B, C, D) is irreducible if it is controllable and observable.

Proof: If the system S(A, B, C, D) in (3.1) is either uncontrollable or unobservable, then by Secs. 3.2.1 to 3.2.3, it is irreducible. It is thus sufficient to prove that if the system is controllable and observable, it is irreducible. This is shown here by contradiction.

Let the n-dimensional system in (3.1) be controllable and observable, and assume that there exists a linear time-invariant system $\bar{S}(\bar{A}, \bar{B}, \bar{C}, \bar{D})$:

$$\bar{S}: \quad \begin{aligned} \dot{\bar{x}}(t) &= \bar{A}\bar{x}(t) + \bar{B}u(t) \\ \bar{y}(t) &= \bar{C}\bar{x}(t) + \bar{D}u(t) \end{aligned} \tag{3.79}$$

of dimension $\bar{n} < n$, which is zero-state equivalent to S(A, B, C, D). Thus by definition of zero-state equivalence,

$$Ce^{At}B + D\delta(t) = \bar{C}e^{\bar{A}t}\bar{B} + \bar{D}\delta(t) \qquad \text{for } t \text{ on } (0, \infty)$$

which implies that

$$CA^kB = \bar{C}\bar{A}^k\bar{B}, \quad k \in 0,\ 1,\ 2,\ \ldots \tag{3.80}$$

From (3.9) and (3.16),

$$RP \triangleq \begin{bmatrix} C \\ CA \\ \vdots \\ CA^{n-1} \end{bmatrix} [B \;\; AB \cdots A^{n-1}B]$$

$$= \begin{bmatrix} CB & CAB \cdots CA^{n-1}B \\ CAB & CA^2B \cdots CA^nB \\ \vdots & \vdots \\ CA^{n-1} & CA^nB \cdots CA^{2(n-1)}B \end{bmatrix}$$

$$\equiv \bar{R}_{n-1}\bar{P}_{n-1} \quad \text{by using (3.80)} \tag{3.81}$$

where $\bar{R}$ and $\bar{P}$ are the observability and state controllability matrices of $\bar{S}(\bar{A}, \bar{B}, \bar{C}, \bar{D})$. Since S is controllable and observable $\rho(R) = \rho(P) = n$ [i.e., $\rho(RP) = n$]. The matrices $\bar{R}_{n-1}$ and $\bar{P}_{n-1}$ are of dimension $\ell\bar{n} \times \bar{n}$ and $\bar{n} \times \bar{n}m$; hence the matrix $\bar{R}_{n-1}\bar{P}_{n-1}$ can at most have rank $\bar{n}$. However, (3.81) implies that $\rho(\bar{R}_{n-1}\bar{P}_{n-1} = n > \bar{n}$, which is a contradiction. This proves the theorem.

If S(A, B, C, D) has a prescribed transfer-function matrix G(s), the dynamic equation in (3.1) is called a realization of G(s). If S(·) is controllable and observable, (3.1) is called an irreducible realization of G(s). The following development shows that all the irreducible realizations of G(s) are equivalent.*

Let S(A, B, C, D) in (3.1) be an irreducible realization of an $\ell \times m$ proper rational matrix G(s). Then $\bar{S}(\bar{A}, \bar{B}, \bar{C}, \bar{D})$ is also an irreducible realization of G(s) if (A, B, C, D) and $(\bar{A}, \bar{B}, \bar{C}, \bar{D})$ are equivalent; that is, there exists a nonsingular constant matrix T such that $\bar{A} = T^{-1}AT$, $\bar{B} = T^{-1}$, $\bar{C} = CT$, and $D = \bar{D}$. Sufficiency follows from Secs. 2.2.3 and 3.1.2. If S(A, B, C, D) and $\bar{S}(\bar{A}, \bar{B}, \bar{C}, \bar{D})$ are realizations of the same G(s), then from (3.79), (3.80), and (3.81),

$$D = \bar{D}; \quad RP = \bar{R}\bar{P}; \quad RAP = \bar{R}\bar{A}\bar{P} \tag{3.82}$$

The irreducibility assumption implies that $\rho(\bar{R}) = n$; hence the matrix $\bar{R}^t\bar{R}$ is nonsingular; hence, from (3.82),

*A lowercase superscript t is used here to denote the transpose to avoid any confusion with the transformation matrix T.

$$\bar{P} = (\bar{R}^t\bar{R})^{-1}\bar{R}^t R P \triangleq T^{-1}P \tag{3.83}$$

Since state controllability and observability are invariant under transformation, the equation above implies that $\rho(T) = n$. Hence T qualifies as an equivalent transformation; since $\rho(P) = n$, it also implies that

$$T^{-1} = (\bar{P}P^t)(PP^t)^{-1} \tag{3.84}$$

Combining (3.82), (3.83), and (3.84), we have

$$R = \bar{R}T^{-1} \quad \text{and} \quad T^{-1}A = \bar{A}T^{-1}$$

which implies that

$$\bar{B} = T^{-1}B; \quad C = \bar{C}T; \quad \bar{A} = T^{-1}AT$$

The assertion above implies that all the reducible realizations of G(s) have the same dimension. Physically, the dimension of an irreducible dynamic equation is the minimal number of integrators or the minimal number of energy storage elements required to generate the given transfer-function matrix.

3.3 ALGEBRAIC INTERPRETATIONS [13-18]

The concepts of controllability and observability are extended here to multivariable systems in polynomial matrix form:

$$\begin{aligned} P(s)\xi(s) &= Q(s)u(s) \\ y(s) &= R(s)\xi(s) + W(s)u(s) \end{aligned} \tag{3.85}$$

where $\xi(s)$ is an r-state polynomial vector, u(s) and y(s) are m-input control and ℓ-output response vectors, and P(s), Q(s), R(s), and W(s) are polynomial matrices of appropriate dimensions, P(s) being an r-square nonsingular matrix. The relationship between $S\{P(s), Q(s), R(s), W(s)\}$ in (3.85) and the state-space representation S(A, B, C, D) in (3.1) (described in Sec. 2.4) and the preceding derivations lead to the following definition.

<u>Definition 3.6: Controllability</u> A state variable $\xi(s)$ is said to be <u>controllable</u> if it is given by a solution $\xi(s)$ in R^r for some $u \in R^m$. The system in (3.85) is said to be completely state controllable if all r components of $\xi \in R^r$ are controllable. The controllable state variables (elements of ξ) are those obtainable by using finite input control over a finite time period. The set of controllable state variables is a subspace of the state space called the <u>controllable subspace</u>.

The following theorems [17] form the basis of an algebraic condition for controllability.

Theorem 3.7 Let

$$\underbrace{(s^{n-1} + p_{n-1}s^{n-1} + \cdots + p_0)}_{p(s)}\xi(s) = \underbrace{(q_{n-1}s^{n-2} + \cdots + q_0)}_{q(s)}u(s) \tag{3.86}$$

where p(s) and q(s) are relatively prime, be the polynomial representation of a system. Then for any $\bar{\xi}(s)$ given by $p(s)\bar{\xi}(s) = 0$, there exists a $u(s) \in R$ such that the unique solution of $\xi(s)$ in (3.86) has the property

$$\xi(t) = \bar{\xi}(t), \quad \forall\, t > t_0$$

Proof of the theorem follows from the fact that p(s) and q(s) have no common factors, and hence the solution of (3.86) is unique.

Theorem 3.8 Let $P_1(s)$ and $Q_1(s)$ be two relatively left prime polynomial matrices of dimensions $(r \times r)$ and $(r \times m)$, respectively, with $P_1(s)$ nonsingular. If there exist two other polynomial matrices $P_2(s)$ and $Q_2(s)$ with the same properties and $P_2(s)$ is nonsingular, such that

$$[P_1(s)]^{-1}Q_1(s) = [P_2(s)]^{-1}Q_2(s) \tag{3.87}$$

then $\{P_1(s), Q_1(s)\}$ and $\{P_2(s), Q_2(s)\}$ are related through a unimodular matrix U(s) such that

$$P_2(s) = U(s)P_1(s) \quad \text{and} \quad Q_2(s) = U(s)Q_1(s) \tag{3.88}$$

Proof of the theorem follows from the definitions and derivations in Secs. 1.5 and 2.4.

Theorem 3.9 The system $S\{P(s), Q(s), R(s), W(s)\}$ in (3.85) is said to be completely state controllable if $\{P(s), Q(s)\}$ are relatively left prime.

A formal proof of the theorem is given in Refs. 12 and 17, but a simpler, and to a certain extent incomplete proof based of Theorems 3.7 and 3.8 is given below.

Let P(s) and Q(s) be relatively prime; then $[P(s)]^{-1}Q(s)$ is a rational matrix, which may be assumed to be proper without any loss of generality. Let M(s) be a Smith-McMillan form of $[P(s)]^{-1}Q(s)$. Thus there are two unimodular matrices $U_1(s)$ and $U_2(s)$ such that

$$[U_1(s)]^{-1}\{[P(s)]^{-1}Q(s)\}U_2(s) = M(s)$$

or

$$[P(s)U_1(s)]^{-1}Q(s)U_2(s) = M(s) \triangleq [\phi(s)]^{-1}[\psi(s)] \tag{3.89}$$

where $\phi(s)$ and $\psi(s)$ are diagonal relatively prime polynomial matrices (Sec. 1.5).

Using Theorem 3.8 from (3.86), there exists a unimodular matrix $U_3(s)$ such that

$$\left.\begin{aligned} U_3(s)P(s)U_1(s) &= \phi(s) = \{\phi_i(s)\} \\ &\text{and} \\ U_3(s)Q(s)U_2(s) &= \psi(s) = \{\psi_i(s)\} \end{aligned}\right\} \quad i \in 1, \ldots, r \tag{3.90}$$

Since $U_{1,2,3}(s)$ are unimodular matrices, by the definitions in (3.89), (3.90) shows that the solution of $\xi(s)$ in

$$P(s)\xi(s) = Q(s)U(s)$$

is equivalent to the solution of

$$\phi_i(s)\xi_i(s) = \psi_i(s)u_i(s), \quad \forall\, i \in 1, \ldots, r \tag{3.91}$$

which has a unique solution for $\xi(s) \in R$, since $\phi(s)$ and $\psi(s)$ are relatively (left) prime (Theorem 3.7); that is, $\xi(s)$ is completely controllable. This completes the proof.

3.3.1 Input-Decoupling Zeros

If $\{P(s), Q(s)\}$ have a common (nonunimodular) left divisor $Z_\ell(s)$ (i.e., they are not relatively left prime), then

$$P(s) = Z_\ell(s)P_1(s) \quad \text{and} \quad Q(s) = Z_\ell(s)Q_1(s) \tag{3.92}$$

where $\{P_1(s), Q_1(s)\}$ are relatively left prime with $Z_\ell(s)$ nonunimodular. If the equation

$$P_1(s)\bar{\xi}(s) = 0 \tag{3.93}$$

has a solution, it follows that $\bar{\xi}(s) \in x^{cs}$ is the controllability subspace of $\{P(s), Q(s)\}$, where $\delta\{P_1(s)\} < \delta\{P(s)\}$, and the zeros of det $\{P_1(s)\}$ are these roots of det $\{P(s)\}$ which are associated with the controllable state variables—elements of $\bar{\xi}(s)$. The zeros of det $\{Z_\ell(s)\}$ correspond to those not appearing in

$$P_1(s)\xi(s) = Q_1(s)u(s) \tag{3.94}$$

(i.e., not "connected" to the input control). Consequently, the zeros of det $\{Z_\ell(s)\}$ may be interpreted as those elements of $\xi(s)$ [not present in $\bar{\xi}(s)$] which are not excited by the input control. These have been termed input-decoupling zeros, as the corresponding "modes" are "decoupled" from the input [12].

Let $S_{id}(s)$ be the Smith form of $Z_\ell(s)$; then there exist two unimodular matrices such that

$$Z_\ell(s) = L_1(s)S_{id}(s)R_1(s) \tag{3.95}$$

and the zeros* of the invariant polynomials of $S_{id}(s)$ are the input-decoupling zeros of $\{P(s), Q(s)\}$. Then by using the properties of polynomial matrices,†

$$[P(s) \quad -Q(s)] = L(s)[I \quad 0]R(s)$$

which gives

$$[P(s) \quad -Q(s)] = \underbrace{L_1(s)S_{id}(s)R_1(s)}_{Z_\ell(s)}\underbrace{L_2(s)[I \mid 0]R_2(s)}_{[P_1(s) \mid -Q_1(s)]}$$

$$= L_1(s)[S_{id}(s) \mid 0]\underbrace{\left[\begin{array}{c|c} R_1(s)L_2(s) & 0 \\ \hline 0 & I \end{array}\right]}_{R_3(s)} R_2(s)$$

$$= L_1(s)[S_{id}(s) \mid 0]R_3(s) \tag{3.96}$$

where $R_3(s)$ is unimodular. Since the Smith form of any polynomial matrix is unique, (3.96) shows that $[S_{id}(s) \mid 0]$ is the Smith form of $[P(s) \mid -Q(s)]$. This leads to the following definition:

*These are also called Smith zeros of $Z_\ell(s)$; idz's are therefore the Smith zeros of the greatest common left divisor of $T_c(s)$ defined in Secs. 2.4.3 and 2.4.4.

†This is an alternative definition of relative primeness [12, 18] and follows from (1.102) and (1.103); since there exists an unimodular matrix U(s) such that $U(s)[P_1(s) \mid -Q_1(s)] = [R(s) \mid 0] = [I \mid 0]R(s)$. If $P_1(s)$ and $Q_1(s)$ are relatively prime, any gcrd R(s) is unimodular (i.e., $[R(s)]^{-1}$ is polynomial). Hence $U(s)[P\ (s) \mid -Q\ (s)]\,[R(s)]^{-1} = [I \mid 0]$.

Definition 3.7: Input-decoupling zero The input-decoupling zeros (idz's) of a polynomial system $S[P(s), Q(s), R(s), W(s)\}$ are the zeros of the invariant polynomials of the Smith form of $[P(s) \mid -Q(s)]$.

If $\delta\{P(s)\} \equiv n$, then from (2.124), the system matrices

$$T(s) = \begin{bmatrix} P(s) & -Q(s) \\ R(s) & W(s) \end{bmatrix} \overset{se}{=} \begin{bmatrix} sI_n - A & -B \\ C & D(s) \end{bmatrix} \tag{3.97}$$

Consequently, the following equivalences may be established:

$$T_c(s) = [P(s) \quad -Q(s)] \overset{se}{=} [sI_n - A \quad -B] \quad \text{and} \quad T_o(s) = \begin{bmatrix} P(s) \\ R(s) \end{bmatrix} \overset{se}{=} \begin{bmatrix} sI_n - A \\ C \end{bmatrix} \tag{3.98}$$

which lead to the following definitions.

Definition 3.8: Input-decoupling zero The input-decoupling zeros (idz's) are the zeros of the invariant polynomials of the Smith form of

$$T_c(s) = [sI - A \mid -B] \tag{3.99}$$

Thus if a system does not have any input-decoupling zeros, the matrices $(sI - A)$ and B are relatively left prime; that is, the Smith form of $T_c(s)$ is $[I \quad 0]$, which is consistent with the definition in Sec. 2.4.4(a) and leads to the following theorem [12].

Theorem 3.10 The polynomial matrices $\{(sI_n - A), B\}$ of dimensions $n \times n$ and $n \times m$ are relatively left prime if any one of the following equivalent conditions:

1. There exists no real or complex nonzero vector α which simultaneously satisfies, for some i,

$$\alpha^*(\lambda_i I_n - A) = 0 \quad \text{and} \quad \alpha^* B = 0 \tag{3.100}$$

where λ_i, $i \in 1, \ldots, n$, are the eigenvalues of A.

2. The $(n \times mn)$ matrix P has rank n, where

$$P = [B \quad AB \quad A^{p-1}B]$$

p being an integer less than or equal to the degree of the minimal polynomial of A (i.e., $p < n$).

3. There exists an $(n \times n)$ polynomial matrix $N_1(s)$ with elements of degree $< p - 2$ and another $(m \times n)$ polynomial matrix $N_2(s)$ with elements of degree $< p - 1$ such that

$$(sI_n - A)N_1(s) + BN_2(s) = I_n$$

where p is the degree of minimal polynomial of A.

<u>Proof:</u> Proof of (1) follows directly from the fact that if $(sI_n - A)$ and B are relatively prime, then the matrix

$$T_c(s) = [sI_n - A \mid -B]$$

has rank n for all s, and no α exists that satisfies (3.100). A similar argument follows for the complex-conjugate transpose α^*.

A simplified algebraic proof of (2) is given here.† From (3.99),

$$T_c(s) \equiv [sI - A \mid -B] = [0 \mid sI_n - A] \left[\begin{array}{c|c} I_n & 0 \\ \hline I_n & -(sI_n - A)^{-1}B \end{array}\right] \tag{3.101}$$

Since

$$(sI_n - A)^{-1} = \sum_{k=1}^{p} \alpha_k \frac{A^{k-1}}{s^k}$$

where α_k are scalars and p is the degree of minimum polynomial of A,

$$(sI_n - A)^{-1}B = [B \cdot A^{k-1}B \cdot A^{p-1}B] \begin{bmatrix} \frac{\alpha_1}{s} \\ \vdots \\ \frac{\alpha_k}{s^k} \\ \vdots \\ \frac{\alpha_{p-1}}{s^{p-1}} \end{bmatrix} \tag{3.102}$$

†Formal proof of (2) using the invariant subspace of A^* is given in Ref. 12, p. 73.

Thus combining the two equations above, we have

$$\rho[T_c(s) = \rho[I_n(sI - A)^{-1}]B = \rho[(sI - A)^{-1}]B = \rho[B \cdot A^{k-1}B \cdot A^{p-1}B]$$

Since $[P(s) \mid -Q(s)]$ and $[sI - A \mid -B]$ are system equivalent, proof of (2) follows from the fact that

$$\rho[P(s) \mid -Q(s)] = \rho[sI - A \mid -B] = \rho[T_c(s)] \tag{3.103}$$

Proof of the last part of the theorem follows from the relative primeness of $(sI_n - A)$ and B [(1.98)] and is left to the reader.

Combination of the preceding definitions of input-decoupling zeros and Theorem 3.10 leads to the following alternative definition of state controllability.

Definition 3.9: State controllability A system S(A, B, C, D) is said to be completely state controllable if it has no input-decoupling zeros.

This definition provides a simple method of checking the state controllability of systems represented in any of the two standard forms

$$\begin{bmatrix} P(s) & -Q(s) \\ R(s) & W(s) \end{bmatrix} \quad \text{or} \quad \begin{bmatrix} sI - A & -B \\ C & D(s) \end{bmatrix}$$

Computation of idz's The input-decoupling zeros of the uncontrollable modes of a system, described in either the polynomial-matrix form or the state-space form, may be identified by using the following results based on the preceding developments.

1. Polynomial-matrix-form representation: The input-decoupling zeros are the roots of $\det\{Z_\ell(s)\} = 0$, where $Z_\ell(s)$ is the non-unimodular greatest common left divisor (gcld) of $T_c(s) = [P(s) \mid -Q(s)]$. [This follows from (3.95) and (3.96), and Definition 3.7.]

Since $Z_\ell(s)$ may be formed by inspection of the rows of $T_c(s)$, this interpretation provides a very simple method of identifying the idz of $S\{P(s), Q(s), R(s), W(s)\}$. Once $Z_\ell(s)$ has been formed, the idz may be obtained directly from the zeros of $\det\{Z_\ell(s)\}$.

2. State-space-form representation: Since the idz's represent the uncontrollable modes of the system S(A, B, C, D), then by Sec. 3.2.1 there is a nonsingular transformation matrix P which transforms the system*

*For notational simplicity, it is assumed that the system is completely controllable and $T_c = P$, $W_c = P^{-1}$ (as in the notation of Sec. 3.2.1). The results, however, also apply for the case where $\rho(T_c) < n$. [P is used as the transformation matrix to avoid confusion with $T_c(s)$, used widely in this section.]

into its controllable form. Since such transformation is done through constant matrices, the resulting controllable form $S_c(A_c, B_c, C_c, D_c)$ is system equivalent to the original system S(A, B, C, D). Thus

$$[sI - A \mid -B] \overset{se}{=} \left[\begin{array}{cc|c} sI - A_c & -A_{12} & -B_c \\ 0 & sI - A_{\bar{c}} & 0 \end{array}\right]$$

$$\overset{se}{=} [P^{-1}(sI - A)P \mid -P^{-1}B] \tag{3.104}$$

where A_c and $A_{\bar{c}}$ denote the controllable and uncontrollable modes of the system.

Therefore,

$$\rho[T_c(s)] \equiv \rho \left\{ P^{-1}[sI - A \mid -B] \begin{bmatrix} P & 0 \\ 0 & I_m \end{bmatrix} \right\} \tag{3.105}$$

Since pre- and postmultiplication of $[sI - A \mid -B]$ represents row and column operations, respectively, the existence of a nonsingular P implies that the right-hand side of (3.104) [i.e., the controllable form of $T_c(s)$] may be derived from T(s) by a finite number of elementary row and column operations. If P is computed by using the derivation in Sec. 3.2, then the controllable form of $T_c(s)$

$$\left[\begin{array}{cc|c} sI - A_c & -A_{12} & -B_c \\ 0 & sI - A_{\bar{c}} & 0 \end{array}\right]$$

may be obtained directly from (3.104), which shows that the input-decoupling zeros are given by the roots of $\det\{sI - A_{\bar{c}}\} = 0$. Thus the idz may also be obtained fairly easily from T(s) in state-space form. Methods of computing idz's are illustrated below [12, 18].

<u>Example 3.9: Controllability, idz, and pole-zero cancellation</u> The system S(A, B, C) in (3.61) is considered here to illustrate the direct relationship betweeen pole-zero cancellation, stat controllability, and input-decoupling zeros. The input-state transfer function of the system is

$$G_1(s) = (sI - A)^{-1}B = \begin{bmatrix} s - 1 & -1 & 0 \\ 0 & s - 1 & 0 \\ 0 & -1 & s - 1 \end{bmatrix}^{-1} \begin{bmatrix} 0 & 1 \\ 1 & 0 \\ 0 & 1 \end{bmatrix} \tag{3.106}$$

$$= \frac{1}{(s-1)^3}\begin{bmatrix} (s-1)^2 & s-1 & 0 \\ 0 & (s-1)^2 & 0 \\ 0 & s-1 & (s-1)^2 \end{bmatrix}$$

which shows that there is one pole-zero cancellation for s = 1, making the system uncontrollable, as seen in Example 3.6. A check for the existence of idz's may be made directly by using the derivations above. By direct substitution of the transformation matrix T_c from Example 3.6 in (3.105), the state controllable form of $T_c(s)$ is given by*

$$P^{-1}[sI - A \mid -B]\begin{bmatrix} P & 0 \\ 0 & I \end{bmatrix}$$

$$= \begin{bmatrix} 1 & 0 & 0 \\ 0 & 1 & 0 \\ 1 & 0 & 1 \end{bmatrix}\left[\begin{array}{ccc|cc} s-1 & -1 & 0 & 0 & -1 \\ 0 & s-1 & 0 & -1 & 0 \\ 0 & -1 & s-1 & 0 & -1 \end{array}\right]\begin{bmatrix} 1 & 0 & 0 & 0 & 0 \\ 0 & 1 & 0 & 0 & 0 \\ 1 & 0 & 1 & 0 & 0 \\ 0 & 0 & 0 & 1 & 0 \\ 0 & 0 & 0 & 0 & 1 \end{bmatrix}$$

$$= \left[\begin{array}{cc|c|cc} s-1 & -1 & 0 & 0 & -1 \\ 0 & s-1 & 0 & -1 & 0 \\ \hline 0 & 0 & s-1 & 0 & 0 \end{array}\right]$$

$$= \left[\begin{array}{c|c|c} sI - A_c & A_{11} & B_c \\ \hline 0 & sI - A_{\bar{c}} & 0 \end{array}\right] \tag{3.107}$$

which is consistent with previous results.

Example 3.10: Computation of idz of T(s)

$$T(s) = \left[\begin{array}{cc|cc} (s+1)(s+2) & 0 & -(s+2) & -(s+2) \\ 0 & (s+1)(s+2) & s & -(s+1) \\ \hline 1 & 0 & 0 & 0 \\ 0 & 1 & 0 & 0 \end{array}\right] \begin{matrix} \Big\} T_c(s) \\ \\ \end{matrix} \tag{3.108}$$

The block diagram of the system transfer function and equivalent signal flow chart are shown in Fig. 3.6. Since only row operations are allowed here to

*An arbitrary third column has been added to obtain a nonsingular P.

derive the gcld of the rows of T(s), the following equivalence is obtained to identify the idz [which are not apparent from (3.108)]:

$$T_c(s) = \underbrace{\begin{bmatrix} s+2 & 0 \\ 0 & 1 \end{bmatrix}}_{Z_\ell(s)} \underbrace{\begin{bmatrix} s+1 & 0 \\ 0 & (s+1)(s+2) \end{bmatrix}}_{P_1(s)} \underbrace{\begin{bmatrix} -1 & -1 \\ s & -(s+1) \end{bmatrix}}_{Q_1(s)}$$

Therefore, the system in (3.108) has one input decoupling zero at s = -2. The system matrix in state-space form, derived from Fig. 3.6, is given by

$$T(s) = \left[\begin{array}{cccc|cc} s+1 & 0 & 0 & 0 & -1 & 0 \\ 1 & s+2 & 0 & 0 & 1 & 0 \\ 0 & 0 & s+1 & 0 & 0 & -1 \\ 0 & 0 & 0 & s+2 & 0 & -1 \\ \hline 1 & 0 & 1 & 0 & 0 & 0 \\ 0 & 1 & 0 & 1 & 0 & 0 \end{array}\right] \quad \text{(first four rows: } T_c(s))$$

which again does not make the existence of any decoupling zero apparent; the following column operation is performed.*

$$T_c(s):\ R2 = R1 + R2 \rightarrow \left[\begin{array}{cccc|cc} s+1 & 0 & 0 & 0 & -1 & 0 \\ s+2 & s+2 & 0 & 0 & 0 & 0 \\ 0 & 0 & s+1 & 0 & 0 & -1 \\ 0 & 0 & 0 & s+2 & 0 & -1 \end{array}\right]$$

$$C1 = C1 - C2 \rightarrow \left[\begin{array}{cccc|cc} s+1 & 0 & 0 & 0 & -1 & 0 \\ 0 & s+2 & 0 & 0 & 0 & 0 \\ 0 & 0 & s+1 & 0 & 0 & -1 \\ 0 & 0 & 0 & s+2 & 0 & -1 \end{array}\right]$$

$$\begin{array}{l}\text{Interchange} \\ \text{of R2 and R4}\end{array} \rightarrow \left[\begin{array}{cccc|cc} s+2 & 0 & 0 & 0 & -1 & 0 \\ 0 & 0 & 0 & s+2 & 0 & -1 \\ 0 & 0 & s+1 & 0 & 0 & -1 \\ 0 & s+2 & 0 & 0 & 0 & 0 \end{array}\right]$$

*In subsequent derivations the notation Ra = Rb + Rc means that row a is replaced by the addition of rows b and c (or multiples of them as indicated); similar notation applies to column operations.

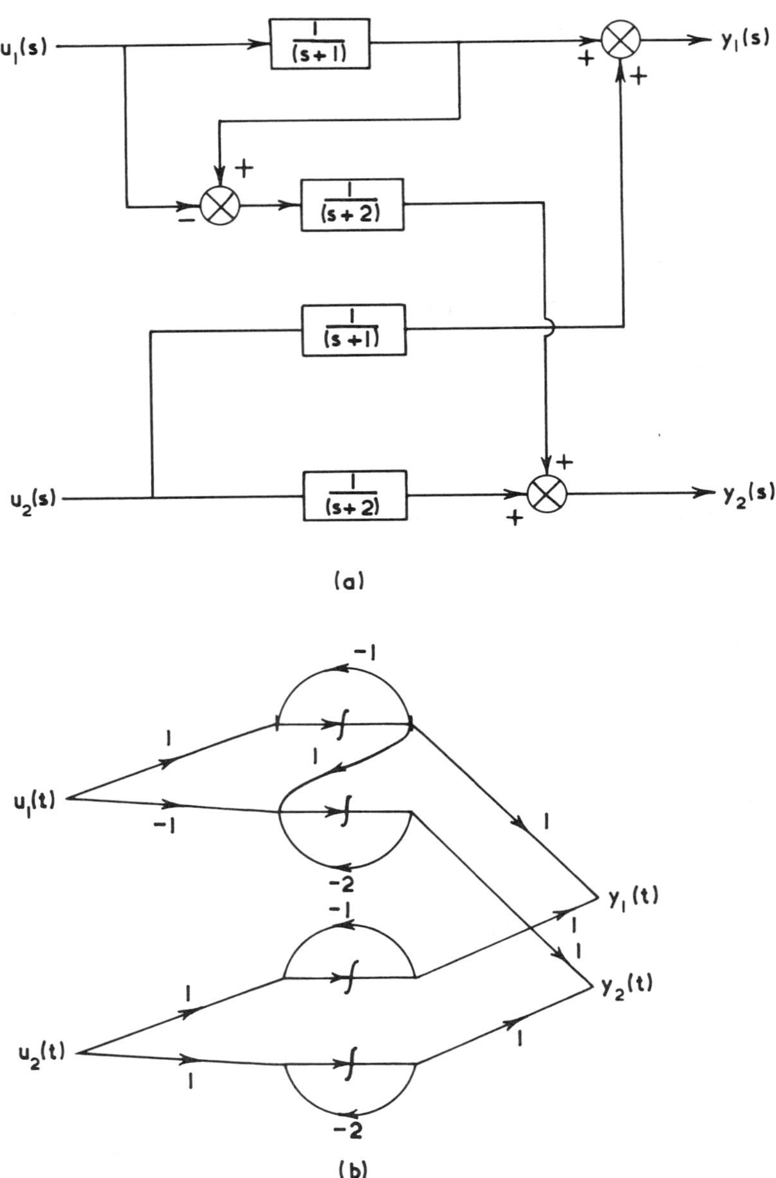

Fig. 3.6 Representation of the system in (3.108): (a) block diagram of $G(s) \equiv [P(s), Q(s), R(s), W(s)]$; (b) flow chart of $S(A, B, C, D)$.

$$\text{Interchange of C2 and C4} \rightarrow \left[\begin{array}{cccc|cc} s+2 & 0 & 0 & 0 & -1 & 0 \\ 0 & s+2 & 0 & 0 & 0 & -1 \\ 0 & 0 & s+1 & 0 & 0 & -1 \\ \hline 0 & 0 & 0 & s+2 & 0 & 0 \end{array}\right]$$

which is in the required form and shows that s = -2 is an uncontrollable mode or idz of the system. Ths is consistent with the observation that in Fig. 3.6b, there is a pole-zero cancellation at s = -2 in the input-state transfer-function matrix, thereby making the system not completely state controllable.

The above row/column operations may be represented by the following pre- and postmultiplications:

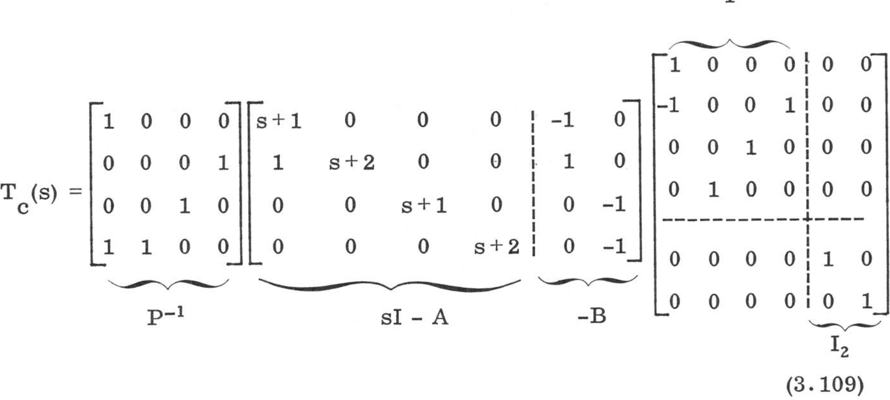

$$T_c(s) = \underbrace{\begin{bmatrix} 1 & 0 & 0 & 0 \\ 0 & 0 & 0 & 1 \\ 0 & 0 & 1 & 0 \\ 1 & 1 & 0 & 0 \end{bmatrix}}_{P^{-1}} \left[\begin{array}{cccc|cc} s+1 & 0 & 0 & 0 & -1 & 0 \\ 1 & s+2 & 0 & 0 & 1 & 0 \\ 0 & 0 & s+1 & 0 & 0 & -1 \\ 0 & 0 & 0 & s+2 & 0 & -1 \end{array}\right] \left[\begin{array}{cccc|cc} 1 & 0 & 0 & 0 & 0 & 0 \\ -1 & 0 & 0 & 1 & 0 & 0 \\ 0 & 0 & 1 & 0 & 0 & 0 \\ 0 & 1 & 0 & 0 & 0 & 0 \\ \hline 0 & 0 & 0 & 0 & 1 & 0 \\ 0 & 0 & 0 & 0 & 0 & 1 \end{array}\right] \quad (3.109)$$

where the first block of the middle matrix is $sI - A$, the second block is $-B$, the upper-left block of the last matrix is P and its lower-right block is I_2.

It can be easily checked that for the system described by (3.108) and Fig. 3.6, the matrix P forms a basis for transformation of (3.108) into a controllable form. Detailed derivations of P (by following the procedure in Sec. 3.2.1) are left to the reader.

3.3.2 Output-Decoupling Zeros

By using the derivations above and the observability properties in Sec. 3.1.2, the following theorem may be states.

<u>Theorem 3.11</u> The system S{P(s), Q(s), R(s), W(s)} is said to be completely observable if {P(s), R(s)} are relatively right prime.

Because of the system similarity in (3.97), this theorem is equivalent to the following definitions.

Definition 3.10 The system S(A, B, C, D) is completely observable if $\begin{bmatrix} sI_n - A \\ C \end{bmatrix}$ are relatively right prime; that is, the rank the ($\ell n \times n$) matrix R is n, where R is given by

$$R = [C \;\; A^T C^T \cdots (A^T)^{n-1} C^T]$$

Definition 3.11 The output-decoupling zeros of a polynomial matrix representation S{P(s), Q(s), R(s), W(s)} are the zeros of invariant polynomials of the Smith form of

$$T_o(s) = \begin{bmatrix} P(s) \\ R(s) \end{bmatrix}$$

A consequence of this definition, as with idz's, is that the odz's are the Smith zeros of the greatest common right divisor of $T_o(s)$. Thus if P(s) and R(s) are not relatively right prime, there exists a polynomial matrix (i.e., non-unimodular) $Z_r(s)$ (right divisor) such that

$$P(s) = P_2(s)Z_r(s) \quad \text{and} \quad R(s) = R_2(s)Z_r(s)$$

and the odz's are the zeros of det $\{Z_r(s)\}$.

Definition 3.12 The output-decoupling zeros are the zeros of the invariant polynomials of the Smith form of

$$T_o(s) = \begin{bmatrix} sI - A \\ C \end{bmatrix}$$

Thus the output-decoupling zeros of a system may be derived through the Smith form of $T_o(s)$ in state space or polynomial form. For a system to be completely observable, there should be no output-decoupling zeros; that is, the Smith form of

$$\begin{bmatrix} P(s) \\ \hline R(s) \end{bmatrix} \quad \text{or} \quad \begin{bmatrix} sI - A \\ \hline C \end{bmatrix}$$

should have the form

$$\begin{bmatrix} I \\ 0 \end{bmatrix}$$

Derivation of odz's These may be derived directly from either the polynomial representation or the state-space form of the system matrix T(s) by using the definitions above and procedures similar to those for idz.

1. T(s) in polynomial form: These are given by the gcrd of $T_o(s)$,

$$T_o(s) = \begin{bmatrix} P(s) \\ R(s) \end{bmatrix} = \begin{bmatrix} P_2(s) \\ R_2(s) \end{bmatrix} Z_r(s) \tag{3.110}$$

Again, $Z_r(s)$ may be found by inspection of the column of $T_o(s)$, and the odz's may be identified through some elementary column operations. The odz's are then the zeros of det $\{Z_r(s)\}$.

2. State-space form: This again follows from a dual argument of idz, since for any observable system (Sec. 3.2.2), there exists a nonsingular transformation matrix* V which transforms the original system S(A, B, C, D) into its observable form $S_o(A_o, B_o, C_o, D_o)$, given by

$$\begin{bmatrix} A_o & 0 \\ A_{21} & A_{\bar{o}} \end{bmatrix}; \quad \begin{bmatrix} B_o \\ B_{\bar{o}} \end{bmatrix}; \quad [C_o \quad 0]; \quad D$$

Again since V is constant,

$$\begin{bmatrix} sI - A \\ \\ C \end{bmatrix} \overset{se}{\equiv} \left[\begin{array}{c|c} sI - A_o & 0 \\ -A_{21} & sI - A_{\bar{o}} \\ \hline C_c & 0 \end{array}\right] \equiv \left[\begin{array}{c} V^{-1}(sI - A)V \\ \\ \hline CV \end{array}\right] \tag{3.111}$$

where A_o and $A_{\bar{o}}$ denote the observable and unobservable modes of the system. Therefore,

$$\rho[T_o(s)] \equiv \rho \left\{ \begin{bmatrix} V^{-1} & 0 \\ 0 & I_\ell \end{bmatrix} \begin{bmatrix} sI - A \\ C \end{bmatrix} [V] \right\} \tag{3.112}$$

Thus the odz's may be computed by transforming $T_o(s)$ into the observable form of (3.111) through a finite number of elementary row and column operations. Alternatively, if V is known, odz's may be identified by using (3.112); the odz's in either case are given by the zeros of det $\{sI - A_{\bar{o}}\}$. The method of computing odz's is illustrated by using the following numerical example.

*For notational convenience V is assumed to have full rank.

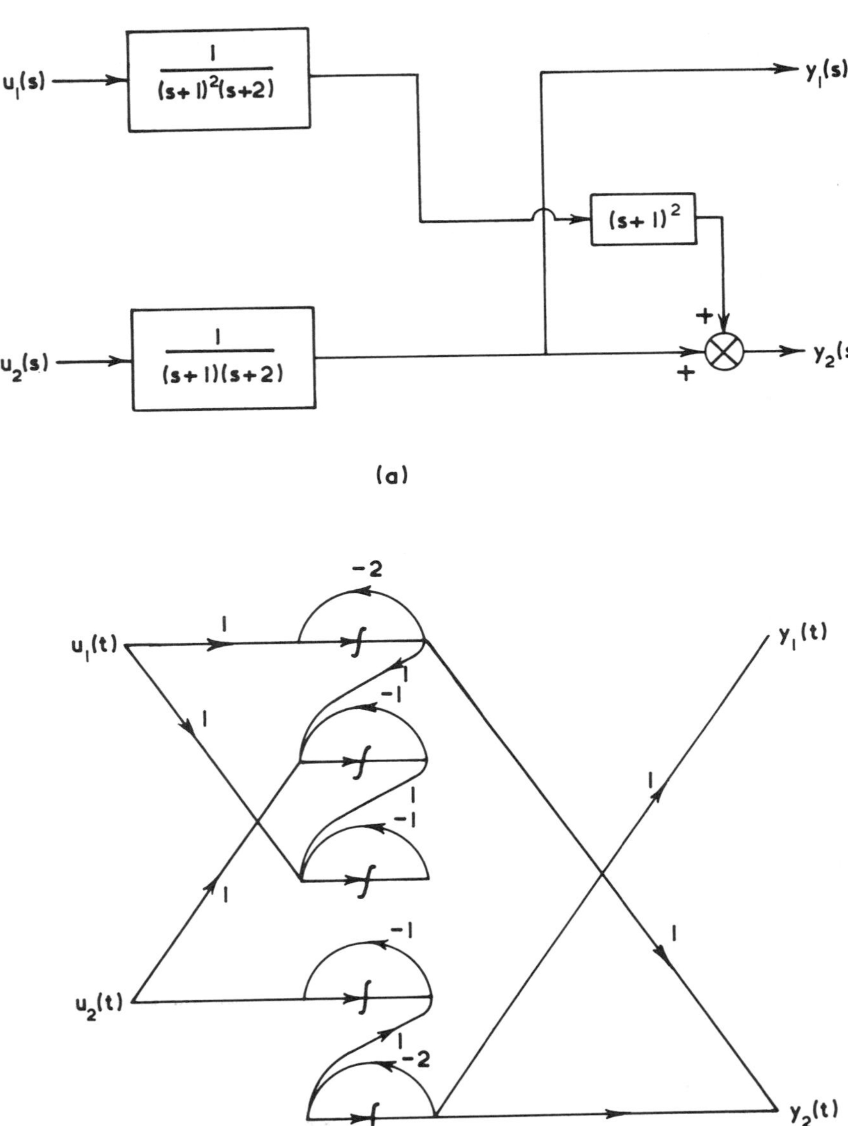

Fig. 3.7 (a) Transfer-function block diagram and (b) signal flow graph of the system in (3.113).

Example 3.11: Computation of the output-decoupling zeros of T(s)

$$T(s) = \left[\begin{array}{cc|cc} (s+1)^2(s+2) & 0 & -1 & 0 \\ 0 & (s+1)(s+2) & 0 & -1 \\ \hline 0 & 1 & 0 & 0 \\ (s+1)^2 & 1 & 0 & 0 \end{array}\right] \qquad (3.113)$$

The block diagram of the transfer function and an equivalent signal flow graph of the system are shown in Fig. 3.7a. The existence of odz's in the system above is apparent since

$$T_o(s) = \begin{bmatrix} (s+1)^2(s+2) & 0 \\ 0 & (s+1)(s+2) \\ 0 & 1 \\ (s+1)^2 & 1 \end{bmatrix} = \begin{array}{c} P_1(s) \left\{ \vphantom{\begin{array}{c} 0 \\ 0 \end{array}} \right. \\ R_1(s) \left\{ \vphantom{\begin{array}{c} 0 \\ 0 \end{array}} \right. \end{array} \left[\begin{array}{cc} s+2 & 0 \\ 0 & (s+1)(s+2) \\ \hline 0 & 1 \\ 1 & 1 \end{array}\right] \underbrace{\begin{bmatrix} (s+1)^2 & 0 \\ 0 & 1 \end{bmatrix}}_{Z_r(s)}$$

The system therefore has (two) output-decoupling zeros, at s = -1, 1. Computation of the odz's from the system matrix may now proceed by using the state-space representation of Fig. 3.7b, from which

$$T(s) = \left[\begin{array}{ccccc|cc} s+2 & 0 & 0 & 0 & 0 & -1 & 0 \\ -1 & s+2 & 0 & 0 & 0 & 0 & -1 \\ 0 & -1 & s+1 & 0 & 0 & -1 & 0 \\ 0 & 0 & 0 & s+1 & -1 & 0 & -1 \\ 0 & 0 & 0 & 0 & s+2 & 0 & 0 \\ \hline 0 & 0 & 0 & 0 & 1 & 0 & 0 \\ 1 & 0 & 0 & 0 & 1 & 0 & 0 \end{array}\right] \qquad (3.114)$$

where the matrix $T_o(s)$ needs the following row/column interchange for transformation into the observable form.

$$T_o(s):\ \text{C3 and C5 interchange} \rightarrow \left[\begin{array}{ccccc} s+2 & 0 & 0 & 0 & 0 \\ -1 & s+1 & 0 & 0 & 0 \\ 0 & -1 & 0 & 0 & s+1 \\ 0 & 0 & -1 & s+1 & 0 \\ 0 & 0 & s+2 & 0 & 0 \\ \hline 0 & 0 & 1 & 0 & 0 \\ 1 & 0 & 1 & 0 & 0 \end{array}\right]$$

$$\text{R3 and R5 interchange} \rightarrow \left[\begin{array}{ccccc} s+2 & 0 & 0 & 0 & 0 \\ -1 & s+1 & 0 & 0 & 0 \\ 0 & 0 & s+2 & 0 & 0 \\ 0 & 0 & -1 & s+1 & 0 \\ 0 & -1 & 0 & 0 & s+1 \\ \hline 0 & 0 & 1 & 0 & 0 \\ 1 & 0 & 1 & 0 & 0 \end{array}\right]$$

which is in the required form, showing that the output-decoupling zeros are at -1, -1. The interchanges above are equivalent to the following pre- and postmultiplications:

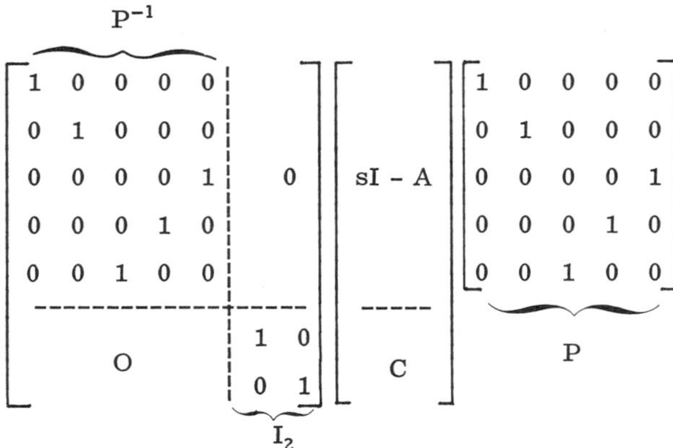

$$\left[\begin{array}{c|c} \overbrace{\begin{matrix} 1 & 0 & 0 & 0 & 0 \\ 0 & 1 & 0 & 0 & 0 \\ 0 & 0 & 0 & 0 & 1 \\ 0 & 0 & 0 & 1 & 0 \\ 0 & 0 & 1 & 0 & 0 \end{matrix}}^{P^{-1}} & 0 \\ \hline O & \underbrace{\begin{matrix} 1 & 0 \\ 0 & 1 \end{matrix}}_{I_2} \end{array}\right] \left[\begin{array}{c} sI - A \\ \hline C \end{array}\right] \underbrace{\left[\begin{matrix} 1 & 0 & 0 & 0 & 0 \\ 0 & 1 & 0 & 0 & 0 \\ 0 & 0 & 0 & 0 & 1 \\ 0 & 0 & 0 & 1 & 0 \\ 0 & 0 & 1 & 0 & 0 \end{matrix}\right]}_{P}$$

where

$$P^{-1}AP = \left[\begin{array}{ccc|cc} -2 & 0 & 0 & 0 & 0 \\ 1 & -1 & 0 & 0 & 0 \\ 0 & 0 & -2 & 0 & 0 \\ \hline 0 & 0 & 1 & -1 & 0 \\ 0 & 1 & 0 & 0 & -1 \end{array}\right] \equiv \left[\begin{array}{c|c} A_o & 0 \\ \hline A_{12} & A_{\bar{o}} \end{array}\right] \tag{3.115}$$

$$CP = \left[\begin{array}{ccc|cc} 0 & 0 & 1 & 0 & 0 \\ 1 & 0 & 1 & 0 & 0 \end{array}\right] \equiv [C_o \;\vdots\; 0]$$

which are consistent with the unobservable modes shown in Fig. 3.7b.

3.3.3 Input/Output Decoupling Zeros [6, 12, 18]

Let the system matrix T(s) be given by using (3.76):

$$T(s) = \begin{bmatrix} P(s) & -Q(s) \\ R(s) & W(s) \end{bmatrix}$$

$$= \begin{bmatrix} sI - A & -B \\ C & D(s) \end{bmatrix} \overset{se}{\equiv} \left[\begin{array}{c|cccc} sI - \bar{A}_{c\bar{o}} & -\bar{A}_{12} & -\bar{A}_{13} & -\bar{A}_{14} & -\bar{B}_{c\bar{o}} \\ 0 & sI - \bar{A}_{co} & -A_{23} & 0 & -\bar{B}_{co} \\ 0 & 0 & sI - \bar{A}_{\bar{c}o} & 0 & 0 \\ 0 & 0 & -\bar{A}_{43} & sI - \bar{A}_{\bar{c}\bar{o}} & 0 \\ \hline 0 & C_{co} & \bar{C}_{\bar{c}o} & 0 & D(s) \end{array}\right] \tag{3.116}$$

has a set of input-decoupling zeros, denoted by {idz}, which are the zeros of the Smith form of [P(s) -Q(s)] or the zeros of $Z_\ell(s)$, where

$$T_c(s) = [P(s) \;\vdots\; -Q(s)] \equiv Z_\ell(s)[P_1(s) \quad -Q_1(s)] \tag{3.117}$$

Let the {idz} be removed* by following the procedure in the preceding section. Then the resulting polynomial system matrix, without any idz, becomes

*{·} denotes the set of "·".

$$T_1(s) = \begin{bmatrix} P_1(s) & -Q_1(s) \\ R(s) & W(s) \end{bmatrix} \tag{3.118}$$

By using the properties of system equivalence, the system matrix in state-space form without any idz is

$$T_1(s) \overset{se}{=} \left[\begin{array}{cc|c} sI - \bar{A}_{c\bar{o}} & -\bar{A}_{12} & -\bar{B}_{c\bar{o}} \\ 0 & sI - \bar{A}_{co} & -\bar{B}_{co} \\ \hline 0 & \bar{C}_{co} & D(s) \end{array}\right] \tag{3.119}$$

which as can be seen is completely controllable. Thus while removing the input-decoupling zeros of T(s) in (3.116), the uncontrollable <u>and</u> unobservable modes are also removed. These modes, given by the zeros of det $\{sI - A_{\bar{c}\bar{o}}\}$ are called the <u>input/output-decoupling zeros</u>. The significance of their name, iodz's for short, follows from the fact that they disappear when either the input-decoupling zeros or the output-decoupling zeros of the system are removed. This is consistent with the definition of iodz given in Sec. 2.4, which is not repeated here. Thus iodz's are equivalent to those modes which are neither controllable nor observable and do not appear in either the input-state or the state-output transfer-function matrices. Therefore,

$$\{\text{iodz}\} \text{ of } T(s) = \begin{cases} \{\text{idz}\} \text{ of } T(s) - \{\text{odz}\} \text{ of } T_1(s) \text{ if idz's are first removed} \\ \{\text{odz}\} \text{ of } T(s) - \{\text{idz}\} \text{ of } T_2(s) \text{ if odz's are first removed} \end{cases}$$

where

$$T_1(s) = \begin{bmatrix} P_1(s) & -Q_1(s) \\ R(s) & W(s) \end{bmatrix} \text{ and } [P(s) \;\; Q(s)] = Z_\ell(s)[P_1(s) \mid -Q_1(s)]$$

$\{P_1(s), Q_1(s)\}$ being relatively left prime

(3.120)

$$T_2(s) = \begin{bmatrix} P_2(s) & -Q(s) \\ R_2(s) & W(s) \end{bmatrix} \text{ and } \begin{bmatrix} P(s) \\ R(s) \end{bmatrix} = \begin{bmatrix} P_2(s) \\ R_2(s) \end{bmatrix} Z_r(s)$$

$\{P_2(s), R_2(s)\}$ being relatively right prime

Thus the iodz's may be obtained directly through computation of the idz's and odz's.

Derivation of input/output-decoupling zeros follows from the combination of methods for the idz's and odz's described earlier, and is outlined below.

1. Polynomial-form representation: From the preceding derivation, idz and odz are the zeros of the gcld and gcrd of $T_c(s)$ and $T_o(s)$, respectively, which may be obtained through the elementary operations below.

$$(1)\ T(s) = \begin{bmatrix} P(s) & -Q(s) \\ R(s) & W(s) \end{bmatrix} \overset{se}{\equiv} \underbrace{\begin{bmatrix} Z_\ell(s) & 0 \\ 0 & I \end{bmatrix}}_{\text{idz}} \underbrace{\begin{bmatrix} P_1(s) & -Q_1(s) \\ R(s) & W(s) \end{bmatrix}}_{T_1(s)}$$

$$\overset{se}{\equiv} \underbrace{\begin{bmatrix} Z_\ell(s) & 0 \\ 0 & I \end{bmatrix}}_{\text{idz}} \underbrace{\begin{bmatrix} \bar{P}(s) & -Q_1(s) \\ R_1(s) & W(s) \end{bmatrix}}_{\bar{T}(s)} \underbrace{\begin{bmatrix} \bar{Z}_r(s) & 0 \\ 0 & I \end{bmatrix}}_{\text{odz's after idz's have been removed}} \tag{3.121a}$$

and

$$(2)\ T(s) = \begin{bmatrix} P(s) & -Q(s) \\ R(s) & W(s) \end{bmatrix} \overset{se}{\equiv} \underbrace{\begin{bmatrix} P_2(s) & -Q(s) \\ R_2(s) & W(s) \end{bmatrix}}_{T_2(s)} \underbrace{\begin{bmatrix} Z_r(s) & 0 \\ 0 & I \end{bmatrix}}_{\text{odz}}$$

$$\overset{se}{\equiv} \underbrace{\begin{bmatrix} \bar{Z}_\ell(s) & 0 \\ 0 & I \end{bmatrix}}_{\text{idz's after odz's have been removed}} \underbrace{\begin{bmatrix} \bar{P}(s) & Q(s) \\ R_2(s) & \end{bmatrix}}_{\bar{T}(s)} \underbrace{\begin{bmatrix} Z_r(s) & 0 \\ 0 & I \end{bmatrix}}_{\text{odz}} \tag{3.121b}$$

where $T_1(s)$ denotes the system matrix when only the idz's have been removed, and $T_2(s)$ denotes the system matrix when only the odz's have been removed; $\bar{T}(s)$ denotes the system matrix when both the idz's and odz's have been removed. Thus comparison of (3.120) and (3.121) shows that if the system has no iodz, $Z_\ell(s) \equiv \bar{Z}_\ell(s)$ and $Z_r(s) \equiv \bar{Z}_r(s)$. If there are iodz's in $T(s)$, then these are those zeros of $Z_\ell(s)$ and $Z_r(s)$ which are not present in $\bar{Z}_\ell(s)$ and $\bar{Z}_r(s)$, respectively.

2. State-space form: This follows from (3.116) and the decomposition theorem (Theorem 3.5). Since the nonsingular transformation matrix

P (= TV) transforms S(A, B, C, D) into the canonical form of (3.116), there exists a finite number of row and column operations which accomplish the transformation in (3.116). Thus by extending (3.105) and (3.112),

$$\rho[T(s)] = \rho\left\{\begin{bmatrix} (PV)^{-1} & 0 \\ 0 & I_\ell \end{bmatrix}\begin{bmatrix} sI - A & -B \\ C & D \end{bmatrix}\begin{bmatrix} PV & 0 \\ 0 & I_m \end{bmatrix}\right\} \tag{3.122}$$

where P and V are transformation matrices to obtain controllable and observable forms, respectively. Therefore, the decoupling zeros of T(s) in state-space form may either be determined through row/column operations or by direct multiplication by using (3.122). A few numerical examples are given below.

Example 3.12: Decoupling zeros from state-space form of T(s)

$$T(s) = \left[\begin{array}{ccc|cc} s+1 & 0 & 0 & 0 & 0 \\ 0 & s+2 & 0 & -1 & -1 \\ 0 & 0 & s+3 & -1 & 0 \\ \hline 1 & 1 & 0 & 0 & 0 \\ 0 & 1 & 0 & 0 & 0 \end{array}\right] \tag{3.123}$$

$$\equiv \overbrace{\left[\begin{array}{ccc|c} s+1 & 0 & 0 & \\ 0 & 1 & 0 & 0 \\ 0 & 0 & 1 & \\ \hline & 0 & & I_2 \end{array}\right]}^{Z_\ell(s)} \overbrace{\left[\begin{array}{ccc|cc} 1 & 0 & 0 & 0 & 0 \\ 0 & s+2 & 0 & -1 & -1 \\ 0 & 0 & 1 & -1 & 0 \\ \hline 1 & 1 & 0 & 0 & 0 \\ 0 & 1 & 0 & 0 & 0 \end{array}\right]}^{\bar{P}(s)} \overbrace{\left[\begin{array}{ccc|c} 1 & 0 & 0 & \\ 0 & 1 & 0 & 0 \\ 0 & 0 & s+3 & \\ \hline & 0 & & I_2 \end{array}\right]}^{\bar{Z}_r(s)}$$

$$\equiv \begin{bmatrix} Z_\ell(s) & 0 \\ 0 & I \end{bmatrix} \underbrace{\begin{bmatrix} \bar{P}(s) & -B \\ C & 0 \end{bmatrix}}_{\bar{T}(s)} \begin{bmatrix} \bar{Z}_r(s) & 0 \\ 0 & I \end{bmatrix} \tag{3.124}$$

$$\det\{T(s)\} = \det\{Z_\ell(s)\}\ \det\{\bar{P}(s)\}\ \det\{\bar{Z}_r(s)\}$$

Hence the following observations may be made:

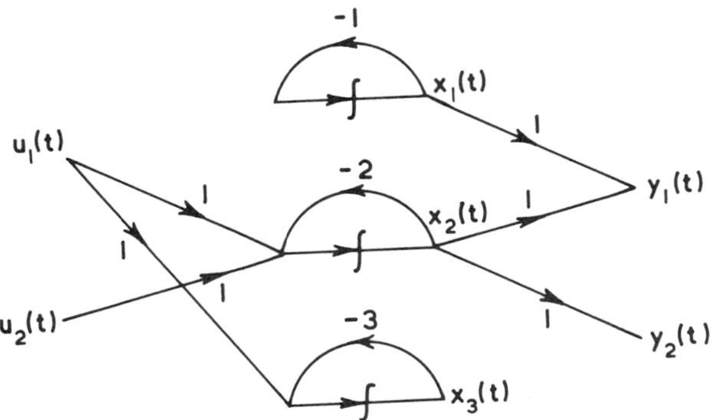

Fig. 3.8 Uncontrollable and unobservable system of Example 3.12.

$$\text{idz} = \{-1\}; \quad \text{odz} = \{-3\}; \quad \text{iodz} = \{\text{none}\}; \quad \text{sz} = \{-2\}$$

Thus the system is neither completely controllable nor completely observable, which is apparent from the signal flow graph in Fig. 3.8 and the following derivation:

$$\begin{aligned} G(s) &= C[\bar{P}(s)]^{-1}B \\ &= \begin{bmatrix} 1 & 1 & 0 \\ 0 & 1 & 0 \end{bmatrix} \begin{bmatrix} 1 & 0 & 0 \\ 0 & \frac{1}{s+2} & 0 \\ 0 & 0 & 1 \end{bmatrix} \begin{bmatrix} 0 & 0 \\ 1 & 1 \\ 1 & 0 \end{bmatrix} \\ &= \frac{1}{s+2} \begin{bmatrix} 1 & 1 \\ 1 & 1 \end{bmatrix} \end{aligned} \tag{3.125}$$

Example 3.13: Input/output-decoupling zero from state-space form T(s)

$$\begin{bmatrix} \dot{x}_1(t) \\ \dot{x}_2(t) \\ \dot{x}_3(t) \end{bmatrix} = \begin{bmatrix} 2 & 1 & 1 \\ 1 & 2 & 1 \\ 0 & 0 & 1 \end{bmatrix} \begin{bmatrix} x_1(t) \\ x_2(t) \\ x_3(t) \end{bmatrix} + \begin{bmatrix} 1 & 1 \\ 1 & 2 \\ -2 & -3 \end{bmatrix} \begin{bmatrix} u_1(t) \\ u_2(t) \end{bmatrix}$$

$$\begin{bmatrix} y_1(t) \\ y_2(t) \end{bmatrix} = \begin{bmatrix} -1 & 1 & 1 \\ 1 & -1 & 1 \end{bmatrix} \begin{bmatrix} x_1(t) \\ x_2(t) \\ x_3(t) \end{bmatrix} \tag{3.126}$$

The system has three poles: -1, -1, -3. The system matrix is

$$T(s) = \left[\begin{array}{ccc|cc} s-2 & -1 & -1 & -1 & -1 \\ -1 & s-2 & -1 & -1 & -2 \\ 0 & 0 & s-1 & 2 & 3 \\ \hline -1 & 1 & 1 & 0 & 0 \\ 0 & -1 & 1 & 0 & 0 \end{array}\right]$$

which does not make the existence of any decoupling zeros apparent. The following row/column operations are performed on T(s) to bring it into the canonical form of (3.116).

$$\left.\begin{array}{l} R1 = -R1 + R2 \\ R2 = -R3 \\ R3 = R1 + R2 + R3 \end{array}\right\} \left[\begin{array}{ccc|cc} -(s-1) & s-1 & 0 & 0 & -1 \\ 0 & 0 & -(s-1) & -2 & -3 \\ s-3 & s-3 & s-3 & 0 & 0 \\ \hline -1 & 1 & 1 & 0 & 0 \\ 1 & -1 & 1 & 0 & 0 \end{array}\right]$$

$$\left.\begin{array}{l} C1 = -C1 + C2 \\ C2 = C1 + C2 - 2C3 \\ C3 = C1 + C2 \end{array}\right\} \left[\begin{array}{ccc|cc} 2(s-1) & 0 & 0 & 0 & -1 \\ 0 & 2(s-1) & 0 & -2 & -3 \\ 0 & 0 & 2(s-3) & 0 & 0 \\ \hline 2 & -2 & 0 & 0 & 0 \\ -2 & -2 & 0 & 0 & 0 \end{array}\right]$$

Therefore,

$$T(s) \overset{se}{=} \begin{array}{l} sI - A_{co}\left\{ \right. \\ \\ \\ \\ \end{array} \underbrace{\left[\begin{array}{cc|c|cc} -2(s-1) & 0 & 0 & 0 & -1 \\ 0 & 2(s-1) & 0 & -2 & -3 \\ \hline 0 & 0 & 2(s-3) & 0 & 0 \\ \hline 2 & -2 & 0 & 0 & 0 \\ -2 & -2 & 0 & 0 & 0 \end{array}\right]}_{C_{co} \quad\quad C_{\bar{c}\bar{o}}} \begin{array}{l} \left.\right\} B_{co} \\ \\ \left.\right\} B_{\bar{c}o} \\ \\ \\ \end{array} \qquad (3.127)$$

which shows that $sI - A_{\bar{c}\bar{o}} \equiv s - 3$; thus $s = 3$ is an iodz of the system. This is not quite clear from the signal flow graph (Fig. 3.9a) of the original system in (3.126). It has, however, been noted in Sec. 3.2 that the uncontrol-

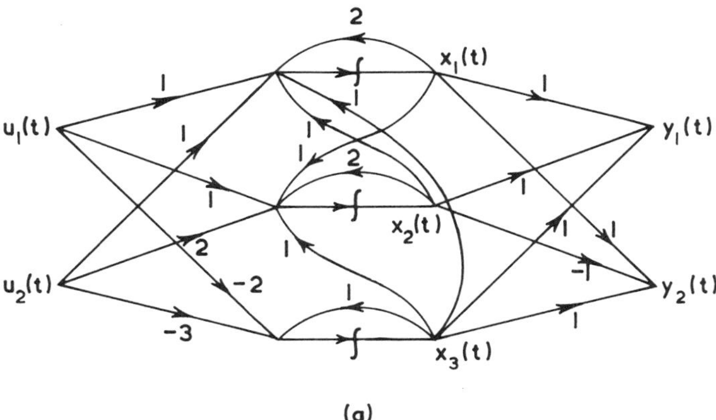

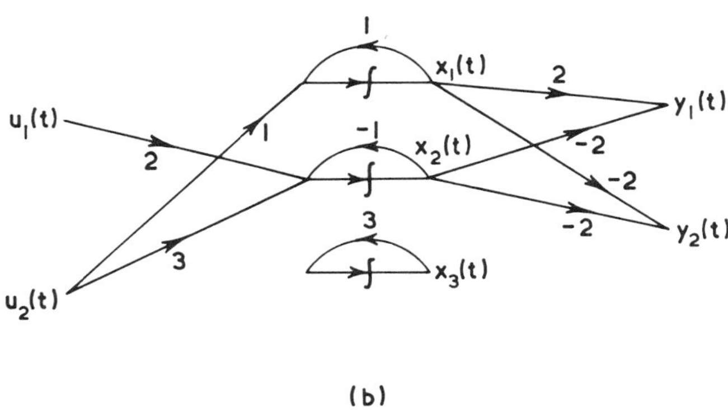

Fig. 3.9 Signal flow graph of (3.126): (a) original system S(A, B, C); (b) normalized system $S(A_n, B_n, C_n)$.

lable and unobservable modes of S(A, B, C, D) may be readily identified through its normalized equations, which for the above yields the signal flow graph in Fig. 3.9b, where the existence of an uncontrollable and unobservable mode is apparent. It is also worth noting here that for systems with known modal matrix, the equivalent form in (3.127) may be directly obtained from the original T(s) by

$$T(s) \overset{se}{=} \begin{bmatrix} M^{-1} & 0 \\ 0 & I \end{bmatrix} \begin{bmatrix} sI - A & -B \\ C & D \end{bmatrix} \begin{bmatrix} M & 0 \\ 0 & I \end{bmatrix}$$

$$\underset{=}{\text{se}} \begin{bmatrix} sI - M^{-1}AM & -M^{-1}B \\ CM & D \end{bmatrix}$$

The modal matrix thus corresponds the row/column operations on T(s). For this system, the modal matrix M derived from A in (3.126) is

$$M = \begin{bmatrix} -1 & 1 & 1 \\ 1 & 1 & 1 \\ 0 & -2 & 0 \end{bmatrix}; \quad M^{-1} = \frac{1}{2}\begin{bmatrix} -1 & 1 & 0 \\ 0 & 0 & -1 \\ 1 & 1 & 1 \end{bmatrix}$$

which correspond to the elementary row (≡ M^{-1}) and column (≡ M) operations performed above.

Example 3.14: Input/output-decoupling zeros of T(s) in polynomial form

$$T(s) = \left[\begin{array}{cccc|cc} (s+1)^2(s+2) & 0 & 0 & 0 & 1 & 0 \\ 0 & (s+1)(s+2)^2 & 0 & 0 & 0 & -1 \\ 0 & 0 & (s+1)(s+2) & 0 & 1 & 0 \\ 0 & 0 & 0 & s+2 & 0 & s+1 \\ \hline -1 & -1 & 0 & 0 & 0 & 0 \\ 0 & 0 & -1 & -1 & 0 & 0 \end{array}\right] \quad (3.128)$$

Existence of any decoupling zeros is not apparent, so a few elementary row/column operations are necessary.

$$\begin{matrix} R1 = R1 - R3 \\ R2 = R2 - R4 \end{matrix} \Big\} \left[\begin{array}{cccc|cc} (s+1)^2(s+2) & 0 & -(s+1)(s+2) & 0 & 0 & 0 \\ 0 & (s+1)(s+2)^2 & 0 & -(s+2) & 0 & -(s+2) \\ 0 & 0 & (s+1)(s+2) & 0 & 1 & 0 \\ 0 & 0 & 0 & s+2 & 0 & s+1 \\ \hline -1 & -1 & 0 & 0 & 0 & 0 \\ 0 & 0 & -1 & -1 & 0 & 0 \end{array}\right]$$

$$
\left.\begin{matrix} C1 = C1 - C2 \\ C3 = C3 - C4 \end{matrix}\right\}
\left[\begin{array}{cccc|cc}
(s+1)^2(s+2) & 0 & -(s+1)(s+2) & 0 & 0 & 0 \\
-(s+1)(s+2)^2 & (s+1)(s+2)^2 & s+2 & -(s+2) & 0 & -(s+2) \\
0 & 0 & (s+1)(s+2) & 0 & 1 & 0 \\
0 & 0 & -(s+2) & s+2 & 0 & s+1 \\
\hline
0 & -1 & 0 & 0 & 0 & 0 \\
0 & 0 & 0 & -1 & 0 & 0
\end{array}\right]
$$

Therefore,

$$
T(s) \equiv \underbrace{\left[\begin{array}{cccc|c}
(s+1)(s+2) & 0 & 0 & 0 & \\
0 & s+2 & 0 & 0 & \\
0 & 0 & 1 & 0 & 0 \\
0 & 0 & 0 & 1 & \\
\hline
 & & 0 & & I_2
\end{array}\right]}_{Z_\ell(s)}
\underbrace{\left[\begin{array}{cccc|cc}
s+1 & 0 & -1 & 0 & 0 & 0 \\
-(s+1)(s+2) & (s+1)(s+2) & 1 & -1 & 0 & -1 \\
0 & 0 & (s+1)(s+2) & 0 & 1 & 0 \\
0 & 0 & -(s+2) & s+2 & 0 & s+1 \\
\hline
0 & -1 & 0 & 0 & 0 & 0 \\
0 & 0 & 0 & -1 & 0 & 0
\end{array}\right]}_{T_1(s)}
\tag{3.129a}
$$

and

$$
T(s) \equiv \underbrace{\left[\begin{array}{cccccc}
s+1 & 0 & -(s+1) & 0 & 0 & 0 \\
-(s+2) & (s+1)(s+2)^2 & 1 & -(s+2) & 0 & -(s+2) \\
0 & 0 & s+1 & 0 & 1 & 0 \\
0 & 0 & -1 & s+2 & 0 & s+1 \\
\hline
0 & -1 & 0 & 0 & 0 & 0 \\
0 & 0 & 0 & -1 & 0 & 0
\end{array}\right]}_{T_2(s)}
\underbrace{\left[\begin{array}{cccc|c}
(s+1)(s+2) & 0 & 0 & 0 & \\
0 & 1 & 0 & 0 & \\
0 & 0 & s+2 & 0 & 0 \\
0 & 0 & 0 & 1 & \\
\hline
 & & 0 & & I_2
\end{array}\right]}_{Z_r(s)}
\tag{3.129b}
$$

giving

$$\{idz\} = -1,\ -2,\ -2 \quad \text{and} \quad \{odz\} = -1,\ -2,\ -2$$

To identify the input/output-decoupling zero(s), the output-decoupling zero(s) of $T_1(s)$ [$\equiv$ $T(s)$ with idz's removed] are to be evaluated; $T_1(s)$ in (3.129a) may be expressed as

$$T_1(s) = \underbrace{\left[\begin{array}{cccc:cc} 1 & 0 & -1 & 0 & 0 & 0 \\ -(s+2) & (s+1)(s+2) & 1 & -1 & 0 & -1 \\ 0 & 0 & (s+1)(s+2) & 0 & 1 & 0 \\ 0 & 0 & -(s+2) & s+2 & 0 & s+1 \\ \hdashline 0 & -1 & 0 & 0 & 0 & 0 \\ 0 & 0 & 0 & -1 & 0 & 0 \end{array}\right]}_{\overline{T}(s)} \left[\begin{array}{c:c} \overbrace{\begin{matrix} s+1 & 0 & 0 & 0 \\ 0 & 1 & 0 & 0 \\ 0 & 0 & 1 & 0 \\ 0 & 0 & 0 & 1 \end{matrix}}^{\overline{Z}_r(s)} & 0 \\ \hdashline 0 & I_2 \end{array}\right] \tag{3.129c}$$

which shows that only one odz is left in $T_1(s)$ if the idz's of T(s) are removed.

Hence for the system in (3.128), $\{iodz\} = -2, -2$. [The same result is obtained if the idz's of $T_2(s)$ in (3.129b) are evaluated.] Thus

$$\{idz\} = -1,\ -2,\ -2 \qquad \{odz\} = -1,\ -2,\ -2 \qquad \{iodz\} = -2,\ -2$$

$$\{sz\} = \text{set of zeros of } P(s) = -1,\ -1,\ -2,\ -1,\ -2,\ -2,\ -1,\ -2,\ -2$$

$$\triangleq \{idz\} + \{odz\} + \{tz\} - \{iodz\}$$

$$\{tz\} = -1,\ -1,\ -2,\ -2,\ -2$$

which are the poles of the transfer-function representation of T(s) as shown in Fig. 3.10.

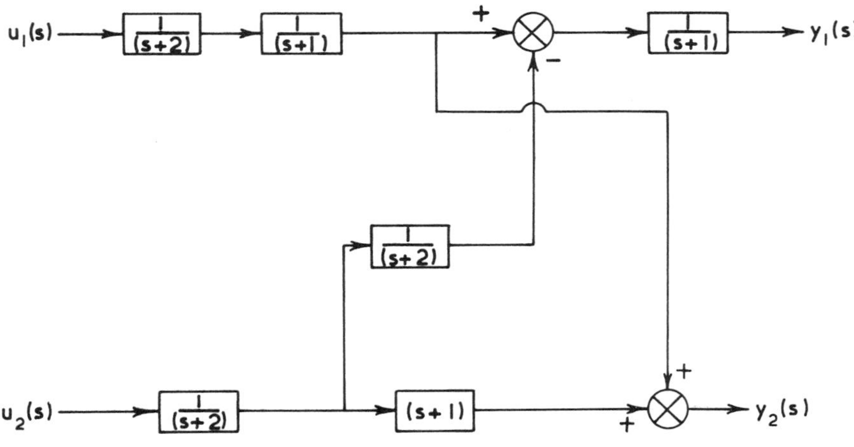

Fig. 3.10 Transfer function of (3.128).

3.3.4 Reduction of Order

The derivations in Sec. 3.2 show that the transfer-function matrix of any system S(A, B, C, D) contains only the controllable and observable modes, and consequently the removal of the uncontrollable and unobservable modes do not have any influence on the transfer function. The derivation in Sec. 3.3.3 also suggest that these modes or decoupling zeros may be identified and removed through the system matrix T(s) expressed in either polynomial form or state-space form. It is also clear from the previous examples that the system order is reduced when the decoupling zeros are eliminated. Since these zeros are absent in the transfer-function matrix, reduction of system order by this method has practical application.

This section develops the notion of least-order system matrix representation by using the concept of input and output decoupling zeros. Let there be two polynomial matrix representation of the same system, given by

$$T_1(s) = \begin{bmatrix} P_1(s) & -Q_1(s) \\ R_1(s) & W_1(s) \end{bmatrix} \quad \text{and} \quad T_2(s) = \begin{bmatrix} P_2(s) & -Q_2(s) \\ R_2(s) & W_1(s) \end{bmatrix} \tag{3.130}$$

where $\delta\{P_1(s)\} = n_1$ and $\delta\{P_2(s)\} = n_2$ with $n_1 > n_2$ [i.e., the representation $T_1(s)$ is of higher order than $T_2(s)$]. Since both $T_1(s)$ and $T_2(s)$ represent the same system, by definition they have the same transfer function,

$$\begin{aligned} G(s) &= R_1(s)P_1^{-1}(s)Q_1(s) + W_1(s) \\ &= R_2(s)P_2^{-1}(s)Q_2(s) + W_1(s) \end{aligned} \tag{3.131}$$

The difference in the orders of $T_1(s)$ and $T_2(s)$ is plausible since there is no unique choice of the polynomial state vector $\xi(s)$ (Sec. 2.4).

Because $T_1(s)$ and $T_2(s)$ have the same transfer function, they are strictly system equivalence (Theorem 2.3), and by Theorem 2.5, there is a transformation of system equivalence which carries $T_1(s)$ into $T_2(s)$. Therefore, the lower-order system $T_2(s)$ is obtainable from $T_1(s)$, which gives the same transfer-function matrix G(s). This leads to the following theorem.

<u>Theorem 3.12</u> If a system $S\{P_1(s), Q_1(s), R_1(s), W_1(s)\}$ of order n_1 $[= \delta\{\det(P_1(s))\}]$ has any decoupling zeros (input or output or input/output), there exists another polynomial-matrix representation of the system $S\{P_2(s), Q_2(s), R_2(s), W_1(s)\}$ which has no decoupling zeros, and is of order $n_2 < n_1$, where $n_2 = \delta\{\det(P_2(s))\}$.

Proof: To prove the theorem, it is assumed that the system has some input decoupling zeros and some output decoupling zeros. Then from (3.120), there exist two nonsingular polynomial matrices $Z_\ell(s)$ and $Z_r(s)$ which contain the input and output decoupling zeros of the system, respectively. The transfer function matrix of $T_1(s)$ is therefore given by

$$\begin{aligned} G_1(s) &= R_1(s)[P_1(s)]^{-1}Q_1(s) + W_1(s) \\ &= R_1(s)[Z_\ell(s)P_1'(s)]^{-1}Z_\ell(s)Q_2(s) + W_1(s) \\ &= R_1(s)[P_1'(s)]^{-1}Q_2(s) + W_1(s) \end{aligned} \tag{3.132}$$

The representation $S[P_1'(s), Q_2(s), R_1(s), W_1(s)\}$ has no input-decoupling zeros. If the output-decoupling zeros of (3.132) are now to be identified, then, the transfer-function matrix in (3.132) becomes

$$\begin{aligned} G_1(s) &\equiv R_2(s)Z_r(s)[P_2(s)Z_r(s)]^{-1}Q_2(s) + W_1(s) \\ &= R_2(s)[P_2(s)]^{-1}Q_2(s) + W_1(s) \\ &\equiv G_2(s) \end{aligned} \tag{3.133}$$

To complete the proof, the following identity obtained directly from (3.132) and (3.133) is used:

$$P_1(s) = Z_\ell(s)P_1'(s) = Z_\ell(s)P_2(s)Z_r(s) \tag{3.134}$$

that is, $\delta\{\det P_1(s)\} = \delta\{\det Z_\ell(s)\} + \delta\{\det P_2(s)\} + \delta\{\det Z_r(s)\}$ or

$$\begin{aligned} n_1 &= n_{idz} + n_2 + n_{odz} \\ &> n_2 \end{aligned} \tag{3.135}$$

where n_{idz} and n_{odz} denote the number of idz's and odz s; (3.135) follows directly from the definition given earlier for the case without any iodz's. If there are any iodz's (3.135) may be modified to

$$\begin{aligned} n_1 &= n_{idz} + n_2 + n_{odz} - n_{iodz} \\ &> n_2 \end{aligned} \tag{3.136}$$

This completes the proof and leads to the following theorem and definition.

Theorem 3.13 The transfer function of any system remains unchanged if its decoupling zeros are removed.

Definition 3.13: Least-order system An m-input/ℓ-output system $S\{P(s), Q(s), R(s)W(s)\}$ is said to be of least order if it has no decoupling zeros.

Thus if $T_1(s)$ is a system matrix with order n_1 with no decoupling zeros and having $G_1(s)$ as its transfer function matrix, then by using the foregoing development, it may be concluded that there exists no polynomial system representation $T_2(s)$, with order $n_2 < n_1$, yielding the same transfer function matrix $G_1(s)$. Thus a least-order system is equivalent to a completely controllable and observable (≡ output controllable) system—which is consistent with the concept of decoupling zeros and state controllability and observability discussed earlier.

There are several methods of computing the least order of a given T(s) or its equivalent G(s); one method, which is closely related to the developments in Sec. 3.2.3, is given below [12].

Theorem 3.14 The least-order $\bar{n}$ of an m-input/ℓ-output system is equal to the rank of the $(\ell p \times mp)$ matrix H given by

$$H = \begin{bmatrix} G_1 & \cdots & G_2 & \cdots & G_p \\ \vdots & & & & \\ G_p & \cdots & G_{p+1} & \cdots & G_{2p-1} \end{bmatrix} \tag{3.137}$$

where G_k, $k \in 1, \ldots, (2p - 1)$ are the coefficients in the expansion of G(s) for large enough $|s|$:

$$G(s) = D(s) + \frac{G_1}{s} + \frac{G_2}{s^2} + \cdots \tag{3.138}$$

and p is the degree of the monic least common denominator of the elements of G(s).

Proof: Let the system matrix T(s) of least order producing G(s) be given by

$$T(s) = \begin{bmatrix} sI_{\bar{n}} - A & B \\ C & D(s) \end{bmatrix}$$

Since, for large enough $|s|$,

$$(sI_{\bar{n}} - A)^{-1} \equiv \frac{I_{\bar{n}}}{s} + \frac{A}{s^2} + \cdots$$

it follows that

$$G(s) = C(sI - A)^{-1}B + D(s) \equiv D(s) + \frac{CB}{s} + \frac{CAB}{s^2} + \cdots \tag{3.139}$$

hence $G_k = CA^{k-1}B$ for $k \in 1, 2, \ldots$

By comparing (3.138) and (3.139), the matrix H in (3.137) may be expressed as

$$H = \underbrace{\begin{bmatrix} CB & CAB & \cdots & CA^{k-1}B \\ \vdots & & & \\ CA^{k-1} & CA^{k}B & \cdots & CA^{2k-1}B \\ CA^{k-1}B & & \cdots & CA^{2k-1}B \end{bmatrix}}_{P} \equiv \underbrace{\begin{bmatrix} C \\ \vdots \\ CA^{k-1} \end{bmatrix}}_{R} [B \;\; AB \;\; A^{k-1}B] \tag{3.140}$$

which, for k = p, has rank $\bar{n}$, the order by the controllable and observable pact of the system (i.e., the least order of the system). This completes the proof. This method of calculating the least order of a given G(s) is illustrated below [19].

Example 3.15: Computation of the least order of G(s)

$$G(s) = \begin{bmatrix} \frac{1}{s+1} & \frac{2}{s+1} \\ -\frac{1}{(s+1)(s+2)} & \frac{1}{s+2} \end{bmatrix} \tag{3.141}$$

The least common denominator of all elements of G(s) is

$$d(s) = (s+1)(s+2) = s^2 + 3s + 2 \rightarrow \text{degree } p = 2$$

A general expansion for the strictly proper part of G(s) is

$$G(s) = \frac{Q_1 s^{p-1} + Q_2 s^{p-2} + \cdots + Q_p}{s^p + a_1 s^{p-1} + a_2 s^{p-2} + \cdots + a_p} \equiv \frac{\sum_{k=1}^{p} Q_k s^{p-k}}{d(s)}$$

which when compared with (3.138) for the same coefficients of s gives [3]

$$Q_1 = G_1$$

$$Q_2 = G_2 + a_1 G_1$$

$$Q_3 = G_3 + a_1 G_2 + a_2 G_1$$

$$\vdots$$

$$Q_p = G_p + a_1 G_{p-1} + \cdots + a_{p-1} G_1$$

$$O = G_{p+j} + a_1 G_{p+j-i} + \cdots + a_p G_j \quad \text{for } j \in 1, 2, \ldots$$

For the system in (3.141),

$$G(s) = \frac{1}{\underset{\substack{\uparrow \\ a_1}}{s^2} + \underset{\substack{\uparrow \\ a_2}}{3s} + 2}\left\{s\underbrace{\begin{bmatrix} 1 & 2 \\ 0 & 1 \end{bmatrix}}_{Q_1} + \underbrace{\begin{bmatrix} 2 & 4 \\ -1 & 1 \end{bmatrix}}_{Q_2}\right\}$$

and

$$G_1 = Q_1 = \begin{bmatrix} 1 & 2 \\ 0 & 1 \end{bmatrix}; \quad G_2 = Q_2 - a_1G_1 = \begin{bmatrix} -1 & -2 \\ -1 & -2 \end{bmatrix}; \quad G_3 = -[a_1G_2 + a_2G_1] = \begin{bmatrix} 1 & 2 \\ 3 & 4 \end{bmatrix}$$

with $Q_3 = 0$

Therefore,

$$H = \begin{bmatrix} G_1 & G_2 \\ G_2 & G_3 \end{bmatrix} = \begin{bmatrix} 1 & 2 & -1 & -2 \\ 0 & 1 & -1 & -2 \\ -1 & -2 & 1 & 3 \\ -1 & -2 & 3 & 4 \end{bmatrix} \rightarrow \rho(H) = 3$$

The system matrix corresponding to (3.141) is

$$T_1(s) = \left[\begin{array}{ccc|cc} s+1 & 0 & 0 & -1 & 0 \\ 0 & s+1 & 0 & 0 & -1 \\ 0 & 0 & s+2 & -1 & -1 \\ \hline 1 & 2 & 0 & 0 & 0 \\ -1 & 0 & 1 & 0 & 0 \end{array}\right] \qquad (3.142)$$

which is of order 3 and has no decoupling zeros (Fig. 3.11a). An alternative representation of the system transfer-function matrix is

$$T_2(s) = \left[\begin{array}{cccc|cc} s+1 & 0 & 0 & 0 & -1 & 0 \\ 0 & s+2 & 0 & 0 & 0 & -1 \\ 0 & 0 & (s+1)(s+2) & 0 & -1 & 0 \\ 0 & 0 & 0 & s+2 & 0 & -1 \\ \hline 1 & 2 & 0 & 0 & 0 & 0 \\ 0 & 0 & -1 & 1 & 0 & 0 \end{array}\right] \delta\{\det T_1(s)\} = 5$$

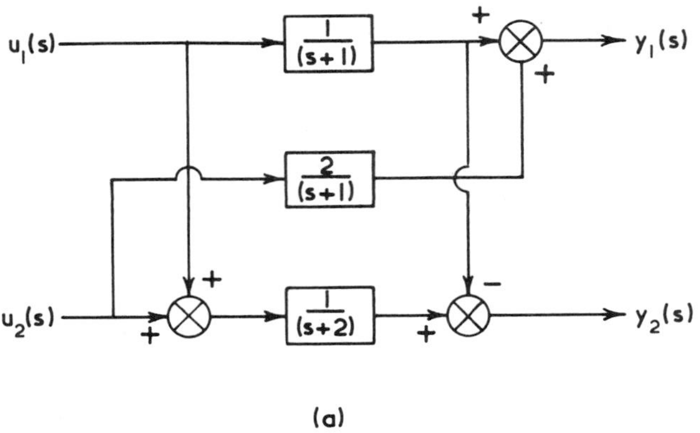

(a)

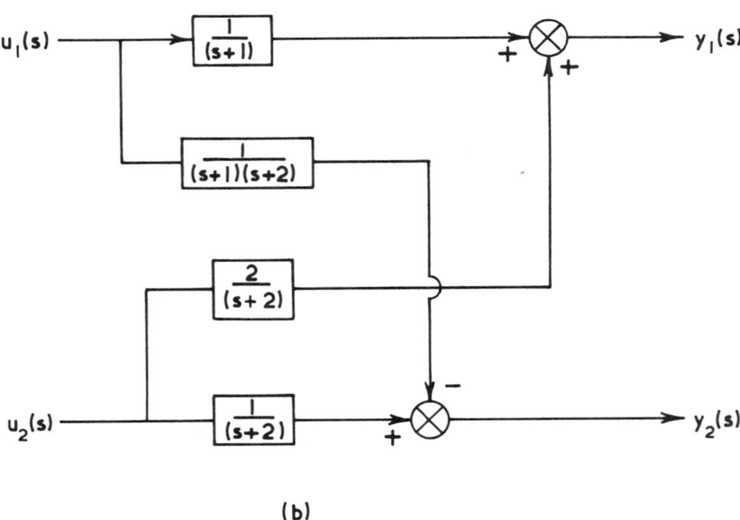

(b)

Fig. 3.11 Block diagram representation of G(s) in (3.141): (a) representation of $T_1(s)$ of order 3; (b) representation with $T_2(s)$ of order 5.

$$\xrightarrow{\text{se}} \left[\begin{array}{cccc|cc} s+1 & 0 & -(s+1)(s+2) & 0 & 0 & 0 \\ 0 & s+2 & 0 & -(s+2) & 0 & 0 \\ 0 & 0 & (s+1)(s+2) & 0 & -1 & 0 \\ 0 & 0 & 0 & s+2 & 0 & -1 \\ \hline 1 & 2 & 0 & 0 & 0 & 0 \\ 0 & 0 & -1 & 1 & 0 & 0 \end{array}\right] \begin{cases} R1 = R1 - R3 \\ R2 = R2 - R4 \end{cases}$$

$$\underset{\text{idz: }(s+2)}{\xrightarrow{\text{se}}} \left[\begin{array}{cccc|cc} 2(s+1) & 0 & -(s+1)(s+2) & 0 & 0 & 0 \\ -(s+2) & (s+2) & -(s+2) & -(s+2) & 0 & 0 \\ 0 & 0 & (s+1)(s+2) & 0 & -1 & 0 \\ 0 & 0 & 0 & s+2 & 0 & -1 \\ \hline 0 & 2 & 0 & 0 & 0 & 0 \\ 0 & 0 & 0 & 1 & 1 & 0 \end{array}\right] \begin{cases} C1 = 2C1 - C2 \\ C3 = \ \ C3 + C4 \end{cases}$$

$$\uparrow \text{ odz: } (s+2) \qquad (3.143)$$

Thus when the decoupling zeros are removed, the degree of det $\{T_2(s)\}$ becomes 3; this is apparent from the two block diagram representations of G(s) in Fig. 3.11b, from which yet another system matrix of order 4 can be formulated [19].

The following theorem on the reduction of order in terms of a system matrix T(s) may now be stated.

Theorem 3.15 Let G(s) be the transfer function of an m-input/ ℓ-output system and $n = \delta\{G(s)\}$. Let $T_1(s)$ be a system matrix of order $n_1 > n$ ($n_1 \equiv \delta\{\det T_1(s)\}$) which gives rise to G(s), and let T(s) contain a total number of n_d decoupling zeros (i.e., $n_d = \{idz\} + \{odz\} - \{iodz\}$). Then there exists a least-order system matrix $T_2(s)$ or order $n = n_1 - n_d$, which also gives rise to the same G(s).

Proof: The second point of theorem follows directly, as the transfer function matrix remains unchanged when the decoupling zeros are removed (Theorem 3.13). The first part is based on Theorem 3.12. To prove this, let $Z_\ell(s)$ and $Z_r(s)$ be the polynomial matrices containing the input- and output-decoupling zeros of $T_1(s)$, respectively. Then $T_1(s)$ may be expressed as [using (3.120)]

$$T_1(s) = \left[\begin{array}{c|c} P_1(s) & -Q_1(s) \\ \hline R_1(s) & W_1(s) \end{array}\right] \equiv \left[\begin{array}{c|c} Z_\ell(s)P_2(s)Z_r(s) & -Z_\ell(s)Q_2(s) \\ \hline R_2(s)Z_r(s) & W_1(s) \end{array}\right]$$

$$\equiv \left[\begin{array}{c|c} Z_\ell(s) & 0 \\ \hline 0 & I_m \end{array}\right] \left[\begin{array}{c|c} P_2(s) & -Q_2(s) \\ \hline R_2(s) & W_1(s) \end{array}\right] \left[\begin{array}{c|c} Z_r(s) & 0 \\ \hline 0 & I_\ell \end{array}\right]$$

$$\underset{\rightarrow}{se} \left[\begin{array}{c|c} P_2(s) & -Q_2(s) \\ \hline R_2(s) & W_1(s) \end{array}\right] = T_2(s) \tag{3.144}$$

Thus $T_1(s)$ and $T_2(s)$ are system equivalent since they are assumed to yield the transfer-function matrix. From (3.144), $\delta\{\det T_1(s)\} = n_1 = n_d + \delta\{\det T_2(s)\}$. Since $T_2(s)$ is assumed to be of least order and by (3.144) gives rise to the transfer-function matrix G(s),

$$\delta\{\det T_2(s)\} = \delta\{G(s)\} = n$$

Therefore,

$$n_1 = n_d + n > n$$

The development above provides an algebraic method, through row/column operations of T(s), of deriving the least order of any systems. It can be shown easily that the result obtained in this way is similar to those given by Sec. 3.2.3. This is not expanded further here.

Finally, by observing the relationship between the zeros of T(s) and poles of G(s) in the previous examples and (3.144), the following theorem may be stated.

Theorem 3.16 The poles of the transfer-function matrix G(s) are equal to the zeros of the corresponding system matrix less all decoupling zeros; that is,

Poles of $G(s)(p_g)$ = systems zeros of T(s)(sz) - [input-decoupling zeros (idz's) + output-decoupling zeros (odz's) - input/output-decoupling zeros (iodz's)] of T(s) (3.145)

Proof of the theorem follows from the definitions and developments in Sec. 3.3.3 and is not considered here. The theorem shows that the poles of

G(s) are the same as the transmission zeros defined earlier; this is illustrated below [21].

Example 3.16: Computation of poles and zeros of G(s) and T(s)

$$T(s) = \left[\begin{array}{ccc|ccc} s^2+2s & 0 & 0 & 1 & 2s-1 & s-1 \\ 0 & s^2+3s+2 & 0 & 0 & 2 & 1 \\ 0 & 0 & 1 & 0 & 1 & 0 \\ \hline 1 & 0 & 0 & 0 & 0 & 0 \\ -s & s^2 & 0 & 0 & 0 & 0 \end{array}\right] \tag{3.146}$$

(a) Computation of poles of G(s)

$$\begin{aligned} G(s) &= R(s)[P(s)]^{-1}Q(s) + W(s) \\ &= \begin{bmatrix} \dfrac{1}{s(s+2)} & \dfrac{2s-1}{s(s+2)} & \dfrac{s-1}{s(s+2)} \\ -\dfrac{1}{s+2} & -\dfrac{s-1}{(s+1)(s+2)} & \dfrac{1}{(s+1)(s+2)} \end{bmatrix} = \frac{N(s)}{d(s)} \end{aligned} \tag{3.147}$$

where

$$d(s) = s(s+1)(s+2) \quad \text{and} \quad N(s) = \begin{bmatrix} s+1 & 2s^2-s-1 & (s-1)(s+1) \\ -s(s+1) & -s(s-1) & s \end{bmatrix}$$

Poles of G(s) are, therefore,

$$p_g = \{0, -1, -2\}$$

(b) Computation of zeros of T(s)

The following row/column operations are performed on T(s):

$$\left.\begin{array}{l} C5 = 3C5 - 5C6 \\ C4 = -C4 + C5 - 2C6 \\ C6 = 3C4 + C6 \end{array}\right\} \rightarrow \left[\begin{array}{ccc|ccc} s^2+2s & 0 & 0 & 0 & s+2 & s+2 \\ 0 & s^2+3s+2 & 0 & 0 & 1 & 1 \\ 0 & 0 & 1 & 1 & 3 & 0 \\ \hline 1 & 0 & 0 & 0 & 0 & 0 \\ -s & s^2 & 0 & 0 & 0 & 0 \end{array}\right]$$

which has one idz at s = 2. Therefore, for this system,

$$\text{idz} = \{-2\}\ ; \quad \text{odz} = \text{none}; \quad \text{iodz} = \{\text{none}\}$$

$$\text{sz} = \{0, -1, -2, -2\} = \{p_g\} + \{\text{idz}\} \equiv \{\text{tz}\} + \{\text{idz}\}$$

Thus the poles of G(s) are the zeros of T(s) after the removal of the decoupling zeros.

To demonstrate the significance of sz, the following state-space representation* of G(s) in (3.147) is considered [21]:

$$A = \begin{bmatrix} 0 & 1 & 0 & 0 \\ 0 & -2 & 0 & 0 \\ 0 & 0 & 0 & 1 \\ 0 & 0 & -2 & -3 \end{bmatrix}, \quad B = \begin{bmatrix} 0 & 2 & 1 \\ 1 & -5 & -3 \\ 0 & 0 & 0 \\ 0 & 2 & 1 \end{bmatrix}, \quad C = \begin{bmatrix} 1 & 0 & 0 & 0 \\ 0 & -1 & -2 & -3 \end{bmatrix} \tag{3.148}$$

the characteristic polynomial of this representation being

$$\det(sI - A) = s(s + 1)(s + 2)^2$$

Thus the zeros of the characteristic polynomial are the same as the system zeros of (3.146). This is consistent with observations in Sec. 2.4.4.

Since the design of most physical systems is based on knowledge of the systems' behavior, a thorough understanding of the structural properties is essential for "good correspondence" between expected (often obtained through simulation) and actual (experimental) responses. A considerable amount of "theoretical" effort is thus directed toward the analysis of the mathematical model by using complicated numerical algorithms, and the results developed in this chapter form the basis of such theoretical analysis. While describing some relatively simple methods of identifying the controllable and observable forms of the system, the direct correspondence between state-space and polynomial representations has been established in this chapter, which is used extensively in subsequent chapters.

*Methods of deriving this representation are considered in Chap. 4.

REFERENCES

1. Ogata, K., State Space Analysis of Control Systems, Prentice-Hall, Englewood Cliffs, N.J., 1967.
2. DeRusso, P. M., Roy, J. R., and Close, C. M., State Variables for Engineers, Wiley, New York, 1965.
3. Chen, C. T., Introduction to Linear System Theory, Holt, Rinehart and Winston, New York, 1970.

4. Barnett, S., Introduction to Mathematical Control Theory, Clarendon Press, Oxford, 1975.
5. Luenberger, D. G., Introduction to Dynamic Systems, Wiley, New York, 1979.
6. Kailath, T., Linear Systems, Prentice-Hall, Englewood Cliffs, N.J., 1980.
7. Kalman, R. E., Falb, P. L., and Arbib, M. A., Topics in Mathematical System Theory, McGraw-Hill, New York, 1969.
8. Zadeh, L. A., and Desoer, C. A., Linear System Theory, McGraw-Hill, New York, 1963.
9. Bucy, R. S., "Canonical forms for multivariable systems," Trans. IEEE, AC-13: 567-569 (1968).
10. Mayne, D. Q., "Computational procedure for the minimal realization of transfer-function matrices," Proc. IEE, 115: 1363-1368 (1968).
11. Luenberger, D. G., "Canonical forms for linear multivariable systems," Trans. IEEE, AC-12: 290-293 (1967).
12. Rosenbrock, H. H., State-Space and Multivariable Theory, Thomas Nelson, London, 1970.
13. Rosenbrock, H. H., "Properties of linear constant systems," Proc. IEE, 117: 1717-1720 (1970).
14. Rosenbrock, H. H., "The transformation of strict system equivalence," Int. J. Control, 25: 11-19 (1977).
15. Rosenbrock, H. H., and Hayton, G. E., "Dynamical indices of a transfer function matrix," Int. J. Control, 20: 177-189 (1974).
16. MacFarlane, A. G. J., "Complex variable design methods," in Modern Approaches to Control System Design, N. Munro (Ed.), Peter Peregrinus, London, 1979.
17. Pernebo, L., "Algebraic control theory for linear multivariable systems," Thesis, Lund Institute of Technology, Lund, Sweden, 1978.
18. Fuhrmann, P. A., "On strict system equivalence and similarity," Int. J. Control, 25: 5-10 (1977).
19. Gilbert, E. G., "Controllability and observability in multivariable control systems," SIAM J. Control, 2: 128-151 (1963).
20. Kalman, R. E., "Mathematical description of linear dynamical systems," SIAM J. Control, 1: 152-192 (1963).
21. Heyman, M., and Thorpe, J. A., "Transfer equivalence of linear dynamical systems," SIAM J. Control, 8: 19-40 (1970).
22. Director, S. W., Introduction to Systems Theory, McGraw-Hill, New York, 1972.
23. McGillem, C. D., and Cooper, G. R., Continuous and Discrete Signal and System Analysis, Holt, Rinehart and Winston, New York, 1974.
24. Brockett, R. W., Finite Dimensional Linear Systems, Wiley, New York, 1970.

4

Transfer-Function Realizations

It was noted in Chap. 2 that the transfer-function representation provides an input/output description of linear systems, while the time-domain representation through state and output equations contains information about the input/output behavior as well as the internal structure of linear (and nonlinear) systems. Because of the analytical advantages associated with the time-domain representations, a large number of synthesis/design techniques are based on state/output equations.* On the other hand, because of practical/operational problems,† the majority of engineering systems are described through input/output response/transfer functions. Thus to establish a bridge between these two complementary aspects of control, a (ideally one-to-one) correspondence between the two representations is necessary. In addition to this, it is usually more efficient to simulate complex systems through a set of integro-differential equations rather than through a transfer-function matrix.‡ This problem of translating system representation from G(s) to S(A, B, C, D), known as the realization problem, has been extensively studied, yielding a number of realization procedures. Some basic aspects of realization theory and a few selected methods are considered in this chapter.

4.1 REALIZATION PROBLEM [1-9]

The derivations in Chap. 3 showed that the relationship between state-space and input/output representation for a proper system is

*A number of synthesis/design techniques based on frequency-domain representations are now available; some of these are considered in Chap. 10.
†Although this is changing due to the advent of new sensing and signal-processing techniques.
‡This situation is also changing due to the availability of a range of high-level block-diagram-orientated languages on a number of digital computer systems.

$$G(s) = C(sI - A)^{-1}B + D$$

Thus for a given set of system matrices A, B, C, and D, the transfer function matrix G(s) may be uniquely determined. The realization problem is the inverse problem where the system matrices A, B, C, and D are to be evaluated for a given G(s). A realization is said to be minimal if the dimension of the square matrix A [or the state vector x(t)] is minimal. It can be shown that every rational proper G(s) has minimal realizations, and all minimal realizations are equivalent. The example below outlines the significance of minimal realization [1].

Example 4.1: Realization concept—one-input/one-output system

$$\begin{aligned} S(A,b,c):\ \dot{x}(t) &= \text{diag}\{a_1, a_2, a_3, a_4\}x(t) + [b_1, b_2, 0, 0]^T u(t) \\ y(t) &= [0, c_2, 0, c_4]y(t) + [0]u(t) \end{aligned} \tag{4.1}$$

The transfer-function matrix of this fourth-order system is

$$\begin{aligned} g(s) &= c(sI - A)^{-1}b \\ &= [0, c_2, 0, c_4]\begin{bmatrix} s - a_1 & & & \\ & s - a_2 & & \\ & & s - a_3 & \\ & & & s - a_4 \end{bmatrix}^{-1}\begin{bmatrix} b_1 \\ b_2 \\ 0 \\ 0 \end{bmatrix} \\ &= c_2 \frac{1}{s - a_2} b_2 \end{aligned} \tag{4.2}$$

The signal flow graph of the state-space representation Fig. 4.1 shows that the state-controllable and state-observable (≡ output controllable) part of the system may be represented by

$$\begin{aligned} \dot{x}_2(t) &= a_2 x_2(t) + b_2(t) \\ y(t) &= c_2 x_2(t) \end{aligned} \tag{4.3}$$

and this defines the passage of control signal from the input terminal to the output terminal, and has a transfer function $c_2 b_2/(s - a_2)$. It can therefore be concluded that (4.3) is an "adequate" realization of g(s) in (4.2), and this is the minimal realization of g(s). The description is (4.1) is a nonminimal realization of g(s).

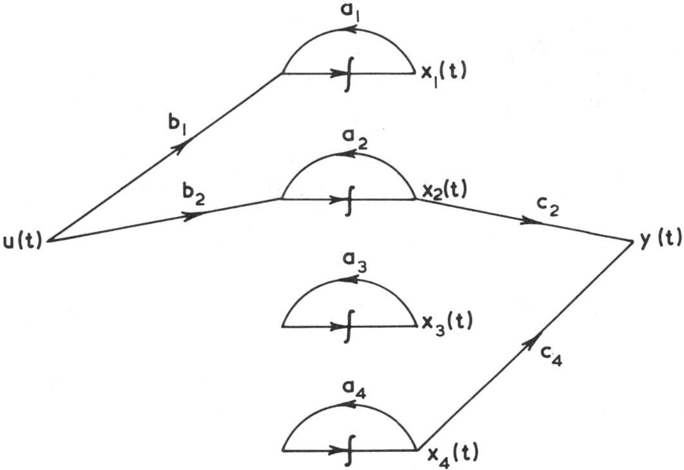

Fig. 4.1 Signal flow graph of Example 4.1.

4.1.1 Concepts and Definitions

As the example illustrates, for every realizable transfer function (matrix), there is an unlimited number of linear time-invariant state and output equations, and there are direct relationships between controllability and observability properties of S(A, B, C, D) and minimal realizations. These lead to the following concepts associated with realization theory [2].

Theorem 4.1

1. The impulse-response matrix of a linear time-invariant system S(A, B, C, D) depends solely on the controllable and observable part of the system $S_{co}(A_{co}, B_{co}, C_{co}, D_{co})$ and is given explicitly by

$$H(t, \tau) = C_{co}\phi_{co}(t, t_0)B_{co} + D_{co} \tag{4.4}$$

where $\phi_{co}(\cdot)$ is the state transition matrix corresponding to S_{co}.
2. Knowledge of the impulse-response matrix $H(t, \tau)$ identifies the completely controllable and completely observable part of the system, and this part (S_{co}) is itself a dynamical system with the smallest dimension among all realizations of $H(t, \tau)$.
3. A realization of S(A, B, C, D) is minimal (≡ irreducible) if and only if at all times it consists of S_{co} alone; thus every irreducible

realization of S is completely controllable and completely observable (i.e., completely output controllable).

4. Any two completely controllable and completely observable realization of S(A, B, C, D) are algebraically equivalent.

Proof of the theorem follows from the canonical decomposition theory (Theorem 3.5) and Fig. 3.5. A consequence of the theorem is given below.

Definition 4.1 A realization

$$\begin{aligned} \dot{x}(t) &= Ax(t) + Bu(t) \\ y(t) &= Cx(t) + Du(t) \end{aligned} \tag{4.5}$$

of a rational transfer function matrix G(s) is said to be minimal or irreducible if not other realization

$$\begin{aligned} \dot{\hat{x}}(t) &= \hat{A}\hat{x}(t) + \hat{B}u(t) \\ y(t) &= \hat{C}\hat{x}(t) + \hat{D}u(t) \end{aligned} \tag{4.6}$$

of G(s) exists with dimension of $\hat{x}(t)$ less than the dimension of x(t), the minimal order of realization being $n_{min} = \delta\{G(s)\}$.

With Theorem 4.1 and the derivations in Sec. 3.2, it is apparent that a minimal realization of G(s) can easily be obtained through the derivation of the controllable and observable canonical forms of (4.4). This procedure, although conceptually simple, requires a significant amount of numerical computation. Before considering the various realization procedures for multivariable systems, some basic results for single-input systems are stated below without detailed derivations [2, 3, 7].

4.1.2 Standard Forms

The various canonical forms of realization of

$$\begin{aligned} g(s) = \frac{N(s)}{\Delta(s)} &= \frac{\beta_1 s^{n-1} + \beta_2 s^{n-2} + \cdots + \beta_n}{s^n + \alpha_1 s^{n-1} + \cdots + \alpha_n} + d \\ &= \bar{g}(s) + d \end{aligned} \tag{4.7}$$

are given here; these are based on the results in Sec. 3.2.

(a) Controllable canonical form

$$\begin{bmatrix} \dot{x}_1(t) \\ \dot{x}_2(t) \\ \vdots \\ \dot{x}_{n-1}(t) \\ \dot{x}_n(t) \end{bmatrix} = \begin{bmatrix} 0 & 1 & 0 & \cdots & 0 \\ 0 & 0 & 1 & \cdots & 0 \\ \vdots & & & & \\ & 0 & 0 & \cdots & 1 \\ -\alpha_n & -\alpha_{n-1} & -\alpha_{n-2} & \cdots & -\alpha_1 \end{bmatrix} \begin{bmatrix} x_1(t) \\ x_2(t) \\ \vdots \\ x_{n-1}(t) \\ x_n(t) \end{bmatrix} + \begin{bmatrix} 0 \\ 0 \\ \vdots \\ 0 \\ 1 \end{bmatrix} u(t) \tag{4.8}$$

$$y(t) = [\beta_n, \beta_{n-1}, \cdots, \beta_2, \beta_1] \begin{bmatrix} x_1(t) \\ x_2(t) \\ \vdots \\ x_{n-1}(t) \\ x_n(t) \end{bmatrix} + du(t)$$

(b) Observable canonical form

$$\begin{bmatrix} \dot{x}_1(t) \\ \dot{x}_2(t) \\ \vdots \\ \dot{x}_{n-1}(t) \\ \dot{x}_n(t) \end{bmatrix} = \begin{bmatrix} 0 & 0 & \cdots & 0 & -\alpha_n \\ 1 & 0 & \cdots & 0 & -\alpha_{n-1} \\ \vdots & & & & \\ 0 & 0 & \cdots & 0 & -\alpha_2 \\ 0 & 0 & \cdots & 1 & -\alpha_1 \end{bmatrix} \begin{bmatrix} x_1(t) \\ x_2(t) \\ \vdots \\ x_{n-1}(t) \\ x_n(t) \end{bmatrix} + \begin{bmatrix} \beta_n \\ \beta_{n-1} \\ \vdots \\ \beta_2 \\ \beta_1 \end{bmatrix} u(t) \tag{4.9}$$

$$y(t) = [0 \ 0 \ \cdots \ 0 \ 1] \begin{bmatrix} x_1(t) \\ x_2(t) \\ \vdots \\ x_{n-1}(t) \\ x_n(t) \end{bmatrix}$$

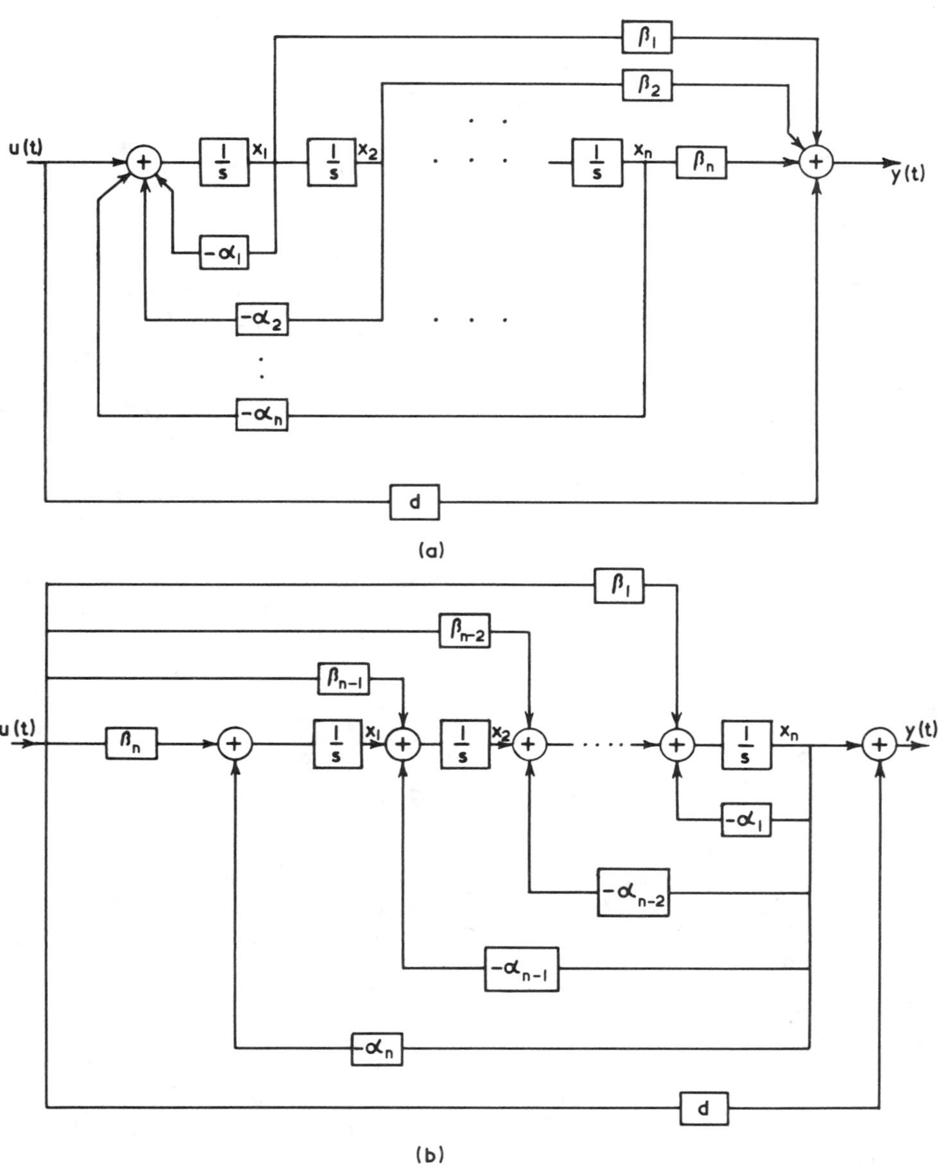

Fig. 4.2 Canonical-form realization of g(s) in (4.7): (a) controllable form; (b) observable form.

The controllable and observable canonical form realizations are shown in Fig. 4.2, and the derivations are illustrated through the following example [3].

Example 4.2

$$g(s) = \frac{4s^3 + 25s^2 + 45s + 34}{2s^3 + 12s^2 + 20s + 16} \equiv \frac{0.5s^2 + 2.5s + 1}{s^3 + 6s^2 + 10s + 8} + 2 \qquad (4.10)$$

Controllable form:

$$\dot{x}(t) = \begin{bmatrix} 0 & 1 & 0 \\ 0 & 0 & 1 \\ -8 & -10 & -6 \end{bmatrix} x(t) + \begin{bmatrix} 0 \\ 0 \\ 1 \end{bmatrix} u(t); \; y(t) = [1 \;\; 2.5 \;\; 0.5]x(t) + 2u(t) \qquad (4.11)$$

Observable form:

$$x(t) = \begin{bmatrix} 0 & 0 & -8 \\ 1 & 0 & -10 \\ 0 & 1 & -6 \end{bmatrix} x(t) + \begin{bmatrix} 1 \\ 2.5 \\ 0.5 \end{bmatrix} u(t); \; y(t) = [0 \;\; 0 \;\; 1]x(t) + 2u(t) \qquad (4.12)$$

(c) Jordan canonical form

The forms above, also known as companion form representations, are convenient in that they can be derived directly from the transfer function g(s). Another method, derived by partial fraction expansion of g(s), is given below.

Distinct eigenvalues of A: In this case the coordinate transformation through the modal matrix (Sec. 2.2) $x(t) = Mz(t)$ transforms the transfer function g(s) into

$$g(s) = c(sI - A)^{-1}b + d = cM(sI - \Lambda)^{-1}M^{-1}b + d \quad (\Lambda = M^{-1}AM) \qquad (4.13)$$

$$= c_n \operatorname{diag}\left\{\frac{1}{s - \lambda_k}\right\} b_n + d = \sum_{k=1}^{n} \frac{c_{nk} b_{nk}}{s - \lambda_k} + d, \quad k \in 1, \ldots, n \qquad (4.14)$$

Thus if g(s) can be expanded as

$$g(s) = \sum_{k=1}^{n} \frac{h_k}{s + \lambda_k} + d$$

the elements of c and b can be directed computed from

$$h_k = c_{nk} b_{nk}; \quad k \in 1, \ldots, n \tag{4.15}$$

This is illustrated below [8].

Example 4.3

$$g(s) = \frac{s^2 + 4s + 5}{s^3 + 6s^2 + 11s + 5} \quad (\lambda = -1, -2, -3)$$

$$= \frac{1}{s+1} - \frac{1}{s+2} + \frac{1}{s+3} \tag{4.16}$$

A normalized representation is therefore

$$z(t) = \begin{bmatrix} -1 & & 0 \\ & -2 & \\ 0 & & -3 \end{bmatrix} z(t) + \begin{bmatrix} \frac{1}{2} \\ -1 \\ \frac{1}{2} \end{bmatrix} u(t); \quad y(t) = [2 \;\; 1 \;\; 2] z(t) \tag{4.17}$$

Repeated eigenvalues of A: The idea of extending the above to the case of repeated roots is outlined through the expansion of g(s) into

$$g(s) = \frac{e_1}{(s - \lambda_1)^3} + \frac{e_2}{(s - \lambda_1)^2} + \frac{e_3}{s - \lambda_1} + \frac{e_4}{s - \lambda_2} + \frac{e_5}{s - \lambda_3} + d \tag{4.18}$$

where g(s) has poles at λ_1, λ_1, λ_1, λ_2, and λ_3.

By using the transformation x(t) = Jx(t) it can be shown (Sec. 1.4) that the derivations above would yield the following realization for (4.18):

$$\begin{bmatrix} z_1(t) \\ z_2(t) \\ z_3(t) \\ z_4(t) \\ z_5(t) \end{bmatrix} = \left[\begin{array}{ccc|cc} \lambda_1 & 1 & 0 & 0 & 0 \\ 0 & \lambda_1 & 0 & 0 & 0 \\ 0 & 0 & \lambda_1 & 0 & 0 \\ \hline 0 & 0 & 0 & \lambda_2 & 0 \\ 0 & 0 & 0 & 0 & \lambda_3 \end{array}\right] z(t) + \begin{bmatrix} 0 \\ 0 \\ 1 \\ 1 \\ 1 \end{bmatrix} u(t)$$

$$y(t) = [e_1 \;\; e_2 \;\; e_3 \;\; e_4 \;\; e_5] z(t) + d \tag{4.19}$$

The representation of g(s) above is called the Jordan canonical form. This is shown in Fig. 4.3a. An alternative form of representation (changes in b and c) is shown in Fig. 4.3b. Realization in Jordan canonical form has two difficulties. First, the denominator of the transfer function must be factored, or the poles of g(s) must be computed. Second, if g(s) has complex poles, A, b, c will have complex elements, and some equivalent transformation with complex-conjugate elements may have to be introduced [3].

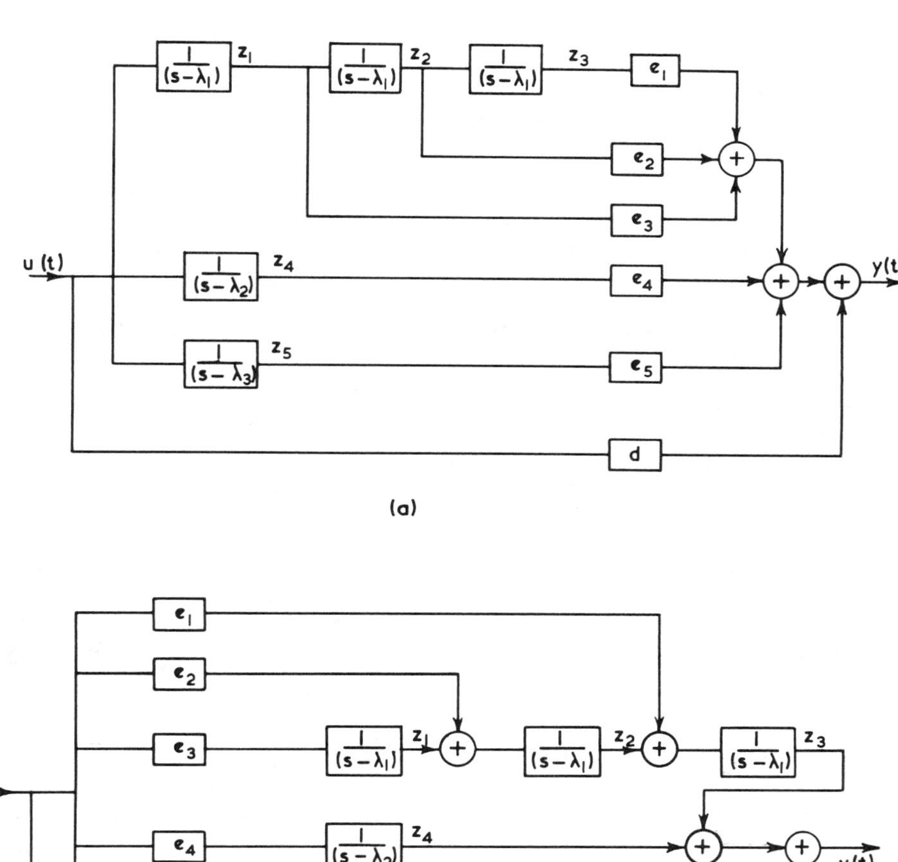

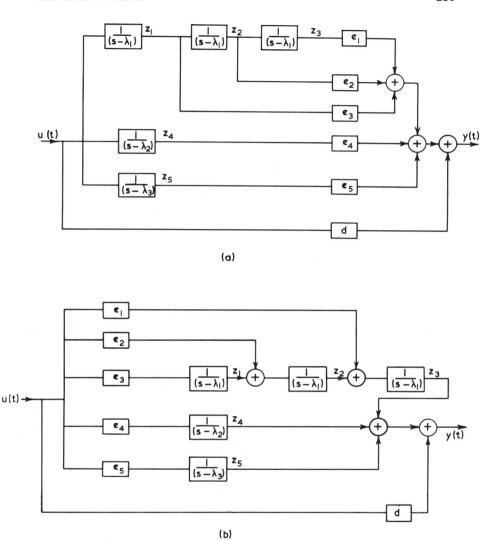

Fig. 4.3 Two different Jordan-canonical-form realizations of (4.18).

In most cases, however, the elements of J, b, and c can be obtained through suitable numerical algorithms [9].

Before the results above for single-input and single-output systems are extended to multivariable systems, the following theorems are stated to form a basis for further analytical developments.

Theorem 4.2 For the single-input/single-output system,

$$\dot{x}(t) = Ax(t) + bu(t); \quad y(t) = cx(t) + du(t) \tag{4.20}$$

where A, b, c, and d are of dimensions $n \times n$, $n \times 1$, $1 \times n$, and 1×1, if the transfer function is given by

$$g(s) = c(sI - A)^{-1}b + d = \frac{N(s)}{\det\{sI - A\}} + d = \frac{N(s)}{\Delta(s)} + d \tag{4.21}$$

then (4.20) is irreducible only if the polynomials N(s) and $\Delta(s)$ do not have nontrivial common factor (i.e., if they are relatively prime). This leads to the following theorem.

Theorem 4.3 The dynamical equation (4.20) is irreducible only if the denominator of the transfer function in (4.21) is equal to the characteristic polynomial of A and the degree of the denominator of the transfer function is equal to the dimension of A.

4.2 NONMINIMAL REALIZATIONS [3,4,6,10,13]

The derivations of the previous sections are directly extended here to multi-input/multi-output systems. Although such realizations are not minimal, they provide a simple procedure for realizing transfer-function matrices, and a basis for more refined methods considered in the following sections.

4.2.1 Block Forms

Let $\alpha(s)$ be the monic least common multiple (lcm) of the denominators of the elements of the proper rational matrix G(s) of order $\ell \times m$; then, from (4.4),

$$G(s) = \{g_{ij}(s)\} = \left\{\frac{\hat{g}_{ij}(s)}{\alpha(s)}\right\} = \frac{\hat{G}(s)}{\alpha(s)} = C(sI - A)^{-1}B + D \tag{4.22}$$

where $\hat{G}(s) = \{\hat{g}_{ij}(s)\}$, $\hat{g}_{ij}(s) = \alpha(s)g_{ij}(s)$, $\forall\, i \in 1, \ldots, \ell;\ j \in 1, \ldots, m$.

If $p = \delta\{\alpha(s)\}$, then $p \le n = \delta\{\det G(s)\}$, with $\hat{G}(s)$ expressed as

$$\hat{G}(s) = \hat{G}_1 s^{p-1} + \hat{G}_2 s^{p-2} + \cdots + \hat{G}_{p-1} s + \hat{G}_p \tag{4.23}$$

where

$$\alpha(s) = s^p + \alpha_1 s^{p-1} + \cdots + \alpha_{p-1} s + \alpha_p$$

and $\hat{G}_k$, $k \in 0, \ldots, p-1$, are $(\ell \times m)$ constant matrices. Then by following the realization for the single-input/single-output systems in (4.8), the controllable realization of an m-input/ℓ-output system may be derived [3, 6]

$$\begin{bmatrix} \dot{x}_1(t) \\ \dot{x}_2(t) \\ \vdots \\ \dot{x}_n(t) \end{bmatrix} = \begin{bmatrix} 0 & I_m & 0 & \cdots & 0 \\ 0 & 0 & I_m & \cdots & 0 \\ 0 & 0 & 0 & \cdots & I_m \\ -\alpha_p I_m & -\alpha_{p-1} I_m & -\alpha_{p-2} I_m & \cdots & -\alpha_1 I_m \end{bmatrix} \begin{bmatrix} x_1(t) \\ x_2(t) \\ \vdots \\ x_n(t) \end{bmatrix} + \begin{bmatrix} 0 \\ 0 \\ \vdots \\ 0 \\ I_m \end{bmatrix} \begin{bmatrix} u_1(t) \\ \vdots \\ u_m(t) \end{bmatrix} \tag{4.24}$$

$$\begin{bmatrix} y_1(t) \\ \vdots \\ y_\ell(t) \end{bmatrix} = [\hat{G}_p \quad \hat{G}_{p-1} \quad \cdots \quad \hat{G}_1] \begin{bmatrix} x_1(t) \\ \vdots \\ x_n(t) \end{bmatrix} + Du(t)$$

where

$$D = \lim_{s \to \infty} G(s)$$

The transfer-function matrix of (4.24) can easily be shown to be

$$G(s) = \frac{\sum_{k=1}^{p} \hat{G}_k s^{p-k}}{\sum_{k=0}^{p} \alpha_k s^{p-k}} \quad \text{with } \alpha_0 = 1 \tag{4.25}$$

This form of realization (called direct realization [6]), however, is not minimal since the order of the A matrix is $p \times m$, which, in most cases, is likely to be greater than n [the degree of G(s)], where $p \leq n$. If the observable form in (4.9) were used, the resulting multivariable observable realization would be as follows:

$$A = \begin{bmatrix} 0 & 0 & \cdots & 0 & -\alpha_p I_\ell \\ I_\ell & 0 & \cdots & 0 & -\alpha_{p-1} I_\ell \\ \vdots & \vdots & & \vdots & \vdots \\ 0 & 0 & \cdots & I_\ell & -\alpha_1 I_\ell \end{bmatrix}; \quad B = \begin{bmatrix} \hat{G}_p \\ \hat{G}_{p-1} \\ \vdots \\ \hat{G}_1 \end{bmatrix}; \quad C^T = \begin{bmatrix} 0 \\ 0 \\ \vdots \\ I_\ell \end{bmatrix}; \quad D = \lim_{s\to\infty} G(s) \tag{4.26}$$

The dimension of the state vector in the observable form is $\ell \times p$, which again is likely to be greater than n, the order of minimal realization (Theorem 4.1). This method of direct (nonminimal) realization is illustrated below.

Example 4.4: Direct realization of G(s)

$$G(s) = \begin{bmatrix} -\dfrac{s}{(s+1)^2} & \dfrac{1}{s+1} \\ \dfrac{2s+1}{s(s+1)} & \dfrac{1}{s+1} \end{bmatrix} \tag{4.27}$$

The monic least common denominator of the denominators of the elements of G(s) is

$$\alpha(s) = s(s+1)^2 = s^3 + 2s^2 + s; \quad \delta\{\alpha(s)\} = p = 3$$

G(s) may therefore be expressed as, from (4.25),

$$G(s) = \frac{1}{\alpha(s)}\left[\sum_{k=1}^{3} \hat{G}_k s^{p-k}\right] = \frac{1}{\alpha(s)}[\hat{G}_1 s^2 + \hat{G}_2 s + \hat{G}_3] \tag{4.28}$$

Then $\hat{G}(s)$ is expanded to transform it into the form (4.23):

$$\hat{G}(s) = \begin{bmatrix} -s^2 & s(s+1) \\ (2s+1)(s+1) & s(s+1) \end{bmatrix} = \begin{bmatrix} -s^2 & s^2+s \\ 2s^2+3s+1 & s^2+s \end{bmatrix}$$

$$\equiv \underbrace{\begin{bmatrix} -1 & 1 \\ 2 & 1 \end{bmatrix}}_{\hat{G}_1} s^2 + \underbrace{\begin{bmatrix} 0 & 1 \\ 3 & 1 \end{bmatrix}}_{\hat{G}_2} s + \underbrace{\begin{bmatrix} 0 & 0 \\ 1 & 0 \end{bmatrix}}_{\hat{G}_3}$$

$$\lim_{s\to\infty} G(s) = [0]$$

The direct realization in controllable form is therefore, from (4.24):

$$A = \begin{bmatrix} 0 & I_2 & 0 \\ 0 & 0 & I_2 \\ -\alpha_3 I_2 & -\alpha_2 I_2 & -\alpha_1 I_2 \end{bmatrix}, \quad \alpha_1 = 2, \ \ \alpha_2 = 1, \ \alpha_3 = 0$$

$$= \begin{bmatrix} 0 & 0 & 1 & 0 & 0 & 0 \\ 0 & 0 & 0 & 1 & 0 & 0 \\ 0 & 0 & 0 & 0 & 1 & 0 \\ 0 & 0 & 0 & 0 & 0 & 1 \\ 0 & 0 & -1 & 0 & -2 & 0 \\ 0 & 0 & 0 & -1 & 0 & -2 \end{bmatrix}; \quad B = \begin{bmatrix} 0 & 0 \\ 0 & 0 \\ 0 & 0 \\ 0 & 0 \\ 1 & 0 \\ 0 & 1 \end{bmatrix}; \ C = [\hat{G}_3 \ \ \hat{G}_2 \ \ \hat{G}_1]$$

$$= \begin{bmatrix} 0 & 0 & 0 & 1 & -1 & 1 \\ 1 & 0 & 3 & 1 & 2 & 1 \end{bmatrix} \tag{4.29a}$$

The observable form of this direct realization, obtained from (4.26), is

$$A = \begin{bmatrix} 0 & 0 & -\alpha_3 I_2 \\ I_2 & 0 & -\alpha_2 I_2 \\ 0 & I_2 & -\alpha_1 I_2 \end{bmatrix} = \begin{bmatrix} 0 & 0 & 0 & 0 & 0 & 0 \\ 0 & 0 & 0 & 0 & 0 & 0 \\ 1 & 0 & 0 & 0 & -1 & 0 \\ 0 & 1 & 0 & 0 & 0 & -1 \\ 0 & 0 & 1 & 0 & -2 & 0 \\ 0 & 0 & 0 & 1 & 0 & -2 \end{bmatrix};$$

$$B = \begin{bmatrix} \hat{G}_3 \\ \hat{G}_2 \\ \hat{G}_1 \end{bmatrix} = \begin{bmatrix} 0 & 0 \\ 1 & 0 \\ 0 & 1 \\ 3 & 1 \\ -1 & 1 \\ 2 & 1 \end{bmatrix}; \quad C = [0 \ \ 0 \ \ I_2] = \begin{bmatrix} 0 & 0 & 0 & 0 & 1 & 0 \\ 0 & 0 & 0 & 0 & 0 & 1 \end{bmatrix} \tag{4.29b}$$

The realization procedure above, although fairly simple in terms of numerical computation, is rather artificial in that the order of the resulting system is excessively high. The method, however, may be refined to derive reduced order realizations; these reduced realizations, though nonminimal, form a realistic basis for minimal realization. The modified method is based on identifying the least common multipliers of the denominators of the m columns (for controllable form) and the ℓ rows (for observable form) of the $(\ell \times m)$ transfer-function matrix G(s), rather than a single lcm for all the $(\ell \times m)$ elements of G(s). The derivations of controllable and observable forms are considered separately [1, 11, 14].

4.2.2 Controllable Companion Form [1, 6, 10, 12]

Let $\{g_j(s)\}$, $j \in 1, \ldots, m$, denote the m columns of the proper rational matrix G(s), $d^j(s)$ be the monic lcm of the denominators of the elements of $g_j(s)$, and let $\delta j = \deg\{d^j(s)\}$ for $j \in 1, \ldots, m$. Then G(s) may be expressed as

$$G(s) = [g_1(s) \cdot g_j(s) \cdot g_m(s)]$$

and any element $g_{ij}(s)$, $i \in 1, \ldots, \ell$, $j \in 1, \ldots, m$, may be expressed as

$$g_{ij}(s) = \frac{\beta_1^{ij} s^{\delta j-1} + \beta_2^{ij} s^{\delta j-2} + \cdots + \beta_{\delta j}}{s^{\delta j} + \alpha_1^j s^{\delta j-1} + \cdots + \alpha_{\delta j}^j} \tag{4.30}$$

Let the coefficients of the numerator and denominator polynomials of (4.30) be represented through the elements of two column vectors β^{ij} and α^j:

$$\beta^{ij} = \begin{bmatrix} \beta_{\delta j}^{ij} \\ \beta_{\delta j-1}^{ij} \\ \vdots \\ \beta_1^{ij} \end{bmatrix} (\delta j \times 1); \qquad \alpha^j = \begin{bmatrix} \alpha_{\delta j}^{j} \\ \alpha_{\delta j-1}^{j} \\ \vdots \\ \alpha_1^{j} \end{bmatrix} (\delta j \times 1) \tag{4.31}$$

By following (4.8) a controllable realization of $g_{ij}(s)$ may be expressed by $S_j(A_j, b_j, c_{ij})$, where

$$A_j = \begin{bmatrix} 0 & 1 & 0 \cdots & 0 \\ 0 & 0 & 1 \cdots & 0 \\ \vdots & & & \\ 0 & 0 & \cdots & 1 \\ -\alpha_{\delta j}^j & -\alpha_{\delta j-1}^j & \cdots & -\alpha_1^j \end{bmatrix} (\delta_j \times \delta_j); \quad b_j = \begin{bmatrix} 0 \\ 0 \\ \vdots \\ 0 \\ 1 \end{bmatrix} (\delta j \times 1);$$

$$c_{ij} = [\beta^{ij}]^T (1 \times \delta j)$$
$$= [\beta^{ij}_{\delta j} \quad \beta^{ij}_{\delta j-1} \quad \cdots \quad \beta^{ij}_1]$$

Defining a vector e_k as the kth column of an identity matrix (i.e., the kth element of e_k is unity while all others are zero), the matrices above may be expressed in compact form as

$$A_j = \begin{bmatrix} e_2^T \\ \vdots \\ e_{\delta j}^T \\ -[\alpha^j]^T \end{bmatrix}; \quad b_j = [e_{\delta j}]; \quad c_{ij} = [\beta^{ij}_{\delta j}]^T \tag{4.32}$$

$$(\text{i.e., } c_{11} = [\beta^{11}_{\delta 1}]^T, \; c_{21} = [\beta^{21}_{\delta 1}]^T, \text{ etc.})$$

Realization of the entire system may now be completed by combining the individual realizations are given above by taking j from 1 to m and is given by $S(A_c, B_c, C_c)$, where

$$A_c = \begin{bmatrix} A_1 & 0 & 0 & 0 \\ 0 & A_2 & 0 & 0 \\ & & \ddots & \\ 0 & 0 & A_j & 0 \\ & & & \ddots \\ 0 & 0 & 0 & A_m \end{bmatrix}_{(N_c \times N_c)} = \text{diag}\,\{A_j\}, \; j \in 1, \ldots, m \tag{4.33a}$$

$$B_c = \begin{bmatrix} b_1 & & & 0 \\ & b_2 & & \\ & & \ddots & \\ & & b_j & \\ & & & \ddots \\ 0 & & & b_m \end{bmatrix}_{(N_c \times m)} = \text{diag}\,\{b_j\}, \; j \in 1, \ldots, m \tag{4.33b}$$

$$C_c = \begin{bmatrix} c_{11} & c_{12} & \cdots & c_{1j} & \cdots & c_{1m} \\ c_{21} & c_{22} & \cdots & c_{2j} & \cdots & c_{2m} \\ \vdots \\ c_{i1} & c_{i2} & \cdots & c_{ij} & \cdots & c_{im} \\ \vdots \\ c_{\ell 1} & c_{\ell 2} & \cdots & c_{\ell j} & \cdots & c_{\ell m} \end{bmatrix}_{(\ell \times N_c)} \tag{4.33c}$$

where any $c_{ij} = [\beta^{ij}_{\delta j}]^T$ is a $(1 \times \delta j)$ row vector and $N_c = \Sigma^m_{j=1} \delta_j$. This realization may be proved to be controllable by extending (4.8) to multi-input/multioutput systems; the method is illustrated below [6].

Example 4.5: Derivation of controllable form of $G(s) = [g_1(s) \;\; g_2(s)]$

$$G(s) = \begin{bmatrix} \dfrac{1}{(s-1)^2} & \dfrac{1}{(s-1)(s+3)} \\ -\dfrac{6}{(s-1)(s+3)^2} & \dfrac{s-2}{(s+3)^2} \end{bmatrix} \tag{4.34}$$

$$g_1(s) = \begin{bmatrix} \dfrac{1}{(s-1)^2} \\ -\dfrac{6}{(s-1)(s+3)^2} \end{bmatrix} = \frac{1}{(s-1)^2(s+3)^2} \begin{bmatrix} (s+3)^2 \\ -6(s-1) \end{bmatrix}$$

$$= \frac{1}{s^4 + 4s^3 - 2s^2 - 12s + 9} \begin{bmatrix} (s^2 + 6s + 9) \\ -6s + 6 \end{bmatrix}; \qquad \delta_1 = 4$$

$$A_1 = \begin{bmatrix} 0 & 1 & 0 & 0 \\ 0 & 0 & 1 & 0 \\ 0 & 0 & 0 & 1 \\ -9 & 12 & 2 & -4 \end{bmatrix}; \quad b_1 = \begin{bmatrix} 0 \\ 0 \\ 0 \\ 1 \end{bmatrix}; \quad \beta^{11}_{\delta 1} = \begin{bmatrix} 9 \\ 6 \\ 1 \\ 0 \end{bmatrix}; \quad \beta^{21}_{\delta 1} = \begin{bmatrix} 6 \\ -6 \\ 0 \\ 0 \end{bmatrix}$$

Thus

$$c_{11} = [\beta^{11}_{\delta 1}]^T = [9 \;\; 6 \;\; 1 \;\; 0]; \qquad c_{21} = [\beta^{21}_{\delta 1}]^T = [6 \;\; -6 \;\; 0 \;\; 0]$$

$$g_2(s) = \begin{bmatrix} \dfrac{1}{(s-1)(s+3)} \\ \\ \dfrac{s-2}{(s+3)^2} \end{bmatrix} = \frac{1}{(s-1)(s+3)^2}\begin{bmatrix} s+3 \\ \\ (s-2)(s-1) \end{bmatrix}$$

$$= \frac{1}{s^3+5s^2+3s-9}\begin{bmatrix} s+3 \\ \\ s^2-3s+2 \end{bmatrix}; \qquad \delta 2 = 3$$

$$A_2 = \begin{bmatrix} 0 & 1 & 0 \\ 0 & 0 & 1 \\ 9 & -3 & -5 \end{bmatrix}; \quad b_2 = \begin{bmatrix} 0 \\ 0 \\ 1 \end{bmatrix}; \quad \beta_{\delta 2}^{12} = \begin{bmatrix} 3 \\ 1 \\ 0 \end{bmatrix}; \quad \beta_{\delta 2}^{22} = \begin{bmatrix} 2 \\ -3 \\ 1 \end{bmatrix}$$

$$c_{12} = [\beta_{\delta 2}^{12}]^T = [3 \;\; 1 \;\; 0]; \qquad c_{22} = [\beta_{\delta 2}^{22}]^T = [2 \;\; -3 \;\; 1]$$

The complete controllable realization is therefore

$$A_c = \begin{bmatrix} A_1 & 0 \\ 0 & A_2 \end{bmatrix} = \left[\begin{array}{cccc|ccc} 0 & 1 & 0 & 0 & & & \\ 0 & 0 & 1 & 0 & & & \\ 0 & 0 & 0 & 1 & & & \\ -9 & 12 & 2 & -4 & & & \\ \hline & & & & 0 & 1 & 0 \\ & & & & 0 & 0 & 1 \\ & & & & 9 & -3 & -5 \end{array}\right];$$

$$B_c = \begin{bmatrix} b_1 & 0 \\ 0 & b_2 \end{bmatrix} \left[\begin{array}{c|c} 0 & \\ 0 & \\ 0 & \\ 1 & \\ \hline & 0 \\ & 0 \\ & 1 \end{array}\right]; \quad C_c = \begin{bmatrix} c_{11} & c_{12} \\ c_{21} & c_{22} \end{bmatrix} = \begin{bmatrix} 9 & 6 & 1 & 0 & 3 & 1 & 0 \\ 6 & -6 & 0 & 0 & 2 & -3 & 1 \end{bmatrix} \tag{4.35}$$

The order of the controllable realization above is $N_c = \delta 1 + \delta 2 = 7$, which can be shown to be less than the direct realization of (4.34) obtained by using Sec. 4.2.1.

4.2.3 Observable Companion Form [1, 6, 10, 12]

The proper rational $(\ell \times m)$ transfer-function matrix G(s) is now expressed in terms of its ℓ rows $g_i(s)$, $i \in 1, \ldots, \ell$,

$$G(s) = \begin{bmatrix} g_1(s) \\ \vdots \\ g_i(s) \\ \vdots \\ g_\ell(s) \end{bmatrix}$$

Let $d^i(s)$ be the monic lcm of the denominators of all the elements of any ith row, and δi be the degree of $d^i(s)$. Then any element $g_{ij}(s)$ may be expressed as

$$g_{ij}(s) = \frac{\gamma_1^{ij} s^{\delta i-1} + \gamma_2^{ij} s^{\delta i-2} + \cdots + \gamma_{\delta i}^{ij}}{s^{\delta i} + \alpha_1^i s^{\delta i-1} + \cdots + \alpha_{\delta i}^i} \tag{4.36}$$

The coefficients of the numerator and denominator polynomials are assumed to be the elements of the following column vectors:

$$\gamma^{ij} = \begin{bmatrix} \gamma_{\delta i}^{ij} \\ \gamma_{\delta i-1}^{ij} \\ \vdots \\ \gamma_1^{ij} \end{bmatrix} (\delta i \times 1); \quad \alpha^i = \begin{bmatrix} \alpha_{\delta i}^{i} \\ \alpha_{\delta i-1}^{i} \\ \vdots \\ \alpha_1^{i} \end{bmatrix} (\delta i \times 1) \tag{4.37}$$

Then by using (4.9) an observable realization of the scalar transfer function $g_{ij}(s)$ in (4.36) may be described by $S_i(A_i, b_{ij}, c_i)$, where

$$A_i = \begin{bmatrix} 0 & 0 & \cdots & 0 & -\alpha_{\delta i}^i \\ 1 & 0 & \cdots & 0 & -\alpha_{\delta i-1}^i \\ 0 & 0 & \cdots & 1 & -\alpha_1^i \end{bmatrix} (\delta i \times \delta i)$$

$$\equiv [e_2 \quad e_3 \quad \cdots \quad e_{\delta i} \quad -\alpha^i]$$

$$b_{ij} = [\gamma^{ij}]; \quad \text{i.e.,}\ b_{11} = [\gamma^{11}_{\delta 1}],\ \ b_{12} = [\gamma^{12}_{\delta 1}],\ \text{etc.}$$

$$c_i = [0 \quad 0 \quad 1] = [e_{\delta i}]^T \quad (1 \times \delta i)$$

where, as before, $e_{\delta i}$ is the unit column vector of dimension δi for any i.

Therefore, if the realization above were extended to all $i \in 1, \ldots, \ell$, the observable realization of G(s) is obtained, which is given by $S(A_o, B_o, C_o)$:

$$A_o = \begin{bmatrix} A_1 & 0 & 0 & 0 & 0 \\ 0 & A_2 & 0 & 0 & 0 \\ 0 & 0 & 0 & A_i & 0 \\ 0 & 0 & 0 & 0 & A_\ell \end{bmatrix}_{(N_o \times N_o)} = \text{diag}\{A_i\}, \quad i \in 1, \ldots, \ell \tag{4.38a}$$

$$B_o = \begin{bmatrix} b_{11} & b_{12} & \cdots & b_{ij} & \cdots & b_{1m} \\ \cdot & \cdot & & \cdot & & \cdot \\ b_{i1} & b_{i2} & \cdots & b_{ij} & \cdots & b_{im} \\ \cdot & \cdot & & \cdot & & \cdot \\ b_{\ell 1} & b_{\ell 2} & \cdots & b_{\ell j} & \cdots & b_{\ell m} \end{bmatrix}_{(N_o \times m)} \tag{4.38b}$$

$$C_o = \begin{bmatrix} c_1 & & & & & \\ & c_2 & & & & \\ & & \ddots & & & \\ & & & c_i & & \\ & & & & \ddots & \\ & & & & & c_\ell \end{bmatrix}_{(\ell \times N_o)} = \text{diag}\{c_i\}; \quad c_i = e_{\delta i}, \quad i \in 1, \ldots, \ell \tag{4.38c}$$

where each $b_{ij} = [\gamma^{ij}_{\delta i}]$ is a $(\delta i \times 1)$ column vector, and $N_o = \Sigma_{i=1}^{\ell} \delta i$. The realization may be shown to be observable. The method is illustrated by using the rows of (4.34).

<u>Example 4.6: Observable realization of</u>

$$G(s) = \begin{bmatrix} g_1(s) \\ g_2(s) \end{bmatrix} \tag{4.39}$$

$$g_1(s) = \begin{bmatrix} \frac{1}{(s-1)^2} & \frac{1}{(s-1)(s+3)} \end{bmatrix} = \frac{1}{(s-1)^2(s+3)}[s+3 \quad s-1]$$

$$= \frac{1}{s^3+s^2-5s+3}[s+3 \quad s-1]; \qquad \delta 1 = 3$$

$$A_1 = \begin{bmatrix} 0 & 0 & -3 \\ 1 & 0 & 5 \\ 0 & 1 & -1 \end{bmatrix}; \quad \gamma_{\delta 1}^{11} = \begin{bmatrix} 3 \\ 1 \\ 0 \end{bmatrix}; \quad \gamma_{\delta 1}^{12} = \begin{bmatrix} -1 \\ 1 \\ 0 \end{bmatrix}; \quad c_1 = [0 \quad 0 \quad 1]$$

$$g_2(s) = \begin{bmatrix} -\frac{6}{(s-1)(s+3)^2} & \frac{s-2}{(s+3)^2} \end{bmatrix} = \frac{1}{(s-1)(s+3)^2}[-6 \quad (s-1)(s-2)]$$

$$= \frac{1}{s^3+5s^2+3s-9}[-6 \quad s^2-3s+2]; \qquad \delta 2 = 3$$

$$A_2 = \begin{bmatrix} 0 & 0 & 9 \\ 1 & 0 & -3 \\ 0 & 1 & -5 \end{bmatrix}; \quad \gamma_{\delta 2}^{21} = \begin{bmatrix} -6 \\ 0 \\ 0 \end{bmatrix}; \quad \gamma_{\delta 2}^{22} = \begin{bmatrix} 2 \\ -3 \\ 1 \end{bmatrix}; \quad c_2 = [0 \quad 0 \quad 1]$$

The observable realization of (4.39) is thus

$$A_o = \begin{bmatrix} A_1 & 0 \\ 0 & A_2 \end{bmatrix} = \left[\begin{array}{ccc|ccc} 0 & 0 & -3 & & & \\ 1 & 0 & 5 & & 0 & \\ 0 & 1 & -1 & & & \\ \hline & & & 0 & 0 & 9 \\ & 0 & & 1 & 0 & -3 \\ & & & 0 & 1 & 5 \end{array}\right]$$

$$B_o = \begin{bmatrix} b_{11} & b_{12} \\ b_{21} & b_{22} \end{bmatrix} = \begin{bmatrix} \gamma_{\delta 1}^{11} & \gamma_{\delta 1}^{12} \\ \gamma_{\delta 2}^{21} & \gamma_{\delta 2}^{22} \end{bmatrix} = \begin{bmatrix} 3 & -1 \\ 1 & 1 \\ 0 & 0 \\ -6 & 2 \\ 0 & -3 \\ 0 & 1 \end{bmatrix} \tag{4.40}$$

$$C_o = \begin{bmatrix} c_1 & 0 \\ 0 & c_2 \end{bmatrix} = \begin{bmatrix} 0 & 0 & 1 & 0 & 0 & 0 \\ 0 & 0 & 0 & 0 & 0 & 1 \end{bmatrix}$$

Order of this realization $N_o = \delta_1 + \delta_2 = 6$ is less than its controllable form. It can be seen that for the example above, the degree of the monic lcm of the denominators of all elements of G(s) is 4. By Sec. 3.3.4, therefore, the system is of fourth order. This suggests that neither of the realizations above is completely controllable and observable. Hence the realizations (which have orders $\Sigma_1^m \delta j$ or $\Sigma_1^\ell \delta i$) are unlikely to be minimal. If, however, the unobservable modes in the controllable realization $S_c(A_c, B_c, C_c)$ [or the uncontrollable modes in the observable realization $S_o(A_o, B_o, C_o)$] are identified and removed, the resulting system can be made completely controllable and observable, and hence irreducible. This can be achieved by applying the canonical transformations of Sec. 3.2 on either $S_c(A_c, B_c, C_c)$ or $S_o(A_o, B_o, C_o)$. Thus by using this two-stage procedure, an irreducible realization of any rational G(s) may be obtained. This is illustrated below [15].

Example 4.7: Irreducible realization of G(s)

$$G(s) = \begin{bmatrix} \frac{s+1}{s^2} & \frac{s+1}{s^2} & \frac{1}{s} \\ 0 & 0 & \frac{1}{s} \end{bmatrix} \qquad (4.41)$$

$$\underbrace{\qquad}_{g_1(s)} \quad \underbrace{\qquad}_{g_2(s)} \quad \underbrace{\qquad}_{g_3(s)}$$

Stage 1: Controllable form of G(s):

$$g_1(s) = \frac{1}{s^2}\begin{bmatrix} s+1 \\ 0 \end{bmatrix} \equiv \frac{1}{s^2 + \alpha_1^1 s + \alpha_2^1}\begin{bmatrix} \beta_{\delta 1}^{11} s + \beta_{\delta 1-1}^{11} \\ \beta_{\delta 1}^{21} s + \beta_{\delta 1-1}^{22} \end{bmatrix}; \quad \delta_1 = 2$$

$$g_2(s) = \frac{1}{s^2}\begin{bmatrix} s+1 \\ 0 \end{bmatrix} \equiv \frac{1}{s^2 + \alpha_1^2 s + \alpha_2^2}\begin{bmatrix} \beta_{\delta 2}^{12} s + \beta_{\delta 2-1}^{12} \\ \beta_{\delta 2}^{22} s + \beta_{\delta 2-1}^{22} \end{bmatrix}; \quad \delta_2 = 2$$

$$g_3(s) = \frac{1}{s}\begin{bmatrix} 1 \\ 1 \end{bmatrix} \equiv \frac{1}{s + \alpha_1^3}\begin{bmatrix} \beta_{\delta 3}^{13} \\ \beta_{\delta 3}^{23} \end{bmatrix}; \quad \delta_3 = 1$$

Therefore, the controllable realization of the above is

$$A_c = \begin{bmatrix} A_1 & & 0 \\ & A_2 & \\ 0 & & A_3 \end{bmatrix} = \left[\begin{array}{cc|cc|c} 0 & 1 & & & \\ -\alpha_2^1 & -\alpha_1^1 & & 0 & \\ \hline & & 0 & 1 & \\ & 0 & -\alpha_2^2 & -\alpha_1^2 & \\ \hline & & & & -\alpha_1^3 \end{array}\right] \equiv \left[\begin{array}{cc|cc|c} 0 & 1 & & & \\ 0 & 0 & & & \\ \hline & & 0 & 1 & \\ & & 0 & 0 & \\ \hline & & & & 0 \end{array}\right];$$

$$B_c = \begin{bmatrix} b_1 & & 0 \\ & b_2 & \\ 0 & & b_3 \end{bmatrix} = \left[\begin{array}{c|c|c} 0 & & \\ 1 & & \\ \hline & 0 & \\ & 1 & \\ \hline & & 0 \\ & & 1 \end{array}\right]; \qquad (4.42)$$

$$C_c = \begin{bmatrix} [\beta_{\delta 1}^{11}]^T & [\beta_{\delta 2}^{12}]^T & [\beta_{\delta 3}^{12}]^T \\ [\beta^{21}{}_{\delta_1}] & [\beta_{\delta 2}^{22}]^T & [\beta_{\delta 3}^{23}]^T \end{bmatrix} = \begin{bmatrix} 1 & 0 & 1 & 0 & 1 \\ 0 & 0 & 0 & 0 & 1 \end{bmatrix}$$

Stage 2:

1. Check for controllability: $\rho[B \quad AB] = 5 \rightarrow$ controllable.
2. Check for observability: only three linearly independent columns of C, hence the observability index is 3. The observable form of (4.41) can be shown to be [15]:

$$A_o = \begin{bmatrix} 0 & 1 & 0 \\ 0 & 0 & 0 \\ 0 & 0 & 0 \end{bmatrix}; \quad B_o = \begin{bmatrix} 0 & 0 & 1 \\ 1 & 1 & 0 \\ 0 & 0 & 1 \end{bmatrix}; \quad C_o = \begin{bmatrix} 1 & 1 & 0 \\ 0 & 0 & 1 \end{bmatrix} \qquad (4.43)$$

of order $n_{co} < n$. Detailed derivations (using Sec. 3.2) are left to the reader.

4.2.4 Normal Form

Since the transfer-function matrix represents the controllable and observable part of the system, the method of direct derivation of state-space representation of Sec. 4.1.2(c) may be extended to multivariable system as well. This is considered here for distinct eigenvalue case, realization for systems with repeated eigenvalues (Jordan form) is considered in the following section.

The transfer-function matrix may in this case be directly extended from (4.14) as

$$G(s) = C_n[sI - \Lambda]^{-1}B_n + D \tag{4.44}$$

where C_n and B_n are the normal forms of C and B, and $\Lambda = \text{diag}\{\lambda_k\}$, $k = 1, \ldots, n$, are the roots of A; (4.44) may thus be expressed as

$$G(s) = \frac{N(s)}{d(s)} = \sum_{k=1}^{n} \frac{G_k}{s - \lambda_k} + D; \quad d(s) = \prod_{k=1}^{n} (s - \lambda_k) \tag{4.45}$$

where

$$G_k = c_{nk} b_{nk}^T \tag{4.46}$$

c_{nk} is the kth column of c_n, b_{nk}^T is the kth row of B_n, and G_k is of dimension $\ell \times m$ [same as G(s)]. The residue matrix G_k in (4.46), being a vector outer product, has rank 1 and can be derived by

$$G_k = \lim_{s \to \lambda_k} (s - \lambda_k)G(s)$$

with

$$D = \lim_{s \to \infty} G(s) \tag{4.47}$$

Once G_k, $k \in 1, \ldots, n$, are computed, c_{nk} may be derived for an arbitrary value of b_{nk}^T, or vice versa.* The resulting realization is therefore not unique, and consequently may not be completely controllable or completely observable. If, however, the uncontrollable or unobservable modes of the system are known, the summation in (4.45) need only be taken up to n_{co}, the number of controllable and observable modes. The order of the controllable and observable realization will then be of order $n_{co} < n$. The method is illustrated below.

Example 4.8: Realization of G(s) through residual matrices

$$G(s) = \begin{bmatrix} \frac{1}{s+1} & \frac{2}{s+1} & \frac{1}{s+4} \\ 0 & \frac{1}{s+2} & \frac{1}{s+4} \\ \frac{1}{s+1} & \frac{2(s+3)}{(s+1)(s+2)(s+3)} & 0 \end{bmatrix} \tag{4.48}$$

*An alternative method of deriving b_{nk}^T and c_{nk} is given in Refs. 3 and 16.

The lcm of the denominators of all the elements of G(s) = d(s) = (s + 1)(s + 2)(s + 3)(s + 4).

Full-order realization:

$$G_1 = \lim_{\lambda \to -1} (s+1)G(s) = \begin{bmatrix} 1 & 2 & 0 \\ 0 & 0 & 0 \\ 1 & 2 & 0 \end{bmatrix} = \underbrace{\begin{bmatrix} 1 \\ 0 \\ 1 \end{bmatrix}}_{c_{n1}} \underbrace{[1 \quad 2 \quad 0]}_{b_{n1}^T}$$

Similarly,

$$G_2 = \underbrace{\begin{bmatrix} 0 \\ 1 \\ -2 \end{bmatrix}}_{c_{n2}} \underbrace{[0 \quad 1 \quad 0]}_{b_{n2}^T}; \quad G_3 = \underbrace{\begin{bmatrix} 0 \\ 0 \\ 0 \end{bmatrix}}_{c_{n3}} \underbrace{[1 \quad 1 \quad 1]}_{b_{n3}^T}; \quad G_4 = \underbrace{\begin{bmatrix} 1 \\ 1 \\ 0 \end{bmatrix}}_{c_{n4}} \underbrace{[0 \quad 0 \quad 1]}_{b_{n4}^T}$$

The realization is, therefore,

$$\Lambda = \text{diag}\{-1, -2, -3, -4\}; \quad B_n = \begin{bmatrix} 1 & 2 & 0 \\ 0 & 1 & 0 \\ 1 & 1 & 1 \\ 0 & 0 & 1 \end{bmatrix}; \quad C_n = \begin{bmatrix} 1 & 0 & 0 & 1 \\ 0 & 1 & 0 & 1 \\ 1 & -2 & 0 & 0 \end{bmatrix} \qquad (4.49)$$

The system is not completely observable (Fig. 4.4).

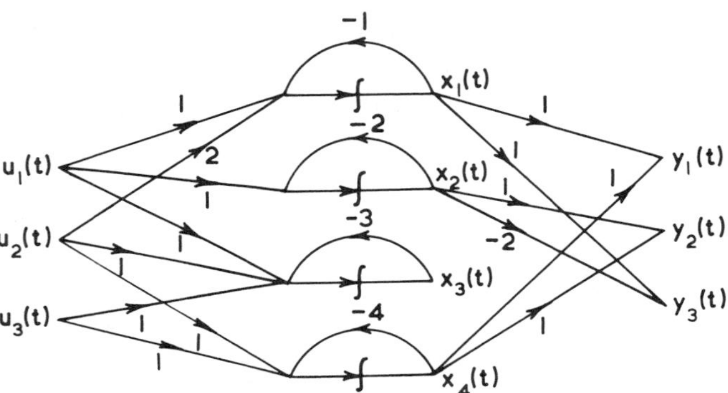

Fig. 4.4 Signal flow graph of full-order realization of (4.48); $x_3(t)$ does not appear at the output.

Reduced-order realization: By inspection, it is seen that $G_{32} \equiv 2/(s+1)(s+2)$. This reduces d(s) to (s + 1)(s + 2)(s + 4). The resulting realization therefore need not contain the pole $\lambda = -3$ and the third row and column of B_n and C_n (Fig. 4.4). Thus the reduced-order realization is

$$\Lambda = \text{diag}\{-1,-2,-4\}; \quad B_n = \begin{bmatrix} 1 & 2 & 0 \\ 0 & 1 & 0 \\ 0 & 0 & 1 \end{bmatrix}; \quad C_n = \begin{bmatrix} 1 & 0 & 1 \\ 0 & 1 & 1 \\ 1 & -2 & 0 \end{bmatrix} \tag{4.50}$$

which is completely controllable and observable and hence a minimal realization of G(s).

<u>Example 4.9: Completely controllable and observable realization with</u>

$$G(s) = \begin{bmatrix} \dfrac{1}{(s+3)(s+4)} & \dfrac{1}{s+1} \\ \dfrac{1}{s+3} & 0 \end{bmatrix} \tag{4.51}$$

The system has three poles at -1, -3, and -4. The residue matrices are therefore

$$G_1 = G_{\lambda_1=-1} = \lim_{s\to-1} (s+1)G(s) = \begin{bmatrix} 0 & 1 \\ 0 & 0 \end{bmatrix};$$

$$G_2 = G_{\lambda_2=-3} = \lim_{s\to-3} (s+3)G(s) = \begin{bmatrix} 1 & 0 \\ 1 & 0 \end{bmatrix};$$

$$G_3 = G_{\lambda_3=-4} = \lim_{s\to-4} (s+4)G(s) = \begin{bmatrix} -1 & 0 \\ 0 & 0 \end{bmatrix}$$

Therefore,

$$G_1 = c_{n1}b_{n1}^T = \begin{bmatrix} 0 & 1 \\ 0 & 0 \end{bmatrix} = \begin{bmatrix} 1 \\ 0 \end{bmatrix}[0 \quad 1]; \quad G_2 = c_{n2}b_{n2}^T = \begin{bmatrix} 1 & 0 \\ 1 & 0 \end{bmatrix} = \begin{bmatrix} 1 \\ 1 \end{bmatrix}[1 \quad 0];$$

$$G_3 = c_{n3}b_{n3}^T = \begin{bmatrix} -1 & 0 \\ 0 & 0 \end{bmatrix} = \begin{bmatrix} 1 \\ 0 \end{bmatrix}[-1 \quad 0]$$

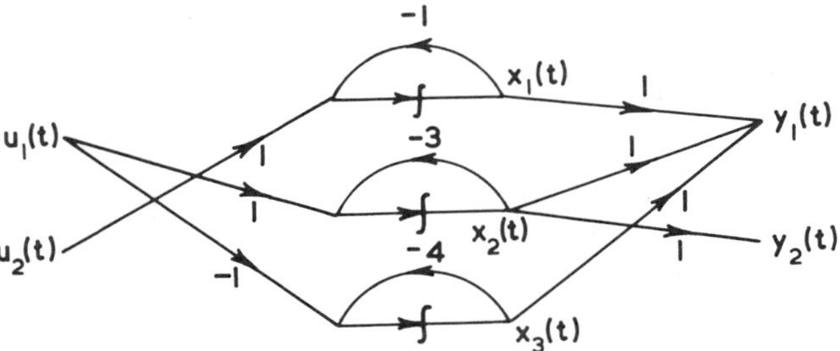

Fig. 4.5 Signal flow graph of minimal realization of (4.51).

The diagonal realization is therefore

$$\Lambda = \text{diag}\{-1, -3, -4\}; \quad B_n = [b_{n1}^T \; b_{n2}^T \; b_{n3}^T] = \begin{bmatrix} 0 & 1 \\ 1 & 0 \\ -1 & 0 \end{bmatrix}; \quad c_n = [c_{n1} \; c_{n2} \; c_{n3}] = \begin{bmatrix} 1 & 1 & 1 \\ 0 & 1 & 0 \end{bmatrix} \tag{4.52}$$

The realization is controllable and observable (Fig. 4.5).

4.3 IRREDUCIBLE REALIZATIONS

The realization methods considered in previous sections, although conceptually simple and convenient for hand calculations, have the disadvantage that they may require further transformations to yield the minimal representations. In the following derivations it is shown that the Smith-McMillan form of a rational transfer-function matrix may be constructively used to obtain a controllable and observable (i.e., minimal) realization. Before the realization procedure is developed, however, the concept of direct sum is first established.

4.3.1 Direct Sum

It has been shown earlier that for any m-input/ℓ-output nth-order system S(A, B, C) with repeated eigenvalues, there exists a similarity transformation which yields the irreducible* form $\bar{S}(\bar{A}, \bar{B}, \bar{C})$ defined by

*Controllable and observable (Sec. 3.2.4).

$$\bar{A} = T^{-1}AT; \quad \bar{B} = T^{-1}B; \quad \bar{C} = CT \tag{4.53}$$

Thus if A has q distinct eigenvalues $\lambda 1, \lambda 2, \ldots, \lambda k, \ldots, \lambda q$, with any λk being repeated nk times $\left(\Sigma_{k=1}^{q} nk = n\right)$, then the triple $(\bar{A}, \bar{B}, \bar{C})$ is given by (using notations slightly different from those in Sec. 3.1.2).

$$\bar{A} = \text{diag}\,\{J^k(\lambda k, nk)\}_{(n\times n)}, \quad k \in 1, \ldots, q;$$

$$\bar{B} = \begin{bmatrix} \bar{B}^1 \\ \vdots \\ \bar{B}^k \\ \vdots \\ \bar{B}^q \end{bmatrix}_{(n\times m)} ; \quad \bar{C} = [\bar{C}^1 \cdots \bar{C}^k \cdots \bar{C}^q]_{(\ell\times n)} \tag{4.54}$$

where $J^k(\lambda k, nk)$ is the Jordan block associated with the eigenvalue λk in A, $k \in 1, \ldots, q$, and has the form

$$J^k(\lambda k, nk) = \begin{bmatrix} \lambda k & 1 & 0 & 0 & 0 \\ 0 & \lambda k & 1 & 0 & 0 \\ \vdots & & & & \\ 0 & 0 & 0 & \lambda k & 1 \\ 0 & 0 & 0 & 0 & \lambda k \end{bmatrix}_{(nk\times nk)} \tag{4.55}$$

In the subsequent derivations the Jordan form in (4.54) is assumed to be irreducible, and consequently the input and the output matrices have full rank [i.e., $\rho(\bar{B}) = m$ and $\rho(\bar{C}) = \ell$]. The submatrices B^k and C^k contain the following m-row vectors (denoted by subscript r) and ℓ-column vectors (denoted by subscript c) [from (3.38) and (3.39)]:

$$\bar{B}^k = nk \underbrace{\begin{bmatrix} b^k_{r1} \\ \vdots \\ b^k_{rj} \\ \vdots \\ b^k_{rnk} \end{bmatrix}}_{m} \leftarrow \left\{ \begin{array}{l} \text{jth row of } \bar{B}^k\text{:} \\ \text{an m-row vector} \end{array} \right. \tag{4.56a}$$

$$\bar{C}^k = \overbrace{[C^k_{c1} \cdot C^k_{cj} \cdot C^k_{cnk}]}^{nk} \}\ \ell \tag{4.56b}$$

↑ jth column of $\bar{C}^k$: an ℓ-column vector

and

$$n = \sum_{k=1}^{q} nk \tag{4.56c}$$

Combining (4.54) and (4.55), the transfer-function matrix of the transformed system may be expressed as

$$G(s) = C(sI - A)^{-1}B = \sum_{k=1}^{q} \bar{C}^k[sI - J^k(\lambda k, nk)]^{-1}\bar{B}^k = \sum_{k=1}^{q} G_k(s) \tag{4.57}$$

where $G_k(s)$ is the transfer-function matrix of the kth subsystem $\bar{S}_k(J^k, \bar{B}^k, \bar{C}^k)$ with $\delta\{G_k(s)\} = nk$. The derivations above show that the vector space x^n is the direct sum* of the q vectors $x^{n1}, \ldots, x^{nk}, \ldots, x^{nq}$. Thus the composite state vector

$$\begin{bmatrix} x^{n1} \\ x^{n2} \\ \vdots \\ x^{nq} \end{bmatrix}$$

according to the definition in Sec. 1.1.4, is the state vector of the composite system $\bar{S}(\cdot)$ formed by the parallel connection of the subsystems $\bar{S}_k(\cdot)$, that is,

$$\begin{aligned} u(t) &= u_k(t) \\ y(t) &= \sum_{k=1}^{q} y_k(t) \end{aligned} \tag{4.58}$$

*In the general definition, neither the subsystems $\bar{S}_k(\cdot)$ nor their direct sum $\bar{S}(\cdot)$ need be in Jordan form. The transformation is used for computional convenience.

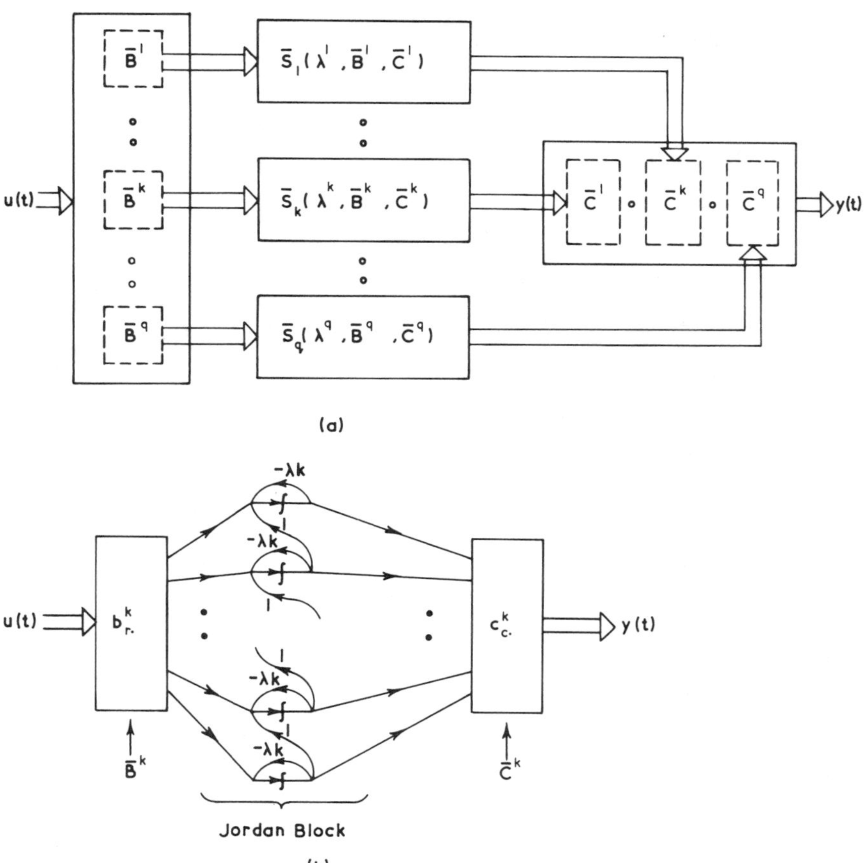

Fig. 4.6 Schematic representation of (a) Jordan form $\bar{S}(J, \bar{B}, \bar{C})$ in (4.54); (b) the kth Jordan block of $S_k(J_k, \bar{B}_k, \bar{C}_k)$ in (4.55) and (4.56).

The resulting representations are shown in Fig. 4.6; consequently, the dynamical system represented by (4.57) and (4.58) may also be expressed as

$$\begin{bmatrix} \dot{x}^{n1} \\ \vdots \\ \dot{x}_{nk} \\ \vdots \\ \dot{x}_{nq} \end{bmatrix} = \begin{bmatrix} J^1(\lambda 1, n1) & & & & \\ & \ddots & & & \\ & & J^k(\lambda k, nk) & & \\ & & & \ddots & \\ & & & & J^q(\lambda q, nq) \end{bmatrix} \begin{bmatrix} x^{n1} \\ \vdots \\ x^{nk} \\ \vdots \\ x^{nq} \end{bmatrix} + \begin{bmatrix} \bar{B}^1 \\ \vdots \\ \bar{B}^k \\ \vdots \\ \bar{B}^q \end{bmatrix} u(t)$$

$$y(t) = [\bar{C}^1 \cdots \bar{C}^k \cdots \bar{C}^q] \begin{bmatrix} x^{n1} \\ \vdots \\ x^{nk} \\ \vdots \\ x^{nq} \end{bmatrix} \tag{4.59}$$

Thus once the transfer function $G_k(s)$ for any subsystem and the associated repeated pole λk are identified, their state-space representations may be derived independently. Another consequence of the above is that the direct sum $\bar{S}(\cdot)$ is controllable (observable) only if each of its constituent subsystems $\bar{S}_k(\cdot)$ is controllable (observable).

4.3.2 Jordan Form [15-18]

The derivations below establish a correspondence between the foregoing diagonal representation in the time domain and the transfer function through the Smith-McMillan form (Sec. 1.5).

It is assumed that the transformed system in (4.57) has a strictly proper $(\ell \times m)$ transfer-function matrix*; then it can be expressed as

$$G(s) = L(s)M(s)R(s) \tag{4.60}$$

where $L(s)$ and $R(s)$ are unimodular matrices of dimensions $\ell \times \ell$ and $m \times m$, respectively, and $M(s)$ is the Smith-McMillan form of $G(s)$ given by

$$M(s) = \begin{array}{cc} \left[\begin{array}{c|c} \text{diag}\left\{\dfrac{\psi_k(s)}{\phi_k(s)}\right\} & 0 \\ \hline 0 & 0 \end{array}\right] & \begin{array}{l} \}\ q, \quad \delta\{\phi_k(s)\} = nk \\ \}\ \ell - q \end{array} \\ \begin{array}{cc} \underbrace{\qquad\qquad\quad}_{q} & \underbrace{\qquad}_{m-q} \end{array} & \end{array}$$

$$= \underbrace{\begin{array}{r} q\{ \\ \ell-q\{ \end{array}\left[\begin{array}{c|c} \text{diag}\{\psi_k(s)\} & 0 \\ \hline \underbrace{0}_{q} & \underbrace{0}_{m-q} \end{array}\right]}_{\Lambda(s)} \underbrace{\left[\begin{array}{c|c} \text{diag}\{\phi_k(s)\} & 0 \\ \hline \underbrace{0}_{q} & \underbrace{I}_{m-q} \end{array}\right]^{-1}}_{\Gamma(s)} \begin{array}{l} \}q, \ k \in 1, \ldots, q \\ \}m-q \end{array} \tag{4.61}$$

*This is generally valid, since the transmission matrix D can be computed directly as $D = \lim_{s \to \infty} G(s)$, and the strictly proper part of $G(s)$ easily identified.

Thus if $L^{cj}(s)$ denotes the jth column of L(s) and $R^{rj}(s)$ the jth row of R(s) for any $j \in 1, \ldots, \ell$, for L(s) and j = 1, ..., m for R(s), then (4.61) becomes

$$G(s) = L(s)[\Lambda(s)][\Gamma(s)]^{-1}R(s) = \{L^{cj}(s)\}[\mathrm{diag}\{\psi_{jk}(s)\}][\mathrm{diag}\{\phi_{jk}(s)\}]^{-1}\{R^{rj}(s)\}$$

$$= \left\{ L^{cj}(s)\psi_{jk}(s)\,\frac{1}{\phi_{jk}(s)}\,R^{rj}(s) \right\} \qquad (4.62)$$

where

(1) $\psi_{jk}(s)$ and $\phi_{jk}(s)$ represent the corresponding diagonal elements of $\Lambda(s)$ and $\Gamma(s)$, defined as below:

$$\psi_{jk}(s) = \begin{cases} \psi_k(s) & \text{for } j = k \in 1, 2, \ldots, q \\ 0 & \text{for } j = (q+1), \ldots, \ell \end{cases}$$

$$\phi_{jk}(s) = \begin{cases} \phi_k(s) & \text{for } j = k \in 1, 2, \ldots, q \\ 1 & \text{for } j = (q+1), \ldots, m \end{cases}$$

Consequently,

$$(2)\quad L^{cj}(s)\psi_{jk}(s) = \begin{cases} L^{cj}\psi_k(s) & \text{for } j = k \in 1, 2, \ldots, q \\ 0 & \text{for } j = (q+1), \ldots, m \end{cases} \qquad (4.63)$$

and since there is only one pole λk associated with any $\phi_k(s)$, $k \in 1, 2, \ldots, q$,

$$(3)\quad \phi_k(s) = (s - \lambda k)^{nk} \qquad (4.64)$$

where $\delta\{\phi_k(s)\} > \delta\{\phi_{k-1}(s)\}$ because of the divisibility property of the invariant polynomials of M(s). Thus, combining (4.62), (4.63), and (4.64), we have*

$$G(s) = \sum_{j=1} \frac{L^{cj}(s)\psi_{jk}(s)R^{rj}(s)}{\phi_{jk}(s)}$$

$$= \sum_{j=k=1}^{q} \frac{L^{cj}\psi_k(s)R^{rj}(s)}{\phi_k(s)} \qquad (4.65)$$

$$= \sum_{j=k=1}^{q} \frac{\bar{L}^{cj}(s)R^{rj}(s)}{\phi_k(s)}$$

where

*The summation being taken over j = 1 to max(m, ℓ).

$$\bar{L}^{cj}(s) = L^{cj}(s)\psi_k(s), \quad j = k = 1, 2, \ldots, q$$

Thus comparison of (4.57) and (4.65) yields*

$$G_k(s) = \frac{\bar{L}^{cj}(s)R^{rj}(s)}{\phi_k(s)} \quad \text{for } k = j = 1, 2, \ldots, q \tag{4.66}$$

To establish a correspondence between (4.54) and (4.65), the following identity is necessary:

$$[sI - J^k(\lambda k, nk)]^{-1} = \frac{1}{(s - \lambda k)^{nk}} \operatorname{adj}\{sI - J^k(\lambda k, nk)\}$$

$$= \frac{1}{\phi_k(s)} \begin{bmatrix} (s-\lambda k)^{nk-1} & (s-\lambda k)^{nk-2} & \cdots & 1 \\ 0 & (s-\lambda k)^{nk-1} & \cdots & s-\lambda k \\ \vdots & & & \\ 0 & 0 & & (s-\lambda k)^{nk-1} \end{bmatrix}$$

$$\equiv \frac{1}{\phi_k(s)} \begin{bmatrix} 1 \\ s-\lambda k \\ \vdots \\ (s-\lambda k)^{nk-1} \end{bmatrix} [(s-\lambda k)^{nk-1} \cdots s-\lambda k \;\; 1] \tag{4.67}$$

Since

$$G_k(s) = \bar{C}^k[sI - J^k(\lambda k, nk)]^{-1}\bar{B}^k \tag{4.68}$$

combination of (4.66) and (4.67) gives

$$\bar{L}^{cj}(s)R^{rj}(s) = \phi_k(s)G_k(s) = \bar{C}^k \underbrace{\begin{bmatrix} 1 \\ s-\lambda k \\ \vdots \\ (s-\lambda k)^{nk-1} \end{bmatrix}}_{\text{column matrix}} \underbrace{[(s-\lambda k)^{nk-1} \cdots s-\lambda k \;\; 1]}_{\text{row matrix}}\bar{B}^k \tag{4.69}$$

*Each $G_k(s)$ is of order $\ell \times m$.

Thus, equating matrices of similar dimensions gives us

$$\bar{L}^{cj}(s) = \bar{C}^{k} \begin{bmatrix} 1 \\ s-\lambda k \\ \vdots \\ (s-\lambda k)^{nk-1} \end{bmatrix}; \quad j = k = 1, 2, \ldots, q \qquad (4.70a)$$

and

$$R^{rj}(s) = [(s-\lambda_k)^{nk-1} \cdots s-\lambda_k \quad 1]\bar{B}^{k}; \quad j = k = 1, 2, \ldots, q \qquad (4.70b)$$

since $\bar{S}_k(J_k, \bar{B}_k, \bar{C}_k)$, $k \in 1, 2, \ldots, q$, are assumed to be controllable and observable, the derivation above provides a procedure for irreducible realization of a given G(s). A few illustrative examples are given below to highlight the main features of this procedure [15, 18].

Example 4.10: System with multiple poles at the origin

$$G(s) = \frac{1}{s^4} \begin{bmatrix} s^3 - s^2 + 1 & 1 & -s^3 + s^2 - 2 \\ 1.5s + 1 & s + 1 & -1.5s - 2 \\ s^3 - 9s^2 - s + 1 & -s^2 + 1 & s^3 - s - s \end{bmatrix} \qquad (4.71)$$

The degree of G(s) can be seen to be 8, and hence the order of irreducible realization (n_{min}) is 8.

Stage 1—Transformation of G(s) into its Smith-McMillan form:

$$G(s) = L(s)M(s)R(s) \equiv L(s)\Lambda(s)[\Gamma(s)]^{-1}R(s)$$

For the system in (4.71),

$$M(s) = \begin{bmatrix} \frac{1}{s^4} & 0 & 0 \\ 0 & \frac{1}{s^3} & 0 \\ 0 & 0 & \frac{1}{s} \end{bmatrix}$$

$$L(s) = \begin{bmatrix} 1 & 0 & 0 \\ 1+s & 1 & 0 \\ 1-s^2 & -2-10s+2s^2 & 1 \end{bmatrix}$$

$$R(s) = \begin{bmatrix} s^3 - s^2 + 1 & 1 & -s^3 + s^2 - 2 \\ -s^3 + s + 0.5 & 0 & s^3 - s + 0.5 \\ 2s^3 - 9s - 5s + 9 & 0 & -2s^3 + 9s^2 + 5s - 9 \end{bmatrix}$$

Therefore,

$$\Lambda(s) = \text{diag}\{\psi_j(s)\} = I_3$$

$$\Gamma(s) = \text{diag}\{\phi_j(s)\} = \text{diag}\{s^4, s^3, s\}$$

Stage 2—Identification of the number of Jordan blocks: For this system, there are three Jordan blocks, $J(0,4)$, $J(0,3)$, and $J(0,1)$, corresponding to $\phi_1(s)$, $\phi_2(s)$, and $\phi_3(s)$ in $\Gamma(s)$.

Stage 3—Realization of the individual blocks: In view of (4.69) and (4.70), the three blocks may be realized independently.

Block 1: $J(0,4)$: $n1 = 4$, $\lambda 1 = 0$

$$\bar{C}^1(sI - J_1)^{-1}\bar{B}^1 = \bar{C}^1 \begin{bmatrix} 1 \\ s \\ s^2 \\ s^3 \end{bmatrix} [s^3 \quad s^2 \quad s \quad 1] \left| \frac{1}{s^4} \right| \bar{B}^1$$

$$\equiv \underbrace{\bar{C}^1 \begin{bmatrix} 1 \\ s \\ s^2 \\ s^3 \end{bmatrix}}_{L^{c1}(s)} \underbrace{\left| \frac{1}{s^4} \right|}_{\frac{1}{\phi_1(s)}} \underbrace{[s^3 \quad s^2 \quad s \quad 1]\bar{B}^1}_{R^{r1}(s)} \tag{4.72}$$

where $L^{c1}(s)$ and $R^{r1}(s)$ are the first column and the first row of $L(s)$ and $R(s)$, respectively, and

$$\bar{C}^1 = \{c^1_{ij}\}$$
$$\bar{B}^1 = \{b^1_{ij}\} \tag{4.73}$$

Combining (4.72) and (4.73) gives us

$$\begin{bmatrix} c^1_{11} & c^1_{12} & c^1_{13} & c^1_{14} \\ c^1_{21} & c^1_{22} & c^1_{23} & c^1_{24} \\ c^1_{31} & c^1_{32} & c^1_{33} & c^1_{34} \end{bmatrix} \begin{bmatrix} 1 \\ s \\ s^2 \\ s^3 \end{bmatrix} = L^{c1}(s) = \begin{bmatrix} 1 \\ 1+s \\ 1-s^2 \end{bmatrix}$$

Comparison of terms of with similar powers of s gives

$$\bar{C}^1 = \begin{bmatrix} 1 & 0 & 0 & 0 \\ 1 & 1 & 0 & 0 \\ 1 & 0 & -1 & 0 \end{bmatrix}$$

Similarly,

$$[s^3 \quad s^2 \quad s \quad 1] \begin{bmatrix} b^1_{11} & b^1_{12} & b^1_{13} \\ b^1_{21} & b^1_{22} & b^1_{23} \\ b^1_{31} & b^1_{32} & b^1_{33} \\ b^1_{41} & b^1_{42} & b^1_{43} \end{bmatrix} = [s^3 - s^2 + 1 \quad 1 \quad -s^3 + s^2 - 2]$$

which yields

$$\bar{B}^1 = \begin{bmatrix} 1 & 0 & -1 \\ -1 & 0 & 1 \\ 0 & 0 & 0 \\ 1 & 1 & -2 \end{bmatrix}$$

<u>Block 2</u>: $J_2(0,3)$, $n2 = 3$, $\lambda 2 = 0$. The procedure is as before and uses the following equations:

$$[\bar{C}^2] \begin{bmatrix} 1 \\ s \\ s^2 \end{bmatrix} = \bar{L}^{c2}(s) \quad \text{and} \quad [s^2 \quad s \quad 1][\bar{B}^2] = R^{r2}(s)$$

giving

$$\bar{C}^2 = \begin{bmatrix} c^2_{11} & c^2_{12} & c^2_{13} \\ c^2_{21} & c^2_{22} & c^2_{23} \\ c^2_{31} & c^2_{32} & c^2_{33} \end{bmatrix} = \begin{bmatrix} 0 & 0 & 0 \\ 1 & 0 & 0 \\ -2 & -10 & 2 \end{bmatrix}$$

$$\bar{B}^2 = \begin{bmatrix} b_{11}^2 & b_{12}^2 & b_{13}^2 \\ b_{21}^2 & b_{22}^2 & b_{23}^2 \\ b_{31}^2 & b_{32}^2 & b_{33}^2 \end{bmatrix} = \begin{bmatrix} 0 & 0 & 0 \\ 1 & 0 & -1 \\ 0.5 & 0 & 0.5 \end{bmatrix}$$

Block 3: $J_3(0,1)$, $n3 = 1$, $\lambda 3 = 0$. The equations are

$$[\bar{C}^3][1] = \bar{L}^{c3}(s) \quad \text{and} \quad [1][\bar{B}^3] = R^{r2}(s)$$

which when equated for similar powers of s give

$$\bar{C}^3 = \begin{bmatrix} c_{11}^3 \\ c_{21}^3 \\ c_{23}^3 \end{bmatrix} = \begin{bmatrix} 0 \\ 0 \\ 1 \end{bmatrix} \quad \text{and} \quad \bar{B}^3 = [b_{11}^3 \;\; b_{12}^3 \;\; b_{13}^3] = [9 \;\; 0 \;\; -9]$$

Comparison of the various blocks of $\bar{B}^j$ and $\bar{C}^j$ suggest that terms marked † in R(s) do not appear in $\bar{B}$ and $\bar{C}$. The minimal realization of G(s) is therefore given by

$$\bar{A} = \text{diag}\,\{J^k(\lambda k, nk)\} = \left[\begin{array}{cccc|ccc|c} 0 & 1 & 0 & 0 & & & & \\ 0 & 0 & 1 & 0 & & 0 & & \\ 0 & 0 & 0 & 1 & & & & \\ 0 & 0 & 0 & 0 & & & & \\ \hline & & & & 0 & 1 & 0 & \\ & 0 & & & 0 & 0 & 1 & \\ & & & & 0 & 0 & 0 & \\ \hline & & & & & & & 0 \end{array}\right] \tag{4.74a}$$

$$\bar{B} = \begin{bmatrix} \bar{B}^1 \\ \bar{B}^2 \\ \bar{B}^3 \end{bmatrix} = \left[\begin{array}{ccc} 1 & 0 & -1 \\ -1 & 0 & 1 \\ 0 & 0 & 0 \\ 1 & 1 & -2 \\ \hline 0 & 0 & 0 \\ 1 & 0 & -1 \\ -0.5 & 0 & 0.5 \\ \hline 9 & 0 & -9 \end{array}\right] \tag{4.74b}$$

$$\bar{C} = [\bar{C}^1 \;\; \bar{C}^2 \;\; \bar{C}^3] = \left[\begin{array}{cccc|ccc|c} 1 & 0 & 0 & 0 & 0 & 0 & 0 & 0 \\ 1 & 1 & 0 & 0 & 1 & 0 & 0 & 0 \\ 1 & 0 & -1 & 0 & -2 & -10 & 2 & 1 \end{array}\right] \quad (4.74c)$$

which can be seen to be controllable $[\rho(\bar{B}) = 3]$ and observable $[\rho(\bar{C}) = 3]$.

A close observation of the derivation above would suggest that realization could almost be derived by inspecting the unimodular matrices L(s) and R(s). This, although a strength of the procedure, may not be possible when the multiple poles are not at the origin ($\lambda k \neq 0$) and the order of multiplicity (nk) is high. The problems of deriving the irreducible realization by hand calculations are highlighted through the following example.

Example 4.11: Irreducible realization of G(s)

$$G(s) = \begin{bmatrix} \dfrac{s+3}{(s+1)^3} & \dfrac{(s+1)(s+3)}{(s+1)^3} \\ \dfrac{s(s+3)}{(s+1)^3} & \dfrac{s(s+3)}{(s+1)^2} - \dfrac{s}{(s+2)^2} \end{bmatrix} \quad (4.75)$$

Stage 1: The Smith-McMillan form of G(s) is given by

$$G(s) = \begin{bmatrix} 1 & 0 \\ s & 1 \end{bmatrix} \begin{bmatrix} \dfrac{s+3}{(s+1)^3} & 0 \\ 0 & \dfrac{s}{(s+2)^2} \end{bmatrix} \begin{bmatrix} 1 & s+1 \\ 0 & -1 \end{bmatrix}$$

$$= \begin{bmatrix} 1 & 0 \\ s & 1 \end{bmatrix} \begin{bmatrix} s+3 & 0 \\ 0 & s \end{bmatrix} \begin{bmatrix} (s+1)^3 & 0 \\ 0 & (s+2)^2 \end{bmatrix}^{-1} \begin{bmatrix} 1 & s+1 \\ 0 & -1 \end{bmatrix}$$

$$= \underbrace{\begin{bmatrix} s+3 & 0 \\ s(s+3) & s \end{bmatrix}}_{\bar{L}(s)} \underbrace{\begin{bmatrix} (s+1)^3 & 0 \\ 0 & (s+2)^2 \end{bmatrix}^{-1}}_{\Gamma(s)} \underbrace{\begin{bmatrix} 1 & s+1 \\ 0 & -1 \end{bmatrix}}_{R(s)}$$

Stage 2: There are two Jordan blocks, J(-1,3) and J(-2,2), corresponding to $\phi_1(s) = (s+1)^3$ and $\phi_2(s) = (s+2)^2$ in $\Gamma(s)$.

Stage 3: Realization of blocks 1 and 2.

Block 1: The equations are

$$\bar{C}^1\begin{bmatrix}1\\ s+1\\ (s+1)^2\end{bmatrix} = \bar{L}^{c1}(s) \rightarrow \begin{bmatrix}c_{11}^1 + c_{12}^1(s+1) + c_{13}^1(s+1)^2\\ c_{21}^1 + c_{22}^1(s+1) + c_{23}^1(s+1)^2\end{bmatrix} = \begin{bmatrix}s+3\\ s(s+3)\end{bmatrix}$$

and

$$[(s+1)^2 \quad s+1 \quad 1]\bar{B}^1 = R^{r1}(s)$$
$$\rightarrow [(s+1)^2 b_{11}^1 + (s+1)b_{21}^1 + b_{31}^1 \mid (s+1)^2 b_{12}^1 + (s+1)b_{22}^1 + b_{32}^1] = [1 \mid s+1]$$

As can be seen, solutions for the elements of $\bar{C}^1$ and $\bar{B}^1$ may not be easy to obtain without the aid of any numerical algorithm. The solutions for this case are

$$\bar{C}^1 = \begin{bmatrix}2 & 1 & 0\\ -2 & 1 & 1\end{bmatrix}; \quad \bar{B}^1 = \begin{bmatrix}0 & 0\\ 0 & 1\\ 1 & 0\end{bmatrix}$$

Block 2: The input/output matrices are derived as above:

$$\bar{C}^2 = \begin{bmatrix}0 & 0\\ -2 & 1\end{bmatrix}; \quad \bar{B}^2 = \begin{bmatrix}0 & -1\\ 0 & 0\end{bmatrix}$$

The realization is therefore given by

$$\bar{A} = \text{diag}\{J^k(\lambda k, nk)\} = \left[\begin{array}{ccc|cc}-1 & 1 & 0 & & \\ 0 & -1 & 1 & & 0\\ 0 & 0 & -1 & & \\ \hline & & & -2 & 1\\ & 0 & & 0 & -2\end{array}\right] \tag{4.76a}$$

$$\bar{B} = \begin{bmatrix}\bar{B}^1\\ \bar{B}^2\end{bmatrix} = \left[\begin{array}{cc}0 & 0\\ 0 & 1\\ 1 & 0\\ \hline 0 & -1\\ 0 & 0\end{array}\right] \tag{4.76b}$$

$$\bar{C} = [\bar{C}^1 \ \bar{C}^2] = \left[\begin{array}{ccc|cc}2 & 1 & 0 & 0 & 0\\ -2 & 1 & 1 & -2 & 1\end{array}\right] \tag{4.76c}$$

For the given system $\delta\{G(s)\} = 5 = \delta\{\Gamma(s)\} = n_{min}$, and therefore the realization above is minimal.

4.3.3 Companion Form [15-18]

The realization procedure in the preceding section, while simple and fairly convenient for hard calculations, has the limitation that any denominator polynomial $\phi_j(s)$ should contain only one root. The concept of this special case is extended below to remove any structural constraint on the Smith-McMillan form of a proper rational G(s). The derivation of a generalized irreducible algorithm uses the rather complex modulo theory of modern algebra.* For the sake of brevity, only the definition of modulo and the basic equations for realization are given, without any detailed derivation of the abstract theory.

Definition 4.2: Modulo Two polynomials $\alpha = \alpha(s)$ and $\beta = \beta(s)$ are said to be congruent modulo the polynomial $\phi = \phi(s)$ [written as $\alpha \equiv \beta \pmod{\phi}$] if α and β have the same remainder after division by ϕ.

Similarly, two polynomial matrices P = P(s) and Q = Q(s) are said to be congruent modulo $\phi = \phi(s)$ ($\equiv$ a polynomial) [written as $P \equiv Q \pmod{\phi}$], provided that the corresponding elements of P and Q have the same remainders after division of ϕ. Thus $P = Q \pmod{\phi}$ only if there exists a polynomial matrix R = R(s) such that $P = Q + \phi R$.

Definition 4.3: Equivalence† Two rational matrices, $G_1 = G_1(s)$ and $G_2 = G_2(s)$, of appropriate dimensions with least common denominators for each of their entries as $\phi_1 = \phi_1(s)$ and $\phi_2 = \phi_2(s)$, respectively, are said to be equivalent if $\phi_1 = \phi_2$ and there exist two unimodular matrices $U_1(s)$ and $U_2(s)$ such that $\phi G_2 = U_1(\phi G_1)U_2 \pmod{\phi}$, where $\phi = \phi_1 = \phi_2$.

Since two linear time-invariant systems are said to be equivalent (Sec. 3.4) if their transfer functions are equivalent, Definition 4.3 forms another definition of equivalence. This alternative definition, together with the derivations in Secs. 4.3.1 and 4.3.2, forms the basis of the realization procedure below.

Since the system S(A, B, C) is assumed to have nonrepeated eigenvalues, its controllable and observable form $\hat{S}(\hat{A}, \hat{B}, \hat{C})$ may be assumed to consist of a set of controllable and observation subsystems $\hat{S}_i(\hat{A}, \hat{B}_i, \hat{C}_i)$ with the corresponding $\hat{A}_i$ being expressed in the companion form:‡

*G. Birkhoff, Survey of Modern Algebra, Macmillan, New York, 1953.
†Called weakly equivalent in Ref. 18.
‡Controllable canonical form; subscript i is used here to distinguish this representation from the canonical form in the previous section.

$$\hat{A}_i = \begin{bmatrix} 0 & 1 & \cdots & 0 \\ 0 & 0 & \cdots & 0 \\ 0 & 0 & \cdots & 1 \\ -\alpha_0^i & -\alpha_1^i & \cdots & -\alpha_{ni-1}^i \end{bmatrix} \tag{4.77}$$

associated with the characteristic (minimal) polynomial

$$\phi_i(s) = s^{ni} + \alpha_{ni-1}^i s^{ni-1} + \alpha_{ni-2}^i s^{ni-2} + \cdots + \alpha_0^i$$

It can then be easily shown that for any $ni < n$ (= the order of $\hat{A}$) [18]

$$\operatorname{adj}[sI - \hat{A}_i] = \begin{bmatrix} 1 \\ s \\ \vdots \\ s^{ni-1} \end{bmatrix} \underbrace{[\gamma_0(s) \;\; \gamma_1(s) \cdots \gamma_{ni-1}(s)]}_{(ni\,-\,1)}$$

$$- \phi_i(s) \left.\underbrace{\begin{bmatrix} 0 & 0 & \cdots & 0 & 0 \\ 1 & 0 & \cdots & 0 & 0 \\ s & 1 & \cdots & 0 & 0 \\ s^{ni-2} & s^{ni-3} & \cdots & 1 & 0 \end{bmatrix}}_{(ni\,-\,1)}\right\} (ni - 1) \tag{4.78}$$

where $\gamma_k(s)$ is a polynomial given by

$$\gamma_k(s) = s^{ni-1-k} + \alpha_{ni-1-k} s^{ni-2-k} + \cdots + \alpha_k s + \alpha_{k+1}; \quad k \in 0, 1, \ldots, ni-1$$

Using the definition of modulo, the equation above may be expressed as

$$\operatorname{adj}[sI - \hat{A}_i] = \begin{bmatrix} 1 \\ s \\ \vdots \\ s^{ni-1} \end{bmatrix} [\gamma_0(s) \;\; \gamma_1(s) \cdots \gamma_{ni-1}(s)] \pmod{\phi_i}$$

The transfer function $G_i(s)$ of the ith subsystem $\hat{S}_i(\hat{A}_i, \hat{B}_i, \hat{C}_i)$ may therefore be expressed as

$$\phi_i(s)G_i(s) = \hat{C}_i \begin{bmatrix} 1 \\ s \\ \vdots \\ s^{ni-1} \end{bmatrix} [\gamma_0(s) \ \ \gamma_1(s) \cdots \gamma_{ni-1}(s)]\hat{B}_i \pmod{\phi_i} \tag{4.79}$$

The definition of modulo is now used to establish a relationship between (4.79) and its Smith-McMillan form. Since G(s) is proper rational, it can be expressed as

$$G(s) = \{G_i(s)\} = L(s)M(s)R(s) = \{L_i^c(s)\} \left[\text{diag} \left\{\frac{\psi_i(s)}{\phi_i(s)}\right\}\right] \{R_i^r(s)\}$$

or

$$G_i(s) = L_i^c(s) \frac{\psi_i(s)}{\phi_i(s)} R_i^r(s); \qquad i \in 1, \ldots, q \text{ [q as in (4.61)]} \tag{4.80}$$

where L_i^c is a column matrix and $R_i^r(s)$ is a row matrix. By the definition of modulo, then from (4.80),

$$\phi_i G_i = L_i^c \left\{\phi_i \left(\frac{\psi_i}{\phi_i}\right)\right\} R_i^r \pmod{\phi_i} \tag{4.81}$$

Thus combining (4.79) and (4.81) gives

$$L_i^c(s)\,(\psi_i(s))\,R_i^r(s) \pmod{\phi_i} \equiv \hat{C}_i \begin{bmatrix} 1 \\ s \\ \vdots \\ s^{ni-1} \end{bmatrix} [\gamma_0(s) \ \ \gamma_1(s) \cdots \gamma_{ni-1}(s)]\hat{B}_i \pmod{\phi_i} \tag{4.82}$$

which may be split into two separate equations:

$$\hat{C}_i \begin{bmatrix} 1 \\ s \\ \vdots \\ s^{ni-1} \end{bmatrix} = L_i^c(s)\psi_i(s) \pmod{\phi_i} \tag{4.83a}$$

$$[\gamma_0(s) \ \ \gamma_1(s) \cdots \gamma_{ni-1}(s)]\hat{B}_i = R_i^r(s) \pmod{\phi_i} \tag{4.83b}$$

Therefore, combining the derivations above and the definition of equivalence, we have

$$\{\phi_i G_i\} \equiv \{L_i^c(s)\}\{\phi_i M_i\}\{R_i^r(s)\} \pmod{\phi_i}$$

or

$$\begin{aligned} \phi G &\equiv L[\phi M]R \pmod{\phi} \\ &\equiv L[M']R \pmod{\phi} \end{aligned} \tag{4.83c}$$

which is consistent, since M is the Smith-McMillan form of G if ϕ is assumed to be the least common multiplier of all the denominators of M [i.e., if $\phi(s)$ is the pole polynomial of G(s) or the minimal polynomial of A].

The significance of (4.83) is that while G is equivalent to its Smith-McMillan form M, ϕG is equivalent to ϕM, where ϕM is obtained by replacing the numerator of each element of M by its remainder after division by the corresponding denominator. This follows from the definition of modulo, a further interpretation of which is as follows: If α/β is any quotient of polynomials with β dividing ϕ, then the remainder after dividing $\phi(\alpha/\beta)$ by ϕ is just $\phi(\alpha'/\beta)$, where α' is the remainder after dividing α by β.

A comparison of the results above and the derivation of the repeated-eigenvalue case would suggest that the use of (4.83) stems from the need of establishing a conceptually simple mechanism of obtaining the inverse of $(sI - \hat{A})$, where $\hat{A} = \text{diag}\{\hat{A}_i\}$, $\hat{A}_i$ being in companion form. As a consequence, the resulting realization represents $\phi(s)M(s)$, which by Definition 4.3 is equivalent to M(s) and hence is equivalent to a realization of G(s). This method of realization is illustrated below [18]. The order of minimal realization in this case is

$$n_{min} = \sum_{i=1}^{q} \delta\{\phi_i(s)\} = \sum_{i=1}^{q} ni$$

Example 4.12: Irreducible realization of G(s)

$$G(s) = \begin{bmatrix} \dfrac{1}{s(s+2)} & \dfrac{2s-1}{s(s+2)} & \dfrac{s-1}{s(s+2)} \\ -\dfrac{1}{s+2} & -\dfrac{s-1}{(s+1)(s+2)} & \dfrac{1}{(s+1)(s+2)} \end{bmatrix} \tag{4.84}$$

Step 1: For this system, $\phi = \phi(s) = s(s+1)(s+2) = s^3 + \underset{\overset{\uparrow}{\alpha_2}}{3}s^2 + \underset{\overset{\uparrow}{\alpha_1}}{2}s$, and the Smith-McMillan form is given by

$$G(s) = \underbrace{\begin{bmatrix} 1 & 0 \\ -s & 1 \end{bmatrix}}_{L(s)} \underbrace{\begin{bmatrix} \frac{1}{s^2+2s} & 0 & 0 \\ 0 & \frac{s^2}{s^2+3s+2} & 0 \end{bmatrix}}_{M(s)} \underbrace{\begin{bmatrix} 1 & 2s-1 & s-1 \\ 0 & 2 & 1 \\ 0 & 1 & 0 \end{bmatrix}}_{R(s)} \tag{4.85}$$

From (4.83),

$$\phi(s)G(s) = L(s)[\phi(s)M(s)]R(s) \pmod{\phi} \tag{4.86}$$

where

$$[\phi(s)M(s)] = \begin{bmatrix} \frac{1}{s^2+2s} & 0 & 0 \\ 0 & \frac{-3s-2}{s^2+3s+2} & 0 \end{bmatrix} \tag{4.87}$$

(It can be seen that (4.87) represents the proper part of M(s)—which provides another way of deriving $[\phi(s)M(s)]$).

Step 1: Once (4.86) is established, realization of the subsystems corresponding to the q subsystems $[G_i(s)]$ follows the same procedure as for repeated eigenvalue case.

Block 1: Second-order system, since $\phi_1(s) = s^2 + 2s \equiv s^2 + \alpha_1^1 s + \alpha_0^1$; therefore,

$$\hat{A}_1 = \begin{bmatrix} 0 & 1 \\ 0 & -2 \end{bmatrix} \rightarrow n1 = 2 \tag{4.88a}$$

and as the system in (4.84) has three inputs and two outputs, the corresponding $\hat{B}^1$ and $\hat{C}^1$ may be assumed to be

$$\hat{B}_1 = \begin{bmatrix} b_{11}^1 & b_{12}^1 & b_{13}^1 \\ b_{21}^1 & b_{22}^1 & b_{23}^1 \end{bmatrix} \quad \text{and} \quad \hat{C}_1 = \begin{bmatrix} c_{11}^1 & c_{12}^1 \\ c_{21}^1 & c_{22}^1 \end{bmatrix} \tag{4.88b}$$

Using (4.83), (4.85), and (4.88), we obtain

$$\begin{bmatrix} c_{11}^1 & c_{12}^1 \\ c_{21}^1 & c_{22}^1 \end{bmatrix} \begin{bmatrix} 1 \\ s \end{bmatrix} = L_1^c \psi_1(s) \pmod{\phi_1} = \begin{bmatrix} 1 \\ -s \end{bmatrix} [1]$$

and equating coefficients of equal powers of s, this gives

$$\hat{C}_1 = \begin{bmatrix} 1 & 0 \\ 0 & -1 \end{bmatrix}$$

Also,

$$[\gamma_0(s) \quad \gamma_1(s)] \begin{bmatrix} b_{11}^1 & b_{12}^1 & b_{13}^1 \\ b_{21}^1 & b_{22}^1 & b_{23}^1 \end{bmatrix} = R_1^r(s) \ (\mathrm{mod}\ \phi_1) = [1 \quad 2s-1 \quad s-1]$$

where

$$\gamma_k(s) = s^{n1-1-k} + \alpha_{n1-1-k}^1 s^{n1-2-k} + \cdots + \alpha_{k+1}^1; \quad k = 0, 1, \ldots, n1-1$$

$$n1 = 2; \quad \gamma_0(s) = s + \alpha_1^1 = s + 2; \quad \gamma_1(s) = 1$$

Therefore, from above, substituting the corresponding values gives us

$$[s+2 \quad 1] \begin{bmatrix} b_{11}^1 & b_{12}^1 & b_{13}^1 \\ b_{21}^1 & b_{22}^1 & b_{23}^1 \end{bmatrix} = [1 \quad 2s-1 \quad s-1]$$

which gives

$$\hat{B}_1 = \begin{bmatrix} 0 & 2 & 1 \\ 1 & -5 & -3 \end{bmatrix}$$

<u>Block 2</u>: $\phi_2(s) = s^2 + 3s + 2 \equiv s^2 + \alpha_1^2 s + \alpha_0^2$; therefore, n2 = 2 and

$$\hat{A}_2 = \begin{bmatrix} 0 & 1 \\ -2 & -3 \end{bmatrix}$$

and let

$$\hat{B}_2 = \begin{bmatrix} b_{11}^2 & b_{12}^2 & b_1^2 \\ b_{21}^2 & b_{22}^2 & b_2^2 \end{bmatrix} \quad \text{and} \quad \hat{C}_2 = \begin{bmatrix} \hat{c}_{11}^2 & \hat{c}_{12}^2 \\ \hat{c}_{21}^2 & \hat{c}_{22}^2 \end{bmatrix}$$

Following the same procedure as above:

(1) $$\begin{bmatrix} c_{11}^2 & c_{12}^2 \\ c_{21}^2 & c_{22}^2 \end{bmatrix}\begin{bmatrix} 1 \\ s \end{bmatrix} = L_2^c\psi_2(s) \pmod{\phi_2} = \begin{bmatrix} 0 \\ 1 \end{bmatrix}[-3s \quad -2]$$

which gives

$$\hat{C}_2 = \begin{bmatrix} 0 & 0 \\ -2 & -3 \end{bmatrix}$$

(2) For this block,

$$\gamma_k(s) = s^{n2-1-k} + \alpha^2_{n1-1-k} s^{n2-2-k} + \cdots + \alpha^2_{k+1}$$

Therefore, for n2 = 2, and k = 0, ... n2 - 1 = 0, 1

$$\gamma_0(s) = s + \alpha_1^2 = s + 3; \quad \text{and} \quad \gamma_1(s) = 1$$

Thus by substituting the above in (4.83), we have

$$[\gamma_0(s) \ \ \gamma_1(s)]\hat{B}_2 = R_2^r(s) \pmod{\phi_2}$$

or

$$[s+3 \quad 1]\begin{bmatrix} b_{11}^2 & b_{12}^2 & b_1^2 \\ b_{21}^2 & b_{22}^2 & b_2^2 \end{bmatrix} = [0 \quad 2 \quad 1]$$

giving

$$\hat{B}_2 = \begin{bmatrix} 0 & 0 & 0 \\ 0 & 2 & 1 \end{bmatrix}$$

This completes the computation since the third column of $[\psi(s)M(s)]$ is null. The complete irreducible realization of G(s) in (4.84) is therefore $\hat{S}(\hat{A}, \hat{B}, \hat{C})$, where

$$\hat{A} = \begin{bmatrix} \hat{A}_1 & 0 \\ 0 & \hat{A}_2 \end{bmatrix} = \left[\begin{array}{cc|cc} 0 & 1 & & \\ 0 & -2 & \multicolumn{2}{c}{0} \\ \hline & & 0 & 1 \\ \multicolumn{2}{c|}{0} & -2 & -3 \end{array}\right] \tag{4.89a}$$

$$\hat{B} = \begin{bmatrix} \hat{B}_1 \\ \hat{B}_2 \end{bmatrix} = \begin{bmatrix} 0 & 2 & 1 \\ 1 & -5 & -3 \\ 0 & 0 & 0 \\ 0 & 2 & 1 \end{bmatrix} \tag{4.89b}$$

$$\hat{C} = [\hat{C}_1 \quad \hat{C}_2] = \begin{bmatrix} 1 & 0 & 0 & 0 \\ 0 & -1 & -2 & -3 \end{bmatrix} \tag{4.89c}$$

[If G(s) were proper, the transmission matrix D would be calculated from $D = \lim_{s \to \infty} G(s)$.]

4.4 REALIZATION FROM MATRIX FRACTIONS [3, 6, 10, 11, 19-25]

The realization procedures described in preceding sections are based on a proper rational matrix G(s). These methods, which are applicable to a majority of engineering systems where transfer functions (matrices) are obtainable from input/output data, are not directly applicable to systems that are described mathematically through a set of differential equations (assuming zero initial conditions)

$$D(s)y(s) = N(s)u(s) \tag{4.90a}$$

where u(s) is the m-input control, y(s) is the ℓ-output variable, and N(s) and D(s) are polynomial matrices of appropriate dimensions, respectively. Furthermore, a realization procedure for (4.90a) provides an alternative method when the available system description is in one of the two forms*

$$y(s) = \begin{cases} N_r(s)[D_r(s)]^{-1}u(s) \\ [D_\ell(s)]^{-1}N_\ell(s)u(s) \end{cases} \tag{4.90b}$$

where N(s) and D(s) are relatively prime and $\delta\{N(s)\} \not> \delta\{D(s)\}$ in each case. The realization methods from this matrix-fraction description of multivariable systems are based on some basic relationships that may be derived from the results of rational matrices outlined in Sec. 1.5; these are considered in the following section. The subsequent sections use these results to develop methods of obtaining canonical-form realizations from a given description of the form in (4.90).

*The dimensions are $N_r(s)$: $\ell \times m$; $D_r(s)$: $m \times m$ and $D_\ell(s)$: $\ell \times \ell$; $N_\ell(s)$: $\ell \times m$.

4.4.1 Matrix-Fraction Description

The derivations here are based on the concept of row/column reduction of polynomial matrices (Sec. 1.5) and on the assumption that for any linear system, the polynomial matrix associated with the output variable has full rank [i.e., $\rho[D(s)] = \min(\ell, m)$] and that it can be reduced to a row-proper form, as in (1.107).

Theorem 4.4 Any m-square polynomial matrix D(s) is equivalent to the following polynomial matrices:

(1) $\Gamma_r \operatorname{diag}\{s^{dri}\} + \operatorname{diag}\{1, s, \ldots, s^{dri-1}\} L_r$

(2) $\Gamma_c \operatorname{diag}\{s^{dcj}\} + L_c \operatorname{diag}\left\{\begin{matrix} 1 \\ s \\ \vdots \\ s^{dcj-1} \end{matrix}\right\}$

where the constant square matrices Γ and L have the following dimensions: Γ_r, Γ_c: (m × m) and nonsingular; L_r: (Σ dri × Σ dri); and L_c(Σ dcj × Σ dcj)— dri and dcj being the highest degrees in any ith row ($i \in 1, \ldots, m$) and jth column ($j \in 1, \ldots, m$) of D(s). The structure of the various diagonal matrices is

$$\operatorname{diag}\{s^{dri}\} = \begin{bmatrix} s^{dr1} & & & & \\ & \ddots & & & \\ & & s^{dri} & & \\ & & & \ddots & \\ & & & & s^{drm} \end{bmatrix}$$

$$\operatorname{diag}\{1, s, \ldots, s^{dri}\} = \begin{bmatrix} \boxed{1, s, \ldots, s^{dr1}} & & & & \\ & \ddots & & & \\ & & \boxed{1, s, \ldots, s^{dri}} & & \\ & & & \ddots & \\ & & & & \boxed{1, s, \ldots, s^{drm}} \end{bmatrix} = \psi_r(s)$$

$$\operatorname{diag}\{s^{dcj}\} = \begin{bmatrix} s^{dc1} & & & & \\ & \ddots & & & \\ & & s^{dcj} & & \\ & & & \ddots & \\ & & & & s^{dcm} \end{bmatrix}$$

$$\operatorname{diag}\left\{\begin{matrix} 1 \\ \vdots \\ s \\ \vdots \\ s^{dcj} \end{matrix}\right\} = \begin{bmatrix} \begin{matrix} 1 \\ s \\ \vdots \\ s^{dc1} \end{matrix} & & & & \\ & \ddots & & & \\ & & \begin{matrix} 1 \\ s \\ \vdots \\ s^{dcj} \end{matrix} & & \\ & & & \ddots & \\ & & & & \begin{matrix} 1 \\ s \\ \vdots \\ s^{dcm} \end{matrix} \end{bmatrix} = \psi_c(s)$$

Proof: Since any given N(s) may be made to be row or column proper through elementary row or column operation:

$$N(s) \equiv N_r(s)U_r(s) \quad \text{row operation} \tag{4.91}$$

$$\equiv U_c(s)N_r(s) \quad \text{column operation} \tag{4.92}$$

where $U_r(s)$ and $U_c(s)$ are unimodular matrices; proof of the theorem then follows directly if the matrices $L_r(s)$ in (1.107) and $L_c(s)$ in (1.105) are expressed as

$$\left.\begin{array}{l} (1)\quad L_r(s) = \text{diag}\{1, s, \ldots, s^{dri-1}\}\bar{L}_r = \psi_r(s)\bar{L}_r \\ (2)\quad L_c(s) = \bar{L}_c \,\text{diag}\left\{\begin{matrix} 1 \\ s \\ \vdots \\ s^{dcj-1} \end{matrix}\right\} = \bar{L}_c\psi_c(s) \end{array}\right\} \text{where } \bar{L}_r \text{ and } \bar{L}_c \text{ are constant matrices} \tag{4.93, 4.94}$$

Since $L_r(s)$ and $L_c(s)$ consists of lower-order terms (i.e., terms of power less than dri and dcj, respectively, for any i, j), it is only necessary to show that they can be broken into forms as in (4.93) and (4.94). Since the rows or columns of N(s) can be rearranged such that dci > ··· > dcj > ··· > dcm, or dri > ··· > dri > drm, this follows directly from (1.104) and the structure of $\psi_c(s)$ and $\psi_r(s)$.

An example of this representation is given below.

Example 4.13: Representation of D(s) as in Theorem 4.4

$$D(s) = \begin{bmatrix} (s+2)^2(s+1) & s+1 \\ 0 & (s+1)^2 \end{bmatrix} \tag{4.95}$$

$$= \underbrace{\left[\begin{array}{c|c} 1 & 0 \\ \hline 0 & 1 \end{array}\right]}_{\Gamma_r} \underbrace{\left[\begin{array}{c|c} s^3 & 0 \\ \hline 0 & s^2 \end{array}\right]}_{\text{diag}\{s^{dri}\}} + \underbrace{\left[\begin{array}{ccc|cc} 1 & s & s^2 & 0 & 0 \\ \hline 0 & 0 & 0 & 1 & s \end{array}\right]}_{\psi_r(s)} \underbrace{\begin{bmatrix} 4 & 1 \\ 8 & 1 \\ 5 & 0 \\ 0 & 1 \\ 0 & 2 \end{bmatrix}}_{\bar{L}_r} \tag{4.96}$$

$$= \underbrace{\left[\begin{array}{c|c} s^3 & 0 \\ \hline 0 & s^2 \end{array}\right]}_{\text{diag}\{s^{dci}\}} \underbrace{\left[\begin{array}{c|c} 1 & 0 \\ \hline 0 & 1 \end{array}\right]}_{\Gamma_c} + \underbrace{\begin{bmatrix} 4 & 8 & 5 & 1 & 1 \\ 0 & 0 & 0 & 1 & 2 \end{bmatrix}}_{\bar{L}_c} \underbrace{\left[\begin{array}{c|c} 1 & 0 \\ s & 0 \\ \hline s^2 & 0 \\ 0 & 1 \\ 0 & s \end{array}\right]}_{\psi_c(s)} \tag{4.97}$$

Having established a method of "expanding" D(s), the general theorem on matrix-fraction description may now be stated.

Theorem 4.5 A strictly proper transfer-function matrix $G_{sp}(s)$ of order $\ell \times m$ may be expressed in either of the two forms:

$$\text{(1)}\quad G_{sp}(s) = [\Gamma(s) + \psi_r(s)P]^{-1}[\psi_r(s)Q] \tag{4.98}$$

$$\text{(2)}\quad G_{sp}(s) = [\bar{Q}\psi_c(s)][\Gamma(s) + \bar{P}\psi_c(s)]^{-1} \tag{4.99}$$

where

$$\Gamma(s) = \text{diag}\,\{s^{ni}\} \text{ for (1)}: \rightarrow \text{dimension } \ell \times \ell$$

$$= \text{diag}\,\{s^{nj}\} \text{ for (2)}: \rightarrow \text{dimension } m \times m$$

$$\psi_r(s) = \text{diag}\,\{1, s, \ldots, s^{ni-1}\} \rightarrow \text{dimension } \ell \times \sum_{i=1}^{\ell} ni$$

$$\psi_c(s) = \text{diag}\begin{Bmatrix} 1 \\ s \\ \vdots \\ s^{nj-1} \end{Bmatrix} \rightarrow \text{dimension } \sum_{j=1}^{m} nj \times m$$

P and Q are constant (independent of s) matrices of dimensions:

$$P\text{: } \sum_{i=1}^{\ell} ni \times \ell; \quad Q\text{: } \sum_{i=1}^{\ell} ni \times m; \quad \bar{P}\text{: } m \times \sum_{j=1}^{m} nj; \quad \bar{Q}\text{: } \ell \times \sum_{j=1}^{m} nj$$

and ni (nj) is the degree of the denominator polynomials of the elements in the ith row (column) of $G_{sp}(s)$.

Proof: Part (1). Let $g_i(s)$ be the monic least common multiple of the denominators of all the elements in the ith row of $G_{sp}(s)$. Then

$$G_{sp}(s) = \begin{bmatrix} \frac{\hat{g}_{11}(s)}{g_1(s)} & \cdots & \frac{\hat{g}_{ij}(s)}{g_1(s)} & & \frac{\hat{g}_{1m}(s)}{g_1(s)} \\ \vdots & & \vdots & & \vdots \\ \frac{\hat{g}_{i1}(s)}{g_i(s)} & \cdots & \frac{\hat{g}_{ij}(s)}{g_i(s)} & & \frac{\hat{g}_{im}(s)}{g_i(s)} \\ \vdots & & \vdots & & \vdots \\ \frac{\hat{g}_{\ell 1}(s)}{g_\ell(s)} & \cdots & \frac{\hat{g}_{\ell j}(s)}{g_\ell(s)} & \cdots & \frac{\hat{g}_{\ell m}(s)}{g_\ell(s)} \end{bmatrix} \tag{4.100}$$

Since ni is the degree of $g_i(s)$, then $\delta\{g_i(s)\} = ni > \delta\{\hat{g}_{ij}(s)\}$ for any $i \in 1, \ldots, \ell$. Let $\Lambda(s)$ be the diagonal matrix defined as

$$\Lambda(s) = \text{diag}\{g_i(s)\}, \quad i \in 1, \ldots, \ell \tag{4.101}$$

where $g_i(s)$ has the form

$$g_i(s) = s^{ni} + p_1^i s^{ni-1} + \cdots + p_{ni}^i \tag{4.102}$$

Then $G_{sp}(s)$ may be expressed as

$$G_{sp}(s) = [\Lambda(s)]^{-1}\hat{G}(s) \tag{4.103}$$

where $\hat{G}(s) = \{\hat{g}_{ij}(s)\}$. For any $i \in 1, \ldots, \ell$, $j \in 1, \ldots, m$, let $\hat{g}_{ij}(s)$ have the form

$$\hat{g}_{ij}(s) = q_1^{ij} s^{ni-1} + q_2^{ij} s^{ni-2} + \cdots + q_{ni}^{ij} \tag{4.104}$$

From (4.102) and (4.104), the denominator and the numerators of the elements of $G_{sp}(s)$ may be expressed as

$$g_i(s) = s^{ni} + \underbrace{[1, s, \ldots, s^{ni-1}]}_{(1 \times ni)} \underbrace{\begin{bmatrix} p^i_{ni} \\ p^i_{ni-1} \\ \vdots \\ p^i_1 \end{bmatrix}}_{(ni \times 1)} = s^{ni} + \psi^i_r(s)p^i \tag{4.105}$$

$$i \in 1, \ldots, \ell; \; j \in 1, \ldots, m$$

$$\hat{g}_{ij}(s) = \underbrace{[1, s, \ldots, s^{ni-1}]}_{(1 \times ni)} \underbrace{\begin{bmatrix} q^{ij}_{ni} \\ q^{ij}_{ni-1} \\ \vdots \\ q^{ij}_1 \end{bmatrix}}_{(ni \times 1)} = \psi^i_r(s)q^{ij} \tag{4.106}$$

where ψ^i_r is a row vector and p^i and $\mathbf{q}^{ij}$ are column vectors of dimensions as indicated. Combining (4.102) and (4.103), we have

$$\Lambda(s) = \text{diag}\,\{s^{ni}\} + \psi_r(s)P \tag{4.107}$$

where

$$\psi_r(s) = \overbrace{\begin{bmatrix} 1, s, \ldots, s^{n1-1} & & & & & 0 \\ & \cdot & & & & \\ & & \cdot & & & \\ & & & s, \ldots, s^{ni-1} & & \\ & & & & \cdot & \\ & & & & & \cdot \\ 0 & & & & & 1, s, \ldots, s^{n\ell-1} \end{bmatrix}}^{\Sigma\, ni \text{ columns}} \Bigg\} \ell \text{ rows}$$

$$= \text{diag}\,\{1, s, \ldots, s^{ni-1}\}, \quad i \in 1, \ldots, \ell$$

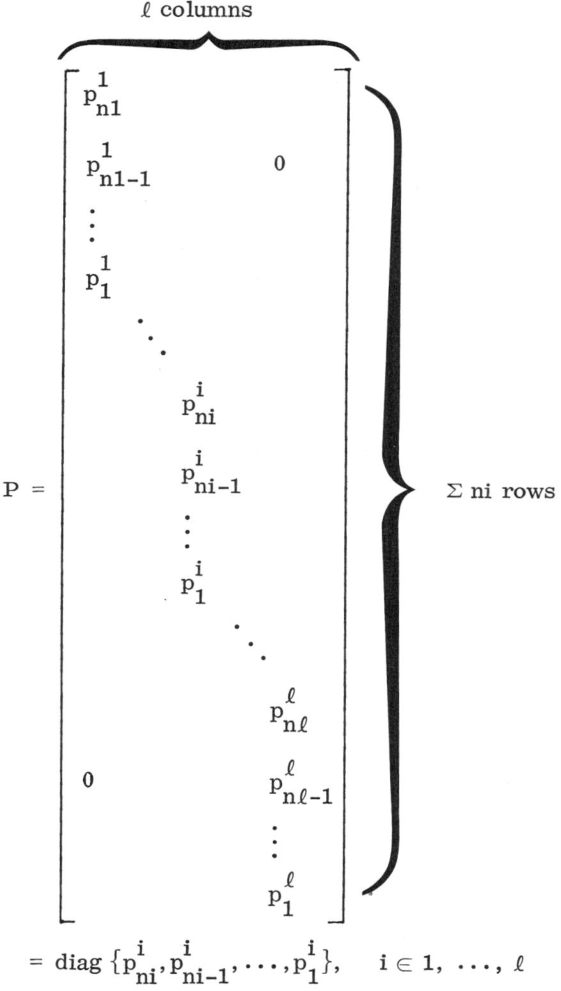

$$= \text{diag}\,\{p^i_{ni}, p^i_{ni-1}, \ldots, p^i_1\}, \qquad i \in 1, \ldots, \ell$$

Similarly, combining (4.104) and (4.106) gives

$$\hat{G} = \psi_r(s)Q \tag{4.108}$$

where $\psi_r(s)$ is as given above and Q is given by

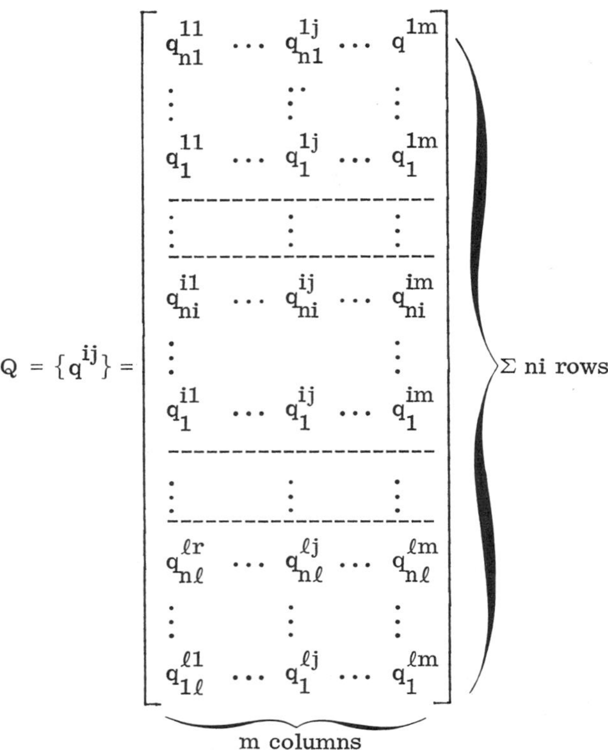

Combination of (4.107) and (4.108) gives

$$G_{sp}(s) = [\Lambda(s)]^{-1}\hat{G}(s) = [\Gamma(s) + \psi_r(s)P]^{-1}[\psi_r(s)Q] \tag{4.109a}$$

Part (2) of the theorem may be proved in a similar way by assuming $g_j(s)$ as the monic least common multiple of denominators of all the elements in the jth column of $G_{sp}(s)$, $j \in 1, \ldots, m$. Then, as in (4.103),

$$G_{sp}(s) = \hat{\bar{G}}(s)[\bar{\Lambda}(s)]^{-1}$$

where

$$\left.\begin{aligned} \bar{\Lambda}(s) &= \text{diag}\,\{g_j(s)\} \\ \hat{\bar{G}}(s) &= \{\hat{\bar{g}}_{ij}(s)\} \end{aligned}\right\} \quad i \in 1, \ldots, \ell;\; j \in 1, \ldots, m \tag{4.109b}$$

Detailed derivation of (4.109b) is left to the reader.

In some cases it may be necessary to perform some permutation (interchange of rows/columns) on the given transfer function (yielding an equivalent system) to obtain the exact forms of (4.109a) and (4.109b). This is illustrated below.

Example 4.14: Matrix-fraction description of G(s)

$$G(s) = \begin{bmatrix} \frac{1}{(s+1)^2(s+2)^2} & \frac{1}{(s+2)^2} \\ -\frac{1}{(s+2)^2} & -\frac{1}{(s+2)^2} \end{bmatrix} \quad (4.110)$$

(1) Left matrix-fraction description (4.98):

$$G(s) = [D_\ell(s)]^{-1}N_\ell(s) = \begin{bmatrix} (s+1)^2(s+2)^2 & 0 \\ 0 & (s+2)^2 \end{bmatrix}^{-1} \begin{bmatrix} 1 & (s+1)^2 \\ -1 & -1 \end{bmatrix}$$

$$D_\ell(s) = \begin{bmatrix} s^4+6s^3+13s^2+12s+4 & 0 \\ 0 & s^2+4s+4 \end{bmatrix}$$

$$= \begin{bmatrix} s^4 & 0 \\ 0 & s^2 \end{bmatrix} + \begin{bmatrix} 6s^3+13s^2+12s+4 & 0 \\ 0 & 4s+4 \end{bmatrix}$$

Therefore,

$$\Gamma(s) = \begin{bmatrix} s^3 & 0 \\ 0 & s^2 \end{bmatrix}$$

and, consequently,

$$\psi_r(s) = \begin{bmatrix} 1 & s & s^2 & s^3 & 0 & 0 \\ 0 & 0 & 0 & 0 & 1 & s \end{bmatrix} \quad (4.111a)$$

giving

$$\begin{bmatrix} 6s^3 + 13s^2 + 12s + 4 & 0 \\ 0 & 4s + 4 \end{bmatrix} = \begin{bmatrix} 1 & s & s^2 & s^3 & 0 & 0 \\ 0 & 0 & 0 & 0 & 1 & s \end{bmatrix} \underbrace{\begin{bmatrix} 4 & 0 \\ 12 & 0 \\ 13 & 0 \\ 6 & 0 \\ 0 & 4 \\ 0 & 4 \end{bmatrix}}_{P} \qquad (4.111b)$$

and

$$N_\ell(s) = \begin{bmatrix} 1 & s^2 + 2s + 1 \\ -1 & -1 \end{bmatrix} = \begin{bmatrix} 1 & s & s^2 & s^3 & 0 & 0 \\ 0 & 0 & 0 & 0 & 1 & s \end{bmatrix} \underbrace{\begin{bmatrix} 1 & 1 \\ 0 & 2 \\ 0 & 1 \\ 0 & 0 \\ -1 & -1 \\ 0 & 0 \end{bmatrix}}_{Q} \qquad (4.111c)$$

(2) Right matrix-fraction description (4.99):

$$G(s) = N_r(s)[D_r(s)]^{-1} = \begin{bmatrix} 1 & 0 \\ -1 & s \end{bmatrix} \begin{bmatrix} 0 & (s+1)^2(s+2) \\ (s+2)^2 & s+2 \end{bmatrix}^{-1}$$

$$D_r(s) = \begin{bmatrix} 0 & s^3 + 4s^2 + 5s + 2 \\ s^2 + 4s + 4 & s + 2 \end{bmatrix}$$

$$= \begin{bmatrix} s^3 & 0 \\ 0 & s^2 \end{bmatrix} \underbrace{\begin{bmatrix} 0 & 1 \\ 1 & 0 \end{bmatrix}}_{\Gamma c} + \begin{bmatrix} 0 & 4s^2 + 5s + 2 \\ 4s + 4 & s + 2 \end{bmatrix}$$

where

$$\Gamma(s) = \text{diag}\,\{s^3, s^2\}\Gamma_c \qquad (\Gamma_c \text{ being a permutation matrix})$$

and

$$\psi_c(s) = \begin{bmatrix} 1 & 0 \\ s & 0 \\ s^2 & 0 \\ 0 & 1 \\ 0 & s \end{bmatrix} \tag{4.112a}$$

giving

$$\begin{bmatrix} 0 & 4s^2+5s+2 \\ 4s+4 & s+2 \end{bmatrix} = \underbrace{\begin{bmatrix} 2 & 5 & 4 & 0 & 0 \\ 2 & 1 & 0 & 4 & 4 \end{bmatrix}}_{\bar{P}} \begin{bmatrix} 1 & 0 \\ s & 0 \\ s^2 & 0 \\ 0 & 1 \\ 0 & s \end{bmatrix} \Gamma_c \tag{4.112b}$$

and

$$N_r(s) = \begin{bmatrix} 1 & 0 \\ -1 & s \end{bmatrix} = \underbrace{\begin{bmatrix} 1 & 0 & 0 & 0 & 0 \\ -1 & 0 & 0 & 0 & 1 \end{bmatrix}}_{\bar{Q}} \begin{bmatrix} 1 & 0 \\ s & 0 \\ s^2 & 0 \\ 0 & 1 \\ 0 & s \end{bmatrix} \tag{4.112c}$$

Combining (4.112a), (4.112b), and (4.112c), we have

$$\begin{aligned} G(s) &= [\bar{Q}\psi_c(s)][\Gamma(s)\Gamma_c + \bar{P}\psi_c(s)\Gamma_c]^{-1} \\ &= [\bar{Q}\psi_c(s)]K[\Gamma(s) + \bar{P}\psi_c(s)]^{-1} \end{aligned}$$

which represent the same system with different numbering of the input/output terminals. The derivations above and the results in Secs. 4.2.2 and 4.2.3 are used in the following sections to develop observable and controllable canonical-form realization methods.

4.4.2 Observable Canonical Form [20-24]

By extending the results in Sec. 3.2, it can be shown that if the output matrix of a purely dynamic m-input/ℓ-output nth-order system S(A, B, C) has full rank, it can be transformed into an observable canonical form $\hat{S}_o(\hat{A}_o, \hat{B}_o, \hat{C}_o)$

through a nonsingular transformation of the state vector, where the matrices A, B, C and $\hat{A}_o$, $\hat{B}_o$, $\hat{C}_o$ have appropriate dimensions and $\hat{S}_o$ has the following structure:

$$\hat{A}_o = \underbrace{\begin{bmatrix} \hat{A}_{o11} & \hat{A}_{o12} & \cdots & \hat{A}_{o1\ell} \\ \hat{A}_{o21} & \hat{A}_{o22} & \cdots & \hat{A}_{o2\ell} \\ \vdots & & & \\ \hat{A}_{o\ell 1} & \hat{A}_{o\ell 2} & \cdots & \hat{A}_{o\ell\ell} \end{bmatrix}}_{n \text{ columns}} \Bigg\} n \text{ rows} \tag{4.113a}$$

with $\hat{A}_{oii}$ in an (ni × ni) observable companion matrix [as in (4.9)]:

$$\hat{A}_{oii} = \underbrace{\begin{bmatrix} 0 & 0 & \cdots & 0 & -p^i_{ni} \\ 1 & 0 & \cdots & 0 & -p^i_{ni-1} \\ 0 & 1 & \cdots & 0 & -p^i_{ni-2} \\ \vdots & & & & \\ 0 & 0 & \cdots & 1 & -p^i_1 \end{bmatrix}}_{ni \text{ columns}} \Bigg\} ni \text{ rows} \tag{4.113b}$$

and $\hat{A}_{oij}$, $i \neq j$, is an (ni × nj) matrix given by

$$\hat{A}_{oij} = \underbrace{\begin{bmatrix} 0 & 0 & \cdots & 0 & -p^i_{nj} \\ 0 & 0 & \cdots & 0 & -p^i_{nj-1} \\ \vdots & & & & \\ 0 & 0 & \cdots & 0 & -p^i_1 \end{bmatrix}}_{nj \text{ columns}} \Bigg\} ni \text{ rows} \tag{4.113c}$$

The observable companion matrix $\hat{A}_{oii}$ has 1's occupying positions below the diagonal elements and the last (ni)th column equal to some nonzero column vector. Elements of this column vector constitute the characteristic polynomial* associated with $\hat{A}_{oii}$. For $i \neq j$, $\hat{A}_{oij}$ is an $(ni \times nj)$ matrix which is identically equal to zero excepth for the final (nj)th column, as shown above, ni, $i \in 1, \ldots, \ell$, being the observability indices, obtained during the transformation procedure. Let αi's be defined as

$$\alpha 0 = 0; \quad \alpha k = \sum_{i=1}^{k} ni, \quad k \in 1, 2, \ldots, \ell$$

that is,

$$\alpha 0 = 0; \ \alpha 1 = n1; \ \alpha 2 = n1 + n2; \ \ldots; \ \alpha i = n1 + n2 + \cdots + ni; \cdots;$$
$$\alpha \ell = n1 + n2 + \cdots + n\ell = n$$

The output matrix $\hat{C}_o$ has the following form:

$$\hat{C}_o = \underbrace{\begin{bmatrix} 0 & \cdots & 0 & 1 & 0 & \cdots & 0 & 0 & 0 & \cdots & 0 \\ 0 & \cdots & 0 & 0 & 0 & \cdots & 0 & 1 & 0 & \cdots & 0 \\ 0 & \cdots & 0 & 0 & 0 & \cdots & 0 & 0 & 0 & \cdots & 0 \\ 0 & \cdots & 0 & 0 & 0 & \cdots & 0 & 0 & 0 & \cdots & 0 \\ 0 & \cdots & 0 & 0 & 0 & \cdots & 0 & 0 & 0 & \cdots & 1 \end{bmatrix}}_{n1 + n2 + \cdots + n\ell = n} \Bigg\} \ell \qquad (4.113d)$$

The input matrix $\hat{B}_o$ is of order $(n \ \ + n2 + \cdots + n\ell = n) \quad m$ and assumes no special form. For notational convenience the matrices above are often expressed as

$$\hat{A}_{oii} = \underbrace{\begin{bmatrix} 0 & 0 & \cdots & 0 & \times \\ 1 & 0 & \cdots & 0 & \times \\ \vdots & & & & \\ 0 & 0 & \cdots & 0 & \times \\ 0 & 0 & \cdots & 1 & \times \end{bmatrix}}_{ni} \Bigg\} ni; \quad \hat{A}_{oij} = \underbrace{\begin{bmatrix} 0 & \cdots & 0 & \times \\ 0 & \cdots & 0 & \times \\ \vdots & & & \\ 0 & \cdots & 0 & \times \\ 0 & \cdots & 0 & \times \end{bmatrix}}_{nj} \Bigg\} ni \qquad (4.114)$$

and

*$\det \{sI - \hat{A}_{oii}\} = s^{ni} + p_1^i s^{ni-1} + \cdots + p_{ni-2}^i s^2 + p_{ni-1}^i + p_{ni}^i$.

$$\hat{C}_o = \underbrace{\begin{bmatrix} 0 & 0 & \cdots & 0 & \overset{\downarrow}{1} & 0 & \cdots & 0 & \overset{\downarrow}{0} & \cdots & \overset{\downarrow}{0} \\ 0 & 0 & \cdots & 0 & 0 & 0 & \cdots & 0 & 1 & \cdots & 0 \\ \vdots \\ 0 & 0 & \cdots & 0 & 0 & 0 & \cdots & 0 & 0 & \cdots & 1 \end{bmatrix}}_{n} \Bigg\} \ell \tag{4.115}$$

where × denotes a nonzero element.

To derive an observable realization procedure, let $\hat{A}_\ell$ be the constant $n \times \ell$ matrix consisting of the ℓ ordered αk columns of $\hat{A}_o$ [marked ↓ in (4.114)], $\hat{C}_\ell$ (= I_ℓ) be the constant $\ell \times \ell$ matrix consisting of the ℓ ordered αk columns of $\hat{C}_o$. The following theorem (called the structure theorem [20-24]) may now be stated.

Theorem 4.6 The strictly proper transfer-function matrix of an m-input/ℓ-output nth-order ($n > \ell$) completely observable system S(A, B, C) may be expressed as

$$C(sI - A)^{-1}B = G_{sp}(s) = \hat{C}_\ell[\Gamma(s) - \psi_r(s)\hat{A}_\ell]^{-1}\psi_r(s)\hat{B}_o$$

where $\Gamma(s)$ and $\psi_r(s)$ are as defined in (4.98).

Proof: Since S(A, B, C) is completely observable, there exists a nonsingular matrix T such that

$$\begin{aligned} G_{sp}(s) = C(sI - A)^{-1}B &\equiv CT[sI - T^{-1}AT]^{-1}T^{-1}B \\ &\equiv \hat{C}_o[sI - \hat{A}_o]^{-1}\hat{B}_o \end{aligned} \tag{4.116}$$

The theorem may thus be proved by showing that

$$\hat{C}_o[sI - \hat{A}_o]^{-1} = \hat{C}_\ell[\Gamma(s) - \psi_r(s)\hat{A}_\ell]^{-1}\psi_r(s) \tag{4.117}$$

which is equivalent to (as $\hat{C}_\ell = I_\ell$)

$$[\Gamma(s) - \psi_r\hat{A}_\ell]\hat{C}_o = \psi_r(s)[sI - \hat{A}_o] \tag{4.118}$$

By direct multiplication and using Theorem 4.5, the right-hand side of the equation above becomes

$$\psi_r(s)[sI - \hat{A}_o] = \text{diag}\{1, s, \ldots, s^{ni-1}\}\begin{bmatrix} sI - A_{o11} & -A_{o12} & \cdots & -A_{o1m} \\ \vdots & & & \\ -A_{o\ell 1} & -A_{o\ell 2} & \cdots & sI - A_{o\ell m} \end{bmatrix}$$

$$\triangleq \begin{bmatrix} g_1(s) & 0 & 0 \\ \vdots & & \\ 0 & g_i(s) & 0 \\ \vdots & & \\ 0 & 0 & g_\ell(s) \end{bmatrix} = \Lambda(s) \tag{4.119}$$

The left-hand side of (4.118) may now be expanded as

$$[\text{diag}\{s^{ni}\} - \text{diag}\{1, s, \ldots, s^{ni-1}\}\underbrace{\begin{bmatrix} \times & \times & \cdots & \times \\ \times & \times & \cdots & \times \\ \vdots & & & \\ \times & \times & \cdots & \times \end{bmatrix}}_{\ell \text{ nonzero columns of } \hat{A}_o}\}\hat{C}_o \tag{4.120}$$

$$\triangleq \text{diag}\{s^{ni}\} - \text{diag}\{-p_1^i s^{ni-1} - p_2^i s^{ni-2} - \cdots - p_{ni}^i\} \tag{4.121}$$

$$= \text{diag}\{g_i(s)\} \qquad i \in 1, \ldots, \ell$$

$$= \Lambda(s)$$

This completes the proof. Since the structure of $\hat{A}_o$ is known a priori, Theorems 4.5 and 4.6 may be combined to obtain the following observable realization theorem.

Theorem 4.7 The observable-form realization of the m-input/ℓ-output strictly proper system described by the polynomial matrix representation

$$\underbrace{[\text{diag}\{s^{ni}\} - \text{diag}\{1, s, \ldots, s^{ni-1}\}P]}_{D(s)}y(s) = \underbrace{[\psi_r(s)Q]}_{N(s)}U(s) \tag{4.122}$$

is $\hat{S}_o(\hat{A}_o, \hat{B}_o, \hat{C}_o)$, where the columns of the $\Sigma\, ni \times \ell$ matrix P form the ℓ nonzero columns of the $\Sigma\, ni$-square matrix $\hat{A}_o$; $\hat{B}_o \equiv Q$ and $\hat{C}_o$ consists of

ℓ rows with 1's in any (αk)th element ($\alpha k = \Sigma_{i=1}^{k}$ ni, k $\in$ 1, 2, ..., ℓ) and all other elements in $\hat{C}_o$ are 0. The order of realization is Σ ni, i $\in$ 1, ..., ℓ, where ni is the degree of the least common multiplier of the elements of the ith row of the polynomial matrix D(s)

The realization procedure* based on the preceding results is outlined through the following illustrative example [23].

Example 4.15: Observable-form realization of

$$\begin{bmatrix} s^2+3s+1 & 2s+3 \\ s^3+3s^2+s & 3s^2+3s+6 \end{bmatrix} \begin{bmatrix} y_1(s) \\ y_2(s) \end{bmatrix} = \begin{bmatrix} 1 & 0 \\ s+1 & s+3 \end{bmatrix} \begin{bmatrix} u_1(s) \\ u_2(s) \end{bmatrix} \tag{4.123}$$

Step 1—Transformation of D(s) into row-proper form: Since the analytical results in Theorems 4.5 and 4.6 are based on the row-proper form of D(s), this step is concerned with deriving an equivalent form of (4.123), where the polynomial on the left is row proper. This is achieved by the postmultiplication of a unimodular matrix $U_\ell(s)$ such that $D_1(s)$ is row proper:

$$\underbrace{U_\ell(s)D(s)}_{D_1(s)}y(s) = \underbrace{U_\ell(s)N(s)}_{N_1(s)}u(s) \tag{4.124}$$

For the system in (4.124),

$$U_\ell(s) = \begin{bmatrix} -s & 1 \\ 1 & 0 \end{bmatrix}$$

giving

$$D_1(s) = \begin{bmatrix} 0 & s^2+6 \\ s^2+3s+1 & 2s+3 \end{bmatrix} \quad \text{and} \quad N_1(s) = \begin{bmatrix} 1 & s+3 \\ 1 & 0 \end{bmatrix} \tag{4.125}$$

Step 2—Transformation of (4.125) into a form such that the elements with highest degrees are in the principal diagonal positions: This is

*A system with strictly proper G(s) [$\delta\{D(s)\} > \delta\{N(s)\}$] is considered here to demonstrate the applicability of (4.117). Although Theorem 4.6 may be extended to proper G(s) [24], such an extension is not essential, since in any numerical algorithm the strictly proper part of G(s) may be easily isolated from any given form. A method of checking the strict properness of $[D(s)]^{-1}N(s)$ is given in step 9.

achieved by appropriate column interchange in $D_1(s)$, that is, through a matrix multiplication of the form

$$\underbrace{KD_1(s)}_{D_0(s)}y(s) = \underbrace{KN_1(s)}_{N_0(s)}u(s) \tag{4.126}$$

The constant matrix K for (4.126) is

$$K = \begin{bmatrix} 0 & 1 \\ 1 & 0 \end{bmatrix}$$

giving

$$D_0 = \begin{bmatrix} s^2+6 & 0 \\ 2s+3 & s^2+3s+1 \end{bmatrix}; \quad N_0(s) = \begin{bmatrix} s+3 & 1 \\ 0 & 1 \end{bmatrix} \tag{4.127}$$

Step 3—Identification of the elements in $D_0(s)$ in (4.126) with highest-degree polynomials and their degrees: For the given system,

$$g_1(s) = s^2 + 6, \qquad n1 = 2$$

$$g_2(s) = s^2 + 3s + 1, \qquad n2 = 2$$

Step 4—Decomposition of the matrix $D_0(s)$:

$$D_0(s) = \begin{bmatrix} g_1(s) & d_{12}(s) \\ d_{21}(s) & g_2(s) \end{bmatrix}$$

$$= \underbrace{\begin{bmatrix} s^2 & 0 \\ 0 & s^2 \end{bmatrix}}_{\Gamma(s)} - \underbrace{\begin{bmatrix} 1 & s & 0 & 0 \\ 0 & 0 & 1 & s \end{bmatrix}}_{\psi_r(s)} \underbrace{\begin{bmatrix} -6 & 0 \\ 0 & 0 \\ -3 & -1 \\ -2 & -3 \end{bmatrix}}_{P}$$

Step 5—Decomposition of the matrix $N_0(s)$ as

$$N_0(s) = \underbrace{\begin{bmatrix} 1 & s & 0 & 0 \\ 0 & 0 & 1 & s \end{bmatrix}}_{\psi_r(s)} \underbrace{\begin{bmatrix} 1 & 1 \\ 3 & 0 \\ 0 & 1 \\ 0 & 0 \end{bmatrix}}_{Q}$$

Step 6—Computation of αk k = 1, 2 and identification of $\hat{C}_\ell$:

$\alpha 1 = n1 = 2$

$\alpha 2 = n1 + n2 = 4$

first column, second element: $\alpha 1 = 2$

$$\hat{C}_o = \begin{bmatrix} 0 & 1 & 0 & 0 \\ 0 & 0 & 0 & 1 \end{bmatrix} \rightarrow \hat{C}_\ell = \begin{bmatrix} 1 & 0 \\ 0 & 1 \end{bmatrix}$$

$\alpha 2 = 4$: second column, fourth element

Step 7—Formulation of $\hat{B}_o$:

$$\hat{B}_o = Q = \begin{bmatrix} 1 & 1 \\ 3 & 0 \\ 0 & 1 \\ 0 & 0 \end{bmatrix}$$

Step 8—Formulation of $\hat{A}_o$:

$$A_o = \left[\begin{array}{cc|cc} 0 & \times & 0 & \times \\ 1 & \times & 0 & \times \\ \hline 0 & \times & 0 & \times \\ 0 & \times & 1 & \times \end{array}\right] = \begin{bmatrix} 0 & -6 & 0 & 0 \\ 1 & 0 & 0 & 0 \\ 0 & -3 & 0 & -1 \\ 0 & -2 & 1 & -3 \end{bmatrix} \tag{4.128}$$

(Columns 1 and 3: known from the structure of $\hat{A}_o$; columns 2 and 4: first column and second column of P.)

Step 9—Computation of the transmission matrix D: Since $g_i(s)$ are monic, D can be directly computed as

$$D = \lim_{s\to\infty} [\operatorname{diag}\{s^{ni}\}]^{-1} N_0(s)$$

$$\lim_{s\to\infty} \begin{bmatrix} \frac{1}{s^2} & 0 \\ 0 & \frac{1}{s^2} \end{bmatrix} \begin{bmatrix} s+3 & 1 \\ 0 & 1 \end{bmatrix} = \begin{bmatrix} 0 & 0 \\ 0 & 0 \end{bmatrix} \tag{4.129}$$

Step 10—Rearrangement of the columns of $\hat{C}_o$ and $\hat{B}_o$: Since the columns of $D_0(s)$ and $N_0(s)$ are interchanged to bring the highest-degree polynomial terms in the principal diagonal positions, rearrangement of the columns of $\hat{C}_o$ and $\hat{B}_o$ is necessary to retain the original input/output numbering, giving

$$C_o = K^{-1}\hat{C}_o = \begin{bmatrix} 0 & 0 & 0 & 1 \\ 0 & 1 & 0 & 0 \end{bmatrix} \tag{4.130}$$

$$B_o = \hat{B}_o K^{-1} = \begin{bmatrix} 1 & 1 \\ 0 & 3 \\ 1 & 0 \\ 0 & 0 \end{bmatrix} \tag{4.131}$$

The observable canonical-form realization of (4.123) is therefore given by $S(A_o, B_o, C_o, D)$, where $A_o = \hat{A}_o$ and B_o, C_o, and D as given in (4.128) to (4.131).

4.4.3 Controllable Canonical Form [20-24]

If the transfer-function matrix representation of an m-input/ℓ-output system is given through a right inverse:

$$y(s) = N(s)[D(s)]^{-1}U(s) \tag{4.132}$$

[as opposed to the left inverse in (4.90b)], a controllable canonical-form realization can be obtained by using Theorem 4.5. As before, the derivation here is based on the results in Sec. 4.4.1.

The controllable canonical form $\hat{S}_c(\hat{A}_c, \hat{B}_c, \hat{C}_c)$ of a completely controllable purely dynamic system has the following form:

$$\hat{A}_c = \underbrace{\begin{bmatrix} A_{c11} & A_{c12} & \cdots & A_{c1m} \\ A_{c12} & A_{c22} & \cdots & A_{c2m} \\ \vdots & & & \\ A_{cm1} & A_{cm2} & \cdots & A_{cmm} \end{bmatrix}}_{\text{n columns}} \Bigg\} \text{ n rows} \tag{4.133}$$

with $\hat{A}_{cii}$ and $\hat{A}_{cij}$ as

$$\hat{A}_{cii} = \underbrace{\begin{bmatrix} 0 & 1 & 0 & \cdots & 0 \\ 0 & 0 & 1 & \cdots & 0 \\ \vdots & & & & \\ 0 & 0 & 0 & \cdots & 1 \\ \times & \times & \times & \cdots & \times \end{bmatrix}}_{nj} \Bigg\} nj; \quad \hat{A}_{cij} = \underbrace{\begin{bmatrix} 0 & 0 & \cdots & 0 \\ 0 & 0 & \cdots & 0 \\ \vdots & & & \\ 0 & 0 & \cdots & 0 \\ \times & \times & \cdots & \times \end{bmatrix}}_{ni} \Bigg\} nj; \quad i \neq j \tag{4.134}$$

where × denotes a nonzero element. That is, $\hat{A}_{cii}$ is a (nj × nj) companion matrix with 1's occupying positions above the diagonal elements, and the last [i.e., (nj)th] row equal to some nonzero row vector. For i ≠ j, $\hat{A}_{cij}$ is an (nj × ni) matrix whose last row is nonzero and all of whose other rows are null. As earlier, let

$$\beta k = \sum_{j=1}^{k} nj, \quad k \in 1, 2, \ldots, m \tag{4.135}$$

The input matrix here has the form

$$\hat{B}_c = \underbrace{\begin{bmatrix} 0 & 0 & \cdots & 0 \\ 0 & 0 & \cdots & 0 \\ \vdots & & & \\ 1 & 0 & \cdots & 0 \\ 0 & 0 & \cdots & 0 \\ \vdots & & & \\ 0 & 1 & \cdots & 0 \\ \vdots & & & \\ 0 & 0 & \cdots & 0 \\ 0 & 0 & \cdots & 1 \end{bmatrix}}_{m} \Bigg\} \Sigma\, nj = n \tag{4.136}$$

$\hat{C}_c$ has no special form in this representation.

Let $\hat{A}_m$ be the $(n \times m)$ matrix consisting of the m ordered βk columns of $\hat{A}_c$ (marked → above), and $\hat{B}_m$ $(= I_m)$ be the $(m \times m)$ matrix consisting of the m ordered βk rows of $\hat{B}_o$. The following theorem may now be stated.

Theorem 4.8 The strictly proper transfer function matrix of an m-input/ℓ-output nth-order $(n > m)$ completely controllable system S(A, B, C) may be expressed as

$$C(sI - A)^{-1}B = G_{sp}(s) = \hat{C}_c\psi_c(s)[\Gamma(s) - \hat{A}_m\psi_c(s)]^{-1}\hat{B}_m \tag{4.137}$$

where ψ_c and $\Gamma(s)$ are as defined in Theorem 4.5.

Theorem 4.9 The controllable canonical-form realization of the m-input/ℓ-output strictly proper system is described by the matrix-fraction representation

$$y(s) = \underbrace{[\bar{Q}\psi_c(s)]}_{N(s)}\underbrace{[\operatorname{diag}\{s^{nj}\} - \bar{P}\operatorname{diag}\begin{bmatrix} 1 \\ s \\ \cdot \\ \cdot \\ \cdot \\ s^{nj} \end{bmatrix}]^{-1}}_{[D(s)]^{-1}}u(s) \tag{4.138}$$

in $\hat{S}_c(\hat{A}_c, \hat{B}_c, \hat{C}_c)$, where the columns of the $m \times \Sigma\, nj$ matrix $\bar{P}$ form the m nonzero columns of the $\Sigma\, nj$-square matrix $\hat{A}_c$, $\hat{C}_c = \bar{Q}$, $\hat{B}_c$ consists of the m rows with 1's in any βkth element $(\beta k = \Sigma_{j=1}^{k} nj,\ k \in 1, \ldots, m)$, and all other elements of $\hat{B}_c$ are 0. The order of realization is $\Sigma\, nj$, $j \in 1, 2, \ldots, m$, where nj is the degree of the lcm of the elements of the jth column of the polynomial matrix D(s).

Detailed proofs of the foregoing theorems are not considered here, but the resulting realization procedure is illustrated below.

Example 4.16: Controllable-form realization of

$$N(s)[D(s)]^{-1} \equiv \begin{bmatrix} s & 0 \\ -s & s^2 \end{bmatrix}\begin{bmatrix} 0 & -(s^3 + 4s^2 + 5s + 2) \\ s^2 + 4s + 4 & s + 2 \end{bmatrix}^{-1} \tag{4.139}$$

Step 1—Transformation of D(s) into column-proper form: This can be easily achieved by postmultiplication of a unimodular matrix $U_r(s)$ such that $D_1(s)$ is column proper, where

$$\underbrace{N(s)U_r(s)}_{N_1(s)}\ [\underbrace{D(s)U_r(s)}_{D_1(s)}]^{-1} \equiv N_1(s)[D_1(s)]^{-1}$$

D(s) in (4.139) is column proper, so this step is omitted $[D(s) = D_1(s)]$.

Step 2—Rearrangement of the rows of $D_1(s)$ to bring the highest-degree polynomials into their principal diagonal positions: This is achieved by a postmultiplying $D_1(s)$ by a matrix L given by

$$L = \begin{bmatrix} 0 & 1 \\ -1 & 0 \end{bmatrix} \qquad [D_1(s) = D(s) \text{ in this case}]$$

which transforms (4.139) into

$$\underbrace{N_1(s)L}_{N_0(s)}[\underbrace{D_1(s)L}_{D_0(s)}]^{-1} = \underbrace{\begin{bmatrix} 0 & s \\ -s^2 & -s \end{bmatrix}}_{N_0(s)} \underbrace{\begin{bmatrix} s^3+4s^2+5s+2 & 0 \\ -(s+2) & s^2+4s+4 \end{bmatrix}}_{D_0(s)}^{-1} \tag{4.140}$$

Step 3—Identification of the elements in $D_0(s)$ with highest-degree polynomials in its columns:

$$g_1(s) = s^3 + 4s^2 + 5s + 2, \qquad n1 = 3$$

$$g_2(s) = s^2 + 4s + 4, \qquad n2 = 2$$

Step 4—Decomposition of $D_0(s)$:

$$D_0(s) = \underbrace{\begin{bmatrix} s^3 & 0 \\ 0 & s^2 \end{bmatrix}}_{\Gamma(s)} - \underbrace{\begin{bmatrix} -2 & -5 & -4 & 0 & 0 \\ 2 & 1 & 0 & -4 & -4 \end{bmatrix}}_{\bar{P}} \underbrace{\begin{bmatrix} 1 & 0 \\ s & 0 \\ s^2 & 0 \\ 0 & 1 \\ 0 & s \end{bmatrix}}_{\psi_c(s)}$$

Step 5—Decomposition of $N_0(s)$:

$$N_0(s) = \underbrace{\begin{bmatrix} 0 & 0 & 0 & 0 & 1 \\ 0 & 0 & -1 & 0 & -1 \end{bmatrix}}_{\bar{Q}} \underbrace{\begin{bmatrix} 1 & 0 \\ s & 0 \\ s^2 & 0 \\ 0 & 1 \\ 0 & s \end{bmatrix}}_{\psi_c(s)}$$

Step 6—Computation of βk, $k \in 1, 2$, and formulation of $\hat{B}_m$:

$\beta 1 = n1 = 3$

$\beta 2 = n1 + n2 = 5$

$$\hat{B}_c = \begin{bmatrix} 0 & 0 \\ 0 & 0 \\ 1 & 0 \\ 0 & 0 \\ 0 & 1 \end{bmatrix} \begin{matrix} \\ \\ \leftarrow \beta 1 = 3 \\ \text{first row} \quad \text{third row} \\ \\ \end{matrix} \rightarrow \hat{B}_m = \begin{bmatrix} 1 & 0 \\ 0 & 1 \end{bmatrix}$$

$\beta 2 = 5 \rightarrow$ (fifth row)

second column   fifth row

Step 7—Formulation of $\hat{C}_c$:

$$\hat{C}_c = \bar{Q} = \begin{bmatrix} 0 & 0 & 0 & 0 & 1 \\ 0 & 0 & 1 & 0 & 1 \end{bmatrix}$$

Step 8—Formulation of $\bar{A}_c$:

$$\hat{A}_c = \left[\begin{array}{ccc|cc} 0 & 1 & 0 & 0 & 0 \\ 0 & 0 & 0 & 0 & 0 \\ \times & \times & \times & \times & \times \\ \hline 0 & 0 & 0 & 0 & 1 \\ \times & \times & \times & \times & \times \end{array}\right] \begin{matrix} \leftarrow ** \\ \leftarrow ** \\ \longleftarrow \text{first row of P} \\ \leftarrow ** \\ \longleftarrow \text{second row of P} \end{matrix} \qquad (4.141a)$$

**rows known from the structure of $\hat{A}_c$

$$= \begin{bmatrix} 0 & 1 & 0 & 0 & 0 \\ 0 & 0 & 1 & 0 & 0 \\ -2 & -5 & -4 & 0 & 0 \\ 0 & 0 & 0 & 0 & 1 \\ 2 & 1 & 0 & -4 & -4 \end{bmatrix}$$

Step 9—Computation of the transmission matrix D:

$$D = \lim_{s \to \infty} N_0(s)[\operatorname{diag}\{s^{nj}\}]^{-1}$$

$$= \lim_{s \to \infty} \begin{bmatrix} 0 & s \\ -s^2 & -s \end{bmatrix} \begin{bmatrix} \frac{1}{s^3} & 0 \\ 0 & \frac{1}{s^2} \end{bmatrix} = \begin{bmatrix} 0 & 0 \\ 0 & 0 \end{bmatrix} \tag{4.141b}$$

Step 10—Rearrangement of the rows of $\hat{C}_o$ and $\hat{B}_o$: Since the columns of D(s) and N(s) were interchanged through L, the original input/output numbering is manifested by reverse transformation; that is,

$$B_c = \hat{B}_c L^{-1} = \begin{bmatrix} 0 & 0 \\ 0 & 0 \\ 0 & -1 \\ 0 & 0 \\ 1 & 0 \end{bmatrix} \tag{4.141c}$$

$$C_c = L^{-1}\hat{C}_c = \begin{bmatrix} 0 & 0 & 1 & 0 & 1 \\ 0 & 0 & 0 & 0 & 1 \end{bmatrix} \tag{4.141d}$$

The controllable canonical-form realization of (4.123) is therefore given by $S_c(A_c, B_c, C_c, D)$ when $A_c = \hat{A}_c$ and B_c, C_c, and D as given above.

Once a realization S(A, B, C) is obtained by using any of the preceding methods, its minimality may be checked by using either the degree or the Markov parameters of G(s). In the former method, as indicated earlier, the order of minimal realization is

$$n_{min} = \delta\{G(s)\}$$

$$= \Sigma\, \delta\{\psi_i(s)\}$$

$\psi_i(s)$ being the invariant polynomials in the Smith-McMillan form of G(s). This relationship is especially suitable for the methods in Sec. 4.3, where the $\psi_i(s)$ are available.

In the Markov parameter method, the order of minimal realization is

$$n_{min} = \rho(M)$$

where M is the infinite (block) Hankel matrix [a special form denoted by H in (3.140)].

$$M = \begin{bmatrix} CB & CAB & CA^2B & \cdots \\ CAB & CA^2B & CA^3B & \cdots \\ CA^2B & CA^3B & CA^4B & \cdots \\ \vdots & & & \end{bmatrix}$$

The criterion above follows from controllability and observability properties, since minimality of S(A, B, C) implies complete state controllability and complete observability. This method is particularly suitable if numerical algorithms are used in the realization procedure.

The problems of translating one form of system representation into another were first identified through the fundamental results that the input/output description reveals only the controllable and observable modes and that this part of the system defines the smallest state-space representation among other representations with the same input/output relationships. These results and associated realization methods (and algorithms) initiated the creation of extensive literature on the subject. Work on this fundamental problem of system theory highlighted the need for using the "right" relationship between the input/output and state-space models for a particular problem. As a result, it is now well established that a transfer function matrix can be used to study stability properties of a time-invariant multivariable system only if the representation of the system being used in the analysis is minimal. The effects of the "hidden" poles and zeros on system stability are analyzed in the following chapter through the properties of decoupling zeros introduced in earlier chapters. Another example of the importance of realization theory in the context of feedback is the linear quadratic control problem. Stability of the closed-loop optimized system can be assured only if optimization is carried out by using a completely controllable and observable model. These are considered in the following chapter.

REFERENCES

1. Mayne, D. Q., "Computational procedure for the minimal realization of transfer-function matrices," Proc. IEE, 115: 1363-1368 (1968).
2. Kalman, R. E., "Mathematical description of linear dynamical systems," SIAM J. Control, 1: 152-192 (1963).
3. Chen, C. T., Introduction to Linear System Theory, Holt, Rinehart and Winston, New York, 1970.
4. Panda, S. P., and Chen, C. T., "Irreducible Jordan form realization of a transfer-function matrix," Trans. IEEE, AC-14: 66-69 (1969).
5. Kuo, Y. L., "On the irreducible Jordan form realization and the degree of a rational matrix," Trans. IEEE, CT-17: 322-332 (1970).
6. Kailath, T., Linear Systems, Prentice-Hall, Englewood Cliffs, N.J., 1980.
7. Fortmann, T. E., and Hitz, K. L., An Introduction to Linear Control Systems, Marcel Dekker, New York, 1977.
8. Blackman, P. F., Introduction to State-Variables Analysis, Macmillan, London, 1977.
9. Melsa, J. L., and Jones, S. K., Computer Programs for Computational Assistance in the Study of Linear Control Theory, McGraw-Hill, New York, 1970.
10. Rosenbrock, H. H., State-Space and Multivariable Theory, Thomas Nelson, London, 1970.
11. Rosenbrock, H. H., "Computation of minimal representations of a rational transfer-function matrix," Proc. IEE, 115: 325-327 (1968).
12. Munro, N., "Minimal realization of transfer-function matrices using the system matrix," Proc. IEE, 118: 1298-1301 (1971).
13. Gueguen, C. J., and Toumire, E., "Comments of 'Irreducible Jordan form realization of a rational matrix'," Trans. IEEE, AC-14: 783-784 (1969).
14. Mayne, D. Q., "An elementary derivation of Rosenbrock's minimal realization algorithm," Trans. IEEE, AC-18: 306-307 (1973).
15. Kalman, R. E., "Irreducible realization and the degree of a rational matrix," SIAM J. Control, 13: 520-544 (1965).
16. Gilbert, E. G., "Controllability-observability in multivariable control systems," SIAM J. Control, 2: 128-151 (1963).
17. Chen, C. T., "Representations of linear time-invariant composite systems," Trans. IEEE, AC-13: 277-283 (1968).
18. Heymann, M., and Thorpe, J. A., "Transfer equivalence of linear dynamical systems," SIAM J. Control, 8: 19-40 (1970).
19. Rosenbrock, H. H., "On linear system theory," Proc. IEE, 114: 1353-1359 (1967).
20. Wolovich, W. A., and Falb, P. L., "On the structure of multivariable systems," SIAM J. Control, 7: 437-451 (1969).
21. MacDuffee, C. C., Theory of Matrices, Chelsea, New York, 1959.

22. Wolovich, W. A., "On the numerators and zeros of rational transfer matrices," Trans. IEEE, AC-18: 544-546 (1973).
23. Wolovich, W. A., "The determination of state-space representation of linear multivariable systems," Automatica, 9: 97-106 (1973).
24. Wolovich, W. A., Linear Multivariable Systems, Springer-Verlag, New York, 1974.
25. Silverman, L. M., "Realization of linear dynamical systems," Trans. IEEE, AC-16: 554-567 (1971).

5
Stability and Control

Stability, like controllability and observability, discussed earlier, is a qualitative property of an engineering system which is not affected by the state of the system or input signals. The importance of stability is emphasized by the fact that almost all workable systems are designed to be "stable." If a system is not stable, it is usually of no use in practice.

Although stability theory is a vast subject, most analytical results are quite limited in their applications. In view of this, and to relate the theoretical criteria to practical synthesis methods, only a selection of definitions, concepts, and criteria associated with linear systems are presented in this chapter. Following the development in the preceding chapters, stability properties are studied here in terms of input/output descriptions as well as state-space representations within the same framework. This rather abstract aspect of system analysis is followed by a brief discussion on some related topics: sensitivity, integrity, and interaction.

5.1 MULTIVARIABLE SYSTEMS

The basic definitions and criteria [1-3] commonly used in the analysis of single-input/single-output systems and outlined briefly in Appendixes 1 and 2 are extended here to multivariable systems represented either through the $(\ell \times m)$ proper rational transfer-function matrix G(s) or by the dynamic equations S(A, B, C, t).* For the sake of generality, a time-varying system is considered during the initial part of the following development.

If $H(t, \tau)$ is the impulse-response matrix of the multivariable system, then from Sec. 2.3,

*The transmission matrix D does not appear in the stability analysis and is assumed to be null in this section. The system with $D \equiv 0$ is known as a purely dynamic system and is represented by a strictly proper G(s).

$$y(t) = \{y_i(t)\} = \int_{-\infty}^{t} H(t, \tau)u(\tau)\, d\tau, \qquad i \in 1, \ldots, \ell \tag{5.1}$$

where

$$H(t, \tau) = \{h_{ij}(t, \tau)\}, \qquad i \in 1, \ldots, \ell; j \in 1, \ldots, m$$

and $h_{ij}(\cdot)$ is the (scalar) impulse response between the jth input control $u_j(t)$ and the ith output variable $y_i(t)$.

As in a single-input/single-output system, a relaxed multivariable system is defined to be <u>BIBO stable</u> only if for any bounded input control, the output response is bounded. Since u(t) and y(t) are vectors in this multivariable case, the notion of boundedness implies that every component of the input and output vector is bounded. By applying the stability criteria derived in Appendix 1 to every pair of input/output terminals [there are $(\ell \times m)$ such pairs], and using the fact that the sum of a finite number of bounded functions is bounded, the following theorem may be formulated.

<u>Theorem 5.1</u> A relaxed multivariable system described by (5.1) is BIBO stable only if there exists a finite number k, such that for every element of $H(\cdot)$,

$$\int_{-\infty}^{t} h_{ij}(t, \tau)\, d\tau < k < \infty \tag{5.2}$$

By appropriate modification a similar constraint may be obtained for time-invariant systems. Theorem 5.1, combined with Theorem A8, leads to the following theorem:

<u>Theorem 5.2</u> A relaxed multi-input/multi-output system represented by a proper rational transfer function matrix G(s) is BIBO stable if the poles of every element $g_{ij}(s)$ of G(s) have negative real parts.

To derive the stability in the time domain, the state response is decomposed into the zero-state response and the zero-input response (Sec. 2.2), giving for time varying systems,

$$\begin{aligned} x(t) &= \phi(t: t_0, x_0, u) \\ &= \phi(t: t_0, 0, u) + \phi(t, t_0, x_0, 0) \end{aligned} \tag{5.3}$$

By following the derivations in Appendix 2, every equilibrium state of

$$\dot{x}(t) = A(t)x(t) \tag{5.4}$$

may be defined to be stable only if there exists some constant k such that

$$\phi(t, t_0) \leq k < \infty \quad \text{for any } t_0 \text{ and } \forall\, t > t_0 \tag{5.5}$$

The following definition for time-domain stability may now be given.

Definition 5.1: Total stability A linear dynamical system is said to be totally stable only if for any initial state and, for any bounded input, the output as well as the state variables are bounded.

Consequently, the conditions for total stability are more stringent than those of bounded-input/bounded-output stability, requiring boundedness of the output as well as of all state variables, and boundedness must hold not only for the zero state but also for any initial state. Since a bounded-input/bounded-output system may not contain state variables which increase indefinitely with time, practical systems should satisfy the criteria of total stability. This leads to the following theorem.

Theorem 5.3 The system S(A, B, C, t) described by

$$\dot{x}(t) = A(t)x(t) + B(t)u(t); \quad y(t) = C(t)x(t) \tag{5.6}$$

is totally stable only if the matrices B(t) and C(t) are bounded in $(-\infty, \infty)$, and the zero-input system is asymptotically stable.

Proof of the theorem follows from (5.2) and is based on the assumption that A(t), B(t), C(t) are continuous functions of t for $-\infty < t < \infty$. For linear time-invariant systems, the zero-state response is characterized by

$$G(s) = C(sI - A)^{-1}B \tag{5.7}$$

which by definition is BIBO stable if all the poles of every element $g_{ij}(s)$ of G(s) have negative real parts. The zero input is controlled by

$$\dot{x}(t) = Ax(t)$$

or

$$x(t) = e^{At}x(t_0) \tag{5.8}$$

where the equilibrium state of (5.8) is a solution of

$$\dot{x}(t) = 0$$

or

$$Ax(t) = 0 \tag{5.9}$$

and has the property that

$$x_{\text{equilibrium}} = x_e = e^{At}x_e, \quad \forall\, t > 0$$

This leads to the following theorem.

Theorem 5.4 The zero state of $\dot{x}(t) = Ax(t)$ is asymptotically stable only if all the eigenvalues of A have negative real parts.

Proof: For the zero state to be asymptotically stable, $\|e^{At}\|$ should be bounded, and should tend to zero as $t \to \infty$; or equivalently from the above, $\|e^{At}\| \to 0$ as $t \to \infty$. Since every entry of e^{At} is of the form $t^k e^{p_i}$, $p_i = (\sigma_i + j\omega_i)t$, $\|e^{At}\| \to 0$ as $t \to \infty$ if all eigenvalues of A have negative real parts.

The eigenvalues of A are the roots of the characteristic equation of A (i.e., $\det\{sI - A\} = 0$). Thus the asymptotic stability of the zero-input response of the system can be easily determined through the characteristic polynomial of A, using the classical Routh-Hurwitz criterion outlined in Appendix 2.

The bounded-input/bounded-output stability of the linear time-invariant system S(A, B, C) is determined by the poles of G(s); since

$$G(s) = C(sI - A)^{-1}B = C\,\frac{\text{adj}\,(sI - A)}{\det\{sI - A\}}\,B \tag{5.10}$$

every pole of G(s) is an eigenvalue of A. Thus, if the zero state of S(A, B, C) is asymptotically stable, the zero-state response of S(·) will also be bounded-input/bounded-output stable. Since the transfer-function matrix G(s) contains only the controllable and observable state variables, the bounded-input/bounded-output stability of the zero-state response in general does not imply asymptotic stability of the zero state.

For a completely controllable and observable system, the characteristic polynomial of A is equal to the characteristic polynomial of G(s) (Sec. 2.3), and consequently every eigenvalue of A is a pole of G(s) and every pole of G(s) is an eigenvalue of A. This leads to the following theorem.

Theorem 5.5 If a linear time-invariant system S(A, B, C) in (5.6) is completely controllable and observable, the following statements are equivalent:

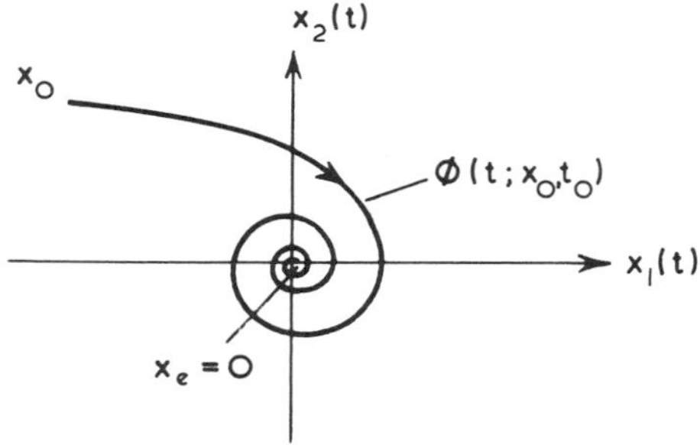

Fig. 5.1 Asymptotically stable system: motion of x(t) on (x_1, x_2) state plane.

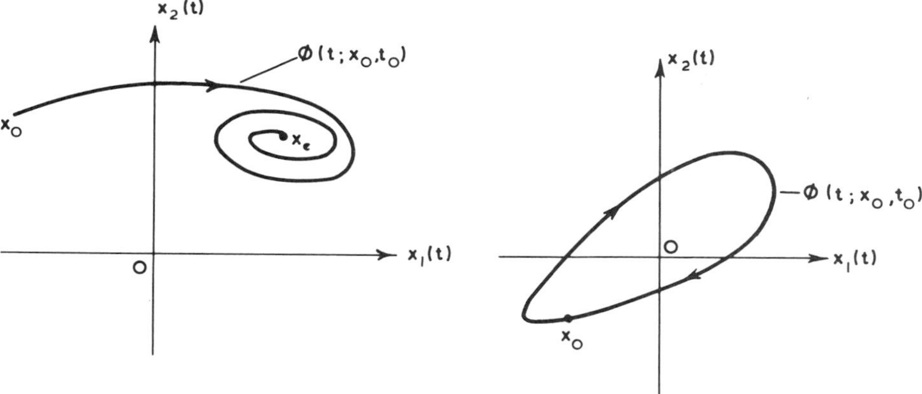

Fig. 5.2 Stable (but not asymptotically stable) system: motion of x(t) on (x_1, x_2) state plane.

1. The system is totally stable.
2. The zero-state response of the system is bounded-input/bounded-output stable.
3. Under zero-input response, the zero state of the system is asymptotically stable.
4. All poles of the transfer-function matrix of the system have negative real parts.
5. All eigenvalues of the matrix A have negative real parts.

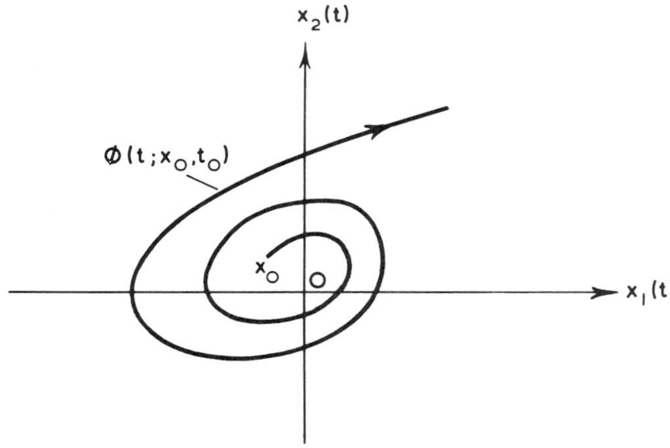

Fig. 5.3 Unstable system: motion of x(t) on (x_1, x_2) state space.

A summary of the foregoing stability definitions and concepts is given below [4].

1. Asymptotic stability—$x(t) \to 0$ as $t \to \infty$ (Fig. 5.1): All eigenvalues of A, distinct or repeated, are on the left half of the complex frequency plane [i.e., $\mathrm{Re}\,[\lambda_i(A)] < 0$, $\forall\, i \in (1, \ldots, n)$].
2. Stability—$x(t) < \infty$ as $t \to \infty$ (Fig. 5.2): All eigenvalues of A are either on the left half of complex frequency plane or on the imaginary axis. Those strictly on the left-half plane may be repeated; those on the imaginary axis must be distinct [i.e., $\mathrm{Re}\,[\lambda_i(A)] < 0$, $\forall\, i \in (1, \ldots, n)$] and if [$\mathrm{Re}\,\lambda_k(A)] = 0$, then $\lambda_j(A) \neq \lambda_k(A)$, $\forall\, j \neq k$.
3. Instability—$x(t) \to \infty$ as $t \to \infty$ (Fig. 5.3): One or more eigenvalues of A are strictly on the right half of the complex frequency plane, or repeated eigenvalues on the imaginary axis. $\mathrm{Re}\,[\lambda_i(A)] > 0$ for some i, $\mathrm{Re}\,[\lambda_j(A)] = \mathrm{Re}\,[\lambda_k(A)] = 0$, or for some j and k, $\mathrm{Im}\,[\lambda_j(A)] = \mathrm{Im}\,[\lambda_k(A)]$.

5.1.1 Feedback Systems [2, 5–8]

This section develops the stability criteria for multivariable systems with unity- and nonunity-feedback configurations, as shown in Figs. 5.4 and 5.5. The proper transfer function $G_o(s)$ in Fig. 5.4 is assumed to have the same number of inputs and outputs (i.e., $m = \ell$), primarily for numerical simplicity; the results may easily be modified for $m \neq \ell$. The closed-loop transfer function matrix in Fig. 5.4 is given by

$$G_c(s) = [I + G_o(s)]^{-1} G_o(s) \equiv \frac{1}{\Delta_c(s)} \{\mathrm{adj}\,[I + G_o(s)]\} G_o(s) \tag{5.11}$$

where

$$\Delta_c(s) = \det \{I + G_o(s)\}$$

In (5.11) and the equations that follow, a subscript o denotes an open-loop system and a c, a closed-loop system.

If $G_o(s)$ is of dimension 1, the stability of the closed-loop system depends on the zeros of $[1 + g(s)]$. For the multivariable case, however,

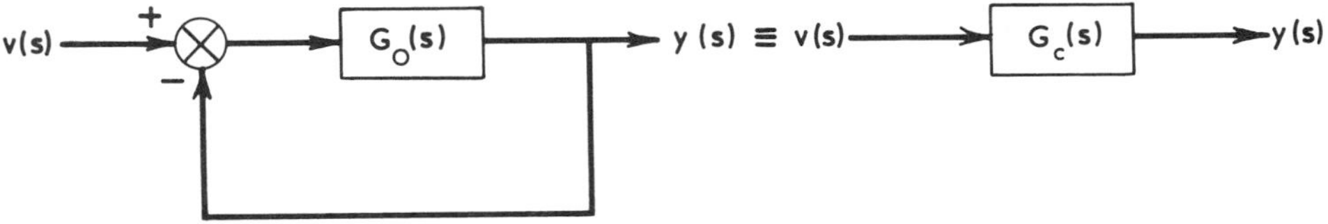

Fig. 5.4 Unity-feedback multivariable system.

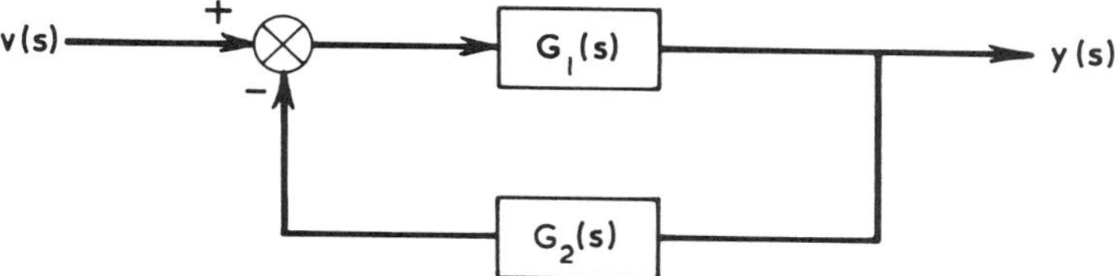

Fig. 5.5 Nonunity-feedback multivariable system.

the stability of the closed-loop system does not depend on the zeros of $[I + G(s)]$ above. This is shown below.

Theorem 5.6 If $S_o(A_o, B_o, C_o, D_o)$ is the irreducible realization of $G_o(s)$, the unity feedback system in Fig. 5.4 is totally stable only if the zeros of the polynomial $\Delta_o(s)\Delta_c(s)$ have negative real parts, where $\Delta_o(s)$ and $\Delta_c(s)$ are the characteristic polynomials of the open- and closed-loop systems, respectively.

Proof: Since S_o is the irreducible realization of $G_o(s)$, it is completely controllable and completely observable; hence the pole polynomial of $G_o(s)$ is equivalent to the characteristic polynomial of A_o. Thus

$$\Delta_o(s) = \det\{sI - A_o\} \equiv \text{lcm of the denominators of all the elements of } G(s)$$

For the closed-loop system in Fig. 5.4, the state and output equations are

$$\begin{aligned} \dot{x}_c(t) &= [A_o - B_o[I+D_o]^{-1}C_o]x_o(t) + B_o[I+D_o]^{-1}v(t) = A_c x_o(t) + B_c v(t) \\ y_c(t) &= [I + D_o]C_o x_o(t) + D_o[I + D_o]^{-1}v(t) = C_c x_o(t) + D_c v(t) \end{aligned} \tag{5.12}$$

Since there is no additional state variable in the closed-loop system, $x_o(t)$ appears in (5.6) and (5.12).

By earlier derivations, the system in (5.12) is stable only if all the eigenvalues of A_c have negative real parts, where the characteristic polynomial of A_c is

$$\begin{aligned} \det\{sI - A_c\} &= \det\{sI - A_o + B_o(I + D_o)^{-1}C_o\} \\ &\equiv \det\{(sI - A_o)[(I + D_o) + C_o(sI - A_o)^{-1}B_o](I + D_o)^{-1}\} \\ &= \underbrace{\det\{sI - A_o\}}_{\Delta_o(s)} \underbrace{\det\{I + G_o(s)\}}_{\Delta_c(s)} \underbrace{\det\{(I + D_o)^{-1}\}}_{K} \equiv K\Delta_o(s)\Delta_c(s) \end{aligned} \tag{5.13}$$

where K is a constant real number, which concludes the proof. Thus if $G_o(s)$ has no poles on the right-half s-plane, the feedback system is totally stable only if the Nyquist plot (Appendix 2) of det $\{I + G_o(s)\}$ does not encircle or go through the origin or, equivalently, the Nyquist plot of [det $\{I + G_o(s)\}$ - 1] does not encircle or go through the critical point (-1 + j0). This is considered in detail in Sec. 5.2.

If for the open-loop system the matrix-fraction representation is

$$G_o(s) = \frac{N_o(s)}{D_o(s)} \equiv N_o(s)[D_o(s)]^{-1} \tag{5.14}$$

where $\delta\{D_o(s)\} \geq \delta\{N_o(s)\}$ and these are co-prime, then for the unity feedback system, the requirement above reduces to the following theorem:

Theorem 5.7 The closed-loop system in Fig. 5.4 is stable if the zeros of the polynomial $\Delta_o(s)$ det $\{I + G_o(s)\}$ have negative real parts, where det $\{I + G_o(\infty)\} \neq 0$.

If $\Delta_c(s)$ in (5.13) is expressed as

$$\Delta_c(s) = \det\{I + G_o(s)\} = \frac{N_c(s)}{D_c(s)} \tag{5.15}$$

where $N_c(s)$ and $D_c(s)$ are polynomials in s, $\delta\{D_c(s)\} \geq \delta\{N_c(s)\}$, then, from (5.11), the closed-loop transfer function becomes

$$G_c(s) = \frac{\operatorname{adj}[I + G_o(s)]}{N_c(s)} G_o(s)D_c(s) \equiv \frac{\operatorname{adj}[I + G_o(s)]\{\hat{g}_{ij}(s)\}}{N_c(s)} \cdot \frac{1}{\Delta_o(s)/D_c(s)}$$

$$\equiv \frac{P(s)}{N_c(s)\hat{N}(s)} \tag{5.16}$$

where

$$G_o(s) = \frac{\{\hat{g}_{ij}(s)\}}{\Delta_o(s)} \quad \text{and} \quad \hat{N}(s) = \frac{\Delta_o(s)}{D_c(s)}$$

The equations above suggest that the stability of the closed-loop system with unity feedback may be ascertained from the three polynomials $\Delta_o(s)$, $N_c(s)$, and $D_c(s)$, defined above. This is stated in the following theorem.

Theorem 5.8 If the zeros $N_c(s)$ and $\hat{N}(s)$ defined in (5.15) and (5.16) are in the left-half s-plane, the system in Fig. 5.4 is stable.

Theorem 5.8 (proof not given here) forms the basis of deriving a necessary and sufficient condition for stability of a system with nonunity feedback shown in Fig. 5.5.

Theorem 5.9 The multivariable feedback system in Fig. 5.5 is stable only if all the zeros of $N(s)$ and $\bar{N}(s)$ have negative real parts where

$$\frac{N(s)}{D(s)} = \det\{I + G_1(s)G_2(s)\} \quad \text{and} \quad \bar{N}(s) = \frac{\Delta_1(s)\Delta_2(s)}{D(s)}$$

$\Delta_1(s)$ and $\Delta_2(s)$ being the characteristic polynomials of $G_1(s)$ and $G_2(s)$, provided that $\det\{I + D_1D_2\} \neq 0$.

Proof: Let the transfer functions $G_1(s)$ and $G_2(s)$ be represented by the two controllable and observable systems $S_1(A_1, B_1, C_1, D_1)$ and $S_2(A_2, B_2, C_2, D_2)$, respectively. Then the state and output equations of the closed-loop system in

$$\dot{\bar{x}}(t) = \begin{bmatrix} \dot{x}_1(t) \\ --- \\ \dot{x}_2(t) \end{bmatrix} = \underbrace{\left[\begin{array}{c|c} A_1 - B_1D_2PC_1 & -B_1C_2 + B_1D_2PD_1C_2 \\ \hline B_2PC_1 & A_2 - B_2PD_1C_2 \end{array}\right]}_{\bar{A}} \underbrace{\begin{bmatrix} x_1(t) \\ --- \\ x_2(t) \end{bmatrix}}_{\bar{x}(t)} + \underbrace{\begin{bmatrix} B_1 - B_1D_2PD_1 \\ \hline B_2PD_1 \end{bmatrix}}_{\bar{B}} v(t)$$

(5.17)

$$\bar{y}(t) = \underbrace{[PC_1 \mid -PD_1C_2]}_{\bar{C}}\bar{x}(t) + \underbrace{PD_1}_{\bar{D}} v(t)$$

where $P = [I + D_1D_2]^{-1}$.

As can be seen, the computation of the characteristic polynomial $\det\{sI - \bar{A}\}$ in general terms is a formidable one, but to prove Theorem 5.9, detailed derivation for the case with $D_1 = D_2 \equiv 0$ is given below.

The characteristic polynomial of the closed-loop system with $D_1 = D_2 \equiv 0$ is given by*

$$\begin{aligned}\Delta_c(s) &= \det\{sI - \bar{A}\} = \det\left\{\begin{bmatrix} sI - A_1 & B_1C_2 \\ -B_2C_1 & sI - A_2 \end{bmatrix}\right\} \\ &= \det\{sI - A_1\}\det\{(sI - A_2) + B_2C_1[sI - A_1]^{-1}B_1C_2\} \\ &= \det\{sI - A_1\}\det\{sI - A_2\}\det\{I + [sI - A_2]^{-1}B_2C_1[sI - A_1]^{-1}B_1C_2\}\end{aligned}$$

*Section 1.2.2 and using the identity $\det[I + AB] = \det[I + BA]$.

$$= \det\{sI - A_1\} \det\{sI - A_2\} \det\{I + C_2[sI - A_2]^{-1}B_2 C_1[sI - A]^{-1} B_1\}$$
$$= \Delta_1(s)\Delta_2(s) \det\{I + G_1(s)G_2(s)\} \tag{5.18}$$

Thus the eigenvalues of $\bar{A}$ are the zeros of the polynomial $\Delta_c(s)$, which may be expressed as

$$\Delta_c(s) = \Delta_1(s)\Delta_2(s) \frac{N(s)}{D(s)} \tag{5.19a}$$
$$\equiv \frac{\Delta_1(s)\Delta_2(s)}{D(s)} N(s) \equiv \bar{N}(s)N(s)$$

when D_1 and D_2 are nonnull, $\Delta_c(s)$ is modified to

$$\Delta_c(s) = \Delta_1(s)\Delta_2(s) \det\{I + G_1(s)G_2(s)\}/\det\{I + D_1D_2\}$$
$$\equiv K\Delta_1(s)\Delta_2(s) \frac{N(s)}{D(s)} \equiv \bar{N}(s)N_1(s);\ K = \text{real constant} \tag{5.19b}$$

which completes the proof.

The basic results presented here suggest ways of determining the closed-loop stability from open-loop parameters. Although these methods are useful in establishing some rules of thumb, as shown by the following example [6], their applications are limited to controllable and observable

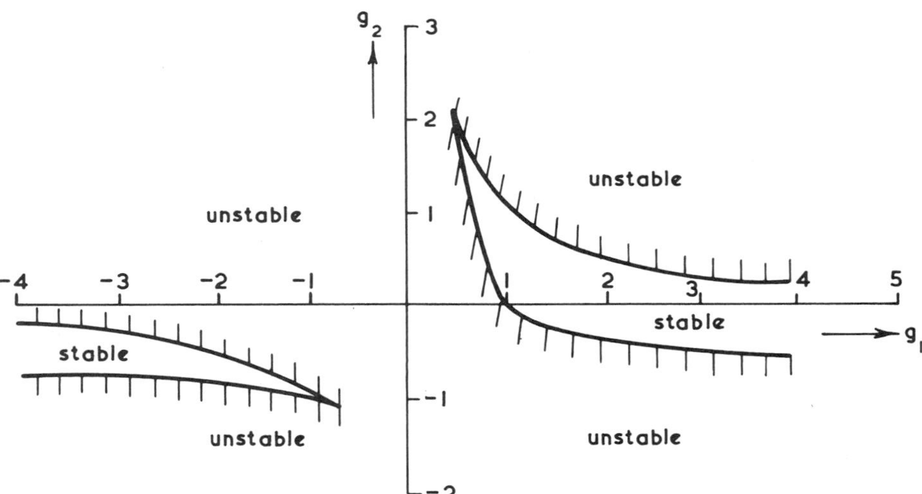

Fig. 5.6 Stability region of the system in (5.20) with unity feedback.

systems. More general methods of stability analysis are considered in Sec. 5.2.

Example 5.1: Assessment of the stability of the unity feedback system with

$$G_o(s) = \begin{bmatrix} \frac{g_1}{s-1} & g_2 \\ g_1 & \frac{g_2}{s+2} \end{bmatrix} \tag{5.20}$$

For $g_i \neq 0$, $i \in 1, 2$; $\Delta_o(s) = s^2 + s - 2$:

$$\Delta_c(s) = \frac{(1 - g_1 g_2)s^2 + (g_1 + g_2 - g_1 g_2 + 1)s + (2g_1 - g_2 + 3g_1 g_2 - 2)}{s^2 + s - 2} \equiv \frac{N_c(s)}{D_c(s)} \tag{5.21}$$

Thus by Theorem 5.8 the closed-loop system is stable for these values of g_1 and g_2 for which the roots of $N_c(s) = 0$ have negative real parts [$\hat{N}(s) = \Delta_o(s)/D_c(s) = 1$ in this case]. The regions of stability for different combinations of g_1 and g_2 are shown in Fig. 5.6.

5.1.2 State Feedback

This section examines briefly the effect of linear state variable feedback (lsvf), or state feedback* for short, on the stability of a system. It has been noted in Sec. 2.2 that given an input control, the dynamics of any system is completely characterized by the behavior of its state vector. Derivations in Appendix 1 also show that asymptotic as well as absolute stability of linear systems may be ascertained through the trajectory of the state vector $x(t) \in X^n$ in the n-dimensional state space. Thus the analysis of the influence of state feedback law of the type

$$u(t) = v(t) + Kx(t) \tag{5.22}$$

on the stability of the system S(A, B, C, D)

$$\begin{aligned} \dot{x}(t) &= Ax(t) + Bu(t) \\ y(t) &= Cx(t) + Du(t) \end{aligned} \tag{5.23}$$

has considerable practical importance [K is a real constant matrix of dimension $m \times n$; $u(t) \in U^m$, $y(t) \in Y^\ell$, $x(t) \in X^n$].

*A special class of state feedback control laws derived through quadratic criterion is considered in Sec. 5.1.4.

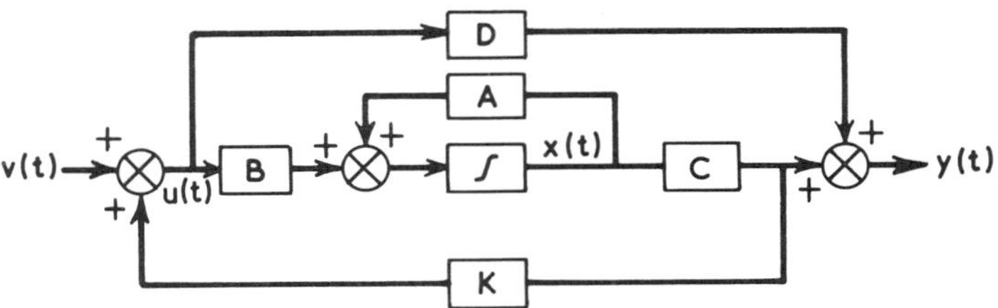

Fig. 5.7 Closed-loop control through state feedback: multivariable system.

Since the stability of the open-loop system in (5.23) may be ascertained from the properties of the matrix A, the closed-loop system, described by Fig. 5.7, is

$$\begin{aligned} \dot{x}(t) &= Ax(t) + B[Kx(t) + v(t)] = (A + BK)\,x(t) + Bv(t) \\ &= \bar{A}x(t) + Bv(t) \\ y(t) &= [C + DK]\,x(t) + Dv(t) = \bar{C}x(t) + Dv(t) \end{aligned} \tag{5.24}$$

The stability characteristics may be analyzed by evaluating the eigenvalues of the matrix

$$\bar{A} = A + BK \tag{5.25}$$

where K is an $(m \times n)$ constant matrix, called the state feedback matrix. The practical significance of state feedback control is apparent from (5.25), which suggests that the eigenvalues or the poles of the closed-loop system (Fig. 5.7) may be modified through the suitable choice of K, provided that all the state variables are available for feedback. Both asymptotic stability and absolute stability may be achieved through state feedback subject to state controllability of the open-loop system (5.23).

For asymptotic stability, the state feedback matrix K is chosen such that for any ith eigenvalue of $\bar{A}$,

$$\text{Re}\,\lambda_i(\bar{A}) < 0 \qquad \text{for any } i \in 1, \ldots, n \tag{5.26}$$

It has been shown in Appendix 2 that the stability of (5.23) can be tested by choosing two matrices P and Q such that

$$PA + A^T P = -Q \tag{5.27}$$

where P and Q are positive definite.

For absolute stability, the elements of K are chosen to satisfy, for any eigenvalue of $\bar{A}$,

$$\text{Re}\ \lambda_i(\bar{A}) < -\alpha, \quad i \in 1, \ldots, n \tag{5.28}$$

where α is a specified positive real number chosen to make the transient response of the system $S(\bar{A}, B, \bar{C}, D)$ in (5.24) decay at least as rapidly as $e^{-\alpha t}$. To derive a condition similar to (5.27) for the closed-loop system, (5.28) is rewritten as

$$\text{Re}\ \lambda_i(\bar{A}) < -\alpha$$

or

$$\text{Re}\ \lambda_i(\bar{A} + \alpha I) < 0 \tag{5.29}$$

which is equivalent to (5.26). Hence there is a condition equivalent to (5.27) for the closed-loop system $\bar{A}$ in replaced by $(\bar{A} + \alpha I)$, yielding

$$P(\bar{A} + \alpha I) + (\bar{A} + \alpha I)^T P = -Q$$

or

$$P\bar{A} + A^T P + 2\alpha I = -Q \tag{5.30}$$

where α is a positive scalar and P and Q are semidefinite matrices. Computation of P and Q in (5.27) and (5.30) may, as indicated in Appendix 2, be simplified by choosing $P = Q = I$.

The development above shows that the eigenvalues of a system can be controlled (i.e., assigned or shifted) through a linear state feedback control law of the type in (5.22). The relationship between this pole assignability and state controllability is outlined through the following two theorems [2].

Theorem 5.10 Single-input/single-output system. If a single-input/single-output system is completely state controllable, then by state feedback the eigenvalues of $(A + bk)$ can be arbitrarily assigned.

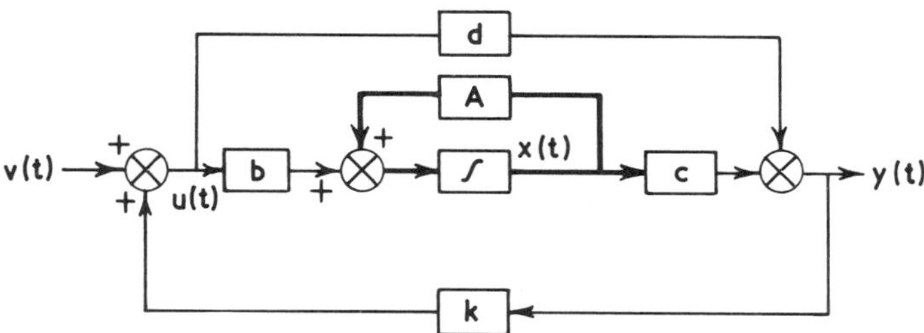

Fig. 5.8 Closed-loop control through state feedback: single-input/single-output system.

Proof: It can be shown that if the system S(A, b, c, d) (Fig. 5.8) is controllable, it can be transformed into the following controllable canonical form:

$$\bar{S}(\bar{A},\bar{b},\bar{c},\bar{d}): \quad \begin{aligned} \dot{z}(t) &= \bar{A}z(t) + \bar{b}u(t) \\ y(t) &= \bar{c}z(t) + \bar{d}u(t) \end{aligned} \tag{5.31}$$

where

$$\bar{A} = M^{-1}AM = \begin{bmatrix} 0 & 1 & \cdots & 0 & 1 \\ 0 & 0 & \cdots & 0 & 0 \\ \vdots & & & & \\ 0 & 0 & \cdots & 0 & 1 \\ -\alpha_n & -\alpha_{n-1} & \cdots & -\alpha_2 & -\alpha_1 \end{bmatrix}; \quad \bar{b} = M^{-1}b = \begin{bmatrix} 0 \\ 0 \\ \vdots \\ 0 \\ 1 \end{bmatrix}$$

$$\bar{c} = cM = [\beta_n \quad \beta_{n-1} \quad \cdots \quad \beta_2 \quad \beta_1]; \quad \bar{d} = d$$

where α_i and β_i's are related to the system transfer function

$$g(s) = \frac{\beta_1 s^{n-1} + \beta_2 s^{n-2} + \cdots + \beta_n}{s^{n-1} + \alpha_1 s^{n-1} + \alpha_2 s^{n-2} + \cdots + \alpha_n} + \bar{d} \tag{5.32}$$

Because of the equivalence transformation, the state feedback law in (5.22) for single-input/single-output systems becomes

$$u(t) = v(t) + \bar{k}z(t), \qquad \text{where } \bar{k} = kM$$

Since

$$\begin{aligned} \det\,[sI - (A + bk)] &= \det\{M^{-1}[sI - (A + bk)]M\} = \det\{sI - M^{-1}(A + bk)M\} \\ &= \det\{sI - (M^{-1}AM + M^{-1}bkM)\} = \det\{sI - (\bar{A} + \bar{b}\bar{k})\} \end{aligned}$$

assigning eigenvalues to (A + bk) is equivalent to assigning eigenvalues to $(\bar{A} + \bar{b}\bar{k})$.

Let the characteristic polynomial of the matrix (A + bk), or equivalently, of $(\bar{A} + \bar{b}\bar{k})$ with desired eigenvalues be

$$s^n + p_1 s^{n-1} + \cdots + p_n$$

If $\bar{k}$ is chosen as

$$\bar{k} = [(\alpha_n - p_n)\ (\alpha_{n-1} - p_{n-1})\ \cdots\ (\alpha_1 - p_1)]$$

then the dynamical equations of the system after the introduction of state feedback control law becomes

$$z(t) = \begin{bmatrix} 0 & 1 & 0 & \cdots & 0 & 0 \\ 0 & 0 & 1 & \cdots & 0 & 0 \\ \vdots & & & & & \\ 0 & 0 & 0 & \cdots & 0 & 1 \\ -p_n & -p_{n-1} & -p_{n-2} & \cdots & -p_2 & -p_1 \end{bmatrix} z(t) + \begin{bmatrix} 0 \\ 0 \\ \vdots \\ 0 \\ 1 \end{bmatrix} v(t)$$

$$y = [\{\beta_n + d(\alpha_n - p_n)\}\ \{\beta_{n-1} + d(\alpha_{n-1} - p_{n-1})\}\ \cdots\ \{\beta_1 + d(\alpha_1 - p_1)\}]\bar{x} + dv(t) \tag{5.33}$$

which has the desired characteristic polynomial and hence the desired eigenvalues.

If a dynamical system is completely state controllable, all its eigenvalues can be arbitrarily assigned by the introduction of state feedback. It has been shown in Sec. 3.2 that if a system is not completely state controllable, the state equation may be transformed into

$$\begin{bmatrix} \dot{\bar{z}}_c(t) \\ \dot{z}_{\bar{c}}(t) \end{bmatrix} = \begin{bmatrix} \bar{A}_c & \bar{A}_{12} \\ 0 & \bar{A}_{\bar{c}} \end{bmatrix} \begin{bmatrix} \bar{z}_c(t) \\ z_{\bar{c}}(t) \end{bmatrix} + \begin{bmatrix} \bar{b}_c \\ 0 \end{bmatrix} u(t) \tag{5.34}$$

where the reduced equation

$$\dot{\bar{z}}_c(t) = \bar{A}_c \bar{z}_c(t) + \bar{b}_1 u(t)$$

is completely state controllable. Because of the form of $\bar{A}_c$, the set of eigenvalues of $\bar{A}$ is the union of the sets of eigenvalues of $\bar{A}_c$ and $\bar{A}_{\bar{c}}$. Due to the form of the input control matrix in (5.34) the matrix $\bar{A}_{\bar{c}}$ is not affected by state feedback control. Hence the eigenvalues of $\bar{A}_{\bar{c}}$ cannot be controlled.

Thus if $\{\bar{A}_c, \bar{b}_c\}$ is completely state controllable, all the eigenvalues of $\bar{A}_{\bar{c}}$ can be arbitrarily assigned. Hence the total number of eigenvalues that are affected by a linear state feedback control law is equal to the order of the controllable subsystem in the canonical form of the system. From this discussion, it is apparent that if the uncontrolled part of the canonical form, $\bar{A}_{\bar{c}}$, contains unstable (positive real parts) eigenvalues, the system cannot be stabilized by state feedback.*

*This is considered later in the context of stabilizability.

To extend the result for single-input/single-output system discussed above to multivariable systems, let the system S(A, B, C, D) defined in (5.23) be completely state controllable. Then the state controllability matrix P has rank n, where

$$P = [B \;\; AB \cdots A^{n-1}B] \quad [b_1 \cdots b_m Ab_1 \cdots Ab_m \cdots A^{n-1}b_1 \cdots A^{n-1}b_m];$$

$$B = \{b_j\}, \quad j = 1, \ldots, m \qquad (5.35)$$

If there exists a b_j such that the submatrix

$$[b_j \;\; Ab_j \cdots A^{n-1}b_j]$$

has rank n, control of the state variables of S(·) can be achieved by using the jth component of u(t) alone (Sec. 3.2.1). If there exists no such b_j, control of the state cannot be achieved by a single component of u(t). With this concept and the development above, the following may be asserted.

Theorem 5.11: Multivariable systems If the system S(A, B, C, D) in (5.23) is completely state controllable, the eigenvalue of (A + BK) can be arbitrarily assigned through the state feedback law in (5.22).

Proof: By introducing a state feedback law of the form

$$u(t) = w(t) + K_1 x(t) \qquad (5.36)$$

to the system S(A, B, C, D) in (5.23), the state equation of the closed-loop system becomes

$$\dot{x}(t) = (A + BK_1)x(t) + Bw(t) \qquad (5.37)$$

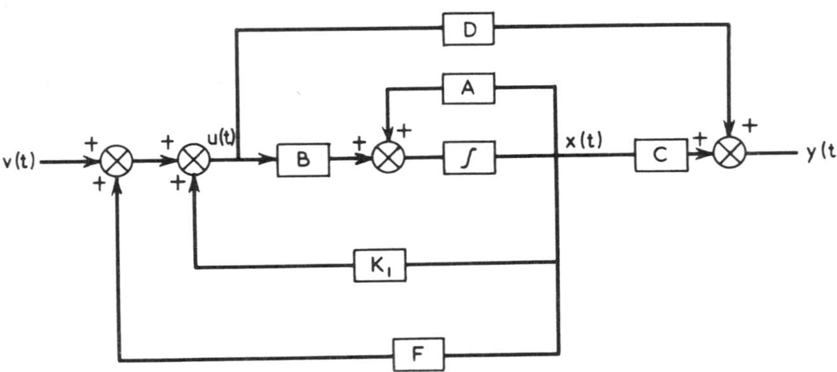

Fig. 5.9 State feedback control of a multivariable system.

Since $\{A, B\}$ is state controllable, the gain matrix K_1 can be chosen so that $\{\bar{A}, b_1\}$ is state controllable, where $\bar{A} = A + BK_1$ and b_1 is the first column of B. Let another state feedback control law (shown in Fig. 5.9) of the form

$$w(t) = v(t) + Fx(t) \tag{5.38}$$

be introduced to the system, where F has the form

$$F = \begin{bmatrix} f_1 \\ 0 \\ \cdot \\ \vdots \\ 0 \end{bmatrix}; \quad f_1 \text{ and } 0 \text{ are } (1 \times n) \text{ row vectors}$$

Combining (5.36) and (5.38), The state equation of the system shown in Fig. 5.9 becomes

$$\dot{x}(t) = (\bar{A} + BF)x(t) + Bv(t) = (\bar{A} + b_1f_1)x(t) + Bv(t) \tag{5.39}$$

Since $\{\bar{A}, b_1\}$ is state controllable, the eigenvalues of $(\bar{A} + b_1f_1)$ can be arbitrarily assigned by a proper choice of f_1 (Theorem 5.10). The state feedback law obtained by combining (5.36) and (5.38) is given by

$$u(t) = v(t) + (K_1 + F)x(t) = v(t) + Kx(t) \tag{5.40}$$

where $K = K_1 + F$ is the required feedback gain matrix, which completes the proof. Methods of computing K_1 and F are considered later.

We have thus established the fact that for a completely state controllable system, a designer has complete freedom in the choice of closed-loop poles and hence of the dynamics of the closed-loop system. The following theorems are needed for further analysis.

Theorem 5.12 The state controllability of a linear multivariable system is invariant under any linear state feedback law.

Proof of the theorem follows from the definition of state controllability and is not expanded here.

Theorem 5.12, and its proof, may also be stated in the following form.

Theorem 5.13 If $\{A, B\}$ is controllable and b_j, $j \in 1, \ldots, m$ are the column vectors of B, then for any j, $j \in 1, \ldots, m$, there exists an $(m \times n)$ real constant matrix K_j such that $\{A + BK_j, b_j\}$ is controllable.

Proof of this theorem is based on the equivalent transformation $x(t) = Mz(t)$ and is omitted here.

Having established the facts that (1) state controllability is invariant under state feedback, and (2) for a completely state controllable system, the designer has total freedom in the choice of closed-loop poles, and hence the dynamics of the closed-loop system, through an appropriate constant $m \times n$ state feedback matrix, it is relevant to raise the question: How much flexibility ($\equiv$ freedom) in the choice of closed-loop poles is lost if the open-loop system is not completely state controllable? This is discussed below.

It has been shown in Sec. 3.2 that by an appropriate choice of a transformation matrix T, any linear time-invariant system may be transformed into its controllable form,

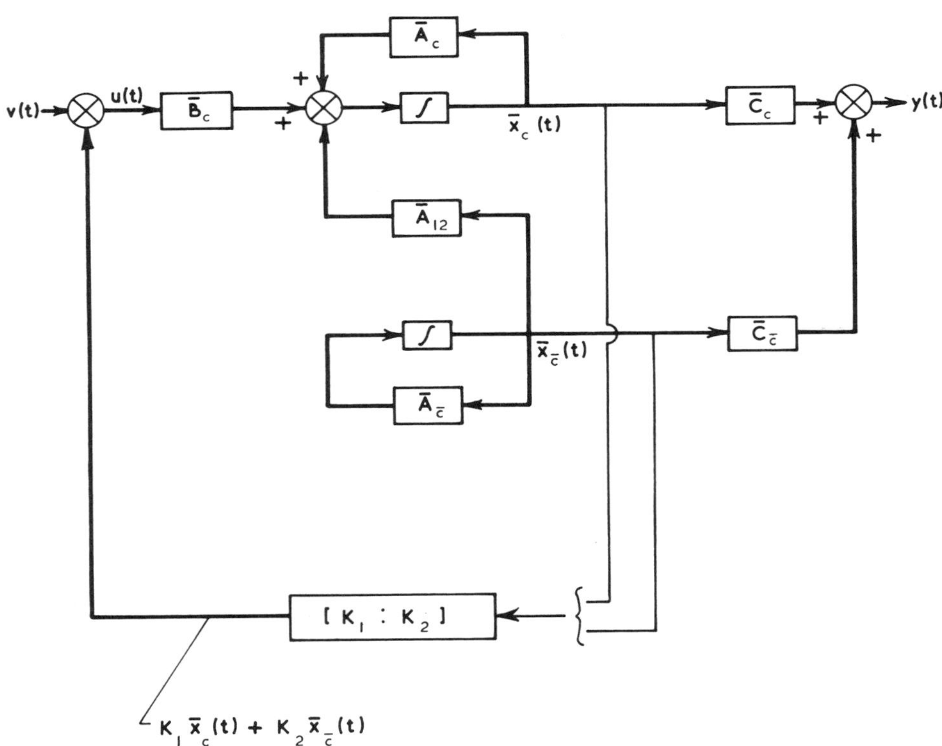

Fig. 5.10 Controllable-form representation of a state feedback system showing that control can be exerted only on the state variables appearing in $\bar{x}_c(t)$. A more general configuration of this representation may be derived from the canonical form [Fig. 3.5 and Eq. (3.77)].

$$\dot{\bar{x}}(t) = \begin{array}{r} n_c\{ \\ n-n_c\{ \end{array}\left[\begin{array}{c} \dot{\bar{x}}_c(t) \\ \hline \dot{\bar{x}}_{\bar{c}}(t) \end{array}\right] = \underset{\begin{array}{cc} \underbrace{\quad}_{n_c} & \underbrace{\qquad}_{n-n_c} \end{array}}{\left[\begin{array}{c|c} \bar{A}_c & \bar{A}_{12} \\ \hline 0 & \bar{A}_{\bar{c}} \end{array}\right]} \left[\begin{array}{c} \bar{x}_c(t) \\ \hline \bar{x}_{\bar{c}}(t) \end{array}\right] + \begin{array}{r} n_c\{ \\ n-n_c\{ \end{array}\left[\begin{array}{c} \bar{B}_c \\ \hline 0 \end{array}\right] u(t) \tag{5.41}$$

$$y(t) = [\bar{C}_c \quad \bar{C}_{\bar{c}}]\bar{x}(t) + Du(t)$$

where the reduced equation

$$\begin{aligned} \dot{x}_c(t) &= \bar{A}_c\bar{x}_c(t) + \bar{B}_c u(t) \\ y(t) &= \bar{C}\bar{x}(t) + Du(t) \end{aligned} \tag{5.42}$$

is completely state controllable, n_c being the controllability index of the system and $x_c(t) \in X^{nc}$. In view of the form of (5.41), it is apparent that the eigenvalues of the submatrix $\bar{A}_{\bar{c}}$ are not affected by the introduction of state feedback (as shown in Fig. 5.10), while all the eigenvalues of the subsystem $\bar{S}(\bar{A}_c, \bar{B}_c, \bar{C}_c)$ of order n_c may be arbitrarily assigned through a constant state feedback matrix. This leads to the following definition.

Definition 5.2: Stabilizability A linear system is said to be stabilizable through state feedback only if all of its uncontrollable modes are stable.

The notion of stabilizability has considerable practical significance, since in many applications, the design requirement may only be to move the poles from the right half of the s-plane to the left. From (5.42) it is apparent that if $\bar{A}_{\bar{c}}$ has any eigenvalues with positive real parts ($\equiv$ unstable poles), the system cannot be stabilized through state feedback.

5.1.3 Output Feedback

Since the state variables, by definition, are chosen from the conceptual viewpoint during the process of establishing a mathematical model of a system, in some cases the state vector may not have any specific physical significance, and hence may not be available at the output terminals or may not even be measurable. In these cases, control through output feedback, rather than by state feedback, would have to be considered. The appropriate control then becomes

$$u(t) = v(t) + Hu(t) \tag{5.43}$$

In view of the close relationship between state controllability and pole assignability, the output feedback control law is often translated into an equivalent state feedback law by using

$$y = Cx(t) \tag{5.44}$$

which then allows use of the state feedback synthesis procedure by the use of the "equivalent" state feedback matrix HC. This method, although applicable to many practical cases, modifies the definition of stabilizability.

Definition 5.3: Stabilizability A linear system is said to be stabilizable by output feedback only if all the uncontrollable and unobservable modes of the system are stable.

Thus a linear system may only be stabilized by output feedback if the modes that are not output controllable are stable. This is not expanded

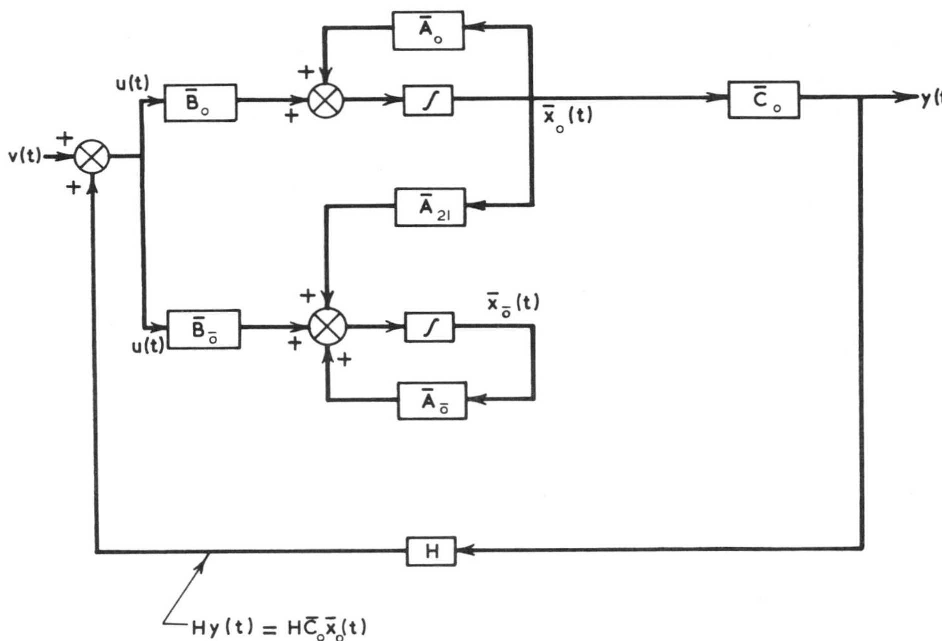

Fig. 5.11 Observable form representation with output feedback; only the modes appearing in $\bar{x}_o(t)$ may be controlled. A more general configuration of this one may be derived from Fig. 3.5 and Eq. (3.77).

further here, but can be proved by using the derivations in Sec. 3.2 (and Fig. 5.11).

Finally, a few comments on the location of zeros. It can be seen that for single-input/single-output systems, the state feedback modifies the location of the closed-loop poles while leaving the location of the zeros unchanged. For the multivariable case, however, the location of the closed-loop poles as well as the zeros are influenced by the state feedback gain matrix, as shown below.

$$G_c(s) = C \frac{\text{adj } (sI - A - BK)}{\det\{sI - A - BK\}} B \equiv \frac{\{\hat{g}_{ij}(s)\}}{\det\{sI - A - BK\}} \tag{5.45}$$

where

$$\hat{g}_{ij}(s) = c_i^r \text{ adj } (sI - A - BK) b_j^c \tag{5.46}$$

where $\hat{g}_{ij}(s)$ is the numerator polynomial of any (i, j)th element of $G_c(s)$. Thus care may be needed in the location of zeros while achieving stabilization by state feedback. This is considered in Chap. 6.

5.1.4 Optimal Control [19-26]

The derivations in the preceding section highlight the practical usefulness of state feedback (provided that the condition for state controllability is satisfied) in the stabilization of linear, and often nonlinear, systems. In addition to asymptotic or absolute stability, state feedback gain matrix K may be chosen to make the zero state converge within an arbitrarily short time; this is achieved by preassigning the closed-loop poles in the far-left-half s-plane. Although, theoretically, there is no limit on the speed of this response, the resulting demand on input energy may be very high, thereby violating the assumption of bounded input control. Thus in practical terms, there are limits on "how far" the closed-loop poles may be moved away from the imaginary axis. In addition, the design requirement may be to control a particular system parameter in a specified manner while simultaneously achieving stability. These considerations lead to the optimization problem, where the speed of response, behavior of a particular variable, and stability may be taken into account simultaneously.

Since the time response of the state variables is sufficient to define the behavior of the system S(A, B, C, D) in (5.23), formulation of the optimization problem is based on establishing a performance criterion in terms of the state vector which gives a qualitative and quantitative measure of the system behavior. There are many criteria that may be used to express how fast an initial state is reduced to its zero state (≡ origin), but the most commonly used one is the quadratic integral criterion*

$$J(x, u, t) = \frac{1}{2} \int_{t_0}^{T} \{[x(t)]^t Q[x(t)]\} \, dt \tag{5.47}$$

Q being a positive-semidefinite symmetric matrix† (Sec. 1.1); the integrand continually measures the deviation of x(t) from its zero state.‡ The weighting matrix Q determines the relative weighting of the individual elements of x(t). The criterion in (5.47) thus indicates the cumulative deviation of the state vector x(t) from the zero state during the time interval $t_0 < t < T$. The choice of the elements of Q permits the flexibility of treating each state variable independently and attaching different importance (weighting) to the various elements of x(t).

Although (5.47) provides a concise criteria for the time response of the state vector, the input control that minimizes J(·) may be unacceptably high. To avoid this possibility the criterion in (5.47) is usually modified to include the input control vector u(t), thereby modifying the performance criterion to:

$$J(x, u, t) = \frac{1}{2} \int_{t_0}^{T} \{[x(t)]^t Q[x(t)] + [u(t)]^t R[u(t)]\} \, dt \tag{5.48}$$

where R is a positive-definite weighting matrix for the input control vector.** The criteria in (5.48) imposes a constraint on the input control while reducing the total value of the integrand. Since the criterion J(x, u, t) is quadratic and contains a factor for input energy (≡ cost), it is known as the <u>quadratic cost function</u> or <u>quadratic performance criterion</u>, and the technique of evaluating a control law that minimizes J(x, u, t) is called <u>linear quadratic control</u> (LQC). [All functions in J(x, u, t) are, in a general analysis, assumed to be continuous in time.]]

The theory of LQC is now well established†† and, despite the difficulty of defining the ideal performance index, has been used successfully in the design of linear and nonlinear systems for over a decade. The influence of control on this cost function is analyzed briefly in this section.

*In the literature, the factor 1/2 in the integrand is sometimes omitted. Presence of 1/2 merely indicates an averaging of the integrand.
†Q is a function of time for time-varying systems.
‡In this section a superscript t is used to denote a transposed vector or matrix.
**R is a function of time for time-varying systems.
††See <u>Trans. IEEE</u>, AC-16 (December 1971): special issue on LQC.

(a) Influence of control

Let the open-loop system S(A, B, C, D) in (5.23) be an asymptotically stable time-invariant system. Then the free system, that is, the uncontrolled system defined by

$$\dot{x}(t) = Ax(t) \tag{5.49}$$

is asymptotically stable and thus there exist two positive-definite symmetric matrices P and Q such that (Appendix 2)

$$PA + A^tP = -Q \tag{5.50}$$

and a suitable Lyapunov function may be chosen as

$$V\{x(t)\} = [x(t)]^tP[x(t)] \tag{5.51}$$

The time derivative of V(·) is

$$\begin{aligned} \frac{d}{dt}[V\{x(t)\}] &= [\dot{x}(t)]^tP[x(t)] + [x(t)]^tP[\dot{x}(t)] \\ &\equiv [x(t)]^t\{PA + A^tP\}[x(t)] \\ &= -[x(t)]^tQ[x(t)] \end{aligned} \tag{5.52}$$

in the absence of input control, the cost function in (5.48) becomes

$$\begin{aligned} J(x, 0, t_0) &= \frac{1}{2}\int_{t_0}^{\infty} [x(t)]^tQ[x(t)] \\ &\equiv -\frac{1}{2}\int_{t_0}^{\infty} \frac{d}{dt}[V\{x(t)\}] \\ &= \frac{1}{2}V\{x(t_0)\} \end{aligned} \tag{5.53}$$

Since the system is assumed to be asymptotically stable, $\lim_{t\to\infty} V\{x(t)\} = 0$. When input control is included, the derivative of the Lyapunov function in (5.51) becomes

$$\frac{d}{dt}[V\{x(t)\}] = [x(t)]^t\{[PA + A^tP][x(t)] + [u(t)^tB^tPx(t) + [x(t)]^tPBu(t)\}$$

or

$$\frac{1}{2}\frac{d}{dt}[V\{x(t)\}] = -\frac{1}{2}[x(t)]^tQ[x(t)] + [u(t)]^tB^tPu(t) \tag{5.54}$$

Let the input control u(t) be realized through the state feedback control law of the form

$$u(t) = Kx(t)$$

where K is chosen as

$$K = -B^t P \tag{5.55}$$

Assuming asymptotic stability of the system with input control, integration of (5.54) gives

$$\frac{1}{2}\int_{t_0}^{T} \frac{d}{dt}[V\{x(t)\}] = -\frac{1}{2}\int_{t_0}^{T} [x(t)]^t Q[x(t)]\, dt - \int_{t_0}^{T} [u(t)]^t K[u(t)]\, dt$$

or

$$-J(x, 0, t_0) = -J(x, u, t_0) + \frac{1}{2}\int_{t_0}^{T} [u(t)]^t R[u(t)]\, dt - \int_{t_0}^{T} [u(t)]^t K[u(t)]\, dt \tag{5.56}$$

when J(x, u, t) corresponds to the cost function with control law (5.55).

Substituting $2\bar{R} = R$ and rearranging (5.56) gives us

$$J(x, 0, t_0) = J(x, u, t_0) + \int_{t_0}^{T} [u(t)]^t [K - \bar{R}]\, [u(t)] \tag{5.57}$$

$$> J(x, u, t) \quad \text{if } [K - \bar{R}] \text{ is positive definite}$$

Thus by introducing a state feedback control law of the type in (5.55), the cost function of the closed-loop system is reduced; that is, an improvement in performance is achieved provided that $[K - \bar{R}]$ is a true positive-definite matrix. Let the system S(A, B, C, D) be unstable in open-loop, and let there be a state feedback law as in (5.22) which stabilizes it. Then, for the closed-loop system, the cost function is given by

$$J(x, u, t)_c = \int_{t_0}^{T} \{[x_c(t)]^t Q[x_c(t)] + [v(t)]^t R[v(t)]\}\, dt \tag{5.58}$$

the closed-loop system being described by

$$\dot{x}_c(t) = (A + BK)x_c(t) + Bv(t)$$

where (A + BK) has all eigenvalues with negative real parts. Then by using a derivation similar to above, it can be shown that the cost function is reduced with the introduction of a stabilizing feedback control law.

(b) Optimal feedback

The optimal feedback control law that minimizes $J(\cdot)$ may be shown for the general linear time-varying case, given by

$$u(t)_{op} = -[R(t)]^{-1}[B(t)]^{t}P(t)x(t) \tag{5.59}$$

where P(t) is the solution of the matrix Riccati equation

$$-\dot{P}(t) = P(t)A(t) + [A(t)]^{t}P(t) - P(t)B(t)[R(t)]^{-1}[B(t)]^{t}P(t) + Q(t)$$

For the time-invariant case, the equation above, letting $T \rightarrow \infty$, becomes

$$0 = PA + A^{t}P - PBR^{-1}B^{t}P + Q \tag{5.60}$$

In the time-invariant case, the optimal control law becomes

$$u(t)_{op} = -R^{-1}B^{t}Px(t) \equiv Kx(t) \tag{5.61}$$

where K is an $(m \times n)$ constant matrix.

Use of the state feedback law in (5.61) results in a system in which all errors approach zero as time approaches infinity; that is, the closed-loop system is asymptotically stable; this is shown analytically below.

(c) Stability

The system in (5.49) with optimal state feedback law as in (5.61) is given by

$$\dot{x}(t) = (A + BK)x(t) = (A - BR^{-1}B^{t}P)x(t) \tag{5.62}$$

Therefore, for time-invariant systems,

$$[\dot{x}(t)]^{t} = [x(t)]^{t}[A - BR^{-1}B^{t}P]^{t}$$

and

$$\begin{aligned}
\frac{d}{dt}\{[x(t)]^{t}P[x(t)]\} &= [\dot{x}(t)]^{t}P[x(t)] + [x(t)]^{t}P[\dot{x}(t)] \\
&= [x(t)]^{t}[A - BR^{-1}B^{t}P]^{t}P[x(t)] + [x(t)]^{t}P[A - BR^{-1}B^{t}P][x(t)] \\
&= [x(t)]^{t}\{[A - BR^{-1}B^{t}P]^{t}P + P[A - BR^{-1}B^{t}P]\}[x(t)] \\
&\equiv [x(t)]^{t}\{PA + A^{t}P - 2PBR^{-1}B^{t}P\}[x(t)] \\
&= [x(t)]^{t}\{-PBR^{-1}B^{t}P - Q\}[x(t)] \qquad \text{from (5.60)} \\
&\equiv [x(t)]^{t}\{-PBR^{-1}B^{t}P - Q\}[x(t)]
\end{aligned} \tag{5.63}$$

Since R is positive definite and Q is positive semidefinite in (5.48), the right-hand side of (5.63) is always negative except when u(t) and x(t) are zero. Furthermore, $[x(t)]^t P[x(t)]$ is positive definite, yet its time derivative is always negative; therefore, $[x(t)]^t P[x(t)]$ may be assumed to be a Lyapunov function and consequently the system controlled by (5.61) is asymptotically stable.

Usually, minimization of the quadratic performance index J(·) is in itself unimportant; however, the theory embodied in the development above provides a systematic method of synthesizing an asymptotically stable feedback controller for large-scale linear dynamic systems.

(d) Controllability and observability [18, 19]

By the definitions in Sec. 3.1 for a completely state controllable system, there exists a finite (≡ bounded) input control u(t) which moves the state vector x(t) to the origin of the state space in a finite time. Once the state vector reaches this origin, it will remain there if no further input control is applied,* and thus the cost of moving the state from any arbitrary point in the state space to the origin (≡ final desired position) is finite. Consequently, the optimal control law necessary for the identical transition will have no cost higher than this finite cost, and therefore $J_{min}(\cdot) < \infty$. The conclusion from this observation is that for $J_{min}(\cdot) < \infty$ the system must be completely state controllable. This also follows from the argument that if the system were controllable and a number of these uncontrollable modes unstable, $J_{min}(\cdot)$ can never be guaranteed to be finite.

In a similar manner, it can be argued that if the system were not completely observable, there will exist some modes that do not contribute to the formation of the performance index J(·). Consequently, if any of these unobservable modes were unstable, there could be no guarantee that $J_{min}(\cdot)$ would always remain finite. This imposes the condition that for $J_{min} < \infty$, the pair (C, A) should be completely observable, or at least its unobservable modes should be stable.† The following observations may therefore be made:

1. For a completely state controllable system, the cost function J(·) is finite.
2. For a completely observable system, the optimal control law $u_{op}(t)$ yields a stable closed-loop system.

Finally, it is worth noting that there always exists an optimal control law even if S(A, B, C, D) is neither completely controllable nor completely observable. However, this is not expanded on here.

*An ideal integrator is controllable in this sense.
†Systems whose unobservable modes are stable are often called <u>detectable</u>.

5.2 GENERALIZED STABILITY CRITERIA

To extend the classical stability criteria to multi-input/multioutput systems it is necessary to establish a relationship between the open- and closed-loop characteristic polynomials. This is achieved by extending the single-input/ single-output configuration [27] in Fig. 5.12a (Appendix 2) to the multivariable case (Fig. 5.12b), by using the concepts of return-difference and return-ratio matrices [28]. In this closed-loop configuration, the following notations are used:*

G(s): $m \times m$ matrix of the open-loop transfer function
K(s): $m \times m$ matrix of the feedforward controller transfer function
H(s): $m \times m$ matrix of the feedback/transducer transfer function
r(s): m-vector of closed-loop reference input transforms
u(s) and y(s) are the m-vector of input and output transforms and G(s), K(s), and H(s) are matrices over the field of rational functions in the complex variable s

The closed-loop transfer-function matrix of the system is given by

$$R(s) = [I_m + G(s)K(s)H(s)]^{-1}G(s)K(s) \tag{5.64}$$

where I_m is an m-square identity matrix.

With all the loops broken as shown in Fig. 5.12b, if a signal $\alpha(s)$ is injected at a, the transformed signal returned at b will be $-G(s)K(s)H(s)\alpha(s)$, and consequently, the difference between the "injected" and "returned" signals is

$$[I_m + G(s)K(s)H(s)]\alpha(s) = [I_m + T(s)]\alpha(s) = F(s)\alpha(s) \tag{5.65}$$

where

$$T(s) = G(s)K(s)H(s) \quad \text{and} \quad F(s) = I_m + T(s) \tag{5.66}$$

The matrix T(s), denoting the loop transfer function of the system with ab broken, is defined as the return-ratio matrix (or return-ratio operator) and the matrix F(s) is defined as the return-difference matrix (or return-difference operator).

To derive an expression between the closed-loop and open-loop characteristic polynomials of Fig. 5.12b, let $Q(s) = G(s)K(s)$ denote the forward transfer function of the system. Then by comparing Fig. 5.12b with Fig. 5.5 and using (5.19b) for the general case,

$$\Delta_r(s) = \Delta_q(s)\Delta_h(s) \det\{F(s)\} \tag{5.67}$$

or

*m-input/m-output systems are considered in subsequent derivations.

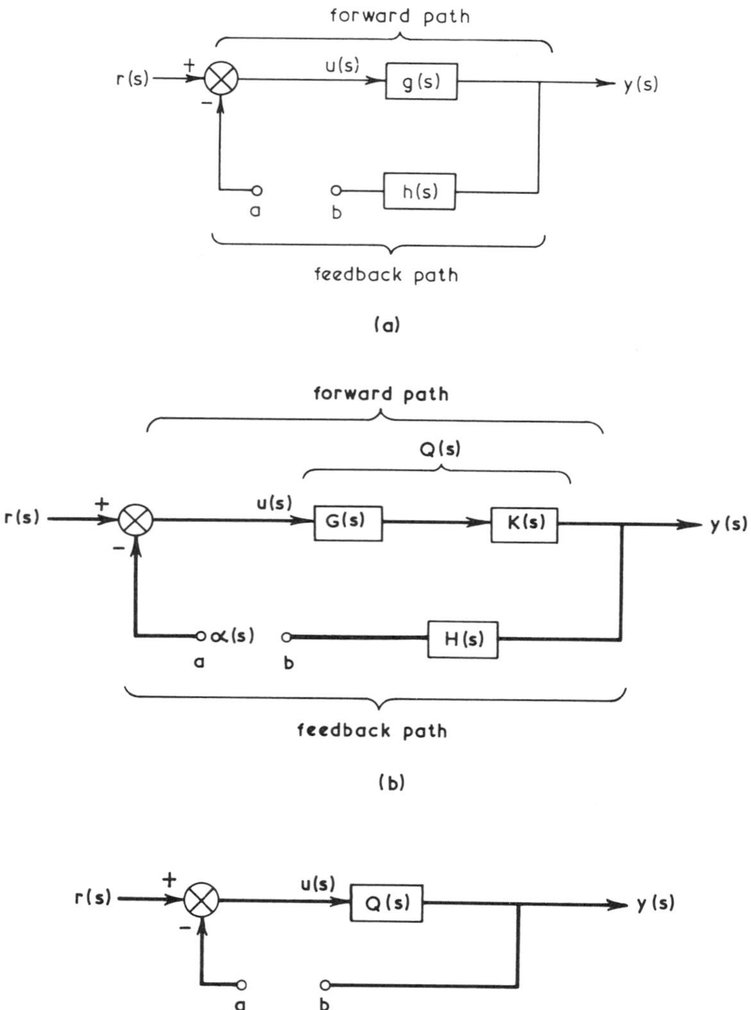

Fig. 5.12 Feedback configurations with breaks for the calculation of return difference: (a) single-input/single-output system; (b) multivariable system with nonunity feedback; (c) multivariable system with unity feedback.

$$\det\{F(s)\} = \frac{\Delta_r(s)}{\Delta_q(s)\Delta_h(s)} \equiv \frac{\phi_c(s)}{\phi_o(s)} \tag{5.68}$$

where $\Delta_q(s)$, $\Delta_h(s)$, and $\Delta_r(s)$ are the characteristic polynomials associated with the forward transfer function Q(s), feedback transfer function H(s), and the closed-loop transfer function R(s). Consequently, $\phi_c(s)$ denotes the closed-loop characteristic polynomial (CLCP) of the system and $\phi_o(s)$ the loop-gain characteristic polynomial* (LGCP) of the system. Thus, combining (5.64) and (5.68), we get

$$\det\{F(s)\} = \frac{\text{CLCP}}{\text{LGCP}} = \frac{\phi_c(s)}{\phi_o(s)} = \frac{\det\{Q(s)\}}{\det\{R(s)\}} \tag{5.69}$$

If the transfer-function matrices in Fig. 5.12b are proper, the identity above becomes

$$\frac{\det\{F(s)\}}{\det\{F(\infty)\}} = \frac{\phi_c(s)}{\phi_o(s)} = \frac{\det\{Q(s)\}}{\det\{R(s)\}} \tag{5.70}$$

where

$$\det\{F(\infty)\} = \det\{I + D_g D_h\} \neq 0 \tag{5.71}$$

$$D_q = \lim_{s\to\infty} Q(s) \quad \text{and} \quad D_h = \lim_{s\to\infty} H(s) \tag{5.72}$$

For the unity feedback system (Fig. 5.12c), the derivations above lead to an elementary multivariable stability criterion in terms of the return-difference matrix F(s); this is outlined below [26-30].

Let $\mathscr{D}$ be the contour in the complex plane described clockwise and consisting of the imaginary axis from $-j\alpha$ to $+j\alpha$, α being the radius of a large semicircle in the right-half plane, large enough to enclose every zero of det $\{Q(s)\}$ and det $\{R(s)\}$ which lies within $\mathscr{D}$ and is centered on the origin. Suppose that $\mathscr{D}$ maps into a closed curve Γ in the complex plane under the mapping det $\{F(s)\}$. Then the system is closed-loop stable if no point within $\mathscr{D}$ maps onto the origin of the complex plane under the mapping det $\{F(s)\}$. Thus the system is closed-loop stable if Γ does not enclose the origin of the complex plane. If $\|\det\{F(s)\}\| \to 1$ as $s \to \infty$, then taking ∞ as arbitrarily large, Γ_f may be assumed to be the locus of $F(j\omega)$. This yields the multivariable Nyquist-type criterion for stability: If det $\{R(s)\}$ maps $\mathscr{D}$ into Γ_c and encircles the origin n_c times clockwise, and det $\{Q(s)\}$ maps $\mathscr{D}$ into Γ_o encircling the origin n_o times clockwise, then the multivariable closed-loop system is stable only if

*Called open-loop characteristic polynomial (OLCP) in the literature [31]. The term loop gain is used here for conformity with definitions in Appendix 2 and since the true open-loop characteristic polynomial is $\Delta_q(s)$ rather than $\Delta_q(s)\Delta_h(s)$. For the case of unity or constant feedback systems, LGCP = OCLP.

$$n_c - n_o = p_o \tag{5.73}$$

where p_o is the number of right-half-plane zeros of the open-loop characteristic polynomial. From (5.69), it therefore follows that if det $\{F(s)\}$ maps $\mathscr{D}$ into Γ_f encircling the origin of the complex plane n_f times in the clockwise direction, then

$$n_f = n_o - n_c \tag{5.74}$$

It thus follows from (5.73) and (5.74) that a necessary and sufficient condition for closed-loop stability is that

$$n_f = -p_o \tag{5.75}$$

where the open-loop stable case follows by setting $p_o = 0$.

The physical significance of the derivations above is that a unity feedback multivariable system, with a set of feedback loops, is stable only if a complex plane mapping of det $\{F(s)\}$ encircles the origin (following the usual direction convention) as many times as there are right-half-plane zeros in the open-loop characteristic polynomial. This result is extended in the following sections.

5.2.1 Characteristic Loci [31,34-42]*

The concept of characteristic loci is based on the notion that, like constant matrices, any square-matrix-valued function G(s) of a complex variable s, with $g_{ij}(s)$ being rational functions, has a set of eigenvalues and a set of associated eigenvectors. This notion stems from the fact that for any given complex frequency s_o, an m-square G(s) becomes a matrix $G(s_o)$ whose elements are complex numbers, and consequently $G(s_o)$ will have a set of complex eigenvalues $g_i(s_o)$, $i \in 1, \ldots, m$. If $\omega_i(s_o)$ is the eigenvector associated with $g_i(s_o)$, then

$$G(s_o)\omega_i(s_o) \triangleq g_i(s_o)\omega_i(s_o) \tag{5.76}$$

or

$$[g_i(s_o)I_m - G(s_o)]\omega_i(s_o) = 0, \quad i \in 1, \ldots, m \tag{5.77}$$

where $g_i(s_o)$ is a root of det $\{g_i(s_o)I_m - G(s_o)\} = 0$.

Thus by generalizing (5.76), if $\omega(s) = \{\omega_i(s)\}$, $i \in 1, \ldots, m$, represents the set of eigenvectors of the rational m-square transfer-function matrix G(s) for any arbitrary s which has a corresponding set of eigenvalues g(s), then

$$G(s)\,\omega(s) \triangleq g(s)\,\omega(s)$$

or

*Derivations in this section follow closely the original work of MacFarlane and co-workers [31].

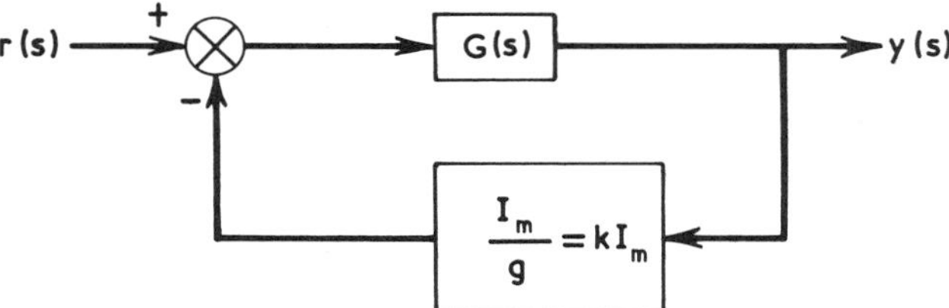

Fig. 5.13 Special form of feedback system.

$$[g(s)I - G(s)]\,\omega(s) = 0 \tag{5.78}$$

where $g(s) = \{g_i(s)\}$, $i \in 1, \ldots, m$, is a set of m eigenvalues corresponding to the roots of

$$\det\{g(s)I_m - G(s)\} = 0 \tag{5.79}$$

To derive the physical significance of g(s), let $g_i(s) = g$, $\forall\, i$ (g being a complex scalar gain). Then (5.79) represents the characteristic polynomial of the rather special m-input/m-output feedback system in Fig. 5.13. Then by the usual notion, the characteristic frequencies of the closed-loop system are the frequencies for which the value of I_m/g is such that the closed-loop system has infinite gain. These characteristic frequencies correspond to the closed-loop poles of the system in Fig. 5.13. Thus corresponding to the feedback gain $g_i(s) = g$, there is a closed-loop pole or a characteristic frequency.

In single-input/single-output systems, Nyquist plots represent the variations of this feedback gain (g) for values of frequencies (ω) around the Nyquist $\mathscr{D}$ contour. In single-input/single-output root loci, the plots of characteristic frequencies are obtained for different gains on the negative real axis. In the multivariable case (Fig. 5.13), there are two independent parameters: complex gains g_i and the complex frequency s. If the elements of the gain vector $g = \{g_i\}$ were expressed as functions of the complex frequency s, the feedback signals in Fig. 5.13 may be considered either as functions of complex frequency (giving frequency loci instead of root loci) or as functions of complex gains (giving gain loci instead of Nyquist plots). This generalization of the feedback gain function forms the basis of an algebraic interpretation of poles and zeros and extension of the classical frequency-domain concept of multi-input/multioutput systems [35, 38]. Some aspects of this generalized frequency-domain analysis are considered here.

The determinant in (5.79) yields a set of m functions of complex variables having the form

$$[g(s)]^m + a_1(s)[g(s)]^{m-1} + \cdots + a_m(s) = 0 \tag{5.80}$$

where any $a_i(s)$, $i \in 1, \ldots, m$, is a rational function in s.

Let these coefficients have a least common denominator $b_0(s)$. Then (5.80) may be rewritten as

$$[g(s)]^m + \frac{b_1(s)}{b_0(s)}[g(s)]^{m-1} + \cdots + \frac{b_m(s)}{b_0(s)} = 0$$

or

$$b_0(s)[g(s)]^m + b_1(s)[g(s)]^{m-1} + \cdots + b_m(s) = 0 \tag{5.81}$$

where any coefficient $b_i(s)$, $i \in 0, \ldots, m$, is a polynomial function. The eigenvector g(s) introduced above, as displayed by (5.80), is a function of a complex variable, and has a value defined by either (5.80) or (5.81). The set of m values of $g_i(s)$, $i \in 1, \ldots, m$, are called the <u>characteristic transfer functions</u> of G(s) or just the <u>characteristic functions</u> of G(s).* This analytic function g(s) satisfies the functional equation

$$\Delta(s,g) = b_0(s)[g(s)]^m + b_1(s)[g(s)]^{m-1} + \cdots + b_m(s) = 0; \quad b_0(s) \not\equiv 0 \tag{5.82}$$

and is called an <u>algebraic function</u>.

If $D_g(s)$ is the discriminant† of (5.79), the following definitions are relevant [38]:

1. A point $s = s_1$ is an ordinary point of g(s) if

$$b_0(s_1) \neq 0 \quad \text{and} \quad D_g(s_1) \neq 0 \tag{5.83}$$

2. A point $s = s_2$ is a critical point of g(s) if either

$$b_0(s_2) = 0 \quad \text{or} \quad D_g(s_2) = 0 \tag{5.84}$$

3. A point $s = s_3$ is called a <u>branch point</u> if it corresponds to a solution of

$$D_g(s) = 0 \tag{5.85}$$

4. The set of values of $s = s_z$ that satisfy

$$b_m(s) = 0 \tag{5.86}$$

*In general, these will be irrational and will be studied in the context of Riemann surfaces [35,40].

†The leading principal minors of a matrix are called the <u>discriminants</u>.

are defined to be the zeros of the algebraic function g(s).

5. The set of values of $s = s_p$ that satisfy

$$b_0(s) = 0 \tag{5.87}$$

are defined to be the poles of the algebraic function g(s).

The set of values of s given by (5.86) and (5.87) may be shown to define the zeros and poles of g(s) even when there are common factors in $b_i(s)$ for $i \in 1, \ldots, m$. A correspondence between these poles and zeros and those defined in Secs. 2.3 and 2.4 may be established directly through (5.79), since

$$\begin{aligned}\det\{g(s)I_m - G(s)\} &\equiv [g(s)]^m - [\operatorname{tr} G(s)][g(s)]^{m-1} + [\Sigma \text{ (principal minors of } G(s) \text{ of order 2)}][g(s)]^{m-2} + \cdots + (-1)^m \det\{G(s)\} \\ &\equiv [g(s)]^m + a_1(s)[g(s)]^{m-1} + a_2(s)[g(s)]^{m-2} + \cdots + a_m(s) \\ &\equiv b_0(s)[g(s)]^m + b_1(s)[g(s)]^{m-1} + b_2(s)[g(s)]^{m-2} + \cdots + b_m(s)\end{aligned} \tag{5.88}$$

Thus b_0 is the least common denominator of all the nonzero sums of the principal minors of the same order of G(s).

By the definitions in Sec. 2.4, for the transfer-function matrix G(s) the pole polynomial p(s) is the least common denominator of all nonzero minors of all orders of G(s). Hence if c(s) is a monic polynomial* whose zeros are those poles of G(s) that do not appear in the polynomial $b_0(s)$, the following identity may be established:

$$\begin{aligned}p(s) &= c(s)b_0(s) \\ a_m(s) &= \frac{b_m(s)}{b_0(s)} = \frac{b_m(s)}{p(s)/c(s)} = \frac{b_m(s)c(s)}{p(s)}\end{aligned} \tag{5.89}$$

Furthermore, since the zero polynomial z(s) of the square matrix G(s) is the numerator of the rational function obtained from $\det\{G(s)\}$, after adjusting $\det\{G(s)\}$ to make its denominator polynomial equivalent to p(s), the zero polynomial becomes

$$z(s) = a_m(s)p(s) = \frac{b_m(s)}{b_0(s)} c(s)b_0(s) = c(s)b_m(s) \tag{5.90}$$

*c(s) contains the decoupling zeros in G(s).

Thus by taking $i \in 1, \ldots, m$, the following general relationships may be written:

$$
\begin{aligned}
p(s) &= c(s)\left[\prod_{i=1}^{m} b_{i0}(s)\right] \\
z(s) &= c(s)\left[\prod_{i=1}^{m} b_{i,ki}(s)\right]
\end{aligned}
\tag{5.91}
$$

where the algebraic function corresponding to the ith characteristic function $g_i(s)$ is given by

$$
\begin{aligned}
\Delta_i(s, g_i) &= [g_i(s)]^{ki} + a_{i1}[g_i(s)]^{ki-1} + \cdots + a_{iki} \\
&= b_{i0}(s)[g_i(s)]^{ki} + b_{i1}(s)[g_i(s)]^{ki-1} + \cdots + b_{i,ki}(s)
\end{aligned}
\tag{5.92}
$$

Two examples to illustrate pole-zero computation and the plotting of the characteristic loci are considered below.

Example 5.2 [33]: Characteristic functions and pole-zero polynomial of

$$
G(s) = \begin{bmatrix} \dfrac{1}{s+1} & 0 \\ -\dfrac{1}{s-1} & \dfrac{1}{s+2} \end{bmatrix}
\tag{5.93}
$$

Thus:

$$
\begin{aligned}
&\text{Pole polynomial:} \quad p(s) = (s+1)(s+2)(s-1) \\
&\text{Zero polynomial:} \quad z(s) = s - 1
\end{aligned}
\tag{5.94}
$$

The characteristic equation for G(s) is

$$
\det\{g(s)I_2 - G(s)\} = \det\left\{\begin{bmatrix} g(s) - \dfrac{1}{s+1} & 0 \\ +\dfrac{1}{s-1} & g(s) - \dfrac{1}{s+2} \end{bmatrix}\right\}
$$

$$
= \underbrace{\left\{g(s) - \frac{1}{s+1}\right\}}_{\Delta_1(s)} \underbrace{\left\{g(s) - \frac{1}{s+2}\right\}}_{\Delta_2(s)}
$$

$$\Delta_1(s) = g_1(s) - \frac{1}{s+1} \equiv g_1(s) + a_{11}(s) \rightarrow a_{11}(s) = -\frac{1}{s+1}$$
$$\Delta_2(s) = g_2(s) - \frac{1}{s+2} \equiv g_2(s) + a_{21}(s) \rightarrow a_{21}(s) = -\frac{1}{s+2} \tag{5.95}$$

Therefore, the coefficients of the algebraic functions in (5.92) are

$$b_{10} = s + 1 \quad \text{and} \quad b_{11} = -1$$
$$b_{20} = s + 2 \quad \text{and} \quad b_{21} = -1 \tag{5.96}$$

Consequently,

$$b_0(s) = b_{10}(s)b_{20}(s) \equiv (s + 1)(s + 2$$

Elimination of the common factors between p(s) and $b_0(s)$ gives

$$c(s) = s - 1$$

The zero polynomial is therefore given by

$$z(s) = c(s)b_{11}(s)b_{21}(s) \equiv s - 1$$

which is consistent with (5.94).

Example 5.3 [39]: Derivation of characteristic loci of G(s)

$$G(s) = \frac{1}{s^2 + 0.6s + 1}\begin{bmatrix} 0.5(s+2) & 4(s+0.1) \\ 0.3(s+8.3) & 3.2(s+1.25) \end{bmatrix} \tag{5.97}$$

The characteristic equation of G(s) is

$$\Delta(s,g) = \det\left\{\begin{bmatrix} g(s) - \dfrac{0.5(s+2)}{s^2+0.6s+1} & -\dfrac{4(s+0.1)}{s^2+0.6s+1} \\ -\dfrac{0.3(s+8.3)}{s^2+0.6s+1} & g(s) - \dfrac{3.2(s+1.25)}{s^2+0.6s+1} \end{bmatrix}\right\}$$
$$= \frac{1}{\Delta(s)}\{[g(s)\,\Delta s - 0.5(s+2)][g(s)\,\Delta(s) - 3.2(s+1.25)] - 1.2(s+0.1)(s+8.3)\} \tag{5.98}$$

where $\Delta(s) = \text{OLCP} = s^2 + 0.6s + 1$. After simplification, (5.98) yields the following algebraic equation:

$$\underbrace{[\Delta(s)]^2}_{b_0(s)}\{g(s)\}^2 + \underbrace{[-3.7(s+1.35)\ \Delta(s)]}_{b_1(s)}\{g(s)\} + \underbrace{[0.4(s^2-12.2s+7.5]}_{b_2(s)} = 0 \quad (5.99)$$

The two complex frequency-dependent roots of g(s) in (5.99) correspond to the characteristic frequency loci (≡ root loci) of the open-loop system in (5.97) shown in Fig. 5.14. These are drawn by choosing specific values of the angular frequency ω_1 ($s_1 = j\omega_1$) and solving (5.99) to obtain the corresponding values of the roots of g(s): $g_1(j\omega_1)$ and $g_2(j\omega_1)$. The procedure is then repeated with a different value of ω_2 ($s_2 = j\omega_2$).

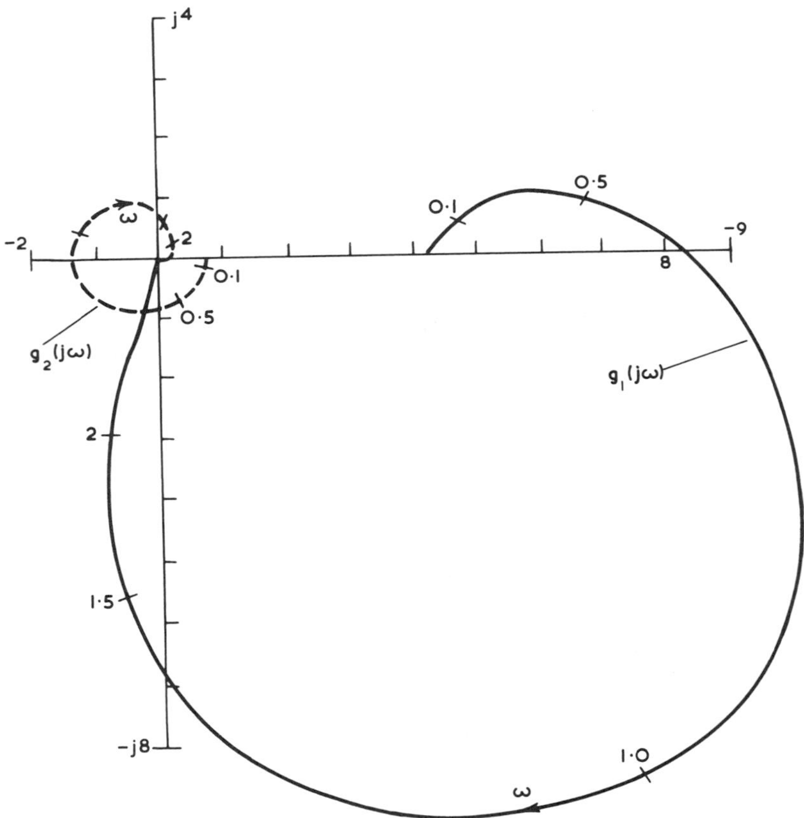

Fig. 5.14 Characteristic frequency loci $g_1(j\omega)$ and $g_2(j\omega)$ corresponding to G(s) in (5.97).

5.2.2 Nyquist Stability Criterion [31, 32, 37–42]*

The concept of characteristic loci introduced in the preceding section is used here as the basis of developing a multivariable Nyquist stability criterion. In the analysis of the stability of multi-input/multioutput systems, each characteristic locus is treated separately according to the Nyquist criterion of single-input/single-output systems (Appendix 2). Then the set of m loci may be pieced together to derive the resulting Nyquist "contours" of the overall system, noting that the contour in this case will be a set of complex manifolds, as opposed to the two-dimensional contours on the complex frequency plane in the single-input/single-output case. These complex manifolds have recently been studies [32, 33] using the concept of Riemann surfaces [40, 41]. This study, details not considered here, concluded that each characteristic locus may be assessed individually by using the standard Nyquist encirclement criterion (Fig. 5.12a).

Number of clockwise encirclements of the origin made by the f(s) locus Γ_f in the f-plane $[f(s) = 1 + g(s)h(s)]$	=	number of zeros of f(s) encircled by the s-plane locus Γ_s in the s-plane - numbers of poles of f(s) encircled by the s-plane locus Γ_s in the s-plane

(5.100a)

and then added up algebraically to ascertain the stability of the complete system (i.e., all m-input/m-output subsystems). That is, for any ith channel, the stability criterion is

$$En_{fi} = n_{fzi} - n_{fpi} \tag{5.100b}$$

and for the complete system, the criterion is

$$\sum_{i=1}^{m} En_{fi} = \sum_{i=1}^{m} n_{fzi} - \sum_{i=1}^{m} n_{fpi} \tag{5.100c}$$

where

$f_i(s)$ = ith characteristic function of the m-square matrix F(s)

En_{fi} = net sum of clockwise encirclements of the origin in the complex plane by the locus† of $f_i(s)$, $f(s) = \{f_i(s)\}$ being the set of m-characteristic functions of $F(s) = I + Q(s)H(s)$

*This section is based on Refs. 33 and 39.

†This is a simpler description. The proper description is that En_{fi} is the net sum of clockwise encirclement of the origin in the complex plane by the set of image curves under $f_i(s)$ of the set of curves formed on the corresponding Riemann surface by piecing together the appropriate set of Nyquist contours when forming the Riemann surface [corresponding to $f_i(s)$] [33].

n_{fzi} = number of right-half s-plane zeros of $f_i(s)$

n_{fpi} = number of right-half s-plane poles of $f_i(s)$

The last two parameters are obtainable from the polynomials similar to (5.91) and the first from graphical plotting of $f_i(s)$ through the solution of

$$\det\{f(s)I_m - F(s)\} = 0, \quad f(s) = \{f_i(s)\}, \quad i \in 1, \ldots, m$$

This idea of using the classical principle of the argument concept for the multivariable stability analysis through characteristic loci has been termed the extended principle of the argument [32] and forms the basis of generalizing the Nyquist stability criterion to multivariable systems. In the derivations below, the relationship between the various characteristic polynomials of the system in Fig. 5.12b is established in two stages: in the first stage, the loop gain (i.e., all loops at a in Fig. 5.12 open) characteristic polynomial $\phi_o(s)$ is evaluated by using

$$\begin{aligned}\phi_o(s) &= \text{denominator polynomial of } \det\{Q(s)H(s)\}\\ &\equiv \text{denominator polynomial of } \det\{G(s)K(s)H(s)\}\end{aligned} \tag{5.101a}$$

and making suitable adjustments in the polynomial above to take into account the pole-zero cancellations [i.e., uncontrollable and unobservable modes (or decoupling zeros)]. Once $\phi_o(s)$ is obtained, the closed-loop characteristic polynomial may then be derived by using (5.70):

$$\phi_c(s) = \frac{\det\{F(s)\}}{\Delta_f(\infty)}\,\phi_o(s), \qquad \text{where } \Delta_f(\infty) = \det\{F(\infty)\} \tag{5.101b}$$

The resulting ϕ_c may again have to be modified to take into account pole-zero cancellations in the closed-loop transfer-function matrix. These derivations are first considered here by using the following terminology:

1. Zeros of a characteristic polynomial $\phi(s) \triangleq$ roots of $\phi(s) = 0$
2. Zero polynomial of a transfer function matrix $A(s) \equiv$ numerator polynomial of $A(s)$
3. Pole polynomial of a transfer function matrix $A(s) \equiv$ denominator polynomial of $A(s)$
4. Thus if $A(s) = z(s)/p(s)$, then
   (a) Poles of $A(s)$ are the zeros of the polynomial $p(s)$.
   (b) Zeros of $A(s)$ are the zeros of the polynomial $z(s)$.

Assuming that the open-loop system has a number of uncontrollable and unobservable modes, the complete (i.e., prior to any numerator-denominator cancellations) pole polynomial of $Q(s)$ may be expressed as

$$\phi_q(s) = p_q(s)p_{dq}(s) \tag{5.102}$$

where $p_q(s)$ represents those poles that appear in Q(s), and $p_{dq}(s)$ represents those poles canceled by an equivalent zero polynomial [i.e., the zeros of the polynomial $p_{dq}(s)$ are the decoupling zeros].

Similarly, for the transfer-function matrix H(s), the pole polynomial is

$$\phi_h(s) = p_h(s)p_{dh}(s) \tag{5.103}$$

and the pole polynomial of the return-ratio matrix T(s) may be expressed as

$$\phi_t(s) = p_q(s)p_h(s)p_{dt}(s) = p(s)p_{dt}(s) \tag{5.104}$$

where $p_{dt}(s)$ represents the polynomial whose zeros are identical to those poles of Q(s) and H(s) that are lost when Q(s)H(s) is formed.* Thus by taking into the cancellations above, the loop-gain pole polynomial (i.e., LGCP) of the system, with all loops broken at a, as in Fig. 5.12b, may be expressed as

$$\phi_o(s) = p(s)p_{dt}(s)p_{dq}(s)p_{dh}(s) = p(s)p_{dt}(s)p_d(s) \tag{5.105}$$

actually appearing in Q(s)H(s)

lost in the product Q(s)H(s)

cancellations within Q(s)

cancellations within H(s)

$p_d(s)$

Thus the zeros of p(s) are the zeros of the $\phi_o(s)$.

To relate (5.105) to det {F(s)}, let

$$\det\{F(s)\} = \frac{z(s)}{p(s)} \tag{5.106}$$

where z(s) and p(s) are related to the characteristic functions [as in (5.92)] $f_i(s)$ of F(s) of the form

$$\Delta_i(s, f_i) = b^f_{i,o}(s)[f_i(s)]^{ki} + b^f_{i,1}(s)[f_i(s)]^{ki-1} + \cdots + b^f_{i,ki}(s) \tag{5.107a}$$

by

*From (5.66) the m-square return-ratio matrix T(s) = Q(s)H(s) and the m-square return-difference matrix F(s) have the same pole polynomial.

$$z(s) = c(s) \prod_{i=1}^{m} b^f_{i,ki}(s) \quad \text{and} \quad p(s) = c(s) \prod_{i=1}^{m} b^f_{io}(s) \tag{5.107b}$$

corresponds to cancellations in det $\{F(s)\}$

Combining (5.105) to (5.107), the closed-loop characteristic polynomial of the system in Fig. 5.12b is

$$\phi_c(s) = \frac{\det\{F(s)\}}{\Delta_f(\infty)} \phi_o(s) \equiv \frac{1}{\Delta_f(\infty)} \frac{z(s)}{p(s)} p(s)p_{dt}(s)p_d(s)$$

$$= \frac{1}{\Delta_f(\infty)} p_{dt}(s)p_d(s)c(s) \prod_{i=1}^{m} b^f_{i,ki}(s) \tag{5.108}$$

and an alternative form of LGCP in (5.105) being

$$\phi_o(s) = p_{dt}(s)p_d(s)c(s) \prod_{i=1}^{m} b^f_{io}(s) \tag{5.109}$$

Before proceeding further, the following parameters (all positive or zero) need to be defined:

- p_o = number of right-half s-plane zeros of LGCP [i.e., the roots of $\phi_o(s) = 0$]
- p_{dt} = number of right-half s-plane roots of $p_{dt}(s) = 0$
- p_d = number of right-half s-plane roots of $p_d(s) = 0$
- c = number of right-half s-plane roots of $c(s) = 0$
- n_{fzi} = number of right-half s-plane zeros associated with the characteristic function $f_i(s)$, $i \in 1, \ldots, m$
- n_{fpi} = number of right-half s-plane poles associated with the characteristic function $f_i(s)$, $i \in 1, \ldots, m$

Then combining the derivations above gives us

$$p_o = \Sigma\, n_{fpi} + p_{dt} + p_d + c \tag{5.110}$$

and for stability, by extending (5.100) over $i \in 1, \ldots, m$, with En denoting clockwise encirclement:

$$\sum_{i=1}^{m} En_{fi} = \sum_{i=1}^{m} n_{fzi} - \sum_{i=1}^{m} n_{fpi} \tag{5.111}$$

or

$$\sum_{i=1}^{m} En_{fi} = \sum_{i=1}^{m} n_{fzi} + p_{dt} + p_d + c - p_o \tag{5.112}$$

where the parameters are either positive or zero. Thus for the closed-loop system in Fig. 5.12b, (5.110) and (5.112) may be used to form a basis for stability tests: as a special case, let the roots of the following four monic polynomial functions be on the negative left half of the s-plane:

$$\begin{aligned} p_{dt}(s) &= 0 \quad \text{i.e.,}\ p_{dt} = 0 \\ p_d(s) &= 0 \quad \text{i.e.,}\ p_d = 0 \\ c(s) &= 0 \quad \text{i.e.,}\ c = 0 \\ \prod_{i=1}^{m} b_{i,ki}^{f} &= 0 \quad \text{i.e.,}\ \Sigma n_{fzi} = 0 \end{aligned} \tag{5.113}$$

The stability condition in (5.111) then reduces to

$$\sum_{i=1}^{m} En_{fi} = -\sum_{i=1}^{m} n_{fpi} \tag{5.114}$$

where

En_{fi} = net sum of clockwise encirclements of the origin in the complex plane by the ith characteristic loci of F(s) corresponding to its ith characteristic function $f_i(s)$

By shifting the origin of the imaginary axis, as in the single-input/single-output case, this condition may be further reduced to

$$\sum En_{ti} = -p_o \tag{5.115}$$

where

$$\begin{aligned} F(s) &= I + Q(s)H(s) = I + T(s) \\ f_i(s) &= 1 + t_i(s) \end{aligned} \tag{5.116}$$

where

En_{ti} = net sum of clockwise encirclements of the critical point (-1 + j0) in the complex plane by the characteristic loci associated with the ith characteristic function $t_i(s)$ of T(s)

This is stated in the following theorem [38].

Theorem 5.14 The closed-loop system in Fig. 5.12b is stable only if

Net sum of counterclockwise encirclements of the critical point (-1+j0) by the set of characteristic loci of the return ratio matrix	=	Total number of right-half s-plane zeros of the loop-gain characteristic polynomial [this is equivalent to "the total number of right-half s-plane poles appearing in Q(s) and H(s)"]

If the system is completely controllable and observable, or if its uncontrollable/unobservable modes are stable (i.e., if the system is detectable), then $p_d = p_{dt} = c \equiv 0$ and the test for stability may be readily accomplished using (5.12b).

Proof of Theorem 5.14 follows from the fact that the characteristic loci of F(s) and T(s) are identical in shapes [because of (5.116)] except that the latter set of loci are obtained from the former by a shift of origin from (0, 0) to (-1 + j0, 0).

The derivation above may be summed up to provide the following multivariable generalization of the Nyquist stability criterion.

Theorem 5.15 The closed-loop system in Fig. 5.12b (without any unstable uncontrollable and unobservable modes) is stable only if

Net sum of counterclockwise encirclements of the critical point (-1 + j0) by the set of characteristic loci $\{t_i(s)\}$ of the return-ratio matrix T(s) [≡ Q(s)H(s)]

- net sum of counterclockwise encirclements of the origin by the set of characteristic loci $\{f_i(s)\}$ of the return difference matrix F(s) [≡ I + Q(s)H(s)]

= total number of right-half s-plane zeros of the loop-gain transfer-function matrix [≡ Q(s)H(s)] [this last item being equivalent to the total number of right-half s-plane zeros in the transfer function Q(s)H(s)]

Proof of the theorem requires that with $p_d = p_q = c = 0$,

$$\sum_{i=1}^{m} En_{ti} - \sum_{i=1}^{m} En_{fi} = -\sum_{i=1}^{m} n_{tzi} \tag{5.117}$$

Since the system is stable in closed-loop,

$$\sum_{i=1}^{m} En_{fi} = \sum_{i=1}^{m} n_{fzi} - \sum_{i=1}^{m} n_{fpi} \equiv -\sum_{i=1}^{m} n_{fpi}$$

and

$$\sum_{i=1}^{m} En_{ti} = \sum_{i=1}^{m} n_{tzi} - \sum_{i=1}^{m} n_{tpi}$$

and consequently,

$$\sum_{i=1}^{m} En_{fi} - \sum_{i=1}^{m} En_{ti} = -\sum_{i=1}^{m} n_{tzi} \tag{5.118}$$

since the pole polynomial of F(s) is the same as the pole polynomial of Q(s)H(s). An alternative form of the theorem is given below, which is followed by an illustrative example.

Theorem 5.16 If for the detectable system in Fig. 5.12b, the ith characteristic loci $q_i(s)$ of Q(s) encircles the critical point (-1 + j0) En_{cqi} times and the origin En_{oqi} times, the system is closed-loop stable if

$$\sum_{i=1}^{m} En_{cqi} - \sum_{i=1}^{m} En_{oqi} = -p_o \tag{5.119}$$

where p_o is the number of right-half s-plane poles in Q(s).

Example 5.4 [33, 39]: Closed-loop stability through generated Nyquist criterion of G(s)

$$G(s) = \begin{bmatrix} \dfrac{s-1}{1.25(s+1)(s+2)} & \dfrac{s}{1.25(s+1)(s+2)} \\[2ex] -\dfrac{6}{1.25(s+2)(s+2)} & \dfrac{s-2}{1.25(s+1)(s+2)} \end{bmatrix} \tag{5.120}$$

The characteristic function for the open-loop system is given by

$$\det\{g(s)I - G(s)\} = \Delta(s, g) = \underbrace{1.56(s+1)(s+2)}_{b_o(s) \equiv p_g(s)}\{g(s)\}^2 - 5(s-1.5)\{g(s)\} + \underbrace{1}_{b_2(s) \equiv z_g(s)} \equiv 0$$

Therefore, the number of right-half s-plane poles = 0 and the number of right-half s-plane zeros = 0.

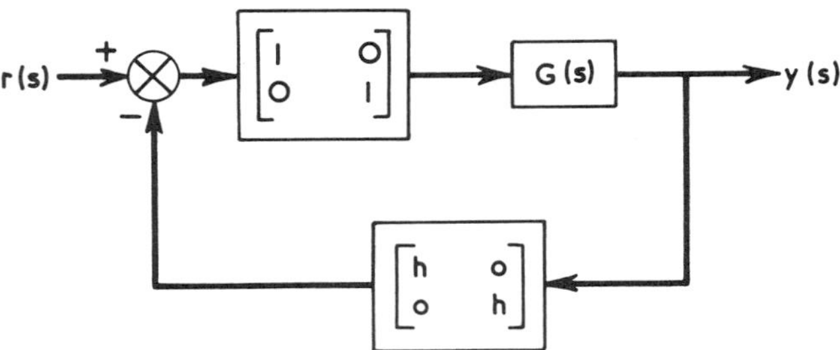

Fig. 5.15 Constant feedback system with G(s) in (5.120).

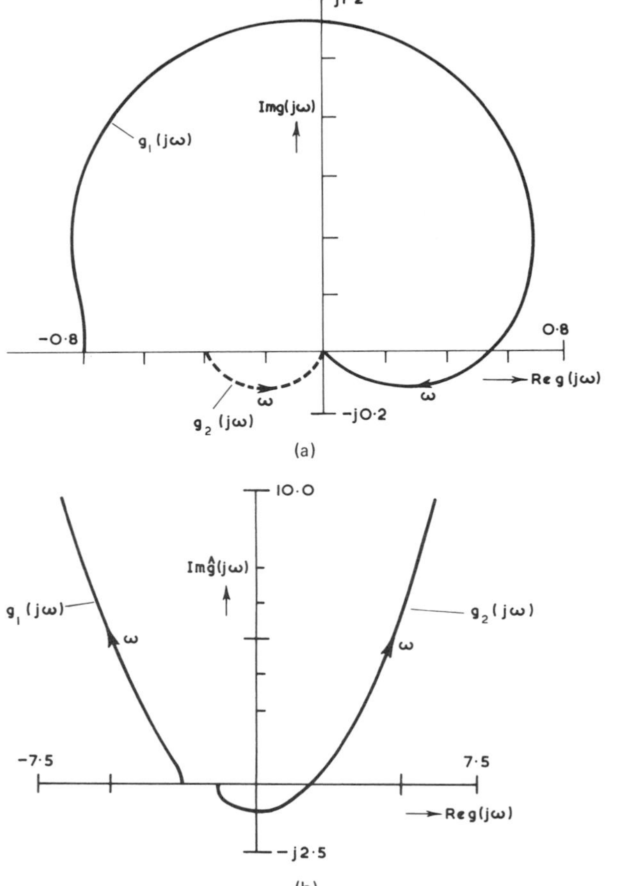

Fig. 5.16 (a) Characteristic loci of Q(s) in Example 5.4. (b) Inverse characteristic gain loci for G(s).

Table 5.1 Effect on the Closed-Loop System of Varying h

Range of critical points on the real axis $-1/h$	Observations on encirclements by $g_1(s)$ or $g_2(s)$	Encirclements by the individual loci CW = +ve	Net encirclement ACW = -ve	Value of h (inverse of range in column 1)	Closed-loop performance[a]
$-\infty$ to -0.8	None	-	0	$1.25 > h > 0$	Stable
-0.8 to -0.4	One CW by $g_1(s)$	$g_1(s)$: +1 $g_2(s)$: 0	+1	$2.5 > h > 1.25$	Unstable
-0.4 to -0.0	One by each in opposite directions	$g_1(s)$: +1 $g_2(s)$: -1	0	$+\infty > h > 2.5$	Stable
$+0.0$ to $+0.53$	Two CW by $g_1(s)$	$g_1(s)$: +2 $g_2(s) = 0$	+2	$-1.88 > h > -\infty$	Unstable
$+0.53$ to $+\infty$	None	-	0	$0 > h > -1.88$	Stable

[a] Stable if $\sum En(g, -1/h) = \sum En(g_i, 0) = 0$.

For the sake of convenience, the closed-loop configuration in Fig. 5.15 is considered, for which

$$K(s) = \begin{bmatrix} 1 & 0 \\ 0 & 1 \end{bmatrix} \quad \text{and} \quad H(s) = \begin{bmatrix} h & 0 \\ 0 & h \end{bmatrix} h \rightarrow \text{constant gain}$$

The characteristic loci of Q(s) ≡ G(s) are shown in Fig. 5.16a. The system is stable in open-loop (no right-half s-plane poles, $p_o = 0$). Let En(g - ∞) be the net sum of encirclements of the point (-∞ + j0) in the clockwise direction, then by taking this point as the critical point, by Theorems 5.14-5.16, the conditions for closed-loop stability become

(1) Theorem 5.15: $\mathrm{En}(g, -\infty) = -z_g$ (5.121)

(2) Theorem 5.16: $\mathrm{En}(g, -\infty) - \mathrm{En}(g, 0) = -p_g$ (5.122)

With the critical point at (-(1/h) + j0), Table 5.1 [39] indicates the effect on the closed-loop system of varying h.

If the values in Table 5.1 were plotted on a two-dimensional gain plane, the stability regions may be clearly identified; this is shown in

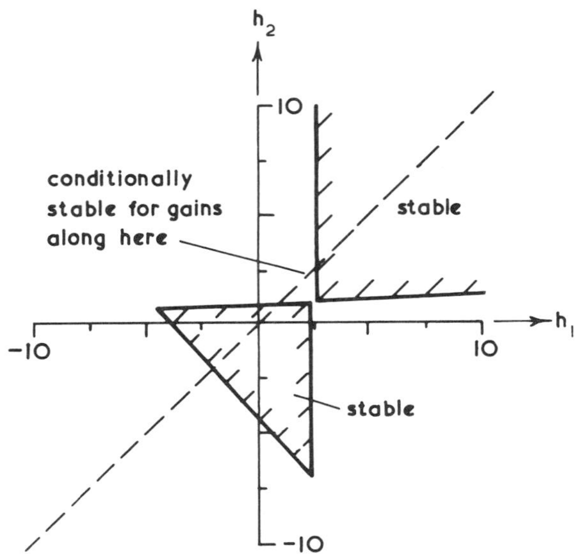

Fig. 5.17 Stability regions with different constant-feedback gains with G(s) in (5.120).

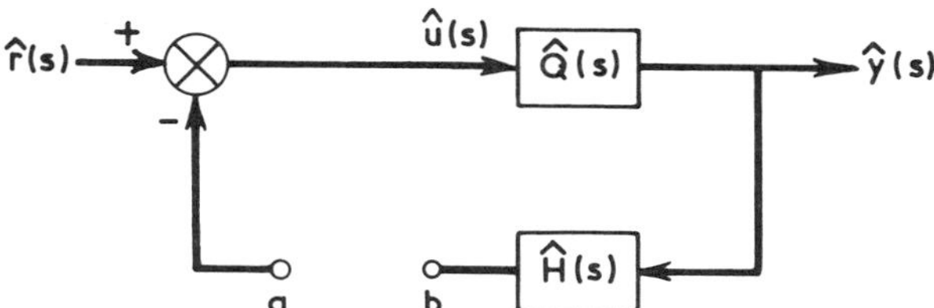

Fig. 5.18 Representation with inverse transfer-function matrices: nonunity feedback (with breaks for the computation of return difference).

Fig. 5.16b. The transition from stability to instability and stability regions as h increases is related to the existence of a branch point at (0.042 + j0). The role of this branch point and the effects of branch points on stability in general have been studied extensively in [31]. Details of these are not considered here.

5.2.3 Inverse Nyquist Criterion [31,32,43]*

The derivations here involve the establishment of a relationship between the open- and closed-loop characteristic polynomials of the m-input/m-output system in Fig. 5.18 using the inverses of the various transfer-function matrices. The following elementary results form the basis of derivations in this section. Let G(s) be the m-square rational transfer-function matrix of the open-loop system in Fig. 5.18. Then its pole and zero polynomials [$p_g(s)$ and $z_g(s)$] may be expressed by using its Smith-McMillan form,†

$$M(s) = L_1(s)G(s)L_2(s) \tag{5.123}$$

$$= \text{diag}\left\{\frac{\alpha_i(s)}{\beta_i(s)}\right\}, \quad i \in 1, \ldots, m \tag{5.124}$$

and

$$p_g(s) = \prod_{i=1}^{m} \beta_i(s), \quad z_g(s) = \prod_{i=1}^{m} \alpha_i(s) \tag{5.125}$$

Then by using (1.127), the inverse of G(s) may be expressed as‡

*Derivations in this section closely follow Ref. 43.

†$\alpha_i(s)$ and $\beta_i(s)$ are used for notational convenience. These are equal to $\psi_i(s)$ and $\phi_i(s)$, respectively, in the Smith-McMillan form in Sec. 1.5.3.

‡A ^ is used in subsequent derivations to denote the inverse of the corresponding matrix.

$$[G(s)]^{-1} = \hat{G}(s) = [L_2(s)]^{-1}[M(s)]^{-1}[L_1(s)]^{-1} = \hat{L}_2(s)\hat{M}(s)\hat{L}_1(s) \tag{5.126}$$

where T is a symmetric constant matrix given by (1.128) and $\hat{M}(s)$ is related to (5.125) by

$$\hat{M}(s) = \text{diag}\left\{\frac{\beta_i(s)}{\alpha_i(s)}\right\}, \quad i \in 1, \ldots, m \tag{5.127}$$

and represents the Smith-McMillan form of $\hat{G}(s)$. Hence the pole and zero polynomials of $\hat{G}(s)$ are

$$p_{\hat{g}}(s) = \prod_{i=1}^{m} \alpha_i(s) \equiv z_g(s)$$

and (5.128)

$$z_{\hat{g}}(s) = \prod_{i=1}^{m} \beta_i(s) = p_g(s)$$

Thus the pole (zero) polynomial of G(s) is equal to the zero (pole) polynomial of $\hat{G}(s)$.

By following the basic concepts introduced in the preceding section, a characteristic function $\hat{g}(s)$ associated with $\hat{G}(s)$ may be defined. This is called the <u>inverse characteristic function</u> and like g(s) [the characteristic function of G(s)] is an algebraic function. However, as the eigenvalues of a matrix are the reciprocals of the eigenvalues of the inverse matrix, the branch points of g(s) and $\hat{g}(s)$ are the same, and consequently they share the same Riemann surface [31,43]. This observation, together with the relationship between pole and zero polynomials of G(s) and $\hat{G}(s)$, forms the basis of the following development.

To derive the relationship between characteristic polynomials and poles and zeros of various inverse transfer-function matrices, the following identities are necessary.

1. From (5.64) for the closed-loop system in Fig. 5.18,

$$[R(s)]^{-1} = [Q(s)]^{-1}[I_m + Q(s)H(s)]$$

that is,

$$\hat{R}(s) = \hat{Q}(s) + H(s) \tag{5.129}$$

2. Consequently, from (5.70),

$$\frac{\det\{F(s)\}}{\Delta_f(\infty)} \triangleq \frac{\text{CLCP}}{\text{LGCP}} \triangleq \frac{\phi_c(s)}{\phi_o(s)} = \frac{\det\{Q(s)\}}{\det\{R(s)\}} \equiv \frac{\det\{\hat{R}(s)\}}{\det\{\hat{Q}(s)\}}$$

that is,

$$\frac{\det\{\hat{R}(s)\}}{\det\{\hat{Q}(s)\}} = \frac{\phi_c(s)}{\phi_o(s)} \tag{5.130}$$

where $\phi_o(s)$ and $\phi_c(s)$ are as defined earlier.

Thus if $p_{\hat{r}}(s)$ and $p_{\hat{q}}(s)$ represent the pole polynomials, and $z_{\hat{r}}(s)$ and $z_{\hat{q}}(s)$ the zero polynomials of $\hat{R}(s)$ and $\hat{Q}(s)$, then by using (5.128),

$$\det\{\hat{R}(s)\} \triangleq \frac{z_{\hat{r}}(s)}{p_{\hat{r}}(s)} \equiv \frac{p_r(s)}{z_r(s)}$$

$$\det\{\hat{Q}(s)\} \triangleq \frac{z_{\hat{q}}(s)}{p_{\hat{q}}(s)} \equiv \frac{p_q(s)}{z_q(s)} \tag{5.131}$$

where $p_r(s)[p_q(s)]$ and $z_r(s)[z_q(s)]$ represent the pole and zero polynomials of $R(s)[Q(s)]$, and consequently, for the inverse system,

$$\frac{\det\{\hat{R}(s)\}}{\det\{\hat{Q}(s)\}} = \frac{z_{\hat{r}}(s)}{p_{\hat{r}}(s)}\,\frac{p_{\hat{q}}(s)}{z_{\hat{q}}(s)} \equiv \frac{z_{\hat{r}}(s)}{p_{\hat{q}}(s)p_h(s)}\,\frac{p_{\hat{q}}(s)}{z_{\hat{q}}(s)}; \quad [\text{since } p_{\hat{r}}(s) = p_{\hat{q}}(s)p_h(s), \text{ from (5.129)}]$$

$$= \frac{z_{\hat{r}}(s)}{p_h(s)p_q(s)} = \frac{p_r(s)}{p_h(s)p_q(s)} \tag{5.132}$$

which is consistent with (5.130). Combining (5.104), (5.105), (5.130), (5.131), and (5.132), we have

$$\phi_c(s) = \frac{\det\{\hat{R}(s)\}}{\det\{\hat{Q}(s)\}}\phi_o(s) = \frac{z_{\hat{r}}(s)}{p_h(s)p_q(s)}\,p_q(s)p_h(s)p_d(s)p_{dt}(s)$$

$$\equiv p_{dt}(s)p_d(s)c(s)\prod_{i=1}^{m} b^{\hat{r}}_{i,ki}(s) \tag{5.133}$$

where the $b^{\hat{r}}_{i,ki}(s)$ are the coefficients of the algebraic functions associated with $\hat{R}(s)$:

$$\Delta_i(s,\hat{r}_i) = b^{\hat{r}}_{i,0}(s)[\hat{r}_i(s)]^{ki} + b^{\hat{r}}_{i,1}(s)[\hat{r}_i(s)]^{ki-1} + \cdots + b^{\hat{r}}_{i,ki}(s)$$

and consequently,

$$z_{\hat{r}}(s) \triangleq c(s)\prod_{i=1}^{m} b^{\hat{r}}_{i,ki}(s) \quad \text{and} \quad p_{\hat{r}}(s) \triangleq c(s)\prod_{i=1}^{m} b^{\hat{r}}_{i,0}(s) \tag{5.134}$$

where $p_{dt}(s)$, $p_d(s)$, and $c(s)$ are as defined in Sec. 5.2.2.

Although the derivations above relate the open-loop characteristic polynomial to the characteristic functions $\hat{r}_i(s)$ of the inverse closed-loop transfer-function matrix, these do not lead to any general statement on the stability of the closed-loop system in terms of $\hat{Q}(s)$, Q(s), and H(s). This is due to the structure of $\hat{R}(s)$ given by (5.129), which, while providing a general structure of the denominator polynomial, does not provide a correlation between the zeros of $\hat{R}(s)$, $\hat{Q}(s)$, and H(s). This has led to the formulation of stability criterion through inverse transfer functions for two specific cases:

1. Unity feedback or feedback through a constant diagonal matrix, $H(s) = I_m$ or $H(s) = \text{diag}\{h_i\}$ [h_i being constants, $i \in 1, \ldots, m$]
2. Diagonally dominant systems, where $\hat{Q}(s)$ and H(s) are chosen such that $\hat{R}(s)$ is diagonally dominant

The unity feedback case is considered here and the diagonal dominance in the following section.

For the unity feedback case (Fig. 5.19), the closed-loop inverse transfer-function matrix in (5.129) becomes

$$\hat{R}(s) = \hat{Q}(s) + I \equiv \hat{T}(s) + I \tag{5.135}$$

where

$$\hat{\phi}_o(s) = p_{dq}(s)p(s) \qquad \text{and} \qquad \det\{\hat{R}(s)\} = \frac{\hat{z}(s)}{\hat{p}(s)} \tag{5.136}$$

$p_{dq}(s)$: cancellations in forming Q(s); $p(s)$: actually appearing in Q(s)

where $\hat{z}(s)$ and $\hat{p}(s)$ are related to the characteristic functions $\hat{r}_i(s)$ of $\hat{R}(s)$ of the form

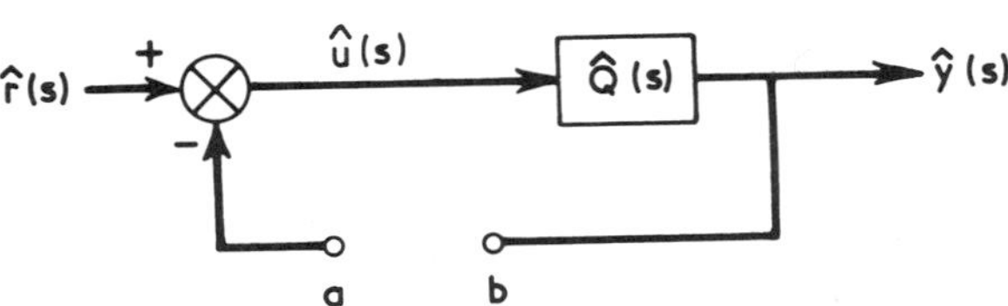

Fig. 5.19 Representation with inverse transfer-function matrices: unity feedback (with breaks for the computation of return difference).

$$\Delta_i(s,\hat{r}_i) = b^{\hat{r}}_{i,0}(s)[\hat{r}_i(s)]^{ki} + b^{\hat{r}}_{i,1}(s)[\hat{r}_i(s)]^{ki-1} + \cdots + b^{\hat{r}}_{i,ki}(s) \quad (5.137a)$$

by

$$\hat{z}(s) = c(s)\prod_{i=1}^{m} b^{\hat{r}}_{i,ki}(s) \quad \text{and} \quad p(s) = c(s)\prod_{i=1}^{m} b^{\hat{r}}_{i,o}(s) \quad (5.137b)$$

corresponds to cancellations in $\hat{R}(s)$

Therefore, combining (5.130), (5.136), and (5.137), the closed-loop characteristic polynomial of (5.136) is given by

$$\hat{\phi}_c(s) = p_{dq}(s)c(s)\prod_{i=1}^{m} b^{\hat{r}}_{i,ki}(s) \quad (5.138)$$

which is of the same form as (5.108) except that in this case, the characteristic functions are associated with the inverse of the closed-loop transfer function $[\hat{R}(s)]$ rather than the return-ratio matrix $[F(s)]$. Furthermore, in this case, due to (5.135), the characteristic functions $\hat{r}_i(s)$ are the same as the characteristic functions associated with $\hat{Q}(s)$, which are the inverse characteristic functions of the return-ratio matrix. Then using the notations

- $\hat{p}_o$ = number of right-half s-plane roots of $\hat{\phi}_o(s) = 0$
- p_{dq} = number of right-half s-plane roots of $p_{dq}(s) = 0$
- c = number of right-half s-plane roots of $c(s) = 0$
- $\hat{n}_{\hat{r}zi}$ = number of right-half s-plane zeros associated with the characteristic functions $\hat{r}_i(s)$ of $\hat{R}(s)$, $i \in 1, \ldots, m$
- $\hat{n}_{\hat{r}pi}$ = number of right-half s-plane poles associated with the characteristic functions $\hat{r}_i(s)$ of $\hat{R}(s)$, $i \in 1, \ldots, m$

and the observations above, the following relations may be derived from (5.110) and (5.111):

$$(1)\quad \hat{p}_o = \sum_{i=1}^{m} \hat{n}_{\hat{r}pi} + p_{dq} + c \quad (5.139)$$

and for stability:

$$(2)\quad \sum_{i=1}^{m} \hat{E}n_{\hat{r}i} = \sum_{i=1}^{m} \hat{n}_{\hat{r}zi} - \sum_{i=1}^{m} \hat{n}_{\hat{r}pi} \quad (5.140a)$$

or

$$(3)\quad \sum_{i=1}^{m} \hat{E}n_{\hat{r}i} = \sum_{i=1}^{m} \hat{n}_{\hat{r}zi} + p_{dq} + c - \hat{p}_o \tag{5.140b}$$

where $\hat{E}n$ denotes clockwise encirclement of the appropriate inverse characteristic loci, and the parameters in the equation above are either zero or positive. Since $\Sigma_{i=1}^{m} \hat{n}_{\hat{r}zi}$ in (5.140) denotes the total number of right-half s-plane zeros of the characteristic function $\hat{r}_i(s)$, (5.140b) leads to the following necessary and sufficient condition for stability, when as a special case $p_{dq} = c = \Sigma_{i=1}^{m} \hat{n}_{\hat{r}zi} \equiv 0$:

$$\sum_{i=1}^{m} \hat{E}n_{\hat{r}i} = -\hat{p}_o \tag{5.141}$$

By shifting the origin of the imaginary axis and by using (5.135), this condition may be reduced to

$$\sum_{i=1}^{n} \hat{E}n_{\hat{q}i} = -\hat{p}_o \tag{5.142}$$

where $\hat{E}n_{\hat{q}i}$ denotes the net sum of encirclements of the critical point $(-1+j0)$ in the complex plane by the ith characteristic loci associated with the ith characteristic function $\hat{q}_i(s)$ of $\hat{Q}(s)$. Formal statement of the inverse Nyquist stability criterion may now be given.

Theorem 5.17 The closed-loop system in Fig. 5.19 is stable only if

Net sum of counterclockwise encirclements of the critical point (-1 + j0) by the set of inverse characteristic loci of the return-ratio matrix [Q(s)]	=	total number of right-half s-plane zeros of Q(s)

Proof of the theorem follows from the fact that (1) the zeros of Q(s) are equivalent to the poles of $\hat{Q}(s)$ [i.e., the zeros of the loop-gain characteristic polynomial $\hat{\phi}_o(s)$]; (2) the inverse characteristic loci of the return-difference matrix [≡ Q(s)] are identical to the inverse characteristic loci of the closed-loop transfer function matrix [R(s)] except that the former set is obtained from the latter by a change of origin from (0, 0) to (-1 + j0); and (3) the characteristic loci of any square transfer-function matrix Q(s) are the same as the inverse characteristic loci of the corresponding inverse transfer-function matrix $\hat{Q}(s)$. Hence the proof follows from (5.142). Thus if the uncontrollable and unobservable modes of G(s) in Fig. 5.19 are stable [i.e., $p_{dq} = 0$], Theorem 5.17 is equivalent to the more general inverse Nyquist stability criterion.

Theorem 5.18 The closed-loop system in Fig. 5.19 is stable if

Net sum of counterclockwise encirclements of the critical point $(-1 + j0)$ by the set of inverse characteristic loci of $Q(s) = G(s)K(s)$ $(\equiv \Sigma\, \hat{E}_{\hat{q}i})$

- net sum of counterclockwise encirclements of the origin by the set of inverse characteristic loci of $Q(s)$ $[\equiv \Sigma\, \hat{E}n_{\hat{r}i}]$

= total number of right-half s-plane poles of G(s) and K(s)

The proof of the theorem requires that

$$\sum_{i=1}^{m} \hat{E}n_{\hat{q}i} - \sum_{i=1}^{m} \hat{E}n_{\hat{r}i} = \sum_{i=1}^{m} n_{qpi}$$

where n_{qpi} is the number of right-half s-plane poles associated with the characteristic function $q_i(s)$ of $Q(s)$.

Since the system has no unstable uncontrollable or unobservable modes, $p_{dq}(s) = 0$ and consequently, from (5.100) for

$$\text{closed loop} \rightarrow \sum \hat{E}n_{\hat{r}i} = \sum \hat{n}_{\hat{r}zi} - \sum \hat{n}_{\hat{r}pi} \tag{5.143}$$

$$\text{open loop} \rightarrow \sum En_{qi} = \sum \hat{n}_{\hat{q}zi} - \sum \hat{n}_{\hat{q}pi} \tag{5.144}$$

where the pole polynomials of $\hat{R}(s)$ and $\hat{Q}(s)$ are identical, and consequently from above,

$$\sum \hat{E}n_{\hat{q}i} - \sum \hat{E}n_{\hat{r}i} = \sum \hat{n}_{\hat{q}zi} - \sum \hat{n}_{\hat{r}zi} \equiv \sum_{i=1}^{n} n_{qpi} \tag{5.145}$$

since for stability of the closed-loop system, $\hat{R}(s)$ cannot have any zeros on the right-half s-plane.

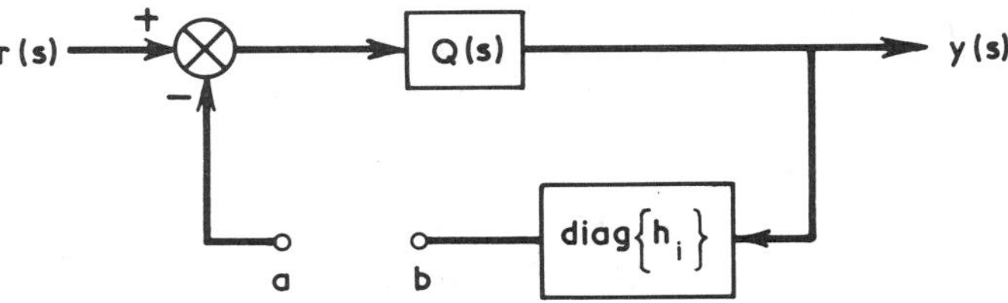

Fig. 5.20 System with constant-diagonal-feedback matrix.

If the feedback configuration in Fig. 5.20 is used, the closed-loop system transfer function becomes

$$\hat{R}(s) = \hat{Q}(s) + \text{diag}\{h_i\} \tag{5.146}$$

and consequently,

$$\hat{r}_i(s) = \hat{q}_i(s) + h_i, \quad i \in 1, \ldots, m \tag{5.147}$$

This implies that the critical point for the ith inverse characteristic loci is moved from (-1 + j0) to ($-h_i$ + j0). In the simplest case where $h_i = h$, $\forall\, i \in 1, \ldots, m$, the critical point is shifted from (-1 + j0) to (-h + j0) for all loci. In both cases the corresponding modifications in Theorems 5.17 and 5.18 may be easily made and are stated in Theorem 5.19. This is followed by an illustrative example.

Theorem 5.19 If for the detectable system in Fig. 5.20, the characteristic loci $\hat{q}_i(s)$ of $\hat{Q}(s)$ encircle the critical point ($-h_i$ + j0) $\hat{E}n_{c\hat{q}i}$ times and the origin $\hat{E}n_{o\hat{q}i}$ times, the closed-loop system is stable only if

$$\sum_{i=1}^{m} \hat{E}n_{c\hat{q}i} - \sum_{i=1}^{m} \hat{E}n_{o\hat{q}i} = -p_o$$

where $\hat{p}_o$ is the number of right-half s-plane poles of Q(s)

Example 5.5: Closed-loop stability through inverse Nyquist criterion The system with loop transfer function in (5.120) is considered here. For the sake of simplicity, and for comparison with the earlier result, the configuration in Fig. 5.15 is analyzed.

The inverse characteristic loci $\hat{q}_1(s)$ and $\hat{q}_2(s)$ of $Q(s) = K(s)G(s) \equiv G(s)$ are shown in Fig. 5.17. The conditions for closed-loop stability in this case, from Theorems 5.17 and 5.18, are:

(1) Theorem 5.17: $En(\hat{q}_1, -\alpha) = -z_q$ (5.148)

(2) Theorem 5.18: $En(\hat{q}, -\alpha) - En(\hat{q}, 0) = -p_q$ (5.149)

where $En(\hat{q}, \cdot)$ represents the net sum of clockwise encirclements of the point (·) by the inverse characteristic loci $\hat{q}(s)$ of Q(s). With these conditions, from Fig. 5.17 with a critical point at (-h, 0), the following observations may be made.

1. Critical point in $(-\infty, -2.5)$: No encirclements of any point in this region; hence the closed-loop system is stable for values of feedback gain $+\infty > h > +2.5$.
2. Critical point in $(-2.5, -1.25)$: One encirclement by $g_1(s)$ in clock-

wise direction, and none by $g_2(s)$; hence the system is unstable for $+2.5 > h > +1.25$.

3. Critical point in (-1.25,+1.88): One encirclement by $g_1(s)$ in clockwise direction and one by $g_2(s)$ in counterclockwise direction; hence net encirclement is zero, and the system is stable for $+1.25 > h > -0.188$.

These observations are similar to those in Table 5.1, which is a consequence of the fact that the characteristic loci of $\hat{G}(s)$ and the inverse characteristic loci of $G(s)$ are identical, as noted earlier.

5.2.4 Diagonal Dominance [44-51]*

The stability criteria in the preceding sections are based on the determination of rather complicated functions of parameters (e.g., feedback gains) which need to be adjusted during the synthesis. It is thus desirable to relate stability conditions to the behavior of the Nyquist plots of individual elements of the various transfer function matrices. This would allow the treatment of the system, to a certain extent, like a set of independent single-input/single-output subsystems or channels. This can be achieved through the concept of diagonal dominance of matrices in the field of complex numbers.

Definition 5.4: Diagonal dominance A square matrix A of order m, in the field of complex numbers, is diagonally dominant (d.d.) if for every $i \in 1, \ldots, m$, and every j in this range except $j \neq i$,

$$|a_{ii}| > \sum_{\substack{j=1 \\ j\neq i}}^{m} |a_{ij}| \tag{5.150a}$$

$$|a_{ii}| > \sum_{\substack{j=1 \\ j\neq i}}^{m} |a_{ji}| \tag{5.150b}$$

$$|a_{ii}| > \sum_{\substack{j=1 \\ j\neq i}}^{m} (|a_{ij}| + |a_{ji}|)/2 \tag{5.150c}$$

If (5.150a) is satisfied, the modulus of every leading diagonal element exceeds the sum of the moduli of the other elements in the same row of A:

*Results presented here are based on original work of Rosenbrock and coworkers [44-48].

then A is said to be diagonally dominant row-wise (d.d.r.). Similarly, if (5.150b) is saisfied, A is said to be diagonally dominant column-wise (d.d.c.). If (5.150c) applies, the matrix is said to have mean diagonal dominance (d.d.m.). When it is not necessary to distinguish between row, column, or mean dominance, A is simply termed to have diagonal dominance.

The significance of the foregoing conditions is that they imply that (1) the matrix A is nonsingular, and (2) the d.d. matrix A remains so whenever any of its diagonal elements is increased in magnitude and whenever any off-diagonal element is decreased in amplitude.

Definition 5.5: Diagonal dominance on a contour A square matrix Z(s) of order m, in the field of rational functions of the complex variable s, is diagonally dominant on a contour $\mathscr{D}$ in the (complex frequency) s-plane if it is either d.d.c., d.d.r., or d.d.m. for every value of s on $\mathscr{D}$.

Although the definition above is applicable to the field of complex rational functions of s, only the field of real rational functions of s is of interest in the design procedure. One useful extension of this definition is that if Z(s) is d.d. on $\mathscr{D}$, then f(s)Z(s) is also d.d. on $\mathscr{D}$, where f(s) is a real function of s such that no zeros or poles of f(s) lie on $\mathscr{D}$.

A consequence of the definitions above and the stability criteria in the preceding section is the following theorem [47].

Theorem 5.20 If a rational transfer function matrix $G(s) = \{g_{ij}(s)\}$, $i, j \in 1, \ldots, m$, is diagonal-dominant for every s on $\mathscr{D}$, then the number of encirclements of the origin by the Nyquist plot of det $\{G(s)\}$ is the sum of the numbers of encirclements by the Nyquist plots of the diagonal elements of G(s).

This theorem, proof not considered here, leads to the following stability criterion for the closed-loop system in Fig. 5.20 [47].

Theorem 5.21 For the closed-loop system in Fig. 5.20, the Nyquist plot of $q_{ii}(s)$ encircle the point $((1/h) + j0)$, En_{qii} times, and for each s on $\mathscr{D}$,

$$\left| q_{ii}(s) + \frac{1}{h_i} \right| > \delta_i(s), \qquad i \in 1, \ldots, m \tag{5.151}$$

where

$$\delta_i(s) = \sum_{\substack{j=1 \\ j \neq i}}^{m} |q_{ij}(s)| \quad \text{or} \quad \delta_i(s) = \sum_{\substack{j=1 \\ j \neq i}}^{m} |q_{ji}(s)| \quad \text{or} \quad \delta_i(s) = \sum_{\substack{j=1 \\ j \neq i}}^{m} [|q_{ij}(s) + q_{ji}(s)|]/2$$

[i.e., $Q(s) = \{q_{ij}(s)\}$ is d.d.]. Then the closed-loop system is asymptotically stable if

$$\sum_{i=1}^{m} En_{qii} = -p_o \tag{5.152}$$

where p_o denotes the number of right-half s-plane poles* in Q(s) and En is taken to be positive in the clockwise direction.

Theorem 5.21 enables one to obtain a subset of the stability region in the space of feedback gains h_i by drawing the Nyquist plot of each $q_{ii}(s)$ surrounded by circles of radius $\delta_i(s)$. Moreover, the bounds obtained for each feedback gain are independent of the values at which the other gains are set. This theorem, combined with previous results, leads to the following generalized inverse Nyquist criterion for diagonally dominant systems.

<u>Theorem 5.22</u> If the Nyquist plot of $\hat{q}_{ii}(s)$, where $[Q(s)]^{-1} = [G(s)K(s)]^{-1} = \{\hat{q}_{ii}(s)\}$, encircles the point $-h_i$ $\hat{E}n_{c\hat{q}ii}$ times, and the origin $\hat{E}n_{oqii}$ times, such that for each s on $\mathscr{D}$,

$$|\hat{q}_{ii}(s) + h_i| > \hat{\delta}_i(s) \quad \text{and} \quad |\hat{q}_{ii}(s)| > \hat{\delta}_i(s), \quad i \in 1, \ldots, m$$

where $\hat{\delta}_i(s)$ is defined in terms of the off-diagonal element of $\hat{Q}(s) = [Q(s)]^{-1}$ in the same way as $\delta_i(s)$ in terms of the off-diagonal elements of Q(s) in Theorem 5.21, the closed-loop system is asymptotically stable only if

$$\sum_{i=1}^{m} En_{cqii} - \sum_{i=1}^{m} \hat{E}n_{oqii} = -p_o \tag{5.153}$$

where p_o is the number of right-half s-plane poles of Q(s). It should be noted that, although different forms of diagonal dominance are allowed, the same definition of $\delta_i(s)$ and $\hat{\delta}_i(s)$ will have to be used for each i at any given point s on $\mathscr{D}$.

The relationship between diagonal dominance and stability may be studied further through the concept of Gershgorin bands or disks [39,44,46, 50]. This is outlined briefly below.

<u>Theorem 5.23</u> Gershgorin theorem. If z is a square matrix of order m in the field of complex numbers, its eigenvalues lie, in the complex plane, within the family of m circles (called <u>Gershgorin circles</u>) having their centers at z_{ii} with radii $\Sigma_{j=1, j\neq i}^{m} |z_{ij}|$. Moreover, since the eigenvalues are unchanged by transposing Z, the eigenvalues of Z^T also lie within the

*That is, p_o is equal to the number of right-half s-plane zeros in the open-loop characteristic polynomial $\phi_o(s)$.

union of the circles having their centers at z_{ii} with radii $\Sigma_{j, j \neq i} |z_{ji}|$, where $i \in 1, 2, \ldots, m$.

The proof of this theorem is omitted here, but a simple example is given below. A few comments are relevant [49]:

1. The first set of circles are sometimes referred to as the row set of Gershgorin circles and the second set as the column set of Gershgorin circles.
2. If Z is a diagonal matrix, the elements z_{ii} are themselves eigenvalues of Z and all radii are zero. The theorem is then a statement of the limits of the area of the complex plane within which the eigenvalues are confined when the nondiagonal elements are added to a diagonal matrix.
3. The theorem does not state that one particular eigenvalue remains within one particular Gershgorin circle as these nondiagonal elements are increased in modulus, for as this occurs, the circles of either set increase in radius and at some stage overlap, allowing an eigenvalue to move from inside one circle to the inside of another.

Example 5.6: Gershgorin circles of Z

$$Z = \begin{bmatrix} 0 & 1 \\ -2 & -3 \end{bmatrix} \tag{5.154}$$

1. Eigenvalues of Z: $\lambda_1 = -1$, $\lambda_2 = -2$.
2. Gershgorin circles: These are drawn with radii 1 and 2 and centers at 0 and -3, respectively (Fig. 5.21).
3. By Theorem 5.23, λ_1 and λ_2 should be within the above two circles, which is consistent with (1) above.

In some cases the eigenvalues may lie within fewer than m circles rather than within all of them.

Definition 5.6: Gershgorin bands [44,49,50] If Z(s) is in the field of real rational functions of s, and s describes some simple contour $\mathscr{D}$ in the s-plane, then in general the center of a Gershgorin circle ($z_{ii}(s)$) and its radius ($\Sigma_{j, j \neq i} |z_{ij}(s)|$ or $\Sigma |z_{ji}(s)|$) will vary with s. As s describes $\mathscr{D}$ once, the assembly of points in the complex plane on or within the successive positions of a particular Gershgorin circle as s goes around $\mathscr{D}$ is called a Gershgorin band (or Gershgorin disk).

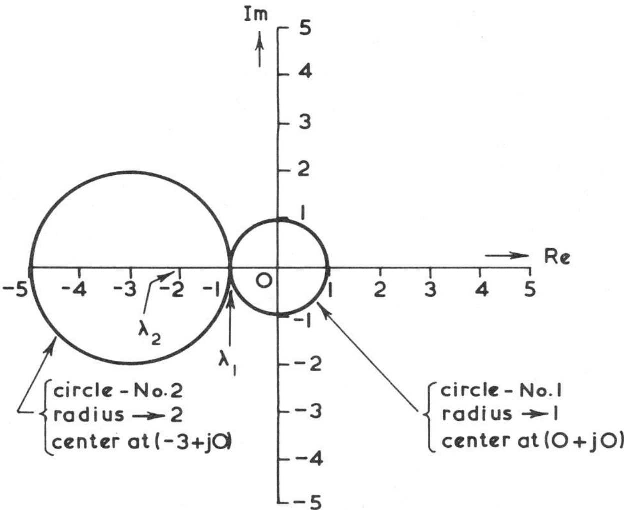

Fig. 5.21 Gershgorin circles of (5.154).

To outline the graphical interpretation of Definition 5.6, let $z_{11}(s)$ and $z_{22}(s)$ be the diagonal elements of a d.d. 2×2 rational matrix Z(s). Let the two graphs of $z_{11}(s)$ and $z_{22}(s)$ be as shown in Fig. 5.22a, plotted with s (= jω) increasing from 0. For some selected values of s, let a family of Gershgorin circles be drawn; these are shown in Fig. 5.22b. (In this case, there are circles with center at s and radii $|z_{12}(s)|$ and $|z_{21}(s)|$ for the two diagonal elements. The two graphs in Fig. 5.22a and b are now superimposed to obtain Fig. 5.22c, with the contours of the circles being joined by dashed lines. The area covered by these lines for each diagonal element of Z(s) (Fig. 5.22d) is the corresponding Gershgorin band. With this basic concept, the following theorem may be noted.

Theorem 5.24 Let $\hat{Q}(s)$ and $\{\hat{Q}(s) + H\}$ be the diagonally dominant on $\mathscr{D}$. Let the image of $\mathscr{D}$ under $\hat{q}_{ii}$ go $\hat{En}_{oi}$ times around the origin and $\hat{En}_{ci}$ times around the critical point $(-h_i, 0)$, $i \in 1, \ldots, m$. If the open-loop system has p_o poles in the complex right-half plane, the system in Fig. 5.20 is asymptotically stable only if

$$\sum_{i=1}^{m} \hat{En}_{ci} - \sum_{i=1}^{m} \hat{En}_{oi} = -p_o \tag{5.155}$$

Theorem 5.24 (proof follows from previous derivation) forms the basis of the following interpretation [46]. First the usual Nyquist plots of $\hat{q}_{ii}$

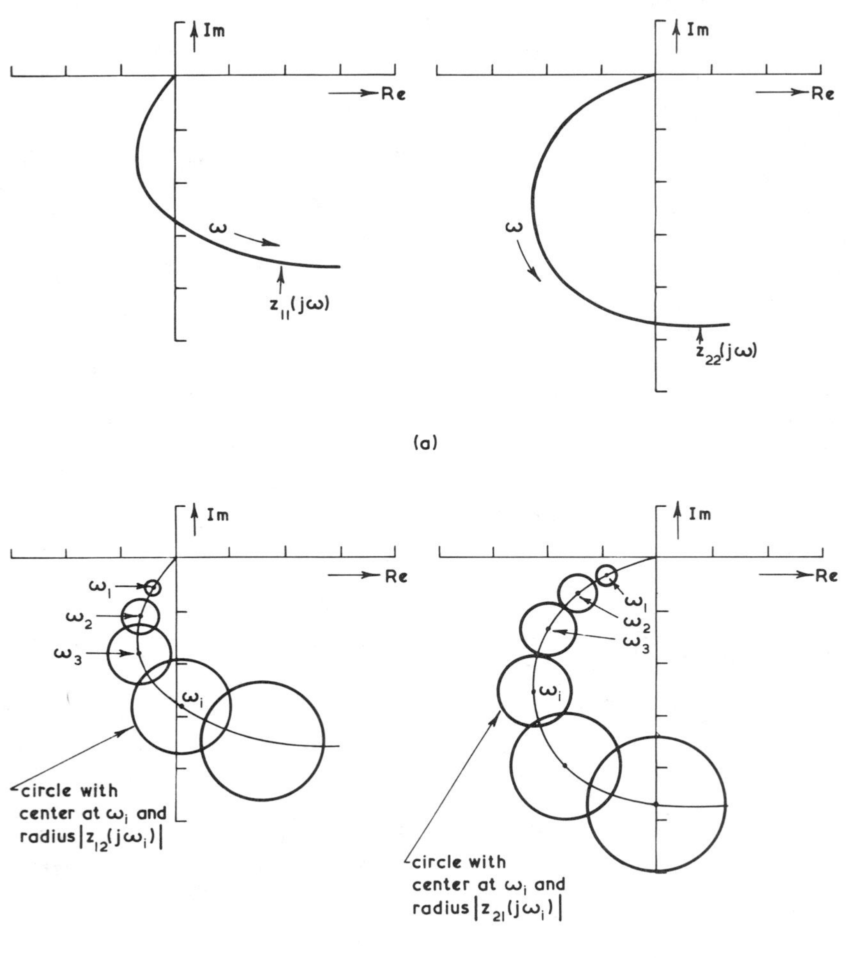

are drawn. On this are superimposed the Gershgorin circles with center $\hat{q}_{ii}(s)$ and radius

$$\hat{\delta}_i(s) = \sum_{\substack{j=1 \\ j\neq i}}^{m} |q_{ij}(s)| \quad \text{or} \quad \sum_{\substack{j=1 \\ j\neq i}}^{m} |\hat{q}_{ji}(s)|$$

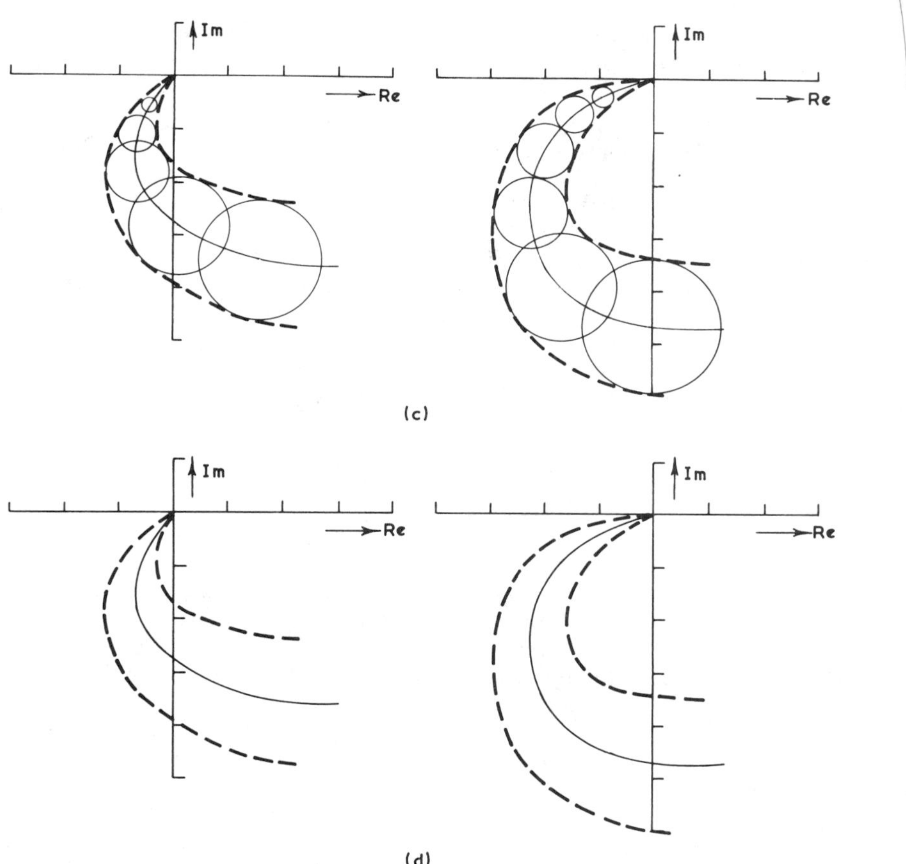

Fig. 5.22 Graphical development of Gershgorin bands for $Z(s) = [z_{ij}(s)]$, $i, j = 1, 2$: (a) complex plane plotting of $z_{11}(j\omega)$ and $z_{22}(j\omega)$; (b) Gershgorin circles with arbitrary centers and radii; (c) contours of the Gershgorin circles; (d) Gershgorin bands.

for a selected value of s on $\mathscr{D}$ (Fig. 5.22). These circles sweep out the Gershgorin band as s goes once round D. If this band does not include the origin or the critical point $(-h_i, 0)$ for all $i \in 1, \ldots, m$, the $\hat{Q}(s)$ and $\hat{Q}(s) + H$ are dominant on $\mathscr{D}$. If $\hat{Q}(s)$ is d.d. on $\mathscr{D}$, then $\hat{E}n_{oi}$ is the number of times the Gershgorin band encircles the origin, while $\hat{E}n_{ci}$ is the number of times it encircles the critical point $(-h_i, 0)$. If the origin or the critical point is inside the Gershgorin band, there is no dominance. Thus the Gershgorin bands may be used as graphical means of testing the diagonal dominance of a given $\hat{Q}(s)$ by using the following criteria:

1. If the Gershgorin bands swept out by the diagonal elements of $\hat{Q}(s)$ exclude the origin, $\hat{Q}(s)$ is d.d. on the contour involved.
2. If the Gershgorin bands swept out by the diagonal elements of $\hat{R}(s)$ [$= \hat{Q}(s) + H$] exclude the points $(-h_i, 0)$, then $\hat{R}(s)$ is d.d. on the contour involved.
3. Consequently, if H is diagonal (or d.d.), the stability of $\hat{R}(s)$ is controlled by the same criterion as those obtained by setting all off-diagonal terms of $\hat{Q}(s)$ to zero.

Thus if there is d.d., (5.155) gives a necessary and sufficient condition for asymptotic stability in terms of encirclements by the Gershgorin bands.

5.3 DESIGN REQUIREMENTS

Although most theoretical analysis of control systems are concerned primarily with stability, in practical application there are other factors which, although not always quantifiable, may have a significant influence on system behavior. Three of such factors—sensitivity, interaction, and integrity—which are considered to be most significant and which may be analyzed together with the system stability are outlined here. The aim of the derivations below is not to present an extensive analysis but rather to include an introductory discussion with a view to outlining their influence on system performance within a broad framework.

5.3.1 Sensitivity [52-57]

Although a certain number of design objectives may be achieved through the modification of the system transfer function in open-loop through a precompensator, feedback configuration is preferred, due primarily to the following factors:

1. Effects of variations in system parameters are reduced in the presence of a feedback loop.
2. Effects of disturbance (unknown and mostly undesirable) inputs may be reduced through feedback.

A very brief account of these two aspects is presented here.

(a) Parameter variation

In his classical study, Bode defined feedback largely in terms of its ability to reduce parameter-variation effects, and defined a sensitivity function as the logarithmic derivative of the system transfer function with respect to a parameter α, given by

$$S_g(\alpha, s) = \frac{\partial[\log_e g(s,x)]}{\partial x} = \frac{d\{g(s,\alpha)\}}{d\alpha} \frac{\alpha}{g(s,\alpha)} \tag{5.156}$$

where $g(s, \alpha)$ is the scalar transfer function and α is a parameter in $g(\cdot)$ which is subject to variation. A good feedback design ensures that $|S(\cdot)| \ll 1$ in the frequency band of interest. If the parameter variation is large, then for scalar transfer function, an equivalent sensitivity function is

$$S_g(\alpha, s) = \frac{\Delta g(s,\alpha)}{g(s,\alpha)} \frac{\alpha}{\Delta\alpha} \tag{5.157}$$

where $\Delta g(s, \alpha)$ is the change in $g(s, \alpha)$ due to a change $\Delta\alpha$ in α, where α and $g(\cdot)$ represent the unchanged ($\equiv$ nominal) values.

If $S_g(\alpha, s)$ is real, the Bode sensitivity function is the ratio of the percentage change in $g(s, \alpha)$ to the percentage change in α. This definition may be easily extended to include the sensitivity of one transfer function $g(\cdot)$ with respect to another $h(\cdot)$, giving a function

$$S_g(h, s) = \frac{\partial[\log_e g\{s, h(s)\}]}{\partial\{h(s)\}} \tag{5.158}$$

The definition above leads to the commonly used expression for sensitivity function of a single-input/single-output closed-loop system (Fig. 5.23).

$$S_g(h, s) = \frac{1}{1 + g(s)k(s)h(s)} \tag{5.159}$$

In the following derivation a similar expression and an equivalent form for a multi-input/multioutput system is established.

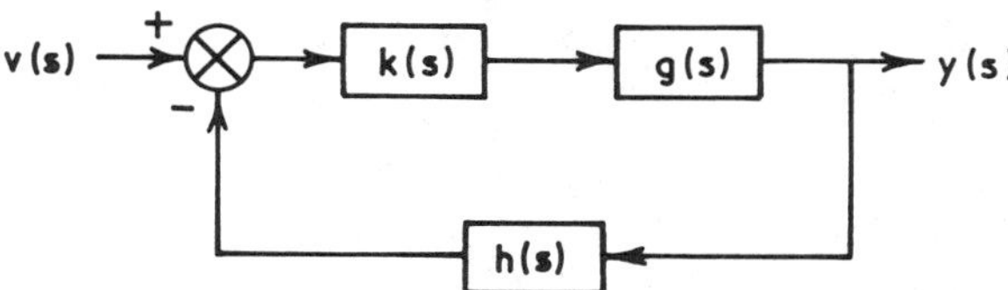

Fig. 5.23 General feedback configuration of a single-input/single-output system.

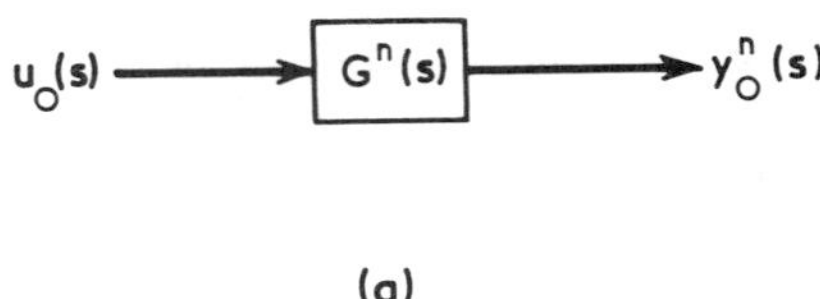

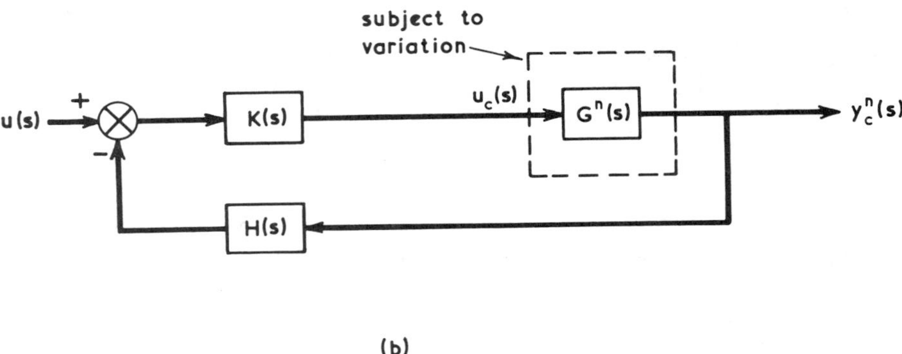

Fig. 5.24 Nominal (without parameter variations) system configurations: (a) open loop; (b) closed loop.

By using the notation in Fig. 5.24 and assuming that only the system transfer function G(s) is subject to variation, the following relationships may be derived for the nominal case:

(1) Nominal open-loop: $y_o^n(s) = G^n(s)u_o(s)$ (5.160)

(2) Nominal closed-loop: $y_c^n(s) = G^n(s)u_c(s);$

$$u_c(s) = K(s)[u(s) - H(s)y_c^n(s)] \tag{5.161}$$

For the sake of simplicity, all transfer functions are assumed to be m-square proper, although the equations and the subsequent results may be easily extended to systems where number of inputs ≠ number of outputs.

Let the system transfer function assume a new matrix $G^v(s)$ due to variations in its internal parameters $[G^n(s) \neq G^v(s)]$. If the compensators K(s) and H(s) were designed for $G^n(s)$, the output of the open-loop system and the closed-loop system [with $G^v(s)$ as transfer-function matrix] are related by the following equations (Fig. 5.25):

(3) With parameter variation in open-loop: $y_o^v(s) = G^v(s)u_o(s)$ (5.162)

(4) With parameter variation in closed-loop: $y_c^v(s) = G^v(s)u_c(s)$;

$$u_c(s) = K(s)[u_s(s) - H(s)y_c^n(s)] \tag{5.163}$$

the input controls $u_o(s)$ and $u_c(s)$ remaining the same in all cases (1) to (4).

The sensitivity matrix S(G, s) is then defined as

$$S(G, s) = \frac{y_c^v(s) - y_c^n(s)}{y_o^v(s) - y_o^n(s)} \equiv \frac{e_c(s)}{e_o(s)}$$

that is,

$$e_c(s) = S(G, s)e_o(s) \tag{5.164}$$

where $e_o(s)$ and $e_c(s)$ are m-vectors representing the difference in output signals due to parameter variations in open-loop and closed-loop systems, respectively. In the absence of any parameter variations, $\|S(G, s)\| < 1$ if $y_o^n(s) = y_c^n(s)$. By the following derivations of the transfer functions of the two closed-loop systems in Figs. 5.24b and 5.25b, it can be shown that

$$e_c(s) \equiv [I + G^n(s)K(s)H(s)]^{-1}e_o(s) \tag{5.165}$$

(a)

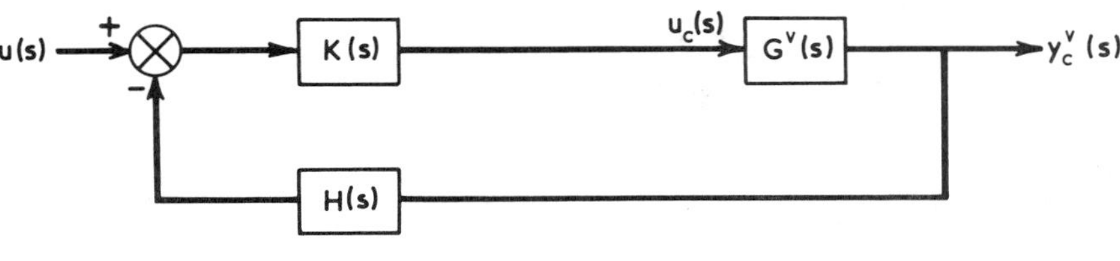

(b)

Fig. 5.25 System configuration with parameter variations: (a) open loop; (b) closed loop.

from which by comparison with (5.164),

$$S(G, s) = [I + G^n(s)K(s)H(s)]^{-1} \equiv [F(s)]^{-1} \tag{5.166}$$

where F(s) is the return-difference matrix of the nominal closed-loop configuration in Fig. 5.24:

$$F(s) = I + G^n(s)K(s)H(s) \tag{5.167}$$

which suggests that even for an m-input/ℓ-output system, the sensitivity matrix S(·) is an m-square matrix. Thus the effect of parameter variations in G(s) may be reduced by choosing the elements of F(jω) such that

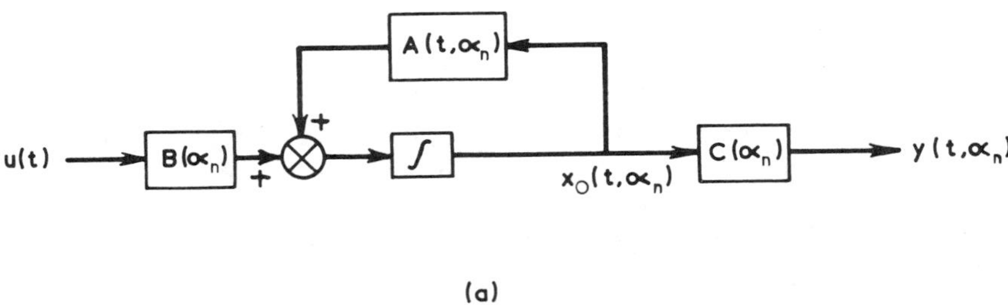

(a)

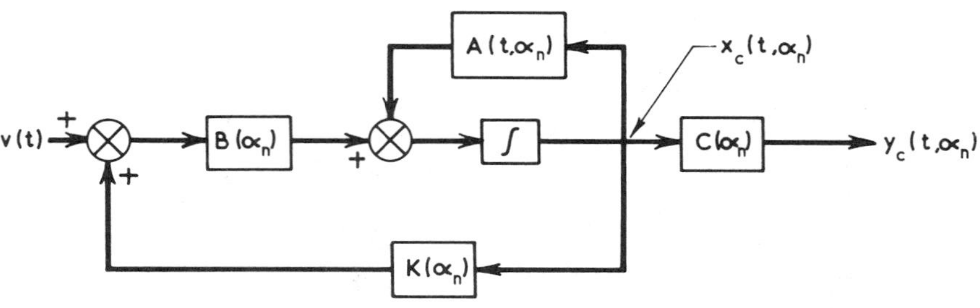

(b)

Fig. 5.26 Nominal representation of $S(A, B, C, \alpha_n, t)$: (a) open-loop system; (b) closed-loop system (≡ nominal open-loop system with nominal state feedback matrix).

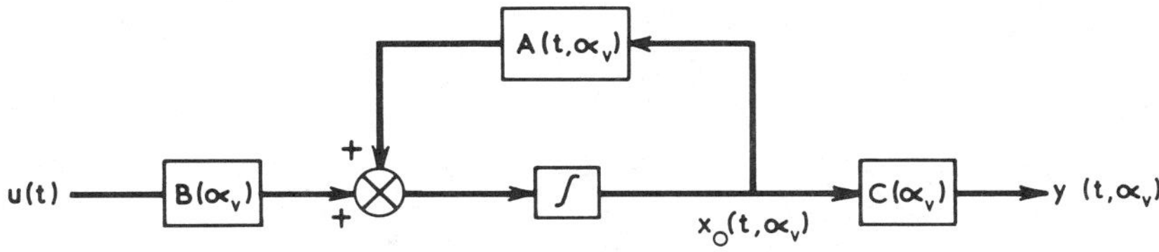

(a)

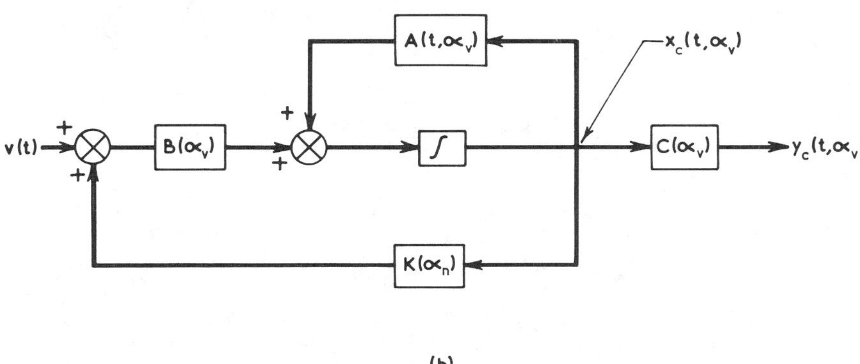

(b)

Fig. 5.27 System configuration with parameter variation: (a) open-loop system; (b) closed-loop system with nominal state feedback matrix.

$$\|F(j\omega)\| = \|I + G^n(j\omega)K(j\omega)H(j\omega)\| \simeq \|G^n(j\omega)K(j\omega)H(j\omega)\| \gg 1 \tag{5.168}$$

over as large a range of $(0, \omega)$ as possible. If the equalities and inequalities above are validated, the feedback loops are said to be <u>tight</u>.

Another equivalent expression for $S(\cdot)$ is

$$S(G, s) = [\Delta R(s)][R^n(s)]^{-1}G^n(s)[\Delta G(s)] \tag{5.169}$$

where

$$\Delta G(s) = G^v(s) - G^n(s) \quad \text{and} \quad \Delta R(s) = R^v(s) - R^n(s)$$

$G^n(s)$ and $R^n(s)$ being the open- and closed-loop transfer function matrices in Fig. 5.24; the expression for $S(\cdot)$ in (5.169) being valid only for m-input/m-output systems with nonsingular $R(s)$.

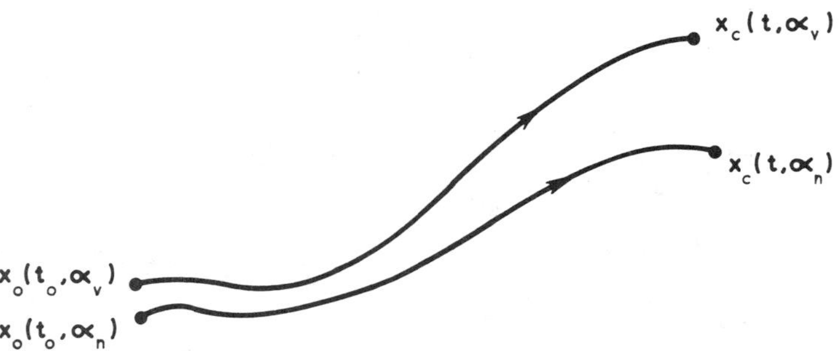

Fig. 5.28 Open-loop to closed-loop trajectories of the state vector.

If the system description were given in state-space form,

$$\dot{x}(t, \alpha) = A(\alpha)x(t, \alpha) + B(\alpha)u(t) ; \qquad y(t, \alpha) = C(\alpha)x(t, \alpha) \tag{5.170}$$

with $x(0, \alpha)$ as the initial state vector, where α is a parameter vector subject to variations. Let a nominal value of $\alpha = \alpha_n$ for which the (desired) nominal state vector is $x(t, \alpha_n)$, and let $K(\alpha_n)$ be a state feedback control law derived by using the representation in (5.170) for $\alpha = \alpha_n$, and as shown in Figs. 5.26 and 5.27; for $\alpha_v = \alpha_n$, the two closed-loop systems are equivalent. The following parameters may be defined for first-order variations in the trajectory changes of $x(\cdot)$ (Fig. 5.28):

$$\begin{aligned} &(1)\ \text{Open-loop system:} \quad \delta x_o(t) \simeq x_o(t_o, \alpha_v) - x_o(t, \alpha_n) \\ &(2)\ \text{Closed-loop system:} \quad \delta x_c(t) \simeq x_c(t, \alpha_v) - x_c(t, \alpha_n) \end{aligned} \tag{5.171}$$

The sensitivity matrix $S(x, \alpha)$ is then defined as

$$S(x, \alpha) = \frac{\delta x_c(t)}{\delta x_o(t)} \tag{5.172}$$

which can be shown to be equivalent to

$$S(x, \alpha) = [I + \phi(t, \alpha_n)BK]^{-1} \tag{5.173}$$

where $\phi(t, \alpha_n)$ is the state transition matrix for the nominal system ($\alpha = \alpha_n$). The matrix above relates the transforms of the open-loop and closed-loop trajectory deviations in the presence of parameter variations, and is equivalent to the expression in (5.167).

An alternative method of expressing the effect of parameter variation on the trajectory of $x(\cdot)$ is through the use of a matrix Z which is related to $K(\alpha)$: a closed-loop system is said to be less sensitive to parameter variation than its nominal open-loop system if

$$\int_0^t [\delta x_c(t)]^T Z[\delta x_c(t)]\, dt < \int_0^t [\delta x_o(t)]^T Z[\delta x_o(t)]\, dt \qquad (5.174)$$

for all $t > 0$ and for a given $Z > 0$.

The relationship between the definition above and the sensitivity matrix $S(\cdot)$ is expressed by the following theorem.

Theorem 5.25 A sufficient condition for the validity of (5.174) is

$$[S(-j\omega)]^T Z[S(j\omega)] - Z < 0, \quad \forall \omega \qquad (5.175)$$

and that the closed-loop system is asymptotically stable. [For single-input/ single-output systems, this condition reduces to $|S(j\omega)| < 1$ with $Z = I$.]

Proof of the theorem is not considered here.

Finally, the relationship between sensitivity and optimal controls (Sec. 5.1) is outlined through the following theorem, stated without proofs.

Theorem 5.26 If the feedback control law $K(\alpha_n)$ in Fig. 5.26 were derived through the minimization of the cost function

$$J = \int_0^\infty \{[x(t, \alpha_n)]^T Q[x(t, \alpha_n)] + [u(t)]^T R[u(t)]\}\, dt$$

the sensitivity inequality in (5.175) is satisfied for all t and for all δx with $Z = [K(\alpha_n)]^T R[K(\alpha_n)]$.

The physical significance of this theorem is that parameters sensitivity may be reduced through an optimal feedback law.

(b) External disturbance [57]

In the presence of a disturbance input (often unknown) signal d(s) (Fig. 5.29), the output of the open-loop system is truly expressed by

$$y(s) = G(s)u(s) + d(s) \qquad (5.176)$$

and the corresponding closed-loop output response becomes

$$y(s) = [F(s)]^{-1}Q(s)v(s) + [F(s)]^{-1}d(s) \qquad (5.177)$$

where F(s) and Q(s) are the return-difference and return-ratio matrices, respectively. The transfer function from d(s) to y(s) is thus $[F(s)]^{-1}$, and

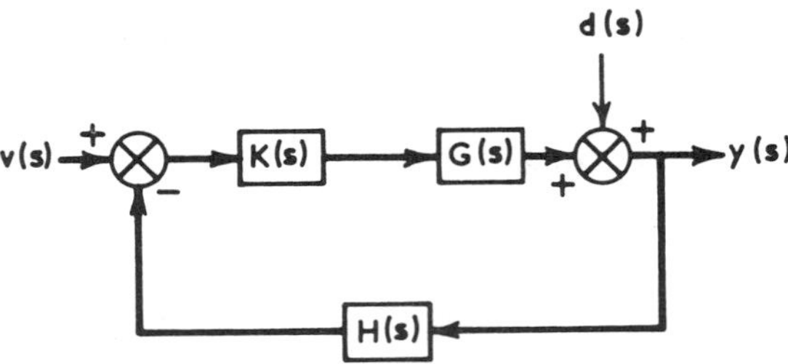

Fig. 5.29 Closed-loop system with disturbance input.

the effect of external disturbances on the output may be reduced by constraining K(s) and H(s) such that for each element of $F(j\omega)$,

$$[F(j\omega)]^{-1} = [I + G(j\omega)K(j\omega)H(j\omega)]^{-1} \ll 1 \tag{5.178}$$

over any frequency spectrum of disturbance $(0, \omega)$. Since it may be tedious to constrain each element of $F(j\omega)$, it is sufficient in most cases to reduce the norm of $[F(j\omega)]^{-1}$, that is,

$$\|[F(j\omega)]^{-1}\| \ll 1 \quad \text{over } (0, \omega) \tag{5.179}$$

Thus in some rather loose sense, the effect of external disturbances may be reduced by making $\|F(j\omega)\| \gg 1$ over $(0, \omega)$, which (as in 5.168) implies that

$$\begin{aligned} \|F(j\omega)\| &= \|I + G(j\omega)K(j\omega)H(j\omega)\| \simeq \|G(j\omega)K(j\omega)H(j\omega)\| \\ &\gg 1 \quad \text{over } (0, \omega) \end{aligned} \tag{5.180}$$

The feedback loops are said to be <u>tight</u> if the constraints above are satisfied.

In some cases complete rejection of a class of disturbances may be required. Let $F(s) = P(s)/\phi_c(s)$, where P(s) is a polynomial matrix and $\phi_c(s)$ is the closed-loop characteristic polynomial. Let d(s) be the Laplace transform of the time-varying external disturbance d(t). Then this disturbance is completely rejected if P(s) can be factorized as

$$P(s) = n(s)P_o(s) \tag{5.181}$$

where n(s) is a polynomial and $P_o(s)$ is a polynomial matrix, the zeros of n(s) being the poles of d(s). [For example, a constant disturbance d(t) = c, d(s) = c/s and for complete rejection of this disturbance P(s) is to be factorized as $P(s) = sP_o(s)$, i.e., n(s) = s.]

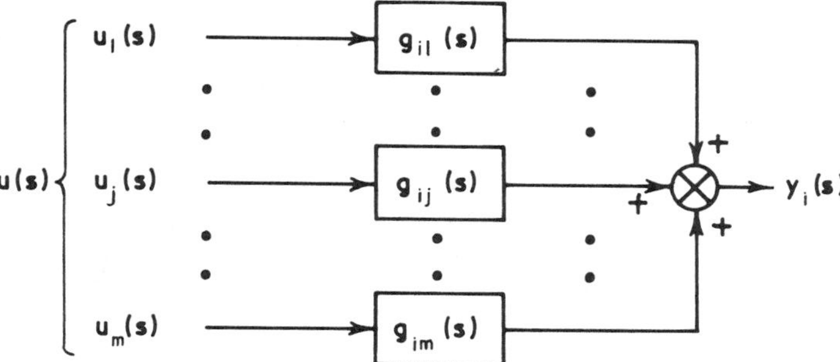

Fig. 5.30 Composite output vector in a multivariable system.

5.3.2 Interaction [58,59]*

In contrast with a single-input/single-output system, any one output of a multivariable system is affected by more than one (usually all) element(s) of the input control vector:

$$y_i(s) = \sum_{j=1}^{m} g_{ij}(s)u_j(s) \tag{5.182}$$

This is depicted by Fig. 5.30, where $g_{ij}(\cdot)$ are the elements of the system transfer function matrix G(s). This is a characterizing feature of multivariable systems and is termed <u>cross-coupling</u> or <u>interaction</u>. In many systems, it is possible that the direct response from any ith input $u_i(s)$ to a corresponding output $y_i(s)$ displays BIBO stability, but not the cross-coupling responses [i.e., from any input $u_j(s)$, $j \neq i$, to the output $y_i(s)$]. This in practice is equivalent to $g_{ii}(s)$ [≡ diagonal term in G(s)] satisfying the stability criterion, while $g_{ij}(s)$, $i \neq j$ [≡ nondiagonal terms in G(s)] do not. The result is that the cross-coupling responses of the system will be unstable, displaying destabilizing effects of interaction. In a similar manner, some systems may display stabilizing effects of interaction. Interaction thus plays an important role in the design of multivariable systems, and because of this, single-loop design without any reference to cross-coupling effects is virtually impossible in multivariable systems.

In time-domain design techniques, this is overcome by designing all input/output loops simultaneously so that the direct as well as the cross-coupling channels satisfy the desired design criterion; whereas in the frequency domain, closed-loop design is often accomplished in a loop-by-loop sequential manner, taking into account all interactions as the process

*Derivations in this section are based on Ref. 58.

unfolds. Another alternative method, of course, is to remove (or significantly remove as is done through diagonal dominance) input/output interaction through a suitable control law, and then design each loop as a single-input/single-output system. Interaction may be completely removed through decoupling design or noninteracting control, and may be achieved either through state/output feedback or through dynamic compensation. Whichever design method is adopted, it is relevant from the synthesis viewpoint to identify (1) the condition for minimization of interaction, or (2) if interaction is preserved, some qualitative measure (≡ interaction index) of its effect on system stability. These two aspects of interaction are outlined briefly here.

(a) Minimization of interaction

Since complete elimination of interaction requires a diagonal transfer function matrix, in the simplest case of unity feedback $[H(s) = I_m]$ the condition for closed-loop noninteraction is

$$Q(s) = G(s)K(s) = \Lambda(s) \tag{5.183}$$

where $\Lambda(s)$ is an m-square matrix. Thus the choice of K(s) that the feedforward compensates is controlled by

$$K(s) = [G(s)]^{-1}\Lambda(s) \tag{5.184}$$

which, however, is not suitable for application, because of computational difficulties and stability requirement of the resulting controller.

If the feedback loops are tight, then, from (5.180), the closed-loop frequency-response matrix is given by

$$R(j\omega) \simeq [G(j\omega)K(j\omega)H(j\omega)]^{-1}G(j\omega)K(j\omega) \tag{5.185}$$

Thus if unity feedback is employed, tight feedback automatically ensures low interaction at low frequency.

At high frequencies, however, $G(j\omega)$ usually tends to zero; from (5.185),

$$\lim_{s\to\infty} R(j\omega) \to \lim_{s\to\infty} G(j\omega)K(j\omega) \tag{5.186}$$

Thus low interaction at high frequencies requires that K(s) be chosen such that $[G(j\omega)K(j\omega)]$, as it tends to a diagonal matrix with $s \to \infty$, should approximate a diagonal form.

A similar conclusion may also be derived from the characteristic loci. Since all characteristic loci should satisfy a stability criterion of Nyquist type, they must have small moduli at high frequencies. Thus since all characteristic loci tend to zero as the modulus of s tends to infinity [equivalent to $G(s) \to 0$ as $s \to \infty$], the only way of removing high-frequency interaction in the closed-loop system is by suitable modification of the characteristic loci of the forward path transfer-function matrix Q(s).

(b) Interaction index

The primary purpose of defining an interaction index is to provide a qualitative (and possibly quantitative) measure of the effect of cross-coupling on the overall stability for a given control configuration. Let the state feedback control law

$$u(t) = v(t) + Hy(t) \tag{5.187}$$

for the time-invariant m-input/ℓ-output system S(A, B, C) be split into two stages:

Stage 1: Feedback loops from one particular output $y_i(t)$ to the input control are implemented as shown in Fig. 5.31a. Let this be termed closure of the ith loop.

Stage 2: Feedback loops from all outputs to all inputs controls are implemented as shown in Fig. 5.31b. Let this be termed closure of all loops.

Let the output feedback $m \times \ell$ matrix H in (5.187) be given by

$$H = \begin{bmatrix} h_{11} & \cdots & h_{1i} & \cdots & h_{1\ell} \\ h_{21} & \cdots & h_{2i} & \cdots & h_{2\ell} \\ \vdots & & & & \\ h_{m1} & \cdots & h_{mi} & \cdots & h_{m\ell} \end{bmatrix} \tag{5.188}$$

(h_{mi} column: ith column of H $\equiv h_i^c$, $i \in 1, \ldots,$)

and the input/output matrices of S(A, B, C) be

$$B = [b_1^c \cdots b_j^c \cdots b_m^c], \qquad C = \begin{bmatrix} c_1^r \\ c_i^r \\ \vdots \\ c_\ell^r \end{bmatrix} \leftarrow \text{ith row of C, } i \in 1, \ldots, \ell$$

(b_j^c: jth column of B, $j \in 1, \ldots, m$)

Combining the above, the following closed-loop state equations may be formulated:

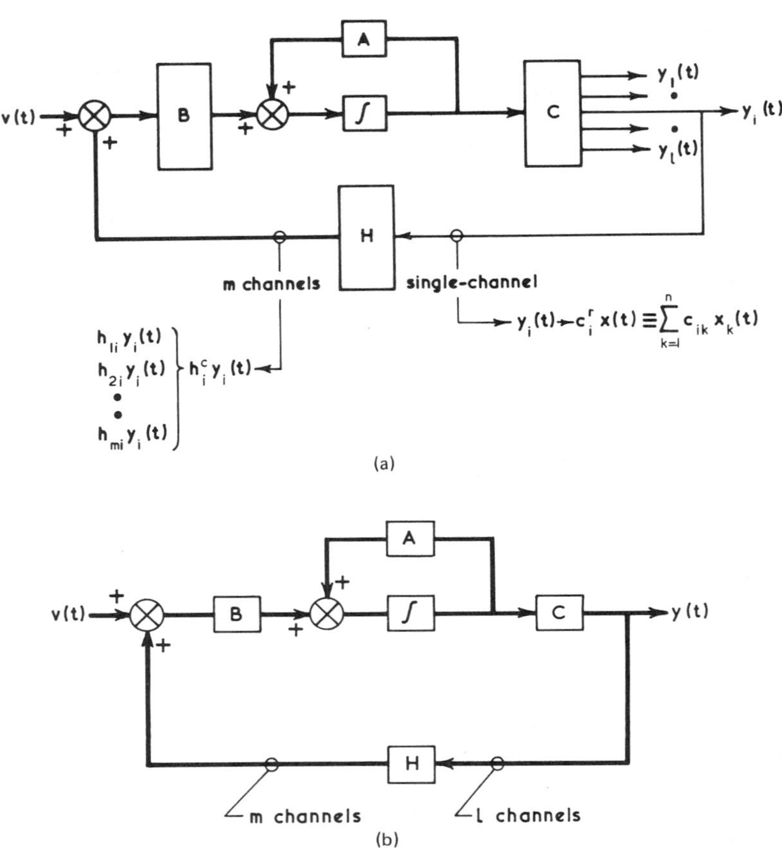

Fig. 5.31 Feedback loops to formulate the interaction index: (a) closure of the ith loop ($i \in 1, \ldots, \ell$)} (b) closure of all loops.

(1) Closure of the ith loop:

$$\dot{x}(t) = Ax(t) + \sum_{j=1}^{m} \{b_j^c h_i^c y_i(t)\} + Bv(t) = Ax(t) + \sum_{j=1}^{m} \{b_j^c h_{ij} c_i^r x(t)\} + Bv(t)$$

$$\equiv \left[A + \sum_{j=1}^{m} b_j^c h_{ij} c_i^r \right] x(t) + Bv(t) \tag{5.189}$$

(2) Closure of all loops:

$$\dot{x}(t) = [A + BHC]x(t) + Bv(t) = \left[A + \sum_{i=1}^{\ell}\left\{\sum_{j=1}^{m} b_j^c h_{ij} c_i^r\right\}\right] x(t) + Bv(t)$$

$$\equiv \left[A + \sum_{j=1}^{m} (b_j^c h_{1j} c_1^r) + \sum_{j=1}^{m} (b_j^c h_{2j} c_2^r) + \cdots + \sum_{j=1}^{m} (b_j^c h_{\ell j} c_\ell^r)\right] x(t) + Bv(t) \tag{5.190}$$

The output equation for both cases is

$$y(t) = Cx(t) \tag{5.191}$$

The effectiveness of the two control configurations in bringing the ith output to the desired value may be measured through the amplitudes of the following error variables:

$$\left.\begin{aligned} &\text{(1) Closure of ith loop: } e_i(t) = \max_{x(t_0)} \int_{t_0}^{\infty} [y_i(t)]^2\, dt \\ &\text{(2) Closure of all loops: } e_{i\ all}(t) = \max_{x(t_0)} \int_{t_0}^{\infty} [y_{i\ all}(t)]^2\, dt \end{aligned}\right\} [x(t_0)]^T[x(t_0)] = 1 \tag{5.192}$$

where $y_i(t)$ and $y_{i\ all}(t)$ represent the output responses for closure of the ith loop and the closure of all loops, respectively, for zero input control. Thus

$$e_i(\infty) \geq 0 \quad \text{and} \quad e_{i\ all}(\infty) \geq 0$$

represent the average deviations of the ith output, for any $i \in 1, \ldots, \ell$, from its desired value with respect to the systems initial state $x(t_0)$. The following definition of this may now be given.

Definition 5.7: Interaction index

$$\text{Interaction index in the ith loop,}\quad I_i = \frac{e_{i\ all}(\infty) - e_i(\infty)}{e_i(\infty)} \equiv \frac{e_{i\ all}(\infty)}{e_i(\infty)} - 1, \quad i \in 1, \ldots, \ell \tag{5.193}$$

Thus

$$-1 < I_i < \infty$$

Since e(t) is positive, the following observations may be made from Definition 5.7:

1. $I_i < 0$ if $e_{i\ all}(\infty) < e_i(\infty)$, that is, if a reduction in average error is achieved by closing all feedback loops for a particular H → favorable (≡ stabilizing) interaction in terms of closed-loop response of the system when all loops are connected.
2. $I_i = 0$ → no interaction in the closed-loop system.
3. $I_i > 0$ if the average error is increased by closing all loops for a particular H → unfavorable (≡ destabilizing) effect when all loops are connected.
4. $I_i \to \infty$ if $e_{i\ all}(\infty) >> e_i(\infty)$, that is, the average error increases as all loops are closed → severe (unstable) interaction will occur when all loops are closed.

5.3.3 Integrity [60]

In designing controllers for multivariable systems, it is important to check the stability of the closed-loop system under the three categories of failure conditions: output sensors, error detection, and actuator failures. Effects of these breakdowns on system stability are examined here by using the concept of a switch matrix.

Definition 5.8: Switch matrix. A switch matrix is a diagonal matrix, denoted by S, with all diagonal elements either 1 (for a normally operating loop) or an arbitrary small quantity ϵ (for a failed loop).

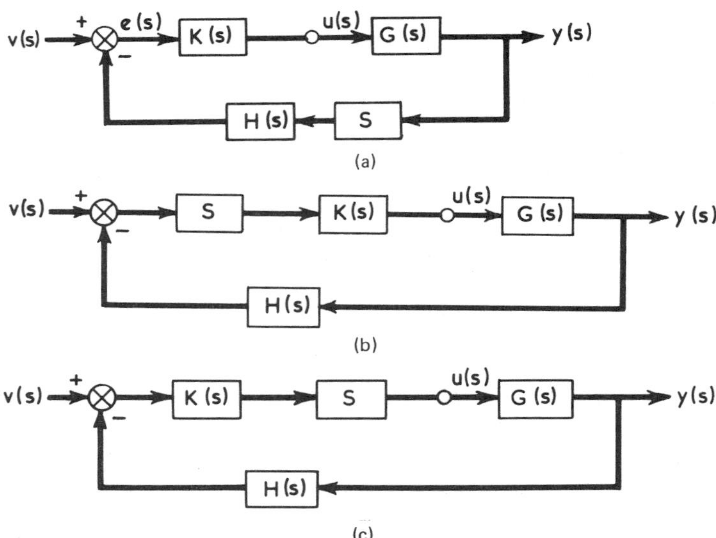

Fig. 5.32 Closed-loop system block diagrams representing three failure conditions: (a) transducer failure; (b) error-detector failure; (c) actuator failure.

The three failure conditions above may then be represented by placing S at appropriate positions in the closed-loop system shown in Fig. 5.32. From these figures, the following three closed-loop transfer functions may be obtained.

Transducer failure:
$$y(s) = [I_m + G(s)K(s)H(s)S]^{-1}G(s)K(s)v(s) \equiv [I_m + T_t(s)S]^{-1}G(s)K(s)v(s) \tag{5.194a}$$

Error-detector failure:
$$y(s) = [I_m + G(s)K(s)SH(s)]^{-1}G(s)K(s)Sv(s) \equiv G(s)K(s)S[I_m + T_e(s)S]^{-1}v(s) \tag{5.194b}$$

Actuator failure:
$$y(s) = [I_m + G(s)SK(s)H(s)]^{-1}G(s)SK(s)v(s) \equiv G(s)S[I_m + T_a(s)S]^{-1}K(s)v(s) \tag{5.194c}$$

where $T_t(s)S$, $T_e(s)S$, and $T_a(s)S$ are the appropriate return-ratio matrices.

Definition 5.9: High integrity. A multivariable feedback system is defined as being of high integrity if it remains stable under all likely failure conditions.

Theorem 5.27 A multivariable feedback system shown in Fig. 5.32 has high integrity against transducer, error-detecting, and actuator failures only if the characteristic loci of the principal submatrices of the matrices $T_t(s)$, $T_e(s)$, and $T_a(s)$ satisfy the stability requirements theorem developed in the previous sections.

Proof: For general representation, let T(s) be defined as

$$\bar{T}(s) = T(s)S \tag{5.195}$$

The subscripts t, a, or e may be added to T(s) and will be used as appropriate in the subsequent analysis. The result of postmultiplication of (5.195) is that all columns of T(s) (corresponding to those columns of S containing an entry ϵ) are multiplied by ϵ. It can easily be shown that by the choice of an appropriate permutation matrix P, $\bar{T}(s)$ may be transformed to

$$T'(s) = P^{-1}\bar{T}(s)P = \begin{bmatrix} T'_{11}(s) & \epsilon T'_{12}(s) \\ T'_{21}(s) & \epsilon T'_{22}(s) \end{bmatrix} \tag{5.196}$$

where $T'_{11}(s)$ is the principal submatrix of $T'(s)$ obtained by deleting all rows and columns of T(s) corresponding to ϵ entries in the associated switch matrix S [$T'_{11}(s)$ and $T'_{22}(s)$ being square by definition]. Thus the nonnull eigenvalues of $T'(s)$ are given by

$$\det\{q(s)I - T'_{11}(s)\} = 0 \tag{5.197}$$

The solution of q(s) from (5.197) thus defines the set of characteristic loci of the return-ratio matrix for those systems that correspond to the failure situations defined by the ϵ-valued diagonal entries in the switch matrix S. The proof is completed by using (5.197) and the stability criteria developed in Sec. 5.2.

One special case is of importance: when $H(s) = kI_m$, k being a scalar. In such a case, it can be seen from (5.194) that $T_t(s)S \equiv T_e(s)S$, and therefore they have the same eigenvalues. Consequently, the eigenvalues of all the corresponding principal submatrices of $T_t(s)$ and $T_e(s)$ will also be the same. That is, the integrity with transducer failure is synonymous with integrity with error-detector failure. From this point of view, it is desirable to choose $H = I_m$ and include all compensating dynamics in the forward path controller K(s). It is perhaps appropriate to indicate here that the ideal combination of high integrity under all failure conditions and good system response may not be achieved simultaneously in most practical design/synthesis. The designer would thus have to identify the "most likely" failure conditions before undertaking design/synthesis.

Classical methods for the design of single-input/single-output systems, which are based on Nyquist plot and root-locus techniques, use the fundamental properties of scalar functions of a complex variable. The need for complicated theory for the stability analysis of multivariable systems is highlighted by the following two-input/two-output system, defined by [26]

$$G(s) = \begin{bmatrix} g_{11}(s) & g_{12}(s) \\ g_{21}(s) & g_{22}(s) \end{bmatrix} \tag{5.198}$$

To identify the problem of single-loop design approach, only the loop between output $y_1(s)$ and input $u_1(s)$ is closed, as shown in Fig. 5.33a. The input/output relationship in the second (open-loop) channel is then given by

$$y_2(s) = g_{21}(s)u_1(s) + g_{22}u_2(s)$$

where

$$u_1(s) = w_1(s) - k_1y_1(s) = w_1(s) - k_1[g_{11}(s)u_1(s) + g_{12}(s)u_2(s)]$$

$$u_1(s) = \frac{w_1(s) - k_1g_{12}(s)u_2(s)}{1 + k_1g_{11}(s)}$$

Combining the equations above gives us

$$y_2(s) = \left[g_{22}(s) - k_1\frac{g_{12}(s)g_{21}(s)}{1 + k_1g_{11}(s)}\right]u_2(s) + \frac{g_{21}(s)}{1+k_1g_{11}(s)}w(s) \tag{5.199}$$

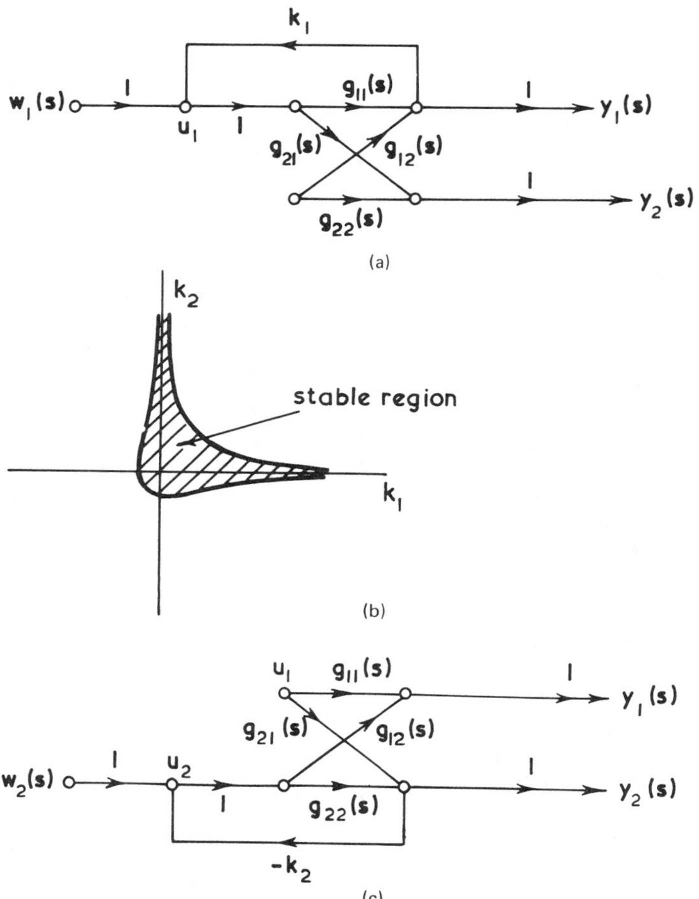

Fig. 5.33 Two-input/two-output system configurations and stability region: (a) system with channel 1 closed; (b) stability region for the system k_1 and k_2 being feedback loop gains in channels 1 and 2, respectively (both loops closed); (c) system with channel 2 closed.

If the loop gain in channel 1 is chosen to be high $[|k_1 g_{11}(s)| >> 1]$, then for $w(s) = 0$ for all s,

$$y_2(s) \simeq \left[\frac{g_{11}(s)g_{22}(s) - g_{12}(s)g_{21}(s)}{g_{11}(s)}\right] u_2(s) \tag{5.200}$$

Thus for sufficiently high gains in closed-loop channel 1, the frequency-response characteristics of the effective transmission between $u_2(j\omega)$ and $y_2(j\omega)$ are given by

$$\frac{g_{11}(j\omega)g_{22}(j\omega) - g_{12}(j\omega)g_{21}(j\omega)}{g_{11}(j\omega)} \qquad (5.201)$$

Difficulties of control arise from the fact that, in addition to the direct transference $g_{22}(j\omega)$ in the second channel, the action of the feedback in the first channel causes an extra transmission term $\{-g_{12}(j\omega)g_{21}(j\omega)/g_{11}(j\omega)\}$ to appear. To illustrate the physical effect of this additional term, let $g_{ij}(s)$, $i, j \in 1, 2$, be all first-order lags; the action of the feedback loop in the first channel is then to add a large amount of additional phase lag between $w_2(j\omega)$ and $y(j\omega)$, the additional signal lagging on the directly transmitted signal (of around an additional 180°). The presence of the controller in channel 1 injecting extra phase lag into the second transference will greatly reduce the amount of loop gain that can be applied in the feedback loop of the controller in the second channel without causing instability. The stable operating region, as a function of both loop gains, will be of the form shown in Fig. 5.33b. This shows that the system will become unstable if high gains are applied in both feedback loops, although it will withstand an arbitrary high feedback gain in any one loop at a time.

If the second channel is closed with the feedback loop shown in Fig. 5.33c with channel 1 in open loop, $y_1(s)$ is given by

$$y_1(s) = \left[g_{11}(s) - k_2 \frac{g_{21}(s)g_{21}(s)}{1 + k_2 g_{22}(s)} \right] u_1(s) + \frac{g_{12}(s)}{1 + k_2 g_{22}(s)} w_2(s) \qquad (5.202)$$

with the same observations on stability criterion being applicable.

The transfer function of channel 2 with channel 1 closed and $w_1(s) = 0$ is given by:

$$\frac{g_{11}(s)g_{22}(s) - g_{12}(s)g_{21}(s)}{g_{11}(s)} = \det\left(\frac{[G(s)]}{g_{11}(s)}\right) \qquad (5.203)$$

Thus, if G(s) has any zeros in the right-half s-plane, the transmission between $w_2(s)$ and $y_2(s)$, with channel 1 feedback loop closed, will become nonminimum phase. This is an extreme form of the problem, since nonminimum phase transference causes difficulties in feedback control schemes. Further studies of more general examples, particularly those in which det [G(s)] vanishes in the right-half s-plane, will show that these effects persist regardless of the particular arrangements of inputs and outputs between which the two feedback loops are connected. They may be avoided only by using the general feedback scheme shown in Fig. 5.34, with off-diagonal elements present in the controller matrix K(s). Most of the difficulties in multiple-loop feedback control stem from the necessity to

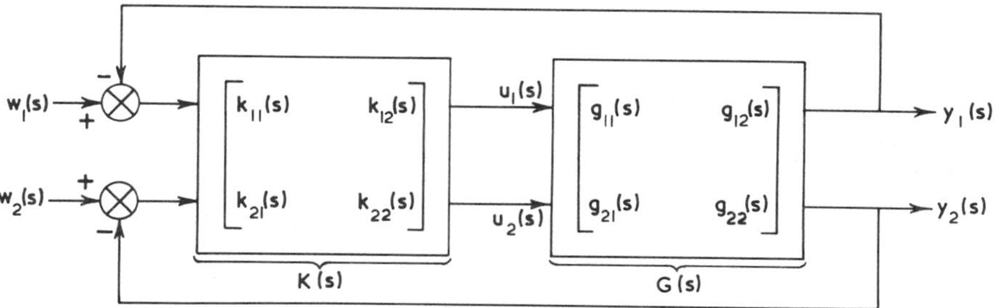

Fig. 5.34 General feedforward compensator configuration.

design for all entries in K(s). There are thus two systematic procedures suitable for multivariable system design:

1. To design all loops together The approach used in most time-domain design techniques is considered in Chaps. 6 to 9.
2. To design loop by loop in a sequential manner, taking account of all interactions as the process unfolds The approach commonly adopted in frequency-domain design techniques is considered in Chap. 10.

REFERENCES

1. Ogata, K., State Space Analysis of Control Systems, Prentice-Hall, Englewood Cliffs, N.J., 1967.
2. Chen, C. T., Introduction to Linear System Theory, Holt, Rinehart and Winston, New York, 1970.
3. Brockett, R. W., Finite Dimensional Linear Systems, Wiley, New York, 1970.
4. Athans, M., Dertovos, M. L., Spann, R. N., and Mason, J. J., Systems, Networks and Computation: Multivariable Methods, McGraw-Hill, New York, 1974.
5. Chen, C. T., "Controllability and observability of composite systems," Trans. IEEE, AC-12: 402-409 (1967).
6. Chen, C. T., "Stability of linear multivariable feedback systems," Proc. IEEE, 56: 821-828 (1968).
7. Chen, C. T., "A proof of the stability of multivariable feedback systems," Proc. IEEE, 56: 2061-2062 (1968).
8. Rosenbrock, H. H., "The stability of multivariable systems," Trans. IEEE, AC-17: 105-106 (1972).
9. Wonham, W. M., "On pole assignment in multi-input controllable linear systems," Trans. IEEE, AC-12: 660-665 (1967).

10. Brockett, R. W., "Poles, zeros and feedback: state space interpretation," Trans. IEEE, AC-10: 129-135 (1965).
11. Kalman, R. E., and Bertram, J. E., "Control system analysis and design via the 'second method' of Lyapunov," ASME J. Basic Eng., 82: 371-393 (1960).
12. La Salle, J. P., and Lefschetz, S., Stability by Lyapunov's Direct Method with Applications, Academic Press, New York, 1961.
13. Chen, C. T., "A note on pole assignment," Trans. IEEE, AC-13: 597-598 (1968).
14. Heymann, M., "Comments on pole assignment in multi-input controllable linear systems," Trans. IEEE, AC-13: 748-749 (1968).
15. Johnson, C. D., and Wonham, W. M., "A note on the transformation to canonical (phase-variable) form," Trans. IEEE, AC-9: 312-313 (1964).
16. Porter, B., Synthesis of Dynamical Systems, Thomas Nelson, London, 1969.
17. Kalman, R. E., Ho, Y. C., and Narendra, K. S., "Controllability of linear dynamical systems," Contrib. Differential Equ., 1: 189-213 (1961).
18. Kailath, T., Linear Systems, Prentice-Hall, Englewood Cliffs, N.J., 1980.
19. Kwakernaak, H., and Sivan, R., Linear Optimal Control Systems, Wiley, New York, 1972.
20. Kalman, R. E., "Contributions to the theory of optimal control," Bol. Soc. Mat. Mexicana, 5: 102-119 (1960).
21. Athans, M., and Falb, P. L., Optimal Control, McGraw-Hill, New York, 1966.
22. Kalman, R. E., Falb, P. L., and Arbib, M. A., Topics in Mathematical System Theory, McGraw-Hill, New York, 1969.
23. Kalman, R. E., "When is a linear control system optimal?" ASME J. Basic Eng., 86: 51-60 (1964).
24. Norton, M., Modern Control Engineering, Pergamon Press, Elmsford, N.Y., 1972.
25. Molinaric, B. P., "The time-invariant linear-quadratic optimal control problem," Automatica, 13: 347-357 (1977).
26. MacFarlane, A. G. J., "Multivariable control system design: a guided tour," Proc. IEE, 117: 1039-1047 (1970).
27. Nyquist, H., "Regeneration theory," Bell Syst. Tech. J., 11: 126-147 (1932).
28. MacFarlane, A. G. J., "Return-difference and return-ratio matrices and their use in analysis and design of multivariable control systems," Proc. IEE, 117: 2037-2059 (1970).
29. Chen, C. T., and Tsang, N. F., "A stability criterion based on the return-difference," Trans. IEEE, E-10: 150-182 (1967).
30. MacFarlane, A. G. J., "A survey of recent results in linear multivariable feedback theory," Automatica, 8: 455-492 (1972).

31. MacFarlane, A. G. J. (Ed.), Complex Variable Methods for Linear Multivariable Feedback Systems, Taylor & Francis, London, 1980.
32. MacFarlane, A. G. J., and Postlethwaite, I., "Extended principle of the argument," Int. J. Control, 27: 49-55 (1978).
33. MacFarlane, A. G. J., and Postlethwaite, I., "Generalised Nyquist stability criterion and multivariable root-loci," Int. J. Control, 25: 81-127 (1977).
34. MacFarlane, A. G. J., and Postlethwaite, I., "Characteristic frequency functions and characteristic gain functions," Int. J. Control, 26: 265-278 (1977).
35. Edmunds, J. M., "Characteristic gains, characteristic frequencies and stability," Int. J. Control, 29: 669-706 (1979).
36. Desoer, C. A., and Wang, Y.-T., "On the generalised Nyquist stability criterion," Trans. IEEE, AC-25: 187-196 (1980).
37. MacFarlane, A. G. J., "Complex-variable-design-methods," in Modern Approaches to Control System Design, N. Munro (Ed.), Peter Peregrinus, London, 1979, Chap. 7.
38. Aroki, M., and Nwokah, O. I., "Bounds for closed-loop transfer functions of multivariable systems," Trans. IEEE, AC-20: 666-670 (1975).
39. MacFarlane, A. G. J., "Relationship between recent developments in linear control theory and classical design techniques," Parts 1-5, Meas. Control, 8 (1975).
40. Bliss, G. A., Algebraic Functions, Colloquium Publications XVI, American Mathematical Society, New York, 1933.
41. Springer, G., Introduction to Rieman Surfaces, Chelsea, New York, 1981.
42. MacFarlane, A. G. J., and Belletrutti, J. J., "The characteristic locus design method," Automatica, 9: 575-588 (1973).
43. Postlethwaite, I., "A generalized inverse Nyquist stability criterion," Int. J. Control, 26: 325-340 (1977).
44. Rosenbrock, H. H., Computer-Aided Control System Design, Academic Press, New York, 1974.
45. Rosenbrock, H. H., "Progress in the design of multivariable control," Meas. Control., 4: 9-11 (1971).
46. Rosenbrock, H. H., "Design of multivariable control systems using the inverse Nyquist array," Proc. IEE, 116: 1929-1936 (1969).
46a. Rosenbrock, H. H., "Inverse Nyquist array design method," in Modern Approaches to Control System Design, N. Munro (Ed.), Peter Peregrinus, London, 1979, Chap. 5.
47. Cook, P. A., "System stability," in Modern Approaches to Control System Design, N. Munro (Ed.), Peter Peregrinus, London, 1979, Chap. 4.
48. McMorran, P. D., "Design of a gas turbine controller using the inverse Nyquist method," Proc. IEE, 117: 2050-2056 (1970).
49. Layton, J. M., Multivariable Control Theory, Peter Peregrinus, London, 1976.

50. Rosenbrock, H. H., and Storey, C., Mathematics of Dynamical Systems, Thomas Nelson, London, 1969.
51. Owens, D. H., Feedback and Multivariable Systems, Peter Peregrinus, London, 1979.
52. Cruz, J. B., and Perkins, W. R., "A new approach to the sensitivity problem in multivariable feedback system design," Trans. IEEE, AC-9: 216-223 (1964).
53. Morgan, B. S., "Sensitivity analysis and synthesis of multivariable systems," Trans. IEEE, AC-11: 506-512 (1966).
54. Perkins, W. R., and Cruz, J. B., "Feedback properties of linear regulators," Trans. IEEE, AC-16: 659-664 (1971).
55. Perkins, W. R., and Cruz, J. B., "The parameter variation problem in state feedback control systems," ASME J. Basic Eng., 87: 120-124 (1965).
56. Kreindler, E., "Closed-loop sensitivity reduction of linear optimal control systems," Trans. IEEE, AC-13: 254-262 (1968).
57. Mayne, D. Q., "Sequential return-difference-design method," in Modern Approaches to Control System Design, N. Munro (Ed.), Peter Peregrinus, London, 1979, Chap. 8.
58. Davison, E. J., "Interaction index for multivariable control systems," Proc. IEE, 117, 459-462 (1970).
59. Bristol, E. H., "On a new measure of interaction for multivariable process control," Trans. IEEE, AC-11: 133-134 (1966).
60. Belletrutti, J. J., and MacFarlane, A. G. J., "Characteristic loci techniques in multivariable-control-system design," Proc. IEE, 118: 1291-1297 (1971).

Bibliography for Part I

Computational Aspects

Akashi, H., Proc. 8th IFAC World Congr., Kyoto, Vol. 1: Control Theory, Session 19: Computational Methods in Control Problems, Pergamon Press, Oxford, 1981, pp. 495-519.

Aplevich, J. D., "Direct computation of canonical forms for linear systems by elementary matrix operations," Trans. IEEE, AC-19: 124-126 (1974).

Basile, G., and Marro, G., "Controlled and conditioned invariant subspaces in linear system theory," J. Optim. Theory Appl., 3: 306-315 (1969).

Cline, A. K., Moler, C. B., Stewart, G. W., and Wilkinson, J. H., "An estimate for the condition number of a matrix," SIAM J. Numer. Anal., 16: 368-375 (1979).

Davison, E. J., Gesing, W., and Wang, S. H., "An algorithm for obtaining the minimal realization of a linear time-invariant system and determining if a system is stabilizable-detectable," Proc. IEEE Control Decis. Conf., 1977, pp. 777-782.

Dongarra, J. J., Bunch, J. R., Moler, C. B., and Stewart, G. W., LINPACK Users Guide, SIAM, Philadelphia, 1979.

Elmqvist, H., "SIMNON—an iterative simulation program for nonlinear systems," Report TFRT-3091, Department of Automatic Control, Lund Institute of Technology, Lund, Sweden, 1975.

Enright, W., "On the efficient and reliable numerical solution of large linear systems of ODE's," Trans. IEEE, AC-24: 905-908 (1979).

Erisman, A. M. (Ed.), Electric Power Problems: The Mathematical Challenge, SIAM, Philadelphia, 1980.

Forsythe, G. E., Malcolm, M. A., and Moler, C. B., Computer Methods for Mathematical Computations, Prentice-Hall, Englewood Cliffs, N.J., 1977.

Garbow, B. S., Boyle, J. M., Dongarra, J. J., and Moler, C. B., Matrix Eigensystem Routines—EISPACK Guide Extension, Springer-Verlag, New York, 1977.

Gantmatcher, F. R., The Theory of Matrices, Vols. 1 and 2, Chelsea, New York, 1959.

Glover, K., "An approximate realization algorithm that directly identifies the system poles," Preprints 7th IFAC World Congr., Helsinki, Vol. 3, 1978, pp. 1789-1796.

Golub, G. H., and Wilkinson, J. H., "Ill-conditioned eigenvalue systems and the computation of the Jordan canonical form," SIAM Rev., 18: 578-619 (1976).

Kleinman, D. L., and Rao Krishna, P., "Extension to the Bartels-Stewart algorithm for linear matrix equation," Trans. IEEE, AC-23: 85-87 (1978).

Kučera, V., "Polynomial equations in control system design," Preprints 7th IFAC World Congress, Helsinki, Vol. 3, 1978, pp. 1815-1822.

Lasdon, L. S., Mitter, S. K., and Warren, A. D., "The conjugate gradient method for optimal control," Trans. IEEE, AC-12: 132-138 (1967).

Laub, A. J., and Moore, B. C., "Calculation of transmission zeros using QZ techniques," Automatica, 14: 557-566 (1978).

Luenberger, D. G., Introduction to Linear and Nonlinear Programming, Addison-Wesley, Reading, Mass., 1973.

Niemi, A. (Ed.), Preprints 7th IFAC World Congr., Helsinki, Vol. 2, Sessions 26A and 26B: Mathematical Systems Theory, Pergamon Press, Oxford, 1978, pp. 1115-1171.

Miller, W., and Wrathal, C., Software for Roundoff Analysis of Matrix Algorithms, Academic Press, New York, 1980.

Moler, C. B., and VanLoan, C. F., "Nineteen dubious ways to compute the exponential of a matrix," SIAM Rev., 20: 801-836 (1978).

Moore, B. C., "Principal component analysis in linear systems: controllability, observability, and model reduction," Trans. IEEE, AC-26: 130-138 (1981).

Paige, C. C., "Properties of numerical algorithms related to computing controllability," Trans. IEEE, AC-26: 17-32 (1981).

Parlett, B. N., The Systematic Eigenvalue Problem, Prentice-Hall, Englewood Cliffs, N.J., 1980.

Polak, E., Computational Methods in Optimization, Academic Press, New York, 1977.

Rosenbrock, H. H., "An automatic method of finding the greatest or least values of a function," Comput. J., 3: 175-184 (1962).

Schoenstadth, A. L., Information Linkage Between Applied Mathematics and Industry, Vol. 2, Academic Press, New York, 1980.

Smith, B. T., Boyle, J. M., Dongarra, J. J., Garbow, B. S., Ikebe, Y., Kama, V. C., and Moler, C. B., Matrix Eigensystem Routines—EISPACK Guide, Springer-Verlag, New York, 1976.

Stewart, G. W., Introduction to Matrix Computation, Academic Press, New York, 1973.

Stoer, J., and Bulirsch, R., Introduction to Numerical Analysis, Springer-Verlag, New York, 1980.

Van-Dooran, P., "The generalized eigenstructure problem: applications in linear system theory," Ph.D. thesis, Catholic University of Leuven, Belgium, 1979.

Vostry, Z., "New algorithm for polynomial spectral factorization with quadratic convergence," Kybernetika, 12: 248-254 (1976).

General Theory

Akashi, H., Proc. 8th IFAC World Congr., Kyoto, Vol. 1: Control Theory, Pergamon Press, Oxford, 1981.

Anderson, B. D. O., "Linear multivariable control systems: a survey," Proc. 5th IFAC World Congr., Paris, 1972.

Atherton, D. P. (Ed.), Multivariable Technological Systems, Preprints 4th IFAC Symp. Multivariable Technol. Syst., New Brunswick, Pergamon Press, Oxford, 1977.

Bengtsson, G., "Output regulation and internal models—a frequency domain approach," Automatica, 13: 333-345 (1977).

Chen, C. T., and Haas, I. J., Elements of Control System Analysis, Prentice-Hall, Englewood Cliffs, N.J., 1968.

Davison, E. J., and Wang, S. H., "Properties and calculation of transmission zeros of linear multivariable systems," Automatica, 10: 643-658 (1974).

Desoer, C. A., and Schulman, J. D., "Zeros and poles of matrix transfer functions and their dynamical interpretation," Trans. IEEE, CAS-21: 3-8 (1974).

Desoer, C. A., and Vidyasagar, M., Feedback Systems: Input-Output Properties, Academic Press, New York, 1975.

Emre, E., and Silverman, L. M., "New criterion and system theoretic interpretations for relatively prime polynomial matrices," Trans. IEEE, AC-22: 239-242 (1977).

Emre, E., Silverman, L. M., and Glover, K., "Generalized dynamic covers for linear systems with applications to deterministic identification and realization problems," Trans. IEEE, AC-22: 26-35 (1977).

Forney, G. D., "Minimal bases of rational vector spaces with application to multi-variable linear systems," SIAM J. Control, 13: 493-520 (1975).

Francis, B. A., and Wonham, W. M., "The internal model principle of control theory," Automatica, 12: 457-465 (1976).

Hermann, R., and Martin, C. F., "Applications of algebraic geometry to systems theory," Trans. IEEE, AC-22: 19-25 (1977).

Heymann, M., "The prime structure of linear dynamical systems," SIAM J. Control, 10: 460-469 (1972).

Horowitz, I. M., and Shaked, U., "Superiority of transfer function over state-variable methods in linear time-invariant feedback system design," Trans. IEEE, AC-20: 84-97 (1975).

Karcarias, N., and Kouvaritakis, B., "The use of frequency transmission concepts in linear multivariable system analysis," Int. J. Control, 28: 197-240 (1978).

Luenberger, D. G., "Dynamic equations in descriptor form," Trans. IEEE, AC-22: 312-321 (1977).

Mayne, D. Q., and Brockett, R. W. (Eds.), Geometric Methods in System Theory, D. Reidel, Dordrecht, The Netherlands, 1973.

MacFarlane, A. G. J., "The development of frequency-response methods in automatic control," Trans. IEEE, AC-24: 250-265 (1979).

MacFarlane, A. G. J., and Karcarias, N., "Relationships between state-space and frequency-response concepts," Preprints 7th IFAC World Congr., Helsinki, Vol. 3, 1978, pp. 1771-1779.

Owens, D. H., "Integrity of multivariable first-order-type systems," Int. J. Control, 23: 827-835 (1976).

Owens, D. H., "Invariant zeros of multivariable systems: a geometric analysis," Int. J. Control, 26: 537-548 (1977).

Owens, D. H., "A note on the orders of the infinite zeros of linear multivariable systems," Int. J. Control, 31: 409-412 (1980).

Patel, R. V., "On the invertibility of linear multivariable systems," Int. J. Control, 22: 683-687 (1975).

Patel, R. V., "On zeros of multivariable systems," Int. J. Control, 21: 559-608 (1975).

Patel, R. V., "On computing the invariant zeros of multivariable systems," Int. J. Control, 24: 145-146 (1976).

Patel, R. V., Sinswat, V., and Fallside, F., "Disturbance zeros in multivariable systems," Int. J. Control, 26: 85-96 (1977).

Rosenbrock, H. H., "Bounds for transfer function matrices," Trans. IEEE, AC-18: 54-56 (1973).

Rosenbrock, H. H., and Pugh, A. C., "Contributions to a hierarchical theory of systems," Int. J. Control, 19: 845-867 (1974).

Rosenbrock, H. H., and Storey, C., Mathematics of Dynamical Systems, Thomas Nelson, London, 1970.

Sain, M. K., "The growing algebraic presence in system engineering: an introduction," Proc. IEEE, 64: 96-111 (1976).

Silverman, L. M., "Inversion of multivariable linear systems," Trans. IEEE, AC-14: 270-276 (1969).

Sinswat, V., and Fallside, F., "Determination of invariant zeros and transmission zeros of all classes of invertible systems," Int. J. Control, 26: 97-114 (1977).

Sontag, E. D., "On the generalized inverses of polynomial and other matrices," Trans. IEEE, AC-25: 514-517 (1980).

Timothy, L. K., and Bona, B. E., State-Space Analysis, McGraw-Hill, New York, 1968.

Wolfe, C. A., and Meditch, J. S., "Theory of system type for linear multivariable servomechanisms," Trans. IEEE, AC-22: 36–46 (1977).
Wolovich, W. A., and Falb, P. L., "On the structure of multivariable systems," SIAM J. Control, 7(3): 437–451 (1969).
Wonham, W. M., Linear Multivariable Control: A Geometric Approach, Springer-Verlag, Heidelberg, West Germany, 1974.
Ozguler, B., Sezer, E., and Huseyin, O., "Relative primeness of multidimensional polynomial matrices," Trans. IEEE, CAS-27: 729-732 (1980).

Controllability, Observability, and Canonical Forms

Bucy, R. S., "Canonical forms for multivariable systems," Trans. IEEE, AC-13: 567-569 (1968).
Brunovsky, P., "A classification of linear controllable systems," Kybernetika, 3: 173-188 (1970).
Davison, E. J., "Connectability and structural controllability of composite systems," Automatica, 13: 109-123 (1977).
Davison, E. J., and Wang, S. H., "New results on the controllability and observability of general composite systems," Trans. IEEE, AC-20: 123-128 (1975).
Ford, D. A., and Johnson, C. D., "Invariant subspaces and the controllability and observability of linear dynamical systems," SIAM J. Control, 6(4): 554-558 (1968).
Gilchrist, J. D., "n-Observability for linear systems," Trans. IEEE, AC-11: 388-395 (1966).
Heymann, M., "A unique canonical form for multivariable systems," Int. J. Control, 12: 913-927 (1970).
Heymann, M., "Controllability subspaces and feedback simulation," Proc. 6th IFAC World Congr., Boston, Vol. 1, 1975, Paper 43.2.
Jurdjevic, V., "Abstract control systems: controllability and observability," SIAM J. Control, 8(3): 424–439 (1970).
Kreindler, E., and Sarachik, P. E., "On the concepts of controllability and observability of linear systems," Trans. IEEE, AC-9, 129-136 (1964).
Lin, C. T., "Structural controllability," Trans. IEEE, AC-19: 201-208 (1974).
MacFarlane, A. G. J., "On the transfer-function matrix and controllability of cascaded systems," Proc. IEE, 114: 1781-1786 (1967).
Mita, T., "On maximal unobservable subspace, zeros and their application," Int. J. Control, 25: 885–899 (1977).
Panda, S. P., and Chen, C. T., "Comments on controllability and observability of composite systems," Trans. IEEE, AC-15: 280-281 (1970).
Popov, V. M., "Invariant description of linear, time-invariant controllable systems," SIAM J. Control, 10: 252-264 (1972).

Silverman, L. M., and Anderson, B. D. O., "Controllability, observability and stability of linear systems," SIAM J. Control, 6(1): 121-129 (1968).
Yoshikawa, T., and Bhattacharyya, S. P., "Partial uniqueness: observability and input identifiability," Trans. IEEE, AC-20: 713-714 (1975).

Realizations

Ackermann, J. E., and Bucy, R. S., "Canonical minimal realization of a matrix of impulse response sequences," Inf. Control, 19: 224-231 (1971).
Audley, D. R., and Pugh, W. J., "On the H-matrix system representation," Trans. IEEE, AC-18: 235-242 (1973).
Bengtsson, G., "Feedback realizations in linear multivariable systems," Trans. IEEE, AC-22: 576-585 (1977).
Lovass-Nagy, V., Miller, R. J., and Powers, D. L., "On the realization by matrix generalized inverses," Int. J. Control, 26: 745-751 (1977).
Roman, J. R., and Bullock, T. E., "Minimal partial realization in a canonical form," Trans. IEEE, AC-20: 529-533 (1975).
Roza, P., and Sinha, N. K., "Minimal realization of a transfer function matrix in canonical forms," Int. J. Control, 21: 273-284 (1975).
Salehi, S. V., "On the state-space realizations of matrix fraction descriptions for multivariable systems," Trans. IEEE, AC-23: 1054-1057 (1978).
Sheih, L. S., Wei, Y. J., and Yates, R., "Minimal realization of transfer function matrices by means of matrix continued fraction," Int. J. Control, 22: 851-859 (1975).
Sinha, N. K., "Minimal realization of transfer function matrices: a comparative study of different methods," Int. J. Control, 22: 627-639 (1975).
Tether, A. J., "Construction of minimal linear state variable models from finite input-output data," Trans. IEEE, AC-15: 427-436 (1970).

Stability, Sensitivity, and Integrity

Aoki, M., "On the feedback stabilizability of decentralized dynamic systems," Automatica, 8: 163-173 (1972).
Barman, J. F., and Katzenelson, J., "A generalized Nyquist-type stability criterion for multivariable feedback systems," Int. J. Control, 20: 593-622 (1974).
Brockett, R. W., "The status of stability theory for deterministic systems," Trans. IEEE, AC-11: 596-605 (1966).
Callier, F. M., and Desoer, C. A., "A stability theorem concerning open-loop unstable convolution feedback systems with dynamical feedbacks," Proc. 6th IFAC World Congr., Boston, Vol. 1, Session 44, Pergamon Press, Oxford, 1975.

Chen, C. F., and Tsay, Y. T., "A general frequency stability criterion for multi-input-output, lumped and distributed-parameter feedback systems," Int. J. Control, 23: 341-359 (1976).

Cook, P. A., "On the stability of interconnected systems," Int. J. Control, 20: 407-415 (1974).

Crossley, T. R., and Porter, B., "Eigenvalue and eigenvector sensitivities in linear systems theory," Int. J. Control, 10: 163-170 (1969).

DeCarlo, R., and Saeks, R., "The encirclement condition—an approach using algebraic topology," Int. J. Control, 26: 279-287 (1977).

DeCarlo, R. A., Murray, J., and Saeks, R., "Multivariable Nyquist theory," Int. J. Control, 25: 657-675 (1977).

Desoer, C. A., and Chan, W. C., "The feedback interconnection of multivariable systems: simplifying theorems for stability," Proc. IEEE, 64: 139-144 (1976).

Desoer, C. A., and Vidyasagar, M., Feedback Systems: Input-Output Properties, Academic Press, New York, 1975.

Eyman, E. D., "An optimal control problem with component failure," Int. J. Control, 25: 589-602 (1977).

Freeman, E. A., "Stability of linear constant multivariable systems," Proc. IEE, 120: 379-384 (1973).

Godbout, L. F., and Jordan, D., "Sensitivity functions for linear dynamic systems," Int. J. Control, 34: 561-576 (1981).

Gori-Giorgi, C., and Grasselli, O. M., "A new approach to the study of parameter insensitivity," Automatica, 11: 181-188 (1975).

Jury, E. I., Inverse and Stability of Dynamic Systems, Wiley-Interscience, New York, 1974.

Jury, E. I., "Stability of multidimensional scalar and matrix polynomials," Proc. IEEE, 66: 1018-1047 (1978).

Kalman, R. E., Falb, P. L., and Arbib, M. A., Topics in Mathematical System Theory, McGraw-Hill, New York, 1969.

Koussiouris, T. G., "A new stability theorem for multivariable systems," Int. J. Control, 32: 435-441 (1980).

Lasalle, J. P., The Stability of Dynamical Systems, SIAM, Philadelphia, 1976.

Mees, A. I., and Atherton, D. P., "The Popov criterion for multiple-loop feedback systems," Trans. IEEE, AC-25: 924-928 (1980).

Narendra, K. S., and Taylor, J. H., Frequency-Domain Criteria for Absolute Stability, Academic Press, New York, 1973.

Popov, V. M., "Dichotomy and stability by frequency-domain methods," Proc. IEEE, 62: 548-562 (1974).

Saeks, R., "On the encirclement condition and its generalization," Trans. IEEE, CAS-22: 780-785 (1975).

Wilkie, D. F., and Van-Schieven, H. M., "On the sensitivity of linear state regulator," Int. J. Control, 12: 709-719 (1970).

Willems, J. L., Stability Theory of Dynamical Systems, Wiley-Interscience, New York, 1970.

Problems for Part I

1. An $m \times n$ matrix A is said to have a right inverse B if AB = I and a left inverse C if CA = I. Show that A has a right inverse if $\rho(A) = m$ and has a left inverse if $\rho(A) = n$.

2. Find a right inverse of $A = \begin{bmatrix} 1 & 3 & 2 & 3 \\ 1 & 4 & 1 & 3 \\ 1 & 3 & 5 & 4 \end{bmatrix}$ if one exists.

3. Show that the submatrix $T = \begin{bmatrix} 1 & 3 & 3 \\ 1 & 4 & 3 \\ 1 & 3 & 4 \end{bmatrix}$ is nonsingular and obtain

$\begin{bmatrix} 7 & -3 & -3 \\ -1 & 1 & 0 \\ 0 & 0 & 0 \\ -1 & 0 & 1 \end{bmatrix}$ as another right inverse of A.

4. Prove that if $|A_{11}| \neq 0$, then $\begin{vmatrix} A_{11} & A_{12} \\ A_{21} & A_{22} \end{vmatrix} = |A_{11}| \cdot |A_{22} - A_{21}A_{11}^{-1}A_{12}|$.

5. Obtain the inverses of the symmetric matrices

$$A = \begin{bmatrix} 1 & 2 & -1 & 2 \\ 2 & 2 & -1 & 1 \\ -1 & -1 & 1 & -1 \\ 2 & 1 & -1 & 2 \end{bmatrix} \quad \text{and} \quad B = \begin{bmatrix} 0 & 1 & 2 & 2 \\ 1 & 1 & 2 & 3 \\ 2 & 2 & 2 & 3 \\ 2 & 3 & 3 & 3 \end{bmatrix}.$$

6. Obtain the inverse of the matrices

$$A = \begin{bmatrix} 1 & 2 & 0 & 0 \\ 0 & 3 & 0 & 0 \\ 0 & 0 & 2 & 1 \\ 0 & 0 & 0 & 3 \end{bmatrix} \quad \text{and} \quad B = \begin{bmatrix} 2 & 5 & 2 & 3 \\ 2 & 3 & 3 & 4 \\ 3 & 6 & 3 & 2 \\ 4 & 12 & 0 & 8 \end{bmatrix}$$

by using the most appropriate method.

7. Find the Jordan form and modal matrices for the following matrices:

$$A = \begin{bmatrix} 1 & 2 \\ -2 & -3 \end{bmatrix} \qquad B = \begin{bmatrix} 5 & 4 & 0 \\ 0 & 1 & 0 \\ -4 & 4 & 1 \end{bmatrix} \qquad C = \begin{bmatrix} 0 & 0 & 1 & 0 \\ 0 & 0 & 0 & 1 \\ 0 & 0 & 0 & 0 \\ 0 & 0 & 0 & 0 \end{bmatrix}$$

8. Find e^A by both Cayley-Hamilton theorem and Sylvester's theorem for the matrix

$$A = \begin{bmatrix} 2 & 1 & 1 \\ 1 & 2 & 1 \\ 1 & 1 & 2 \end{bmatrix}.$$

9. Derive the Smith-forms of the following matrices:

(i) $$\begin{bmatrix} (s+2) & (s+1) & (s+3) \\ s(s+1)^2 & s(s^2+s+1) & s(s+1)(2s+1) \\ (s+1)(s+2) & (s+1)^2 & 3(s+1)^2 \end{bmatrix}$$

(ii) $$\begin{bmatrix} (s+1) & (2s^2+s-1) & (s^2-1) \\ -s(s+1) & -s(s-1) & s \end{bmatrix}$$

(iii) $$\begin{bmatrix} (s^2+1) & s(s^2+1) & s(2s^2-s+1) \\ (s-1) & (s^2+1) & (s^2-2s+1) \\ s^2 & s^3 & (2s^3-2s^2+1) \end{bmatrix}$$

(iv) $$\begin{bmatrix} s & (s-1) & (s+2) \\ s(s+1) & s^2 & s(s+2) \\ s(s-2) & (s-1)(s-2) & (s^2+s-3) \end{bmatrix}$$

$$(v)\quad \begin{bmatrix} \frac{(s^2+1)}{(s-2)} & \frac{(s^2+3s+3)}{(s-1)} & \frac{(s^2+4s-2)}{(s+2)} & \frac{(s^2+3)}{(s-2)} \\ \frac{(3s+1)}{s(s+2)} & \frac{(4s+3)}{(s^2+6s+4)} & \frac{2(s+1)}{(s^2+6s-1)} & \frac{(3s+2)}{(s^2+2s+3)} \end{bmatrix}$$

10. Derive the Smith-McMillan forms of the following rational matrices:

$$(i)\quad \frac{1}{\Delta(s)}\begin{bmatrix} (s+1) & (2s^2+s-1) & (s^2-1) \\ -s(s+1) & -s(s-1) & s \end{bmatrix},\quad \Delta(s) = s(s+1)(s+2)$$

$$(ii)\quad \begin{bmatrix} \frac{1}{(s+1)^2} & \frac{1}{(s+1)(s+2)} \\ \frac{1}{(s+1)(s+2)} & \frac{(s+3)}{(s+2)^2} \end{bmatrix}$$

$$(iii)\quad \frac{1}{\Delta(s)}\begin{bmatrix} 1 & -1 \\ (s^2+s-4) & (2s^2-s-8) \\ (s+2)(s-2) & 2(s+2)(s-2) \end{bmatrix},\quad \Delta(s) = (s+1)(s+2)$$

$$(iv)\quad \frac{1}{(s-1)(s+1)(s+2)}\begin{bmatrix} (s-1)(s+2) & 0 & (s-1)^2 \\ -(s+1)(s+2) & (s-1)(s+1) & (s^2-1) \end{bmatrix}$$

11. Derive the state and output equations of the two interconnected tanks shown in Fig. PI.1, and hence derive the transfer function between input u(s) and output y(s)

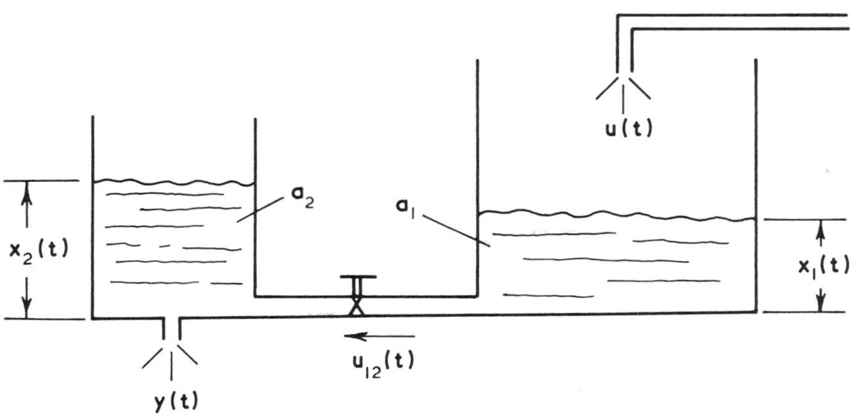

Fig. PI.1

Derive the impulse and step responses of the system with $a_1 = 0.5$, $a_2 = 2.0$, $r_{12} = 2.5$ and $r_2 = 0.5$.

12. Derive the state and output equations of the systems described by the signal flow graphs in Figure PI.2.

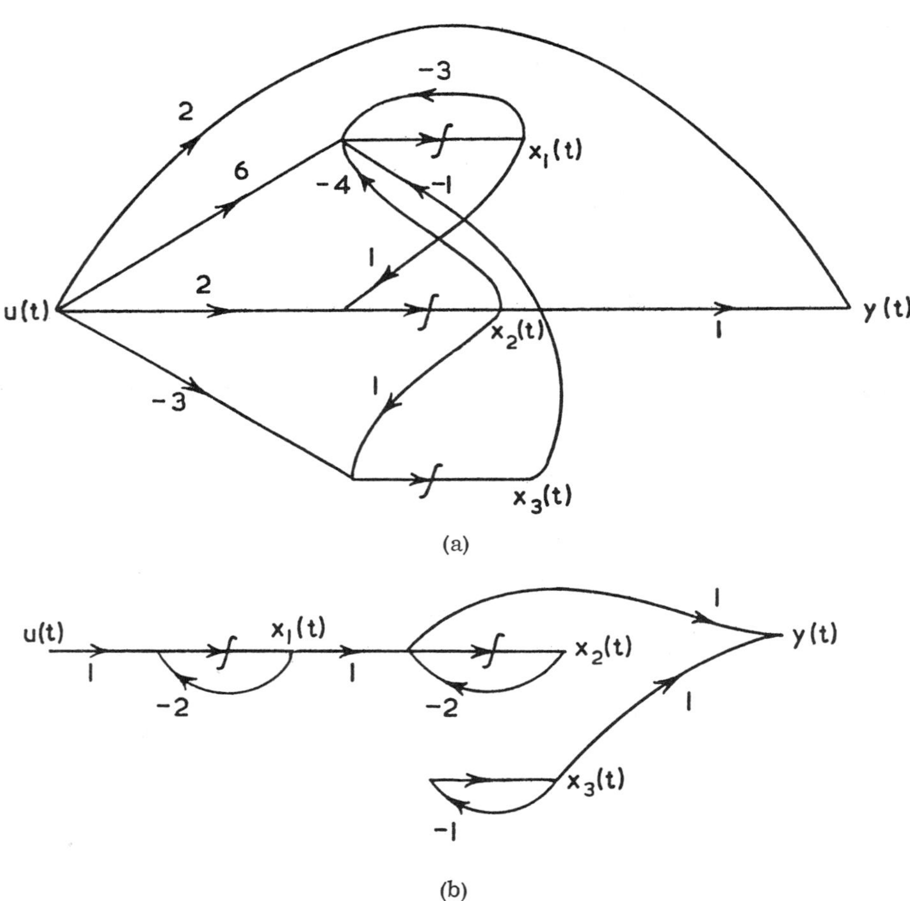

Fig. PI.2

13. Find the impulse response matrix H(t) for the system S(A, B, C):

$$A = \begin{bmatrix} 1 & 0 & 0 \\ 0 & 3 & 0 \\ 0 & 0 & -5 \end{bmatrix}, \quad B = \begin{bmatrix} 1 & 0 \\ 0 & 1 \\ 1 & 1 \end{bmatrix}, \quad C = \begin{bmatrix} 1 & 1 & 0 \\ 0 & 1 & 1 \\ 0 & 0 & 1 \end{bmatrix}$$

14. Find the transfer function matrix and draw the signal flow graphs for the systems described below:

(i) $\ddot{y}_1(t) + 3\dot{y}_1(t) + 2y_1(t) = 2u_1(t) + u_2(t)$

$\dot{y}_2(t) + 2y_2(t) + \dot{y}_1(t) = -2u_1(t) + u_2(t)$

(ii) $\dot{y}_1(t) + y_1(t) = u_1(t) + 2u_2(t)$

$\ddot{y}_2(t) + 3\dot{y}_2(t) + 2y_1(t) = u_2(t) - u_1(t)$

15. Derive the state equations of the systems shown in Figure PI.3.

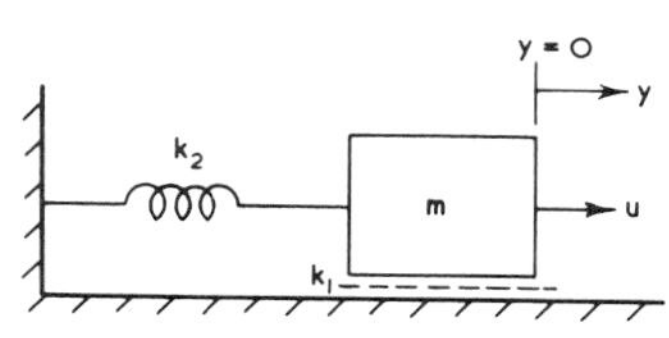

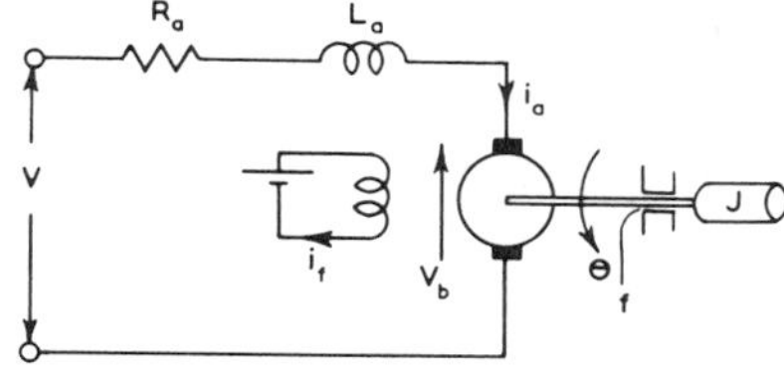

(a) a mechanical system

(b) an armature-controlled d.c. motor

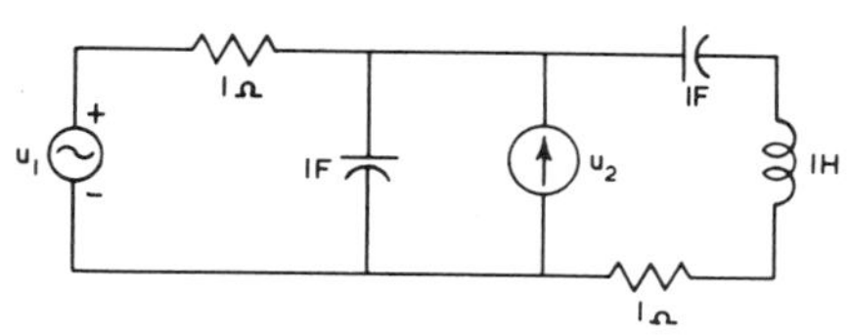

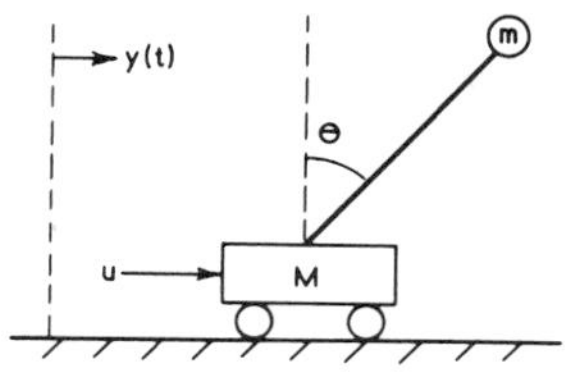

(c) a network with voltage and current sources

(d) an inverted pendulum

Fig. PI.3

16. Comment on the state controllability of the string of four moving vehicles shown in Figure PI.4. The equations of motion for each vehicle are

$$\ddot{y}_i(t) = -\frac{a_i}{m_i} v_i(t) + \frac{1}{m_i} f_i(t).$$

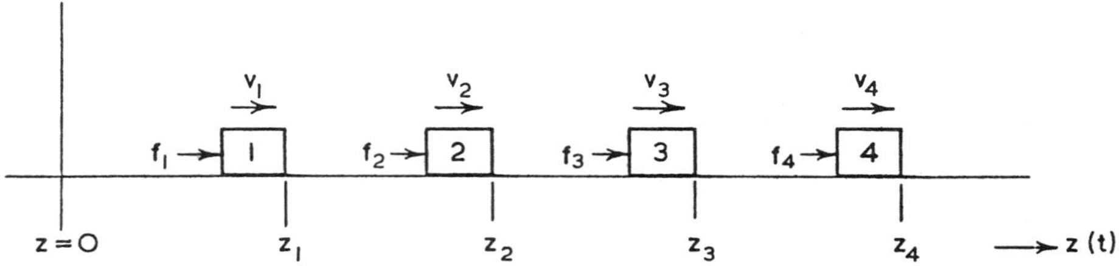

Fig. PI.4

17. Obtain realizations for the following transfer function matrices:

(i) $$\begin{bmatrix} \dfrac{(s+1)}{(s^2+3s+2)} & \dfrac{(s+1)}{(s^2+4s+3)} \end{bmatrix}$$

(ii) $$\begin{bmatrix} \dfrac{1}{(s+1)} & \dfrac{1}{(s+3)} \\[2ex] \dfrac{1}{s} & \dfrac{2}{(s+2)} \end{bmatrix}$$

(iii) $$\begin{bmatrix} \dfrac{s}{(s+1)^2} & \dfrac{1}{s^2} \\[2ex] \dfrac{1}{(s+1)} & \dfrac{1}{(s+2)} \end{bmatrix}$$

(iv) $$\begin{bmatrix} \dfrac{1}{(s+2)^2} & \dfrac{1}{s} \\[2ex] \dfrac{1}{(s+1)} & \dfrac{1}{(s+1)^2} \end{bmatrix}$$

(v) $$\frac{(s-2)}{(s^4-1)} \begin{bmatrix} (3s^2-5s-2) & (2s^2-5s-1) \\ (4s^2-2s+4) & (3s^2-2s-5) \end{bmatrix}$$

(vi) $$\frac{1}{(s^2-1)(s+2)} \begin{bmatrix} (s^2+6) & s^2+s+4 \\ 2s^2-7s-2 & s^2-5s-2 \end{bmatrix}$$

(vii) $$\begin{bmatrix} \dfrac{1}{(s+1)^2} & \dfrac{1}{(s+1)(s+2)} \\[2ex] \dfrac{1}{(s+1)(s+2)} & \dfrac{(s+3)}{(s+2)^2} \end{bmatrix}$$

(viii) $$\begin{bmatrix} \dfrac{1}{(s+1)} & \dfrac{1}{(s+1)} \\[2ex] \dfrac{1}{(s+1)} & \dfrac{1}{(s+1)} \end{bmatrix}$$

18. Determine the inverse of the following transfer function matrices:

(i) $\begin{bmatrix} \frac{1}{(s+1)} & 1 \\ -1 & \frac{2}{(s+2)} \end{bmatrix}$ (ii) $\begin{bmatrix} \frac{1}{(s+1)^2} & \frac{1}{(s+1)(s+2)} \\ \frac{1}{(s+1)(s+2)} & \frac{(s+3)}{(s+2)^2} \end{bmatrix}$

19. Determine if the following transfer function matrices are diagonally dominant and plot their inverse Nyquist arrays:

$$\begin{bmatrix} \frac{1}{(s+1)} & 1 \\ -1 & \frac{2}{(s+2)} \end{bmatrix}$$

(ii) $\frac{1}{(s+1)(2s+1)(s+2)}\begin{bmatrix} (2s+1)(s+2) & 0.75(s+1)(s+2) \\ 0.75(s+1)(s+2) & 2(s+1)(2s+1) \end{bmatrix}$

20. Derive the characteristic functions of the system with the following transfer function matrices:

(i) $\begin{bmatrix} \frac{1}{(s+1)} & \frac{1}{(s+3)} \\ \frac{1}{s} & \frac{2}{(s+2)} \end{bmatrix}$ (ii) $\begin{bmatrix} \frac{1}{(s+1)} & \frac{1}{(s+1)} \\ \frac{1}{(s+1)} & \frac{1}{(s+1)} \end{bmatrix}$

II
SYNTHESIS

Chapter 6: STATE FEEDBACK CONTROL

This chapter presents a number of methods of assigning poles and zeros followed by outlines of model-reference adaptive and linear quadratic control.

Chapter 7: OUTPUT FEEDBACK CONTROL

Pole assignment through feedback and the extension of classical proportional-plus-integral and proportional-plus-derivative to multivariable systems are presented here. An overview of the stability aspect of lower-order model-following control and dynamic compensation is given at the end.

Chapter 8: DESIGN WITH STATE OBSERVER

This chapter first introduces the concept of an observer and then develops a number of methods of computing observers for multivariable systems. The synthesis method through observer is considered briefly in the final section.

Chapter 9: DECOUPLING CONTROL

Methods for deriving state and output feedback decoupling control laws are developed in this chapter.

Chapter 10: GENERALIZED DESIGN TECHNIQUES

This chapter presents a selection of design methods based on the extension of classical control techniques to multivariable systems.

Bibliography and Problems.

6

State Feedback Control

It was shown earlier that in the absence of input control [i.e., $u(t) \equiv 0$], the solution of the state equation of S(A, B, C) has the form, called the <u>zero-input (state) response</u>,

$$x(t) = \sum_{i=1}^{n} \alpha_i e^{\lambda_i t} \omega_i \tag{6.1}$$

where ω_i is the eigenvector of A associated with its eigenvalue λ_i, $i \in 1, \ldots, n$ (λ_i's are assumed to be distinct). Thus the nature of the response characteristics of any linear time-invariant free system is determined by the eigenvalues λ_i. If λ_i's are not real, the response in (6.1) will be oscillatory, corresponding to pairs of complex-conjugate eigenvalues; the decrements and frequencies of these oscillatory modes being controlled by the real and imaginary parts of the corresponding eigenvalue. Thus if a feedback control law of the form

$$u(t) = v(t) + Fx(t) \tag{6.2}$$

is introduced, the free-state response of the closed-loop system*

$$\dot{x}(t) = [A + BF]x(t) + Bu(t); \quad y(t) = Cx(t) \tag{6.3}$$

will have its time responses determined by the eigenvalues of (A + BF). As indicated in Sec. 5.1, if S(A, B, C) is completely state controllable, the matrix (A + BF) may be assigned any set of eigenvalues through an appropriate choice of F. Since

$$y(t) = Cx(t) \equiv C \sum \alpha_i e^{\lambda_i t} \omega_i \tag{6.4}$$

*Referred to as the system $S(\overline{A + BF}, B, C)$ in the subsequent derivations.

by using (6.1), known as the state feedback control law,* the zero-input response of the system may be controlled through the state feedback matrix F. A few commonly used methods of "shifting" the eigenvalues (or the poles) of a system are considered in this chapter. The analytical derivations here assume that all the state variables [i.e., the complete state vector x(t)] of the system being controlled are available for feedback.

6.1 FULL-RANK FEEDBACK [1-7]

The general matrix method of changing the eigenvalues of a linear time-invariant system S(A, B, C) is considered here. The mechanism of pole shifting for single-input/single-output systems through constant feedback of state variables is developed first. This is followed by a general method of shifting any number of poles by using the normalized forms of state equations introduced earlier.

(a) Single-input/single-output system†

The state equation in this case is

$$\dot{x}(t) = Ax(t) + bx(t) \tag{6.5}$$

where b is a n-column vector and u(t) is a scalar input control. Let the real eigenvalues of A be λ_i with associated eigenvectors ω_i, $i \in 1, \ldots, n$. Also let the eigenvectors of A^T by γ_i^T for $i \in 1, \ldots, n$. Then

$$A\omega_i \triangleq \lambda_i\omega_i \quad \text{and} \quad A^T\gamma_i \triangleq \lambda_i\gamma_i^T, \qquad i \in 1, \ldots, n \tag{6.6}$$

and if the eigenvectors are assumed to have been normalized, then

$$\gamma_j^T\omega_i \triangleq \delta_{ji} = \begin{Bmatrix} 0, & i \neq j \\ 1, & i = j \end{Bmatrix} \quad \begin{matrix} i \in 1, \ldots, n \\ j \in 1, \ldots, n \end{matrix} \tag{6.7}$$

Since the complete n-state vector is available for feedback, the control input v(t) given by

$$v(t) = r\sum_{i=1}^{n} f_i x_i(t) \equiv rfx(t) + u(t) \equiv \bar{f}x(t) + u(t) \tag{6.8}$$

*Or a linear state-variable feedback (lsvf) control law.

†The case with distinct eigenvalues is considered here for analytical convenience.

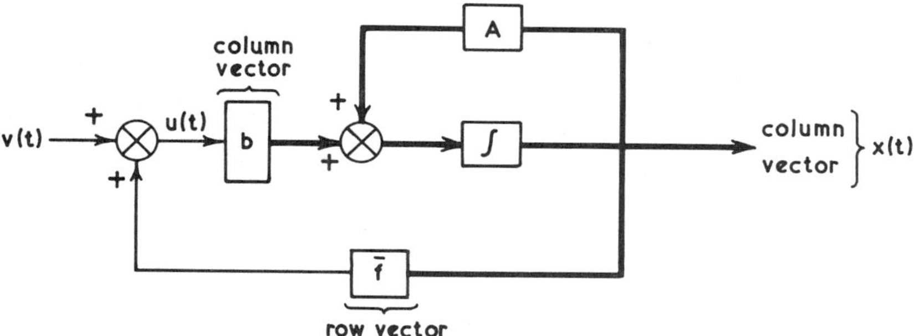

Fig. 6.1 Shifting of one eigenvalue through state feedback.

may be generated through the constant m-row vector $\bar{f}$ as shown in Fig. 6.1, where r is a constant. The closed-loop system then becomes

$$\dot{x}(t) = [A + b\bar{f}]x(t) + v(t) \tag{6.9}$$

Let $\lambda_j(\omega_j)$ be the single real eigenvalue (associated eigenvector) of A, which is to be "moved" to a desired (prespecified) location $\bar{\lambda}_j(\bar{\omega}_j)$ through the state feedback law in (6.8), while the other eigenvalues (and eigenvectors) remain unchanged. To fulfill this requirement, the arbitrary feedback row vector $\bar{f}$ should satisfy the constraints

$$(A + b\bar{f})\omega_i = A\omega_i + brf\omega_i \triangleq \begin{Bmatrix} \lambda_i\omega_i, & i \neq j \\ \bar{\lambda}_j\bar{\omega}_j, & i = j \end{Bmatrix} \quad \begin{matrix} i \in 1, \ldots, n \\ j \in 1, \ldots, n \end{matrix} \tag{6.10}$$

which is satisfied only if feedback row vector f is chosen to be equal to γ_j^T. To analyze the choice of r, let the input column vector b and the arbitrary desired eigenvector $\bar{\omega}_j$ be expanded as*

$$b = \sum_{i=1}^{n} g_i\omega_i \quad \text{and} \quad \bar{\omega}_j = \sum_{i=1}^{j} k_i\omega_i \tag{6.11}$$

Then from (6.10),

*$\omega_1, \ldots, \omega_i, \ldots, \omega_n$ are linearly independent since $\lambda_1, \ldots, \lambda_i, \ldots, \lambda_n$ are distinct.

$$[A+brf]\bar{\omega}_j = \left[A+\left(\sum_{i=1}^{n} g_i\omega_i\right) r\gamma_j^T\right]\left(\sum_{i=1}^{n} k_i\omega_i\right) = \sum_{i=1}^{n} k_i\lambda_i\omega_i + r\left(\sum_{i=1}^{n} g_i\omega_i\right)k_j$$

using (6.7)

$$\triangleq \bar{\lambda}_j\bar{\omega}_j = \bar{\lambda}_j\left(\sum_{i=1}^{n} k_i\omega_i\right) \tag{6.12}$$

which yields

$$\sum_{i=1}^{n} k_i\lambda_i\omega_i + r\left(\sum_{i=1}^{n} g_i\omega_i\right)k_j = \bar{\lambda}_j\left(\sum_{i=1}^{n} k_i\omega_i\right)$$

from which for any $i, j \in 1, \ldots, n$,

$$k_i\lambda_i\omega_i + r(g_i\omega_i)k_j = \bar{\lambda}_j(k_i\omega_i)$$

or

$$k_i\lambda_i + rg_ik_i = \bar{\lambda}_jk_i \tag{6.13}$$

Consequently, for $i = j$,

$$k_j = \frac{rg_jk_j}{\bar{\lambda}_j - \lambda_j} \rightarrow r = \frac{\bar{\lambda}_j - \lambda_j}{g_j} \tag{6.14}$$

which suggests that the desired eigenvalue shift is achieved through a structural modification in the input column vector b, providing that the constant gain r in Fig. 6.1 is chosen as given by (6.14). If λ_j is real, so is the vector γ_j^T and consequently, the control law in (6.8) incorporating γ_j^T as the feedback row vector is physically realizable. The method of deriving the necessary control law is illustrated below.

Example 6.1: Eigenvalue shift through state feedback. The following single-input/single-output second-order system is considered here:

$$\dot{x}(t) = \underbrace{\begin{bmatrix} -3 & 2 \\ 2 & 0 \end{bmatrix}}_{A} x(t) + \underbrace{\begin{bmatrix} 1 \\ 1 \end{bmatrix}}_{b} u(t) \tag{6.15}$$

The system is completely state controllable, and hence both eigenvalues can be assigned freely through an appropriate state feedback control law. The first stage is to derive eigenvalues and eigenvectors of A and A^T:

$$A:\ \lambda_1 = 1,\ \omega_1 = \frac{1}{5}\begin{bmatrix}1\\2\end{bmatrix};\quad \lambda_2 = -4,\ \omega_2 = \frac{1}{5}\begin{bmatrix}2\\-1\end{bmatrix} \tag{6.16}$$

$$A^T:\ \lambda_1 = 1,\ \gamma_1 = \begin{bmatrix}1\\2\end{bmatrix};\quad \lambda_2 = -4,\ \gamma_2 = \begin{bmatrix}2\\-1\end{bmatrix}$$

with $\gamma_1^T\omega_1 = 1$, $\gamma_1^T\omega_2 = 0$, $\gamma_2^T\omega_1 = 0$, and $\gamma_2^T\omega_2 = 1$, as in (6.7). The second stage is to "expand" the input matrix b as in (6.11):

$$b = \begin{bmatrix}1\\1\end{bmatrix} \equiv \frac{3}{5}\begin{bmatrix}1\\2\end{bmatrix} + \frac{1}{5}\begin{bmatrix}2\\-1\end{bmatrix} \equiv \underset{\substack{\uparrow\\ g_1}}{3}\omega_1 + \underset{\substack{\uparrow\\ g_2}}{1}\omega_2 \tag{6.17}$$

In the third stage, the value of the gain constant r_i is computed for any specified eigenvalue. Let the eigenvalue $\lambda_1 = 1$ be required to be moved to $\bar{\lambda}_1 = -4$. Then from (6.14),

$$r = \frac{\bar{\lambda}_1 - \lambda_1}{g_1} = \frac{-4 - 1}{3} = -\frac{5}{3} \tag{6.18}$$

The necessary feedback control law is now derived directly from (6.8):

$$u(t) = r\gamma_1^T x(t) + v(t) = -\frac{5}{3}[1\ \ 2]\begin{bmatrix}x_1(t)\\x_2(t)\end{bmatrix} + v(t) = \begin{bmatrix}\underset{\substack{\uparrow\\ \bar{f}_1}}{-\frac{5}{3}} & \underset{\substack{\uparrow\\ \bar{f}_2}}{-\frac{10}{3}}\end{bmatrix}\begin{bmatrix}x_1(t)\\x_2(t)\end{bmatrix} + v(t)$$

$$= [\bar{f}_1\ \ \bar{f}_2]x(t) + v(t) \tag{6.19}$$

For the corresponding closed-loop system,

$$A + brf = \begin{bmatrix}-3 & 2\\2 & 0\end{bmatrix} + \begin{bmatrix}1\\1\end{bmatrix}[-\frac{5}{3}\ \ -\frac{10}{3}] = \begin{bmatrix}-\frac{14}{3} & -\frac{4}{3}\\ \frac{1}{3} & -\frac{10}{3}\end{bmatrix} \tag{6.20a}$$

$$\det(sI - \overline{A + brf}) = (s + \frac{14}{3})(s + \frac{10}{3}) + \frac{4}{9} = (s + 4)(s + 4) \tag{6.20b}$$

If ℓ number of real eigenvalues are to be changed simultaneously $(1 < \ell < n)$, the input control in the method above needs to be resolved into ℓ components:

$$u(t) = \sum_{k=1}^{\ell} u_k(t), \quad \text{where } u_k(t) = \begin{cases} [\bar{f}^k]x(t) & \text{for } k \in 1, \ldots, \ell - 1 \\ [\bar{f}^\ell]x(t) + v(t) & \text{for } k = \ell \end{cases} \tag{6.21}$$

For $k = 1$, the procedure above yields

$$\dot{x}(t) = Ax(t) + bu(t) = (A + b[\bar{f}^1])\, xt \tag{6.22}$$

which moves the first eigenvalue from λ_1 to some prespecified value $\bar{\lambda}_1$. Thus by taking k sequentially, the closed-loop state equation assumes the form

$$\begin{aligned} \dot{x}(t) &= A + b[u_1(t) + \cdots + u_k(t) + \cdots u_\ell(t)] \\ &= A + b\{[\bar{f}^1] + \cdots + [\bar{f}^k] + \cdots + [\bar{f}^\ell]\} + v(t) \\ &= A + b \sum_{k=1}^{\ell} [\bar{f}^k] + v(t) \end{aligned} \tag{6.23}$$

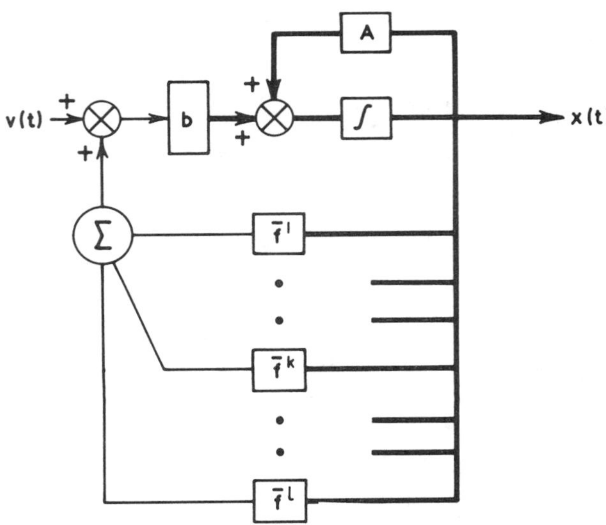

Fig. 6.2 Shifting of several eigenvalues through state feedback.

where $\bar{f}^k = r^k f^k$, and f^k is chosen to be equal to γ_k^T, the kth eigenvector of A^T. The closed-loop system in (6.23) therefore has eigenvalues at $\bar{\lambda}_1, \ldots, \bar{\lambda}_\ell, \lambda_{\ell+1}, \ldots, \lambda_n$. The constant r^k is computed from (6.14). The closed-loop configuration in this case is shown in Fig. 6.2.

By combining (6.14) and (6.17), it can be shown that the structure of the control law required to move an arbitrary number of ℓ poles from $\lambda_1, \ldots, \lambda_\ell$ to $\bar{\lambda}_1, \ldots, \bar{\lambda}_\ell$, while leaving the location of the remaining $\lambda_{\ell+1}, \ldots, \lambda_n$ poles unaltered, will be real even if complex poles are being moved.

(b) Multivariable system

The preceding theory of state feedback control can be readily modified for multivariable systems by partitioning the $(n \times m)$ input matrix B into m column vectors

$$B = [b_1, b_2, \ldots, b_m] \tag{6.24}$$

and then writing the state equation as

$$\dot{x}(t) = Ax(t) + b_1 u_1(t) + b_2 u_2(t) + \cdots + b_m u_m(g) \tag{6.25}$$

where $u_1(t), u_2(t), \ldots, u_m(t)$ are the m elements of the input control vector u(t). Thus by generating each of these input controls in the same manner as in (6.23) with the corresponding column of B, the desired changes in eigenvalues may be obtained. Thus if ℓ eigenvalues of A are to be modified, the required control input is

$$u_j(t) = \sum_{k=1}^{\ell} [\bar{f}^{kj}]x(t) + v_j(t), \qquad j \in 1, \ldots, m; \quad \ell < n \tag{6.26}$$

and the closed-loop system is given by

$$\dot{x}(t) = \left\{ A + \sum_{j=1}^{m} b_j \sum_{k=1}^{\ell} [\bar{f}^{kj}] \right\} x(t) + Bv(t) \tag{6.27}$$

where $v(t) = [v_1(t), \ldots, v_j(t), \ldots, v_m(t)]^T$. Numerical computation in this case is based on considering each (A, b_j) as a single-input/single-output system, the corresponding $\bar{f}^{\ell j}$, for any ℓ, being computed by using (6.8) to (6.14). Once all values of j have been considered, the "total" feedback matrix is derived as $\bar{F} = \Sigma_{j=1}^{m} b_j \Sigma_{k=1}^{\ell} \bar{f}^{kj}$. The method above, although conceptually simple, does not form a suitable basis for developing a numerical algorithm, due mainly to the fact that each column in B is to be considered separately to derive the corresponding row of $\bar{F}$. A more direct approach is possible if either the controllable form or the normalized form of S(A, B, C) is used in the formulation of the state feedback control law. The former method was outlined in Sec. 5.1.2 (Theorem 5.10) for single-input/single-output systems, which can easily be extended to the multivariable case. An alternative method of deriving the state feedback matrix through normalization is outlined below.

Let the system S(A, B, C) have n distinct (real or complex) eigenvalues $\lambda_1, \lambda_2, \ldots, \lambda_n$, given by the diagonal matrix

$$\Lambda = \operatorname{diag}\{\lambda_1, \lambda_2, \ldots, \lambda_n\} \tag{6.28}$$

Let F be the $m \times n$ real constant state feedback matrix in (6.2) which moves the eigenvalues of the closed-loop system $S(\overline{A + BF}, B, C)$ to some preassigned locations denoted by the diagonal matrix

$$\operatorname{diag}\{\pi_1, \ldots, \pi_i, \ldots, \pi_n\} \tag{6.29}$$

The method of coordinate transformation presented in Sec. 2.2.4 can be used to develop a matrix method of evaluating the value of this feedback gain matrix F necessary to shift the eigenvalues λ_i to π_i through complete state feedback.

Let M be the modal matrix of the system, that is, the matrix consisting of the eigenvectors associated with the eigenvalues of A. Then through the coordinate transformation of the form

$$x(t) = Mz(t)$$

the normalized forms of (6.1) and (6.2) become

$$z(t) = M^{-1}AMz(t) + M^{-1}Bu(t) = \Lambda z(t) + B_n u(t) \tag{6.30}$$

$$u(t) = v(t) + FMz(t) = v(t) + F_n z(t) \tag{6.31}$$

Combining (6.30) and (6.31), the normalized form of the closed-loop system in (6.3) becomes

$$\dot{z}(t) = [\Lambda + B_n F_n]z(t) + v(t) \triangleq \pi z(t) + v(t) \tag{6.32}$$

Thus

$$\pi = \Lambda + B_n F_n \rightarrow B_n F_n = [\pi - \Lambda] \tag{6.33}$$

The equation above may therefore be used to compute the feedback gain matrix F_n necessary to alter the modes of the normalized system in (6.30) through state feedback. The structure of this matrix is given by

$$F_n = [B_n^T B_n]^{-1} B_n^T [\pi - \Lambda] \tag{6.34}$$

Since the system is completely state controllable [i.e., $\rho(B_n) = m$], it is always possible to compute F_n from (6.34) for a given π.

By inverse transformation, the feedback F necessary to move the eigenvalues of (A + BF) to π is given by

$$F = F_n M^{-1} = [B_n^T B_n]^{-1} B_n^T [\pi - \Lambda] M^{-1} \tag{6.35}$$

The derivation for distinct eigenvalues above is applicable for shifting any number of real or imaginary eigenvalues of the system. If the system has repeated eigenvalues, the method is still applicable, with Λ being replaced by the Jordan form J, but the numerical computation and the specification of the desired eigenvalues π may be more involved. This is illustrated through examples below (without any discussions on the details of the analytical constraints).

Example 6.2: Eigenvalue shift through coordinate transformation

$$\dot{x}(t) = \begin{bmatrix} -3 & 2 \\ 2 & 0 \end{bmatrix} x(t) + \begin{bmatrix} 1 & 0 \\ 1 & 1 \end{bmatrix} u(t) \tag{6.36}$$

For this system,

$$\Lambda = \begin{bmatrix} 1 & 0 \\ 0 & -4 \end{bmatrix}; \quad M = \begin{bmatrix} 1 & 1 \\ 2 & -1 \end{bmatrix}; \quad M^{-1} = \frac{1}{5}\begin{bmatrix} 1 & 2 \\ 2 & -1 \end{bmatrix} \tag{6.37}$$

$$B_n = M^{-1}B = \frac{1}{5}\begin{bmatrix} 1 & 2 \\ 2 & -1 \end{bmatrix}\begin{bmatrix} 1 & 0 \\ 1 & 1 \end{bmatrix} = \frac{1}{5}\begin{bmatrix} 3 & 2 \\ 1 & -1 \end{bmatrix}; \quad B_n^{-1} = \begin{bmatrix} 1 & 2 \\ 1 & -3 \end{bmatrix} \tag{6.38}$$

$$F_n = B_n^{-1}[\pi - \Lambda] = \begin{bmatrix} 1 & 2 \\ 1 & -3 \end{bmatrix}\begin{bmatrix} \pi_1 - 1 & 0 \\ 0 & \pi_2 + 4 \end{bmatrix} = \begin{bmatrix} \pi_1 - 1 & 2(\pi_2 + 4) \\ \pi_1 - 1 & -3(\pi_2 + 4) \end{bmatrix}$$

$$F = F_n M^{-1} = \frac{1}{5}\begin{bmatrix} \pi_1 - 1 & 2(\pi_2 + 4) \\ \pi_1 - 1 & -3(\pi_2 + 4) \end{bmatrix}\begin{bmatrix} 1 & 2 \\ 2 & -1 \end{bmatrix} = \begin{bmatrix} -1 & -2 \\ -1 & -2 \end{bmatrix} \quad \text{for } \pi_{1,2} = -4 \tag{6.39}$$

With the closed-loop system,

$$A + BF = \begin{bmatrix} -3 & 2 \\ 2 & 0 \end{bmatrix} + \begin{bmatrix} 1 & 0 \\ 1 & 1 \end{bmatrix} \begin{bmatrix} -1 & -2 \\ -1 & -2 \end{bmatrix} = \begin{bmatrix} -4 & 0 \\ 0 & -4 \end{bmatrix} \tag{6.40}$$

6.2 UNITY-RANK STATE FEEDBACK [8-11]

The state feedback matrix F derived in the preceding section will, in general, have full rank (i.e., $\rho[F] = m$). Computation of such an F, which provides complete freedom in the choice of closed-loop poles, usually involves a significant amount of numerical computation. An alternative method, used primarily to reduce computational work, is to impose some structural constraints on F; one such condition which yields no loss of flexibility in the choice of desired eigenvalues of the closed-loop system $S(\overline{A + BF}, B, C)$ is is the unity rank F; that is, there are $(m + n + 1)$ independent elements in F as opposed to $(m \times n)$ when F has full rank.*† Such a constraint transforms the closed-loop system in (6.3) into an equivalent single-input system, thereby introducing considerable simplicity in the derivation of the required state feedback control law in (6.2); this is developed here.

Assuming zero initial state, Laplace transformation of the state equation of $\overline{S}(\overline{A+BF}, B, C)$ gives

$$X(s) = [sI - A - BF]^{-1}BV(s) = \frac{\text{adj } (sI - A - BF)}{\det \{sI - A - BF\}} BV(s) \equiv G_1(s)V(s) \tag{6.41}$$

since F has rank 1, it may be expressed as

$$F = p \times q \tag{6.42}$$

where p is an m-column vector and q an n-row vector.

The characteristic polynomial of the closed-loop system in (6.41) then may be expressed in the form

$$\det \{sI - A - BF\} = \det \{sI - A - Bpq\} = \det \{sI - A\} - q \text{ adj } (sI - A)Bp$$

that is,

$$q \text{ adj } (SI - A)Bp = \det \{sI - A\} - \det \{sI - A - BF\}$$

*An equivalent concept with output feedback is considered in Sec. 7.1.

†n arbitrary elements in any one row and (m - 1) arbitrary multipliers for this row to give the other rows.

$$= [(\alpha_1 - \beta_1) \cdots (\alpha_i - \beta_i) \cdots (\alpha_n - \beta_n)] \begin{bmatrix} s^{n-1} \\ \vdots \\ s^i \\ \vdots \\ 1 \end{bmatrix} = \{(\alpha_i - \beta_i)\}\{s^{n-i}\}, \quad i \in 1, \ldots, n \tag{6.43}$$

where $\det(sI - A) = \Sigma_{i=1}^{n} \alpha_i s^{n-i}$ and $\det(sI - A - BF) = \Sigma_{i=1}^{n} \beta_i s^{n-i}$ are the characteristic polynomials of A and (A + BF), respectively.

Let M be the modal matrix of the open-loop system S(A, B, C). Then, if $\Lambda = \text{diag}\{\lambda_1, \lambda_2, \ldots, \lambda_n\}$ denotes the open-loop eigenvalues,

$$\begin{aligned} \text{adj}(sI - A) &\triangleq [sI - A]^{-1} \det\{sI - A\} = M[sI - \Lambda]^{-1} M^{-1} \det\{sI - A\} \\ &= M \, \text{diag}\left\{\frac{1}{s - \lambda_i}\right\} M^{-1} \det\{sI - A\} \\ &= M \, \text{diag}\left\{\frac{\det\{sI - \Lambda\}}{s - \lambda_i}\right\} M^{-1}, \quad i \in 1, \ldots, n \end{aligned} \tag{6.44}$$

or

$$\begin{aligned} \text{adj}(sI - A)Bp &= M \, \text{diag}\left\{\frac{\det\{sI - A\}}{s - \lambda}\right\} M^{-1}Bp = M \, \text{diag}\left\{\frac{\det\{sI - \Lambda\}}{s - \lambda_i}\right\} B_n P \\ &= M \, \text{diag}\left\{\frac{\det\{sI - A\}}{s - \lambda_i}\right\} \{b_{nj}\} p, \quad i \in 1, \ldots, n; \; j \in 1, \ldots, m \end{aligned} \tag{6.45}$$

where b_{nj} is the jth column of the normalized input control matrix B_n.

Since $\det\{sI - A\}/(s - \lambda_i)$ is a polynomial of degree (n - 1), (6.45) may be expressed by

$$\text{adj}(sI - A)Bp = M\psi' \begin{bmatrix} s^{n-1} \\ \vdots \\ s^i \\ \vdots \\ 1 \end{bmatrix} = M\psi'\{s^{n-i}\}, \quad i \in 1, \ldots, n \tag{6.46}$$

where $\psi' \equiv \{\psi_{ij}\}$ i, j $\in$ 1, ..., n, is an n-square matrix, ψ_{ij} being the coefficient of s^{n-j} in the polynomial $\det\{sI - \Lambda\}/(s - \lambda_i)\, b_{nj}$. ψ' can be shown to be nonsingular if the open-loop system is stable and controllable. Combining (6.43), (6.45), and (6.46),

$$\{(\alpha_i - \beta_i)\} = qM\psi' = q\psi \quad (\psi = M\psi') \tag{6.47}$$

The unity-rank state feedback matrix F, from (6.42), is thus given by

$$F = p \cdot q = p\{(\alpha_i - \beta_i)\}\psi^{-1} \tag{6.48}$$

where p is a vector not orthogonal to any b_{nj}, $j = 1, \ldots, m$, and ψ is dependent on the open-loop system parameters and the feedback row vector. The development above thus leads to the following theorem.

Theorem 6.1 For a completely state controllable system all closed-loop poles may be assigned arbitrarily by a unity state feedback gain matrix F such that $F = p\{(\alpha_i - \beta_i)\}\psi^{-1}$, where p is a column vector not orthogonal to B_n, and α_i, β_i are the coefficients of the characteristic polynomials of the open- and closed-loop systems, respectively, ψ is as given in (6.45) and (6.46), which may be formulated in terms of the unknown n elements of the column vector p.

The design procedure requires the expansion of det $\{sI - A\}$ to obtain α and the desired closed-loop characteristic polynomial to assign values to β, and the evaluation of adj (sI - A) and M so that with any p, ψ may be obtained from (6.45) and (6.46) (and, preferably, the computation of B_n to identify the orthogonal column vector of B_n).

For a given system S(A, B, C), and known (desired) closed-loop characteristic polynomial $(\Sigma_{i=1}^{n} \beta_{n-1} s^i)$, ψ is a function of the n elements of the column vector p. The scale of p does not affect F, for if p is multiplied by some constant, so is ψ, and since F contains the factors p and ψ^{-1}, it will remain unchanged. Only the direction of p thus affects F, and this direction is determined by the (m - 1) mutual ratios of the elements of p. In fact, of the (n + m - 1) elements of F, n have been used to control the n coefficients of the closed-loop characteristic polynomial; any selected p determines ψ and q uniquely; hence an infinite number of (p, q) pairs exist, all giving the same polynomial det $\{sI - A - BF\}$. The remaining (m - 1) elements in the state feedback matrix of F may therefore be used to modify the numerator of the closed-look transfer function matrix (specifying the zeros of the system). The effects of varying p may be studied through design algorithms by describing F as explicit functions of p. Examples to illustrate this method of design are given below.

The derivation above is based on the transformation of a multivariable system into a single-input system through the control law

$$u(t) = v(t) + pqx(t)$$

where p is assumed to be a known arbitrary matrix. The closed-loop system then becomes

$$\dot{x}(t) = [A + Bpq]\, x(t) + v(t) = [A + bq]\, x(t) + v(t) \tag{6.49}$$

The unknown vector q then is given by, from (6.47),

$$q = \{(\alpha_i - \beta_i)\}\, \psi^{-1}$$

where ψ is an n-square constant matrix obtained from the coefficients of adj (sI − A). It can easily be shown that the equivalent single-input system in (6.49) is completely controllable for any arbitrary choice of p, in which case ψ is nonsingular. The following example illustrates the computation of q.

Example 6.3 [10]: Pole assignment through unity-rank state feedback The system considered here is

$$\dot{x}(t) = \begin{bmatrix} 0 & 1 \\ 0 & 0 \end{bmatrix} x(t) + \begin{bmatrix} 1 & 0 \\ 0 & 1 \end{bmatrix} u(t) \tag{6.50}$$

The design objective is to find a unity-rank stage feedback matrix F such that the closed-loop poles are at $\lambda_{1,2} = -1 \pm j$. The first state of the design is to compute the right-hand side of (6.43):

$$\begin{aligned} \det\{sI - A\} - \det\{(sI - A - BF)\} &= s^2 - (s + 1 + j)(s + 1 - j) = -2s - 2 \\ &\equiv (\alpha_1 - \beta_1)s + (\alpha_2 - \beta_2)1 \\ &= [(\alpha_1 - \beta_1) \,\vdots\, (\alpha_2 - \beta_2)] \begin{bmatrix} s \\ 1 \end{bmatrix} \end{aligned} \tag{6.51}$$

The next stage is to formulate the left side of (6.43) assuming an arbitrary (to be assigned by the designer) form of $p = [p_1 \;\; p_2]^T$:

$$\begin{aligned} q \operatorname{adj}(sI - A)Bp &= q \begin{bmatrix} s & 1 \\ 0 & s \end{bmatrix} \begin{bmatrix} p_1 \\ p_2 \end{bmatrix} = [q_1 \;\; q_2] \begin{bmatrix} p_1 s + p_2 \\ p_2 s \end{bmatrix} \\ &= [q_1 \;\; q_2] \begin{bmatrix} p_1 & p_2 \\ p_2 & 0 \end{bmatrix} \begin{bmatrix} s \\ 1 \end{bmatrix} = q\psi \begin{bmatrix} s \\ 1 \end{bmatrix} \end{aligned} \tag{6.52}$$

Combining (6.51) and (6.52) and using (6.43) and (6.47),

$$q = \{(\alpha_i - \beta_i)\}\, \psi^{-1} = \frac{1}{p_2^2} [-2 \;\; -2] \begin{bmatrix} 0 & p_2 \\ p_2 & -p_1 \end{bmatrix} = \frac{1}{p_2^2} [-2p_2 \;\; -(2p_2 - 2p_1)] \tag{6.53}$$

where p_1 and p_2 are arbitrary such that $\det\{\psi\} \neq 0$.

The final stage is to compute the unity-feedback matrix

$$F = pq = \begin{bmatrix} p_1 \\ p_2 \end{bmatrix} \begin{bmatrix} -\frac{2}{p_2} & -\frac{2}{p_2}\left(1 - \frac{p_1}{p_2}\right) \end{bmatrix} = \begin{bmatrix} -2\alpha & -2\alpha(1-\alpha) \\ -2 & -2(1-\alpha) \end{bmatrix}, \quad \alpha = \frac{p_1}{p_2} \tag{6.54}$$

Design is now completed by assigning specific values to the elements of p. If $p_1 = 1$, $p_2 = 2$, then from (6.54),

$$F = \begin{bmatrix} -1 & -\frac{1}{2} \\ -2 & -1 \end{bmatrix} \tag{6.55}$$

It is apparent that (6.51) gives an infinite choice of F to place the poles at the desired positions, and that F is dependent only on the ratio of p_1/p_2. This result can be generalized to conclude that for a specified set of closed-loop poles, the unity-rank state feedback matrix, F, depends only on the relative magnitude of the elements of the (arbitrary) column vectors in (6.54). Hence F is unique for particular ratios between the elements of p. This generalization, proof of which is omitted here, is illustrated through the following example.

Example 6.4 [8]: Pole assignment through unity-rank state feedback

The system considered is

$$\dot{x}(t) = \begin{bmatrix} -1 & 2 \\ 1 & -3 \end{bmatrix} x(t) + \begin{bmatrix} 1 & 1 & 0 \\ -2 & 0 & 1 \end{bmatrix} u(t) \tag{6.56}$$

Let the desired location of the closed-loop poles be $\lambda_{1,2} = -1 \pm j$. For the open-loop system,

$$G_1(s) = (sI - A)^{-1}B = \frac{1}{s^2 + 4s + 1} \begin{bmatrix} s-1 & s+3 & 2 \\ -(2s+1) & 1 & s+1 \end{bmatrix} \tag{6.57}$$

Let $p = [p_1, p_2, p_3]^T$; then the equivalent single-input system is

$$g(s) = (sI - A)^{-1}Bp = \frac{1}{s^2 + 4s + 1} \begin{bmatrix} (p_1 + p_2)s - p_1 + 3p_2 + 2p_3 \\ (-2p_1 + p)s - p_1 + p_2 + p_3 \end{bmatrix} \tag{6.58}$$

From (6.43) and the equations above, using $q = [q_1 \;\; q_2]$, we have

$$q_1[(p_1 + p_2)s - p_1 + 3p_2 + 2p_3] + q_2[(-2p_1 + p_3)s - p_1 + p_2 + p_3]$$
$$= (s^2 + 4s + 1) - (s + 1 + j)(s + 1 - j) \tag{6.59}$$

Equating coefficients of similar powers of s, (6.59) gives

$$\begin{bmatrix} -p_1 + 3p_2 + 2p_3 & -p_1 + p_2 + p_3 \\ p_1 + p_2 & -2p_1 + p_3 \end{bmatrix} \begin{bmatrix} q_1 \\ q_2 \end{bmatrix} = \begin{bmatrix} -1 \\ 2 \end{bmatrix} \tag{6.60}$$

which is of the form $\psi q = \{(\alpha_i - \beta_i)\}$, where $\det\{\psi\} = 3p_1^2 - p_2^2 + 2p_3^2 - 6p_1p_2 - 6p_1p_3 + 2p_2p_3$ which should be nonzero for (6.60) to have a solution and for the closed-loop system to be controllable. The controllability condition is then

$$3p_1^2 - p_2^2 + 2p_3^2 - 6p_1p_2 - 6p_1p_3 + 2p_2p_3 \neq 0$$

or

$$\Delta(r) = 3r_1^2 - r_2^2 + 2 - 6r_1r_2 - 6r_1 + 2r_2 \neq 0 \tag{6.61}$$

where

$$r_1 = \frac{p_1}{p_3} \quad \text{and} \quad r_2 = \frac{p_2}{p_3}$$

The value of F, subject to the condition above, is given by

$$F = pq = \frac{1}{\Delta(r)} \begin{bmatrix} \alpha r_1 & \beta r_1 \\ \alpha r_2 & \beta r_2 \\ \alpha & \beta \end{bmatrix} \tag{6.62}$$

where

$$\alpha = 4r_1 - 2r_2 - 3 \quad \text{and} \quad \beta = -(r_1 - 7r_2 - 4)$$

This feedback matrix moves the closed-loop poles to $\lambda_{1,2} = -1 \pm j$ for all values of r_1 and r_2 provided that $\Delta(r)$ in (6.61) is nonzero.

6.3 ZERO ASSIGNMENT [11, 12]

For single-input/single-output systems, the zeros are invariant under constant gain feedback and thus the closed-loop responses can be specified completely. In multivariable systems, however, a class of state feedback matrices exists for a given set of desired closed-loop poles. Since the transfer function matrix of a closed-loop system with state feedback [from (6.41)] is

$$G(s) = C[sI - A - BF]^{-1}B \equiv C\frac{\operatorname{adj}[sI - A - BF]}{\det\{sI - A - BF\}}B \tag{6.63}$$

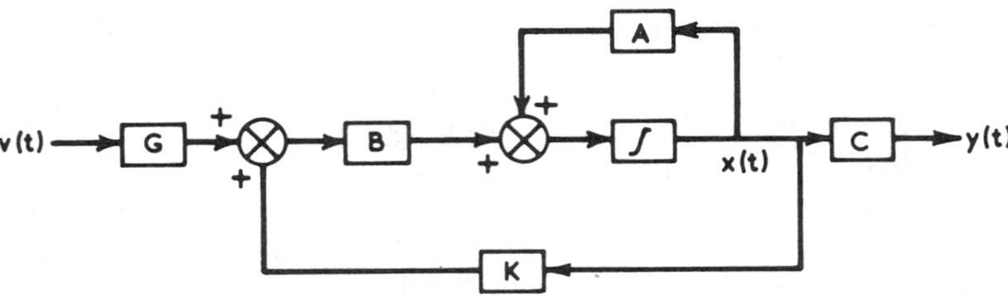

Fig. 6.3 Closed-loop control for zero assignment.

the zeros, as well as the poles, of the multivariable system are affected by linear state feedback control laws. The design methods presented in the preceding sections consider only the location of poles.* Hence for a given set of closed-loop poles, the resulting state feedback gain matrix may yield undesirable zeros. This section presents a method of assigning the location of some zeros of linear time-invariant multivariable systems through a constant gain feedforward matrix while leaving its poles unchanged.

The control law considered here has the form

$$u(t) = Fx(t) + Gv(t) \tag{6.64}$$

where G is an m-square real constant matrix and F is the state feedback matrix for the system S(A, B, C). The state and the output equations of the closed-loop system, shown in Fig. 6.3, then become

$$\dot{x}(t) = (A + BF)x(t) + BGv(t); \quad y(t) = Cx(t) \tag{6.65}$$

The transfer function of the closed-loop system may then become

$$\hat{G}(s) = C\,\frac{\text{adj}\,(sI - A - BF)}{\det\{sI - A - BF\}}\,BG = \frac{N(s)}{\Delta(s)}\,G \tag{6.66}$$

where $N(s) = C\,\text{adj}\,(sI - A - BK)B$ is the $m \times m$ closed-loop numerator polynomial matrix without the feedforward matrix ($G \equiv I$) and $\Delta(s)$ is the closed-loop characteristic polynomial.

It is seen from (6.63) and (6.66) that the feedforward gain matrix G, placed outside the state feedback loop, does not affect $\Delta(s)$, and thus the location of the closed-loop poles controlled by F are unchanged by the inclusion of G in the control law in (6.64). The feedforward controller G,

*Although some limited flexibility in choosing the zeros is inherent in the unity feedback system, as outlined in Sec. 6.2.

however, postmultiplies the numerator polynomial matrix N(s) and thus performs elementary column operations on N(s) thereby affecting the closed-loop zeros. From (6.66) any ith output $y_i(t)$ is related to any jth input $v_j(t)$ by

$$y_i(s) = \frac{n_{ir}(s)}{\Delta(s)} g_{jc}(s)v_j(s) = \frac{\hat{n}_{ij}(s)}{\Delta(s)} v_j(s) \tag{6.67}$$

where $n_{ir}(s)$ is the ith row of N(s) and $g_{jc}(s)$ is the jth column of G, and the (i, j)th element of $\hat{N}(s)G$ is

$$\hat{n}_{ij}(s) = \sum_{k=1}^{m} n_{ik}(s)g_{kj}, \tag{6.68}$$

which shows that the numerator of the transfer function relating the output $y_i(s)$ to the control input $v_j(s)$ is a linear combination of the numerators of transfer functions relating the output $y_i(s)$ to the m control input $v_1(s) \cdots v_m(s)$.

Let the polynomial $n_{ik}(s)$ the (i, k)th element of N(s) in (6.66) be

$$n_{ik}(s) = \alpha^i_{1k} s^{n-1} + \cdots + \alpha^i_{nk}$$

where α's are constants. Then (6.68) becomes

$$\begin{aligned}\hat{n}_{ij}(s) &= g_{1j}(\alpha^i_{11}s^{n-1} + \cdots + \alpha^i_{n1}) + \cdots + g_{mj}(\alpha^i_{1m}s^{n-1} + \cdots + \alpha^i_{nm}) \\ &= (\alpha^i_{11}g_{1j} + \cdots + \alpha^i_{1m}g_{mj})s^{n-1} + \cdots + (\alpha^i_{n1}g_{1j} + \cdots + \alpha^i_{nm}g_{mj}) \\ &\equiv \beta_1 s^{n-1} + \cdots + \beta_n \end{aligned} \tag{6.69}$$

Thus the coefficients of numerator polynomial $\hat{n}_{ij}(s)$ are linear functions of the elements of the jth column of G. These variables can now be chosen so that the transfer function between $y_i(s)$ and $v_j(s)$ has zeros at some specified locations in the complex s-plane. For example, the requirement to have a zero at $s = s_0$ in the $y_i(s)$ to $v_j(s)$ transfer function is

$$\hat{n}_{ij}(s_0) = n_{i1}(s_0)g_{1j} + \cdots + n_{im}(s_0)g_{mj} = 0 \tag{6.70}$$

Similarly, if the value of the coefficient of s^{n-k} in the numerator is specified as β^{ij}_k, then from (6.69),

$$\hat{n}_{ij} = \beta_1^{ij} s^{n-1} + \cdots + \beta_k^{ij} s^{n-k} + \cdots + \beta_n^{ij} \tag{6.71}$$

where

$$\alpha_{k1}^{i} g_{1j} + \cdots + \alpha_{km}^{i} g_{mj} = \beta_k^{ij}$$

Consequently, the specification of one zero position or one numerator coefficient of the transfer function between $y_i(s)$ and $v_j(s)$ gives one linear equation in the unknown elements of jth column of G. Then a set of any r ($\geq$ m) linear simultaneous equations in the m unknown elements $g_{1j}, \ldots, g_{mj}$ may be obtained by relating the r zeros and the numerator coefficients of some or all the transfer functions relating the m outputs to the jth input control, that is,

$$\frac{y_1(s)}{v_j(s)}, \ldots, \frac{v_m(s)}{v_j(s)}$$

These r simultaneous equations may be expressed as

$$Wg_{jc} = \beta \tag{6.72}$$

where W is a constant $r \times m$ matrix and β is a constant r-column vector, both obtained from combining (6.70) and (6.71), where $r \geq m$. Two cases are of interest:

1. Computation of G when r = m: In this case (6.72) can be solved exactly to obtain

$$g_{jc} = W^{-1}\beta \tag{6.73}$$

Therefore, m specifications per input can be met exactly. Of these m, the maximum number of zero specification is equal to (m - 1), since each zero specification introduces one zero element in the vector β, and since β must be nonzero so that $g_{jc} \neq 0$, there must be at one nonzero coefficient specification. The total number of zeros that can always be located exactly using a constant gain feedforward matrix is thus m(m - 1).

2. Computation of G when r > m: In this case (6.72) does not in general have an exact solution and the best approximate solution that minimizes $\| g_{jc} - \beta \|^2$ is given by

$$g_{jc} = (W^T W)^{-1} W^T \beta \tag{6.74}$$

The r specifications are not met exactly; however, the more significant specifications can be weighted relative to the others by multiplying the corresponding rows of W and β by appropriate weighting factors. Again at least one specification per input must be a nonzero coefficient specification so that $\beta \neq 0$. Alternatively, r zeros can be specified per input, provided that one of them is specified approximately by setting the residual of the numerator polynomial to an arbitrarily small value.

The derivations above show that each column of the feedforward controller matrix corresponds to the numerators of transfer functions relating the outputs to the appropriate input control, and is determined from the specifications of r zeros and numerator coefficients of these transfer functions. In this way the controller matrix G is designed one column at a time. The design procedure for r = m is illustrated below.

Example 6.5 [12]: Zero assignment The method of assigning zeros described above is illustrated below for a system with transfer function matrix:

$$G(s) = \frac{1}{s^2+2s+2}\begin{bmatrix} s+1 & -0.5 \\ -s+3 & 2s+2.5 \end{bmatrix} \tag{6.75}$$

Let the desired transfer function matrix be

$$\hat{G}(s) = \frac{1}{s^2+2s+2}\begin{bmatrix} a_1(s+5) & b_1 s - 0.1 \\ a_2 s + 0.1 & b_2(s+3) \end{bmatrix} \tag{6.76}$$

where a and b may assume arbitrary values.

The transfer function of the modified system, from (6.67), is

$$\begin{bmatrix} y_1(s) \\ y_2(s) \end{bmatrix} = \frac{1}{\Delta(s)}\begin{bmatrix} \hat{n}_{11}(s) & \hat{n}_{12}(s) \\ \hat{n}_{21}(s) & \hat{n}_{22}(s) \end{bmatrix}\begin{bmatrix} v_1(s) \\ v_2(s) \end{bmatrix} \tag{6.77}$$

with

$$\hat{n}_{ij}(s) = \sum_{k=1}^{2} n_{ik}(s) g_{kj}, \quad \Delta(s) = s^2 + 2s + 2$$

Let the constant feedforward controller in the feedforward path be $G = \{g_{ij}\}$, j = 1, 2. Then, from (6.68),

$$\hat{N}(s) = N(s)G = \begin{bmatrix} n_{11}(s) & n_{12}(s) \\ n_{21}(s) & n_{22}(s) \end{bmatrix}\begin{bmatrix} g_{11} & g_{12} \\ g_{21} & g_{22} \end{bmatrix}$$

$$= \begin{bmatrix} s+1 & -0.5 \\ -s+3 & 2s+2.5 \end{bmatrix} \begin{bmatrix} g_{11} & g_{12} \\ g_{21} & g_{22} \end{bmatrix} \tag{6.78}$$

$$= \begin{bmatrix} sg_{11} + (g_{11} - 0.5g_{21} & sg_{12} + (g_{12} - 0.5g_{22} \\ s(2g_{21} - g_{11}) + (3g_{11} + 2.5g_{21}) & s(2g_{22} - g_{12}) + (3g_{11} + 2.5g_{22}) \end{bmatrix} \equiv \Delta(s)\hat{G}(s) \tag{6.79}$$

Comparing the elements of (6.77) and (6.79) and using (6.70), we have:

$$\left.\begin{array}{l} \hat{n}_{11}(s) \rightarrow \text{zero at } s=-5\text{:} \quad -5g_{11} + g_{12} - 0.5g_{22} = 0 \\ \hat{n}_{21}(s) \rightarrow \text{constant gain:} \quad 3g_{11} + 2.5g_{21} = 0.1 \end{array}\right\} \rightarrow \begin{bmatrix} g_{11} \\ g_{21} \end{bmatrix} = \frac{1}{170} \begin{bmatrix} -1 \\ 8 \end{bmatrix} \tag{6.80}$$

$$\left.\begin{array}{l} \hat{n}_{12}(s) \rightarrow \text{constant gain:} \quad g_{12} - 0.5g_{22} = -0.1 \\ \hat{n}_{22}(s) \rightarrow \text{zero at } s=-8\text{:} \quad -8(2g_{22} - g_{12}) + 3g_{12} + 2.5g_{22} = 0 \end{array}\right\} \rightarrow \begin{bmatrix} g_{12} \\ g_{22} \end{bmatrix} = \frac{1}{8} \begin{bmatrix} -1.35 \\ -1.1 \end{bmatrix}$$

The resulting transfer-function matrix is

$$\hat{G}(s) = \frac{1}{\Delta(s)} \begin{bmatrix} -58.82(s+5) \times 10^{-2} & -0.17s - 0.1 \\ 0.1s + 0.1 & -0.85(s+8) \end{bmatrix} \tag{6.81}$$

As seen by the example above, the method, although effective, may not in general yield exact zero assignment, due to the presence of fewer variables (g_{ij}) than the number of equations. Numerical methods may, in such cases, have to be used according to (6.74). Another limitation of this relatively straightforward method is the dimension of the input vector. This limitation can be severe when the number of inputs is low and the order of the system is high. However, in cases where the number of undesirable zeros that need to be moved is small, the number of parameters in G can be sufficient.

6.4 LINEAR OPTIMAL CONTROL [13-21]

The quadratic performance index introduced in Sec. 5.1.4 is often used to obtain a closed-form solution for the control which under a reasonable set of constraints yields a stable system. For time-invariant systems, this is achieved by solving the matrix Riccati equation

$$PBR^{-1}B^TP - PA - A^TP - Q = 0 \tag{6.82}$$

where Q and R are the weighting matrices in the performance index

$$J = \frac{1}{2} \int_{t_0}^{t} \{[x(t)]^T Q[x(t)] + [u(t)]^T R[u(t)]\} \, dt \tag{6.83}$$

the resulting optimal feedback law being

$$u(t) = \underbrace{-R^{-1}B^T P}_{K} \, x(t) + v(t) \tag{6.84}$$

Thus for a specified set of Q and R, the state feedback matrix* K may be derived by first solving (6.82) for P [by using one of the established algorithms (Refs. 19 to 21), and then by using (6.84)]. Although this method has been widely used, it has one serious limitation—an absence of any a priori knowledge of the relationship between the performance index and the response of the optimal system. The result is that the design is carried out in an iterative manner through (1) choice of some initial Q and R, and (2) plotting of the resulting time responses. This "direct" design procedure is continued until a satisfactory response is obtained. This problem is partly overcome in the method developed here by expressing the characteristic polynomial of the closed-loop system in terms of the elements of the weighting matrices Q and R. The resulting analysis, although not any simpler in computational term, provides a basis for studying the effects of increasing or decreasing the elements of Q and R on the performance of the system. Once the trends of improvement or deterioration in either stability or system response are identified, the designer can identify a range of preferred weighting matrices and hence performances indices. By using the results and notations in Sec. 5.2, if $\phi_o(s)$ and $\phi_c(s)$ denote the characteristic polynomials of the open-loop system S(A, B, C) and the closed-loop system $S(\overline{A + BK}, B, C)$, where K is given by (6.84), the determinant of the return-difference matrix F(s) of the system (Sec. 5.2) is

$$\det\{F(s)\} = \frac{\phi_c(s)}{\phi_o(s)} = \frac{\det\{sI - A - BK\}}{\det\{sI - A\}} \tag{6.85}$$

To derive a frequency-domain condition of optimality, (6.82) is taken through the following stages [14]:

*K is used here to avoid any confusion with F(s), used subsequently to denote the return-difference matrix.

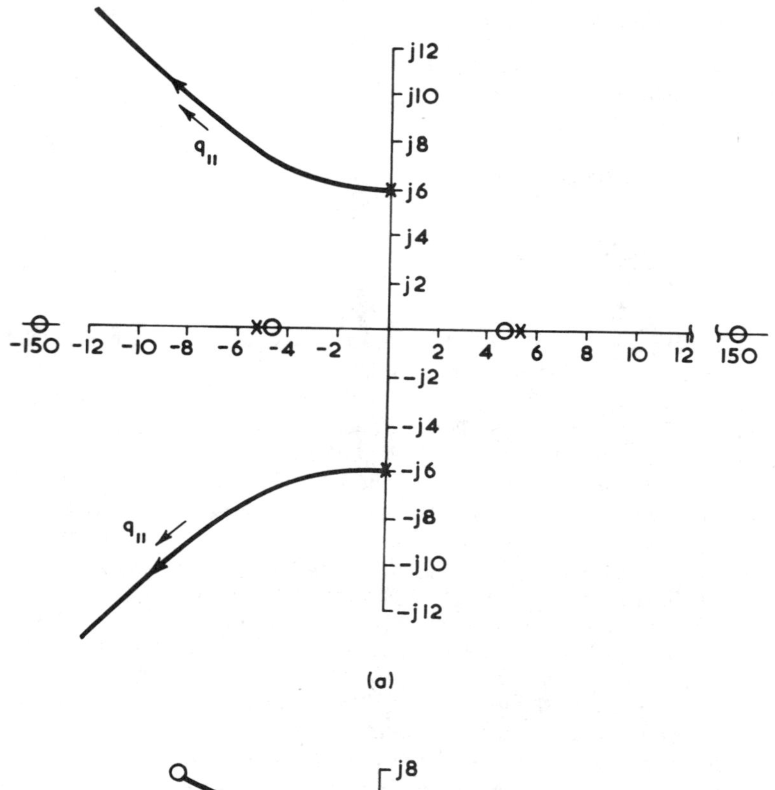
j12
j10
j8
j6
j4
j2
-150
-12
-10
-8
-6
-4
-2
2
4
6
8
10
12
150
-j2
-j4
-j6
-j8
-j10
-j12
q_{11}
q_{11}

(a)

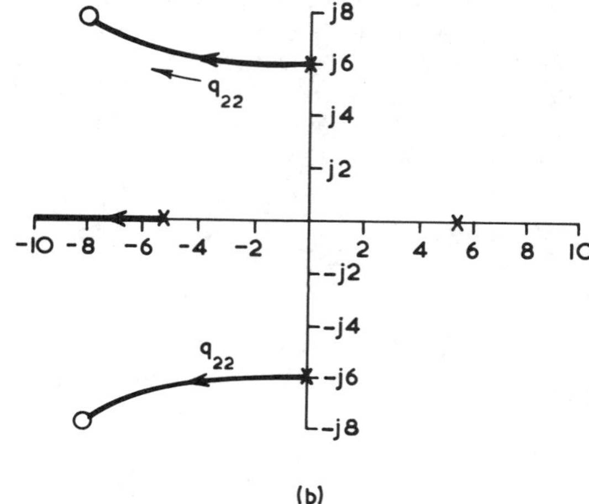
j8
j6
j4
j2
-10
-8
-6
-4
-2
2
4
6
8
10
-j2
-j4
-j6
-j8
q_{22}
q_{22}

(b)

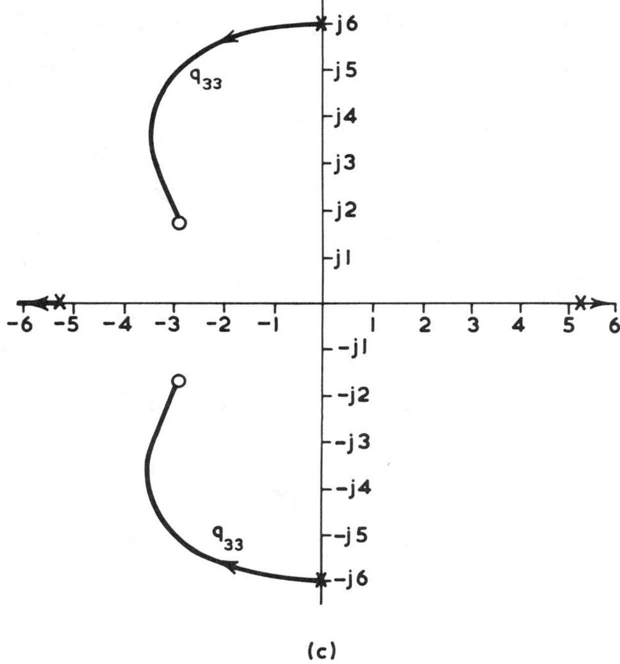

Fig. 6.4 Effect of $Q = \text{diag}\{q_{11}, q_{22}, q_{33}\}$ on root loci: (a) $q_{22} = q_{33} = 0$ with increasing q_{11}; (b) $q_{11} = q_{33} = 0$ with increasing q_{22}; (c) $q_{11} = q_{22} = 0$ with increasing q_{33}.

$$Q = -PA - A^TP + PBR^{-1}B^TP = P[sI - A] + [-sI - A^T]P + PBR^{-1}B^TP$$

Pre- and post-multiplying both sides by $B^T[-sI - A]$ and $[sI - A]^{-1}B$ respectively:

$$B^T[-sI - A^T]^{-1}Q[sI - A]^{-1}B + R = B^T[-sI - A^T]^{-1}\{P[sI - A]$$

$$+ [-sI - A^T]P + PBR^{-1}B^TP\}[sI - A]^{-1}B + R$$

$$= B^T[-sI - A^T]^{-1}PB + B^TP[sI - A]B\ B^T[-sI - A^T]^{-1}PBR^{-1}B^TP[sI - A]^{-1}B + R$$

$$\rightarrow\ [G(-s)]^T Q[G(s)] + R = B^T[-sI - A^T]^{-1}PB + RR^{-1}B^T P[sI - A]B$$

$$+ B^T[-sI - A^T]^{-1}PBR^{-1}RR^{-1}B^T P[sI - A]^{-1}B + R$$

$$= B^T[-sI - A^T]^{-1}K^T R + RK[sI - A]^{-1}B$$

$$+ B^T[-sI - A^T]^{-1}K^T RK[sI - A]^{-1}B + R$$

$$= \{I + B^T[-sI - A^T]K^T\}R\{I + K[sI - A]^{-1}B\}$$

$$= [F(-s)]^T R(F(s)]$$

$$\rightarrow\ R^{-\frac{1}{2}}\{[G(s)]^T Q[G(s)] + R\}\, R^{-\frac{1}{2}} = R^{\frac{1}{2}}[F(-s)]^T R[F(s)]R^{-\frac{1}{2}}$$

$$\rightarrow\ \det\{R^{-\frac{1}{2}}[G(-s)]^T Q[G(s)]R^{-\frac{1}{2}} + I\} = \det\{R^{-\frac{1}{2}}[F(-s)]^T R[F(s)]R^{-\frac{1}{2}}\}$$

$$\rightarrow\ \det\{R^{-\frac{1}{2}}[G(-s)]^T Q[G(s)]R^{-\frac{1}{2}} + I\} = \det\{[F(-s)]^T\}\det\{F(-s)\}$$

$$= \frac{\phi_c(-s)\phi_c(s)}{\phi_o(-s)\phi_o(s)}$$

$$\rightarrow\ \phi_c(-s)\phi_c(s) = \phi_o(-s)\phi_o(s)\det\{I + R^{-\frac{1}{2}}[G(-s)]^T Q[G(s)]R^{-\frac{1}{2}}\} \tag{6.86}$$

which gives an explicit relationship between the characteristic polynomials and the elements of the weighting matrices in the performance index. Thus if the location of the poles of the optimal system is specified, the elements of Q and R become the design parameters. Alternatively, (6.86) may be used to plot a family of root loci for a range of values of Q and R to analyze their effect on the closed-loop poles. A numerical example is considered below.

Example 6.6 [13]: Pole placement through optimal criterion The three-input/three-output third-order system considered here is described by

$$A = \begin{bmatrix} 1 & 2 & -3 \\ -4 & 5 & -0.6 \\ 7 & 8 & -0.9 \end{bmatrix};\quad B = \begin{bmatrix} 0.1 & 0 & 2 \\ 0 & 10 & 0 \\ 0 & 0 & 100 \end{bmatrix};\quad C = \begin{bmatrix} 1 & 0 & 0 \\ 0 & 1 & 0 \\ 0 & 0 & 1 \end{bmatrix} \tag{6.87}$$

with $\phi_o(s) = (s - 5.3)(s + 0.124 - j5.89)(s + 0.124 + j5.89)$, which shows that the open-loop system has one pole on the right-half of the s-plane.

For numerical convenience, the input weighting matrix R in (6.83) is assumed to be an identity matrix. The condition of optimality then becomes, from (6.86),

$$\phi_c(-s)\phi_c(s) = \phi_o(-s)\phi_o(s) \det\{I + [G(-s)]^T Q[G(s)]\}$$

where

$$G(s) = C[sI - A]^{-1}B = \frac{1}{\phi_o(s)} \begin{bmatrix} s^2 - 4.1s + 0.3 & 2s - 22.2 & -3s + 13.8 \\ -4s - 7.8 & s^2 - 0.1s + 20.1 & -0.6s + 12.6 \\ 7s - 67 & 8s + 6 & s^2 - 6s + 13 \end{bmatrix} \tag{6.88}$$

Choosing $Q = \text{diag}\{q_1, q_2, q_3\}$, the movement of the roots of $\phi_c(-s)\phi_c(s)$ may be derived by solving (6.86) for s. These were derived in Ref. 13 and are shown in Fig. 6.4 for various values of q_1, q_2, and q_3, from which the choice in Fig. 6.4c is made due to a fair amount of damping introduced in the closed-loop system. With this choice, the state weighting matrix becomes

$$Q = \begin{bmatrix} 0 & 0 & 0 \\ 0 & 0 & 0 \\ 0 & 0 & 0.125 \end{bmatrix}$$

for which the solution of the matrix Riccati equation in (6.82) is

$$P = \begin{bmatrix} 7.23 & -4.29 & -0.61 \\ -4.29 & 7.65 & 0.335 \\ -0.6116 & 0.335 & 0.3662 \end{bmatrix} \times 10^{-2}$$

$$\rightarrow K = -B^T P = \begin{bmatrix} -7.23 \times 10^{-3} & 4.29 \times 10^{-3} & 0.61 \times 10^{-3} \\ 4.29 \times 10^{-1} & -7.65 \times 10^{-1} & -0.335 \times 10^{-1} \\ 46.7 \times 10^{-2} & -24.92 \times 10^{-2} & -35.4 \times 10^{-2} \end{bmatrix}$$

6.5 ADAPTIVE CONTROL [22-25]

In some applications, the objective of state feedback control is to make the closed-loop system $S(\overline{A + BF}, B, C)$ follow a specified m-input/n-state model (or reference) state vector $x_m(t)$, defined by*

$$\dot{x}_m(t) = A_m x_m(t) + Bu(t) \tag{6.90}$$

rather than pole shifting considered earlier. This is commonly achieved by using a method known as model-reference (or model-following) adaptive control, shown schematically in Fig. 6.5. The control algorithm here is based on generating an appropriate BF such that the error vector

$$e(t) = x_m(t) - x(t) \tag{6.91}$$

tends to zero as t increases [i.e., $x(t) \rightarrow x_m(t)$ as $t \rightarrow \infty$]. Combining (6.3), (6.90), and (6.91), the dynamics of the error state in Fig. 6.5 may be expressed as

$$\dot{e}(t) = A_m x_m(t) - [A + BF]x(t) \equiv A_m e(t) + [A_m - \overline{A + BF}]x(t) \tag{6.92}$$

Thus if the matrix $\hat{F} = BF$ were generated such that $A_m = A + \hat{F}$ and the dynamics of the reference model chosen such that the "reduced" system

$$\dot{e}(t) = A_m e(t) \tag{6.93}$$

were asymptotically stable, the design objectives would be fulfilled. The dynamic equations required to satisfy these constraints are derived here by using Lyapunov's direct method (Appendix 2).

On the grounds of practical constraints, the model in (6.90) is assumed to be asymptotically stable. Consequently, the free system in (6.93) is asymptotically stable, and hence there exists a positive-definite symmetric matrix P such that

$$A_m^T P + PA_m = -I \tag{6.94}$$

which is derived through the choice of a Lyapunov function

$$V[e(t)] = -[e(t)]^T P[e(t)] \tag{6.95}$$

which is positive semidefinite and tends to zero as time increases. To

*For convenience, the input matrix in the model is chosen to be the same as the system matrix. The general case of $B_m \neq B$ may be considered within the same framework.

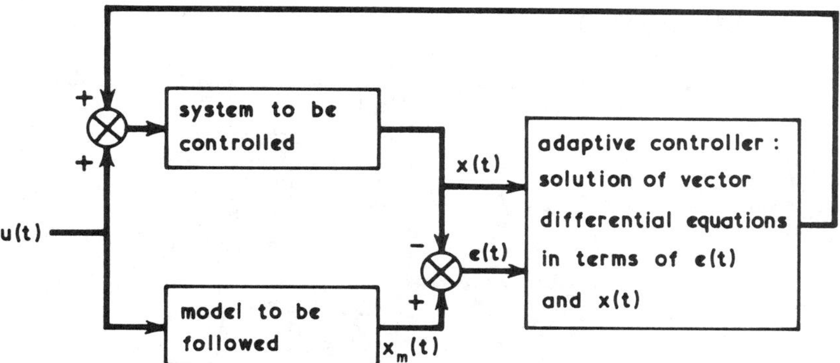

Fig. 6.5 Schematic of a model-reference adaptive scheme.

analyze the stability of the "error-generating" system in (6.93), let the Lyapunov function be chosen as

$$V(e, \hat{F}) = [e(t)]^T P[e(t)] + [\hat{F}]^T Q[\hat{F}] \tag{6.96}$$

where $Q = \{q_{ij}\}$, i, $j \in 1, \ldots, n$ is a positive-semidefinite matrix. If $\hat{F}$ is denoted by $\{\hat{f}_{ij}\}$, i, $j \in 1, \ldots, n$, then

$$[\hat{F}]^T Q[\hat{F}] = \{\hat{f}_{ij}\}^T \{q_{ij}\}\{\hat{f}_{ij}\} = \sum_{i=1}^{n} \sum_{j=1}^{n} q_{ij}\hat{f}_{ij}^2 \tag{6.97}$$

and combination of (6.94), (6.95), (6.96), and (6.97) gives

$$\frac{d}{dt}\{V(e, \hat{F})\} = \underbrace{\left[\frac{de(t)}{dt}\right]^T P[e(t)] + [e(t)]^T P\left[\frac{de(t)}{dt}\right]}_{R} + 2\sum_{i=1}^{n}\sum_{j=1}^{n} q_{ij}\hat{f}_{ij}\frac{d\hat{f}_{ij}}{dt} \tag{6.98}$$

where

$$\begin{aligned} R &= [A_m e(t) + \hat{F}x(t)]^T P[e(t)] + [e(t)]^T P[A_m e(t) + \hat{F}x(t)] \\ &= [e(t)]^T [A_m^T P + PA_m][e(t)] + [\hat{F}x(t)]^T P[e(t)] + [e(t)]^T P[\hat{F}x(t)] \\ &= -[e(t)]^T [e(t)] + 2\sum_{i=1}^{n}\sum_{j=1}^{n} x_j [e(t)]^T p_{ic}\hat{f}_{ij} \end{aligned}$$

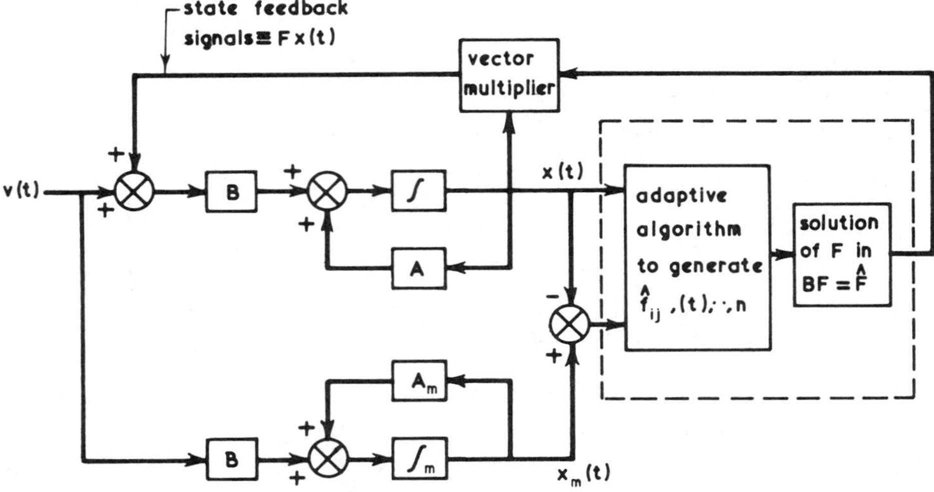

Fig. 6.6 Generation of adaptive state feedback control law.

where p_{ic} is the ith column of P. Thus

$$\frac{d}{dt}\{V(e, \hat{F})\} = -[e(t)]^T[e(t)]^T + 2\sum_{i=1}^{n}\sum_{j=1}^{n}\left\{q_{ij}\frac{d\hat{f}_{ij}}{dt} + x_j[e(t)]^T p_{ic}\right\}\hat{f}_{ij} \tag{6.99}$$

Thus to ensure stability of the composite system while satisfying the constraint $A_m = A + BF \equiv A + \hat{F}$, with the choice of the Lyapunov function as in (6.95), the elements of $\hat{F}$ become

$$q_{ij}\frac{d\hat{f}_{ij}}{dt} = -x_j(t)[e(t)]^T p_{ic}, \qquad i, j \in 1, \ldots, n \tag{6.100}$$

which can be generated by using x(t) and $x_m(t)$ as input control vectors by using the configuration shown in Fig. 6.6, where the elements of F may be derived by solving

$$BF = \hat{F} \tag{6.101}$$

for a given B. If the system S(A, B, C) is completely controllable [i.e., $\rho(B) = m$], then (6.101) has a unique solution for F. Alternatively, the "best" solution for F may be evaluated through the minimization of $\|BF - \hat{F}\|^2$. An illustrative example for a completely controllable system is given below.

Example 6.7: Model following through adaptive state feedback The system in the previous example (6.87) is considered here to illustrate the method of deriving a model reference adaptive state feedback controller. Let the model this system is required to follow as $t \to \infty$ be

$$\begin{bmatrix} \dot{x}_{1m}(t) \\ \dot{x}_{2m}(t) \\ \dot{x}_{3m}(t) \end{bmatrix} = \underbrace{\begin{bmatrix} 0 & 1 & 0 \\ -1 & -2 & 0 \\ 0 & 0 & -1 \end{bmatrix}}_{A_m} \underbrace{\begin{bmatrix} x_{1m}(t) \\ x_{2m}(t) \\ x_{3m}(t) \end{bmatrix}}_{x_m(t)} + \underbrace{\begin{bmatrix} 0.1 & 0 & 2 \\ 0 & 10 & 0 \\ 0 & 0 & 100 \end{bmatrix}}_{B} \begin{bmatrix} u_1(t) \\ u_2(t) \\ u_3(t) \end{bmatrix} \tag{6.102}$$

The design procedure starts with the computation of P in (6.94). Since the model is chosen to be asymptotically stable, there exists a positive-definite symmetric P, which in this case is obtained by solving [20]

$$\begin{bmatrix} 0 & -1 & 0 \\ 1 & -2 & 0 \\ 0 & 0 & 1 \end{bmatrix} \begin{bmatrix} p_{11} & p_{12} & p_{13} \\ p_{21} & p_{22} & p_{23} \\ p_{31} & p_{32} & p_{33} \end{bmatrix} + \begin{bmatrix} p_{11} & p_{12} & p_{13} \\ p_{21} & p_{22} & p_{23} \\ p_{31} & p_{32} & p_{33} \end{bmatrix} \begin{bmatrix} 0 & 1 & 0 \\ -1 & -2 & 0 \\ 0 & 0 & 1 \end{bmatrix} = \begin{bmatrix} -1 & 0 & 0 \\ 0 & -1 & 0 \\ 0 & 0 & -1 \end{bmatrix}$$

$$\to P = \frac{1}{2} \begin{bmatrix} 3 & 1 & 0 \\ 1 & 1 & 0 \\ 0 & 0 & 1 \end{bmatrix} \tag{6.103}$$

$$\uparrow p_{ic} \quad \uparrow p_{2c} \quad \uparrow p_{3c}$$

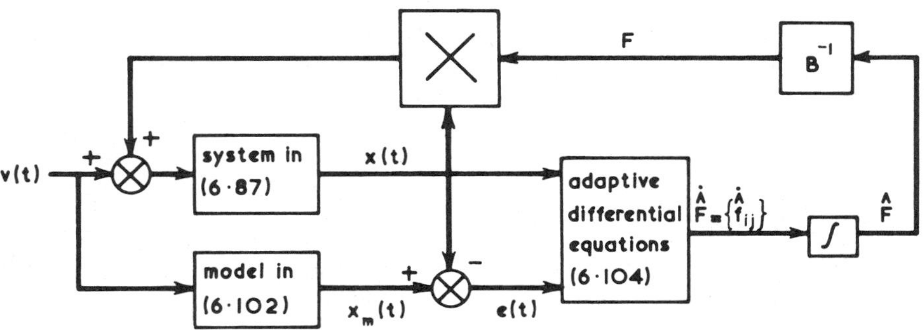

Fig. 6.7 Adaptive state feedback law for Example 6.7.

The control laws for each of the 3 × 3 adaptive state feedback channel may now be derived directly from (6.100):

$$\frac{d\hat{f}_{ij}}{dt} = -\frac{x_j(t)}{q_{ij}}\{[e(t)]^T p_{ic}\}, \quad i, j \in 1, 2, 3$$

which gives

$$\dot{\hat{f}} = -\frac{x_1(t)}{2q_{11}}\left\{[e_1(t) \;\; e_2(t) \;\; e_3(t)]\begin{bmatrix}3\\1\\0\end{bmatrix}\right\} = -\frac{x_1}{2q_{11}}\underbrace{\{3e_1(t) + e_2(t)\}}_{\alpha_1(t)} \tag{6.104a}$$

$$\dot{\hat{f}}_{12} = -\frac{x_2(t)}{2q_{12}}\alpha_1(t) \quad \text{and} \quad \dot{\hat{f}}_{13} = -\frac{x_3(t)}{2q_{12}}\alpha_1(t)$$

$$\dot{\hat{f}}_{21} = -\frac{x_2(t)}{2q_{21}}\left\{[e_1(t) \;\; e_2(t) \;\; e_3(t)]\begin{bmatrix}1\\1\\0\end{bmatrix}\right\}$$

$$= -\frac{x_2}{2q_{21}}\underbrace{\{e_1(t) + e_2(t)\}}_{\alpha_2(t)} \tag{6.104b}$$

$$\dot{\hat{f}}_{22} = -\frac{x_2(t)}{2q_{22}}\alpha_2(t) \quad \text{and} \quad \dot{\hat{f}}_{23} = -\frac{x_2(t)}{2q_{23}}\alpha_2(t)$$

$$\dot{\hat{f}}_{31} = -\frac{x_3(t)}{2q_{31}}\left\{[e_1(t) \;\; e_2(t) \;\; e_3(t)]\begin{bmatrix}0\\0\\1\end{bmatrix}\right\}$$

$$= -\frac{x_3(t)}{2q_{31}}\underbrace{\{e_3(t)\}}_{\alpha_3(t)} \tag{6.104c}$$

$$\dot{\hat{f}}_{32} = -\frac{x_3(t)}{2q_{32}}\alpha_3(t) \quad \text{and} \quad \dot{\hat{f}}_{33} = -\frac{x_3(t)}{2q_{33}}\alpha_3(t)$$

Once the time-dependent variables $\hat{f}_{ij}$ are derived by solving (6.104), the appropriate state feedback equivalent is obtained through the inversion

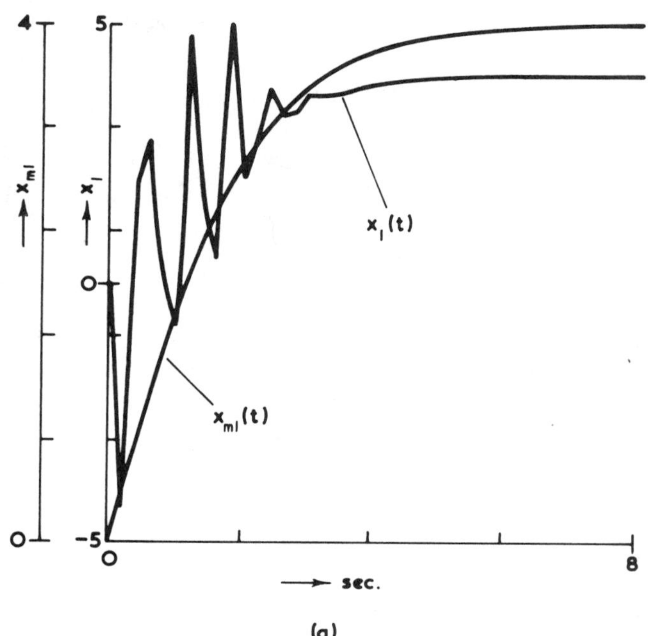

(a)

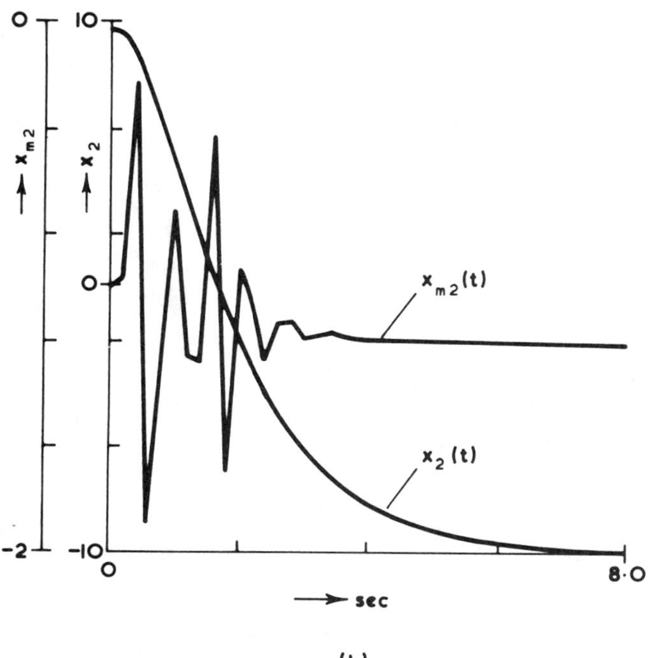

(b)

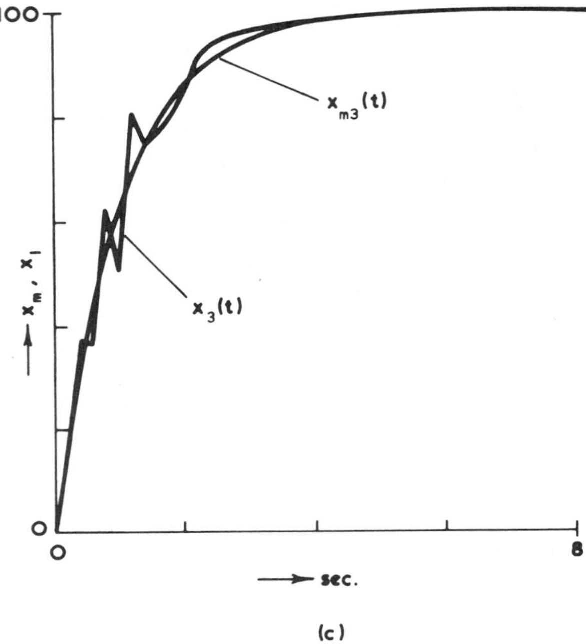

Fig. 6.8 Responses of the three state variables with adaptive controller for $u_1 = u_2 = 0$, $u_3 = 1.0$.

$$F = [B^T B]^{-1} B^T \hat{F} = B^{-1}\hat{F}, \text{ where } \hat{F} \to A_m - A = \begin{bmatrix} -1 & -1 & 3 \\ 3 & -7 & 0.6 \\ -7 & -8 & -0.1 \end{bmatrix} \text{ as } t \to \infty \tag{6.105}$$

The adaptive-loop configuration and some step responses are shown in Figs. 6.7 and 6.8, respectively.

An extensive literature on state feedback control theory now exists, and although a considerable proportion of this deals with abstract mathematics, this has led to the development of a large number of algorithms for the solution of state feedback matrices for given operating requirements: pole assignment, linear quadratic control, adaptive control, and many others not covered here.

The other area of theoretical development that has made a considerable contribution to the understanding of many structural properties of multivariable systems is that of geometric state-space theory. Because of the extent of abstraction needed to present even an overview of this area, this aspect is considered to be beyond the scope of this introductory volume. Readers seeking an introduction to this important area are advised to see Refs. 26 and 27, and then Ref. 28.

REFERENCES

1. Crossley, T. R., and Porter, B., "Synthesis of aircraft modal control systems," Aeronaut. J., 72: 687-701 (1968).
2. Crossley, T. R., and Porter, B., "Synthesis of aircraft modal control systems having real or complex eigenvalues," Aeronaut. J., 73: 138-142 (1969).
3. Porter, B., and Carter, J. D., "Design of multi-loop modal control systems for plants having complex eigenvalues," Trans. Inst. Meas. Control, 1: T61-T68 (1968).
4. Porter, B., Synthesis of Dynamical Systems, Thomas Nelson, London, 1969.
5. Fortman, T. E., and Hitz, K. L., An Introduction to Linear Control Systems, Marcel Dekker, New York, 1977.
6. Luenberger, D. G., Introduction to Dynamic Systems, Wiley, New York, 1979.
7. Wiberg, D. M., State Space and Linear Systems, Schaum's Outline Series, McGraw-Hill, New York, 1971.
8. Fallside, F., and Seraji, H., "Design of multivariable systems using unity-rank feedback," Int. J. Control, 17: 351-364 (1973).
9. Layton, J. M., Multivariable Control Theory, Peter Peregrinus, London, 1976.
10. Fallside, F., and Seraji, H., "Direct design procedure for multivariable feedback systems," Proc. IEE, 118: 797-801 (1971).
11. Fallside, F., "Pole assignment," in Modern Approaches to Control System Design, Peter Peregrinus, London, 1979.
12. Seraji, H., "Design of cascade controllers for zero assignment in multivariable systems," Int. J. Control, 21: 485-496 (1975).
13. Tyler, J. S., "The use of a quadratic performance index to design multivariable control systems," Trans. IEEE, AC-11: 84-92 (1966).
14. MacFarlane, A. G. J., "Return-difference and return-ratio matrices and their use in analysis and design of multivariable feedback control systems," Proc. IEE, 117: 2037-2049 (1970).

15. Fallside, F., and Seraji, H., "Design of optimal systems by a frequency-domain technique," Proc. IEE, 117: 2017-2024 (1970).
16. Kwakernaak, H., and Sivan, R., Linear Optimal Control Systems, Wiley, New York, 1972.
17. Norton, M., Modern Control Engineering, Pergamon Press, Elmsford, N.Y., 1972.
18. Molinaric, B. P., "The time-invariant linear-quadratic optimal control problem," Automatica, 13: 347-357 (1977).
19. Potter, J. E., "Matrix quadratic solutions," SIAM J. Appl. Math., 14: 496-501 (1966).
20. Casti, J., and Ljung, L., "Some analytic and computational results for operator Riccati equations," SIAM J. Control, 13: 817-826 (1975).
21. Russell, D. L., Mathematics of Finite-Dimensional Control Systems, Marcel Dekker, New York, 1979.
22. Porter, B., and Tatnall, M. L., "Performance characteristics of multi-variable model-reference adaptive systems synthesized by Liapunov's direct method," Int. J. Control, 10: 241-257 (1969).
23. Tyler, J. S., "The characteristics of model-following systems as synthesized by optimal control," IEEE Trans., AC-9: 485-498 (1964).
24. Smith, R. A., "Matrix equation XA + BX = C," SIAM J. Appl. Math., 16: 198-201 (1968).
25. Fallside, F., and Seraji, H., "Direct design procedure for multivariable feedback systems," Proc. IEE, 118: 797-801 (1971).
26. Halmos, P. R., Finite-Dimensional Vector Spaces, Van Nostrand Reinhold, New York, 1958.
27. Wonham, W. M., "Geometric state-space theory in linear multivariable control: a status report," Preprints 7th IFAC World Congr., Helsinki, Vol. 3, 1978, pp. 1781-1788.
28. Wonham, W. M., Linear Multivariable Control: A Geometric Approach, Springer-Verlag, New York, 1974.

7
Output Feedback Control

State feedback control, considered in Chap. 6, although a very powerful tool in the analysis and design of linear systems, is restricted in its application to most physical systems.* This is due to the fact that the output variables $y_1(t) \cdots y_m(t)$ rather than the complete n-state vector $x(t) = [x_1(t), \ldots, x_n(t)]^T$ are available for measurement and feedback in the majority of systems encountered by designers.† For the m-input/m-output nth-order system S(A, B, C), it is thus likely that physically realizable feedback control signals will have to be derived from a linear combination of the available outputs. A linear control law of this type, called a linear (or proportional) output feedback law, may be formulated as

$$u(t) = Hy(t) + v(t) \tag{7.1}$$

where the constant m-square matrix H is the output feedback gain matrix. [For time-varying systems, H becomes H(t), a function of time.] This chapter presents a brief account of a number of methods of designing linear multivariable systems through output feedback (Fig. 7.1).

7.1 EXTENSION OF STATE FEEDBACK [1-6]

It was seen in Chap. 6 that for a completely state controllable system S(A, B, C) with state feedback control law

$$u(t) = Fx(t) + v(t) \tag{7.2}$$

the eigenvalue spectrum of the state feedback system

*When the state variables are not available for feedback, a state feedback law can be implemented through the use of observers. This is considered in Chap. 8.

†m-input/m-output systems are considered here for mathematical convenience; most of the results can easily be modified for systems where number of inputs ≠ number of outputs.

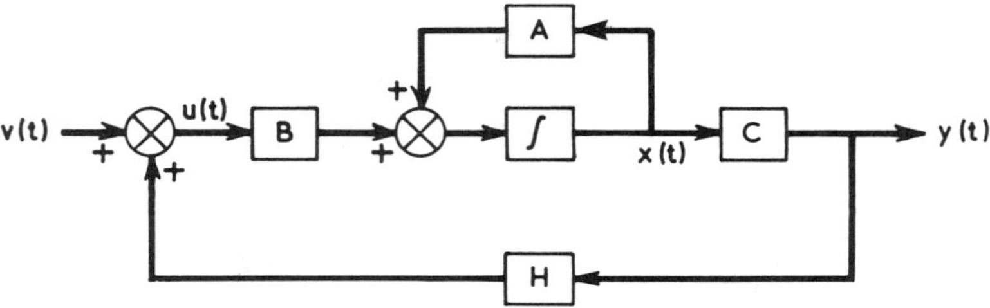

Fig. 7.1 Closed-loop control through output feedback.

$$\begin{aligned} \dot{x}(t) &= (A + BF)x(t) + Bv(t) \\ y(t) &= Cx(t) \end{aligned} \tag{7.3}$$

can be assigned arbitrarily by appropriate choice of the feedback matrix F. This section derives the conditions under which the eigenvalues of the output feedback system (Fig. 7.1)

$$\begin{aligned} \dot{x}(t) &= (A + BHC)x(t) + Bu(t) \\ y &= Cx(t) \end{aligned} \tag{7.4}$$

can be specified arbitrarily through appropriate choice of H. Comparing (7.3) and (7.4), it is clear that the eigenvalues obtained by state feedback control are identical to those obtained by output feedback if

$$H = FC \tag{7.5}$$

Thus for a given F, (7.5) gives an output feedback control law that would yield the same set of eigenvalues were state feedback control law (7.2) implemented.

Of the several methods available for solving (7.5), only two are presented here through the following theorems [2, 6] and numerical examples.

Theorem 7.1 A necessary and sufficient condition for all of the poles described by (7.4) to be arbitrarily assigned by using constant output feedback is that at least one of the set of state feedback matrices F, which achieves the same pole placement, and one of the g-inverses of C satisfy the consistency relationship $FC^gC = F$, where C^g is the generalized inverse* of the $(m \times n)$ matrix C given by $CC^gC = C$.

*The $(m \times n)$ output matrix C is considered to have rank m. A generalized inverse is also known as a pseudo-inverse.

It is to be noted here that from the definition of g-inverses,* C^g is not unique but represents one of a set of g-inverses of C given by

$$C_a^g = C^g + V - C^g C V C C^g \tag{7.6}$$

where C_a^g denotes any g-inverse of C and V is an arbitrary $(n \times m)$ matrix.

Proof: For a particular F and C^{g_1}, let (7.5) not be satisfied (i.e., $FC^gC \neq F$. Then for sufficiency from (7.6)

$$\begin{aligned} FC_a^{g_1}C &= F[C^g + V - C^gCVCC^g]C = FC^gC + FVC - FC^gCVCC^gC \\ &= FC^gC + FVC - FC^gCVC \end{aligned} \tag{7.7}$$

To prove necessity, it has to be shown that the right-hand side of (7.7) does not equal F, that is, the equation

$$FC^gC + FVC - FC^gCVC = F \tag{7.8}$$

does not have a nontrivial solution for the $(n \times m)$ matrix V. We rewrite (7.8) as

$$F(I_n - C^gC)VC = F(I_n - C^gC) \tag{7.9}$$

which is of the form $PXQ = R$. A necessary and sufficient condition for this equation to have a solution for X is that [2, 3]

$$PP^gRQ^gQ = R \tag{7.10}$$

Thus for (7.9) to have a solution for V, the following equation must be satisfied:

$$[F(I - C^gC)][F(I_n - C^gC)]^g[F(I_n - C^gC)]C^gC = F(I_n - C^gC)$$

that is,

$$F(I_n - C^gC)C^gC = F(I_n - C^gC)$$

Hence the solution for V in (7.9) exists only if

$$F(I_n - C^gC) = 0 \rightarrow F = FC^gC \tag{7.11}$$

which completes the proof.

$*CC^gC \triangleq C$.

Computation of generalized inverse [1, 5]

Given an $(m \times n)$ matrix C with rank $r \leq \min\{p, q\}$ with $p > q$, it can be shown that if the matrix C is augmented on the right by a unity matrix of order p, that is

$$\tilde{C} = [C \mid I_p] \tag{7.12}$$

and if the augmented matrix $\tilde{C}$ is reduced using elementary row operations of the form

$$\bar{C} = \left[\begin{array}{cc|c} I_r & \bar{C}_{12} & \bar{C}_{13} \\ 0 & 0 & \bar{C}_{23} \end{array}\right] = [E \mid K] \tag{7.13}$$

then the first q rows of K will be a g-inverse of C (i.e., $C^g = K_q$).

If $p < q$ as is the case in S(A, B, C), the algorithm above may be carried out on C^T, and the required g-inverse is given by K_q^T, which is based on the identity $[C^g]^T = [C^T]^g$.

If the consistency condition of Theorem 7.1 is not satisfied, for a particular g-inverse C^g and a state feedback matrix F, it may be necessary to compute other C^g-inverses. Once the generalized inverse of C is computed, the output feedback matrix that would yield the same closed-loop poles as the state feedback matrix F is computed as

$$H = FC^g \tag{7.14}$$

Example 7.1: Output feedback through extension of state feedback The open-loop system is defined by

$$\dot{x}(t) = \begin{bmatrix} 0 & 1 & 0 \\ -2 & 3 & 0 \\ 5 & 1 & 3 \end{bmatrix} x(t) + \begin{bmatrix} 0 & 0 \\ 1 & 3 \\ 0 & 1 \end{bmatrix} u(t); \quad y(t) = \begin{bmatrix} 0 & 0 & 7 \\ 7 & 9 & 0 \end{bmatrix} x(t) \tag{7.15}$$

with open-loop poles at 1, 2, 3. A suitable state feedback matrix (Sec. 6.1) that would move the closed-loop poles to -3, -3, -4 is given by

$$F = \begin{bmatrix} -7 & -9 & 21 \\ 0 & 0 & -7 \end{bmatrix} \tag{7.16}$$

The preceding derivations are used here to compute an output feedback matrix which would yield the same closed-loop poles.

The operations in (7.12), (7.13) and (7.14) are applied on C^T to compute C^g ($p = m$, $q = n$):

$$\tilde{C}^T = [C^T \mid I_3] = \left[\begin{array}{cc|ccc} 7 & 0 & 1 & 0 & 0 \\ 9 & 0 & 0 & 1 & 0 \\ 0 & 7 & 0 & 0 & 1 \end{array}\right]$$

$$\to \left[\begin{array}{cc|ccc} 1 & 0 & -\frac{1}{7} & \frac{1}{9} & \frac{1}{7} \\ 0 & 1 & 0 & \frac{1}{9} & 0 \\ \hline 7 & 0 & 0 & 0 & 1 \end{array}\right] \tag{7.17}$$

Thus from (7.12), (7.13), and (7.14),

$$K_n^T = \begin{bmatrix} -\frac{1}{7} & \frac{1}{9} & \frac{1}{7} \\ 0 & \frac{1}{9} & 0 \end{bmatrix} \quad \text{and} \quad C^g = K_n = \begin{bmatrix} -\frac{1}{7} & 0 \\ \frac{1}{9} & \frac{1}{9} \\ \frac{1}{7} & 0 \end{bmatrix} \tag{7.18}$$

To test for the consistency condition (7.11),

$$FC^gC = \begin{bmatrix} -7 & -9 & 21 \\ 0 & 0 & -7 \end{bmatrix} \begin{bmatrix} -\frac{1}{7} & 0 \\ \frac{1}{9} & \frac{1}{9} \\ \frac{1}{7} & 0 \end{bmatrix} \begin{bmatrix} 0 & 0 & 7 \\ 7 & 9 & 0 \end{bmatrix} = \begin{bmatrix} -7 & -9 & 21 \\ 0 & 0 & -7 \end{bmatrix} \equiv F$$

From (7.14) and (7.18), the output feedback matrix is given by

$$H = FC^g = \begin{bmatrix} -7 & -9 & 21 \\ 0 & 0 & -7 \end{bmatrix} \begin{bmatrix} -\frac{1}{7} & 0 \\ \frac{1}{9} & \frac{1}{9} \\ \frac{1}{7} & 0 \end{bmatrix} = \begin{bmatrix} 3 & -1 \\ -1 & 0 \end{bmatrix} \tag{7.19}$$

For the closed-loop system with H as above,

$$A + BHC = \begin{bmatrix} 0 & 1 & 0 \\ -9 & -6 & 0 \\ 5 & 1 & -4 \end{bmatrix}$$

which has eigenvalues at -4, -3, -3, as specified. It is worth noting here that the choice of C^g is not unique. An alternative structure of C^g is

$$C^g = \begin{bmatrix} 0 & 0 \\ 0 & \frac{1}{9} \\ \frac{1}{7} & 0 \end{bmatrix} \tag{7.20}$$

for which

$$FC^gC = F \quad \text{and} \quad H = FC^g = \begin{bmatrix} 3 & -1 \\ -1 & 0 \end{bmatrix}$$

Another method of deriving H for a given K is based on the following theorem [6].

Theorem 7.2 The equation F = HC has a solution for $H \in R^{m\times m}$, with $m < n$, only if rank $\begin{bmatrix} F \\ \hline C \end{bmatrix} = m$, where rank (C) = m.

Proof: If there exists a matrix H such that HC = F, where*

$$C = [\hat{C}_1 \mid \hat{C}_2] \quad \text{and} \quad F = [\hat{F}_1 \mid \hat{F}_2] \tag{7.21}$$

$\hat{C}_1 \in R^{m\times m}$, $\hat{C}_2 \in R^{m\times(n-m)}$, $\hat{F}_1 \in R^{m\times m}$, $\hat{F}_2 \in R^{m\times(n-m)}$, then it is evident that

$$H\hat{C}_1 = \hat{F}_1 \quad \text{and} \quad H\hat{C}_2 = \hat{F}_2$$

or

$$\hat{F}_2 = \hat{F}_1\hat{C}_1^{-1}\hat{C}_2 \tag{7.22}$$

so that

$$\begin{bmatrix} F \\ C \end{bmatrix} = \begin{bmatrix} \hat{F}_1 & \hat{F}_1\hat{C}_1^{-1}\hat{C}_2 \\ \hat{C}_1 & \hat{C}_2 \end{bmatrix}$$

*Since $\rho(C) = m < n$, the columns of C can be arranged without any loss of generality to obtain (7.21); a slightly different approach to obtain similar results is adopted in Sec. 9.2.

and therefore that

$$\rho\begin{bmatrix} F \\ C \end{bmatrix} = \rho\begin{bmatrix} I_m & -\hat{F}_1\hat{C}_1^{-1} \\ 0 & I_m \end{bmatrix}\begin{bmatrix} \hat{F}_1 & \hat{F}_1\hat{C}_1^{-1}\hat{C}_2 \\ \hat{C}_1 & \hat{C}_2 \end{bmatrix} = \rho\begin{bmatrix} 0 & 0 \\ \hat{C}_1 & \hat{C}_2 \end{bmatrix}$$

Conversely, if there exist matrices $F \in R^{m\times n}$ and $C \in R^{m\times n}$ so that

$$\rho\begin{bmatrix} F \\ C \end{bmatrix} = m$$

then from (7.21),

$$m = \rho\begin{bmatrix} F \\ C \end{bmatrix} = \rho\begin{bmatrix} I_m & -\hat{F}_1\hat{C}_1^{-1} \\ 0 & I_m \end{bmatrix}\begin{bmatrix} \hat{F}_1 & \hat{F}_2 \\ \hat{C}_1 & \hat{C}_2 \end{bmatrix} = \rho\begin{bmatrix} 0 & \hat{F}_2 - \hat{F}_1\hat{C}_1^{-1}\hat{C}_2 \\ \hat{C}_1 & \hat{C}_2 \end{bmatrix}$$

It follows from the above that $\hat{F}_2 = \hat{F}_1\hat{C}_1^{-1}\hat{C}_2$, which implies that there exists a matrix H such that

$$H\hat{C}_1 = \hat{F}_1 \quad \text{and} \quad H\hat{C}_2 = \hat{F}_2$$

and therefore that $HC = F$.

This method of deriving the class of output feedback matrix from a given state feedback is illustrated below.

Example 7.2: Output feedback through extension of state feedback
The system in Example 7.1 is considered here. For this system

$$\rho\begin{bmatrix} F \\ C \end{bmatrix} = \rho\begin{bmatrix} -7 & -9 & 21 \\ 0 & 0 & -7 \\ 0 & 0 & 7 \\ 7 & 9 & 0 \end{bmatrix} = 2 = m < n$$

and therefore a corresponding output feedback matrix exists as indicated by Theorem 7.2. An appropriate partitioning of C in (7.15) after some elementary column operations, is

$$C = [\hat{C}_1 \mid \hat{C}_2] = \left[\begin{array}{cc|c} 7 & 0 & 0 \\ 0 & 7 & 9 \end{array}\right] \tag{7.23}$$

and the corresponding partitioning of F is

$$F = [\hat{F}_1 \mid \hat{F}_2] = \left[\begin{array}{cc|c} 21 & -7 & -9 \\ -7 & 0 & 0 \end{array}\right] \tag{7.24}$$

The corresponding output feedback matrix is, from (7.22),

$$H = \hat{F}_1\hat{C}_1^{-1} = \begin{bmatrix} 21 & -7 \\ -7 & 0 \end{bmatrix}\begin{bmatrix} \frac{1}{7} & 0 \\ 0 & \frac{1}{7} \end{bmatrix} = \begin{bmatrix} 3 & -1 \\ -1 & 0 \end{bmatrix} \tag{7.25}$$

which as indicated earlier moves the closed-loop poles to the desired locations.

7.2 UNITY-RANK OUTPUT FEEDBACK [7-14]

When the equivalence between the state feedback matrix F and a corresponding output feedback matrix H in the preceding section cannot be established, a simple approach is to extend the concept of unity state feedback of Sec. 6.2 to output feedback*

$$u(t) = v(t) - Hy(t) \tag{7.26}$$

where

$$H = pq$$

with p an m-column vector and q an m-row vector. With this constraint on H, the equations of the resulting closed-loop system in Fig. 7.2 may be expressed as

$$\dot{x}(t) = Ax(t) + Bu(t) \tag{7.27a}$$

$$y(t) = qCx(t) \tag{7.27b}$$

$$u(t) = -py(t) + v(t) \tag{7.27c}$$

which suggests that a unity-rank output feedback reduces the multivariable system S(A, B, C) to an equivalent single-output system† S(A, B, qC) with an output feedback control law of the same form as in (7.27c), for which p can be easily computed. To derive the elements of p and q, as in state feedback case, the closed-loop characteristic polynomial is expressed as

*For conformity of derivations, negative feedback gain is used in this and the following sections.

†As opposed to a single-input system with unity-rank state feedback, as in Sec. 6.2.

$$\det\{sI - A + BHC\} = \det\{sI - A\} + qC \operatorname{adj}(sI - A)Bp$$

or

$$qC \operatorname{adj}\{sI - A\} Bp = [\alpha_1 - \beta_1 \cdots \alpha_i - \beta_i \cdots \alpha_n - \beta_n] \begin{bmatrix} s^{n-1} \\ \vdots \\ s^{n-i} \\ \vdots \\ 1 \end{bmatrix} = \underbrace{\{\alpha_i - \beta_i\}}_{\text{row vector}} \underbrace{\{s^{n-i}\}}_{\text{column vector}} \quad (7.28)$$

where α_i and β_i are the coefficients of the characteristic polynomials of the closed- and open-loop systems, as follows:

$$\begin{aligned} \det\{sI - A + BHC\} &= s^n + \alpha_1 s^{n-1} + \cdots + \alpha_i s^{n-i} + \cdots + \alpha_n \\ \det\{sI - A\} &= s^n + \beta_1 s^{n-1} + \cdots + \beta_i s^{n-i} + \cdots + \beta_n \end{aligned} \quad (7.29)$$

Following the same procedure as in Sec. 6.2, it can be shown that with unity-rank output feedback,

$$qC \operatorname{adj}\{sI - A\} Bp \equiv qC\psi \begin{bmatrix} s^{n-1} \\ \vdots \\ s^{n-i} \\ \vdots \\ 1 \end{bmatrix} \equiv qC\psi\{s^{n-i}\} \equiv q\phi\{s^{n-i}\}, \; i \in 1, \ldots, n \quad (7.30)$$

where ψ, and hence ϕ, contains the elements of p. Elements of ψ are obtained through first multiplying adj (•) by B and p and then arranging the

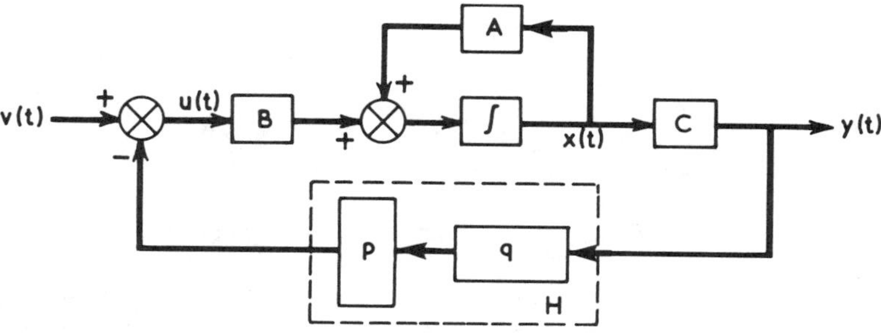

Fig. 7.2 Unity-rank output feedback.

elements with descending powers of s^{n-i}, $i = 1, 2, \ldots, n$. Thus comparison of (7.28) and (7.30) gives

$$q\phi = \{\alpha_i - \beta_i\}, \quad i \in 1, \ldots, n \tag{7.31}$$

where ϕ is an $(m \times n)$ matrix.

In the special case when $m = n$, (7.31) represents n equations in n unknowns, and can be solved uniquely for q as

$$q = \{\alpha_i - \beta_i\}\phi^{-1} \tag{7.32}$$

ϕ being nonsingular when the system is completely state controllable and completely observable.

In the case where $m < n$, (7.31) represents an overdefined set that has a consistent solution only if the solution of m equations satisfies the remaining $(n - m)$ equations. The solutions of the design problem in this case is based on the following derivations.

Let the various matrices in (7.31) be partitioned such that it becomes

$$q\phi = q[\phi_u \mid \phi_\ell] = [\alpha_u - \beta_u \mid \alpha_\ell - \beta_\ell] \tag{7.33}$$

where ϕ_u is a nonsingular $(m \times m)$ matrix and ϕ_ℓ the submatrix containing the remaining $(n - m)$ columns of ϕ, $\alpha_u(\beta_u)$ and $\alpha_\ell(\beta_\ell)$ being the corresponding submatrices of α and β. From (7.33), for nonsingular ϕ_u,

$$q = (\alpha_u - \beta_u)\phi_u^{-1} \tag{7.34}$$

For the overdetermined case, for (7.32) to be consistent, the solution given by (7.34) should satisfy (7.33). Thus

$$q\phi_\ell = \alpha_\ell - \beta_\ell$$

or

$$[\alpha_u - \beta_u]\phi_u^{-1}\phi_\ell = \alpha_\ell - \beta_\ell \tag{7.35}$$

which places $(n - m)$ conditions on the coefficient α_i of $\det\{sI - A - BHC\}$ to be satisfied together with controllability and observability. These constitute the necessary and sufficient conditions for pole assignment using unity-rank output feedback.

The condition of (7.35) can also be computerized to determine whether a chosen set of closed-loop poles can be attained exactly, or within a specified error. The minimization of the error function

$$e = \sum_{i=1}^{n} [\alpha_i - \beta_i]^2 \tag{7.36}$$

may be used in this context. Design examples [7] are given below to illustrate the method above.

Example 7.3: Output feedback through unity rank H, for m = n The method above is illustrated through the system

$$\dot{x}(t) = \begin{bmatrix} -1 & 0 \\ 0 & -4 \end{bmatrix} x(t) + \begin{bmatrix} 0 & 1 \\ 1 & 0 \end{bmatrix} u(t); \quad y(t) = \begin{bmatrix} 1 & 0 \\ 1 & 1 \end{bmatrix} x(t) \tag{7.37}$$

The design objective is to place the closed-loop poles at -2 and -8 using a unity-rank output feedback.

The output feedback matrix is chosen as

$$H = pq = \begin{bmatrix} p_1 \\ p_2 \end{bmatrix} [q_1 \;\; q_2] \tag{7.38}$$

From (7.28), (7.29), and (7.30),

$$qC \operatorname{adj}(sI - A)Bp = q \begin{bmatrix} p_2 s + 4p_2 \\ (p_1 + p_2)s + p_1 + 4p_2 \end{bmatrix} = q \underbrace{\begin{bmatrix} p_2 & 4p_2 \\ p_1 + p_2 & p_1 + 4p_2 \end{bmatrix}}_{\phi} \begin{bmatrix} s \\ 1 \end{bmatrix} \tag{7.39a}$$

and

$$\det\{sI - A + BHC\} - \det\{sI - A\} = (s^2 - 10s + 16) - (s^2 + 5s + 4) = [-15 \;\; 12] \begin{bmatrix} s \\ 1 \end{bmatrix} \tag{7.39b}$$

Combining the above and (7.32) gives us

$$q = \frac{1}{3p_1p_2} [-15 \;\; 12] \begin{bmatrix} -(p_1 + 4p_2) & 4p_2 \\ p_1 + p_2 & -p_2 \end{bmatrix} = \frac{1}{p_1p_2} [9p_1 + 24p_2 \;\vdots\; -24p_2] \tag{7.40}$$

Combining (7.39) and (7.40), we have

$$H = pq = \frac{1}{p_1p_2} \begin{bmatrix} p_1 \\ p_2 \end{bmatrix} [9p_1 - 24p_2 \;\; -24p_2] = \frac{1}{p_1p_2} \begin{bmatrix} 9p_1^2 + 24p_1p_2 & -24p_2^2 \\ 9p_1p_2 + 24p_2^2 & -24p_2 \end{bmatrix}$$

$$= \begin{bmatrix} 9\dfrac{p_1}{p_2} + 24 & -24\dfrac{p_2}{p_1} \\ 9 + 24\dfrac{p_2}{p_1} & -24\dfrac{p_2}{p_1} \end{bmatrix} = \begin{bmatrix} -7 & -24 \\ -7 & -24 \end{bmatrix} \quad \text{for } p_1 = p_2 \tag{7.41}$$

Example 7.4: Output feedback through unity rank H for $m < n$ The system considered here is described by

$$\dot{x}(t) = \begin{bmatrix} 1 & 1 & -2 \\ 2 & 0 & -2 \\ 1 & 2 & 1 \end{bmatrix} x(t) + \begin{bmatrix} 1 & 0 \\ 0 & 0 \\ 0 & 1 \end{bmatrix} u(t); \quad y = \begin{bmatrix} 1 & 0 & 0 \\ 0 & 1 & 0 \end{bmatrix} x(t) \tag{7.42}$$

which is controllable and observable, with open-loop poles at -1, $1.5 \pm j2.4$. Two methods of computing an appropriate output feedback matrix, which would move the closed-loop system poles to -6, $-12 \pm j5$, are considered.

Method 1: The output feedback matrix here is obtained in terms of the elements of the column vector p, as in Example 7.3, for exact location of closed-loop poles.

From (7.31) and using the procedure similar to Example 7.3,

$$qC \operatorname{adj}(sI - A)Bp = [q_1 \quad q_2] \begin{bmatrix} -p_1 & -(p_1 + 2p_2) & 4p_1 - 2p_2 \\ 0 & 2(p_1 - p_2) & -(4p_1 + 2p_2) \end{bmatrix} \begin{bmatrix} s^2 \\ s \\ 1 \end{bmatrix}$$

$$= (s+6)(s+12+j5)(s+12-j5) - \det\{sI - A\}$$

$$= (s^3 + 30s^2 + 265s + 1014) - (s^3 - 2s^2 + 5s + 8)$$

$$[q_1 \quad q_2] \Big[\underbrace{\begin{matrix} p_1 & -(p_1 + 2p_2) \\ 0 & 2(p_1 - p_2) \end{matrix}}_{\phi_u} \;\Big|\; \underbrace{\begin{matrix} 4p_1 - 2p_2 \\ -(4p_1 + 2p_2) \end{matrix}}_{\phi_\ell} \Big] = [\underbrace{32 \quad 260}_{\alpha_u - \beta_u} \;|\; \underbrace{1006}_{\alpha_\ell - \beta_\ell}]$$

from which the consistency condition in (7.35) becomes

$$[32 \quad 260] \begin{bmatrix} \dfrac{1}{p_1} & \dfrac{p_1 + 2p_2}{2p_1(p_1 - p_2)} \\ 0 & \dfrac{1}{2(p_1 - p_2)} \end{bmatrix} \begin{bmatrix} 4p_1 - 2p_2 \\ -(4p_1 + 2p_2) \end{bmatrix} = 1006$$

from which $p_2/p \simeq 3.7$ is an admissible solution. Thus for $p_1 = 1$,

$$p = \begin{bmatrix} 1 \\ 3.7 \end{bmatrix} \quad \text{and} \quad q = [32 \quad -98]$$

$$H = pq = \begin{bmatrix} 1 \\ 3.7 \end{bmatrix} [32 \quad -98] = \begin{bmatrix} 32 & 98 \\ 118.4 & -362.6 \end{bmatrix} \tag{7.43}$$

Method 2: The output feedback matrix H is computed for a specified value of p to illustrate the procedure of approximate closed-loop placement. p is chosen as

$$p = \begin{bmatrix} 1 \\ 3 \end{bmatrix}$$

From (7.28) and (7.29),

$$qC \operatorname{adj}(sI - A)BP = [q_1 \quad q_2] \begin{bmatrix} 1 & -7 & -2 \\ 0 & -4 & -10 \end{bmatrix} \begin{bmatrix} s^2 \\ s \\ 1 \end{bmatrix}$$

$$= (s^3 + a_3 s^2 + a_2 s + a_1) - (s^3 - 2s^2 + 5s + 8) \tag{7.44}$$

where $\det\{R(s)\} = s^3 + a_3 s^2 + a_2 s + a_1$ is the characteristic polynomial of the closed-loop system. From (7.44),

$$\begin{bmatrix} 1 & 0 \\ -7 & -4 \\ -2 & -10 \end{bmatrix} \begin{bmatrix} q_1 \\ q_2 \end{bmatrix} = \begin{bmatrix} a_3 + 2 \\ a_2 - 5 \\ a_1 - 8 \end{bmatrix} \tag{7.45}$$

From the first two equations above,

$$q_1 = a_3 + 2 \quad \text{and} \quad q_2 = \frac{-[(a_2 - 5) + 7(a_3 + 2)]}{4}$$

Substituting these values into the third equation of (7.45), we have

$$-2(a_3 + 2) + \frac{10[(a_2 - 5) + 7(a_3 + 2)]}{4} = (a_1 - 8)[2 \quad -5 \quad -31] \begin{bmatrix} a_1 \\ a_2 \\ a_3 \end{bmatrix} = 53 \tag{7.46}$$

For the desired closed-loop poles -6, $-12 \pm 5j$, the characteristic polynomial is

$$\det\{R(s)\} = s^3 + 30s^2 + 265s + 1014$$

which does not satisfy the consistency condition in (7.46). By using (7.36), the minimization of

$$f = \left\{ [2 \quad -5 \quad -31] \begin{bmatrix} a_1 \\ a_2 \\ a_3 \end{bmatrix} - 53 \right\}^2$$

would provide a set of mean attainable poles. These are -3.7 and -5 ± j15.8, for which the vector q is given by [7]

$$q = [15.17 \quad -102.7]$$

The corresponding output feedback matrix is

$$H = pq = \begin{bmatrix} 15.17 & -102.7 \\ 45.51 & -308.1 \end{bmatrix} \tag{7.47}$$

7.3 PROPORTIONAL-PLUS-ERROR INTEGRAL CONTROL*

The design methods in previous sections have a major deficiency in that they attempt to locate the closed-loop poles at preassigned positions, and if steady-state errors are to be reduced in addition, the feedback loop gains will have to be increased. In some feedback control problems it may be necessary to provide inherent zero state-state error, as required by the true servomechanism system. In classical design methods, steady-state errors are eliminated by incorporating error-integral feedback action, while the desirable system response is obtained through proportional feedback gain. This method of proportion-plus-integral synthesis procedure is extended here to multivariable systems. The design algorithm presented in this section makes use of the concept of cyclicity as well as the properties of state controllability and observability. The notion of cyclicity and its properties are described briefly below [12].

1. <u>Cyclicity of a space with respect to a matrix operator</u>: Let the n-square matrix A be a linear operator in the n-dimensional vector space R^n. Then R^n is said to be cyclic with respect to A only if there exists a vector p in R^n such that the set of n vectors p, Ap, ..., $A^{n-1}p$ span the entire space R^n, that is, if

$$\rho[p \quad Ap \cdots A^{n-1}p] = n$$

When R^n is noncyclic with respect to A, then

*Derivations in this section closely follow Refs. 12 and 13.

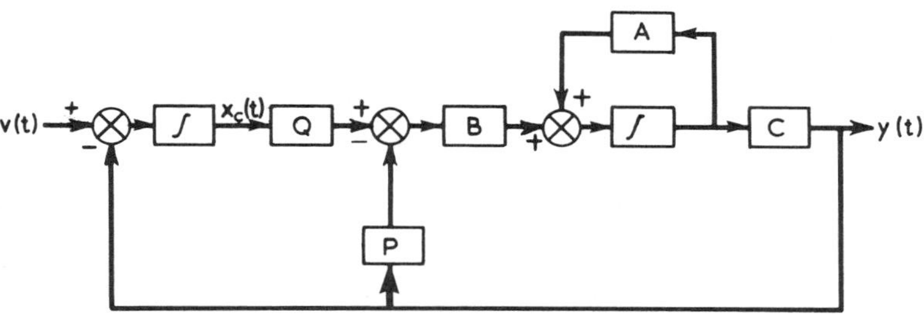

Fig. 7.3 Proportional-plus-error integral control.

$$\rho[p \quad Ap \cdots A^{n-1}p] < n \quad \forall\, p \text{ in } R^n$$

2. Criterion for cyclicity: A necessary and sufficient condition for cyclicity of a system matrix A is that the rational matrix $\phi(s) = (sI - A)^{-1}$ is irreducible (i.e., has no cancellations). In other words, all n^2 elements of the numerator polynomial matrix adj (sI - A) do not have any factors in common with the polynomial det $\{sI - A\}$.

3. Cyclicity in feedback design: The simplest approach of pole assignment of the controllable and observable linear system is through unity rank output feedback matrix H = pq, as discussed in Sec. 7.2. Such an output feedback reduces the multivariable system to an equivalent single-input system S(A, Bp, C) or an equivalent single-output system S(A, B, qC). For the existence of an appropriate p or q such that S(A, Bp, C) is controllable or S(A, B, qC) is observable, the system matrix A must be cyclic. It can be shown [12] that if S(A, B, C) is controllable and observable, any arbitrary choice of q will make S(A, Bq, C) controllable.

If A is noncyclic but S(A, B, C) is controllable and observable, there exists an output feedback matrix H_1 such that $(A - BH_1C)$ is cyclic. Since controllability and observability are invariant under output feedback, the modified system $S(\overline{A - BH_1C}, B, C)$ is controllable and observable. The unity rank feedback matrix $H_2 = pq$ can be then evaluated to assign specified values to the poles of this closed-loop system. The total output feedback matrix for the original system is then $H = H_1 + H_2 = H_1 + pq$.

Analysis of S(A, B, C) with proportional-plus-integral control may now be considered. The controller structure is schematically shown in Fig. 7.3, from which the control law may be derived as

$$u(t) = Qx_c(t) - Py(t) = Qx_c(t) - PCx(t) \tag{7.48}$$

$$= -Py(t) + Q \underbrace{\int [v(t) - y(t)]\,dt}_{\dot{x}_c(t)} = [-P \quad Q]\begin{bmatrix} y(t) \\ x_c(t) \end{bmatrix} \tag{7.49}$$

where P and Q are constant matrices, and $x_c(t)$ is a state vector equal to the integration of the error in the closed-loop system. Necessary conditions for arbitrary pole placement using output feedback are that the augmented system $\bar{S}(\bar{A}, \bar{B}, \bar{C})$ be both controllable and observable. In order to derive a simple and explicit relationship between the closed-loop poles and the controller parameters P and Q, these matrices are initially assumed to have unity rank. The controllability, observability, and cyclicity of the augmented system in Fig. 7.3 are stated below without proof.

Controllability: The augmented system in Fig. 7.3 is state controllable only if

$$\rho[B \mid AB \cdots \mid A^{n-1}B] = n \quad \text{and} \quad \rho\begin{bmatrix} A & B \\ C & 0 \end{bmatrix} = n + m \tag{7.50a}$$

Observability: The augmented system in Fig. 7.3 is observable if the original system S(A, B, C) is observable, that is, if

$$\rho[C^T \quad C^T A^T \quad \cdots \quad C^T (A^T)^{n-1}] = n \tag{7.50b}$$

Cyclicity: Conditions for the augmented system matrix in Fig. 7.3 to be cyclic is that the polynomial matrix $(sI - \bar{A})^{-1}$ be irreducible, where

$$\bar{A} = \begin{bmatrix} A & 0 \\ -C & 0 \end{bmatrix}$$

Synthesis procedure [13]

The controller design algorithm presented below places (3m - 1) poles of the (n + m)th-order closed-loop system in Fig. 7.3 at any specified locations $\lambda_1, \lambda_2, \ldots, \lambda_{3m-1}$ in the complex plane. The design is carried out in three stages.

Stage I: In this stage the augmented system $\bar{S}(\bar{A}, \bar{B}, \bar{C})$ is made cyclic by applying the feedback law

$$u(t) = \hat{Q}x_c(t) + u_c(t) \quad \text{with } \dot{x}_c(t) = v(t) - y(t) \tag{7.51}$$

where $\hat{Q}$ is any arbitrary (m × m) matrix of full rank and $\hat{u}_c(t)$ is the m-control input; the equation of the closed-loop system (Fig. 7.4a) then becomes

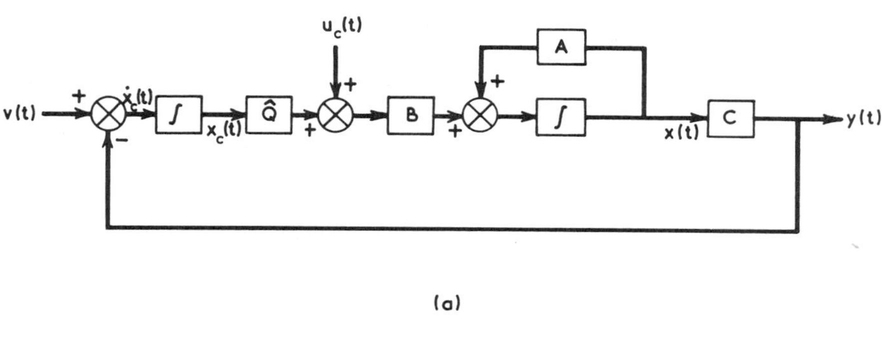

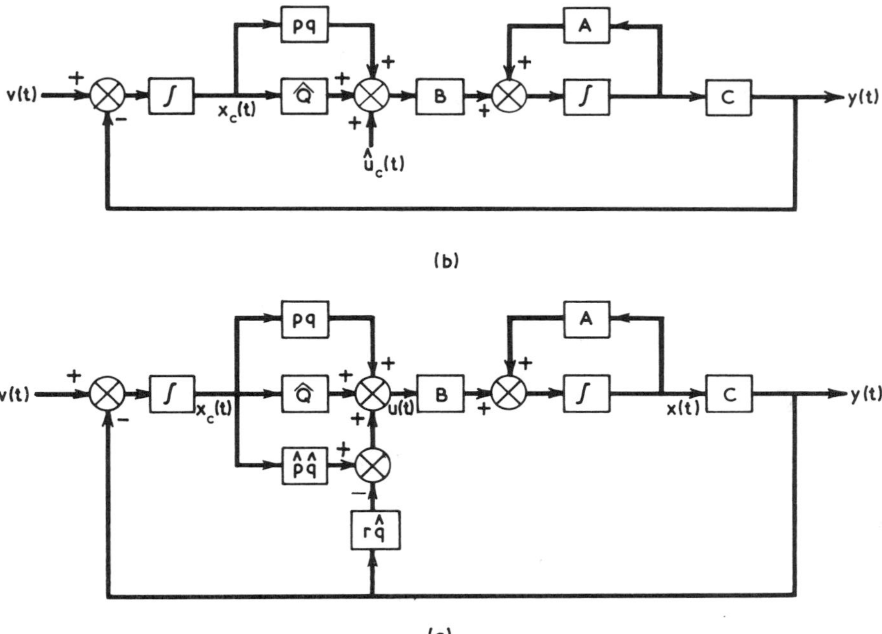

Fig. 7.4 Three stages of proportional-plus-integral synthesis procedure: (a) stage I, control law in (7.51); (b) stage II, control law in (7.54); (c) stage III, control law in (7.59).

$$\begin{bmatrix} \dot{x}(t) \\ \dot{x}_c(t) \end{bmatrix} = \underbrace{\begin{bmatrix} A & B\hat{Q} \\ -C & 0 \end{bmatrix}}_{\bar{A}_1} \begin{bmatrix} x(t) \\ x_c(t) \end{bmatrix} + \begin{bmatrix} 0 \\ I \end{bmatrix} v(t) + \underbrace{\begin{bmatrix} B \\ 0 \end{bmatrix}}_{\bar{B}} u_c(t); \begin{bmatrix} y(t) \\ x_c(t) \end{bmatrix} = \underbrace{\begin{bmatrix} C & 0 \\ 0 & I \end{bmatrix}}_{\bar{C}} \begin{bmatrix} x(t) \\ x_c(t) \end{bmatrix} \quad (7.52)$$

The new modified system matrix $\bar{A}_1$ has distinct poles and is therefore cyclic, with the transfer-function matrix of the system being

$$y(s) = \frac{1}{\det(sI - \bar{A}_1)} \bar{C} \operatorname{adj}(sI - \bar{A}_1)\bar{B}u_c(s) \equiv \frac{1}{\Delta_1(s)} H_1(s)u_c(s)$$

and

$$x_c(s) = \frac{1}{s}\left[v(s) - \frac{1}{\Delta_1(s)} H_1(s)u_c(s)\right]$$

where $H_1(s) = [C \mid 0] \operatorname{adj}(sI - \bar{A})\bar{B}$ and $\Delta_1(s) = \det(sI - \bar{A}_1)$.

Stage II: In this stage $(m - 1)$ poles of the system $\bar{S}(\bar{A}_1, \bar{B}, \bar{C})$ are placed at the distinct specified locations $\lambda_1, \lambda_2, \ldots, \lambda_{m-1}$ by applying the unity-rank feedback control law

$$u_c(t) = pqx_c(t) + \hat{u}_c(t) \quad (7.54)$$

where p and q are $(m \times 1)$ and $(1 \times m)$ vectors, respectively, and $u_c(t)$ is the m-control input; the modified closed-loop characteristic polynomial can be expressed as (Sec. 7.2)

$$\Delta_2(s) = \Delta_1(s) + \frac{1}{s} qH_1(s)p \quad (7.55)$$

In order to place $(m - 1)$ poles at $\lambda_1, \ldots, \lambda_{m-1}$, the vector p is specified arbitrarily such that $(\bar{A}_1, \bar{B}p)$ is controllable and the vector q is found by solving the set of $(m - 1)$ linear equations

$$\Delta_1(\lambda_i) + \frac{1}{\lambda_i} qH_1(\lambda_i)p = \Delta_2(\lambda_i) = 0, \quad i \in 1, \ldots, m - 1 \quad (7.56)$$

The equations of the closed-loop system (Fig. 7.4b) are given by

$$\begin{bmatrix} \dot{x}(t) \\ \dot{x}_c(t) \end{bmatrix} = \underbrace{\begin{bmatrix} A & B(\hat{Q}+pq) \\ -C & 0 \end{bmatrix}}_{\bar{A}_2} \begin{bmatrix} x(t) \\ x_c(t) \end{bmatrix} + \underbrace{\begin{bmatrix} B \\ 0 \end{bmatrix}}_{\bar{B}} \hat{u}_c(t) + \begin{bmatrix} 0 \\ I \end{bmatrix} v(t); \quad y(t) = [C \quad 0] \begin{bmatrix} x(t) \\ x_c(t) \end{bmatrix} \quad (7.57)$$

and $(m - 1)$ poles of this system are at $\lambda_1, \lambda_2, \ldots, \lambda_{m-1}$. The transfer function matrix of the system is

$$\begin{bmatrix} y(s) \\ x_c(s) \end{bmatrix} = \frac{1}{\det(sI - \bar{A}_2)} \bar{C} \operatorname{adj}(sI - \bar{A}_2)\bar{B}u_c(s) = \begin{bmatrix} H_2(s)/\Delta_2(s) \\ \frac{1}{s}\{v(s) - H_2(s)/\Delta_2(s)\} \end{bmatrix} u_c(s)$$

where $H_2(s) = [C \mid 0] \operatorname{adj}(sI - \bar{A}_2)\bar{B}$ and $\Delta_2(s) = \det(sI - \bar{A}_2)$. (7.58)

Stage III: In this final stage, the (m - 1) poles of the system $(\bar{A}_2, \bar{B}, \bar{C})$ at $\lambda_1, \lambda_2, \ldots, \lambda_{m-1}$ are preserved and 2m additional poles are placed at specified locations $\lambda_m, \ldots, \lambda_{3m-1}$ by applying the unity-rank feedback law (Fig. 7.4c)

$$\hat{u}_c(t) = -r\hat{q}y(t) + \hat{p}\hat{q}x_c(t) \tag{7.59}$$

where $\hat{p}$ and r are (m × 1) vectors and $\hat{q}$ is an (1 × m) vector. The resulting closed-loop characteristics polynomial can then be expressed as

$$\Delta_3(s) = \Delta_2(s) + \hat{q}H_2(s)r + \frac{1}{s}\hat{q}H_2(s)\hat{p} \tag{7.60}$$

To preserve the (m - 1) poles at $\lambda_1, \lambda_2, \ldots, \lambda_{m-1}$ in the closed-loop system irrespective of r and $\hat{p}$, from (7.60) the vector $\hat{q}$ must satisfy the equation*

$$\hat{q}H_2(\lambda_i) = 0, \quad i = 1, \ldots, m-1 \tag{7.61}$$

Once $\hat{q}$ is determined, the (m × 1) vectors r and $\hat{p}$ which place 2m additional poles at $\lambda_m, \ldots, \lambda_{3m-1}$ are found by solving the set of 2m linear equations

$$\Delta_2(\lambda_i) + \hat{q}H_2(\lambda_i)r + \frac{1}{\lambda_i}\hat{q}H_2(\lambda_i)\hat{p} = 0, \quad i \in 1, \ldots, 3m-1 \tag{7.62}$$

The total proportional-plus-error integral control law is

$$u(t) = -r\hat{q}y(t) + [\hat{Q} + pq + \hat{p}\hat{q}] \int \{v(t) - y(t)\}\, dt \equiv -Py(t) + Q \int \{v(t) - y(t)\}\, dt$$

where P = rq is the unity-rank (m × m) proportional feedback matrix and Q is the (m × m) integral feedback matrix of full rank m. The equations of the augmented (n + m)th-order closed-loop system is

$$\begin{bmatrix} \dot{x}(t) \\ x_c(t) \end{bmatrix} = \begin{bmatrix} A - BPC & BQ \\ -C & 0 \end{bmatrix} \begin{bmatrix} x(t) \\ x_c(t) \end{bmatrix} + \begin{bmatrix} 0 \\ I \end{bmatrix} v(t) \quad \text{and} \quad y(t) = [C \mid 0] \begin{bmatrix} x(t) \\ x_c(t) \end{bmatrix} \tag{7.63}$$

*This follows from the fact that $\operatorname{adj}(\lambda_i - \bar{A}_2)\{i \in 1, \ldots, m-1\}$ have unity rank and consequently each $H_2(\lambda_i)$ contains only one independent column such that (7.61) is satisfied.

and the (3m - 1) closed-loop poles are at specified locations $\lambda_1, \lambda_2, \ldots, \ldots, \lambda_{3m-1}$. The design algorithm is illustrated below through a numerical example [13].

Example 7.5: Output feedback through proportional-plus-integral control

$$S(A,B,C):\ \dot{x}(t) = \begin{bmatrix} 1 & 3 & 2 \\ 0 & 1 & 2 \\ 0 & 0 & 1 \end{bmatrix} x(t) + \begin{bmatrix} 1 & 0 \\ 2 & 0 \\ 1 & 1 \end{bmatrix} u(t); \quad y(t) = \begin{bmatrix} 1 & 0 & 0 \\ 0 & 1 & 0 \end{bmatrix} x(t) \tag{7.64}$$

Let the desired location of poles with a proportional-plus-integral controller be -1, -2, -3, -4, -5. The open-loop system S(A, B, C) is controllable and observable and cyclic; the design algorithm above can therefore be applied.

Stage I: Since the system is cyclic, the feedback control law (7.51) with $Q = -I_2$ is applied initially to obtain the augmented system $S(\bar{A}_1, \bar{B}, \bar{C})$ in (7.53):

$$\bar{A}_1 = \begin{bmatrix} A & B\hat{Q} \\ -C & 0 \end{bmatrix} = \begin{bmatrix} 1 & 3 & 2 & -1 & 0 \\ 0 & 1 & 2 & -2 & -1 \\ 0 & 0 & 1 & -1 & -1 \\ -1 & 0 & 0 & 0 & 0 \\ 0 & -1 & 0 & 0 & 0 \end{bmatrix}; \quad \bar{B} = \begin{bmatrix} B \\ 0 \end{bmatrix} = \begin{bmatrix} 1 & 0 \\ 2 & 0 \\ 1 & 1 \\ 0 & 0 \\ 0 & 0 \end{bmatrix};$$
$$\bar{C} = \begin{bmatrix} C & 0 \\ 0 & I \end{bmatrix} = \begin{bmatrix} 1 & 0 & 0 & 0 & 0 \\ 0 & 1 & 0 & 0 & 0 \\ 0 & 0 & 0 & 1 & 0 \\ 0 & 0 & 0 & 0 & 1 \end{bmatrix} \tag{7.65}$$

Stage II: One pole of the system in (7.65) is placed at -1 by using the unity rank feedback law (7.54):

$$u_c(s) = pqx_c(t) + \hat{u}_c(t) \quad \text{with } p = \begin{bmatrix} 0 \\ 1 \end{bmatrix} \quad q = [q_1 \quad 0]$$

which using (7.56) yields

$$\Delta_1(-1) + \frac{1}{-1} qH_1(-1)p = -20 - 2q_1 = 0$$

Thus $q_1 = -10$ and the resulting system is $\bar{S}(\bar{A}_2, \bar{B}, \bar{C})$, where

$$\bar{A}_2 = \begin{bmatrix} A & B(\hat{Q}+pq) \\ -C & 0 \end{bmatrix} = \begin{bmatrix} 1 & 3 & 2 & -1 & 0 \\ 0 & 1 & 2 & -2 & 0 \\ 0 & 0 & 1 & -11 & -1 \\ -1 & 0 & 0 & 0 & 0 \\ 0 & -1 & 0 & 0 & 0 \end{bmatrix} \tag{7.66}$$

Stage III: The pole of the system $\bar{S}(\bar{A}_2, \bar{B}, \bar{C})$ at -1 is preserved and four additional poles are placed at -2, -3, -4, and -5 by applying the feedback law in (7.59). To preserve the pole at -1, the vector $\hat{q}$ must satisfy (7.61):

$$\hat{q}H_2(-1) = \hat{q}\begin{bmatrix} -8 & 2 \\ 24 & -6 \end{bmatrix} = 0 \rightarrow \hat{q}\begin{bmatrix} 2 \\ -6 \end{bmatrix} = 0 \rightarrow \hat{q} = [3 \;\; 1]$$

and the closed-loop characteristic polynomial is given by (7.60) with $r = [r_1 \;\; r_2]^T$ and $\hat{p} = [\hat{p}_1 \;\; \hat{p}_2]^T$:

$$\begin{aligned} \Delta_3(s) &= (s+1)[s^4 + (5r_1 - 4)s^3 + (11r_1 + 8r_2 + 5\hat{p}_1 + 6)s^2 + (-14r_1 + 2r_2 + 11\hat{p}_1 + 8\hat{p}_2 - 35)s \\ &\quad + (-14\hat{p}_1 + 2\hat{p}_2 - 2)] \\ &\equiv (s+1)(s+2)(s+3)(s+4)(s+5) = (s+1)(s^4 + 14s^3 + 71s^2 + 154s + 120) \end{aligned}$$

which by equating the coefficient of equal powers of s gives the following set of linear equations to solve for the elements of r and $\hat{p}$:

$$\begin{bmatrix} 5 & 0 & 0 & 0 \\ 11 & 8 & 5 & 0 \\ -14 & 2 & 11 & 8 \\ 0 & 0 & -14 & 2 \end{bmatrix}\begin{bmatrix} r_1 \\ r_2 \\ \hat{p}_1 \\ \hat{p}_2 \end{bmatrix} = \begin{bmatrix} 18 \\ 65 \\ 189 \\ 122 \end{bmatrix} \rightarrow \begin{bmatrix} r_1 \\ r_2 \\ \hat{p}_1 \\ \hat{p}_2 \end{bmatrix} = \begin{bmatrix} 3.6 \\ 5.598 \\ -3.878 \\ 33.857 \end{bmatrix}$$

The parameters of the total proportional-plus-integral control law is thus given by

$$P = r\hat{q} = \begin{bmatrix} 10.8 & 3.6 \\ 16.795 & 5.598 \end{bmatrix} \quad \text{and} \quad Q = \hat{Q} + pq + \hat{p}\hat{q}\begin{bmatrix} -12.633 & -3.878 \\ 91.571 & 32.857 \end{bmatrix} \tag{7.67}$$

Combining (7.64) and (7.67), the characteristic polynomial of the closed-loop system can be shown to have its poles at the desired locations. Step responses of the open- and closed-loop systems are shown in Fig. 7.5.

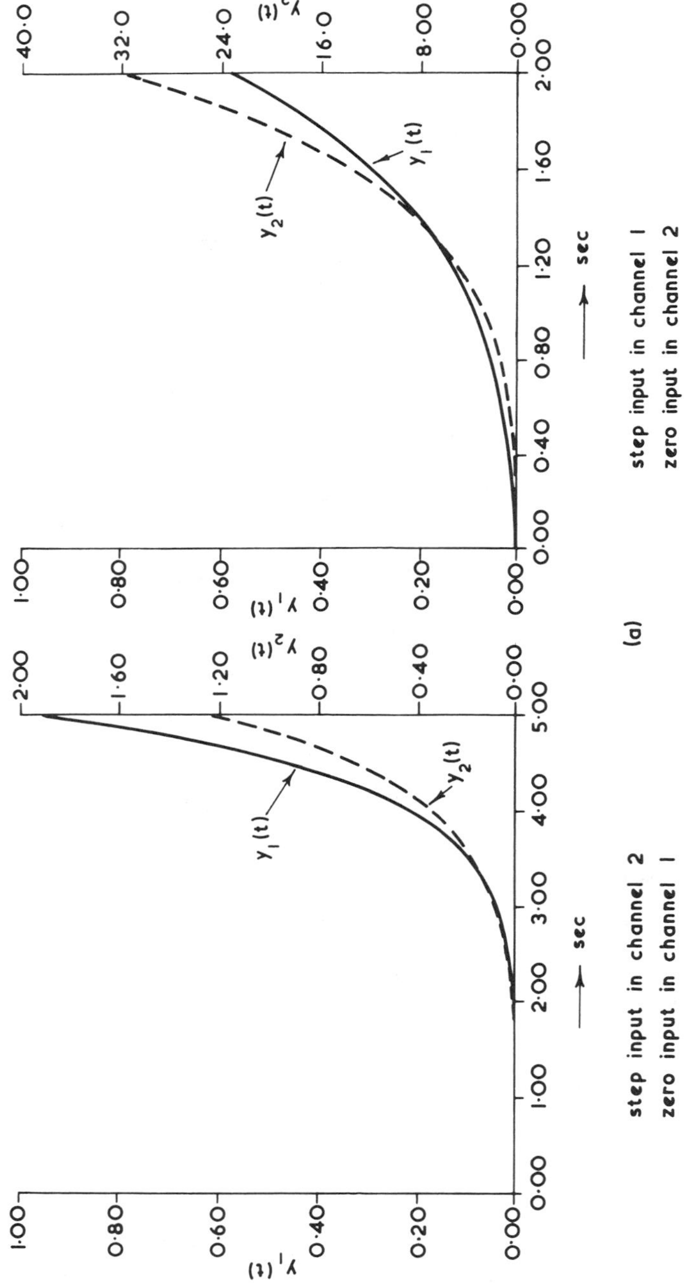

$y_1(t)$
$y_2(t)$
0·00 0·40 0·80 1·20 1·60 2·00
0·00 0·20 0·40 0·60 0·80 1·00
0·00 8·00 16·0 24·0 32·0 40·0
sec
step input in channel 1
zero input in channel 2
0·00 1·00 2·00 3·00 4·00 5·00
0·00 0·40 0·80 1·20 1·60 2·00
step input in channel 2
zero input in channel 1

(a)

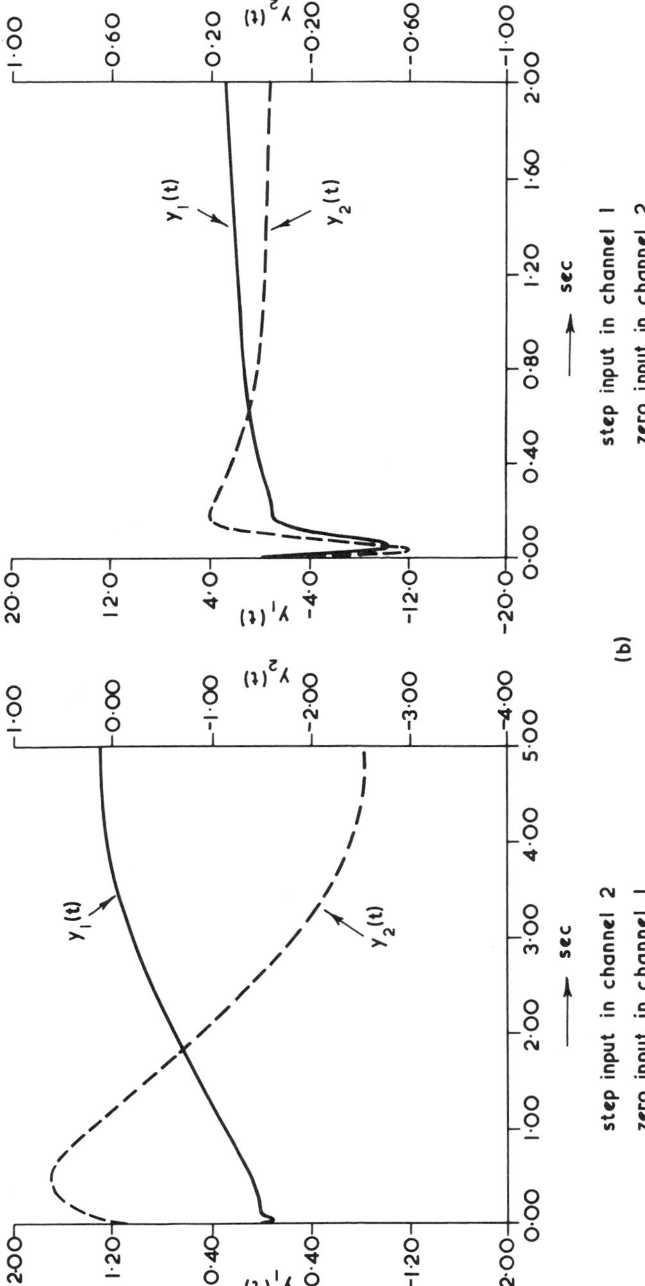

Fig. 7.5 Proportional-plus integral control, step responses of the system in Example 7.5: (a) open-loop system; (b) closed-loop system.

7.4 PROPORTIONAL-PLUS-DERIVATIVE CONTROL*

The analytical results of the preceding section are used here to outline a design procedure for evaluating proportional-plus-derivative output feedback control laws for pole placement. In this two-stage design method, a number of closed-loop poles are first assigned by a constant output-feedback matrix. These poles are then preserved and a number of additional poles are placed using proportional and derivative output feedback. The preservation of poles is achieved by using the fact that in multivariable systems with distinct poles, the numerator transfer-function matrix has unity rank at the system poles.

The system S(A, B, C) considered here is assumed to be completely state controllable and observable and the proportional-plus-derivative feedback control law is formulated as

$$u(t) = v(t) - Py(t) - Q\frac{dy(t)}{dt} \tag{7.68}$$

where v(t) is the closed-loop control input, and P and Q are real constant m-square matrices. The closed-loop system, shown in Fig. 7.6, has the following state equation:

$$\dot{x}(t) = (I+BQC)^{-1}(A-BPC)x(t) + (I+BQC)^{-1}Bv(t); \quad \det(I-BQC) \neq 0 \text{ in general} \tag{7.69}$$

The closed-loop system is of of or n, and thus the derivative term does not introduce any new poles to the system. In the following derivation, a method of evaluating the matrices P and Q which assign the poles of the system is (7.69) at arbitrarily specified locations. These matrices are constructed in two stages.

Stage I: The (m × m) unity-rank proportional output feedback matrix $P_1 = hk$ is applied to the open-loop system S(A, B, C) to obtain the closed-loop characteristic polynomial (Sec. 7.2):

$$\Delta_1(s) = \Delta_o(s) + kH_o(s)h \tag{7.70}$$

where $H_o(s) = C\,\text{adj}\,(sI - A)B$ and $\Delta_o(s) = \det(sI - A)$.

The (1 × m) vector k is specified arbitrarily such that (A, kC) is observable and the (m × 1) vector h is determined to place the (m - 1) closed-loop poles at the distinct specified locations $\lambda_1, \ldots, \lambda_{m-1}$ by solving the (m - 1) linear equations:

$$\Delta_o(\lambda_i) + kH_o(\lambda_i)h = 0, \quad i \in 1, \ldots, m-1 \tag{7.71}$$

*Derivations in this section are based on Ref. 14.

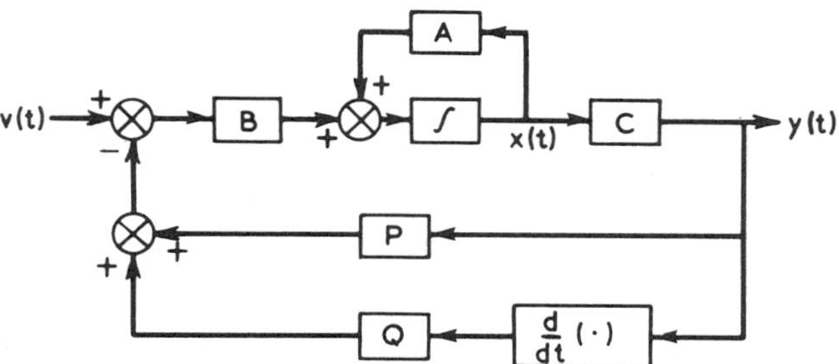

Fig. 7.6 Proportional-plus-derivative control.

The resulting closed-loop system (A_1, B, C), where $A_1 = A - BP_1C$, has $(m - 1)$ poles at $\lambda_1, \ldots, \lambda_{m-1}$.

Stage II: The $(m \times n)$ unity-rank proportional and derivative output feedback matrices $P_2 = \hat{h}\hat{k}$ and $Q = \hat{h}r$ are applied to the system $S(A_1, B, C)$, where $\hat{h}$, $\hat{k}$, and r are $(m \times 1)$, $(1 \times m)$, and $(1 \times m)$ vectors, respectively. This unity-rank structure reduces the problem to the single-input system $S(A, B\hat{h}, C)$ with the closed-loop characteristic polynomial of

$$\Delta_2(s) = \frac{1}{1 + rCB\hat{h}}[\Delta_1(s) + \hat{k}H_1(s)\hat{h} + srH_1(s)\hat{h}] \tag{7.72}$$

where $H_1(s) = C\ \mathrm{adj}\ (sI - A_1)B$ and $\Delta_1(s) = \det\ (sI - A_1)$. The vector $\hat{h}$ is used to preserve the $(m - 1)$ poles of $S(A_1, B, C)$ at $\lambda_1, \ldots, \lambda_{m-1}$ in the closed-loop system by solving

$$H_1(\lambda_i)\hat{h} = 0, \qquad i \in 1, \ldots, m - 1 \tag{7.73}$$

Once $\hat{h}$ is computed, the number of poles in the single-input system $(A_1, B\hat{h}, C)$ that can be placed arbitrarily by the proportional and derivative feedback vectors $\hat{k}$ and r is determined:

$$n_1 = \rho \left\{ \begin{bmatrix} C & \vdots & \underbrace{B\hat{h},\ A_1B\hat{h},\ \ldots,\ A_1^{n-1}B\hat{h}}_{\phi_1} \\ CA_1 & \vdots & \end{bmatrix} \right\}$$

The vectors $\hat{k}$ and r which place n_1 additional poles at the desired locations $\lambda_m, \ldots, \lambda_{m+n_1-1}$ are obtained by solving the n_1 linear equation

$$\Delta_1(\lambda_i) + \hat{k}H_1(\lambda_i)\hat{h} + \lambda_i rH_1(\lambda_i)\hat{h} = 0, \qquad i = m, \ldots, m + n_1 - 1 \tag{7.74}$$

The total proportional feedback matrix is given by

$$P = [h \quad \hat{h}] \begin{bmatrix} k \\ \hat{k} \end{bmatrix} = [hk + \hat{h}\hat{k}] \tag{7.75}$$

(which is of rank 2) and the total derivative feedback matrix

$$Q = \hat{h}r \tag{7.76}$$

(which is of unity rank) place $\alpha = m + n_1 - 1$ poles of the system (A, B, C) at the specified locations. It can be proved that $\rho(\phi_1) = n - m + 1$, and that if

$$\rho \begin{bmatrix} C \\ CA_1 \end{bmatrix} \equiv \rho \begin{bmatrix} C \\ CA \end{bmatrix} = \alpha$$

then

$$n_1 = \rho \left[\begin{array}{c|c} C & \phi_1 \\ CA_1 & \end{array}\right] < \min(\alpha, \ n - m + 1)$$

An illustrative numerical example is given below [14].

<u>Example 7.6: Output feedback through proportional-plus-derivative control</u>

$$\dot{x}(t) = \begin{bmatrix} 0 & 1 & 0 & 0 & 0 \\ 0 & 0 & 1 & 0 & 0 \\ 0 & 0 & 0 & 1 & 0 \\ 0 & 0 & 0 & 0 & 1 \\ 1 & 0 & 0 & 0 & 0 \end{bmatrix} x(t) + \begin{bmatrix} 0 & 1 \\ 0 & 0 \\ 0 & 0 \\ 0 & 0 \\ 1 & 0 \end{bmatrix} u(t); \quad y(t) = \begin{bmatrix} 1 & 0 & 0 & 0 & 0 \\ 0 & 0 & 1 & 0 & 0 \end{bmatrix} x(t) \tag{7.77}$$

The closed-loop poles desired locations are -1, -2, -3, $-1 \pm j$.

Stage I: The transfer function of the system is

$$\frac{H_o(s)}{\Delta_o(s)} = \frac{1}{s^5 - 1} \begin{bmatrix} 1 & s^4 \\ s^2 & s \end{bmatrix} \tag{7.78}$$

The proportional feedback matrix $P_1 = hk$ to place $(m - 1) = 1$ pole at -1 is first calculated. Assuming that $k = [2 \quad 0]$ from (7.71) and (7.78), we have

$$\Delta_o(-1) + kH_o(-1)h = -2 + [2 \quad 0] \begin{bmatrix} 1 & 1 \\ 1 & -1 \end{bmatrix} \begin{bmatrix} h_1 \\ h_2 \end{bmatrix} = -2 + 2h_1 + 2h_2 = 0$$

With $k_2 = 0$, $k_1 = 1$, giving

$$P_1 = \begin{bmatrix} 1 \\ 0 \end{bmatrix} [2 \quad 0] = \begin{bmatrix} 2 & 0 \\ 0 & 0 \end{bmatrix} \tag{7.79}$$

The resulting closed-loop system $S(\overline{A - BP_1C}, B, C)$ has the transfer-function matrix

$$\frac{H_1(s)}{\Delta_1(s)} = \frac{1}{1+s^5} \begin{bmatrix} 1 & s^4 \\ s^2 & -s \end{bmatrix} \tag{7.80}$$

with one closed-loop pole at -1.

Stage II: The proportional and derivative matrices $P_2 (= \hat{h}\hat{k})$ and $Q(= \hat{h}r)$ are to be found. These matrices preserve the pole of $S(\overline{A - BP_1C}, B, C)$ at -1 and place four additional poles at -2, -3, $-1 \pm j$. To preserve the pole at -1, from (7.73) and (7.80), $\hat{h}$ must satisfy the equations

$$H_1(-1)\hat{h} = \begin{bmatrix} 1 & 1 \\ 1 & 1 \end{bmatrix} \hat{h} = 0 \rightarrow [1 \quad 1]\hat{h} = 0 \rightarrow [1 \quad 1] \begin{bmatrix} \hat{h}_1 \\ \hat{h}_2 \end{bmatrix} = 0$$

Choosing $\hat{h}_1 = 1$ arbitrarily,

$$\hat{h} = \begin{bmatrix} 1 \\ -1 \end{bmatrix}$$

In the single-input system $S(\overline{A - BP_1C}, B\hat{h}, C)$, the number of poles that can be placed by $\hat{h}$ and r is given by

$$n_1 = \rho \left\{ \left[\begin{array}{c|c} C & \\ & \phi \\ C(A - BP_1C) & \end{array} \right] \right\} = 4$$

and the closed-loop characteristic polynomial from (7.72) is given by

$$\Delta_2(s) = \frac{1}{1-r_1} \left\{ (s^5+1) + [\hat{k}_1 \quad \hat{k}_2] \begin{bmatrix} 1 & s^4 \\ s^2 & -s \end{bmatrix} \begin{bmatrix} 1 \\ -1 \end{bmatrix} + s[r_1 \quad r_2] \begin{bmatrix} 1 & s^4 \\ s^2 & -s \end{bmatrix} \begin{bmatrix} 1 \\ -1 \end{bmatrix} \right\}$$

$$= \frac{1}{1-r_1} \{ (1-r_1)s^5 - \hat{k}_1 s^4 + r_2 s^3 + (\hat{k}_2 + r_2)s^2 + (\hat{k}_2 + r_1)s + (\hat{k}_1 + 1) \}$$

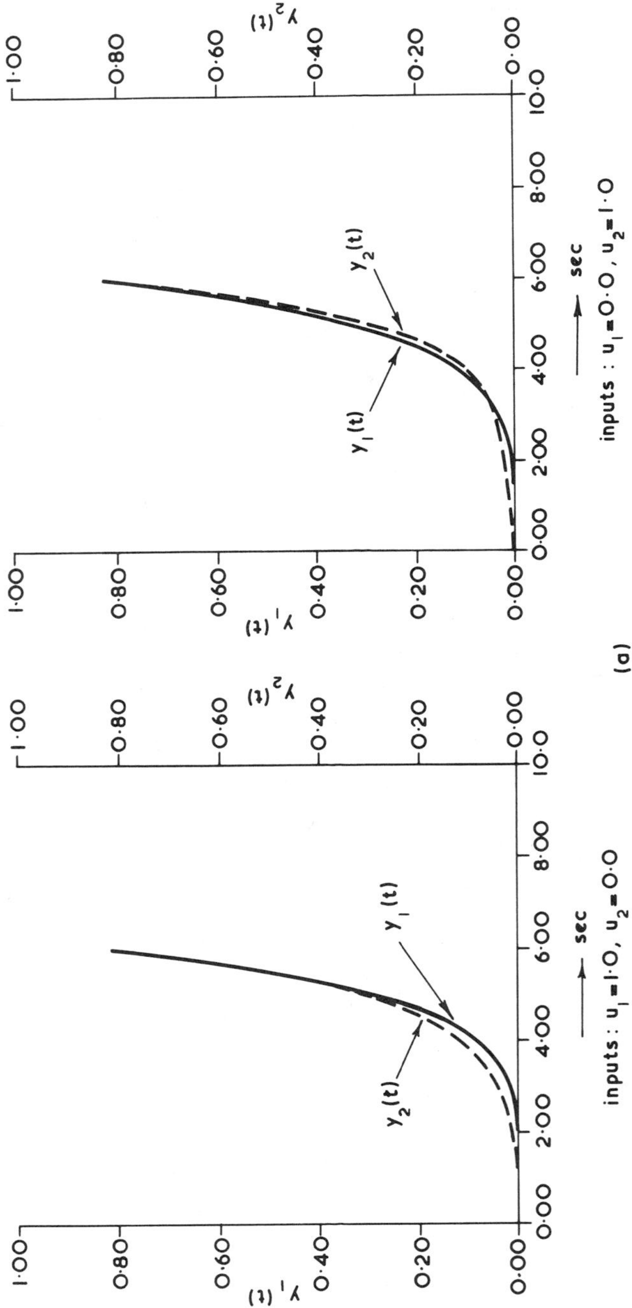

$y_1(t)$
$y_2(t)$
0·00
0·20
0·40
0·60
0·80
1·00
0·00
2·00
4·00
6·00
8·00
10·0
sec
inputs : u_1 = 1·0, u_2 = 0·0
inputs : u_1 = 0·0 , u_2 = 1·0
(a)

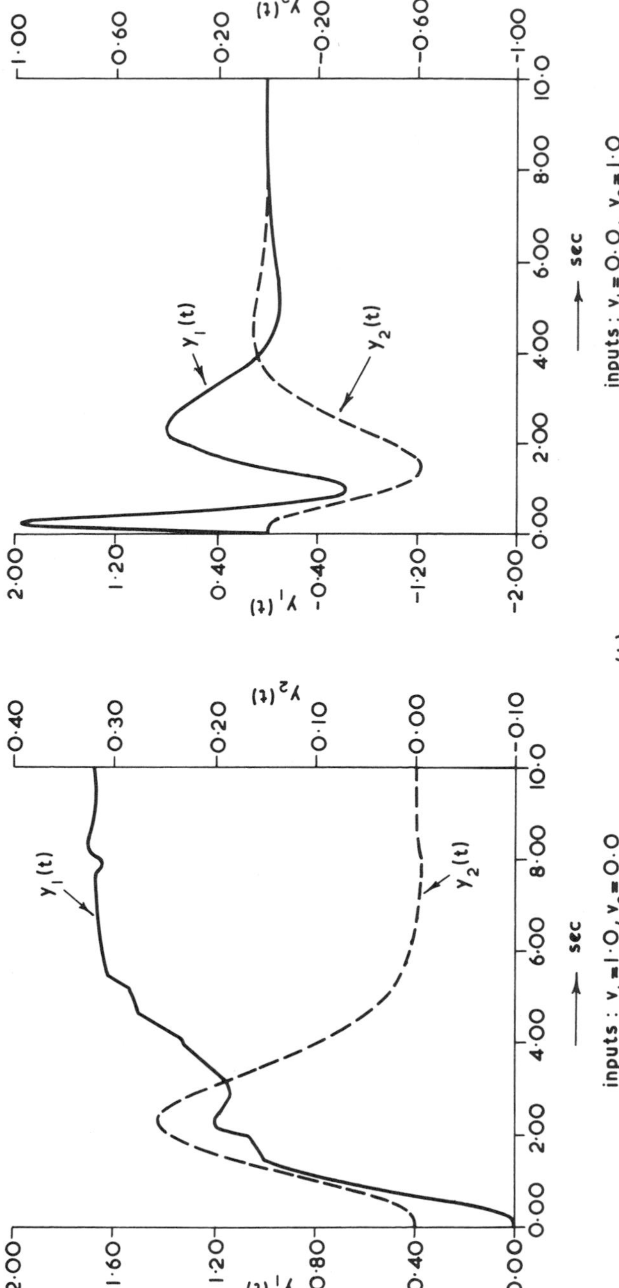

Fig. 7.7 Proportional-plus-integral feedback, step responses of the system in Example 7.6: (a) open-loop system; (b) closed-loop system.

$$= (s+1)\left\{ s^4 + \frac{r_1 - \hat{k}_1 - 1}{1 - r_1} s^3 + \frac{\hat{k}_1 - r_1 + r_2 + 1}{1 - r_1} s^2 + \frac{r_1 + \hat{k}_2 - k_1 - 1}{1 - r_1} s + \frac{\hat{k}_1 + 1}{1 - r_1} \right\}$$

$$\equiv (s+1)(s+2)(s+3)(s+1+j)(s+1-j) = (s+1)(s^4 + 7s^3 + 18s^2 + 22s + 12) \tag{7.81}$$

which by comparing coefficients of similar powers of s gives

$$\begin{bmatrix} -1 & 0 & 8 & 0 \\ 1 & 0 & 17 & 1 \\ -1 & 1 & 23 & 0 \\ 1 & 0 & 12 & 0 \end{bmatrix} \begin{bmatrix} k_1 \\ k_2 \\ r_1 \\ r_2 \end{bmatrix} = \begin{bmatrix} 8 \\ 17 \\ 23 \\ 11 \end{bmatrix} \rightarrow \begin{bmatrix} k_1 \\ k_2 \\ r_1 \\ r_2 \end{bmatrix} = \begin{bmatrix} -0.4 \\ 0.75 \\ 0.95 \\ 1.25 \end{bmatrix} \tag{7.82}$$

Therefore, from (7.75) and (7.76),

$$P_2 = \underbrace{\begin{bmatrix} 1 \\ -1 \end{bmatrix}}_{\hat{h}} \underbrace{[-0.4 \quad 0.75]}_{\hat{k}} = \begin{bmatrix} -0.4 & 0.75 \\ 0.4 & -0.75 \end{bmatrix}$$

and

$$Q = \underbrace{\begin{bmatrix} 1 \\ -1 \end{bmatrix}}_{\hat{h}} \underbrace{[0.95 \quad 1.25]}_{r} = \begin{bmatrix} 0.95 & 1.25 \\ -0.05 & -1.25 \end{bmatrix} \tag{7.83}$$

Thus, combining (7.79) and (7.83) the required proportional and derivative output feedback matrices for the system (7.103) are

$$P = P_1 + P_2 = \begin{bmatrix} 1.6 & 0.75 \\ 0.4 & -0.75 \end{bmatrix} \quad \text{and} \quad Q = \begin{bmatrix} 0.95 & 1.25 \\ -0.95 & -1.25 \end{bmatrix} \tag{7.84}$$

Step responses of the open- and closed-loop systems are shown in Fig. 7.7.

7.5 MODEL-FOLLOWING CONTROL*

The model reference adaptive controller design technique through state feedback developed in Sec. 6.5 requires computation of the Lyanpunov function to ascertain stability and accurate model following. When the state variables of the system are not accessible and the system is controllable and observable, model following may be achieved through output feedback

*This section follows closely to Ref. 15.

provided that the open-loop system satisfies certain conditions. This section develops a relatively simple algebraic method of evaluating an output feedback controller and the necessary condition for implicit model following.

In implicit model following, the output variables of the open-loop system are modified through feedback control so as to approximate the dynamics of a given model. The linear time-invariant multivariable system S(A, B, C) is said to follow a given asymptotically stable model

$$\dot{\eta}(t) = L\eta(t) \tag{7.85}$$

implicitly if $y(t) \equiv \eta(t)$, that is, if the output vector y(t) satisfies the equation

$$\dot{y}(t) = Ly(t) \tag{7.86}$$

as closely as possible. Combining (7.85) and (7.86), the condition for implicit model following may be expressed as

$$\dot{y}(t) - Ly(t) = 0 \tag{7.87}$$

$$[CB]u(t) = [LC - CA]x(t) \tag{7.88}$$

If this equation is to hold for all x(t) through a proper choice of the control input u(t), the vectors in [CB] must be the same vectors as $[\mathscr{L}C - CA]$. The value of the necessary control function may then be obtained by taking the generalized inverse (Sec. 7.2) of $CB\{[CB]^g\}$, and is given by

$$u(t) = [CB]^g[LC - CA]\,x(t) \tag{7.89}$$

The condition to be satisfied by the open-loop system S(A, B, C) for implicit model following of (7.85) may now be obtained by eliminating u(t) from (7.88) and (7.89):

$$[(CB)(CB)^g - I]\,[LC - CA]\,x(t) = 0 \tag{7.90}$$

If CB is nonsingular, this condition will always be satisfied, and the resulting closed-loop system obtained by using (7.89) will be given by

$$\dot{x}(t) = [A + B(CB)^g[LC - CA]]\,x(t); \qquad y(t) = Cx(t) \tag{7.91}$$

Condition for stability:

Since $\rho(C) = m$, by choosing a coordinate transformation

$$z(t) = Tx(t) \tag{7.92}$$

where

$$T = \begin{bmatrix} C \\ \\ E \end{bmatrix}$$

E being an arbitrary $(n - m) \times n$ matrix such that T is nonsingular, the open-loop system S(A, B, C) may be transformed into an equivalent system represented by

$$\begin{aligned} \dot{z}(t) &= TAT^{-1}z(t) + TBu(t) = \bar{A}z(t) + \bar{B}u(t) \\ y(t) &= CT^{-1}z(t) = \bar{C}z(t) \end{aligned} \tag{7.93}$$

where the output matrix $\bar{C}$ has the form $\bar{C} = [I_m \mid 0]$.

The transformed state equation may now be written in the following partitioned form:

$$\dot{x}(t) = \begin{matrix} m\{ \\ n-m\{ \end{matrix} \begin{bmatrix} \dot{y}(t) \\ --- \\ \dot{w}(t) \end{bmatrix} = \begin{matrix} m\{ \\ n-m\{ \end{matrix} \overbrace{\begin{bmatrix} \bar{A}_{11} & \bar{A}_{12} \\ \bar{A}_{21} & \bar{A}_{22} \end{bmatrix}}^{m \quad n-m} \begin{bmatrix} y(t) \\ --- \\ w(t) \end{bmatrix} \begin{matrix} \}m \\ \\ \}n-m \end{matrix} + \begin{bmatrix} \bar{B}_1 \\ -- \\ \bar{B}_2 \end{bmatrix} \begin{matrix} \}m \\ \\ \}n-m \end{matrix} u(t);$$

$$\bar{y} = [I_m \mid 0] \begin{bmatrix} y(t) \\ --- \\ w(t) \end{bmatrix} \tag{7.94}$$

Combining the condition for perfect model following, the transformed system in (7.93) becomes

$$[\bar{B}_1\bar{B}_1^g - I][(L - \bar{A}_{11}) \mid -\bar{A}_{12}] = 0 \tag{7.95}$$

and the feedback control takes the form

$$u(t) = [CB]^g[LC - CA]x(t) \equiv [\bar{C}\bar{B}]^g[L\bar{C} - \bar{C}\bar{A}]x(t) = H\, y(t) + H\, w(t) \tag{7.96}$$

where $H_1 = (\bar{B}_1)^g(L - \bar{A}_{11})$ and $H_2 = -(\bar{B}_1)^g\bar{A}_{12}$. Combining (7.94) and (7.96), the closed-loop model following system becomes

$$\begin{bmatrix} \dot{y}(t) \\ --- \\ \dot{w}(t) \end{bmatrix} = \begin{bmatrix} \bar{A}_{11} + \bar{B}_1H_1 & \bar{A}_{12} + \bar{B}_1H_2 \\ \bar{A}_{21} + \bar{B}_2H_1 & \bar{A}_{22} + \bar{B}_2H_2 \end{bmatrix} \begin{bmatrix} y(t) \\ --- \\ w(t) \end{bmatrix}$$

which when combined with (7.95) reduces to

$$\begin{bmatrix} \dot{y}(t) \\ --- \\ \dot{w}(t) \end{bmatrix} = \begin{bmatrix} L & 0 \\ \bar{A}_{21} + \bar{B}_2\bar{B}_1^g(L - \bar{A}_{11}) & \bar{A}_{22} - \bar{B}_2\bar{B}_1^g\bar{A}_{12} \end{bmatrix} \begin{bmatrix} y(t) \\ --- \\ w(t) \end{bmatrix} \tag{7.97}$$

If S(A, B, C) is assumed to be asymptotically stable, the necessary and sufficient condition for the system represented by (7.91) is to be asymptotically stable is that the eigenvalues of the (n - m) × (n - m) matrix $[\bar{A}_{22} - \bar{B}_2\bar{B}_1^g A_{12}]$ have negative real parts.

In order to express the stability condition of (7.97) in terms of the system matrices A, B, and C, let T^{-1} be partitioned as

$$T^{-1} = [\hat{C} \mid \hat{E}] \tag{7.98}$$

where $\hat{C}$ and $\hat{E}$ are (n × m) and n × (n - m) matrices, respectively; (7.93) then gives

$$\bar{A} = \begin{bmatrix} \bar{A}_{11} & \bar{A}_{12} \\ \bar{A}_{21} & \bar{A}_{22} \end{bmatrix} \equiv \begin{bmatrix} C \\ E \end{bmatrix} A[\hat{C} \;\; \hat{E}] = \begin{bmatrix} CA\hat{C} & CA\hat{E} \\ EA\hat{C} & EA\hat{E} \end{bmatrix}; \quad \bar{B} = \begin{bmatrix} \bar{B}_1 \\ \bar{B}_2 \end{bmatrix} \equiv \begin{bmatrix} C \\ E \end{bmatrix} B = \begin{bmatrix} CB \\ EB \end{bmatrix} \tag{7.99}$$

Therefore,

$$[\bar{A}_{22} - \bar{B}_2\bar{B}_1^g\bar{A}_{12}] = EA\hat{E} - EB(CB)^g CA\hat{E} = E[I - B(CB)^g C]A\hat{E} \tag{7.100}$$

which is independent of the model parameters, and expresses the structural constraint on the open-loop system parameters for the model-following system in (7.97).

Example 7.7 [15]: Output feedback through implicit model following
The synthesis method above is illustrated through the derivation of the model-following control law of

$$\dot{x}(t) = \begin{bmatrix} -3 & 0 & 2 & -2 \\ 1 & -4 & 2 & -2 \\ 1 & -3 & 1 & -2 \\ 1 & 0 & 0 & -4 \end{bmatrix} x(t) + \begin{bmatrix} 3 & 2 \\ 2 & 2 \\ 1 & 1 \\ 1 & 0 \end{bmatrix}; \quad y(t) = \begin{bmatrix} 0 & 0 & 0 & 1 \\ 0 & 0 & 1 & -1 \end{bmatrix} x(t) \tag{7.101}$$

For this system (CB) is nonsingular, and hence the condition for implicit model following (7.90) is satisfied irrespective of the dynamics of the model parameters.

Let the model be chosen to

$$\dot{\eta}(t) = \begin{bmatrix} 0 & 1 \\ -2 & -2 \end{bmatrix} \eta(t) \tag{7.102}$$

A suitable choice of the transformation matrix T is

$$T = \begin{bmatrix} 0 & 0 & 0 & 1 \\ 0 & 0 & 1 & -1 \\ \hline 0 & 1 & 0 & 0 \\ 1 & 0 & 0 & 0 \end{bmatrix} \tag{7.103}$$

where the last two rows ≡ E in (7.92)] of T are to be linearly independent of the two rows of C in (7.101). The inverse of T is given by

$$T^{-1} = \left[\begin{array}{cc|cc} 0 & 0 & 0 & 1 \\ 0 & 0 & 1 & 0 \\ 1 & 1 & 0 & 0 \\ 1 & 0 & 0 & 0 \end{array}\right] = [\hat{C} \mid \hat{E}] \tag{7.104}$$

Combining the foregoing state equation of the system in (7.93) becomes

$$\dot{z}(t) = \bar{A}z(t) + \bar{B}u(t) = \left[\begin{array}{cc|cc} -4 & 0 & 0 & 1 \\ 3 & 1 & -3 & 0 \\ \hline 0 & 2 & -4 & 1 \\ 0 & 2 & 0 & -3 \end{array}\right] x(t) + \left[\begin{array}{cc} 1 & 0 \\ 0 & 1 \\ \hline 2 & 2 \\ 3 & 2 \end{array}\right] u(t) \tag{7.105}$$

$$y(t) = \bar{C}z(t) = \begin{bmatrix} 1 & 0 & 0 & 0 \\ 0 & 1 & 0 & 0 \end{bmatrix} z(t)$$

The control law necessary for (7.101) to follow (7.102) is, from (7.89), given by

$$u(t) = \begin{bmatrix} -1 & 0 & 1 & 3 \\ 0 & 3 & -3 & -2 \end{bmatrix} x(t) \tag{7.106}$$

Using the partitioned matrices in (7.105), the control law above can also be expressed as, from (7.96),

$$u(t) = \begin{bmatrix} 4 & 1 \\ -5 & -3 \end{bmatrix} y(t) + \begin{bmatrix} 0 & 1 \\ -3 & 0 \end{bmatrix} w(t) \tag{7.107}$$

where

$$w(t) = Ex(t) = \begin{bmatrix} 0 & 1 & 0 & 0 \\ 1 & 0 & 0 & 0 \end{bmatrix} x(t)$$

Test for stability: From (7.97) and (7.105),

$$\bar{A}_{22} - \bar{B}_2\bar{B}_1^g\bar{A}_{12} = \begin{bmatrix} -4 & 1 \\ 0 & -3 \end{bmatrix} - \begin{bmatrix} 2 & 2 \\ 3 & 2 \end{bmatrix}\begin{bmatrix} 1 & 0 \\ 0 & 1 \end{bmatrix}\begin{bmatrix} 0 & 1 \\ -3 & 0 \end{bmatrix} = \begin{bmatrix} 2 & -1 \\ 6 & -6 \end{bmatrix} \tag{7.108}$$

which has eigenvalues at 5.162 and -1.162, and consequently the system in (7.101) cannot be made to follow a second-order system.

It is apparent that the synthesis procedure above fails for a given open-loop system when the stability constraint is not satisfied. In such cases it may still be possible to achieve zero error as required by (7.87) by enlarging the class of control laws to include delta functions. This is outlined below [16].

The control law u(t) for model following is expressed as a sum of ordinary and delta functions:

$$u(t, \tau) = u_1(t) + u_\delta(\tau)(t - \tau) \tag{7.109}$$

It is apparent that with the control law above, perfect matching will be absent at the moment the impulse occurs. Adding the delta function above at time t_+, the derivative of the output vector at time t_+ becomes

$$\dot{y}(t_+) = C\dot{x}(t) = CAx(t) + CBu_1(t) + CABu_\delta(t)$$

When combined with (7.87), this gives the modified condition for zero error:

$$CBu_1(t) + CABu_\delta(t) - [LC - CA]\,x(t) = 0 \tag{7.110}$$

Let a new control vector $\bar{u}(t)$ be defined such that

$$\bar{u}(t) \equiv \underbrace{(CB)^g(CB)\bar{u}(t)}_{u_1(t)} + \underbrace{[I - (CB)^g CB]\bar{u}(t)}_{u_\delta(t)} \tag{7.111}$$

Then substituting the corresponding values in (7.110), the input control for implicit model following is given by

$$\underbrace{[CB + CAB[I - (CB)^g CB]}_{R}u(t) = [LC - CA]\,x(t) \tag{7.112}$$

or

$$\bar{u}(t) = R^g[LC - CA]\,x(t) \tag{7.113}$$

R^g being the generalized inverse of R.

Thus eliminating $u_1(t)$ and $u_\delta(t)$ from (7.110) through using (7.113), the modified condition for model following becomes

$$CBR^g[LC - CA]x(t) + CABR^g[LC - CA]x(t) - [LC - CA]x(t) = 0$$

or

$$\{[CB + CAB]R^g - I\}(LC - CA) = 0 \tag{7.114}$$

Thus if (7.114) is satisfied, (7.113) gives the control law that would drive the error $[\dot{y}(t) - Ly(t)]$ to zero, although (7.86) and (7.88) will not hold at time t since the effect of the delta function is not felt until time t_+. At this time $(t+)$ a step change occurs in $\dot{y}(t)$ so as to satisfy (7.86). Implicit model following is thus obtained for at least a time interval that is short in comparison to the fastest time constant of S(A, B, C). As soon as the difference between $\dot{y}(t)$ and $Ly(t)$ exceeds a certain chosen threshold, another delta function is applied; the weighting factor may be suitably chosen by using (7.111). This second delta function restores the equality of (7.86). In this way, implicit model following can be obtained by the application of delta functions whenever the threshold value is exceeded. It is apparent that the smaller the chosen threshold value, the closer will be the spacing of the delta functions, each function having correspondingly smaller strength (amplitude). It can be shown [17] that the required delta functions may be approximately generated by multiplying $u_\delta(t)$ in (7.111) by a large gain constant. This is not considered in detail here. A numerical example and a method of computation of the control law in (7.113) may be found in Ref. 17.

7.6 FEEDBACK COMPENSATION [9]

The methods of control considered in the preceding section are based on constant gain (≡ proportional) feedback matrices. The advantage of such methods of control is that access to the state variables of the system under control is not required. However, two serious limitations of proportional output feedback control methods are that only a limited number of poles of the system may be arbitrarily specified and the flexibility in the choice of closed-loop responses may be lost. In cases where there are fewer outputs than state variables and design requirements other than pole assignment are to be met, the addition of a compensator state vector may yield an extra degree of freedom to the designer. It is shown in this section that the feedback compensator of multivariable systems may be reduced to an equivalent constant feedback design considered in earlier sections. The basic results are presented briefly below.

The feedback compensator dynamics (Fig. 7.8) are described by

$$S_c(A_c, B_c, C_c, D_c): \quad \begin{aligned} \dot{x}_c(t) &= A_c x_c(t) + B_c y(t) \\ y_c(t) &= C_c x(t) + D_c y(t) \end{aligned} \tag{7.115}$$

where $x_c(t)$ is the compensator state vector of order n_c and $y_c(t)$ is the compensator output vector of order m. A_c, B_c, C_c, and D_c are real constant matrices of dimensions $(n_c \times n_c)$, $(n_c \times m)$, $(m \times n_c)$, and $(m \times m)$, respectively. The feedback control law for the system in Fig. 7.9 is

$$u(t) = v(t) + y_c(t) \tag{7.116}$$

and the state equation of the compensated closed-loop system becomes

$$\begin{bmatrix} \dot{x}(t) \\ \dot{x}_c(t) \end{bmatrix} = \begin{bmatrix} A + BD_c C & BC_c \\ B_c C & A_c \end{bmatrix} \begin{bmatrix} x(t) \\ x_c(t) \end{bmatrix} + \begin{bmatrix} B \\ 0 \end{bmatrix} v(t)$$

or

$$\bar{x}(t) = \bar{A}\bar{x}(t) + \bar{B}v(t) \tag{7.117}$$

The output equation of the overall system remains unchanged.

The composite system in (7.117) is of order $(n + n_c)$ with the state vector

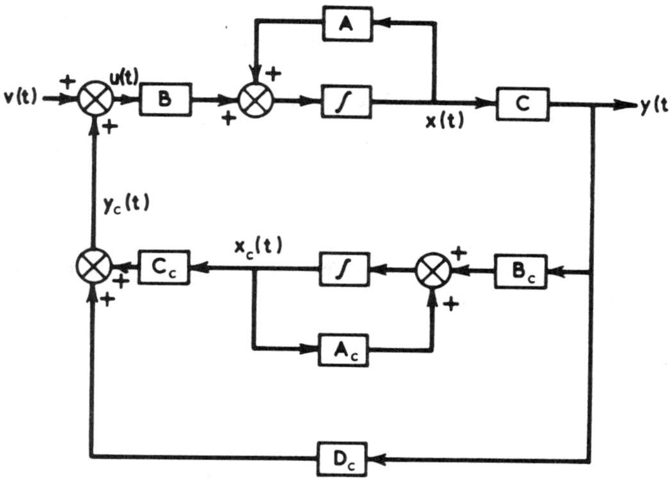

Fig. 7.8 Feedback compensation.

$$\bar{x}(t) = \begin{bmatrix} x(t) \\ x_c(t) \end{bmatrix}$$

and the m-input control vector v(t). The closed-loop system matrix is

$$\bar{A} = \begin{bmatrix} A + BD_cC & BC_c \\ B_cC & A_c \end{bmatrix} = \begin{bmatrix} A & 0 \\ 0 & 0 \end{bmatrix} + \begin{bmatrix} B & 0 \\ 0 & I \end{bmatrix} \begin{bmatrix} D_c & C_c \\ B_c & A_c \end{bmatrix} \begin{bmatrix} C & 0 \\ 0 & I \end{bmatrix}$$

$$= \hat{A} - \hat{B} \underbrace{\begin{bmatrix} D_c & C_c \\ B_c & A_c \end{bmatrix}}_{\hat{H}} \hat{C} \tag{7.118}$$

where $\hat{A}$ is $(n + n_c)$ square, $\hat{B}$ is $(n + n_c) \times (m + n_c)$, and $\hat{C}$ is $(m + n_c) \times (n + n_c)$, which is of similar structure to the closed-loop matrix of an auxiliary system $(\hat{A}, \hat{B}, \hat{C})$ in (7.117) and (7.118), with the $(m + n_c) \times (m + n_c)$ constant output feedback matrix $\bar{H}$ given by

$$\hat{H} = \begin{bmatrix} D_c & C_c \\ B_c & A_c \end{bmatrix} \tag{7.119}$$

Once the compensator structure is established, the dynamic compensator design problem thus reduces to an equivalent constant output feedback design problem. The submatrices of $\hat{H}$ give the required compensator matrices A_c, B_c, C_c, and D_c. Detailed derivations for the design of compensators for multivariable systems are not considered here, and may be found in Refs. 18 and 19.

Derivations included in this chapter demonstrate the use of output feedback for pole assignment and the extension of the classical proportional-plus-integral and proportional-plus-derivative controls. The extent of control that can be exerted through output feedback is dependent on the controllability and observability properties of the system. References to some relevant work in this area are listed in the bibliography.

REFERENCES

1. Munro, N., and Vardulakis, A., "Pole-shifting using output feedback," Int. J. Control, 18: 1267-1273 (1973).
2. Patel, R. V., "Comment on 'Pole-shifting using output feedback'," Int. J. Control, 20: 171-172 (1974).

3. Patel, R. V., "On output feedback pole assignability," Int. J. Control, 20: 955-959 (1974).
4. Rao, C. R., and Mitra, S. K., Generalized Inverse Matrices and Its Applications, Wiley, New York, 1971.
5. Pringle, R. M., and Rayner, A. A., Generalised Inverse Matrices with Applications to Statistics, Griffin Statistical Monographs and Courses, Charles Griffin, London, 1971.
6. Porter, B., "Eigenvalue assignment in linear multivariable systems by output feedback," Int. J. Control, 25: 483-490 (1977).
7. Fallside, F., and Seraji, H., "Pole-shifting procedure for multivariable systems using output feedback," Proc. IEE, 118: 1648-1654 (1971).
8. Layton, J. M., Multivariable Control Theory, Peter Peregrinus, London, 1976.
9. Fallside, F., "Pole assignment," in Modern Approaches to Control System Design, N. Munro (Ed.), Peter Peregrinus, London, 1979, Chap. 11.
10. Fallside, F., and Seraji, H., "Direct design procedure for multivariable feedback systems," Proc. IEE, 118: 797-801 (1971).
11. Jameson, A., "Design of single-input systems for specified roots using output feedback," Trans. IEEE, AC-15: 345-348 (1970).
12. Seraji, H., "Cyclicity of linear multivariable systems," Int. J. Control, 21: 497-504 (1975).
13. Seraji, H., "Design of proportional-plus-integral controllers for multivariable systems," Int. J. Control, 29: 49-63 (1979).
14. Seraji, H., "Pole-placement in multivariable systems using proportional derivative output-feedback," Int. J. Control, 31: 195-207 (1980).
15. Kudva, P., and Gourishankar, "On the stability problem of multivariable model-following systems," Int. J. Control, 24: 801-805 (1976).
16. Tyler, J. S., "The characteristics of model-following systems as synthesized by optimal control," IEEE Trans., AC-9: 485-498 (1964).
17. Erzberger, H., "Analysis and design of model following control systems by state space techniques," Preprints JACC, 1968, pp. 572-581.
18. Brasch, F. M., and Pearson, J. B., "Pole placement using dynamic compensators," Trans. IEEE, AC-15: 34-43 (1970).
19. Pearson, J. B., and Ding, C. Y., "Compensator design for multivariable linear systems," Trans. IEEE, AC-14: 130-134 (1969).

8
Design With State Observer

It was shown in Chap. 6 that the introduction of state feedback in a state controllable system allows complete freedom in the choice of the closed-loop structure of a linear system. Some of this freedom is lost if control action were exerted through the feedback of output variables, as examined in Chap. 7. Since the knowledge of the state vector, not the output, determines the future response if the future input is known, reconstruction of the state vector from linear combinations of its elements (≡ output variables) would allow considerable design flexibility.

The purpose of this chapter is to present methods of reconstructing or estimating the state vector from output variables. In particular, the aim is to evaluate a functional F.

$$\hat{x}(t) = F[y(\tau), \; t_0 \leq \tau < t] \tag{8.1}$$

such that $\hat{x}(t) \simeq Tx(t)$ presents the reconstructed state of the system S(A, B, C, D), where t_0 is the initial time of observation and T a constant transformation matrix. It should be noted that the reconstructed state $\hat{x}(t)$ is a function of the past observation of $y(\tau)$, $t_0 < \tau < t$, and does not depend on future observations of $y(\tau)$, $\tau > t$.

The "simulator" that reconstructs $\hat{x}(t)$ in (8.1) is called a <u>state estimator</u> or an <u>observer</u>. The observer is a dynamic system in which, with increasing time, the output approaches the state that is to be reconstructed. Once the observer has been evaluated, the synthesis procedures in Chap. 6, which assume knowledge of the complete state vector, may be used by replacing the actual state with the reconstructed state.

Let the observer state $\hat{x}(t)$ be defined by the following dynamic simulator (Fig. 8.1):

$$\frac{d\hat{x}(t)}{dt} = L\hat{x}(t) + Mu(t) + Ny(t) \tag{8.2}$$

where $\hat{x}(t)$ is an $\hat{n}$-state vector,* u(t) and y(t) are m-control and ℓ-output

*In the general theory $\hat{x}(t)$ is assumed to be an arbitrary $\hat{n}$-state vector, In practical terms, however, two cases are important: $\hat{n} \leq n$; these are considered separately.

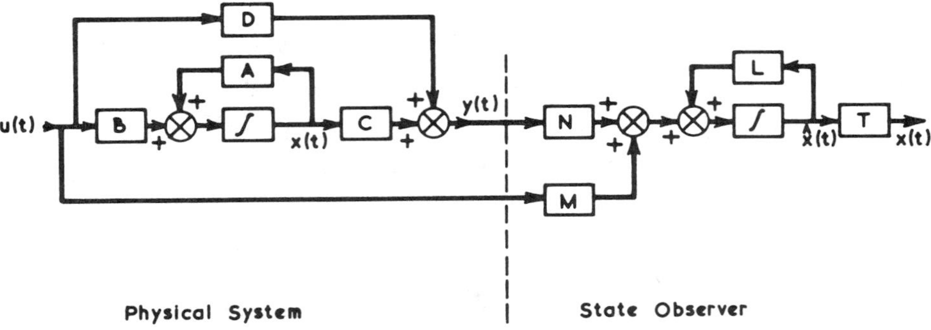

Fig. 8.1 Dynamic simulator for state estimation.

vectors, respectively. L, M, and N are real constant matrices of dimensions $(\hat{n} \times \hat{n})$, $(\hat{n} \times m)$, and $(\hat{n} \times \ell)$, respectively, with $\hat{n} < n$.

Definition 8.1: Observer The system in (8.2) is an observer for the system S(A, B, C, D) if for every initial state $x(t_0)$ of $S(\cdot)$ there exists an initial state $\hat{x}(t_0)$ and a linear transformation matrix T such that

$$\hat{x}(t_0) = Tx(t_0) \text{ implies } \hat{x}(t) = Tx(t), \quad \forall\, t > t_0 \tag{8.3}$$

Definition 8.2: Full-order observer The n-dimensional system in (8.2) is a full-order observer for the n-dimensional system S(A, B, C, D) if

$$\hat{x}(t_0) = Tx(t_0) \text{ implies } \hat{x}(t) = Tx(t), \quad \forall\, t > t_0$$

and the dimension of $\hat{x}(t)$ is the same as that of the state x(t) (i.e., if $\hat{n} = n$).

If the order of $\hat{x}(t)$ is less than that of x(t), the observer is called a reduced-order observer (i.e., $\hat{n} < n$). If the simulator in (8.2) produces a state such that $\hat{x}(t) = x(t)$, $\forall\, t$, with $T = I_n$, it is known as an identity observer. Some basic results related to the structure of these observers are briefly considered in the following two sections. These are followed by synthesis methods for reduced-order observers and their effect on closed-loop control.

8.1 FULL-ORDER OBSERVER [1-5]

Following the postulations above, a reconstruction error vector e(t) may be defined as

$$e(t) = \hat{x}(t) - Tx(t) \simeq 0 \tag{8.4}$$

Combining (8.2) and (8.4) gives us

$$\dot{e}(t) = \dot{\hat{x}}(t) - T\dot{x}(t) = L\hat{x}(t) + Mu(t) + Ny(t) - TAx(t) - TBu(t)$$
$$= L[\hat{x}(t) - Tx(t)] + [NC - TA + LT]x(t) + [M + ND - TB]u(t) \quad (8.5)$$

Thus if the observer and the system matrices were related by the equations

$$TA - LT = NC \quad (8.6a)$$

and

$$TB - ND = M \quad (8.6b)$$

the dynamic equation of the error vector in (8.5) would reduce to $\dot{e}(t) = Le(t)$, which has the solution

$$e(t) = e^{Lt} e(t_0) \quad (8.7a)$$

or

$$\hat{x}(t) - Tx(t) = e^{Lt}[\hat{x}(t_0) - Tx(t_0)] \quad (8.7b)$$

where $\hat{x}(t_0)$ and $x(t_0)$ denote the initial states of $\hat{x}(t)$ and $x(t)$, respectively.

Hence if (8.6) can be satisfied through suitable choice of the real constant matrices L, M, and N, and if the eigenvalues of L are chosen to have negative real parts, the difference between $\hat{x}(t)$ and $Tx(t)$ will merely consist of the transient terms on the right-hand side of (8.7b), and once these have died away, $\hat{x}(t)$ will equal $Tx(t)$. To this extent $\hat{x}(t)$ is an approximation of $Tx(t)$

The constraints in (8.6a) yield $\hat{n}n$ scalar equations for the $\hat{n}n$ elements of T, which may be solved for any arbitrary choice of N, provided that L is chosen such that A and L do not have any common eigenvalue [1, 2]. Thus choosing N and selecting L to have eigenvalues with sufficiently negative real values which are different from those of A, the transformation matrix T may be evaluated. It should be noted at this stage that the condition in (8.6b) is dependent on B and D, and hence these two matrices must be specified even when the system is free [i.e., $u(t) \equiv 0$].

8.2 IDENTITY OBSERVER [1-5]

The derivation in the preceding section to produce an approximation to $Tx(t)$ is useful to obtain $x(t)$ if T has an inverse (i.e., if T is a square nonsingular matrix). Rather than produce a vector $Tx(t)$ and then pass it through an inverse gain transformation matrix ($= T^{-1}$), a more direct approach is to use an identity observer in the development above. This simplifies (8.6) to

$$L = A - NC \tag{8.8a}$$

and

$$M = B - ND \tag{8.8b}$$

Hence, from (8.7),

$$e(t) = \hat{x}(t) - x(t) = e^{(A-NC)t}[\hat{x}(t_0) - x(t_0)] \tag{8.8c}$$

It is shown later that the eigenvalues of (A - NC) in (8.8a) may be chosen arbitrarily (provided that they are real or complex, with the complex eigenvalues occurring in conjugate complex pairs) only if the pair $\{C, A\}$ is completely observable (Sec. 3.1). This result is first developed for single-input/single-output systems, and then extended to multivariable systems [5].

It was shown in Sec. 4.1 that if S(A, b, c, d) is completely observable, then by an equivalent transformation of the form

$$z(t) \equiv \bar{x}^o(t), \qquad \text{where } x(t) = V_o\bar{x}^o(t) \tag{8.9}$$

it can be transformed into the observable canonical form

$$\bar{S}^o: \ \dot{z}(t) = \bar{A}z(t) + \bar{b}u(t); \qquad y(t) = \bar{c}z(t) + du(t) \tag{8.10}$$

$$\bar{A} = V_o^{-1}AV_o = \begin{bmatrix} 0 & 0 & \cdots & 0 & -\alpha_n \\ 1 & 0 & \cdots & 0 & -\alpha_{n-1} \\ \vdots & & & & \\ 0 & 0 & \cdots & 0 & -\alpha_2 \\ 0 & 0 & \cdots & 1 & -\alpha_1 \end{bmatrix}; \ \bar{b} = V_o^{-1}b = \begin{bmatrix} \beta_n \\ \beta_{n-1} \\ \vdots \\ \beta_2 \\ \beta_1 \end{bmatrix} \tag{8.11}$$

$$\bar{c} = cV_o = [0 \quad 0 \quad 0 \quad 1]$$

where α_i and β_i are related to the system transfer function

$$g(s) = \frac{\beta_1 s^{n-1} + \beta_2 s^{n-2} + \cdots + \beta_{n-1}s + \beta_n}{s^n + \alpha_1 s^{n-1} + \cdots + \alpha_{n-1}s + \alpha_n} + d \tag{8.12}$$

and for notational convenience, the vector z(t) is used here to represent the observable form of x(t). Since the system is assumed to be completely observable, z(t) is an n-dimensional state vector.

Using the notions of (8.2) and (8.8), the identity observer for the single-input/single-output system in (8.10) may be expressed as

$$\mathring{z}(t) = \bar{L}\hat{z}(t) + \bar{m}u(t) + \bar{n}y(t) = (\bar{A} - \bar{n}\bar{c})\hat{z}(t) + (\bar{b} - \bar{n}d)u(t) + \bar{n}y(t) \tag{8.13}$$

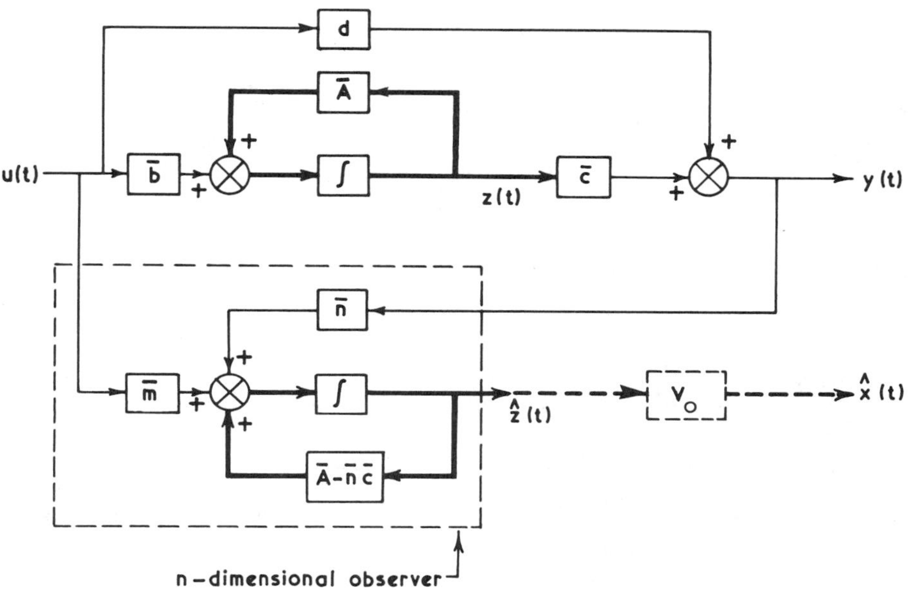

Fig. 8.2 Full-order observer of a single-input/single-output system through transformation (observable form).

where $\bar{L} \equiv (\bar{A} - \bar{n}\bar{c})$ and $\bar{n}$ denotes the parameters of the observer for the normalized state vector z(t) (Fig. 8.2).*

Since the objective here is to evaluate the observer parameters such that (8.13) is asymptotically stable, let the characteristic polynomial of the observer be chosen as

$$p_{\hat{z}}(s) = s^n + p_1 s^{n-1} + \cdots + p_{n-1}s + p_n \tag{8.14}$$

which gives the set of desired eigenvalues. Assuming that (8.13) is in observable canonical form, this choice implies that

$$\bar{L} = \bar{A} - \bar{n}\bar{c} = \begin{bmatrix} 0 & 0 & \cdots & 0 & -p_n \\ 1 & 0 & \cdots & 0 & -p_{n-1} \\ \vdots & & & & \vdots \\ 0 & 0 & \cdots & 1 & -p_1 \end{bmatrix} \tag{8.15}$$

*For notational uniformity n is used in this chapter to denote the system order and the observer parameter for single-input/single-output systems; the exact use is clear in the context.

which by using $\bar{A}$ and $\bar{c}$ in (8.11) gives

$$\bar{n}\bar{c} = \begin{bmatrix} 0 & 0 & \cdots & p_n - \alpha_n \\ 0 & 0 & \cdots & p_{n-1} - \alpha_{n-1} \\ \vdots & & & \\ 0 & 0 & \cdots & p_1 - \alpha_1 \end{bmatrix} \rightarrow \bar{n} = \begin{bmatrix} p_n - \alpha_n \\ p_{n-1} - \alpha_{n-1} \\ \vdots \\ p_1 - \alpha_1 \end{bmatrix} \tag{8.16}$$

By combining the equations above, the observer in (8.13) becomes

$$\dot{\hat{z}}(t) = (\bar{A} - \bar{n}\bar{c})\hat{z} + (\bar{b} - \bar{n}d)u(t) + \bar{n}y(t) \tag{8.17}$$

where $\hat{z}(t)$ is an estimate of the observable state vector $z(t)$ and not of the original system state vector $x(t)$. Consequently, to obtain the required estimate $\hat{x}(t)$ of $x(t)$, the observer in (8.17) will have to be modified through the inverse transformation of (8.9). The resulting configuration is shown by dashed lines in Fig. 8.2.

If the system whose state variables are to be estimated is completely observable, (8.13) suggests that the eigenvalues of the observer can be arbitrarily chosen. If these are chosen to have negative real parts, then regardless of the initial state of the observer, the output $\hat{z}(t)$ will approach the real state $z(t)$ [$\equiv V_o^{-1}x(t)$] asymptotically. Thus for a completely observable system, the transient response of

$$e_{\hat{z}}(t) = \hat{z}(t) - z(t) = e^{(\bar{A}-\bar{n}\bar{c})t}\{\hat{z}(t_0) - z(t_0)\} \tag{8.18}$$

may be made to decay as rapidly as required through the choice of $\bar{n}$. It is to be noted, however, that if this rate of decay is made too rapid, making the eigenvalues of $(\bar{A} - \bar{n}\bar{c})$ too negative in their real parts, the observer will act as an approximate differentiator, accentuating high-frequency noise. A practical compromise is to choose $\bar{n}$ such that the eigenvalues of $(\bar{A} - \bar{n}\bar{c})$ are only slightly more negative than other eigenvalues present in $\bar{S}(\bar{A}, \bar{b}, \bar{c}, d)$.

A correspondence between the estimates of $x(t)$ and $z(t)$ may be obtained by substituting $\hat{x}(t) = V_o z(t)$ in (8.13):

$$\begin{aligned} \hat{x}(t) &= V_o(\bar{A} - \bar{n}\bar{c})V_o^{-1}\hat{x}(t) + V_o(\bar{b} - \bar{n}d)u(t) + V_o\bar{n}y(t) \\ &\equiv (A - nc)\hat{x}(t) + (b - nd)u(t) + ny(t) \end{aligned} \tag{8.19}$$

and consequently,

$$e_{\hat{x}}(t) = \hat{x}(t) - x(t) = e^{(A-nc)t}\{\hat{x}(t_0) - x(t_0)\} \tag{8.20}$$

which indicates that the coordinate transformation does not have any influence on the observer response. The derivation above leads to the following theorem.

Theorem 8.1 If a single-input/single-output time-invariant linear system is completely observable, an asymptotic state estimator with any eigenvalues may be constructed.

Example 8.1 [5]: Identity observer—single-input/single-output system The single-input/single-output system of order 3 described by

$$\dot{x}(t) = \begin{bmatrix} 1 & 2 & 0 \\ 3 & -1 & 1 \\ 0 & 2 & 0 \end{bmatrix} x(t) + \begin{bmatrix} 2 \\ 1 \\ 1 \end{bmatrix} u(t); \quad y(t) = [0 \quad 0 \quad 1]x(t) \tag{8.21}$$

is used here to illustrate the observer design method above.

The first stage of the synthesis procedure is to evaluate the equivalent observable canonical form in (8.10); from Sec. 3.2, this is given by

$$\begin{aligned} \dot{\bar{z}}(t) &= \bar{A}z(t) + \bar{b}u(t) = \begin{bmatrix} 0 & 0 & -2 \\ 1 & 0 & 9 \\ 0 & 1 & 0 \end{bmatrix} z(t) + \begin{bmatrix} 3 \\ 2 \\ 1 \end{bmatrix} u(t) \\ y(t) &= \bar{c}z(t) = [0 \quad 0 \quad 1]z(t) \end{aligned} \tag{8.22}$$

The transfer function of the system is

$$g(s) = \frac{\beta_1 s^2 + \beta_2 s + \beta_3}{s^3 + \alpha_1 s^2 + \alpha_2 s + \alpha_3} = \frac{s^2 + 2s + 3}{s^3 - 9s + 2}$$

The transformation used here is $x(t) = V_o z(t)$, where

$$V_o = \frac{1}{6}\begin{bmatrix} 1 & 1 & 7 \\ 0 & 3 & 0 \\ 0 & 0 & 6 \end{bmatrix} \quad \text{and} \quad V_o^{-1} = \begin{bmatrix} 6 & -2 & -7 \\ 0 & 2 & 0 \\ 0 & 0 & 1 \end{bmatrix} \tag{8.23}$$

In the second stage of design, the desired eigenvalues of an asymptotically stable observer are specified. Let these be at -3, -4, -5; the corresponding characteristic polynomial of $(\bar{A} - \bar{n}\bar{c})$ is then given by

$$\det\{sI - (\bar{A} - \bar{n}\bar{c})\} = (s+3)(s+4)(s+5) = s^3 + 12s^2 + 47s + 60$$

From (8.14) to (8.16), the column vector $\bar{n}$ is thus to be chosen as

$$\bar{n} = \begin{bmatrix} p_3 - \alpha_3 \\ p_2 - \alpha_2 \\ p_1 - \alpha_1 \end{bmatrix} = \begin{bmatrix} 60 - 2 \\ 47 + 9 \\ 12 - 0 \end{bmatrix} = \begin{bmatrix} 58 \\ 56 \\ 12 \end{bmatrix} \tag{8.24}$$

From (8.8), (8.14) to (8.17), and (8.24), the dynamic observer (8.22) is therefore given by

$$\begin{aligned} \dot{\hat{z}}(t) &= (\bar{A} - \bar{n}\bar{c})\hat{z}(t) + \bar{n}y(t) + \bar{b}u(t) \\ &= \begin{bmatrix} 0 & 0 & -60 \\ 1 & 0 & -47 \\ 0 & 1 & -12 \end{bmatrix} \hat{x}(t) + \begin{bmatrix} 58 \\ 56 \\ 12 \end{bmatrix} y(t) + \begin{bmatrix} 3 \\ 2 \\ 1 \end{bmatrix} u(t) \end{aligned} \tag{8.25}$$

($\bar{m} = \bar{b}$ since $\bar{d} = 0$.)

The dynamic observer for x(t) may now be obtained using (8.25) together with the configuration of Fig. 8.2, with V_o^{-1} as given above. Alternatively, (8.19) may be used, in which case the observer equation is given by

$$\dot{\hat{x}}(t) = \begin{bmatrix} 1 & 2 & -33 \\ 3 & -1 & -27 \\ 0 & 2 & -12 \end{bmatrix} \hat{x}(t) + \begin{bmatrix} 33 \\ 28 \\ 12 \end{bmatrix} y(t) + \begin{bmatrix} 2 \\ 1 \\ 1 \end{bmatrix} u(t) \tag{8.26}$$

The results above may be extended to multivariable systems through the following theorem.

Theorem 8.2 If a linear time-invariant multivariable system is completely observable, the eigenvalues of its full-order identity observer can be arbitrarily located in the complex frequency plane.

Proof: Combining (8.2) and (8.8), the equations of an identity observer of full order may be expressed as

$$\dot{\hat{x}}(t) = (A - NC)\hat{x}(t) + (B - ND)u(t) + Ny(t) \tag{8.27}$$

Thus the characteristic polynomial of the observer system is*

$$\det [sI - (A - NC)] \equiv \det [sI - (A^t - C^tN^t)] \tag{8.28}$$

so that the eigenvalues of (A − NC) are identical to those of $(A^t - C^tN^t)$. However, from the condition of pole assignability, the eigenvalues of $(A^t - C^tN^t)$ can be arbitrarily located by choosing appropriate real constant

*A superscript t is used to denote transposed matrices.

matrix N only if the pair $\{A^t, C^t\}$ is completely controllable. This, from the duality property (Theorem 3.2), is equivalent to the observability of the pair $\{A, C\}$.

Thus when the system S(A, B, C) is completely observable, there exists an asymptotically stable full-order identity state observer, and the system S(·) is said to be completely reconstructable. If the system S(A, B, C) is not completely observable, a matrix N may still be found such that the observer is asymptotically stable, although the system is not completely reconstructable; such systems are known to be detectable. Although the stages of numerical computation of L, M, and N for an identity full-order observer are straightforward, the generality associated with (8.8a) may make the choice of prespecified eigenvalues of L difficult, since the specification of det $\{sI - L\}$ does not impose any constraint on the structure of L (except that $A \not\equiv L$). This problem of "nonuniqueness" may be overcome by deriving the observer for the observable or the normalized form of the system, and then including the appropriate inverse transformation. The use of an observable form for multivariable systems, based on an extension of the single-input/single-output systems, is considered later. The use of normalization in the derivation of full-order identity observer is outlined below.

Let P be the modal matrix of the systems S(A, B, C). Then the normal equations of S(·) are*

$$\begin{aligned} \dot{x}_n(t) &= P^{-1}APx_n(t) + P^{-1}Bu(t) \equiv Jx_n(t) + B_n u(t) \\ y(t) &= CPx_n(t) \equiv C_n x_n(t) \end{aligned} \tag{8.29}$$

where $x_n(t)$ is the normal state vector given by $x(t) = Px_n(t)$.

Let $\hat{x}_n(t)$ be the estimated state of the normalized state vector. Then the observer dynamics may be described by

$$\dot{\hat{x}}_n(t) = L_n \hat{x}_n(t) + M_n u(t) + N_n y(t) \tag{8.30}$$

where, by using (8.8),

$$L_n + N_n C_n = J \quad \text{and} \quad B_n = M_n \tag{8.31}$$

For computational convenience, L_n may be chosen to be of the same structure as J, with appropriate block-diagonal elements to give the desired dynamic response. The matrix N_n may then be computed by using generalized inversion of C_n:

$$N_n = [J - L_n]C_n^t[C_n C_n^t]^{-1} \tag{8.32}$$

*D is excluded here for convenience.

which is always valid since the system S($\cdot$) is assumed to be completely observable (i.e., $\rho[C_n] = m$). This method is illustrated below for the distinct eigenvalue case.

Example 8.2: Identity observer through normalization

$$S(A, B, C): \; A = \begin{bmatrix} 2 & -2 & 3 \\ 1 & 1 & 1 \\ 1 & 3 & -1 \end{bmatrix}; \quad B = \begin{bmatrix} 1 & 3 \\ 1 & 1 \\ 1 & 1 \end{bmatrix}; \quad C = \begin{bmatrix} 1 & 1 & 1 \\ 0 & 1 & 0 \end{bmatrix} \tag{8.33a}$$

For this system (Example 1.5),

$$P = \begin{bmatrix} -1 & 11 & 1 \\ 1 & 1 & 1 \\ 1 & -14 & 1 \end{bmatrix} \quad \text{and} \quad P^{-1} = \frac{1}{30}\begin{bmatrix} -15 & 25 & -10 \\ 0 & 2 & -2 \\ 15 & 3 & 12 \end{bmatrix}$$

giving

$$J = \begin{bmatrix} 1 & 0 & 0 \\ 0 & -2 & 0 \\ 0 & 0 & 3 \end{bmatrix}; \quad B_n = \begin{bmatrix} 0 & -1 \\ 0 & 0 \\ 1 & 2 \end{bmatrix}; \quad C_n = \begin{bmatrix} 1 & -2 & 3 \\ 1 & 1 & 1 \end{bmatrix} \tag{8.33b}$$

Thus $\rho[C_n] = 2 = m$ and the system is completely observable; thus a full-order asymptotically stable observer may be evaluated by using (8.30) and (8.31), which for the choice $L_n = \text{diag}\{-2, -3, -4\}$ is

$$\dot{\hat{x}}_n(t) = \underbrace{\begin{bmatrix} -2 & 0 & 0 \\ 0 & -3 & 0 \\ 0 & 0 & -4 \end{bmatrix}}_{L_n} \hat{x}_n(t) + \underbrace{\begin{bmatrix} 0 & -1 \\ 0 & 0 \\ 1 & 2 \end{bmatrix}}_{M_n} u(t) + \underbrace{\frac{1}{38}\begin{bmatrix} 3 & 36 \\ -8 & 18 \\ 49 & 56 \end{bmatrix}}_{N_n} y(t) \tag{8.34}$$

giving an estimate of $x_n(t)$ in (8.29). The required estimate $\hat{x}(t)$ of the original state x(t) may then be obtained through appropriate inverse transformation given by $\hat{x}_n(t) \equiv P^{-1}\hat{x}(t)$, as in the case of single-input/ single-output systems Fig. 8.2 and (8.19).

8.3 REDUCED-ORDER OBSERVER [2-13]

The full-order observers considered earlier are unnecessarily complex since some of the elements of the state vector x(t) are already available through the system output vector y(t). For some systems it may therefore be uneconomic to reconstruct all these elements in the observer vector $\hat{x}(t)$. Some basic theory related to the reduced-order observer is given here. The effect of state feedback through reduced-order observers and synthesis methods are considered in the following sections.

(a) Single-input/single-output system

The derivation is based on the estimation of the observable canonical form (8.10), where the output vector is equivalent to the nth component of the transformed state vector [i.e., $y(t) = z_n(t)$]. It is thus necessary to estimate only the first (n - 1) components of z(t) [i.e., $z_1(t)$, $z_2(t)$, ..., $z_{n-1}(t)$] by using the input u(t) and output y(t) through a configuration shown in Fig. 8.3. As in the preceding section, let this reduced-order observer be evaluated such that it has the characteristic polynomial

$$q_{\hat{z}}(s) = s^{n-1} + q_1 s^{n-2} + \cdots + q_{n-2} s + q_{n-1} \tag{8.35}$$

giving the desired set of eigenvalues. This reduced-order observer for z(t) is chosen to be in the general form as in (8.13):

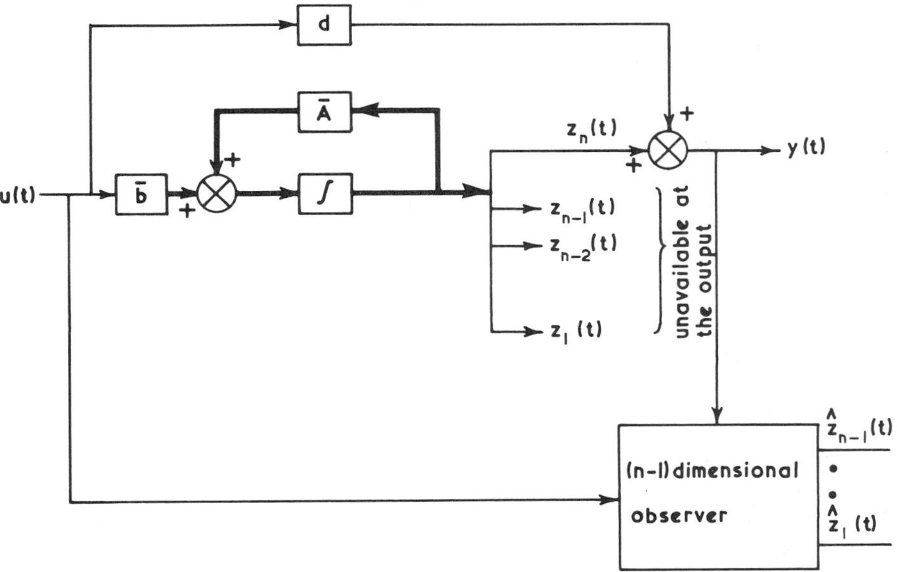

Fig. 8.3 Configuration for the reduced-order observer for single-input/single-output systems.

$$\dot{\hat{z}}(t) = \bar{L}\hat{z}(t) + \bar{m}u(t) + \bar{n}y(t) \equiv \bar{L}\hat{z}(t) + (\bar{T}\bar{b} - \bar{n}d)u(t) + \bar{n}y(t) \tag{8.36}$$

Assuming $\bar{L}$ in (8.36) to be in observable form, (8.35) implies

$$\bar{L} = \left.\underbrace{\begin{bmatrix} 0 & \cdots & 0 & -q_{n-1} \\ 1 & \cdots & 0 & -q_{n-2} \\ \vdots & & & \\ 0 & \cdots & 0 & -q_2 \\ 0 & \cdots & 1 & -q_1 \end{bmatrix}}_{(n-1)\text{ columns}}\right\} (n-1)\text{ rows} \tag{8.37}$$

The unknown parameter $\bar{n}$ in (8.36) is to be derived from (8.6), which in this case becomes

$$\bar{T}\bar{A} - \bar{L}\bar{T} = \bar{n}\bar{c} \tag{8.38}$$

and requires the knowledge of the transformation matrix $\bar{T}$, which for the sake of analytical convenience is chosen as an augmented version of $\bar{L}$ in (8.37):

$$\bar{T} = \left.\overbrace{\begin{bmatrix} \begin{matrix} 1 \\ 0 \\ \vdots \\ 0 \\ 0 \end{matrix} & \underbrace{\begin{matrix} 0 & \cdots & 0 & -q_{n-1} \\ 1 & \cdots & 0 & -q_{n-2} \\ & & & \\ 0 & \cdots & 0 & -q_2 \\ 0 & \cdots & 1 & -q_1 \end{matrix}}_{\bar{L}} \end{bmatrix}}^{n\text{ columns}}\right\} (n-1\text{ rows} \tag{8.39}$$

Using $\bar{A}$ and $\bar{c}$ as in (8.11) and (8.12), this choice of $\bar{T}$ yields

$$\bar{n}\bar{c} = \left.\begin{bmatrix} \underbrace{\begin{matrix} 0 & 0 & \cdots & 0 \\ 0 & 0 & \cdots & 0 \\ \vdots & & & \\ 0 & 0 & \cdots & 0 \\ 0 & 0 & \cdots & 0 \end{matrix}}_{\substack{(n-1)\text{ null} \\ \text{columns}}} & \vdots & \underbrace{\begin{matrix} -q_{n-1}(q_1 - \alpha_1) - \alpha_n \\ -q_{n-2}(q_1 - \alpha_1) - (\alpha_{n-1} + q_{n-1}) \\ \vdots \\ -q_2(q_1 - \alpha_1) - (\alpha_3 + q_3) \\ -q_2(q_1 - \alpha_1) - (\alpha_2 + q_2) \end{matrix}}_{\text{nonnull column}} \end{bmatrix}\right\} (n-1)\text{ rows}$$

$$\rightarrow \bar{n} = \begin{bmatrix} -q_{n-1}(q_1 - \alpha_1) - \alpha_n \\ -q_{n-2}(q_1 - \alpha_1) - (\alpha_{n-1} + q_{n-1}) \\ \vdots \\ -q_2(q_1 - \alpha_1) - (\alpha_3 + q_3) \\ -q_1(q_1 - \alpha_1) - (\alpha_2 + q_2) \end{bmatrix} \tag{8.40}$$

and by using $\bar{b}$ in (8.11) and (8.12),

$$\bar{T}\bar{b} - \bar{n}d = \begin{bmatrix} 1 & 0 & \cdots & 0 & -q_{n-1} \\ 0 & 1 & \cdots & 0 & -q_{n-2} \\ \vdots & & & & \\ 0 & 0 & \cdots & 0 & -q_2 \\ 0 & 0 & \cdots & 1 & -q_1 \end{bmatrix} \begin{bmatrix} \beta_n \\ \beta_{n-1} \\ \vdots \\ \beta_2 \\ \beta_1 \end{bmatrix} - \begin{bmatrix} -q_{n-1}(q_1 - \alpha_1) - \alpha_n \\ -q_{n-2}(q_1 - \alpha_1) - (\alpha_{n-1} + q_{n-1}) \\ \vdots \\ -q_2(q_1 - \alpha_1) - (\alpha_3 + q_3) \\ -q_1(q_1 - \alpha_1) - (\alpha_2 + q_2) \end{bmatrix} \tag{8.41}$$

In the special case when $d \equiv 0$, (8.41) becomes

$$\bar{T}\bar{b} = \begin{bmatrix} \beta_n - q_{n-1}\beta_1 \\ \beta_{n-1} - q_{n-2}\beta_1 \\ \vdots \quad \vdots \\ \beta_3 - q_2\beta_1 \\ \beta_2 - q_1\beta_1 \end{bmatrix} \tag{8.42}$$

Combining the equations above, the observer for the transformed (n - 1) state vector z(t) in (8.11) and (8.12) is given by (8.36). To obtain the required estimate $\hat{x}(t)$ of the original state vector x(t) of S(A, b, c, d), the transformation in Fig. 8.4 may be used, which is always valid since the transformation matrix $\bar{T}_1$ is derived from (8.39) as

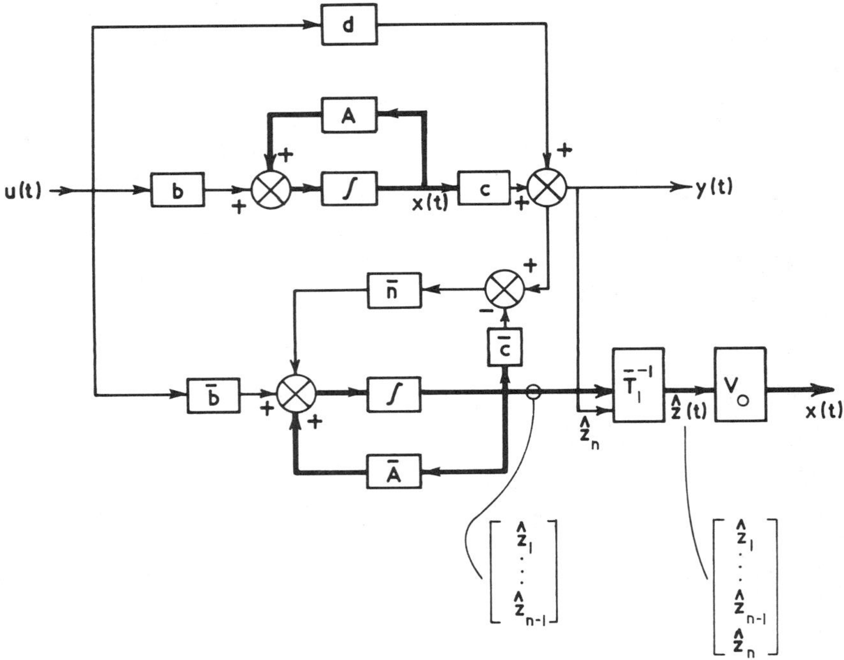

Fig. 8.4 Reduced-order (n - 1)-dimensional observer: single-input/single-output system

$$\bar{T}_1 = \begin{bmatrix} 1 & 0 & \cdots & 0 & -q_{n-1} \\ 0 & 1 & \cdots & 0 & -q_{n-2} \\ \vdots & & & & \\ 0 & 0 & \cdots & 0 & -q_2 \\ 0 & 0 & \cdots & 1 & q_1 \\ \cdots & \cdots & \cdots & \cdots & \cdots \\ 0 & 0 & \cdots & 0 & 1 \end{bmatrix} \begin{matrix} \left.\vphantom{\begin{matrix}1\\0\\ \vdots\\0\\0\end{matrix}}\right\} \bar{T} \\ \leftarrow \text{added row} \end{matrix} \quad \text{(n columns)} \tag{8.43}$$

is nonsingular. Asymptotic stability of the observer in (8.36) is assured due to the arbitrary choice of $q_{\hat{z}}(s)$ in (8.35) with the transient response of the error vector being controlled by the elements of $\bar{L}$ in (8.37). The development above thus provides a constructive procedure of deriving the reduced-order observer of a completely observable system. This derivation is summed up in the following theorem.

Theorem 8.3 For any completely observable single-input/single-output system of order n, there exists an asymptotically stable observer of (n - 1). An illustrative example is given below.

Example 8.3: Reduced-order observer—single-input/single-output system The system in Example 8.1 is considered here to illustrate the method of deriving an (n - 1)th-order asymptotically stable observer. Let the eigenvalues of the required reduced-order [≡ (n - 1) = 2] observer be at -3 and -4. Then the desired characteristic polynomial [from (8.35)] and $\bar{T}_1$ [from (8.43)] are:

$$q_{\hat{z}}(s) = \det\{sI - (\bar{A} - \bar{n}\bar{c})\} = s^2 + 7s + 12 \equiv s^2 + q_1 s + q_2 \tag{8.44}$$

$$\bar{T}_1 = \left[\begin{array}{ccc} & \bar{T} & \\ \hline 0 & 0 & 1 \end{array}\right] = \left[\begin{array}{c|cc} 1 & & \\ 0 & \bar{L} & \\ \hline 0 & 0 & 1 \end{array}\right] = \left[\begin{array}{c|cc} 1 & 0 & -q_2 \\ 0 & 1 & -q_1 \\ \hline 0 & 0 & 1 \end{array}\right] \equiv \begin{bmatrix} 1 & 0 & -12 \\ 0 & 1 & -7 \\ 0 & 0 & 1 \end{bmatrix} \tag{8.45}$$

The parameters of the reduced-order observer are, therefore, from (8.22), (8.40), (8.41), and (8.42),

$$\bar{n} = \begin{bmatrix} -q_2(q_1 - \alpha_1) - \alpha_3 \\ -q_1(q_1 - \alpha_1) - (\alpha_2 + q_2) \end{bmatrix} = \begin{bmatrix} -86 \\ -28 \end{bmatrix} \quad \text{and} \quad \bar{T}\bar{b} = \begin{bmatrix} \beta_3 - q_2\beta_1 \\ \beta_2 - q_1\beta_1 \end{bmatrix} = \begin{bmatrix} -9 \\ -5 \end{bmatrix} \tag{8.46}$$

The reduced-order observer is therefore given by (8.36):

$$\dot{\hat{z}}(t) = \begin{bmatrix} 0 & -12 \\ 1 & -7 \end{bmatrix} \hat{z}(t) + \begin{bmatrix} -9 \\ -5 \end{bmatrix} u(t) + \begin{bmatrix} -86 \\ -28 \end{bmatrix} y(t) \tag{8.47}$$

and from Fig. 8.4, the appropriate inverse transformation to obtain the estimate of the original system $\hat{x}(t)$ from above being

$$\hat{x}(t) = V_0 \bar{T}_1^{-1} \begin{bmatrix} \hat{z}(t) \\ y(t) \end{bmatrix} = \frac{1}{6} \begin{bmatrix} 1 & 1 & 7 \\ 0 & 3 & 0 \\ 0 & 0 & 6 \end{bmatrix} \begin{bmatrix} 1 & 0 & 12 \\ 0 & 1 & 7 \\ 0 & 0 & 1 \end{bmatrix} \begin{bmatrix} \hat{z}_1(t) \\ \hat{z}_2(t) \\ y(t) \end{bmatrix}$$

$$\equiv \frac{1}{6} \begin{bmatrix} 1 & 1 & 26 \\ 0 & 3 & 21 \\ 0 & 0 & 6 \end{bmatrix} \begin{bmatrix} \hat{z}_1(t) \\ \hat{z}_2(t) \\ y(t) \end{bmatrix} \tag{8.48}$$

(b) Multivariable system

In this case, since the output variables $y_1(t), \ldots, y_m(t)$ are derived by linear combinations of the n state variables $x_1(t), \ldots, x_n(t)$, by extending the development of the single-input/single-output system, it may be concluded that the reduced-order observer need only be of dimension (n - m) provided that the original system S(A, B, C, D) is completely observable (i.e., if the m rows of the output matrix C are linearly independent). This is stated through the following theorem.

Theorem 8.4 If the nth-order m-output system S(A, B, C, D) is completely observable, there exists an (n - m)-dimensional asymptotically stable observer.

Proof: Since direct transmission matrix D does not influence the stability of the estimator, it is not included in the following derivations, although the final results may easily be modified for $D \neq 0$. The derivations [3] here are based on direct computation of the observer parameters (i.e., without any transformation to the canonical form). The proof is followed by a synthesis procedure through observable form and normalization.

Assuming that C has full rank, an (n - m)-dimensional vector p(t) is introduced, where

$$p(t) = C_1 x(t) \tag{8.49}$$

such that

$$\rho \begin{bmatrix} C \\ \\ C_1 \end{bmatrix} = n$$

Thus combining (8.49) and the output equation of S(A, B, C),

$$y(t) = Cx(t) \tag{8.50}$$

the n-state vector of S(·) may be expressed as

$$\begin{aligned} x(t) &= \begin{bmatrix} C \\ --- \\ C_1 \end{bmatrix}^{-1} \begin{bmatrix} y(t) \\ --- \\ p(t) \end{bmatrix} \equiv [E_1 \mid E_2] \begin{bmatrix} y(t) \\ \\ p(t) \end{bmatrix} \\ &\equiv E_1 y(t) + E_2 p(t) \end{aligned} \tag{8.51}$$

where the partitioned matrices E_1 and E_2 are defined for convenience.

Combining the equations above, the following expressions may be derived:

$$\dot{y}(t) = C\dot{x}(t) = CAx(t) + CBu(t) \equiv CAE_1y(t) + CAE_2p(t) + CBu(t) \tag{8.52a}$$

$$\dot{p}(t) = C_1\dot{x}(t) = C_1Ax(t) + C_1Bu(t) \equiv C_1AE_1y(t) + C_1AE_2p(t) + C_1Bu(t) \tag{8.52b}$$

Since the input control u(t) and the output variable y(t) are available from S(·), the vector $\dot{y}(t)$ can be derived externally; (8.49) may be used to generate the (n - m)-vector p(t). Once p(t) is derived, the (n - m) set of differential equations of (8.52b) may be realized. Since p(t) is a state vector not available for measurement, a full-order [i.e., (n - m)-dimensional] identity observer may be constructed on the basis of (8.52b) by following Theorem 8.2. Let this reconstructed vector of p(t) be denoted by $\hat{p}(t)$. If $\hat{x}(t)$ denotes the complete reconstructed state vector of the original system S(·), then from (8.51), the desired reconstructed state vector is

$$\hat{x}(t) = E_1y(t) + E_2\hat{p}(t) \tag{8.53}$$

The only constraint in the reconstruction of $\hat{p}(t)$ on the basis of (8.52b) is that the pair $\{CAE_2, C_1AE_2\}$ is observable. This is always true since the submatrices C, C_1 and E_1, E_2 have been chosen to be linearly independent. Derivation of the observer from (8.52b) is based on the requirement that $\hat{p}(t)$ should be as close as possible to p(t), $\forall t$. However, since p(t) is to be used in evaluating $\hat{x}(t)$, and is dependent on the output y(t) of S(·), the error in the estimation of p(t) should be no more than the error in the estimation of $\hat{x}(t)$. This yields

$$\hat{p}(t) - p(t) = \hat{N}[y(t) - C\hat{x}(t)] \tag{8.54}$$

where $\hat{N}$ is a constant gain matrix. Combining (8.52) and (8.54), we have

$$\begin{aligned}\dot{\hat{p}}(t) &= \dot{p}(t) + \hat{N}[\dot{y}(t) - C\dot{\hat{x}}(t)] \\ &\equiv \underbrace{[C_1A - \hat{N}CA]}_{(L)}E_2p(t) + \underbrace{[C_1A - \hat{N}CA]}_{(N)}E_1y(t) + \hat{N}\dot{y}(t) + \underbrace{[C_1 - \hat{N}C]}_{(M)}Bu(t)\end{aligned} \tag{8.55}$$

where the bracketed terms underneath correspond to the estimator variables in (8.2) for $T = I_{n-m}$.

Equation (8.55), although valid in theoretical terms, may not be acceptable in practical realization due to the requirement of generating $\dot{y}(t)$ from the system output equation. This constraint may be removed by choosing another variable q(t) as

$$q(t) = \hat{p}(t) - \hat{N}y(t) \tag{8.56}$$

which when substituted in (8.55) gives

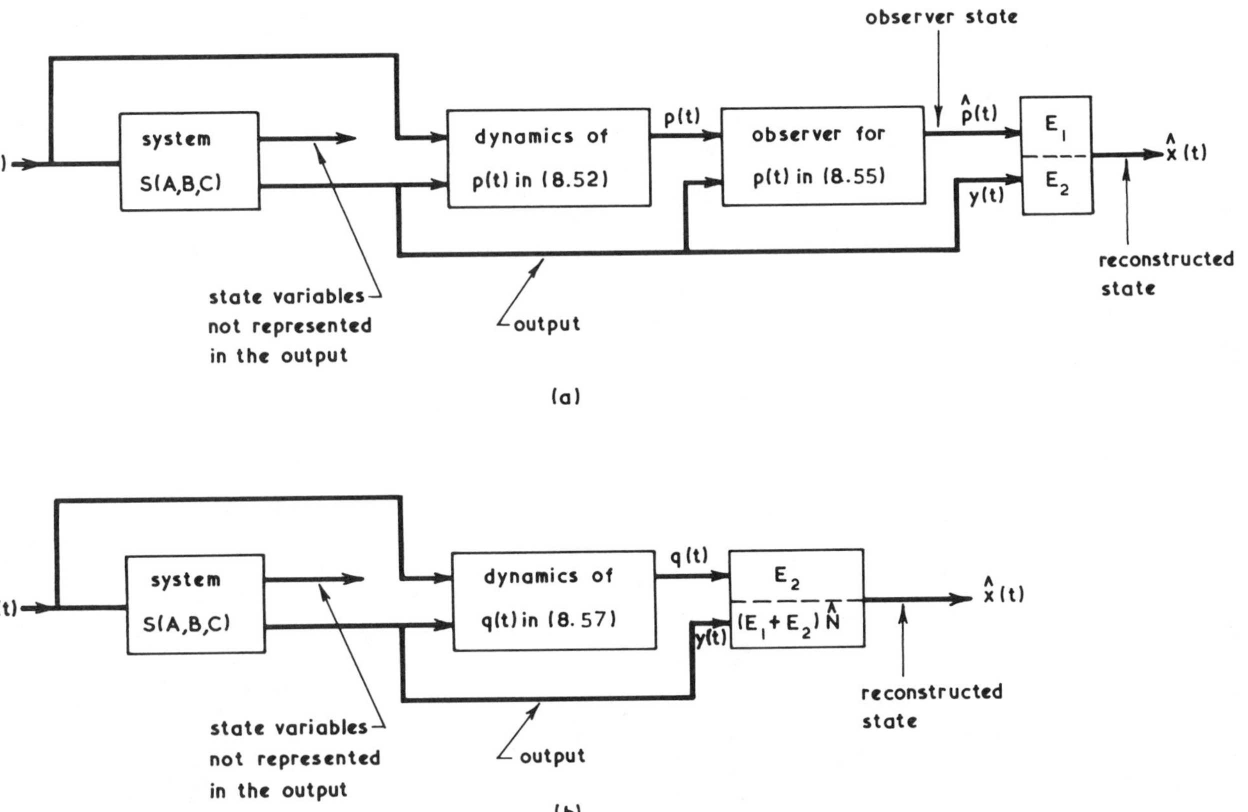

Fig. 8.5 Structure of reduced-order observers without transformation: (a) through the generation of p(t) and its estimation $\hat{p}(t)$; (b) through the generation of q(t).

$$\dot{q}(t) = [C_1A - \hat{N}CA]E_2q(t) + [\{C_1A - \hat{N}CA\}E_2\hat{N} + \{C_1A - \hat{N}CA\}E_1]y(t) + [C_1 - \hat{N}C]Bu(t) \tag{8.57}$$

and does not contain $\dot{y}(t)$. The estimated state vector is then obtained by first solving (8.57) for q(t) and then using (8.56) and (8.53), and has the form

$$\hat{x}(t) = [E_1 + E_2]\hat{N}y(t) + E_2q(t) \tag{8.58}$$

where q(t) is an (n - m) state vector to be constructed externally, as shown in Fig. 8.5. This completes the proof of the theorem.

(c) Synthesis through observable form [8-12]

The derivations above provide a two-stage procedure for the evaluation of the reduced-order observer parameters. The matrices N and M are first computed for desired observer dynamics and a suitable choice of C_1 from (8.55). In the second stage, the (n - m)-order state vector q(t) in (8.57) is generated by using u(t) and y(t) as input controls. The estimated state $\hat{x}(t)$ is then given by (8.58), with E_1 and E_2 as defined in (8.51). This procedure, although fairly simple in conceptual terms, may not be attractive for numerical computation, due mainly to the absence of a simple structural relationship between the observer dynamics and the dynamics of the original system S(·) [this is because of the general form of (8.55)]. This may be overcome if the method of deriving the reduced-order observers for single-input/single-output systems, considered earlier, is extended to multivariable systems. An alternative proof of Theorem 8.4 and a synthesis procedure based on an observable transformation is given below.

By using the notations as in (8.9) and (8.10), the observable form of a multivariable system S(·) may be expressed as

$$\dot{z}(t) = \bar{A}z(t) + \bar{B}u(t); \quad y(t) = \bar{C}z(t) \tag{8.59}$$

where S(·) is assumed to be completely observable [i.e., z(t) is the n-observable state vector] and the matrices $\bar{A}$ and $\bar{C}$ have the following special form (from Secs. 3.2 and 4.4):

$$\bar{A} = V_o^{-1}AV_o = \begin{bmatrix} \bar{A}_{11} & \cdots & \bar{A}_{1j} & \cdots & \bar{A}_{1m} \\ \vdots & & & & \\ \bar{A}_{il} & \cdots & \bar{A}_{ij} & \cdots & \bar{A}_{im} \\ \vdots & & & & \\ \bar{A}_{m1} & & \bar{A}_{mj} & \cdots & \bar{A}_{mm} \end{bmatrix} \text{ with } A_{ii} = \begin{bmatrix} 0 & 0 & \cdots & 0 & -\alpha^i_{ni} \\ 1 & 0 & \cdots & 0 & -\alpha^i_{ni-1} \\ \vdots & & & & \\ 0 & 0 & \cdots & 0 & -\alpha^i_2 \\ 0 & 0 & \cdots & 1 & -\alpha^i_1 \end{bmatrix} (ni \times ni) \tag{8.60a}$$

$$
A_{ij} = \begin{bmatrix} 0 & 0 & \cdots & 0 & \times \\ 0 & 0 & \cdots & 0 & \times \\ \vdots & & & & \\ 0 & 0 & \cdots & 0 & \times \\ 0 & 0 & \cdots & 0 & \times \end{bmatrix} \quad (ni \times nj)
$$

↑ nonnull column

$$
\bar{C} = CV_o = \left[\underbrace{\begin{matrix} 0 & \cdots & 0 & 1 \\ 0 & \cdots & 0 & 0 \\ \vdots & & & \\ 0 & \cdots & 0 & 0 \end{matrix}}_{\bar{C}_1} \;\vdots\; \underbrace{\begin{matrix} 0 & \cdots & 0 & 0 \\ 0 & \cdots & 0 & 1 \\ \vdots & & & \\ 0 & \cdots & 0 & 0 \end{matrix}}_{\bar{C}_i} \;\vdots\; \underbrace{\begin{matrix} 0 & \cdots & 0 & 0 \\ 0 & \cdots & 0 & 0 \\ \vdots & & & \\ 0 & \cdots & 0 & 1 \end{matrix}}_{\bar{C}_m}\right] \tag{8.60b}
$$

where $\bar{C}_i$ is a submatrix of order $m \times ni$ with its ith row equal to the ith row of I_m and all other rows being null, $i \in 1, \ldots, m$. $\bar{B} = V_o^{-1}B$ has no special form. The integer ni denotes the degree of the characteristic polynomial of the subsystem $\bar{S}_i(\bar{A}_{ii}, \bar{B}_i, \bar{C}_i)$:

$$
p_i(s) = s^{ni} + \alpha_1^i s^{ni-1} + \cdots + \alpha_{ni-1}^i s + \alpha_{ni}
$$

no (= maximum ni) is the observability index of S(A, B, C), with $n = \Sigma_{i=1}^{m} ni$.

Thus, in the multivariable case, the effect of the transformation in (8.9) is to convert the system S(A, B, C) into a set of m single-input/single-output subsystems $\bar{S}^i(A_{ii}, B_i, C_i)$ of order ni, $i \in 1, \ldots, m$:

$$
\begin{bmatrix} \dot{z}_1^i \\ \dot{z}_2^i \\ \vdots \\ \dot{z}_{ni-1}^i \\ \dot{z}_{ni}^i \end{bmatrix} = \begin{bmatrix} 0 & \cdots & 0 & -\alpha_{ni}^i \\ 1 & \cdots & 0 & -\alpha_{ni-1}^i \\ \vdots & & & \\ 0 & \cdots & 0 & -\alpha_2^1 \\ 0 & \cdots & 1 & -\alpha_1^i \end{bmatrix} \begin{bmatrix} z_1^i \\ z_2^i \\ \vdots \\ z_{ni-1}^i \\ z_{ni}^i \end{bmatrix} + \begin{bmatrix} b_1^i \\ b_2^i \\ \vdots \\ b_{ni-1}^i \\ b_{ni}^i \end{bmatrix} u_i(t), \tag{8.61}
$$

$$y_i(t) = [0 \;\; 0 \;\; \cdots \;\; 0 \;\; 1] \begin{bmatrix} z_1^i \\ z_2^i \\ \vdots \\ z_{ni-1}^i \\ z_{ni}^i \end{bmatrix}$$

Thus by using Theorem 8.3, an asymptotically stable observer of order $(ni - 1)$ may be constructed for (8.61) for any $i \in 1, \ldots, m$. Hence the total order of state estimator for all these m subsystems is

$$\sum_{i=1}^{m} (ni - 1) = n - m$$

which proves Theorem 8.4.

Thus if $\hat{z}(t)$ denotes the $(n - m)$ estimator state vector related to z(t) by

$$\hat{z}(t) = \bar{T}z(t) \tag{8.62}$$

the observer dynamics may be expressed by extending (8.36):

$$\dot{\hat{z}}(t) = \bar{L}\hat{z}(t) + \bar{M}u(t) + \bar{N}y(t) \tag{8.63a}$$

where

$$\bar{M} = \bar{T}\bar{B}$$
$$\bar{T}\bar{A} - \bar{L}\bar{T} = \bar{N}\bar{C} \tag{8.63b}$$

The synthesis procedure may be developed by extending the single-input/single-output method through first choosing $\bar{L}$ as [from (8.37)]

$$\bar{L} = \begin{bmatrix} \bar{L}_1 & & 0 \\ & \bar{L}_i & \\ 0 & & \bar{L}_m \end{bmatrix} \quad \text{with } \bar{L}_i = \underbrace{\begin{bmatrix} 0 & \cdots & 0 & -r_{ni-1}^i \\ 1 & \cdots & 0 & -r_{ni-2}^i \\ \vdots & & & \\ 0 & \cdots & 0 & -r_2^i \\ 0 & \cdots & 1 & -r_1^i \end{bmatrix}}_{(ni-1)\text{ columns}} \Bigg\} (ni - 1) \text{ rows} \tag{8.64}$$

where the parameters of $\bar{L}_i$ are chosen to achieve the desired dynamics of the ith subsystem of (8.63).

The next stage is to formulate the transformation matrix $\bar{T}$ as [from (8.39)]:

$$\bar{T} = \begin{bmatrix} \bar{T}_1 & & 0 \\ & \bar{T}_i & \\ 0 & & \bar{T}_m \end{bmatrix} \quad \text{with } \bar{T}_i = \overbrace{\begin{bmatrix} 1 & 0 & \cdots & 0 & -r^i_{ni-1} \\ 0 & 1 & \cdots & 0 & -r^i_{ni-2} \\ \vdots & & & & \\ 0 & 0 & \cdots & 0 & -r^i_2 \\ 0 & \underbrace{0 & \cdots & 1 & -r^i_1}_{\bar{L}_i} \end{bmatrix}}^{ni \text{ columns}} \left.\vphantom{\begin{bmatrix}1\\0\\\vdots\\0\\0\end{bmatrix}}\right\} (ni - 1) \text{ rows}$$

$$\equiv [I_{ni-1} \quad r^i] \tag{8.65}$$

where I_{ni-1} is an (ni - 1)-square identity matrix and r^i is a (ni - 1) column vector specifying the desired characteristic polynomial associated with $\bar{L}_i$:

$$r_i(s) = s^{ni-1} + r^i_1 s^{ni-2} + \cdots + r^i_{ni-2} s + r^i_{ni-1} \tag{8.66}$$

Thus if $\bar{N}$ in (8.63) is expressed as a collection of m column vectors,

$$\bar{N} = [\bar{n}_1 \ \cdots \ \bar{n}_i \ \cdots \ \bar{n}_m] \tag{8.67}$$

Combining (8.39), (8.63), (8.64), and (8.65) and (8.66) gives

$$\bar{N}\bar{C}_i = \bar{T}_i\bar{A}_{ii} - \bar{L}_i\bar{T}_i, \qquad i \in 1, \ldots, m \tag{8.68}$$

which may be solved for the (n - m) × m elements of $\bar{N}$.

Combination of the equations above defines the observer for the transformed system in (8.59). Once $\hat{z}(t)$ is obtained, the estimate $\hat{x}(t)$ for the original system x(t) may be obtained by using an inverse transformation of the form in Fig. 8.6, which may be expressed as

$$\hat{x}(t) = V_o \begin{bmatrix} \bar{T} \\ \bar{C} \end{bmatrix}^{-1} \begin{bmatrix} \hat{z}(t) \\ y(t) \end{bmatrix} \equiv V_o[E_1 \ \ E_2] \begin{bmatrix} \hat{z}(t) \\ y(t) \end{bmatrix} \quad \text{with } \begin{bmatrix} \bar{T} \\ \bar{C} \end{bmatrix}^{-1} = [E_1 \ \ E_2]$$

$$\equiv F_1\hat{z}(t) + F_2 y(t) \tag{8.69}$$

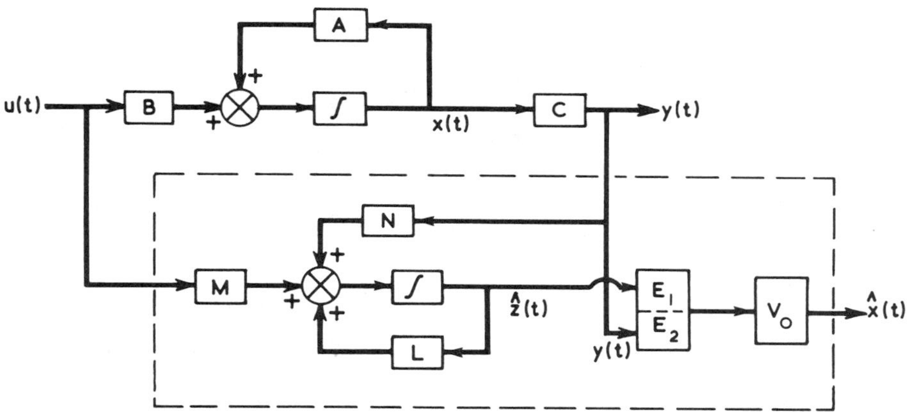

Fig. 8.6 Reduced-order observer through transformation.

where $\hat{z}(t)$ is derived by solving (8.63). An illustrative example is given below [12].

Example 8.4: Reduced-order observer—multivariable system The system is described by

$$S(A, B, C): \; A = \begin{bmatrix} -2 & 1 & 0 & 0 \\ 0 & -2 & 1 & 0 \\ 0 & 0 & -1 & 1 \\ -1 & 0 & 0 & 0 \end{bmatrix}; \quad B = \begin{bmatrix} 0 & 0 \\ 0 & 0 \\ 0 & 1 \\ 1 & 0 \end{bmatrix}; \quad C = \begin{bmatrix} 1 & 0 & 0 & 0 \\ 0 & 0 & 1 & 0 \end{bmatrix} \tag{8.70}$$

For this system, $n_1 = n_2 = 2$; and a suitable choice

$$V_o = \begin{bmatrix} 0 & 1 & 0 & 0 \\ 1 & -2 & 0 & 0 \\ 0 & 0 & 0 & 1 \\ 0 & 0 & 1 & 0 \end{bmatrix} \quad \text{with } V_o^{-1} = \begin{bmatrix} 2 & 1 & 0 & 0 \\ 1 & 0 & 0 & 0 \\ 0 & 0 & 0 & 1 \\ 0 & 0 & 1 & 0 \end{bmatrix} \tag{8.71}$$

gives

$$\bar{A} = \left[\begin{array}{cc|cc} 0 & -4 & 0 & 1 \\ 1 & -4 & 0 & 0 \\ \hline 0 & -1 & 0 & 0 \\ 0 & 0 & 1 & -1 \end{array}\right] = \left[\begin{array}{c|c} \bar{A}_{11} & \bar{A}_{12} \\ \hline \bar{A}_{21} & \bar{A}_{22} \end{array}\right] \equiv \left[\begin{array}{cc|cc} 0 & -\alpha_2^1 & \multicolumn{2}{c}{\multirow{2}{*}{$\bar{A}_{12}$}} \\ 1 & -\alpha_1^1 & & \\ \hline \multicolumn{2}{c|}{\multirow{2}{*}{$\bar{A}_{21}$}} & 0 & -\alpha_2^2 \\ & & 1 & -\alpha_1^2 \end{array}\right]$$

$$\bar{C} = \begin{bmatrix} 0 & 1 & 0 & 0 \\ 0 & 0 & 0 & 1 \end{bmatrix} = [C_1 \mid C_2] \quad \text{and} \quad \bar{B} = \begin{bmatrix} 0 & 0 \\ 0 & 0 \\ 1 & 0 \\ 0 & 1 \end{bmatrix} \tag{8.72}$$

Let the dynamics of the reduced-order (n - m = 2) observer be specified by

$$\bar{L} = \text{diag}\,\{\bar{L}_1, \bar{L}_2\} = \begin{bmatrix} -r_1^1 & 0 \\ 0 & -r_1^2 \end{bmatrix} = \begin{bmatrix} -3 & 0 \\ 0 & -3 \end{bmatrix} \tag{8.73}$$

Then from (8.65) and (8.66),

$$\bar{T} = \text{diag}\,\{\bar{T}_1, \bar{T}_2\} = \left[\begin{array}{c|c|c|c} 1 & \overbrace{-r_1^1}^{\bar{L}_1} & 0 & 0 \\ \hline 0 & 0 & 1 & \underbrace{-r_1^2}_{\bar{L}_2} \end{array}\right] = \begin{bmatrix} 1 & -3 & 0 & 0 \\ 0 & 0 & 1 & -3 \end{bmatrix} \tag{8.74}$$

Consequently, from (8.63),

$$\bar{M} = \bar{T}\bar{B} = \begin{bmatrix} 0 & 0 \\ 1 & -3 \end{bmatrix} \tag{8.75}$$

and

$$\bar{N}\bar{C} = \begin{bmatrix} \bar{n}_{11} & \bar{n}_{12} \\ \bar{n}_{21} & \bar{n}_{22} \end{bmatrix} \begin{bmatrix} 0 & 1 & 0 & 0 \\ 0 & 0 & 0 & 1 \end{bmatrix} = \bar{T}\bar{A} - \bar{L}\bar{T} = \begin{bmatrix} 0 & -1 & 0 & 1 \\ 0 & -1 & 0 & -6 \end{bmatrix}$$

$$\rightarrow \bar{N} = \begin{bmatrix} -1 & 1 \\ -1 & -6 \end{bmatrix} \tag{8.76}$$

The observer in (8.63) is thus

$$\begin{bmatrix} \dot{\hat{z}}_1(t) \\ \dot{\hat{z}}_2(t) \end{bmatrix} = \begin{bmatrix} -3 & 0 \\ 0 & -3 \end{bmatrix} \begin{bmatrix} z_1(t) \\ z_2(t) \end{bmatrix} + \begin{bmatrix} 0 & 0 \\ 1 & -3 \end{bmatrix} u(t) + \begin{bmatrix} -1 & 1 \\ -1 & -6 \end{bmatrix} y(t) \tag{8.77}$$

where $\hat{z}(t)$ is related to $z(t)$ by (8.62). From (8.71) to (8.74),

$$V_o \begin{bmatrix} \bar{T} \\ \\ \bar{C} \end{bmatrix}^{-1} = \begin{bmatrix} 0 & 1 & 0 & 0 \\ 1 & -2 & 0 & 0 \\ 0 & 0 & 0 & 1 \\ 0 & 0 & 1 & 0 \end{bmatrix} \begin{bmatrix} 1 & -3 & 0 & 0 \\ 0 & 0 & 1 & -3 \\ \hline 0 & 1 & 0 & 0 \\ 0 & 0 & 0 & 1 \end{bmatrix}^{-1} = \begin{bmatrix} \underbrace{\begin{matrix} 0 & 0 \\ 1 & 0 \\ 0 & 0 \\ 0 & 1 \end{matrix}}_{F_1} & \vdots & \underbrace{\begin{matrix} 1 & 0 \\ 1 & 0 \\ 0 & 1 \\ 0 & 3 \end{matrix}}_{F_2} \end{bmatrix} \tag{8.78}$$

and hence the estimated state vector $\hat{x}(t)$ of the original system is given by

$$\begin{bmatrix} \hat{x}_1(t) \\ \hat{x}_2(t) \\ \hat{x}_3(t) \\ \hat{x}_4(t) \end{bmatrix} = \begin{bmatrix} 0 & 0 \\ 1 & 0 \\ 0 & 0 \\ 0 & 1 \end{bmatrix} \begin{bmatrix} \hat{z}_1(t) \\ \hat{z}_2(t) \end{bmatrix} + \begin{bmatrix} 1 & 0 \\ 1 & 0 \\ 0 & 1 \\ 0 & 3 \end{bmatrix} \begin{bmatrix} y_1(t) \\ y_2(t) \end{bmatrix} \tag{8.79}$$

Step responses of the original state $x(t)$ in (8.70) and those of the estimated state $\hat{x}(t)$ obtained by combining (8.77) and (8.79) according to Fig. 8.6 are shown in Fig. 8.7 to illustrate the convergence of the estimated state to its actual behavior.

(d) Synthesis through normalization [10]

In some cases it may be more convenient to derive the modal form of $S(\cdot)$ rather than its observable form in (8.59). In such cases, derivations similar to those in (8.29) to (8.32) may be more suitable; these are generalized here for reduced-order observers.

Let $\hat{z}(t)$ be the reduced-order $(n - m)$ estimated state vector of the normalized system $S_n(\cdot)$ in (8.29), described by

$$\dot{\hat{z}}(t) = L_n \hat{z}(t) + M_n u(t) + N_n y(t) \tag{8.80}$$

Here the reduced-order estimated vector $\hat{z}(t)$ is related to the corresponding real-state vector by

$$\hat{z}(t) = Tz_n(t) \tag{8.81}$$

The error vector may therefore be expressed as

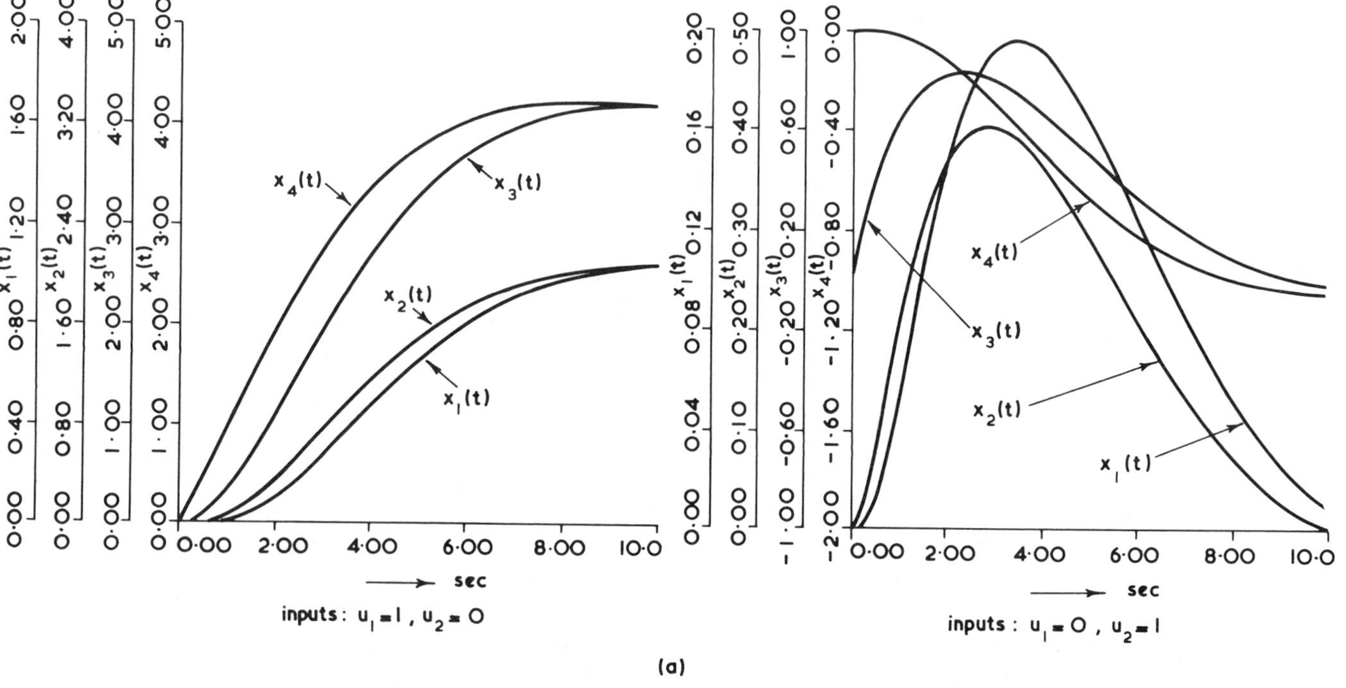

(a)

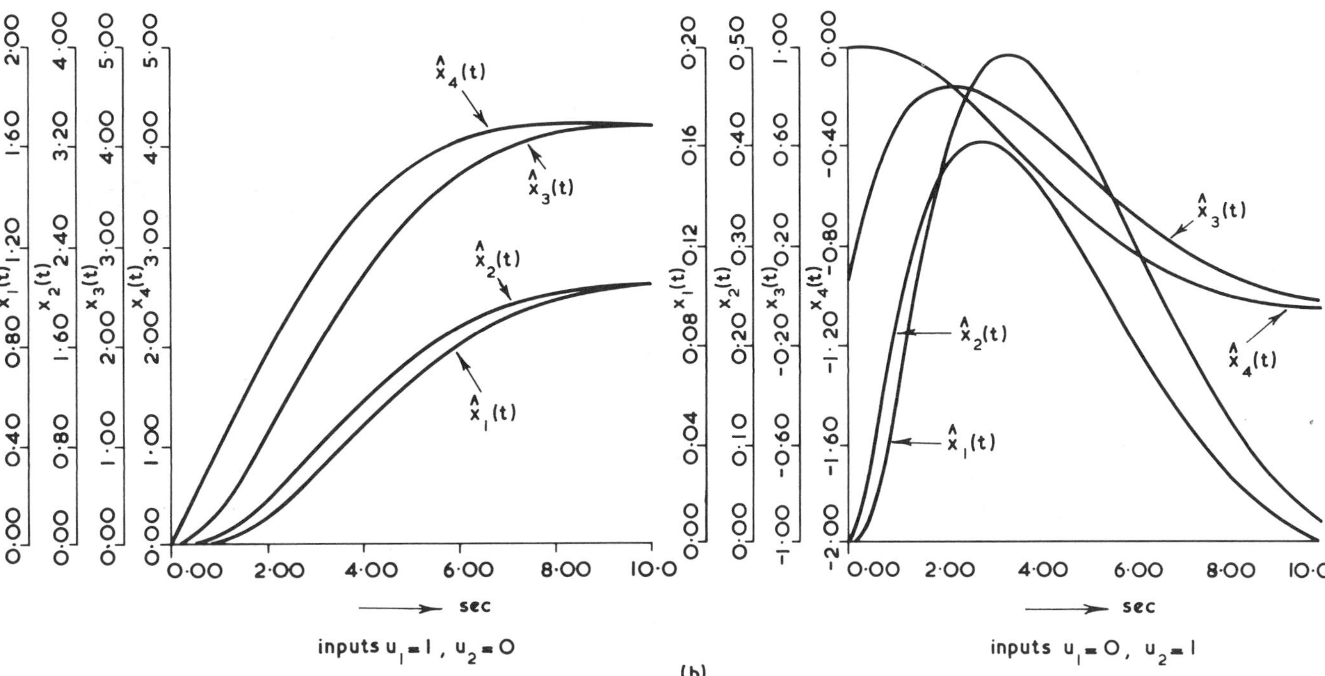

Fig. 8.7 Step responses of (a) the original state x(t) and (b) the observed state $\hat{x}(t)$ of the system in Example 8.4.

$$\dot{\hat{e}}_{\hat{z}}(t) = \dot{\hat{z}}(t) - T\dot{z}_n(t) \equiv L_n[\hat{z}(t) - Tz_n(t)] + [M_n - TB_n]u(t) + [L_nT + N_nC_n - TJ]z_n(t) \tag{8.82}$$

which yields the following conditions for zero-estimation error in steady state [i.e., for $e_{\hat{z}}(t) \to 0$ as $t \to \infty$]:

$$M_n = TB_n$$

and

$$L_nT + N_nC_n = TJ \tag{8.83}$$

The response of the observed reduced-order ($\equiv n - m$) state is given by

$$\hat{z}(t) = Tz_n(t) + e^{L_n t}[\hat{z}(t_0) - Tz_n(t_0)] \equiv Tz_n(t) + \eta(t) \tag{8.84}$$

where $\eta(t) \to 0$ as $t \to \infty$ if the observer in (8.80) and (8.81) is symptotically stable. Once $\hat{z}(t)$ is obtained by solving the equations above, the estimate $\hat{x}(t)$ of the original system state vector $x(t)$ may be obtained by using the configuration in Fig. 8.8, which gives

$$\hat{x}(t) = P\begin{bmatrix} T \\ -- \\ C_n \end{bmatrix}^{-1}\begin{bmatrix} \hat{z}(t) \\ y(t) \end{bmatrix} = P[E_1 \mid E_2]\begin{bmatrix} \hat{z}(t) \\ y(t) \end{bmatrix} = F_1\hat{z}(t) + F_2y(t) \tag{8.85}$$

where

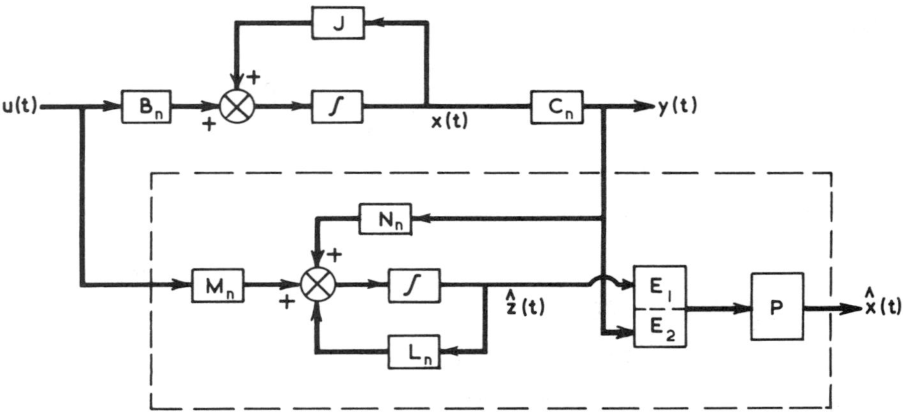

Fig. 8.8 Reduced-order observer through normalization.

$$\begin{bmatrix} T \\ -- \\ C_n \end{bmatrix}^{-1} = [E_1 \quad E_2] \tag{8.86}$$

which exists only if $\{C, A\}$ is completely observable.

The equations above may now be combined to compute the estimator parameters L_n, M_n, and N_n. The elements of the matrix L_n are to be specified by the requirements of exponential decay rate of the error vector in (8.82). In the special cases of (1) full-order observer, T is an n-square nonsingular matrix, (2) full-order identity observer, $T = I_n$, and in view of (8.86) it may not always be possible to evaluate a reduced-order identity observer. Let L_n be chosen as

$$L_n = \text{diag}\,\{\gamma_k\}, \quad k \in 1, 2, \ldots, \ell \tag{8.87}$$

where $\ell = n$ for a full-order observer and $\ell = (n - m)$ for a reduced-order observer. The elements γ_k are chosen to achieve real and imaginary eigenvalues of (8.80) which are different from the eigenvalues of $S(\cdot)$.

Let the $(\ell \times m)$ matrix N_n have the following structure:

$$N_n = \{n_{ki}\} = \begin{bmatrix} n_{11} & \cdots & n_{1i} & \cdots & n_{1m} \\ \vdots & & & & \\ n_{k1} & \cdots & n_{ki} & \cdots & n_{km} \\ \vdots & & & & \\ n_{\ell 1} & \cdots & n_{\ell i} & \cdots & n_{\ell m} \end{bmatrix} \tag{8.88}$$

Then by denoting

$$C_n = \{\bar{c}_{ij}\} = \begin{bmatrix} \bar{c}_{11} & \cdots & \bar{c}_{1j} & \cdots & \bar{c}_{1n} \\ \vdots & & & & \\ \bar{c}_{i1} & \cdots & \bar{c}_{ij} & \cdots & \bar{c}_{in} \\ \vdots & & & & \\ \bar{c}_{m1} & \cdots & \bar{c}_{mj} & \cdots & \bar{c}_{mn} \end{bmatrix} \text{ and } T = \{t_{kj}\} = \begin{bmatrix} t_{11} & \cdots & t_{ij} & \cdots & t_{1n} \\ \vdots & & & & \\ t_{i1} & \cdots & t_{ij} & & \\ \vdots & & & & \\ t_{\ell 1} & \cdots & t_{\ell j} & \cdots & t_{\ell n} \end{bmatrix} \tag{8.89}$$

from (8.83),

$$\{n_{ki}\}\{\bar{c}_{kj}\} = \{t_{kj}\}\,\text{diag}\,\{\lambda_j\} - \text{diag}\,\{\gamma_k\}\{t_{kj}\} \equiv \{\lambda_j t_{kj}\} - \{\gamma_k t_{kj}\} \tag{8.90}$$

and consequently,

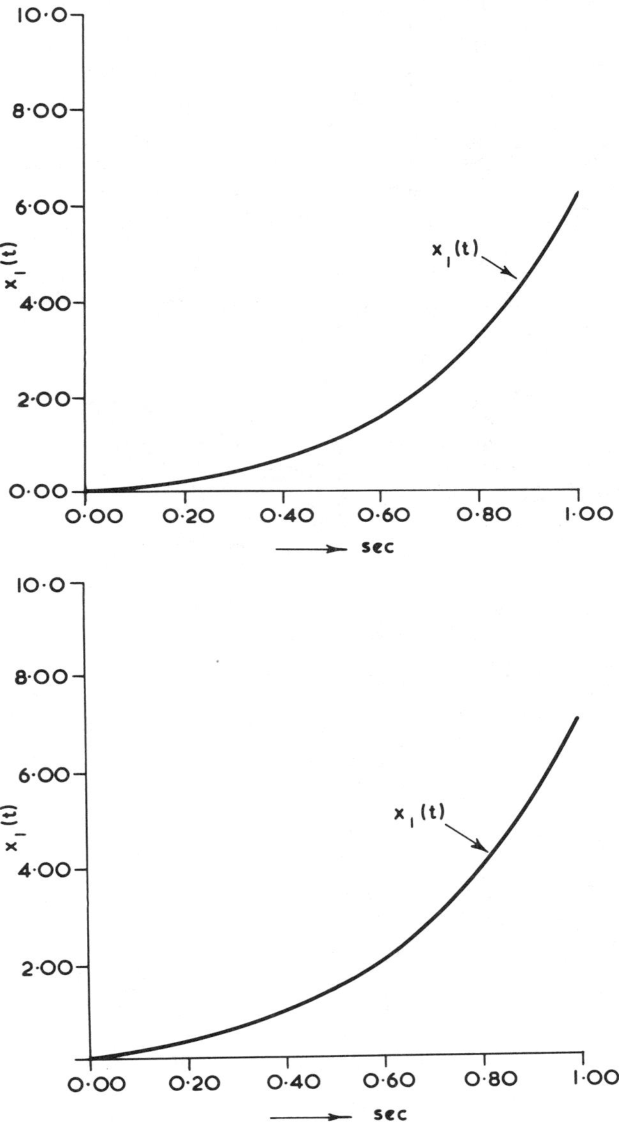

Fig. 8.9 Step responses of (a) the original state x(t) and (b) the observed state $\hat{x}(t)$ of Example 8.5 with reduced-order state estimator.

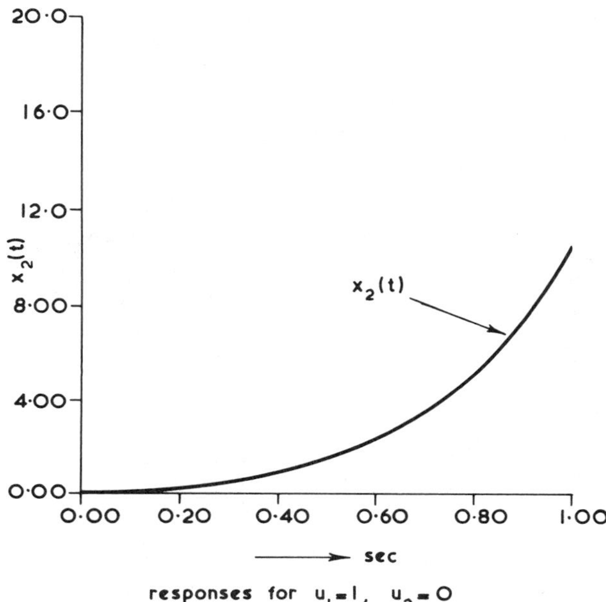

responses for $u_1 = 1$, $u_2 = 0$

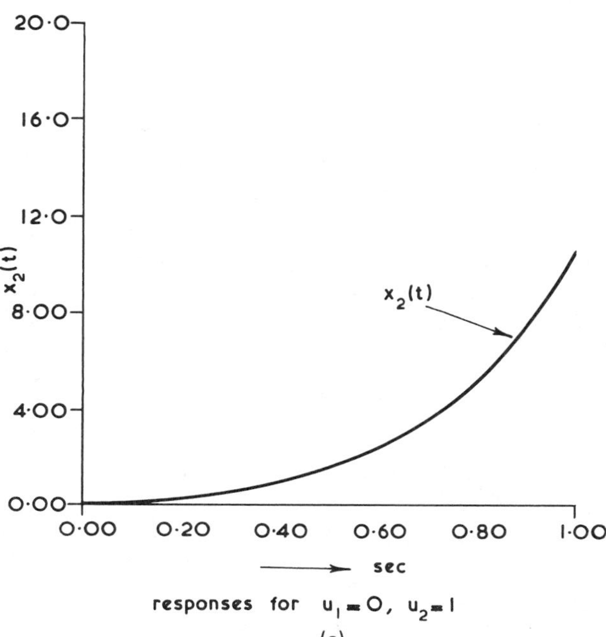

responses for $u_1 = 0$, $u_2 = 1$

(a)

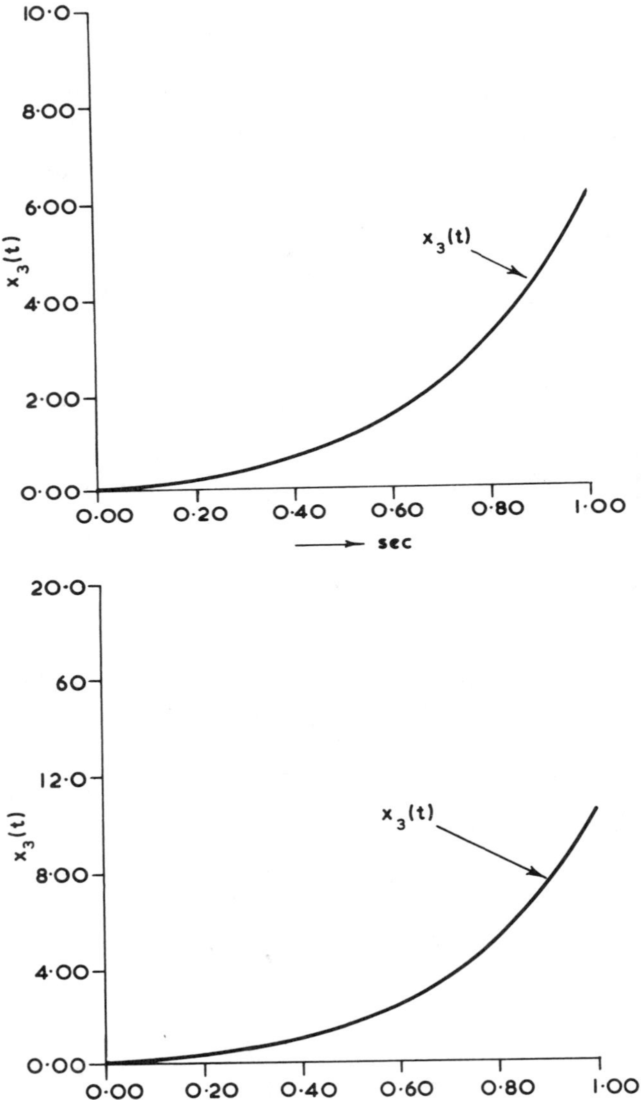
10·0
8·00
6·00
4·00
2·00
0·00
x3(t)
x3(t)
0·00
0·20
0·40
0·60
0·80
1·00
sec
20·0
60
12·0
8·00
4·00
0·00
x3(t)
x3(t)
0·00
0·20
0·40
0·60
0·80
1·00
sec

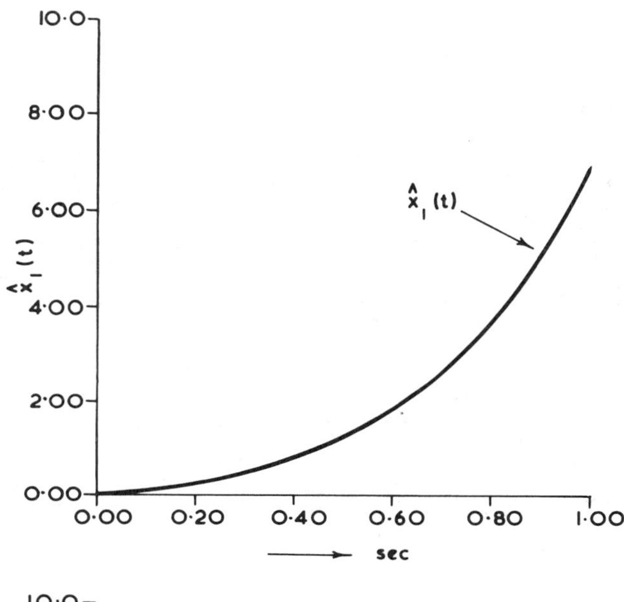
10.0
8.00
6.00
4.00
2.00
0.00
$\hat{x}_1(t)$
$\hat{x}_1(t)$
0.00
0.20
0.40
0.60
0.80
1.00
sec

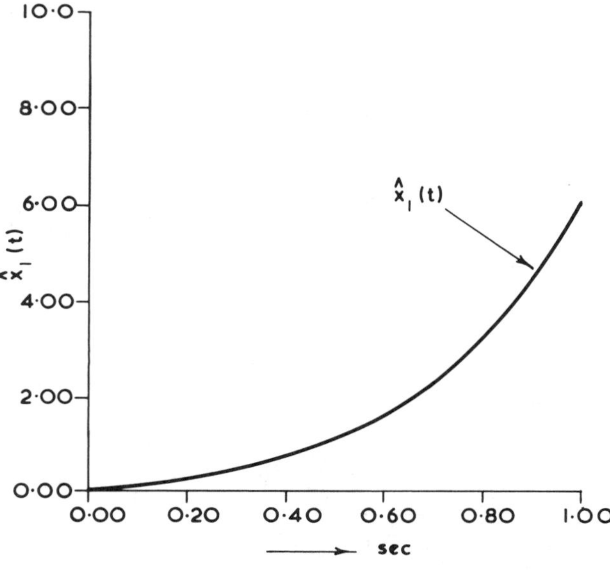
10.0
8.00
6.00
4.00
2.00
0.00
$\hat{x}_1(t)$
$\hat{x}_1(t)$
0.00
0.20
0.40
0.60
0.80
1.00
sec

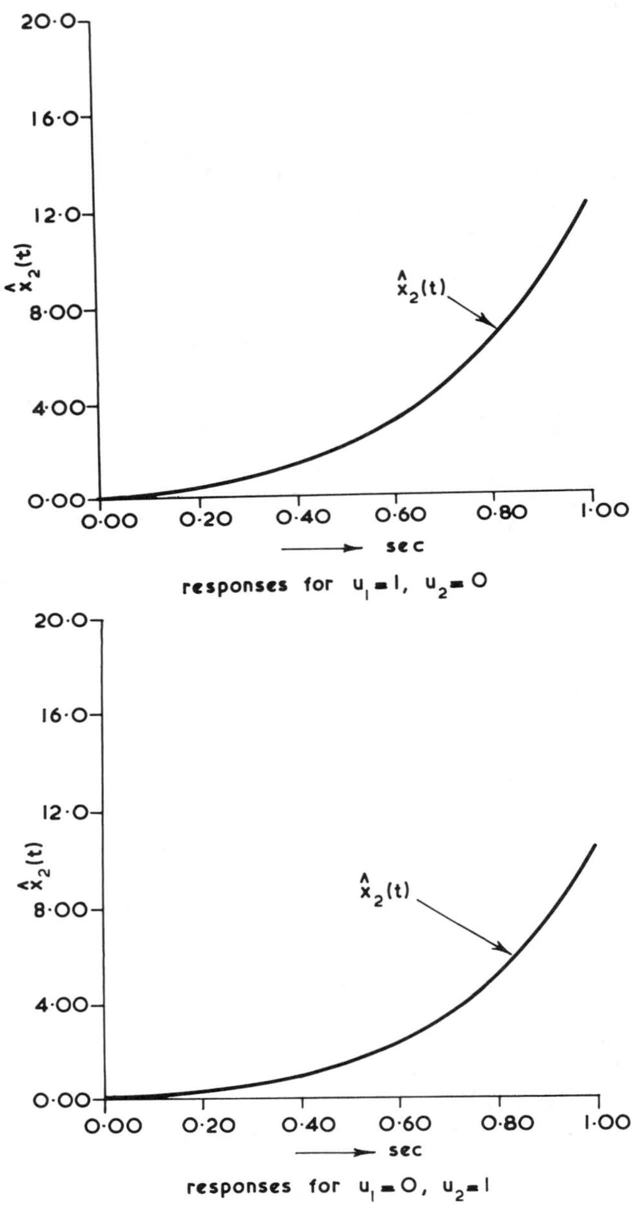
20·0
16·0
12·0
8·00
4·00
0·00
$\hat{x}_2(t)$
$\hat{x}_2(t)$
0·00
0·20
0·40
0·60
0·80
1·00
sec
responses for $u_1 = 1$, $u_2 = 0$
20·0
16·0
12·0
8·00
4·00
0·00
$\hat{x}_2(t)$
$\hat{x}_2(t)$
0·00
0·20
0·40
0·60
0·80
1·00
sec
responses for $u_1 = 0$, $u_2 = 1$

(b)

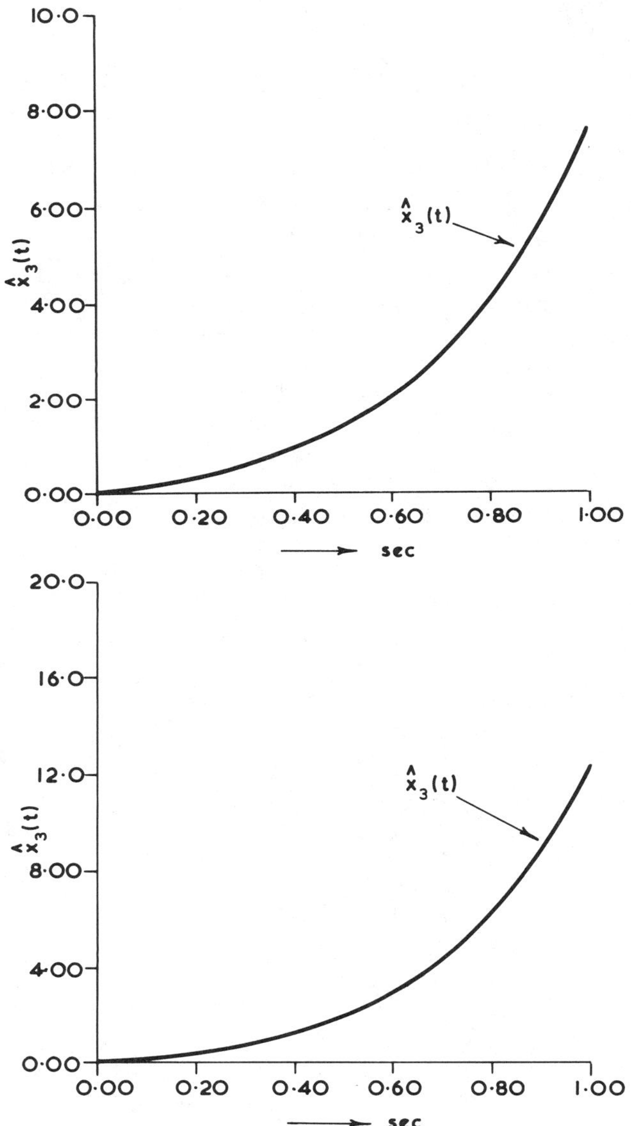
10·0
8·00
6·00
4·00
2·00
0·00
$\hat{x}_3(t)$
$\hat{x}_3(t)$
0·00 0·20 0·40 0·60 0·80 1·00
sec
20·0
16·0
12·0
8·00
4·00
0·00
$\hat{x}_3(t)$
$\hat{x}_3(t)$
0·00 0·20 0·40 0·60 0·80 1·00
sec

$$\{t_{ij}\} = \sum_{k=1}^{\ell} \frac{n_{ki}\bar{c}_{kj}}{\lambda_j - \gamma_i}, \quad \lambda_j \neq \gamma_i; \begin{cases} i = 1, 2, \ldots, m \\ j = 1, 2, \ldots, n \\ k = 1, 2, \ldots, \ell \end{cases} \tag{8.91}$$

Thus if the elements of N_n in (8.88) are chosen such that $\Sigma_k n_{ki}c_{ij} \neq 0$, $\forall$ i, j, k, then (8.91) may be used to derive the corresponding element t_{ij}. The following example demonstrates the method above.

Example 8.5: Reduced-order observer—multivariable system The system in Example 8.2 is considered here to illustrate the method of deriving reduced-order observer through modal transformation.

Since the system in (8.33) has two independent outputs, the required state observer will be a first-order dynamical system with state equation of the form

$$\begin{aligned} \dot{\hat{z}}(t) &= \gamma_1\hat{z}_1(t) + [m_{11} \;\; m_{12}]u(t) + [n_{11} \;\; n_{12}]y(t) \\ \hat{z}(t) &= Tz_n(t), \qquad \text{with } T \equiv [t_{11} \;\; t_{12} \;\; t_{13}] \end{aligned} \tag{8.92}$$

where γ_1, n_{11}, and n_{12} may be chosen arbitrarily: with $n_{11} = 1$, $n_{12} = 2$, and $\gamma_1 = -2$, from (8.83), (8.90), and (8.91):

$$[n_{11} \;\; n_{12}] \begin{bmatrix} 1 & -2 & 3 \\ 1 & 1 & 1 \end{bmatrix} = [t_{11} \;\; t_{12} \;\; t_{13}] \begin{bmatrix} 1 & 0 & 0 \\ 0 & -2 & 0 \\ 0 & 0 & 3 \end{bmatrix} + 2[t_{11} \;\; t_{12} \;\; t_{13}]$$

$$\rightarrow [t_{11} \;\; t_{12} \;\; t_{13}] = [1 \;\; -1 \;\; 1] \tag{8.93}$$

Thus

$$\begin{bmatrix} T \\ -- \\ C_n \end{bmatrix}^{-1} = \begin{bmatrix} 1 & -1 & 1 \\ \hline 1 & -2 & 3 \\ 1 & 1 & 1 \end{bmatrix}^{-1} = \frac{1}{4}\begin{bmatrix} 5 & -2 & 1 \\ -2 & 0 & 2 \\ -3 & 2 & 1 \end{bmatrix} \tag{8.94}$$

where the first column is F_1 and the last two columns are F_2.

From (8.33) and (8.83),

$$[m_{11} \;\; m_{12}] = [1 \;\; -1 \;\; 1] \begin{bmatrix} 0 & -1 \\ 0 & 0 \\ 1 & 2 \end{bmatrix} = [1 \;\; 1] \tag{8.95}$$

Combining the equations above, the first-order observer for the transformed system in (8.33) is given by [$\hat{z}(t)$ being a one-dimensional vector]

$$\hat{z}(t) = \gamma_1\hat{z}(t) + [n_{11}\ \ n_{12}]u(t) + [m_{11}\ \ m_{12}]y(t)$$

$$= -2\hat{z}(t) + [1\ \ 2]\begin{bmatrix} u_1(t) \\ u_2(t) \end{bmatrix} + [1\ \ 1]\begin{bmatrix} y_1(t) \\ y_2(t) \end{bmatrix} \tag{8.96}$$

The complete state vector of the original system in steady state is therefore given by from Example 8.2 and (8.85)

$$\begin{bmatrix} \hat{x}_1(t) \\ \hat{x}_2(t) \\ \hat{x}_3(t) \end{bmatrix} = P\begin{bmatrix} T \\ C_n \end{bmatrix}^{-1}\begin{bmatrix} \hat{z}(t) \\ y(t) \end{bmatrix} \equiv \frac{1}{4}\begin{bmatrix} -30 \\ 0 \\ 30 \end{bmatrix}\hat{z}(t) + \frac{1}{4}\begin{bmatrix} 4 & 22 \\ 0 & 4 \\ 0 & -26 \end{bmatrix}\begin{bmatrix} y_1(t) \\ y_2(t) \end{bmatrix} \tag{8.97}$$

Figure 8.9 shows the convergence of the step responses of the estimated state vector $\hat{x}(t)$ in (8.97) and the original state vector x(t) in (8.33).

8.4 CONTROL THROUGH OBSERVER [7-13]

The derivations in the previous sections provide various methods of constructing a state observer for a given linear time-invariant system S(A, B, C). If the constructed state vector $\hat{x}(t)$ is to be used to realize state feedback control laws considered in Chap. 6, it is essential to study the effect of the feedback of $\hat{x}(t)$ on the closed-loop system. Let the control law necessary for a specified set of closed-loop poles be

$$u(t) = Kx(t) + v(t) \tag{8.98}$$

In terms of the transformed state vector, the control law above takes the form

$$u(t) = \begin{cases} KV_o\bar{z}(t) + v(t) \rightarrow \text{for observable form in (8.59)} \\ KPz_n(t) + v(t) \rightarrow \text{for normal form in (8.29)} \end{cases}$$
$$\equiv \bar{K}z(t) + v(t) \rightarrow \text{in the general case} \tag{8.99}$$

which when combined with (8.69) or (8.85) gives*

$$u(t) = \bar{K}[F_1\hat{z}(t) + F_2y(t)] + v(t) \tag{8.100}$$

where F_1 and F_2 are related to the relevant transformation matrices V_o, P, and $\bar{T}$ or T. For convenience, subsequent derivations here use the general

*These derivations are based on the assumption that the observer is asymptotically stable [i.e., $\hat{z}(t) \rightarrow Tz(t)$ with $t \rightarrow \infty$].

notation $\bar{A}$, $\bar{B}$, $\bar{C}$ to denote both the observable and normal forms of the original system. Thus if the observer is of dimension ℓ, where

$$\ell = \begin{cases} n \text{ for full-order observer} \\ n - m \text{ for reduced-order observer} \end{cases}$$

then the equations of the closed-lop system in Fig. 8.10 with "observer state feedback" are:

$$\begin{aligned} &(1) \quad \dot{z}(t) = \bar{A}\dot{z}(t) + \bar{B}[\bar{K}F_1\hat{z}(t) + \bar{K}F_2y(t) + v(t)] \\ &(2) \quad \dot{\hat{z}}(t) = \bar{L}\hat{z}(t) + T\bar{B}[\bar{K}F_1\hat{z}(t) + \bar{K}F_2y(t) + v(t)] + \bar{N}y(t) \\ &(3) \quad y(t) = \bar{C}z(t) \end{aligned} \tag{8.101}$$

which may be expressed in matrix form as

$$\begin{bmatrix} \dot{z}(t) \\ \dot{\hat{z}}(t) \end{bmatrix} = \underbrace{\begin{bmatrix} \bar{A}+\bar{B}\bar{K}F_2\bar{C} & \bar{B}\bar{K}F_1 \\ T\bar{B}\bar{K}F_2\bar{C}+\bar{N}\bar{C} & \bar{L}+T\bar{B}\bar{K}F_1 \end{bmatrix}}_{(n+\ell)\text{-square}} \underbrace{\begin{bmatrix} z(t) \\ \hat{z}(t) \end{bmatrix}}_{\substack{(n+\ell) \\ \text{state} \\ \text{vector}}} + \underbrace{\begin{bmatrix} \bar{B} \\ T\bar{B} \end{bmatrix}}_{\substack{(n+\ell)\times m \\ \text{input} \\ \text{matrix}}} v(t) \text{ and } y(t) = \underbrace{[\bar{C} \;\; 0]}_{\substack{m\times(n+\ell) \\ \text{output} \\ \text{matrix}}} \tag{8.102}$$

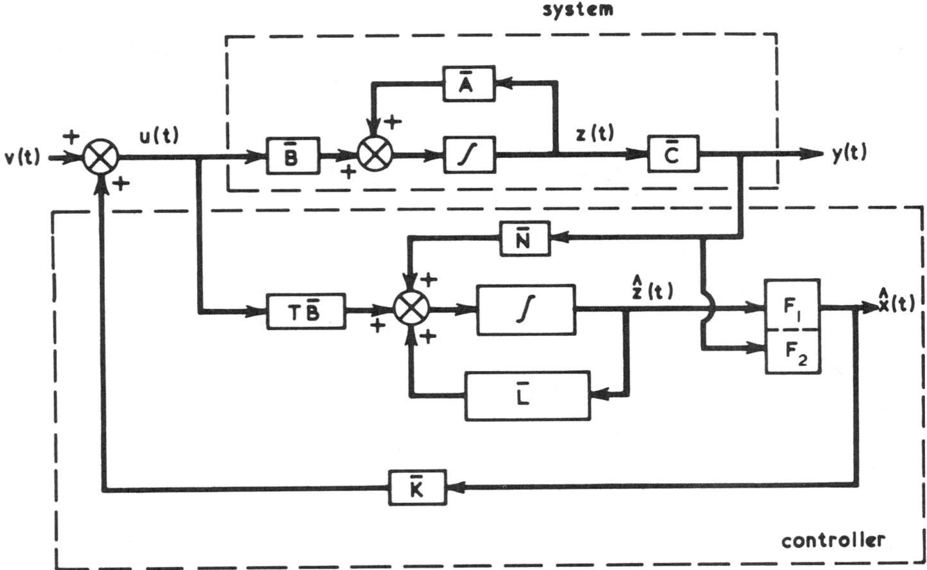

Fig. 8.10 Controller design using state observer.

In order to derive the transfer-function matrix of the composite system described by (8.102), the following transformation of the $(n + \ell)$-state vector $\tilde{z}(t)$ is used:

$$\tilde{z}(t) = \begin{bmatrix} I_n & 0 \\ -T & I_\ell \end{bmatrix} \begin{bmatrix} z(t) \\ \hat{z}(t) \end{bmatrix} \tag{8.103}$$

to obtain an algebraically equivalent system $\tilde{S}(\tilde{A}, \tilde{B}, \tilde{C})$ defined by

$$\tilde{A} = \begin{bmatrix} I_n & 0 \\ -T & I_\ell \end{bmatrix} \begin{bmatrix} \bar{A} + \bar{B}\bar{K}F_2\bar{C} & \bar{B}\bar{K}F_1 \\ T\bar{B}\bar{K}F_2\bar{C} + NC & \bar{L} + T\bar{B}\bar{K}F_1 \end{bmatrix} \begin{bmatrix} I_n & 0 \\ T & T_\ell \end{bmatrix} = \begin{bmatrix} \bar{A} + \bar{B}\bar{K} & \bar{B}\bar{K}F_1 \\ 0 & \bar{L} \end{bmatrix} \tag{8.104}$$

$$\tilde{B} = \begin{bmatrix} I & 0 \\ T & I_\ell \end{bmatrix} \begin{bmatrix} \bar{B} \\ T\bar{B} \end{bmatrix} = \begin{bmatrix} \bar{B} \\ 0 \end{bmatrix}; \quad \tilde{C} = [\bar{C} \;\; 0]$$

Thus by using the identity

$$\begin{bmatrix} P_1 & P_2 \\ 0 & P_4 \end{bmatrix}^{-1} = \begin{bmatrix} P_1^{-1} & -P_1^{-1}P_2P_4^{-1} \\ 0 & P_4^{-1} \end{bmatrix}$$

the transfer-function matrix of the system in Fig. 8.10 is given by

$$\tilde{G}(s) = \tilde{C}[sI - \tilde{A}]^{-1}\tilde{B} = [\bar{C} \;\; 0][sI - \tilde{A}]^{-1} \begin{bmatrix} \bar{B} \\ 0 \end{bmatrix}$$

$$= \bar{C}[sI - (\bar{A} + \bar{B}\bar{K})]^{-1}\bar{B} \tag{8.105a}$$

and

$$\det\{sI - \tilde{A}\} = \det\{sI - (\bar{A} + B\bar{K})\} \det\{sI - \bar{L}\} \tag{8.105b}$$

which suggests that although the order of the closed-loop system in Fig. 8.11 is increased by ℓ, there is no difference (except transient errors) between the performance of the closed-loop system with observer state feedback (Fig. 8.11a) and that of the closed-loop system with feedback of the real state (Fig. 8.11b). Consequently, the design of the state feedback and the design of the state observer may be carried out independently. This conclusion applies to both cases with $\ell = n$ and $\ell = n - m$. This is summarized in the following theorem.

Theorem 8.5: Separation property For a system S(A, B, C) with the observer in (8.2) and (8.6) and feedback control law in (8.99), the charac-

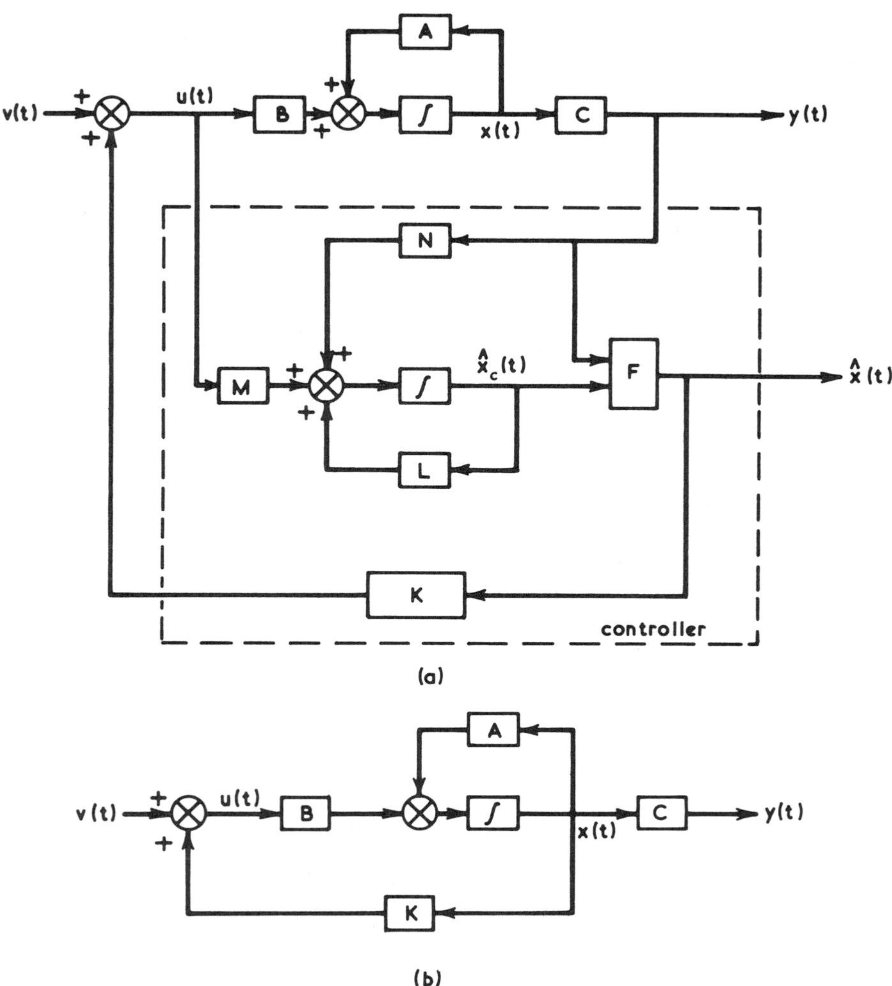

Fig. 8.11 State feedback (a) with estimated state variable feedback and (b) with actual state variables of the system.

teristic polynomial of the closed-loop system can be factored as det $\{sI - A - BK\}$ det $\{sI - L\}$.

The results above may be combined with Theorem 8.4 to obtain the following theorem.

Theorem 8.6 Corresponding to any nth-order completely controllable and completely observable system S(A, B, C) having m linearly independent outputs, a dynamic feedback system of order (n - m) can be constructed such that the (2n - m) eigenvalues of the composite system take any preassigned values.

It is perhaps worth noting that the foregoing result for a linear time-invariant system illustrates the key factor in observer design: The general stability characteristics of a given system are not affected by the inclusion of an asymptotically stable observer. Thus for nonlinear as well as time-invariant systems, an observer can provide appropriate estimates.

As long as the system S(A, B, C) is both controllable and observable, for any desired pole shifting, the state feedback matrix K can be evaluated. The matrix $\bar{L}$ may then be freely chosen to make the observer have any desired dynamics, without disturbing the assigned poles of the system S(A, B, C). In general, the matrix K is introduced in the observer subsystem and the resulting block in the dashed lines in Fig. 8.11 is called the controller or regulator. Once the matrices K, L, and NC have been chosen, the controller is looked upon as a dynamic system which can be realized in any form the designer may choose, canonical or not. The synthesis method is illustrated below.

Example 8.6: Control through observer state feedback The system in Example 8.2 is considered here. Let it be required to shift the eigenvalues $\lambda_1 = 1$ and $\lambda_3 = 3$ to $\lambda_1 = -2$ and $\lambda_3 = -3$ through a state feedback control law using the reduced-order observer derived above.

Let the control law that yields the foregoing eigenvalue shift for the normalized system in (8.33) be

$$u(t) = \bar{K}z(t)$$

Then from Sec. 6.1, the desired locations of the eigenvalues in this case are related to $\bar{K}$ by

$$B_n\bar{K} = \text{diag}\{-2, -2, -3\} - \text{diag}\{1, -2, 3\}$$

which gives

$$\bar{K} = \underbrace{\begin{bmatrix} 5 & -2 \\ -2 & 1 \end{bmatrix}}_{(B_n^T B_n)^{-1}} \underbrace{\begin{bmatrix} 0 & 0 & 1 \\ -1 & 0 & 2 \end{bmatrix}}_{B_n^T} \begin{bmatrix} -3 & & \\ & 0 & \\ & & -6 \end{bmatrix} = \begin{bmatrix} -6 & 0 & -6 \\ 3 & 0 & 0 \end{bmatrix} \qquad (8.106)$$

Combining (8.97), (8.100), and (8.106), the required observer state feedback control is given by

$$u(t) = \bar{K}F_1\hat{z}(t) + \bar{K}F_2 y(t) = \begin{bmatrix} -6 & 0 & -6 \\ 3 & 0 & 0 \end{bmatrix} \frac{1}{4} \begin{bmatrix} -30 \\ 0 \\ 30 \end{bmatrix} \hat{z}(t)$$

$$+ \begin{bmatrix} -6 & 0 & -6 \\ 3 & 0 & 0 \end{bmatrix} \frac{1}{4} \begin{bmatrix} 4 & 22 \\ 0 & 4 \\ 0 & -26 \end{bmatrix} y(t)$$

or

$$\begin{bmatrix} u_1(t) \\ u_2(t) \end{bmatrix} \equiv \begin{bmatrix} 0 \\ -\frac{45}{2} \end{bmatrix} \hat{z}(t) + \begin{bmatrix} -6 & 6 \\ 3 & \frac{33}{2} \end{bmatrix} \begin{bmatrix} y_1(t) \\ y_2(t) \end{bmatrix} \qquad (8.107)$$

The theory of observers for linear systems is now well established, and a number of numerical algorithms for the derivation of full- and reduced-order observers are now available. Some multivariable design methods using output or incomplete state feedback are based on an implied method of realizing the required state feedback control law through observers. In cases where the available signals contain a significant proportion of noise, the observer design algorithms derived here may not be applicable. The method of estimating state variables with noise-corrupted signals is also well established, known as the Kalman-Bucy filter [14], but is not considered here. The importance of observers lies in the fact that any controllable system may be designed to achieve any closed-loop eigenvalues in a two-stage design procedure—the first being the observer design and the second, state feedback.

REFERENCES

1. Layton, J. M., Multivariable Control Theory, Peter Peregrinus, London, 1976.
2. Luenberger, D. G., "An introduction to observers," Trans. IEEE, AC-16: 596-602 (1971).
3. Luenberger, D. G., "Observers for multivariable systems," Trans. IEEE, AC-11: 190-197 (1966).
4. Luenberger, D. G., "Observing the state of a linear system," Trans. IEEE, MIL-8: 74-80 (1964).
5. Chen, C. T., Introduction to Linear System Theory, Holt, Rinehart and Winston, New York, 1970.

6. Fortmann, T. E., and Williamson, D., "Design of low-order observers for linear feedback control laws," Trans. IEEE, AC-17: 301-308 (1972).
7. Kwakernaak, H., and Sivan, R., Linear Optimal Control, Wiley-Interscience, New York, 1972.
8. Cumming, S. D. G., "Design of observers of reduced dynamics," Electron. Lett., 5: 213-214 (1969).
9. Padulo, L., and Arbib, M. A., System Theory, Hemisphere, Washington, D.C., 1974.
10. Crossley, T. R., and Porter, B., "State observer and their applications," in Modern Approaches in Control System Design, N. Munro (Ed.), Peter Peregrinus, London, 1979, Chap. 13.
11. Newmann, M. M., "Design algorithms for minimal order Luenberger observers," Electron. Lett., 5: 390-392 (1969).
12. Munro, N., "Computer-aided-design procedure for reduced-order observers," Proc. IEE, 120: 319-324 (1973).
13. Luenberger, D. G., Introduction to Dynamic Systems, Wiley, New York, 1979.
14. Kalman, R. E., and Bucy, R. S., "New results in linear filtering and prediction theory," ASME J. Basic Eng., 83: 95-108 (1961).

9
Decoupling Control

In linear multivariable systems, a change in any input will usually result in changes in all output variables. Such systems are characterized by coupling or interaction as outlined in Sec. 5.3. It may be useful for certain applications to obtain a system in which such interaction between controls does not occur. The design of multivariable systems so that they become noninteracting or decoupled has received considerable attention in recent years. The design objective of noninteracting (or decoupled) systems is to obtain a system in which each input affects only one output. The primary advantage of such a design is that once noninteraction is achieved, the system is reduced to a number of single-input/single-output channels (subsystems) to which the well-established design techniques may be applied.

In frequency-domain analysis the one-to-one correspondence between the input and output variables may be achieved by constraining the closed-loop transfer function to be diagonal, with diagonal elements not equal to zero, except at a finite set of points on the complex frequency plane. In the time-domain analysis, however, the implication of noninteraction is not so apparent and can only be expressed through a complicated mathematical relation. The constraints and the design procedure for decoupling using state and output feedback control laws are considered in this chapter.

9.1 STATE FEEDBACK DECOUPLING [1–4]

The problem of decoupling linear time-invariant multivariable systems S(A, B, C) is considered here and a synthesis procedure developed. In order to derive the basic concepts of decoupling, a rather special class of systems is first considered: those that are decoupled into m first-order channels or subsystems by a linear state feedback control law

$$u(t) = Fx(t) + Gw(t) \tag{9.1}$$

where w(t) is the new control input, F is the state feedback matrix of order $m \times n$, and G is the input control matrix of order $m \times m$, as shown in

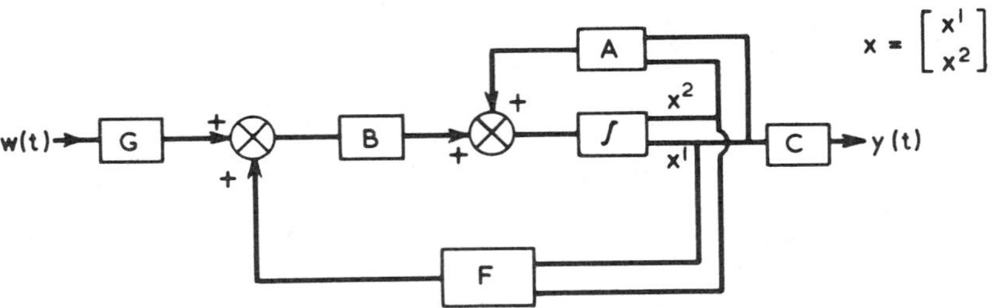

Fig. 9.1 State feedback decoupling problem posed by Morgan.

Fig. 9.1. Although such systems are rare in practice, the analytical results form the basis for decoupling theory in the time domain [1].

Let the differential equations of a linear multivariable system that is decoupled into m first-order subsystems be expressed as

$$\dot{y}(t) = M_o y(t) + \Gamma w(t) \equiv M_o Cx(t) + \Gamma w(t) \tag{9.2}$$

where $\Gamma = \text{diag}\,\{\gamma_i\}$, such that $\Pi_{i=1}^{m} \gamma_i \neq 0$; that is, any ith output is affected by the ith input, and M_o is a diagonal matrix. The characteristic polynomial of this decoupled system is

$$\det\,(sI - M_o)$$

If the decoupled structure above were to be obtained by using a state feedback control law in (9.1), the closed-loop differential equations

$$\dot{y}(t) = C\dot{x}(t) = C(A + BF)x(t) + CBGw(t)$$

and (9.3)

$$\dot{y}(t) = M_o Cx(t) + \Gamma w(t)$$

should be equivalent. This gives the following criteria:

$$M_o C = C(A + BF) \rightarrow F = (CB)^{-1}[M_o C - CA]$$

and (9.4)

$$CBG = \Gamma \rightarrow C = (CB)^{-1}\Gamma$$

Nonsingularity of CB is thus the primary requirement for the existence of the pair of matrices F and G, called the <u>decoupling pair</u>. The control law (9.1) formed out of these two matrices is termed as the state feedback decoupling control law. The important points to be noted from (9.4) are:

The elements of M_o and Γ, specifying the poles and gains of the m decoupled channels, respectively, can be chosen freely while preserving input/output noninteraction. To illustrate the role of M_o and Γ, the following two-input/two-output third-order system is considered.

Example 9.1: State feedback decoupling

$$\dot{x}(t) = \begin{bmatrix} 1 & 1 & 0 \\ 0 & -2 & 0 \\ 0 & 1 & 3 \end{bmatrix} x(t) + \begin{bmatrix} 1 & 1 \\ -1 & 1 \\ -1 & 1 \end{bmatrix} u(t); \quad y(t) = \begin{bmatrix} 1 & 0 & 0 \\ 0 & 0 & 1 \end{bmatrix} x(t) \tag{9.5}$$

for which

$$CB = \begin{bmatrix} 1 & 1 \\ -1 & 1 \end{bmatrix} \rightarrow CB^{-1} = \frac{1}{2}\begin{bmatrix} 1 & -1 \\ 1 & 1 \end{bmatrix}$$

Using (9.4), with $M_o = \text{diag}\{m_{01}, m_{02}\}$ and $\Gamma = \text{diag}\{\gamma_1, \gamma_2\}$, the matrices F and G which decouple (9.5) are given by

$$F = \begin{bmatrix} m_{01} - 1 & -1 & 0 \\ 0 & -1 & m_{02} - 3 \end{bmatrix} \quad \text{and} \quad G = \frac{1}{2}\begin{bmatrix} \gamma_1 & -\gamma_1 \\ \gamma_2 & \gamma_2 \end{bmatrix} \tag{9.6}$$

The open- and closed-loop signal flow graphs are shown in Fig. 9.2. The significance of nonsingularity of CB is that the degree of the polynomial giving the zeros of S(·) is (n - m) [i.e., 1, for (9.5)], whereas the degree of characteristic polynomial of the system is n. This implies that by suitable choice of a state feedback control law, the (n - m) zeros may be canceled and the result is a reduction in the degrees of the numerator and denominator polynomials of the determinant of the closed-loop transfer function matrix. If all zeros are canceled, the degree of the numerator polynomial giving the zeros is (n - m) - (n - m) = 0, and the degree of the characteristic polynomial giving the poles is n - (n - m) = m, and hence there are only m poles to attribute to the m subsystems for which all zeros are canceled by the decoupling pair (F, G). This is demonstrated by the closed-loop transfer function of (9.5) and (9.6):

$$R(s) = C[sI - A - BF]^{-1}BG$$

$$= \begin{bmatrix} \dfrac{(s+2)(s-m_{02})}{(s+2)(s-m_{01})(s-m_{02})} & 0 \\ 0 & \dfrac{(s+2)(s-m_{01})}{(s+2)(s-m_{01})(s-m_{02})} \end{bmatrix} \tag{9.7}$$

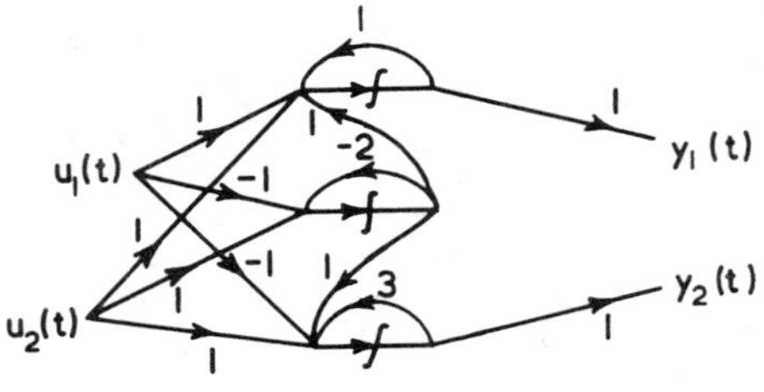

(a)

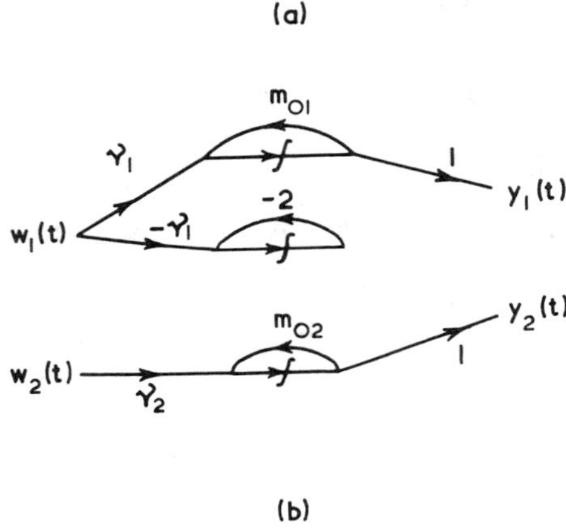

(b)

Fig. 9.2 Signal flow graphs of the open-loop and closed-loop (decoupled) system in (9.5): (a) open-loop system (9.5); (b) closed-loop system [(9.5) and (9.6)].

from which it is apparent that the elements m_{01} and m_{02}, which specify the poles of the decoupled system, should be assigned negative values for closed-loop stability.

Due to the constraint of m first-order decoupled channels, and consequently nonsingular CB, the method above has limited applications. In order to derive a more widely applicable decoupling control law, let there be an arbitrary integer d such that CB, CAB, ..., $CA^{d-1}B$ are all null matrices and CA^dB is a nonsingular matrix [2]. With this assumption, successive differential of the closed-loop equation*

*This system is subsequently referred to as S(A, B, C, F, G) or $s(\overline{A + BF}, BG, C)$.

$$\dot{x}(t) = Ax(t) + B[Fx(t) + Gw(t)]; \quad y(t) = Cx(t) \tag{9.8}$$

gives*

$$
\begin{aligned}
y(t) &= Cx(t) \\
y^{(1)}(t) &= Ax(t) \\
&\vdots \\
y^{(d-1)}(t) &= CA^{d-1}x(t) \\
y^{(d)}(t) &= CA^{d}x(t) \\
y^{(d+1)}(t) &= C(A + BF)^{d+1}x(t) + C(A + BF)^{d}BGw(t) && (9.9) \\
&\vdots \\
y^{(n)}(t) &= C(A + BF)^{n}x(t) + C(A + BF)^{n-1}BGw(t) + C(A + BF)^{n-2}BGw^{(1)}(t) \\
&\quad + \cdots + C(A + BF)^{d}BGw^{(n-d-1)}(t) && (9.10)
\end{aligned}
$$

The set of equations above show that the closed-loop system $S(A, B, C, F, G)$ could be decoupled into m subsystems, each of order d, if F and G were chosen such that

$$C(A + BF)^{d+1} \quad \text{is nonnull}$$

and

$$C(A + BF)^{d}BG \quad \text{is diagonal}$$

In such a case it should be possible to express the closed-loop differential equation of the decoupled system as [using (9.9)]

$$
\begin{aligned}
y^{(d+1)}(t) = {} & M_d y^{(d)}(t) + M_{d-1} y^{(d-1)}(t) + \cdots + M_o y(t) + \Gamma_{d-1} w^{(d-1)}(t) \\
& + \cdots + \Gamma_o w(t)
\end{aligned}
$$

* $y^{(k)}(t) \equiv d^{k}y(t)/dt^{k}$.

$$= \left[\sum_{k=0}^{d} M_k CA^k\right] x(t) + \left[\sum_{k=0}^{d} \Gamma_k \overset{(k)}{w}(t)\right] \tag{9.11}$$

where M_k and Γ_k, $k \in 0, \ldots, d$ are diagonal matrices subject to the condition that not all Γ_k's are singular. Combining (9.9) and (9.11) gives us

$$\Gamma_k \equiv 0, \quad \forall K \quad 0, 1, \ldots, (d-1); \quad \Gamma_o = C(A + BF)^d BG;$$

$$\sum_{k=0}^{d} M_k CA^k = C(A + BF)^{d+1} \tag{9.12}$$

Since $CA^kB = 0$, $\forall k \in 0, \ldots, (d-1)$, (9.12) gives

$$\Gamma_o = CA^dBG \rightarrow G = (CA^dB)^{-1}\Gamma_o$$

$$\sum_{k=0}^{d} M_k CA^k = CA^{d+1} + CA^dBF \rightarrow F = (CA^dB)^{-1}\left\{\sum_{k=0}^{d} M_k CA^k - CA^{d+1}\right\} \tag{9.13}$$

The derivations above show that if an integer d exists for a system S(A, B, C), there exists a pair (F, G) which would decouple the system. The matrices $M_o, M_1, \ldots, M_d$ and Γ_o have the same significance as before and as shown by the closed-loop decoupled differential equation

$$\overset{(d+1)}{y}(t) = M_d \overset{(d)}{y}(t) + M_{d-1} \overset{(d-1)}{y}(t) + \cdots + M_o y(t) + \Gamma_o w(t) \tag{9.14}$$

The dynamics of the decoupled subsystem may, therefore, be controlled through the choice of the elements of M_k's and Γ_o.

The two-input/two-output fourth-order system S(A, B, C) defined below [2] is considered here to illustrate the derivations above.

Example 9.2: State feedback decoupling

$$A = \begin{bmatrix} 0 & 1 & 0 & 0 \\ 0 & -1 & 0 & 0 \\ 0 & 1 & 0 & 1 \\ 0 & 0 & 0 & -2 \end{bmatrix}; \quad B = \begin{bmatrix} 0 & 0 \\ 1 & 0 \\ 0 & 0 \\ 0 & 1 \end{bmatrix}; \quad C = \begin{bmatrix} 1 & 0 & 0 & 0 \\ 0 & 0 & 1 & 0 \end{bmatrix} \tag{9.15}$$

For this system

$$CB = 0; \quad CAB = \begin{bmatrix} 1 & 0 \\ 1 & 1 \end{bmatrix}$$

Thus d = 1. From (9.13), the corresponding state feedback decoupling pair (F, G) is given by

$$F = [CAB]^{-1}[M_oC + M_1CA - CA^2]$$

$$= \begin{bmatrix} 1 & 0 \\ -1 & 1 \end{bmatrix} \left\{ \begin{bmatrix} m_{01} & 0 \\ 0 & m_{02} \end{bmatrix} \begin{bmatrix} 1 & 0 & 0 & 0 \\ 0 & 0 & 1 & 0 \end{bmatrix} \right.$$

$$\left. + \begin{bmatrix} m_{11} & 0 \\ 0 & m_{12} \end{bmatrix} \begin{bmatrix} 0 & 1 & 0 & 0 \\ 0 & 1 & 0 & 1 \end{bmatrix} - \begin{bmatrix} 0 & -1 & 0 & 0 \\ 0 & -1 & 0 & -2 \end{bmatrix} \right\}$$

$$= \begin{bmatrix} m_{01} & m_{11} + 1 & 0 & 0 \\ -m_{01} & -m_{11} + m_{13} & m_{02} & m_{12} + 2 \end{bmatrix}$$

$$G = [CAB]^{-1}\Gamma = \begin{bmatrix} 1 & 0 \\ -1 & 1 \end{bmatrix} \begin{bmatrix} \gamma_1 & 0 \\ 0 & \gamma_2 \end{bmatrix} \tag{9.16}$$

The transfer-function matrix of the closed-loop system is

$$R(s) = C[sI - A - BF]^{-1}BG = \begin{bmatrix} \dfrac{s^2 - m_{12}s - m_{02}}{\Delta(s)} & 0 \\ 0 & \dfrac{s^2 - m_{11}s - m_{01}}{\Delta(s)} \end{bmatrix} \tag{9.17}$$

where $\Delta(s) = (s^2 - m_{11}s - m_{01})(s^2 - m_{12}s - m_{02})$.

Thus the decoupling control law gives rise to unobservable modes in the closed-loop system (defined earlier as output-decoupling zeros). Since the number of modes connected in the closed-loop input/output subsystems is $n = m(d + 1)$, it follows that the number of unobservable modes in the decoupled system is $\xi = n - m(d + 1)$. Since these modes are canceled in (9.17), it may be concluded that the degree of minimal polynomial of the decoupled system is $n - \xi = m(d + 1)$.

Since in general ξ is not equal to zero, the type number* of the different input/output channels will be changed by the application of the decoupling control law using state feedback. Consequently, the response of the decoupled channels or subsystems of $S(\overline{A + BF}, BG, C)$ will be, in general, different from those of the original system $S(A, B, C)$.

The extension above, while applicable to a wider class of system, is restricted, as it is based on the assumption that each decoupled subsystem has the same order. In most physical systems, however, the distribution of the order of the subsystems would be expected to be random, and consequently there would be no single integer (d) that would satisfy the equation

$$CA^kB \equiv \begin{cases} 0, \ \forall k \in 0, 1, \ldots, (d-1) \\ \text{nonsingular for } k = d \end{cases}$$

but this integer should vary from one channel to another.

In order to extend the results above to a general decoupling synthesis procedure, let there exist an integer di such that C_iB, C_iAB, ..., $C_iA^{di-1}B$ are all null row vectors and $C_iA^{di}B$ is a nonnull vector. This definition of di, for any $i \in 1, \ldots, m$, may be expressed as [3]

$$di = \begin{cases} \text{minimum } k \text{ for which } C_iA^kB \neq 0 \\ (n-1) \text{ if } C_iA^kB = 0, \ \forall k \end{cases} \quad \begin{matrix} k \in 1, \ldots, (n-1); \\ i \in 1, \ldots, m \end{matrix} \tag{9.18}$$

where C_i is the ith row of C. Let $\delta = \max di$, $i \in 1, \ldots, m$. With the definitions above and the derivations in (9.9) to (9.12), it can be concluded that the closed-loop system in (9.8) is decoupled into m subsystems, the order of the ith one being di, if

$$C_i(A + BF)^{di}B = \text{a nonnull row vector}$$

and

$$C_i(A + BF)^{di}BF = e_i\text{: the ith row of an m-square identity matrix}$$

The differential equation of the ith subsystem is then of the form

$$y_i^{(di+1)}(t) = m_{di,i}\, y^{(di)}(t) + m_{(di-1),i}\, y^{(di-1)}(t) + \cdots + m_{o,i}y_i(t) + \gamma_{o,i}w_i(t)$$

$$i \in 1, \ldots, m \tag{9.19}$$

where $m_{di,i}$, $m_{(di-1),i}$, ..., $m_{o,i}$ and $\gamma_{o,i}$ ($\neq 0$) are scalars.

*A feedback system is said to be of type m according to the number m of open-loop poles at the origin.

Comparing the equation above with the ith row of (9.11), for decoupling the following conditions are satisfied for any ith row:

$$\sum_{k=0}^{di} m_{k,i} C_i A^k = C_i (A + BF)^{(di+1)}$$

and (9.20)

$$\gamma_{o,i} = C_i (A + BF)^{di} BG$$

Using the definition of di in (9.18), the equations above become

$$\left.\begin{aligned} (C_i A^{di} B) F &= \sum_{k=0}^{di} m_{k,i} C_i A^k - C_i A^{di+1} \\ &\text{and} \\ (C_i A^{di} B) G &= \gamma_{o,i} \end{aligned}\right\} \quad i \in 1, \ldots, m \tag{9.21}$$

Thus if the value of i is taken from 1 to m, the following two sets of equations are obtained:

$$\begin{aligned} \{C_i A^{di} B\} F &= \left\{ \sum_{k=0}^{di} m_{k,i} C_i A^k \right\} - \left\{ C_i A^{di+1} \right\} \\ \{C_i A^{di} B\} G &= \text{diag}\, \{\gamma_{o,i}\} \end{aligned} \tag{9.22}$$

$$B^* F = \sum_{k=0}^{\delta} M_k C A^k - A^* \quad \text{and} \quad B^* G = \Gamma_o \tag{9.23}$$

where $A^* = \{C_i A^{di+1}\}$, $B^* = \{B_i^*\} = \{C_i A^{di} B\}$, and $M_k = \text{diag}\,\{m_{1,k}, \ldots, m_{m,k}\}$, $k \in 0, \ldots, \delta$.

Thus if the rows of B^* were such that $\det(B^*) \neq 0$, there would exist a state feedback decoupling pair given by

$$\begin{aligned} F &= (B^*)^{-1} \left(\sum_{k=0}^{\delta} M_k C A^k - A^* \right) \\ G &= (B^*)^{-1} \Gamma_o \end{aligned} \tag{9.24}$$

The derivation above leads to the formulation of the following theorems on state feedback decoupling [3,4].

Theorem 9.1 If for a linear system defined S(A, B, C) there is a set of integers di, $i \in 1, \ldots, m$, as defined by (9.18), then a necessary and sufficient condition for the existence of a state feedback decoupling control is that $B^* = \{C_i A^{di} B\}$, $i \in 1, \ldots, m$, be nonsingular. A few comments on the results are relevant [4].

(a) Significance of di

The differential equation in (9.19) suggest that there are (di + 1) state variables between the ith input and ith output channels of the decoupled system. Since no additional state variables are introduced by state feedback, (di + 1) is also the number of state variables in an open-loop system in the shortest forward path between the ith input terminal and ith output terminal.

(b) Significance of B*

If $G_i(s)$ is the ith row of the transfer function of the system S(•), then

$$G_i(s) = C_i(sI - A)^{-1}B = \frac{C_i B s^{n-1} + C_i R_1 B s^{n-2} + \cdots + C_i R_{di} B s^{n-di-1} + \cdots + C_i R_{n-1} B}{\det(sI - A)} \tag{9.25}$$

where

$$\det(sI - A) = s^n + \alpha_i s^{n-1} + \cdots + \alpha_n$$

$$R_1 = A + \alpha_1 I$$

$$R_2 = A^2 + \alpha_1 A + \alpha_2 I$$

$$\vdots$$

$$R_{di} = A^{di} + \alpha_1 A^{di-1} + \cdots + \alpha_{di} I$$

$$R_{n-1} = A^{n-1} + \alpha_1 A^{n-2} + \cdots + \alpha_{n-1} I$$

By using the definition of di,

$$G_i(s) = \lim_{|s| \to \infty} \frac{B_i^*}{s^{di+1}} \quad \text{or} \quad B_i^* = \lim_{|s| \to \infty} s^{(di+1)} G_i(s) \tag{9.26}$$

so that B_i^* is the high-frequency gain of the ith subsystem, and B* may be called the high-frequency gain matrix. A more significant interpretation is that the nonsingularity of B* is a sufficient condition for constructing an arbitrary control law to change any initial output $y(t_0)$ to any final output $y(t_f)$ in a finite time interval $t_0 < t < t_f$.

(c) Integrator decoupled system

In the expression of the state feedback matrix F in (9.22) to (9.24), M_k's specify the poles of the decoupled subsystem. If these are chosen as null matrices (i.e., $M_k \equiv 0$, $\forall k$), the decoupling pair is given by

$$F^* = -(B^*)^{-1}A^* \quad \text{and} \quad G = (B^*)^{-1}\Gamma_o$$

and the resulting decoupled system will have all poles at the origin. Such systems are called <u>integrator decoupled systems</u> and may be used to form a basis of decoupling synthesis through coordinate transformation.

(d) Pole-zero cancellation

The total number of poles that appear in the input/output structure of the decoupled system is $\Sigma_{i=1}^{m}$ (di + 1) < n. Since there are no zeros in the decoupled subsystems (9.19) (due to the property of di), the remaining [n - $\Sigma_{i=1}^{n}$ (di + 1)] poles have been canceled by suitable zeros created by the decoupling pair (F, G). In general, n - $\Sigma_{i=1}^{m}$ (di + 1) > 0, and if these cancellations occur on the right half of the complex frequency plane, the resulting decoupled system may not be stable. This point and the decoupling synthesis procedure are illustrated below through a numerical example.

<u>Example 9.3: State feedback decoupling</u> The following system is used here to demonstrate the synthesis procedure when $C_iA^{di}B \neq 0$, $i \in 1, \ldots, m$.

$$A = \begin{bmatrix} 0 & 1 & 1 & 0 \\ 0 & 0 & 1 & 1 \\ 0 & 0 & 0 & 0 \\ 0 & 1 & 0 & 1 \end{bmatrix}; \quad B = \begin{bmatrix} 0 & 0 \\ 1 & 0 \\ 1 & 0 \\ 0 & 1 \end{bmatrix}; \quad C = \begin{bmatrix} 1 & 0 & 0 & 0 \\ 0 & 0 & 0 & 1 \end{bmatrix} \tag{9.27}$$

For this system,

$$C_1B = [0 \;\; 0]; \quad C_1AB = [2 \;\; 0]; \quad C_2B = [0 \;\; 1]$$

Therefore,

$$d_1 = 1; \quad d_2 = 0; \quad \delta = d_1 = 1; \quad B^* = \begin{bmatrix} C_1AB \\ C_2B \end{bmatrix} = \begin{bmatrix} 2 & 0 \\ 0 & 1 \end{bmatrix} \tag{9.28}$$

The number of state variables associated with the shortest forward path of the first subsystem (u_1 to y_1 is $d_1 + 1 = 2$ and of the second subsystem (u_2 to y_2) is $d_2 + 1 = 1$, so that the total number of poles that can be arbitrarily assigned while simultaneously decoupling (9.27) is

$$\eta = (d_1 + 1) + (d_2 + 1) = 3$$

With $M_0 = \text{diag}\{m_{01}, m_{02}\}$, $M_1 = \text{diag}\{m_{11}, 0\}$, and $\Gamma_0 = \text{diag}\{2, 1\}$, the state feedback decoupling pair from (9.22) to (9.24), (9.27), and (9.28) is given by

$$F = \frac{1}{2}\begin{bmatrix} m_{01} & m_{11} & m_{11} - 1 & -1 \\ 0 & -2 & 0 & 2m_{02} - 2 \end{bmatrix}; \quad G = \begin{bmatrix} 1 & 0 \\ 0 & 1 \end{bmatrix} \tag{9.29}$$

The differential equations of the two channels of the closed-loop system are

$$\begin{aligned} \ddot{y}_1(t) - m_{11}\dot{y}_1(t) - m_{01}y_1(t) &= 2w_1(t) \\ \dot{y}_2(t) - m_{02}y_2(t) &= w_2(t) \end{aligned} \tag{9.30}$$

which has the transfer function matrix

$$R(s) = C[sI - \overline{A + BF}]^{-1}BG = \begin{bmatrix} \dfrac{2s}{s(s^2 - m_{11}s - m_{01}} & 0 \\ 0 & \dfrac{1}{s - m_{02}} \end{bmatrix} \tag{9.31}$$

showing that there is a pole-zero cancellation in the first subsystem of $S(\overline{A + BF}, BG, C)$, which is consistent with analytical derivations.

The preceding analysis of state feedback decoupling is based on successive differentiation of the input-state-output equations, and then constraining the cross-coupling terms to zero. A more direct approach is possible using the controllability and observability criteria (Chap. 3). This is considered in Appendix 3 but the principal results relating these concepts in decoupling are stated below [5–7].

Theorem 9.2 A system can be completely decoupled by proportional state feedback only if it is completely output controllable.

This result not only simplifies decoupling analysis, but also provides a direct method of identifying the decouplability of $S(\cdot)$ through either the signal flow graph or the structure of A, B, and C. A synthesis procedure and illustrative example are given in Appendix 3.

9.2 OUTPUT FEEDBACK DECOUPLING

The derivations in the preceding sections assume the availability of the complete state vector for feedback. In practice, as indicated earlier, it is seldom the case that all state variables are individually available. When the state variables are not available, there are two possible ways of realizing a decoupling control law. One is to use a dynamic compensator and the other is to extend the state feedback decoupling laws to include output variables. The latter approach is considered here.

Decoupling analysis of linear time-invariant systems by output feedback can be based on two approaches: in one the analysis of state feedback decoupling of the preceding section is extended to output feedback, and in the other a class of output feedback control laws which are equivalent to state feedback decoupling control laws are derived (along the lines of Sec. 7.1). These two cases are briefly considered here.

9.2.1 Extension of State Feedback Analysis [8]

The output feedback control considered here, to decouple S(A, B, C), is given by

$$u(t) = Hy(t) + Gw(t) \tag{9.32}$$

The closed-loop system with output feedback then becomes†

$$S(A, B, C, H, G): \ \dot{x}(t) = (A + BHC)x(t) + BGw(t); \ \ y(t) = Cx(t) \tag{9.33}$$

From the derivations in the preceding section, the system S(A, B, C, H, G) is said to be decoupled, with $G = B^{*-1}$, if

$$C(A + BHC)^k BB^{*-1} = \text{diagonal}, \quad \forall k \in 0, \ldots, n-1$$

or

$$C_i(A + BHC)^k BB_j^{*-1} = \begin{cases} \alpha_{i,j}, & i = j \\ 0, & i \neq j \end{cases} \qquad i, j \in 1, \ldots, m \tag{9.34}$$

where C_i is the ith row of C and B_j^* is the jth column of B*. Following the previous derivation and the definition of di, it can be shown that

$$C_i(A+BHC)^k BB_j^{*-1} = \begin{cases} 0 & \text{for } k = 0, \ldots, (di-1) \\ C_i A^k BB_j^{*-1} & \text{for } k = di \\ C_i A^{di}(A+BHC)^{k-di} BB_j^{*-1} & \text{for } k \in di+1 \end{cases} \tag{9.35}$$

For k = di + d1 + 1, from (9.35),

† This is referred to as S(A, B, C, H, G).

$$C_i(A+BHC)^{di+d1+1}BB_j^{*-1} = \begin{cases} C_iA^{di}(A+BHC)^{d1+1}BB_j^{*-1} \\ = C_iA^{di+d1+1}BB_1^{*-1} + C_iA^{d1}BH_1 & \text{for } j = 1 \\ \\ C_iA^{di+d1+1}BB_j^{*-1} & \text{for } j \neq 1 \end{cases} \tag{9.36}$$

where H_j and B_j^* are the jth columns of H and B^*, respectively, $j \in 1, \ldots, m$.

Thus for any $i \in 1, \ldots, m$ and $j = 1$,

$$C_iA^{di+d1+1}BB_1^{*-1} + C_iA^{di}BH_1 = \begin{cases} \alpha_{1,d1+1} & \text{for } i = 1 \\ \\ 0 & \text{for } i \neq 1 \end{cases}$$

or

$$C_iA^{di}BH_1 = \alpha_{1,d1+1} - C_iA^{di+d1+1}BB_1^{*-1}$$

that is,

$$B_i^*H_1 = \alpha_{1,d1+1} - C_iA^{di+d1+1}BB_1^{*-1} \quad \text{for } i \in 1, \ldots, m$$

or

$$B^*H_1 = \begin{bmatrix} \alpha_{1,d1+1} \\ \\ 0 \\ \\ 0 \end{bmatrix} - \begin{bmatrix} C_1A^{d1+d1+1} \\ \vdots \\ C_iA^{di+d1+1} \\ \vdots \\ C_mA^{dm+d1+1} \end{bmatrix} BB_1^{*-1} \tag{9.37}$$

or

$$H_1 = (B^*)^{-1}[\gamma_1 - A_{d1+1}^*BB_1^{*-1}], \text{ where } A_{d1+1}^* = \{A^*A^{d1+1}\}$$

which yields

$$\begin{aligned} H &= B^{*-1}\{\Lambda - \underbrace{[A_{d1+1}^*BB_1^{*-1} \cdots A_{dm+1}^*BB_m^{*-1}]}_{A^{**}}\} \\ &= B^{*-1}\{\Lambda - A^{**}\} \end{aligned} \tag{9.38}$$

where $\Lambda = \{\gamma_i\} = \text{diag}\,[\alpha_{1,d1+1}, \ldots, \alpha_{i,di+1}, \ldots, \alpha_{m,dm+1}]$.

The derivation above shows that a decoupling H exists only if B* is nonsingular. The following theorem may now be stated.

Theorem 9.3 A necessary and sufficient condition for the existence of a decoupling pair (H, G) for S(A, B, C) are:

1. B* is nonsingular.
2. $H = -B^{*-1}[A^*_{d1+1}BB^{*-1}_1, \ldots, A_{dm+1}BB^{*-1}_m]$.

The proof of the theorem follows from the derivation above and the following lemma.

Lemma 9.1 If the closed-loop system S(A, B, C, H, G) can be decoupled by (H, G), where $H = G(\Lambda - M)$, Λ being an m-square diagonal matrix, then S(A, B, C, H, G) must necessarily be decoupled by the pair (-GM, G).

Proof: With the pair $[G(\Lambda - M), G]$, the closed-loop system becomes

$$\dot{x}(t) = [A + BG(\Lambda - M)C]x(t) + BGw(t); \quad y(t) = Cx(t) \tag{9.39}$$

Since decoupling is unaffected by the new control law of the form

$$w(t) = v(t) - \Lambda y(t)$$

the input/output interaction properties of the system in (9.39) are similar to those of

$$\begin{aligned} \dot{x}(t) &= [A + BG(\Lambda - M)C]\,x(t) + BG[v(t) - \Lambda y(t)] = [A - BGMC]\,x(t) + BGv(t) \\ y(t) &= Cx(t) \end{aligned} \tag{9.40}$$

which proves the lemma.

The synthesis procedure is illustrated below.

Example 9.4: Output feedback decoupling

$$S(A, B, C): \; A = \begin{bmatrix} 1 & 1 & 0 \\ 0 & 1 & 0 \\ 0 & 0 & 1 \end{bmatrix}; \quad B = \begin{bmatrix} 0 & 1 \\ 1 & 0 \\ 1 & 0 \end{bmatrix}; \quad C = \begin{bmatrix} 1 & -1 & 1 \\ 0 & 1 & 0 \end{bmatrix} \tag{9.41}$$

for this system,

$$d1 = d2 = \delta = 0, \quad CB = B^* = \begin{bmatrix} 0 & 1 \\ 1 & 0 \end{bmatrix}; \quad B^{*-1} = \begin{bmatrix} 0 & 1 \\ 1 & 0 \end{bmatrix} \equiv [B^{*-1}_1 \mid B^{*-1}_2]$$

Consequently, from (9.38),

$$A^{**} = [CA^{d1+1}BB_1^{*-1} \mid CA^{d2+1}BB_2^{*-1}] = \begin{bmatrix} 1 & 1 \\ 0 & 1 \end{bmatrix}$$

and

$$H = B^{*-1}[\Lambda - A^{**}] = \begin{bmatrix} 0 & 1 \\ 1 & 0 \end{bmatrix}\begin{bmatrix} \lambda_1 - 1 & -1 \\ 0 & \lambda_2 - 1 \end{bmatrix} = \begin{bmatrix} 0 & \lambda_2 - 1 \\ \lambda_1 - 1 & -1 \end{bmatrix}$$

With $G = B^{*-1}$ and H as given above, the transfer-function matrix of closed-loop systems is given by

$$R(s) = C[sI - A - BF]^{-1} = \begin{bmatrix} \dfrac{(s-1)(s-\lambda_2)}{(s-1)(s-\lambda_1)(s-\lambda_2)} & 0 \\ 0 & \dfrac{(s-1)(s-\lambda_1)}{(s-1)(s-\lambda_1)(s-\lambda_2)} \end{bmatrix} \tag{9.42}$$

showing that the parameters λ_1 and λ_2 may be freely chosen to achieve the desired response of the decoupled system. It is to be noted here that, in general, as in the case of state feedback, stability and decoupling may not be achieved simultaneously, requiring the addition of compensation to achieve the desired stable response.

9.2.2 Equivalence Between Decoupling F and H [9]

The derivation here is aimed at establishing an equivalent between the state feedback decoupling law given by (9.22) to (9.24) and the output feedback law in (9.32). The purpose of seeking such an equivalence is to achieve decoupling by using output feedback such that the closed-loop system would have the same performance of the system decoupled by state feedback. The derivation is based on the one-to-one correspondence between state and output feedback control laws expressed through the following lemma (Sec. 7.1).

Lemma 9.2 There exists a solution for $F = HC$, where F is the state feedback decoupling matrix in (9.22) to (9.24) and H is the output feedback matrix in (9.32), if

$$\rho \begin{bmatrix} F \\ C \end{bmatrix} = m$$

Proof of this lemma follows from Theorem 7.2 since $\rho(C) = m$ for B^* to be nonsingular. The condition for the existence of an H that is equivalent to the decoupling F is stated by the following theorem.

<u>Theorem 9.4</u> If an m-input/m-output system S(A, B, C) can be decoupled by state feedback, an equivalent output feedback decoupling control law can be evaluated through a one-to-one correspondence between the state and output feedback matrices if the matrix T has rank m, where

$$T = \{T_i\}, \quad \text{where } T_i = \underbrace{\begin{bmatrix} C_i \\ C_i A \\ \vdots \\ C_i A^{di} \\ C_i A^{di+1} \end{bmatrix}}_{m \text{ columns}} \quad (di+1) \text{ rows} \quad \text{for } i \in 1, \ldots, m$$

where di is as defined earlier.

Proof: Since C has rank m, the constraint for the consistency of F = HC can be expressed as

$$\rho \begin{bmatrix} F \\ C \end{bmatrix} = \rho \begin{bmatrix} (B^*)^{-1} \left[\sum_{k=0}^{\delta} M_k C A^k - A^* \right] \\ \hline C \end{bmatrix} = \rho[P] = m$$

where

$$\rho(P) = \rho \left[\begin{array}{c|c} (B^*)^{-1} & 0 \\ \hline 0 & I_m \end{array} \right] \underbrace{\begin{bmatrix} \sum_{k=0}^{\delta} M_k C A^k - A^* \\ \hline C \end{bmatrix}}_{R} = \rho(R)$$

and

$$R = \begin{bmatrix} -I_m & M_\delta & \cdots & M_k & \cdots & M_o \\ 0 & 0 & \cdots & 0 & \cdots & I_m \end{bmatrix} \underbrace{\begin{bmatrix} A^* \\ CA^\delta \\ \vdots \\ CA^k \\ \vdots \\ C \end{bmatrix}}_{T}$$

Since it is only necessary to compute the rows C_iA^k for k = di + 1, some of the rows in any block CA^k will be null for conformity of the above if $\delta \neq di$, $\forall\, i \in 1, \ldots, m$. These rows can be discounted in evaluating the column rank of T. By definition, all M_k's are diagonal matrices, with M_0 being nonsingular ($k \nless 0$, $di \geq 0$). The equation above thus suggests that the rank condition for F = HC is satisfied if

$$\text{column rank (T)} = m \tag{9.43}$$

as the rows of C are linearly independent. Rearranging the rows of T and by eliminating the null rows, (9.43) can be expressed as

$$\rho(T) = \rho\begin{bmatrix} \{C_i\} \\ \vdots \\ \{C_iA^{di}\} \\ \{C_iA^{di+1}\} \end{bmatrix} = \rho\begin{bmatrix} \vdots \\ \hline C_i \\ \vdots \\ C_iA^{di} \\ C_iA^{di+1} \\ \hline \vdots \end{bmatrix} \left.\vphantom{\begin{matrix} C_i \\ \vdots \\ C_iA^{di} \\ C_iA^{di+1}\end{matrix}}\right\} di + 1 \text{ rows} \quad = m \tag{9.44}$$

Equation (9.44) provides an alternative condition for the existence of an equivalence between the state and output feedback decoupling control laws which does not require a priori knowledge of F. Once this condition is satisfied, the decoupling problem is to seek a solution of

$$F = HC \tag{9.45}$$

where F corresponds to the state feedback decoupling matrix (9.22) to (9.24). The corresponding equivalent output feedback matrix that will decouple S(•) is then derived through the following partitioning (Sec. 7.1)

$$[\underbrace{\hat{F}_1}_{m} \mid \underbrace{\hat{F}_2}_{n-m}] = H[\underbrace{\hat{C}_1}_{m} \mid \underbrace{\hat{C}_2}_{n-m}] \tag{9.46}$$

such that $\hat{C}_1$ is nonsingular, which can always be derived through elementary column operations on C, since $\rho[C] = m$. The matrix F is then partitioned such that the m columns of C forming $\hat{C}_1$ correspond to the respective m column in $\hat{F}_1$. Partitioning in this way, as seen in Sec. 7.1, will not be unique, but it will be sufficient to place any nonsingular group of m columns in the position C_1. Thus if (9.44) is satisfied, the equivalent output feedback decoupling matrix may be derived by

$$H = \hat{F}_1[\hat{C}_1]^{-1}, \quad \text{where } \hat{F}_1[\hat{C}_1]^{-1}\hat{C}_2 = \hat{F}_2 \tag{9.47}$$

The method is demonstrated below.

Example 9.5: Output feedback decoupling when $F \equiv HC$ The following system is used to demonstrate the derivation of H from a decoupling F using (9.47):

$$S(A, B, C): \; A = \begin{bmatrix} 1 & 1 & 0 \\ 0 & 1 & 0 \\ 0 & 0 & 1 \end{bmatrix}; \quad B = \begin{bmatrix} 0 & 1 \\ 1 & 0 \\ 1 & 0 \end{bmatrix}; \quad C = \begin{bmatrix} 1 & 1 & -1 \\ 0 & 1 & 0 \end{bmatrix} \tag{9.48}$$

For this system $d1 = d2 = \delta = 0$ and

$$B^* = \begin{bmatrix} 0 & 1 \\ 1 & 0 \end{bmatrix}$$

Thus a state feedback decoupling pair (F, G) exists. From (9.44) the rank of the matrix T is given by

$$\rho(T) = \rho \begin{bmatrix} C \\ CA \end{bmatrix} = \rho \begin{bmatrix} 1 & 1 & -1 \\ 0 & 1 & 0 \\ 1 & 2 & -1 \\ 0 & 1 & 0 \end{bmatrix} = 2 = m$$

Hence state feedback decoupling can be extended to output feedback pling. From (9.22) to (9.24), the state feedback decoupling pair is

$$F = \begin{bmatrix} 0 & m_{02}-1 & 0 \\ m_{01}-1 & m_{01}-2 & -m_{01}+1 \end{bmatrix}; \quad G = \begin{bmatrix} 0 & 1 \\ 1 & 0 \end{bmatrix} \quad \text{with } \Gamma = I, \; M_o = \operatorname{diag}\{m_{01}, m_{02}\}$$

Using the partitioning of (9.46) we have

$$\left[\underbrace{\begin{matrix} 0 & m_{02}-1 \\ m_{01}-1 & m_{01}-2 \end{matrix}}_{\hat{F}_1} \;\middle|\; \underbrace{\begin{matrix} 0 \\ -m_{01}+1 \end{matrix}}_{\hat{F}_2}\right] = H \left[\underbrace{\begin{matrix} 1 & 1 \\ 0 & 1 \end{matrix}}_{\hat{C}_1} \;\middle|\; \underbrace{\begin{matrix} -1 \\ 0 \end{matrix}}_{\hat{C}_2}\right]$$

The output feedback decoupling pair (H, G) is, from (9.47),

$$H = \begin{bmatrix} 0 & m_{02}-1 \\ m_{01}-1 & -1 \end{bmatrix} \quad \text{and} \quad G = \begin{bmatrix} 0 & 1 \\ 1 & 0 \end{bmatrix} \tag{9.49}$$

<u>Example 9.6: Decoupling for the case $F \not\equiv HC$</u> The system below illustrates the case when the consistency condition for F = HC is not satisfied.

$$S(A, B, C): \; A = \begin{bmatrix} -1 & 1 & 0 \\ 0 & -4 & 1 \\ 0 & -2 & -1 \end{bmatrix}; \quad B = \begin{bmatrix} 1 & 0 \\ 1 & 1 \\ 0 & -1 \end{bmatrix}; \quad C = \begin{bmatrix} 1 & 1 & 0 \\ 0 & 1 & -1 \end{bmatrix} \tag{9.50}$$

For this system, $d1 = d2 = \delta = 0$ and

$$\rho(T) = \rho \begin{bmatrix} CA \\ C \end{bmatrix} = \begin{bmatrix} -1 & -3 & 1 \\ 0 & -2 & 2 \\ 1 & 1 & 0 \\ 0 & 1 & -1 \end{bmatrix} = 3 < m$$

and hence state feedback decoupling cannot be extended to output feedback decoupling.

To illustrate the consistency requirement derived above, state feedback decoupling matrix is evaluated by using (9.22) to (9.24):

$$F = (B^*)^{-1}[M_oC - A^*] = \Big[\underbrace{\begin{matrix} 2m_{01}+2 & 2m_{01}-m_{02}+4 \\ -m_{01}-1 & -m_{01}+2m_{02}+1 \end{matrix}}_{\hat{F}_1} \;\Big|\; \underbrace{\begin{matrix} m_{02} \\ -2m_{02}-3 \end{matrix}}_{\hat{F}_2}\Big] \tag{9.51}$$

Corresponding to the partitioning of F above, the output matrix C is partitioned as

$$C = \Big[\underbrace{\begin{matrix} 1 & 1 \\ 0 & 1 \end{matrix}}_{\hat{C}_1} \;\Big|\; \underbrace{\begin{matrix} 0 \\ -1 \end{matrix}}_{\hat{C}_2}\Big] \tag{9.52}$$

From the equations above, for any arbitrary value of $M_o = \text{diag}\{m_{01}, m_{02}\}$,

$$\hat{F}_1[\hat{C}_1]^{-1}\hat{C}_2 = \frac{1}{3}\begin{bmatrix} m_{02} - 2 \\ -2m_{02} - 2 \end{bmatrix} \neq \hat{F}_2 = \begin{bmatrix} m_{02} \\ -2m_{02} - 3 \end{bmatrix} \tag{9.53}$$

Hence state feedback decoupling cannot be extended to decoupling by using output variables. The system in (9.50) is considered in the following section to illustrate a synthesis procedure of extending state feedback decoupling to output feedback in conjunction with a cascade compensator. The matrix G for state feedback decoupling of (9.50) is given by

$$G = (B^*)^{-1}\Gamma = \frac{1}{3}\begin{bmatrix} 2 & -1 \\ -1 & 2 \end{bmatrix} \quad \text{with } \Gamma = I \tag{9.54}$$

9.3 PRECOMPENSATION AND DECOUPLING

The scope of dynamic compensation in the decoupling analysis of multivariable systems is considered here. The derivations presented below are concerned with (1) systems† that cannot be decoupled by state feedback (i.e., systems with singular B*) and (2) systems with nonsingular B* but for which F = HC does not have a solution for H [i.e., (9.43) is not satisfied].

9.3.1 Precompensation for State Feedback Decoupling [4, 12, 13]

The system considered here has a set of integers di defined in (9.18), but the resulting B* is singular [i.e., $\rho(B^*) < m$]. The derivation below shows that if the system is output controllable, a cascade compensator may be evaluated which when precompensate S(A, B, C) (Fig. 9.3) makes the augmented system state feedback decouplable system, the augmented system in Fig. 9.3 being represented by

$$\bar{S}(\bar{A}, \bar{B}, \bar{C}): \;\; \dot{\bar{x}}(t) = \bar{A}\bar{x}(t) + \bar{B}\bar{u}(t); \quad \bar{y}(t) = \bar{C}\bar{x}(t) \tag{9.55}$$

where

$$\bar{A} = \underbrace{\begin{bmatrix} A & BC_c \\ 0 & A_c \end{bmatrix}}_{n+n_c} \Big\} n+n_c; \quad \bar{B} = \underbrace{\begin{bmatrix} BD_c \\ B_c \end{bmatrix}}_{m} \Big\} n+n_c; \quad \bar{C} = \underbrace{[C \;\; 0]}_{n+n_c} \} m$$

The state vector of the augmented system is

†Systems with det {B*} = 0 are said to have weak inherent coupling [4].

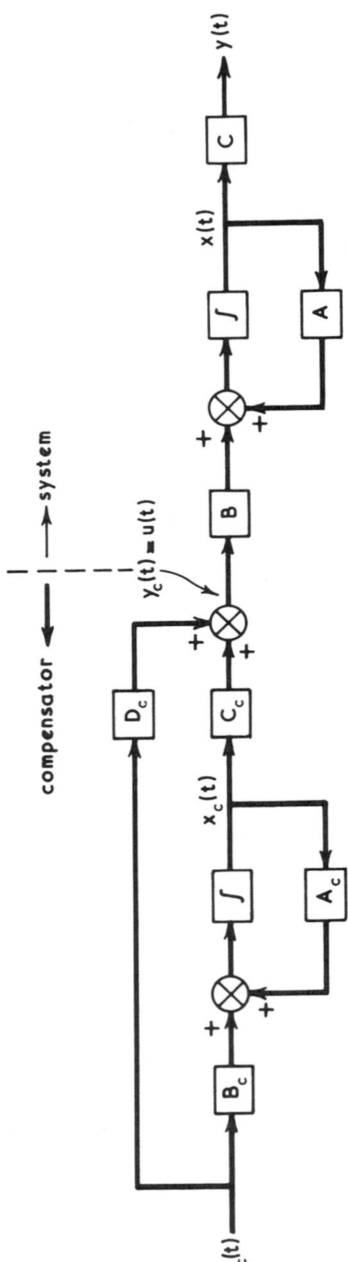

Fig. 9.3 Precompensation for state feedback decoupling.

$$\bar{x}(t) = \begin{bmatrix} x(t) \\ x_c(t) \end{bmatrix}$$

and for the original system $\rho(B^*) = \rho\{C_i A^{di} B\} < m$. Combining (9.18) and (9.55), the following definitions are introduced for $\bar{S}(\bar{A}, \bar{B}, \bar{C})$:

$$\bar{d}i = \begin{cases} \min k \text{ for } \bar{C}_i \bar{A}^k \bar{B} \neq 0 \\ (n+n_c-1) \text{ if } \bar{C}_i \bar{A}^k \bar{B} = 0, \ \forall k \in 0 \end{cases} \quad \begin{matrix} i \in 1, \ldots, m \\ j \in 1, \ldots, (n+n_c-1) \end{matrix} \tag{9.56}$$

$$\bar{\delta} = \max \bar{d}i$$

and consequently

$$\bar{B}_i^* = \bar{C}_i \bar{A}^{\bar{d}i} \bar{B} = [C_i \quad 0] \begin{bmatrix} A & BC_c \\ 0 & A_c \end{bmatrix}^{\bar{d}i} \begin{bmatrix} BD_c \\ B_c \end{bmatrix}$$

$$= [C_i A^{\bar{d}i} B \quad C_i A^{\bar{d}i-1} B \cdots C_i A^{di} B] \begin{bmatrix} D_c \\ C_c B_c \\ \vdots \\ C_c A_c^{\bar{d}i-di-1} B_c \end{bmatrix}, \quad i \in 1, \ldots, m \tag{9.57a}$$

Thus

$$\bar{B}^* = \{B_i^*\} = \underbrace{\begin{bmatrix} C_1 A^{\bar{d}1} B & \cdots & C_1 A^{d1} B & 0 & & & \cdots & 0 \\ & & & \ddots & & & & \\ C_i A^{\bar{d}i} B & \cdots & & & C_i A^{di} B & 0 & \cdots & 0 \\ & & & & & \ddots & & \\ C_m A^{\bar{d}m} B & \cdots & & & & C_m A^{dm} B & \cdots & 0 \end{bmatrix}}_{R} \underbrace{\begin{bmatrix} D_c \\ C_c B_c \\ \vdots \\ C_c A_c^{q-1} B_c \end{bmatrix}}_{T} \tag{9.57b}$$

where $q = \bar{\delta} - \delta$ and 0 is an $1 \times m$ null row vector. This leads to the following theorem.

Theorem 9.5 An output controllable (or invertible) system with singular B^* can always be modified by dynamic precompensation so that the augmented system has nonsingular $\bar{B}^*$.

Proof: From (9.57), the rank of $\bar{B}^*$ is equal to the least rank of the two matrices R and T on the right-hand side, and since T consists of parameters of the compensator $S_c(A_c, B_c, C_c, D_c)$, the rank of T can always be made m, so that the rank of $\bar{B}^*$ is equal to that of R. It can easily be seen from (9.57) that R is a submatrix of the output controllability matrix Q given by

$$Q = [CB \mid CAB \cdots CA^{n-1}B]$$

so that linear independence of the rows of R implies linear independence of the rows of Q as well. Consequently, if Q has rank m, the rows R can always be made independent by suitable choice of pi such that

$$\bar{d}i < di + pi < n - 1, \qquad i \in 1, \ldots, m$$

[i.e., $di < n - 1$ (therefore, $pi > 0$). Hence the proof.

Compensator order: It can be seen from (9.57) the order of the compensator necessary to make $\bar{B}^*$ nonsingular is related in the values of pi, $i \in 1, \ldots, m$, with $\bar{\delta} - \delta = \max pi$.

1. Necessary values of pi: Supposing that $pi = 1, \forall i$, (9.56) and (9.57) give

$$\bar{B}^* = A^*BD_c + B^*C_cB_c \tag{9.58}$$

which may be made to have rank m by suitable choice of C_c and B_c.
2. Sufficient values of pi: Let pi be given by

$$\begin{aligned} pi &= 1, & i &\in 1, \ldots, \overline{m - r} \\ &= 1 + j, & i &\in \overline{m - r + 1}, \ldots, m, \quad j = 1, 2, \ldots, r \ (0 < r < m) \end{aligned} \tag{9.59}$$

where $(m - r)$ is the number of linearly independent rows in B^* arranged to occupy the first $(m - r)$ positions in $\bar{B}^*$ in (9.57). By appropriate numbering of input/output terminals such that $d_1 < d_2 < \cdots < d_m$, this implies that the pi values given in (9.59) are sufficient to give linear independence of the last r rows of R such that R has rank m.

The significance of the minimal value of $pi = 1, \forall i$, is that the compensator introduces at least one state in each of the m-input/m-output channels in Fig. 9.3. Furthermore, no compensator state can appear in more than one input/output channel of the compensator, as this would imply identical rows in C_c. The least order of $S_c(\cdot)$ necessary is thus m.

The sufficient values of pi in (9.59) imply that the compensator dynamics would be interconnected in such a way that

$$\sum_{i=1}^{m} pi = \sum_{i=1}^{m-r} pi + \sum_{i=m-r+1}^{m} pi = m + \sum_{j=1}^{r} j, \quad 0 < r < m \tag{9.60}$$

As the maximum value of r is (m - 1), it is always possible to construct a compensator of order m such that (9.60) is satisfied while maintaining the rank of T in (9.57) as m. The computation produced is illustrated below.

Example 9.7: State feedback decoupling through compensation The system considered is represented by

$$S(A, B, C): \; A = \begin{bmatrix} -1 & 0 & 0 \\ 0 & -2 & 0 \\ 0 & 0 & -3 \end{bmatrix}; \quad B = \begin{bmatrix} 1 & 0 \\ 1 & 1 \\ 1 & 0 \end{bmatrix}; \quad C = \begin{bmatrix} 1 & 1 & 0 \\ 0 & 1 & 1 \end{bmatrix};$$

$$m = 2, \quad n = 3 \tag{9.61}$$

The output controllability matrix is

$$Q = [CB \,\vdots\, CAB \,\vdots\, CA^2B] = \left[\begin{array}{cc|cc|cc} 2 & 1 & -3 & -2 & 5 & 4 \\ 2 & 1 & -5 & -2 & 13 & 4 \end{array}\right] \tag{9.62}$$

which has rank 2. From above, by inspection

$$d_1 = d_2 = \delta = 0; \quad B^* = \begin{bmatrix} 2 & 1 \\ 2 & 1 \end{bmatrix}; \quad \rho(B^*) = 1 < m$$

The system therefore has weak inherent coupling. The stages of designing a precompensator by using Theorem 9.5 is illustrated below [12]. By using the property of $\bar{d}i$, from (9.56), for the augmented system $\bar{S}(\bar{A}, \bar{B}, \bar{C})$ (Fig. 9.3),

$$\bar{C}_i \bar{A}^{\bar{d}i-1} \bar{B} = C_i A^{di} B D_c = 0 \quad \text{or} \quad B^* D_c = 0 \tag{9.63}$$

The feedforward matrix D_c for the compensator is therefore given by

$$\begin{bmatrix} 2 & 1 \\ 2 & 1 \end{bmatrix} D_c = 0 \quad \text{or} \quad D_c = \begin{bmatrix} 1 & 1 \\ -2 & -2 \end{bmatrix} \tag{9.64}$$

For di = 0, $\forall$ i, from (9.57b), and the order of compensator is $\eta_o = m = 2$, and

$$\bar{B}^* = CABD_c + CBC_cB_c$$

since A_c does not appear in the equation above, its choice is arbitrary. Let $\bar{B}^*$ be chosen as

$$\begin{bmatrix} 1 & 0 \\ -1 & -2 \end{bmatrix}$$

and C_c as

$$\begin{bmatrix} 1 & 0 \\ 0 & 1 \end{bmatrix}$$

Then, from the equation above,

$$\begin{bmatrix} 1 & 0 \\ -1 & -2 \end{bmatrix} = \begin{bmatrix} -3 & -2 \\ -5 & -2 \end{bmatrix}\begin{bmatrix} 1 & 1 \\ -2 & -2 \end{bmatrix} + \begin{bmatrix} 2 & 1 \\ 2 & 1 \end{bmatrix}\begin{bmatrix} 1 & 0 \\ 0 & 1 \end{bmatrix} B_c = \begin{bmatrix} -1 & -1 \\ 2 & 1 \end{bmatrix} \tag{9.65}$$

Let A_c be chosen as

$$\begin{bmatrix} \alpha_{11} & \alpha_{12} \\ \alpha_{21} & \alpha_{22} \end{bmatrix}$$

Then the augmented system is represented by, from (9.55),

$$\bar{S}(\bar{A}, \bar{B}, \bar{C}): \ \bar{A} = \begin{bmatrix} -1 & 0 & 0 & 1 & 0 \\ 0 & -2 & 0 & 1 & 1 \\ 0 & 0 & -3 & 1 & 0 \\ 0 & 0 & 0 & \alpha_{11} & \alpha_{12} \\ 0 & 0 & 0 & \alpha_{21} & \alpha_{22} \end{bmatrix}; \quad \bar{B} = \begin{bmatrix} 1 & 1 \\ -1 & -1 \\ 1 & 1 \\ -1 & -1 \\ 2 & 1 \end{bmatrix};$$

$$\bar{C} = \begin{bmatrix} 1 & 1 & 0 & 0 & 0 \\ 0 & 1 & 1 & 0 & 0 \end{bmatrix} \tag{9.66}$$

for which

$$\bar{B}^* = \begin{bmatrix} 1 & 0 \\ -1 & -2 \end{bmatrix}$$

$\bar{d}i = 1, \forall i$, and the state feedback decoupling pair $(\bar{F}, \bar{G})$ may now be calculated using (9.22) to (9.24).

9.3.2 Precompensation for Output Feedback Decoupling [9-11]

The system considered here has a set of integers di as given by (9.18) and a nonsingular B*, but for which the rank condition in (9.44) is not satisfied. The following derivations show that such a system can be suitably compensated such that the augmented systems state feedback matrix will satisfy the condition in (9.47) so that an equivalent output feedback decoupling matrix of the compensated system may be derived. The m-input/m-output purely dynamic cascade compensator $S_c(A_c, B_c, C_c)$ described by

$$\begin{aligned} \dot{x}_c(t) &= A_c x_c(t) + B_c u_c(t) \\ y_c(t) &= C_c x_c(t) \end{aligned} \tag{9.67}$$

where $x_c(t)$ is the n-compensator state, is considered here. The aim of compensation is to modify the state feedback decoupling pair $(\bar{F}, \bar{H})$ of the overall system (Fig. 9.4) $\bar{S}(\bar{A}, \bar{B}, \bar{C})$:

$$\begin{aligned} \dot{\bar{x}}(t) &= \bar{A}\bar{x}(t) + \bar{B}\bar{u}(t) \\ \bar{y}(t) &= \bar{C}\bar{x}(t) \end{aligned} \tag{9.68}$$

where

$$\bar{A} = \begin{bmatrix} A & BC_c \\ 0 & A_c \end{bmatrix}; \quad \bar{B} = \begin{bmatrix} 0 \\ B_c \end{bmatrix}; \quad \bar{C}[C : 0]; \quad \bar{x}(t) = \begin{bmatrix} x(t) \\ x_c(t) \end{bmatrix}$$

such that an equivalent output feedback decoupling matrix $\bar{H}$ can be evaluated from the augmented equation

$$\bar{F} = \bar{H}\bar{C} \tag{9.69}$$

In order to derive the decoupling pair $(\bar{F}, \bar{G})$ for the augmented system in (9.68), the following definitions, along the lines of (9.18), are introduced.

$$\text{dci} = \left\{ \begin{array}{l} \min k\text{:} \; C_{ci}A_c^k B_c = 0 \;\; \text{for } k = 0, 1, \ldots, (n_c - 1) \\ (n_c - 1) \text{ if } C_{ci}A_c^k B_c = 0 \;\; \forall k \end{array} \right\} \begin{array}{l} \beta = \text{minimum } d_{ci}, \\ i \in 1, \ldots, m \end{array} \tag{9.70}$$

$$\bar{d}i = \left\{ \begin{array}{l} \min k\text{:} \; \bar{C}\bar{A}^k\bar{B} = 0 \;\; \text{for } k = 0, 1, \ldots, (n + n_c - 1) \\ (n + n_c - 1) \text{ if } \bar{C}\bar{A}^k\bar{B} = 0 \;\; \forall k \end{array} \right\} \quad \bar{\delta} = \text{maximum } \bar{d}i$$

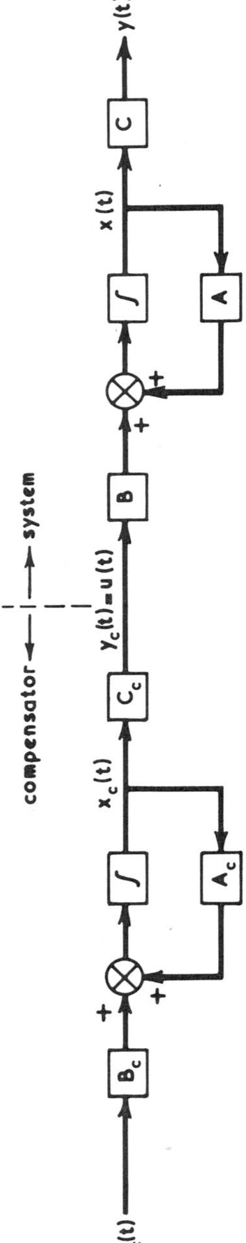

Fig. 9.4 Precompensation for output feedback decoupling.

$$\bar{A}^* = \bar{C}_i \bar{A}^{\bar{d}i+1} \bar{B} \quad \text{and} \quad A_c^* = C_{ci} A_c^{dci+1} B_c$$

From (9.18), (9.69), and (9.70) it can be shown that

$$\bar{C}_i \bar{A}^k \bar{B} = \begin{cases} 0 & \text{for } k = 0, \ldots, (di + dci) \\ C_i A^{di} B C_c A_c^{dci} B_c & \text{for } k = di + dci + 1 \\ 0 & \text{for } k \in (di + dci + 2) \end{cases}$$

Thus

$$\bar{d}i = di + dci + 1; \quad \beta = dci; \quad \bar{\delta} = \delta + \beta + 1; \quad \bar{B}^* = B^* B_c^* \tag{9.71}$$

Furthermore,

$$\bar{A}^* = [A^* A^{\beta+1} \mid A^k A^{\beta} B C_c + \cdots + A^* B C_c A_c^{\beta} + B^* A_c^*] \tag{9.72}$$

Combining the equations, the decoupling pair $(\bar{F}, \bar{G})$ for $\bar{S}(\bar{A}, \bar{B}, \bar{C})$ is given by

$$\begin{aligned} \bar{F} &= (\bar{B}^*)^{-1} \left[\sum_{k=0}^{\bar{\delta}} M_k \bar{C} \bar{A}^k - \bar{A}^* \right] \\ &= (B^* B_c^*)^{-1} \left[\sum_{k=0}^{\delta+\beta+1} M_k C A^k - A^* A^{\beta+1} \,\middle|\, \sum_{\substack{k=0 \\ p=1}}^{\substack{p=k-1 \\ k=\delta+\beta+1}} M_k C A^{k-1} B C_c A_c^{p-1} - (A^* B C_c + B^* C_c A_c) \right] \\ &\equiv [\tilde{F} \mid \tilde{E}] \end{aligned} \tag{9.73}$$

and

$$\bar{G} = \Gamma (B^* B_c^*)^{-1} \quad (\Gamma = \text{diagonal})$$

The partition of $\bar{F}$ corresponds to that of $\bar{C}$ into $[C : 0]$. It is apparent from (9.73) that the compensation has introduced significant differences between $F = HC$ for the original system and $\bar{F} = \bar{H}\bar{C}$ for the augmented system. These differences are capable of ensuring consistency of the latter when the former equation is inconsistent. It is to be noted that the second partitioned submatrix in $\bar{F}$ (i.e., $\tilde{E}$) must be null to ensure the overall consistency of $\bar{F} = \bar{H}\bar{C}$. The compensator parameters appear in the first as well as the second submatrices of $\bar{F}$, and must clearly be constrained such that the latter is null. This requirement forms the basis of design procedure described below through numerical examples.

Extending the interpretation of di to β, the number of state variables ($\equiv n_c$) introduced through $S_c(A_c, B_c, C_c)$ is $m(\beta + 1)$. Since there are $(n - m)$

columns in $\tilde{E}$, the minimum values of β for mathematical consistencies is $\beta_o = n - m - 1$. This minimum value β_o is based on analytical requirements, and in general will impose numerical constraints on a number of elements of M_k, $k \in (\delta + 1), \ldots, (\delta + \beta + 1)$, to satisfy the decoupling requirements discussed above. If some of these M_k elements need to be positive, by significance of M's in Sec. 9.1, the resulting decoupled channels will have poles on the right-hand s-plane.

Example 9.8: Output feedback decoupling through compensation The system in (9.50) is considered here to illustrate the synthesis procedure of extending the class of state feedback decoupling matrix F to achieve consistency of (9.45) and (9.46).

The order of the necessary compensator is $n_c = m(\beta + 1)$, where $\beta = n - m - 1$, thus $n_c = 2$. From (9.50) and (9.73), the state feedback decoupling matrix of the compensated system is

$$\bar{F} = [B^*B_c^*]^{-1}[M_o C + M_1 CA - CA.A \quad : \quad M_1 CBC_c - A^*BC_c - B^*A_c]$$

Since $\bar{F} = \bar{H}\bar{C} = \bar{H}\,[C \mid 0]$, from the above, for consistency of (9.69),

$$\bar{H}\bar{C} = [B^*B_c^*]^{-1}[M_o C + M_1 A - CA^2] \tag{9.74}$$

and

$$0 = M_1 CBC_c - A^*BC_c - B^*A_c \tag{9.75}$$

where $M_o = \text{diag}\{m_{01}, m_{02}\}$ and $M_1 = \text{diag}\{m_{11}, m_{12}\}$. Choosing $B_c = C_c = I$, from above,

$$\bar{H}C = \frac{1}{3}\left[\underbrace{\begin{matrix} 2m_{01} - 2m_{11} - 2 & 2m_{01} - m_{02} - 6m_{11} + 2m_{12} - 14 \\ -m_{01} + m_{11} + 1 & -m_{01} + 2m_{02} + 3m_{11} - 4m_{12} + 1 \end{matrix}}_{\hat{\tilde{F}}_1} \;\middle|\; \underbrace{\begin{matrix} m_{02} + 2m_{11} - 2m_{12} + 4 \\ -2m_{02} - m_{11} + 4m_{12} + 4 \end{matrix}}_{\hat{\tilde{F}}_2}\right]$$

To test the consistency of $\bar{F} = \bar{H}C$ according to (9.45) and (9.46), the following identity will have to be satisfied:

$$\hat{\tilde{F}}_1[\hat{C}_1]^{-1}\hat{C}_2 = \hat{\tilde{F}}_2$$

or

$$\begin{bmatrix} m_{02} + 4m_{11} - 2m_{12} + 12 \\ -2m_{02} - 2m_{11} + 4m_{12} \end{bmatrix} = \begin{bmatrix} m_{02} + 2m_{11} - 2m_{12} + 4 \\ -2m_{02} - m_{11} + 4m_{12} + 4 \end{bmatrix}$$

which gives $m_{11} = -4$ while m_{12} remains arbitrary. When this constraint is

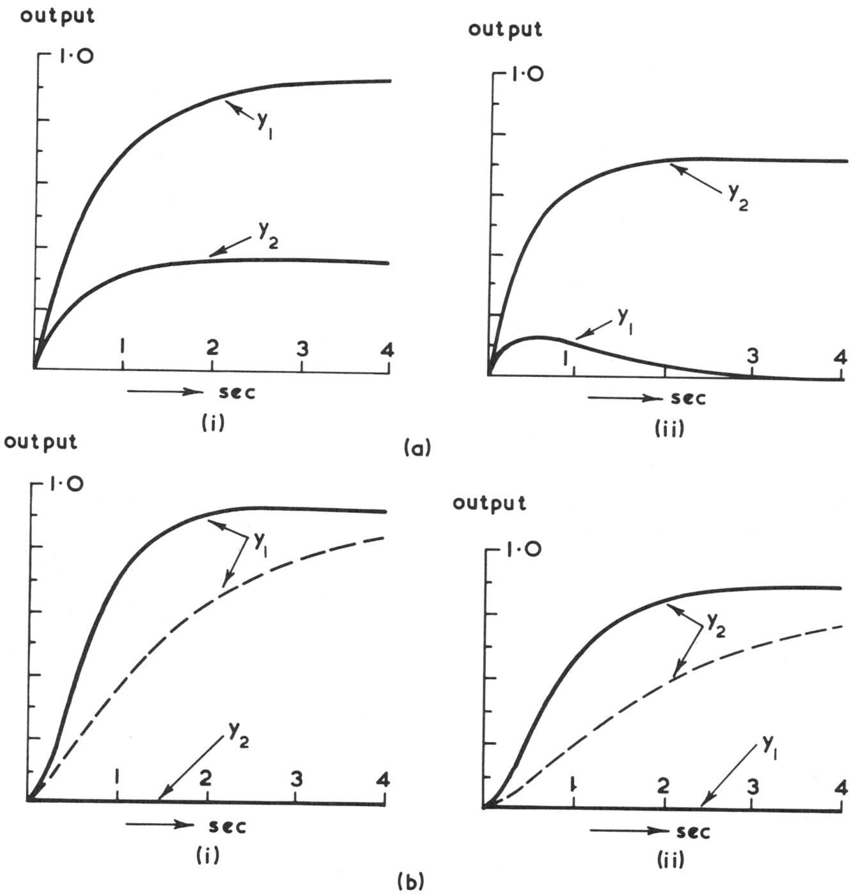

Fig. 9.5 Step responses of the (a) open-loop system in (9.50) and (b) closed-loop system with (9.76) and (9.77). (i) $u_1(t) = 1.0$, $u_2(t) = 0$; (ii) $u_1(t) = 0$, $u_2(t) = 1.0$ [$m_{01} = m_{11} = m_{12} = -4$, $m_{02} = -1$; $\cdots$ $m_{01} = -1$, $m_{02} = m_{12} = -4$, $m_{11} = -2.5$].

inserted in (9.75), the compensator dynamics matrix A_c is formulated as

$$A_c = \frac{1}{3}\begin{bmatrix} -(m_{12} + 10) & -2(m_{12} + 2) \\ 2(m_{12} + 4) & 4(m_{12} + 2) \end{bmatrix} \tag{9.76}$$

and the corresponding $\bar{H}$ is given by

$$\bar{H} = \frac{1}{3}\begin{bmatrix} 2(m_{01}+3) & -m_{02}+2m_{12}+4 \\ -(m_{01}+3) & 2(m_{02}-2m_{12}-4) \end{bmatrix} \tag{9.77}$$

while $\bar{G} = G$ and as given in (9.54).

To demonstrate the effectiveness of the control above, step responses of the system S(A, B, C) with state feedback (9.51) and output feedback (9.76) decoupling control laws are shown in Fig. 9.5.

9.4 DECOUPLING THROUGH MATRIX FRACTION [14-18]*

In many applications, the state-space representation of the system to be decoupled may not be available. Although the transformation of the system transfer-function matrix T(s) into S(A, B, C) may be derived by using any of the realization methods in Chap. 4, a mechanism of deriving an appropriate decoupling output feedback matrix from a given T(s) has computational advantages. This section outlines a fairly simple method of using the matrix-fraction description of a strictly proper T(s) to derive an output feedback decoupling control law.

The derivations here are based on the following preliminary results.

1. The open-loop m-square transfer-function matrix T(s) may be expanded as

$$T(s) = N(s)[D(s)]^{-1} \tag{9.78}$$

where N(s) and D(s) are right co-prime [≡ relatively right prime (Sec. 1.5)].

2. There exists a set of integers di, $i \in 1, \ldots, m$,

$$di = \begin{cases} 0 & \text{if } T_{ir}(s) \equiv 0 \\ \min k: \lim_{s\to\infty} s^{k+1} T_{ir}(s) \neq 0, & k = 0, 1, \ldots \end{cases}$$

such that the m-square matrix B* is nonsingular, where the m-row vectors of B* is defined as

$$B^*_{ir} = \lim_{s\to\infty} s^{di+1} T_{ir}(s), \quad i \in 1, \ldots, m \tag{9.79}$$

*Derivations here follow Ref. 14 closely. T(s) is used here to denote a transfer-function matrix to distinguish it from the constant gain matrix G used in this section.

$T_{ir}(s)$ being the ith row vector of T(s). Both of these conditions are satisfied if T(s) is strictly proper and the system is output controllable—which are consistent with the conditions for state feedback decoupling.

3. With the assumptions above, and using results derived earlier, the following polynomial matrix may be readily formed:

$$R(s) = B^* D(s)[N(s)]^{-1}\psi(s) \tag{9.80a}$$

where

$$\psi(s) = \text{diag}\,\{\psi_i(s)\} \tag{9.80b}$$

and $\psi_i(s)$ being the greatest common divisor of the ith row of N(s), $i \in 1, \ldots, m$.

4. The transfer-function matrix of a closed-loop decoupled system may be expressed as

$$\bar{T}(s) = \text{diag}\left\{\gamma_i \frac{z_i(s)}{p_i(s)}\right\}, \qquad i \in 1, \ldots, m \tag{9.81}$$

where $\Pi_{i=1}^{n} \gamma_i \neq 0$, and $z_i(s)$ and $p_i(s)$ are relatively prime monic polynomials.

The following theorem may now be stated [14].

Theorem 9.6 A necessary and sufficient condition that the system with transfer-function matrix $T(s) \equiv C[sI - A]^{-1}B$ can be decoupled by the constant-output feedback control law

$$u(s) = Hy(s) + Gv(s) \tag{9.82}$$

in that the closed-loop transfer-function matrix has the form of (9.81) and all the following three conditions hold:

1. B^* as defined in (9.79) is nonsingular.
2. $[\psi(s)]^{-1}N(s)$ in (9.80a) for the given T(s) is unimodular.
3. $\psi_j(s)$, $j \in 1, \ldots, m$, are constant multipliers of the off-diagonal elements of $R(s) = \{r_{ij}(s)\}$ [i.e., if $r_{ij}(s)/\psi_j(s)$ is constant (independent of s) for $i, j \in 1, \ldots, m$, $i \neq j$].

A constructive proof of the theorem is given here. The transfer-function matrix of the closed-loop system may be expressed as

$$\begin{aligned}\bar{T}(s) = T(s)[I + HT(s)]^{-1}G &\equiv N(s)[D(s)]^{-1}[1 + HN(s)[D(s)]^{-1}]^{-1}G \\ &= N(s)[G^{-1}D(s) + G^{-1}HN(s)]^{-1}\end{aligned} \tag{9.83}$$

Since N(s) and D(s) are assumed to be relatively right prime, so are N(s) and $[G^{-1}D(s) + G^{-1}HN(s)]$: Let the closed-loop system in (9.83) be decoupled. Then $\bar{T}(s)$ may be expressed in the form (9.81)

$$\bar{T}(s) = \Gamma Z(s)[P(s)]^{-1} \tag{9.84}$$

where

$$\Gamma = \text{diag}\{\gamma_i\}; \quad \prod_{i=1}^{m} \gamma_i \neq 0; \quad Z(s) = \text{diag}\{z_i(s)\}; \quad P(s) = \text{diag}\{p_i(s)\}, \quad i \in 1, \ldots, m$$

Z(s) and P(s) are relatively right prime. Thus by combining (9.83) and (9.84) and using the definition of relative primeness, there exists an unimodular matrix U(s) such that

$$N(s) = \Gamma Z(s)U(s)$$

and (9.85)

$$G^{-1}D(s) + G^{-1}HN(s) = P(s)U(s)$$

Eliminating U(s) from above, we have

$$G^{-1}D(s) + G^{-1}HN(s) = P(s)[\Gamma Z(s)]^{-1}N(s)$$

$$\rightarrow G^{-1}D(s)[N(s)]^{-1}Z(s)\Gamma + G^{-1}HZ(s)\Gamma = P(s) \qquad [\Gamma Z(s) \equiv Z(s)\Gamma]$$

$$\rightarrow \underbrace{B^*D(s)[N(s)]^{-1}Z(s)}_{\substack{R(s)\\ \text{nondiagonal}}} + \underbrace{B^*H\,Z(s)}_{\substack{\bar{H}\\ \text{diagonal}}} = \underbrace{P(s)}_{\text{diagonal}} \qquad [G = B^{*-1}\Gamma] \qquad [Z(s) \rightarrow \equiv \psi(s)] \tag{9.86}$$

$$\rightarrow r_{ij}(s) + \bar{h}_{ij}\psi_j = \begin{Bmatrix} p_i(s) & i = j \\ 0 & i \neq j \end{Bmatrix} \quad i, j \in 1, \ldots, m \tag{9.87}$$

Thus with the choice $Z(s) = \psi(s)$, all three conditions in the theorem are satisfied. The necessary decoupling pair is given by

$$G = B^{*-1}\Gamma \quad \text{and} \quad H = B^{*-1}\bar{H} \tag{9.88}$$

where $\Gamma = \text{diag}\{\gamma_i\}$ is chosen arbitrarily. The method of computing B^{*-1} and $\bar{H}$ from a given T(s) is shown below.

<u>Example 9.9 [14]: Output feedback decoupling through matrix fractions</u> The two-input/two-output system represented by

$$T(s) = \begin{bmatrix} \dfrac{s+1}{s^2-3s+2} & 0 \\ \dfrac{s+1}{s^3-4s^2+s+2} & \dfrac{s^2-3s-2}{s^3-4s^2+s+2} \end{bmatrix} \tag{9.89}$$

is considered here.

Step 1—Computation of d_i and B^* from (9.79):

$$B^*_{1r} = \lim_{s\to\infty} sT_{1r}(s) = [1 \quad 0]; \qquad B^*_{2r} = \lim_{s\to\infty} sT_{2r}(s) = [0 \quad 1]$$

$$\rightarrow d_1 = d_2 = 0, \qquad B^* = \begin{bmatrix} 1 & 0 \\ 0 & 1 \end{bmatrix} \tag{9.90}$$

Thus condition (1) of Theorem 9.6 is satisfied.

Step 2—Matrix-fraction expansion of T(s):

$$T(s) = \begin{bmatrix} s^2-1 & s^3-4s^2+s+6 \\ s+1 & s^2-3s+2 \end{bmatrix} \begin{bmatrix} s^3-4s^2+s+2 & s^4-8s^3+9s^2-8s-2 \\ 0 & 4 \end{bmatrix}^{-1}$$

$$\left.\begin{aligned} \psi_1(s) &= s+1 \\ \psi_2(s) &= 1 \end{aligned}\right\} \psi(s) = \begin{bmatrix} s+1 & 0 \\ 0 & 1 \end{bmatrix} \tag{9.91}$$

Hence

$$[\psi(s)]^{-1}N(s) = \begin{bmatrix} s-1 & s^3-5s+6 \\ s+1 & s^2-3s-2 \end{bmatrix} \tag{9.92}$$

which is unimodular, and hence condition (2) of Theorem 9.6 is satisfied.

$$R(s) = B^*D(s)[N(s)]^{-1}\psi(s) = \begin{bmatrix} s^2-3s-2 & 0 \\ -(s+1) & s-1 \end{bmatrix} \tag{9.93}$$

Step 3—Existence of decoupling H from (9.87):

$$r_{12}(s) = 0 \rightarrow \bar{h}_{12} = 0 \rightarrow \text{constant}$$

$$r_{21}(s) = -(s+1) = -\bar{h}_{21}\psi_1(s) \rightarrow \bar{h}_{21} = -1 \rightarrow \text{constant}$$

which satisfies condition (3) of Theorem 9.6. Let $\bar{h}_{11}$ and $\bar{h}_{22}$ be chosen as 8 and 5, respectively. Then, from (9.87),

$$p_1(s) = r_{11}(s) + \bar{h}_{11}\psi_1(s) = (s^2 - 3s - 2) + 8(s + 1) = s^2 + 5s + 6$$

$$p_2(s) = r_{22}(s) + \bar{h}_{22}\psi_2(s) = (s - 1) + 5 = s + 4$$

Step 4—Computation of (H, G) from (9.88):

$$H = B^{*-1}\bar{H} = \begin{bmatrix} 8 & 0 \\ -1 & 5 \end{bmatrix}; \quad G = B^{*-1}\Gamma = \begin{bmatrix} \gamma_1 & 0 \\ 0 & \gamma_2 \end{bmatrix} \tag{9.94}$$

and the closed-loop system, for $\gamma_1 = \gamma_2 = 1$, is

$$\bar{T}(s) = T(s)[I + HT(s)]^{-1}G = \begin{bmatrix} \dfrac{s+1}{s^2+5s+6} & 0 \\ 0 & \dfrac{1}{s+4} \end{bmatrix} \equiv \begin{bmatrix} \dfrac{z_1(s)}{p_1(s)} & 0 \\ 0 & \dfrac{z_2(s)}{p_2(s)} \end{bmatrix} \tag{9.95}$$

The closed-loop decoupled system above is stable. It is perhaps worth noting here that the decoupling pair (H, G) derived by the method above may not always result in a stable closed-loop system. Once decoupling is achieved, however, the individual subsystems may be stabilized by using any of the single-input/single-output compensation techniques.

Despite the attractions of decoupling, it should be noted that the internal stability of a system is likely to be reduced when input/output noninteraction is achieved. An alternative approach to compensation for output feedback decoupling is to include an observer and accomplish the overall synthesis for decoupling by using the procedure outlined in Chap. 8.

REFERENCES

1. Morgan, B. S., "The synthesis of linear multivariable systems by state variable feedback," Preprints JACC, Stanford Univ., 1964, 468–472.
2. Rekasius, Z. V., "Decoupling of multivariable systems by means of state feedback," Preprints 3rd Annu. Allerton Conf. Circuits System Theory, Monticello, Ill., 1965, pp. 439–448.
3. Falb, P. L., and Wolovich, W. A., "Decoupling in the design and synthesis of multivariable control systems," Trans. IEEE, AC-12: 651–659 (1967).
4. Gilbert, E. G., "The decoupling of multivariable systems by state feedback," SIAM J. Control, 1: 50–61 (1969).
5. Sinha, P. K., "Controllability, observability and decoupling of multivariable systems," Int. J. Control, 26: 603–620 (1977).

6. Mufti, I. H., "On the observability of decoupled systems," Trans. IEEE, AC-14: 75-77 (1969).
7. Mufti, I. H., "A note on the decoupling of multivariable systems," Trans. IEEE, AC-14: 415-416 (1969).
8. Howze, J. W., "Necessary and sufficient conditions for decoupling using output feedback," Trans. IEEE, AC-18: 44-46 (1973).
9. Sinha, P. K., "A new condition for output feedback decoupling of multivariable systems," Trans. IEEE, AC-24: 476-478 (1979).
10. Hazlerigg, A. D. G., and Sinha, P. K., "Noninteracting control by output feedback and dynamic compensation," Trans. IEEE, AC-23: 76-79 (1978).
11. Wade, R. S., "Decoupling of linear time-invariant multivariable control systems," M.Sc. thesis, University of Sussex, 1970.
12. Sinha, P. K., "Dynamic compensation for state feedback decoupling of multivariable systems," Int. J. Control, 24: 673-684 (1978).
13. Cremer, M., "A pre-compensator of minimal order for decoupling a linear multivariable system," Int. J. Control, 15: 1089-1103 (1972).
14. Wang, S. H., and Davison, E. J., "Design of decoupled control systems: a frequency domain approach," Int. J. Control, 21: 529-536 (1975).
15. Wolovich, W. A., "Output feedback decoupling," Trans. IEEE, AC-20: 148-149 (1975).
16. Bayoumi, M. M., and Duffield, T. L., "Output feedback decoupling and pole placement in linear time-invariant systems," Trans. IEEE, AC-22: 142-143 (1977).
17. El-Bagoury, M. A., and Bayoumi, M. M., "Decoupling of multivariable systems by using output feedback," Trans. IEEE, AC-22: 146-149 (1977).
18. Wolovich, W. A., "On the design of non-interactive, left-invertible systems," Int. J. Control, 28: 165-186 (1978).

10
Generalized Design Techniques

The success of the classical frequency-domain approaches, based on Nyquist-Bode frequency-response techniques, is due mainly to the fact that they allow the designer to exert direct control over the properties of the resulting system without any predefined mathematical constraints. Multivariable extension of these methods, however, is not straightforward, due mainly to the presence of input/output cross-coupling terms in the system transfer-function matrix. A substantial amount of effort has been directed over the past decade to understanding the effect of interactions in multivariable systems. As a result, a fairly comprehensive theory for the analysis of systems with matrix transfer functions is now available. An overview of the generalized stability concepts was presented in Chap. 5; these are used here to present a brief summary of the more established design methods which are based on the generalized frequency-response concept. In view of the enormous amount of literature that has been produced in the frequency-domain analysis, adaptation of a single framework is considered to be essential for a chapter that is aimed at providing an introduction to four recently developed design methods rather than outlining the state of the art.

The problem considered here is that of an m-input/m-output multivariable system, represented by the m-square transfer-function matrix G(s) which is to be controlled by the addition of a controller, defined by the m-square transfer-function matrix K(s) and by unity-feedback loops as shown in Fig. 10.1. The following assumptions are made in the analytical development:

1. Elements of G(s) and K(s) are rational polynomial functions of the transform variable s.
2. G(s) represents a stable system (this condition is not essential, but assumed for convenience).
3. det $\{G(s)\}$ and det $\{K(s)\}$ are not identically zero.
4. Zeros of det $\{G(s)\}$ and det $\{K(s)\}$ lie on the left half of the s-plane.
5. Poles of K(s) and G(s) lie on the left half of the s-plane.
6. Poles and zeros do not occur on the imaginary axis.

These are referred to as "conditions" in subsequent sections.

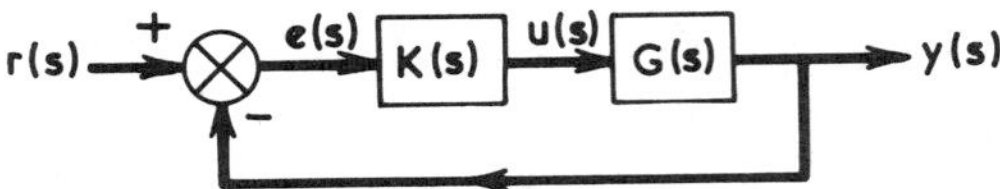

Fig. 10.1 Unity-feedback multivariable system.

10.1 INVERSE-NYQUIST-ARRAY (INA) TECHNIQUE*

This method, originally proposed by Rosenbrock, uses the concept of diagonal dominance (Sec. 5.2.4) to develop a simple relationship between the open- and closed-loop transfer-function matrices of the unity-feedback system with a feedforward (cascade) compensator (Fig. 10.1). The analytical results associated with this design technique is based on the premise that if the open-loop transfer-function matrix G(s) has diagonal dominance over a particular frequency range, the stability of the closed-loop system can be inferred from the generalized inverse Nyquist criterion (Sec. 5.2.3) applied to the Gershgorin bands (Sec. 5.2.4) swept out by the diagonal elements.

By using the notation $(\hat{\cdot})$ to denote the inverse of $(\cdot)$, the transfer-function matrix of the closed-loop system in Fig. 10.1 is given by

$$R(s) = [I_m + G(s)K(s)]^{-1}G(s)K(s) \equiv [I_m + Q(s)]^{-1}Q(s) \tag{10.1}$$

$$\hat{R}(s) = I_m + \hat{Q}(s) \tag{10.2}$$

To analyze the general case where not all loops are closed, an $(m \times m)$ diagonal matrix F is substituted in place of I_m in (10.2), which modifies the inverse closed-loop transfer-function matrix to

$$\hat{R}(s) = F + \hat{Q}(s) \tag{10.3}$$

The structure of F is such that $f_{ii} = 1$ or 0, depending on whether the ith loop is closed (1) or open (0). For example, for $m = 4$, with first and third loops closed, F assumes the form

$$F = \begin{bmatrix} 1 & 0 & 0 & 0 \\ 0 & 0 & 0 & 0 \\ 0 & 0 & 1 & 0 \\ 0 & 0 & 0 & 0 \end{bmatrix} \tag{10.4}$$

Since the mechanism of the transformation from open- to closed-loop form

*References 1 to 3 are followed closely in this section.

is important in this method, the transfer function matrix $\hat{T}(s)$ is introduced:

$$\begin{aligned}\hat{T}(s) &= F + \hat{Q}(s) \\ &= \hat{Q}(s) \rightarrow \text{open-loop system } F \equiv 0 \\ &= \hat{R}(s) \rightarrow \text{closed-loop system } F = I_m \end{aligned} \tag{10.5}$$

The inverse transfer function is not used here to examine the effect of compensation in the feedback path, but because it allows an easy transition from an open-loop to a closed-loop transfer function matrix. If dynamics is introduced in the feedback path, some of the convenience of the method will be lost.

The set of m^2 diagrams representing the elements $\hat{q}_{ij}(j\omega)$ of $\hat{Q}_{ij}(j\omega)$ $(i, j \in 1, \ldots, m)$ is the inverse Nyquist array (INA). The INA allows the elements of $\hat{T}(j\omega)$ to be obtained in an elementary way, since

$$\hat{t}_{ij}(j\omega) = \hat{q}_{ij}(j\omega) \quad \text{for } i \neq j$$

and

$$t_{ii}(j\omega) = \begin{cases} \hat{q}_{ii}(j\omega) & \text{if the ith loop is open} \\ 1 + \hat{q}_{ii}(j\omega) & \text{if the ith loop is closed} \end{cases} \tag{10.6}$$

To obtain the Nyquist diagram for $\hat{t}_{ii}(j\omega)$, it is thus only necessary to shift the origin of the Nyquist diagrams for $\hat{q}_{ii}(j\omega)$ to the point $(-1 + j0)$. To establish a stability criteria of the closed-loop system from its open-loop parameters, the following results, stated without proof, are needed.

Theorem 10.1 If the nonsingular rational polynomial transfer-function matrix $K(s)$ satisfies conditions (4) and (5), it can be written as the product $K_aK_b(s)K_c(s)$, where

K_a is a permutation matrix.

$K_b(s)$ is the result of elementary column operations, each consisting of the addition of a multiple $\alpha_{ij}(s)$ of column i to column j, α_{ij} being a rational polynomial function with all its poles in the left half of the s-plane, and $\det\{K_b(s)\} = 1$.

$K_c(s)$ is a nonsingular diagonal matrix. All poles and all zeros of the principal diagonal elements of $K_c(s)$ are on the left half of the s-plane.

Theorem 10.2 If the nonsingular rational transfer-function matrix $G(s)$ satisfies conditions (4) and (5), then K_a and $K_b(s)$, as defined in Theorem 10.1, can be found such that $G(s)K_aK_b(s)$ is nonsingular, diagonal, and has all the poles and zeros of its principal diagonal elements in the left half of the s-plane.

Proof of this theorem follows from Theorem 10.1 since G(s) can be expressed as $G_c(s)G_b(s)G_a$, where G_a is the permutation matrix. The choice $K_a = G_a^{-1}$ and $K_b(s) = G_b^{-1}(s)$ then gives the desired result. The following stability condition, which follows from derivations in Secs. 5.2.3 and 5.2.4 and referred to as a stability test at various stages of synthesis, may now be given.

Theorem 10.3 Let $\det\{\hat{Q}(s)\}$ map the Nyquist contour $\mathscr{D}$ (section onto $\hat{\Gamma}_o$ which encircles the origin n_o times in a counterclockwise direction) and let $\det\{\hat{T}(s)\}$ map $\mathscr{D}$ onto $\hat{\Gamma}_c$, encircling the origin n_c times in the same direction. Then, subject to conditions (1) to (6), the closed-loop system defined by (10.5) is asymptotically stable only if $\det\{\hat{T}(s)\}$ has no finite imaginary zeros and $(n_c - n_o) = 0$.

Proof of Theorem 10.3 follows from Theorem 5.22. This theorem is inconvenient to use, since it requires computation of the determinant of $\hat{Q}(s)$ and $\hat{T}(s)$. An extended version of the theorem is used in the design, giving a simpler but only sufficient criterion.

Theorem 10.4 Let the elements $\hat{q}_{ii}(s)$ of $\hat{Q}(s)$ map $\mathscr{D}$ onto $\hat{\Gamma}_{oi}$, and the elements $\hat{t}_{ii}(s)$ onto $\hat{\Gamma}_{ci}$. Let $\hat{\Gamma}_{oi}$ and $\hat{\Gamma}_{ci}$ encircle the origin n_{oi} and n_{ci} times, respectively, in the counterclockwise direction. Let $\hat{d}_i(s)$ be defined as

$$\hat{d}_i(s) = \sum_{\substack{j=1 \\ j\neq i}}^{m} \hat{q}_{ij}(s) \tag{10.7}$$

Then:

(1) A sufficient condition for asymptotic stability for the open-loop (forward path) system is that

$$n_o = \sum_{i=1}^{m} n_{oi} \tag{10.8}$$

if

$$|\hat{q}_{ii}(s) - \hat{d}_i(s)| > 0 \quad \text{for all values of s on } \mathscr{D} \text{ and } \forall\, i \in 1, \ldots, m \tag{10.9}$$

and

(2) A sufficient condition for asymptotic stability of the closed-loop system in (10.5) is that

$$n_c = \sum_{i=1}^{m} n_{ci} \tag{10.10}$$

if

$$|\hat{t}_{ii}(s) - \hat{d}_i(s)| > 0 \quad \text{for all values of s on } \mathscr{D} \text{ and } \forall\, i \in 1, \ldots, m \tag{10.11}$$

Controller structure [1]

Since the objective is to design a suitable K(s), it is desirable to know what structure is adequate to describe a general K(s), which satisfies the conditions (1) to (6) stated earlier. By Theorem 10.1, K(s) can be written as

$$K(s) = K_a K_b(s) K_c(s) \tag{10.12}$$

where the three matrices K_a, $K_b(s)$, and $K_c(s)$ have the following properties:

K_a is a permutation matrix. It therefore represents a preliminary renumbering of the inputs to G(s), which usually will be done so that the new input i affects mainly the new input i. The matrix $\hat{K}_a = K_a^{-1}$ is another permutation matrix.
$K_b(s)$ has a determinant $\det\{K_b(s)\} = 1$ and represents a sequence of elementary column operations. Each such operation consists of the addition of a multiple $\alpha_{ij}(s)$ of column i to column j, the matrix G(s) being operated on. Here α_{ij} is a polynomial function with its denominator either 1 or a polynomial with all its zeros in the left half of the left half of the s-plane. The matrix $\hat{K}_b(s) = [K_b(s)]^{-1}$ can be expressed as a corresponding sequence of row matrices; when m = 4, for example, $K_{b1}(s)$ may take the form

$$K_{b1}(s) = \begin{bmatrix} 1 & 0 & \frac{1}{1+s} & 0 \\ 0 & 1 & 0 & 0 \\ 0 & 0 & 1 & 0 \\ 0 & 0 & 0 & 1 \end{bmatrix}$$

representing the addition of 1/(1 + s) times the column 1 of G(s) to column 3 of G(s), the inverse being

$$\hat{K}_{b1}(s) = \begin{bmatrix} 1 & 0 & -\frac{1}{1+s} & 0 \\ 0 & 1 & 0 & 0 \\ 0 & 0 & 1 & 0 \\ 0 & 0 & 0 & 1 \end{bmatrix}$$

which represents the subtraction from row 1 of $\hat{G}(s)$ of 1/(1 + s) times row 3 of $\hat{G}(s)$.

$K_c(s)$ is diagonal, with poles and zeros of the nonzero entries being on the left half of the s-plane. Thus if $K_c(s) = \text{diag}\{k_i(s)\}$, $i \in 1, \ldots, m$, then $\hat{K}_c(s) = [K_c(s)]^{-1} = \text{diag}\{1/k_i(s)\} = \text{diag}\{\hat{k}_i(s)\}$, where $\hat{k}_i(s)$ has all its poles and zeros in the left half of the s-plane. Also, if $|k_i(s_o)| >> 1$, $|\hat{k}_i(s_o)| << 1$.

The structure of the compensator K(s) corresponding to the description above is shown in Fig. 10.2 for m = 3. The matrix $K_b(s)$ modifies the interaction properties of the system, while $K_c(s)$ represents independent single-channel controllers; its m loops $[k_i(s)]$ are called the m <u>principal loops</u>. The importance of decomposition is that a general K(s) can be obtained by successive application of K_a, $K_b(s)$, and $K_c(s)$.

Since $\hat{t}_{ii}(s) = \hat{q}_{ii}(s)$ or $1 + \hat{q}_{ii}(s)$ from (10.6) for the ith loop open or closed, the expressions for $\hat{t}_{ii}(s)$ can be formulated fairly easily, and once (10.11) is checked, the encirclements contributions to n_c may be read from the loci for $\hat{q}_{ii}(s)$ considering the origin as moved to (-1 + j0). Thus when (10.9) and (10.11) are satisfied, n_o and n_c may be obtained by counting the number of encirclements of the loci of diagonal terms of $\hat{Q}(s)$ and $\hat{T}(s)$; Theorem 10.3 may then be applied to test stability.

A narrower sufficient condition for stability under the conditions mentioned earlier and (10.9) and (10.11) is that

$$n_{oi} = n_{ci} \quad \forall\, i \in 1, \ldots, m \tag{10.13}$$

This is a form of Nyquist criterion for single-input/single-output systems. Thus given (10.9) and (10.11), the system is closed-loop stable if each of the $[\hat{q}_{ii}(s)]^{-1}$ (which are similar to ordinary transfer functions) shows closed-loop stability. Either Routh-Hurwitz or root-locus approaches may be used in this context.

The INA method thus yields a two-stage design procedure: first the evaluation of a $\hat{K}(s)$ such that $\hat{K}(s)\hat{G}(s)$ is diagonally dominant, and then the application of single-loop compensation to obtain appropriate dynamics to satisfy stability criteria.

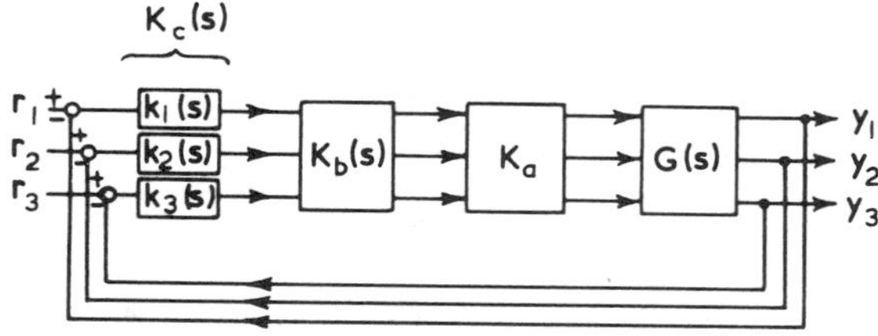

Fig. 10.2 Structure of the closed-loop system with K(s) as in (10.12).

Design procedure

A suite of computer programs is now available which allows the evaluation of $\hat{G}(s)$, $\hat{K}(s)$, and other related matrices in algebraic or numerical form through iterative displays, with a view to designing multivariable systems by using the INA method [4-5]. The salient stages of the design algorithm are outlined below.

Since the basic design criteria are based on the inverse matrices $\hat{G}(s)$, the first step is to compute this matrix $\hat{G}(s) = [G(s)]^{-1}$. The next step is to ascertain whether to operate with the commutation matrix $\hat{K}_a$ or to replace it by I_m (i.e., no commutation of rows); this has to be decided by inspection. If in any row a particular element appears to be greater in modulus on the contour $\mathscr{D}$ than the others, it should be brought into the principal diagonal position by commutation. If no such evidence of dominance exists, a choice of $\hat{K}_a = I_m$ will suffice. The matrix $\hat{K}_a\hat{G}(s)$ is thus obtained.

The next stage is the premultiplication of $\hat{K}_b(s)$ is carried out as above, by a number of elementary row operations on $\hat{K}_a\hat{G}(s)$, with a view to obtaining diagonal dominance (normally row-wise) on the contour $\mathscr{D}$. This task is probably best carried out by seeking to make one row at a time diagonally dominant, each row being left unchanged in subsequent processes (with purely numerical value of K_b as far as possible). The establishment, at any trial stage, of the diagonal dominance of a row requires evaluation of the moduli of all elements at a number of values of s using (10.9). If m is very high, this diagonal dominance can be checked through the digital display of the Nyquist plot for each element in turn, as shown by the illustrative design example (see also Fig. 5.22). At the end of this stage the designer would have a diagonally dominant $\hat{N}(s) = \hat{K}_b(s)\hat{K}_a\hat{G}(s)$. To satisfy the stability requirement, however, the choice of $K_b(s)$ would be restricted such that the diagonal elements of $\hat{K}_b(s)\hat{K}_a\hat{G}(s)$ map $\mathscr{D}$ into loci which encircle the origin and the point $(-1 + j0)$ an equal number of times, as required in Theorem 10.3.

The final stage is to choose the elements of $\hat{K}_c(s) = \text{diag}\{\hat{k}_i(s)\}$. This multiplication $\hat{K}_c(s)\hat{N}(s)$ may affect the diagonal dominance of the rows of $[\hat{N}(s) + F]$, if $F \neq I$, this matrix now becoming $[\hat{K}_c(s)\hat{N}(s) + F]$, as in (10.5). To satisfy the encirclemen criteria of stability, the zeros and the poles of each of $\hat{k}_i(s)$ will have to lie in the left half of the s-plane. It therefore follows that the mappings of $\mathscr{D}$ through $\hat{n}_{ii}(s)$ and through $\hat{k}_i(s)\hat{n}_{ii}(s)$ make an equal number of encirclements of the origin, but the mappings of $[1 + \hat{n}_{ii}(s)]$ and $[1 + k_i(s)\hat{n}_{ii}(s)]$ may make different number of encirclements of the origin, since the two functions, although they have identical number of poles in the right half of the s-plane, may differ in their right-half s-plane zeros.

It is thus useful to choose $K_b(s)$ such that $\hat{N}(s)$ and $[\hat{N}(s) + F]$ are both diagonally dominant and such that the diagonal elements of $\hat{N}(s)$ meet the encirclement criterion. Then various $k_i(s)$ may be used to modify the dynamic properties of the principal diagonal loops of the system. The design method is illustrated below.

Design example 10.1 [1]: INA technique The design procedure based on the INA method is illustrated below with the open-loop transfer function given by

$$G(s) = \begin{bmatrix} \dfrac{1-s}{(1+s)^2} & \dfrac{2-s}{(1+s)^2} \\ \\ \dfrac{1-3s}{3(1+s)^2} & \dfrac{1-s}{(1+s)^2} \end{bmatrix} \tag{10.14}$$

which satisfies conditions (1) to (6) given earlier. The inverse of G(s) is

$$\hat{G}(s) = [G(s)]^{-1} = \begin{bmatrix} 3(1-s)(1+s) & -3(2-s)(1+s) \\ \\ -(1-3s)(1+s) & 3(1-s)(s+2) \end{bmatrix} \tag{10.15}$$

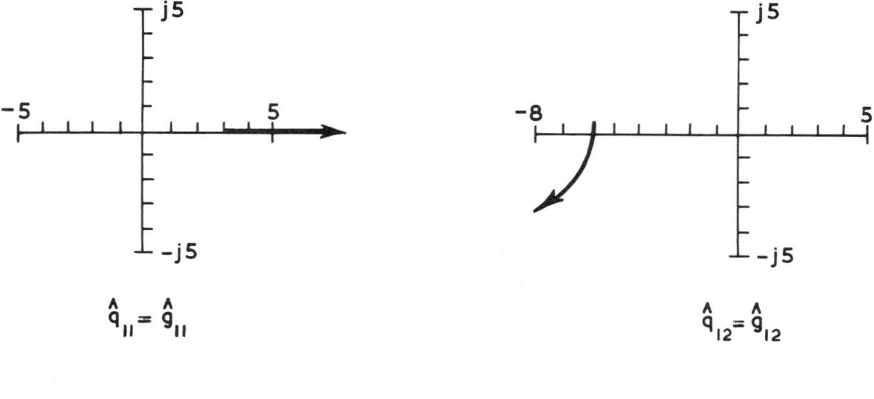

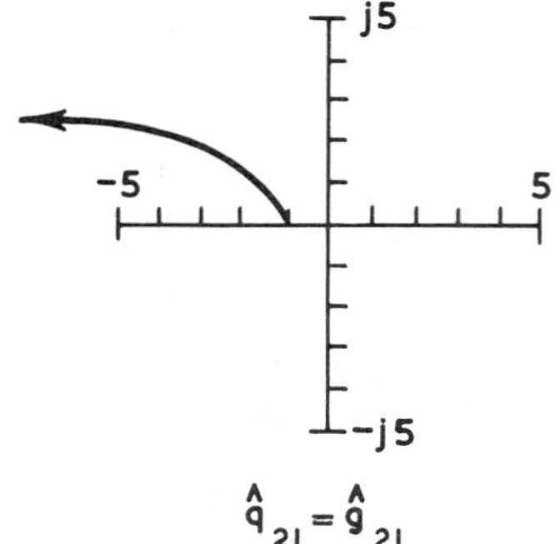

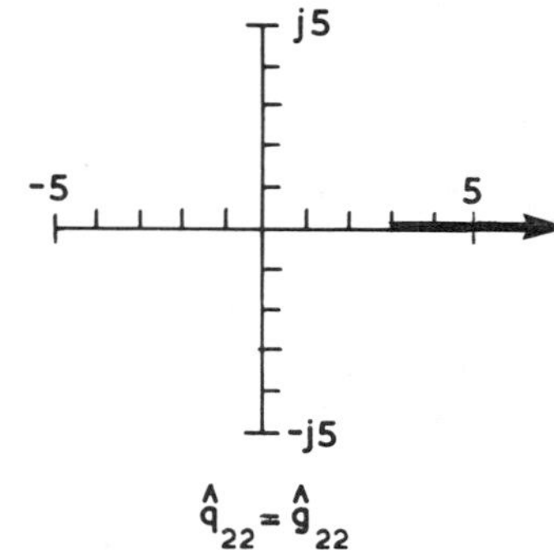

Fig. 10.3 Inverse Nyquist array for (10.15).

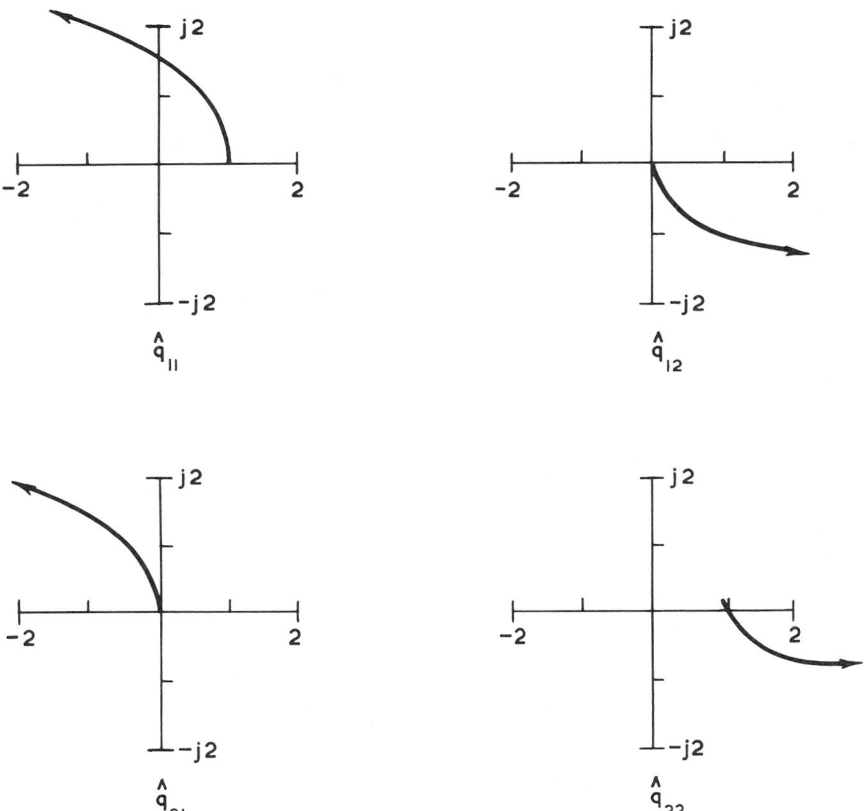

Fig. 10.4 Inverse Nyquist array for (10.16).

The INA for $\hat{G}(s)$ are shown in Fig. 10.3. The next stage of the design is to bring $\hat{Q}(s)$ to diagonal dominance. One first attempt toward this diagonalization is to choose $\hat{K}_a = G(0)$. For physical systems it can be assumed that G(0) is nonsingular, since singularity of G(s) would imply that arbitrary steady-state outputs may be obtained without imposing any constraints on the input controls. For the system above, the procedure for diagonalization of $\hat{Q}(s)$ [and hence Q(s)] is illustrated below.

$\hat{q}_{11}$ $\hat{q}_{12}$

$\hat{q}_{21}$ $\hat{q}_{22}$

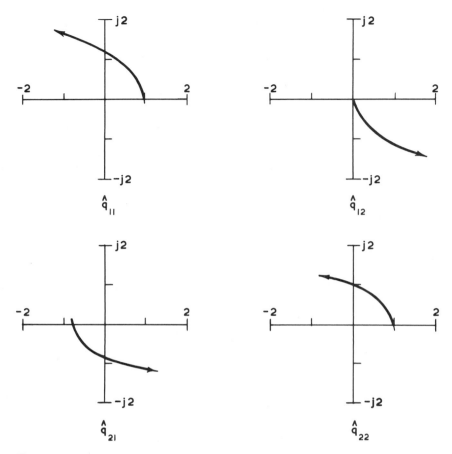

Fig. 10.5 Inverse Nyquist array for (10.18).

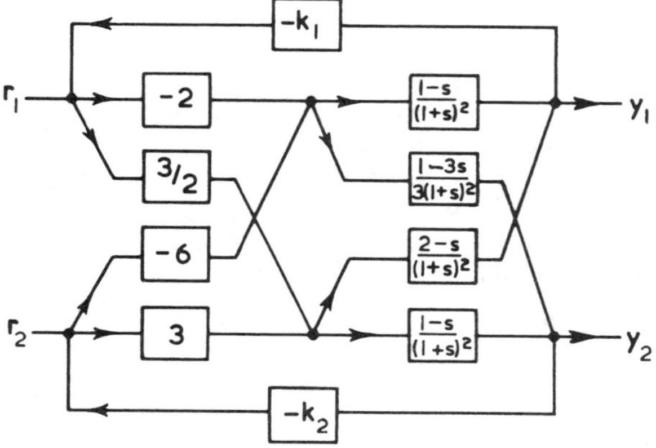

Fig. 10.6 Single-loop compensation for the given system with Q(s) in (10.18).

$$\hat{Q}'(s) = \hat{K}_a\hat{G}(s) = \begin{bmatrix} 1 & 2 \\ \frac{1}{3} & 1 \end{bmatrix} \begin{bmatrix} 3(1-s)(1+s) & -3(2-s)(1+s) \\ -(1-3s)(1+s) & 3(1-s)(1+s) \end{bmatrix}$$

$$= \begin{bmatrix} (1+3s)(1+s) & -3s(1+s) \\ 2s(1+s) & (1-2s)(1+s) \end{bmatrix} \tag{10.16}$$

from which the INA of Fig. 10.4 is obtained. This shows that the diagonal dominance criterion of (10.9) is satisfied by the first row of $\hat{Q}'(s)$, whereas further modifications would be required to achieve diagonal dominance for the second row. Thus the second row of $\hat{Q}'(s)$ need only be altered in the second stage of the design procedure [to obtain $\hat{K}_b(s)$]. Let $\hat{K}_b(s)$ be chosen as

$$K_b = \begin{bmatrix} 1 & 0 \\ -\alpha & 1 \end{bmatrix} \tag{10.17}$$

Then simple calculations show that if the second row of

$$\hat{Q}''(s) = G(0)\hat{K}_b\hat{G}(s)$$

were to be diagonally dominant [i.e., $\hat{q}''_{22}(s) > \hat{q}''_{21}(s)$], then $2/3 < \alpha < 1$. Choosing $\alpha = 5/6$ and the open-loop (forward path) diagonal, $\hat{Q}(s)$ then becomes

$$\hat{Q}(s) = \hat{K}_b(s)\hat{K}_a\hat{G}(s) = \begin{bmatrix} 1 & 0 \\ -\frac{5}{6} & 1 \end{bmatrix} \begin{bmatrix} (1+3s)(1+s) & -3s(1+s) \\ 2s(1+s) & (1-2s)(1+s) \end{bmatrix}$$

$$= (1+s)\begin{bmatrix} 1+3s & -3s \\ -\frac{5+3s}{6} & \frac{2+s}{2} \end{bmatrix} \tag{10.18}$$

which is diagonally dominant; the INA of (10.18) are shown in Fig. 10.5. Since $\hat{Q}(s)$ is now diagonally dominant, design may now proceed on the basis of a single-loop approach, as shown in Fig. 10.6. The compensator matrix $K_c(s)$ has not been included here, although it can be incorporated to improve dynamic response, the compensator without $K_c(s)$ being

$$K(s) = K = K_aK_b = \begin{bmatrix} -2 & -6 \\ \frac{3}{2} & 3 \end{bmatrix} \quad \text{and} \quad Q(s) = G(s)K = \begin{bmatrix} \frac{2+s}{2(1+s)^2} & \frac{3s}{(1+s)^2} \\ \frac{5+3s}{6(1+s)^2} & \frac{1+3s}{(1+s)^2} \end{bmatrix} \tag{10.19}$$

Closed-loop transfer functions

When some of the principal loops are open and some closed, the transfer function between input $r_i(s)$ and output $y_i(s)$ from (10.5) is

$$t_{ii}(s) = \frac{\hat{T}_{ii}(s)}{\det\{\hat{T}(s)\}}$$

The inverse Nyquist diagram for this path is obtained from

$$[t_{ii}(s)]^{-1} = \frac{\det\{\hat{T}(s)\}}{\hat{T}_{ii}(s)}$$

which can be expanded to give

$$[t_{ii}(s)]^{-1} = \begin{cases} \hat{t}_{ii}(s) + \sum_{\substack{j=1\\ j\neq i}}^{m} \hat{t}_{ij}(s)\hat{T}_{ij}(s)/\hat{T}_{ii}(s) & \text{if the ith principal loop is open} \quad (10.20) \\ \hat{t}_{ii}(s) + \sum_{\substack{j=1\\ j\neq i}}^{m} \hat{t}_{ji}(s)\hat{T}_{ji}(s)/\hat{T}_{ii}(s) & \text{if the ith principal loop is closed} \quad (10.21) \end{cases}$$

With increasing feedback gains k_1 and k_2, the system will approximate over an increasingly wide frequency band to a diagonal system. At high frequencies, the effect of interaction may be quantified by using the equations above.

$$[t_{11}(s)]^{-1} = (1+s)(1+3s) - \frac{3s(1+s)^2[(5+3s)/6]}{k_2+(1+s)[(2+s)/2]} \quad \text{with the first loop open} \quad (10.22)$$

$$[t_{22}(s)]^{-1} = (1+s)[(s+s)/2] - \frac{3s(1+s)^2[(5+3s)/6]}{k_1+(1+s)(1+3s)} \quad \text{with the second loop open} \quad (10.23)$$

With knowledge of the equations above, design may proceed using classical synthesis procedure to improve the response of the two channels.

10.2 CHARACTERISTIC-LOCI METHOD*

This method of design of multivariable systems is based on the frequency-response loci associated with a set of characteristic transfer functions

*This section closely follows Refs. 6 to 8.

introduced in Sec. 5.2.1. The definitions and concepts associated with this technique are first outlined; the design procedure is then described.

Let Z(s) be an (m × m) matrix-valued function of a complex frequency variable s whose elements are rational functions of s. For any specific value of the complex frequency $s = s_o$, the corresponding matrix $Z(s_o)$ will be a matrix of complex numbers and will have a set of eigenvalues $\{q_i(s_o), i \in 1, \ldots, m\}$ which are a set of complex numbers. Thus extending the concept of eigenvalues, it may be stated that the eigenvalues of Z(s) are nodes [q(s)] of the (appropriate) characteristic equation formulated as

$$\det \{q(s)I_m - Z(s)\} \triangleq \Delta(q, s) = 0 \tag{10.24}$$

$\Delta(q, s)$ is thus a polynomial in q(s), which in general may not be expressible as a product of factors linear in q(s); the matrix Z(s) will not normally have eigenvalues which are rational functions, $\Delta(q, s)$ usually being expressed in the form

$$\Delta(q, s) = \prod_{i=1}^{m} \Delta_i(q, s) \tag{10.25}$$

where the factors $\Delta_i(q, s)$, $i \in 1, \ldots, m$, are polynomials which are irreducible over the field of rational functions in s. Let this irreducible factor be of the form

$$\Delta_i(q, s) = q_i^{t_i}(s) + a_{i1}a_i^{t_{i-1}}(s) + \cdots + a_{it_1} q(s), \qquad i \in 1, \ldots, m$$

with t_i is the degree of the ith irreducible polynomial and the coefficient a_{ij} ($i \in 1, \ldots, m$, $j \in 1, \ldots, t_i$) are rational functions in s. Let b_{io} be the least common denominator of those coefficients a_{ij}, $j \in 1, \ldots, t_i$. Then (10.25) may be expressed as

$$b_{io}(s)q_i^{t_i} + b_{i1}(s)q_i^{t_i-1} + \cdots + b_{it_i}(s) = 0, \qquad i \in 1, \ldots, m \tag{10.26}$$

where the coefficient $b_{ij}(s)$ ($i \in 1, \ldots, m$, $j \in 1, \ldots, t_i$) are polynomials in s, the functions of complex variable $q_i(s)$ defined by (10.26) being algebraic functions and the set of algebraic functions $q_i(s)$, $i \in 1, \ldots, m$, being the characteristic functions of Z(s).

For any specific value of $s = s_o$, the algebraic function in (10.26) defines an m-degree polynomial which has m-distinct roots, say s_{01}, s_{02}, ..., s_{0m}. Those m roots depend continuously on s, defining m distinct continuous single-valued functions $q_1(s)$, ..., $q_m(s)$ which are characteristic functions of Z(s). Thus the characteristic functions $q_i(s)$, $i \in 1, \ldots, m$, may be considered to be m branches of the algebraic function q(s), and these m values may be determined along any specified path in the s-plane. By substituting $s = j\omega$, this specified path is chosen to be a portion of the imaginary axis. Then with an appropriate variation of the frequency variable ω, a set of loci corresponding to the values of the branches $q_1(j\omega)$, $q_2(j\omega)$, ..., $q_m(j\omega)$ may be obtained by using the following stages:

1. Choosing a value of the angular frequency $\omega = \omega_o$
2. Computing the complex matrix $Z(j\omega_o)$
3. Computing the eigenvalues of $Z(j\omega)$ which are a set of complex numbers denoted by $q_i(j\omega_o)$, $i \in 1, \ldots, m$
4. Plotting the numbers $q_i(j\omega_o)$ on the complex plane

By repeating the stages above for different values of ω, a set of loci on the complex plane may be obtained. The set of loci are called the characteristic loci of Z(s) and are denoted by $q_i(j\omega)$, $i = 1, 2, \ldots, m$. In practice, these loci can be fairly quickly drawn through numerical algorithms (see Ref. 31, Chap. 5).

Design method

The stability analysis in this design synthesis is based on the encirclement theorem considered in Sec. 5.2.1, whereas the overall performance is judged by the integrity theorem outlined in Sec. 5.3. The derivation given below considers unity-feedback configuration; that is, dynamic compensation is included in the feedforward path through K(s), as shown in Fig. 10.1. For reasons of realizability, the m-square elements of K(s) are assumed to be rational functions of s and that $\det\{K(s)\} \not\equiv 0$, in addition to the conditions stated earlier. The actual structure of the controller matrix is derived from the property of a nonsingular matrix in Theorem 10.1, that is, that any nonsingular matrix can be expressed as the product of elementary transformations. Thus

$$K(s) = \prod_{\ell=1}^{n} K_\ell(s) \tag{10.27}$$

where each $K_\ell(s)$, $\ell \in 1, \ldots, n$ (n being an arbitrary number), is an elementary transformation matrix and n is the number of such matrices. The $(m \times m)$ matrix $K_\ell(s)$ is considered to be of the following two types:

Type A: on-diagonal entry:

$$K_\ell(s) = \begin{bmatrix} 1 & 0 & \cdots & 0 & 0 \\ 0 & 1 & \cdots & 0 & 0 \\ 0 & 0 & k_{ii}(s) & 0 & 0 \\ 0 & 0 & \cdots & 0 & 0 \\ 0 & 0 & \cdots & 1 & 0 \\ 0 & 0 & \cdots & 0 & 1 \end{bmatrix}, \quad i \in 1, \ldots, m \tag{10.28}$$

where each on-diagonal entry $k_{ii}(s)$ is a rational function of s, having all its poles and zeros in the left half of the s-plane.

Type B: off-diagonal entry:

$$K_\ell(s) = \begin{bmatrix} 1 & 0 & 0 & 0 & \cdots & 0 \\ 0 & 1 & 0 & 0 & \cdots & 0 \\ 0 & 0 & 1 & k_{ij}(s) & \cdots & 0 \\ 0 & 0 & 0 & 1 & \cdots & 0 \\ 0 & 0 & 0 & 0 & \cdots & 1 \end{bmatrix}, \quad i, j \in 1, \ldots, m \tag{10.29}$$

where the off-diagonal entry $k_{ij}(s)$ is also a rational function of s having all its poles on the left half of the s-plane; its zeros may lie on the right half of the s-plane.

Postmultiplying the open-loop system matrix G(s) by a type A matrix corresponds to the elementary column operation of multiplying all the elements of the ith column of G(s) by $k_{ii}(s)$. Forming the same product with a type B matrix corresponds to the elementary column operation of adding $k_{ij}(s)$ times the ith column of G(s) to the jth column of G(s).

Since unity feedback in all loops is assumed, the system return-ratio becomes the forward-path transference matrix

$$Q(s) = G(s)K(s) \tag{10.30}$$

Thus for stability it is necessary to investigate the effect of the elementary transformation matrices on the characteristic loci of the system. On the basis of the knowledge of these effects, the system may be compensated such that Theorems 5.15 and 5.16 are satisfied.

Let $q_i(s)$ and $g_i(s)$, $i \in 1, \ldots, m$, be the characteristic transfer functions of the return difference [Q(s)] and the system [G(s)] matrices. Then by definition,

$$\det\{Q(s)\} = \det\{G(s)K(s)\} = \prod_{i=1}^{m} q_i(s)$$

and (10.31)

$$\det\{G(s)\} = \prod_{i=j}^{m} g_j(s)$$

Taking logarithms and separating magnitude and phase from (10.31) gives us

$$\sum_{i=1}^{m} \log |q_i(j\omega)| = \log [\det\{K(j\omega)\}] + \sum_{i=1}^{m} \log |g_i(j\omega)|$$

and (10.32)

$$\sum_{i=1}^{m} \text{phase}\{q_i(j\omega)\} = \text{phase}[\det\{K(j\omega)\}] + \sum_{i=1}^{m} \text{phase}\{g_i(j\omega)\}$$

If K(s) is a type A transformation matrix (10.28), then

$$\det\{K(j\omega)\} = k_{ii}(j\omega) \tag{10.33}$$

Consequently, for a type A compensator,

$$\sum_{i=1}^{m} \log|q_i(j\omega)| = \log[\det\{k_{ii}(j\omega)\}] + \sum_{i=1}^{m} \log|g_i(j\omega)|$$

and (10.34)

$$\sum_{i=1}^{m} \text{phase}\{q_i(j\omega)\} = \text{phase}\{k_{ii}(j\omega)\} + \sum_{i=1}^{m} \text{phase}\{g_i(j\omega)\}$$

which shows that the gain and phase shift contributed by the controller term $k_{ii}(j\omega)$ is shared among some or all of the characteristic loci $g_i(j\omega)$ to form the new characteristic loci $q_i(j\omega)$.

If the controller K(s) is chosen to be a type B elementary transformation matrix, then

$$\det\{K(j\omega)\} = 1 \tag{10.35}$$

Consequently

$$\sum_{i=1}^{m} \log|q_i(j\omega)| = \sum_{i=1}^{m} \log|g_i(j\omega)|$$

and (10.36)

$$\sum_{i=1}^{m} \text{phase}\{q_i(j\omega)\} = \sum_{i=1}^{m} \text{phase}\{g_i(j\omega)\}$$

which indicates that any increase in the gain or phase margin in one or some of the characteristic loci $q_i(j\omega)$ results in a corresponding decrease in the net gain and phase margins in one or some of the remaining characteristic loci, and vice versa.

It may be worth noting that the controller K(s) may assume forms other than type A and type B. For example, one other form is

$$K(s) = k(s)I_m \tag{10.37}$$

a scalar matrix controller that multiplies the system characteristic loci by scalar k(s), that is,

$$q_i(j\omega) = k(j\omega)g_i(j\omega), \qquad i \in 1, \ldots, m \tag{10.38}$$

Other forms of K(s) have been indicated earlier and are not considered here.

The derivations above form the basis of a computer-aided design procedure [6,7], where the characteristic loci of G(s) are determined and displayed in the form of Nyquist or Bode diagrams. The stability and integrity are then assessed through the characteristic loci of the appropriate submatrices. If compensation is required, an elementary transformation matrix is introduced, choice of the type of K(s) being dependent on the nature of desired modifications of the characteristic loci. For example, if more phase advance (lead) is necessary to improve stability, type A elementary transformation matrix would be chosen. If phase advance in some of the loci is to be exchanged with phase lag in others, a type B matrix would be appropriate. The actual form of $k_{ii}(s)$ or $k_{ij}(s)$ would be governed by the requirement of improved integrity or reduced cross-coupling (interaction). With these design guidelines, an initial controller is implemented, and the process is repeated with the new set of characteristic loci until the desired design specifications are achieved. Thus it is the cumulative effect of the individual $K_\ell(s)$ that leads to the overall desired performance. The procedure is fairly flexible in that the designer has the freedom of choosing the structure of K(s). The design process is illustrated below.

Design example 10.2 [8]: Characteristic-loci method The two-input/two-output system in Example 10.1 with transfer function matrix

$$G(s) = \begin{bmatrix} \dfrac{1-s}{(1+s)^2} & \dfrac{2-s}{(1+s)^2} \\ \\ \dfrac{1-3s}{3(1+s)^2} & \dfrac{1-s}{(1+s)^2} \end{bmatrix} \tag{10.39}$$

is considered here to illustrate the characteristic-loci design method. The characteristic loci of G(s) are given by the roots $g_1(s)$ and $g_2(s)$ of the characteristic equation

$$\det\{g(s)I - G(s)\} = 0$$

$$\rightarrow (1+s)^4 g(s) - 2(1-s)(1+s)^2 g(s) + [(1-s)^2 - (1-2s)(1-3s)] = 0 \tag{10.40}$$

The characteristic loci are the two roots $g_1(s)$ and $g_2(s)$ of the algebraic equation (10.40). These are shown in Fig. 10.7, where $q_1(j\omega)$ passes through the critical point $(-1 + j0)$ on the complex frequency plane; thus $p_o = 0$ (Theorem 5.16). This is also apparent from

$$\det\{G(s)\} = \frac{1}{3(1+s)^3}$$

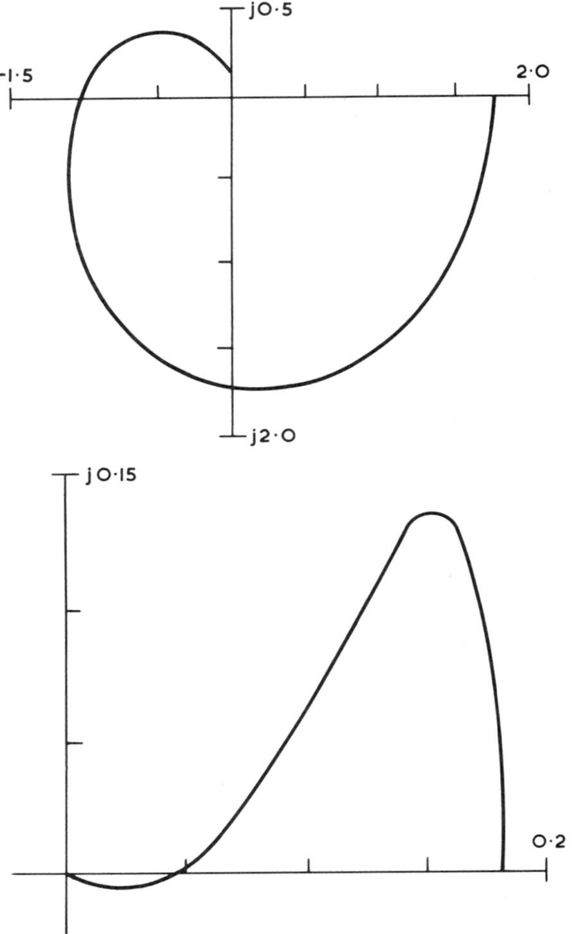

Fig. 10.7 Characteristic loci of the system in (10.39).

The following synthesis is carried out with a view to employing high loop gains to achieve fast transient response. A first step of design, in general, is to remove steady-state interaction by employing a feedforward controller $K_1(s) = -3G^{-1}(0)$, in this case where

$$G^{-1}(0) = \frac{1}{3}\begin{bmatrix} -3 & 6 \\ 1 & -3 \end{bmatrix} \tag{10.41}$$

The return-ratio matrix then becomes

$$Q_1(s) = G(s)K_1(s) = \begin{bmatrix} \frac{1-2s}{(1+s)^2} & \frac{3s}{(1+s)^2} \\ \frac{-2s}{(1+s)^2} & \frac{1+3s}{(1+s)^2} \end{bmatrix} \tag{10.42}$$

which has nonminimum phase (zero at s = +0.5) in one principal minor. For

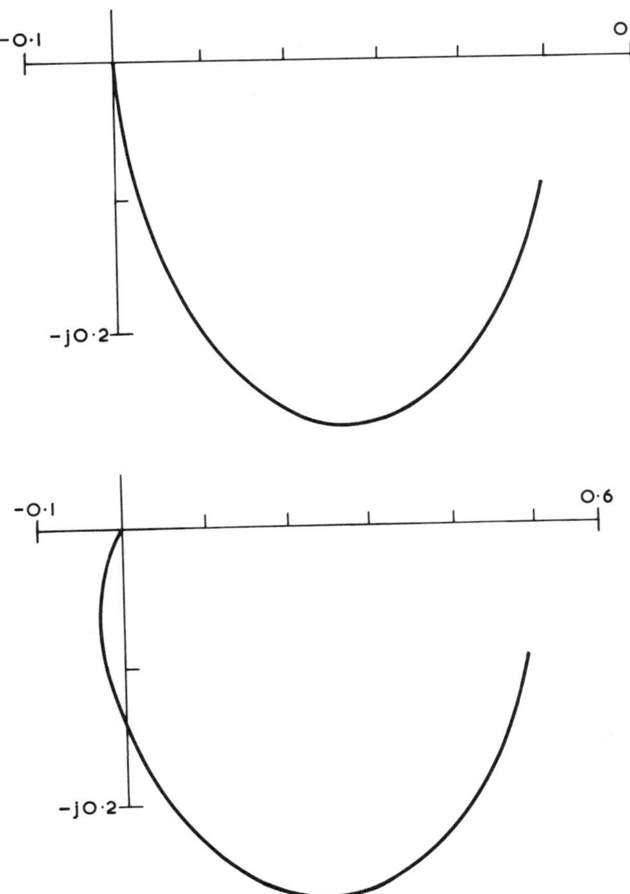

Fig. 10.8 Characteristic loci of (10.43).

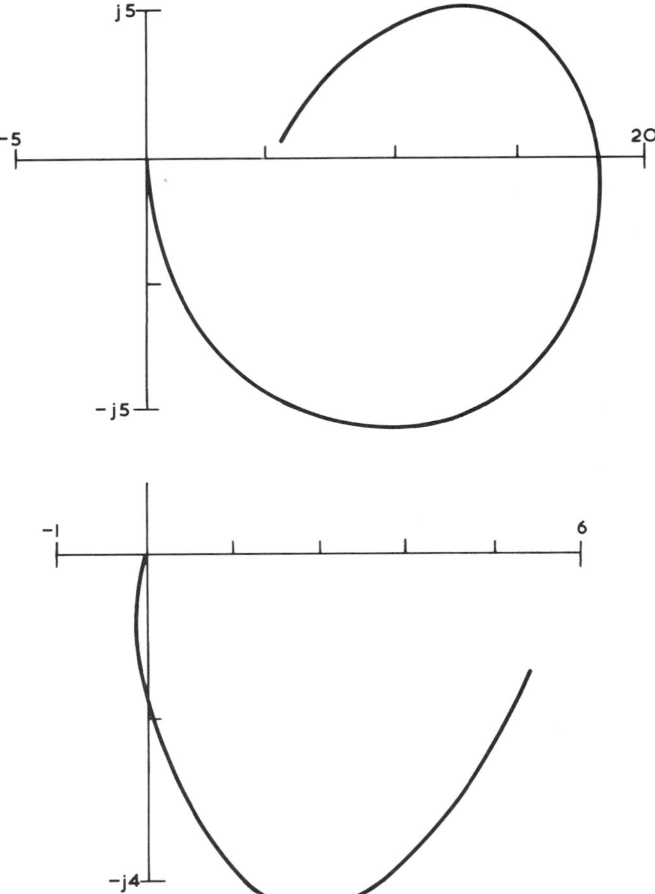

Fig. 10.9 Characteristic loci (10.44).

stability, it is apparent that for the element $(1 - 2s)/(1 + s)^2$, maximum allowable gain is 1, although this can be slightly improved by introducing some phase advance in each loop. Let this be introduced through

$$K_2(s) = \frac{1+s}{10+s} I_2$$

The maximum gain that can be allowed while sustaining stability then becomes 5.5, and let $K_3(s)$ be chosen such that

$$K_3(s) = 5I_2$$

The resulting return-difference matrix thus becomes

$$Q_2(s) = G(s)K_1(s)K_2(s)K_3(s) = \frac{5}{(1+s)(10+s)}\begin{bmatrix} 1-2s & 3s \\ -2s & 1+3s \end{bmatrix} \tag{10.43}$$

The characteristic loci of $Q_2(s)$ are shown in Fig. 10.8. Since there is still one zero on the right hand of the s-plane in (10.43), further modification would be desirable, but the characteristic loci suggest that the system is

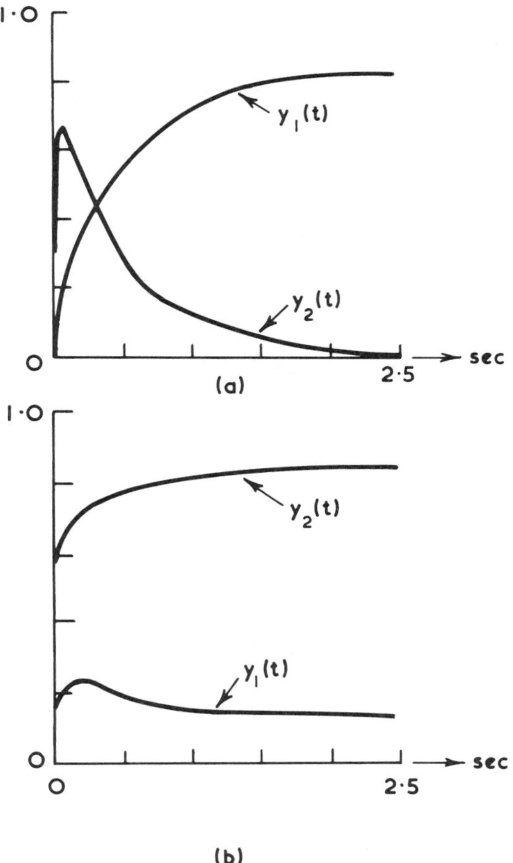

Fig. 10.10 Step responses of the closed-loop system with step input in one channel at a time: (a) step input in channel 1; (b) step input in channel 2.

stable and capable of sustaining high loop gains, although the system will display slow transient responses. To improve the dynamic response and to eliminate the nonminimum phase in $Q_2(1,1)$, the following alternative choice of $K_3'(s)$ may be made:

$$K_3'(s) = 50\begin{bmatrix} 1 & 0 \\ 1 & 1 \end{bmatrix}$$

The resulting return-ratio matrix then becomes

$$Q_3(s) = G(s)K_1(s)K_2(s)K_3'(s) \equiv G(s)K(s)$$

$$= \frac{1}{(1+s)^2}\begin{bmatrix} 1-s & 2-s \\ \dfrac{1-3s}{3} & 1-s \end{bmatrix} \frac{50(s+1)}{s+10}\begin{bmatrix} -3 & -6 \\ 2 & 3 \end{bmatrix} = \frac{50}{(1+s)(10+s)}\begin{bmatrix} 1+s & 3s \\ 1+s & 1+3s \end{bmatrix} \tag{10.44}$$

The resulting characteristic loci are shown in Fig. 10.9.

The closed-loop transfer function of the system with return ratio in (10.44) and unity feedback is given by

$$R(s) = [I+Q(s)]^{-1}Q(s) = \frac{50}{\Delta(s)}\begin{bmatrix} s^3+12s^2+71s+60 & 3s(s^2+11s+10) \\ s^3+12s^2+21s+10 & 3s^3+34s^2+91s+60 \end{bmatrix} \tag{10.45}$$

where $\Delta(s) = s^4 + 222s^3 + 2441s^2 + 5820s + 3600$.

The responses for unit step input to each of the two channels are shown in Fig. 10.10, which indicates the effect, incorporating high gain in $K_3'(s)$. The steady-state errors may be eliminated by introducing another compensator $K_4(s)$ of the type $K_4(s) = s + (1/s)I_2$.

10.3 SEQUENTIAL-RETURN-DIFFERENCE METHOD*

This method of design due to Mayne follows a procedure similar to the INA technique in that it attempts to reduce the multivariable design problem into the classical design of a set of single-input/single-output channels through an updating mechanism. Its main feature is a simple means of studying the effect of closing a particular set of loops on the overall transfer-function matrix of the system. The simplest closed-loop configuration is

*References 9 to 11 are closely followed in this section.

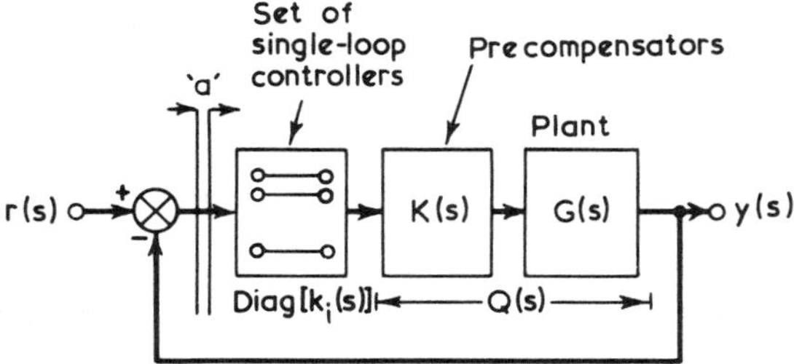

Fig. 10.11 System configuration in sequential return difference method (unity feedback).

shown in Fig. 10.11, where G(s) and K(s) satisfy the conditions mentioned earlier, with K(s) having the form

$$K(s) = K_a K_b(s) \tag{10.46}$$

where K_a is a commutation matrix which alters the order of the columns in G(s) with det $\{K_a\} = 1$, and $K_b(s)$ represents a sequence of elementary column operations in $G(s)K_a$ with det $\{K_b(s)\} = 1$ as in Theorem 10.1. The general closed-loop configuration is shown in Fig. 5.12 where H(s) has properties similar to $K_c(s)$ in Theorem 10.1, that is

$$H(s) = \text{diag}\{h_1(s), \ldots, h_i(s), \ldots, h_m(s)\} \tag{10.47}$$

except that it is in the feedback path, which is the basic structural difference between the INA and the sequential-return-difference (SRD) methods. The breakpoint in the closed-loop system is chosen as a convenient place to refer to the calculation of the return-difference matrix and derive some basic results.

Let $H^\ell(s)$ denote the m-square diagonal matrix* where the first ℓ elements are nonzero, the remaining $(m - \ell)$ elements being zero.

$$H^\ell(s) = \text{diag}\{h_1(s), \ldots, h_\ell(s), \underbrace{0, \ldots, 0}_{m-\ell}\} \tag{10.48}$$

The closed-loop transfer function† of the system in Fig. 10.12 then becomes

$$R^\ell(s) = [I_m + G(s)K(s)H^\ell(s)]^{-1}G(s)K(s) = [I_m + Q(s)H^\ell(s)]^{-1}Q(s) = [F^\ell(s)]^{-1}Q(s) \tag{10.49}$$

where‡

*$H^0(s) = 0$; $H^1(s) = \text{diag}\{h_1(s), 0, \ldots, 0\}$, where only the first loop is closed; $H^m(s) = H(s)$.

†$R^0(s) = Q(s)$; $R^m(s) = R(s)$.

‡$F^0(s) = I_m$; $F^1(s) = I_m + Q(s)H^1(s)$; $F^m(s) = F(s)$.

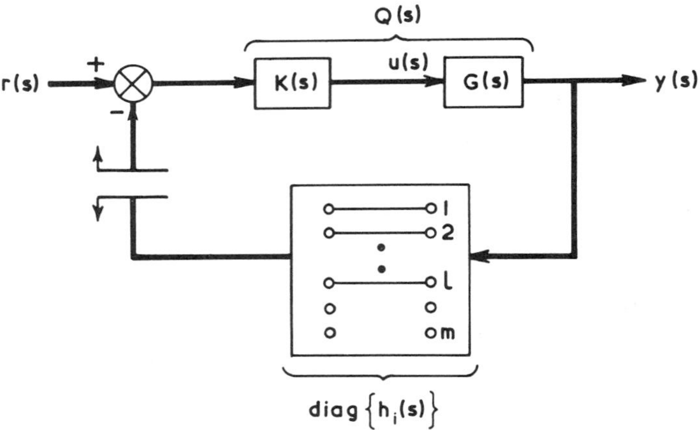

Fig. 10.12 System configuration with first ℓ loops closed (nonunity feedback).

$$Q(s) = G(s)K(s) \quad \text{and} \quad F^{\ell}(s) = I_m + Q(s)H^{\ell}(s) \tag{10.50}$$

so that Q(s) represents the open-loop compensated system in Fig. 10.12 and $F^{\ell}(s)$ represents the corresponding return-difference matrices with the first ℓ loops being closed.

Let a matrix $Q^{\ell}(s) = \{q_{ij}^{\ell}(s)\}$, i, j $\in$ 1, ..., m, be defined such that

$$Q^{\ell}(s) = [F^{\ell}(s)]^{-1}Q(s) \quad \text{with} \quad F^{0}(s) = I_m, \quad \ell \in 1, \ldots, m \tag{10.51}$$

Then the element $q_{ij}^{\ell}(s)$ denotes the (i, j)th element of $Q^{\ell}(s)$. If the first $(\ell - 1)$ loops of H(s) are closed, the scalar return-difference is given by

$$f_{\ell}(s) = 1 + h_{\ell}(s)q_{\ell\ell}^{\ell-1}(s) \tag{10.52}$$

To derive a recursive formula to compute the return-difference matrix of closing ℓ-loops from the return-difference matrix with $(\ell - 1)$ loops closed, (10.50) is used.*

*$q_{cj}(s)$ = jth column of Q(s) and e_{jr} = jth row of I_m. The importance of this scalar return-difference function $f_{\ell}(s)$ arises from the fact that when the element $h_{\ell}(s)$ is added to $H^{\ell-1}(s)$, the feedback $[+h_{\ell}(s)]$ is introduced from the ℓth element of the output vector y(s) to the ℓth element of the input control r(s) (Fig. 10.12). These two variables are at that stage, related by the forward scalar transfer function $q_{\ell\ell}^{\ell-1}(s)$.

$$F^{\ell}(s) = I_m + Q(s)H^{\ell}(s) = I_m + \sum_{j=1}^{\ell} h_j(s)q_{cj}(s)e_{jr}$$

$$= I_m + \sum_{j=1}^{\ell-1} h_j(s)q_{cj}(s)e_{jr} + h_{\ell}(s)q_{c\ell}(s)e_{\ell r} = F^{\ell-1}(s) + h_{\ell}(s)q_{c\ell}(s)e_{\ell r}$$

$$= F^{\ell-1}(s)\{I_m + [F^{\ell-1}(s)]^{-1}h_{\ell}(s)q_{c\ell}(s)e_{\ell r}\} = F^{\ell-1}(s)\{I_m + h_{\ell}(s)q_{c\ell}^{\ell-1}e_{\ell r}\}$$

$$= F^{\ell-1}(s)\bar{F}^{\ell}(s) \tag{10.53}$$

where $\bar{F}^{\ell}(s) = I_m + h_{\ell}(s)q_{c\ell}^{\ell-1}e_{\ell r}$ is the return-difference matrix, prior to closing loop ℓ, after loops 1, 2, ..., $(\ell - 1)$ have been closed. Thus

$$\det\{F^{\ell}(s)\} = \det\{F^{\ell-1}(s)\}\{1 + h_{\ell}(s)q_{c\ell}^{\ell-1}e_{\ell r}\}$$

$$= \det\{F^{\ell-1}(s)\}f_{\ell}(s)$$

$$= \prod_{j=1}^{\ell} f_j(s) \quad [\det\{F^0(s)\} = |I| = 1] = \det\{F^{\ell}(s)\} \tag{10.54}$$

Thus the system in Fig. 10.12, with loops 1, 2, ..., ℓ closed, is stable if the zeros of $f_1(s)$, $f_2(s)$, ..., $f_{\ell}(s)$ are on the left-half of the s-plane, that is, if the scalar return difference $f_j(s)$, seen in designing loops 1, 2, ..., ℓ, satisfies the usually Nyquist criteria. This result leads to the formulation of the following theorem [9].

<u>Theorem 10.5</u> Let $\{f_{\ell}(s),\ \ell = 1, 2, \ldots, m\}$ be the scalar return-difference set, as defined above, which maps the usual Nyquist contour $\mathscr{D}$ on a set of loci $\{\Gamma_{\ell},\ \ell = 1, 2, \ldots, m\}$ in the complex plane. Then if each of the loci satisfies the usual Nyquist stability criterion for return differences (i.e., Γ_{ℓ} does not encircle or pass through the origin of the s-plane), the set of systems obtained by the successive closing of the various loops through $h_{\ell}(s)$; $\ell = 1, 2, \ldots, m$, are all stable.

The foregoing conditions of stability for sequential loop closure may also be stated as (by using Theorem 5.15 or Theorem 5.16)

<u>Theorem 10.6</u> The closed-loop system in Fig. 10.12 is asymptotically stable if the sum of the encirclements of $(-1 + j0)$ by the Nyquist plots of $h_{\ell}(s)q_{\ell\ell}^{\ell-1}$ is $-p_o$, where $q_{\ell\ell}^{\ell-1}(s)$ is the ℓth diagonal element of $Q^{\ell-1}(s)$.

In the case where $h_{\ell}(s)$ are constants, a similar result may be derived involving the encirclement of $(-1/h_{\ell} + j0)$ by $q_{\ell\ell}^{\ell-1}(s)$ or $(-h_{\ell} + j0)$ and the origin by $[q_{\ell}^{\ell-1}(s)]^{-1}$.

Finally, before the design procedure is described, a recursive relationship for the return ratio of the closed-loop system is needed. This is obtained by taking the inverse of (10.53) and multiplying both sides by Q(s); for any $\ell \in 1, \ldots, m$,

$$[F^{\ell}(s)]^{-1}Q(s) = [I_m + h_{\ell}(s)q_{c\ell}^{\ell-1}(s)e_{\ell r}]^{-1}[F^{\ell-1}(s)]^{-1}Q(s)$$

that is,

$$Q^{\ell}(s) \equiv [I_m + h_{\ell}(s)q_{c\ell}^{\ell-1}(s)e_{\ell r}]^{-1}Q^{\ell-1}(s) = \left[I_m - \frac{h_{\ell}(s)q_{c\ell}^{\ell-1}(s)e_{\ell r}}{1 + h_{\ell}q_{\ell\ell}^{\ell-1}(s)}\right]Q^{\ell-1}(s)$$

$$= \left[1 - \frac{h_{\ell}(s)q_{c\ell}^{\ell-1}(s)e_{\ell r}}{f_{\ell}(s)}\right]Q^{\ell-1}(s)$$

Thus

$$Q^{\ell}(s) = Q^{\ell-1}(s) - \frac{h_{\ell}(s)q_{c\ell}^{\ell-1}(s)q_{\ell r}^{\ell-1}}{f_{\ell}(s)} \tag{10.55}$$

which relates $Q^{\ell}(s)$ in terms of $Q^{\ell-1}(s)$.

Design method

The derivations above may be used to select the feedback compensator H(s) by following the stages which are based on alternate use of (10.52) and (10.55):

1. Computation of $Q^0(s) = Q(s)$.
2. Computation of $f_{\ell}(s) = 1 + h_{\ell}(s)q_{\ell\ell}^{\ell-1}(s)$, for $\ell = 1$.

3a. If $\ell = m$, the design procedure is concluded.

3b. If $\ell \neq m$, the following computations are carried out.

4. $\tilde{f}(s) = h_{\ell}(s)/f_{\ell}(s)$.
5. $Q^{\ell}(s) = Q^{\ell-1}(s) - \tilde{f}_{\ell}(s)q_{c\ell}^{\ell-1}(s)q_{\ell r}^{\ell-1}(s)$.
6. ℓ is increased to $(\ell + 1)$ and the stages above are repeated.

The stages above indicate how the "sequential" design of H(s) is accomplished by increasing the value of ℓ starting with $\ell = 1$ and $Q^0(s)$ [= Q(s)], which is computed from (10.50) assuming that K(s) is known. This gives the basis for the design of $H^1(s)$ ($\ell = 1$), whose structure is to be chosen to be as simple as possible, preferably numeric, subject to the constraint that the zeros of $f_1(s)$ ($\ell = 1$) in (10.52) lie on the left-half of the s-plane (Theorem 10-5). If such a $h_1(s)$ can be found, the corresponding $f_1(s)$ would determine the structure of $Q^{\ell}(s)$ in (10.55). The procedure is repeated with $\ell = 2$ in (10.52) and (10.53), until $\ell = m$ is reached.

The feedforward compensating matrix $K(s) = K_a K_b(s)$ is incorporated to improve the stability properties of any particular (or all) loops of the system in Fig. 10.12. The permutation matrix K_a is chosen so that other diagonal elements of $G(s)K_a$ are, preferably, greater in amplitude than over elements in the same row or column for a specified range of s. This may not always be possible. After this initial stage, the constant feedforward controller matrix K_b is found sequentially by considering K_b as the product of m factors.

$$K_b = \prod_{\ell=1}^{m} K_{b\ell}$$

in which each $K_{b\ell}$ is selected immediately before selecting h_ℓ, and each $K_{b\ell}$ has the partitioned form

$$K_{b\ell} = \begin{bmatrix} I_{\ell-1} & 0 \\ 0 & Z_\ell \end{bmatrix}, \quad \ell \in 1, \ldots, m \tag{10.56}$$

where Z_ℓ is a $(m - \ell + 1)$ square matrix representing a sequence of elementary operators, performed on whatever matrix $K_{b\ell}$ multiplies. Then $\det\{K_{b\ell}\} = 1$, $\forall k$, $Z_m = I$, $K_{b1} = Z_1$, and $K_{bm} = I_m$. The return-difference matrix is therefore given by

$$F^{\ell}(s) = I + GK_aK_b(s)H^{\ell}(s) = I + GK_a[K_{b1}, K_{b2}, \ldots, K_{bm}]H^{\ell}(s)$$

$$\equiv I + GK_a[K_{b1}, K_{b2}, \ldots, K_{b\ell}]H^{\ell}(s) \tag{10.57}$$

so that only $K_{b1}, \ldots, K_{b\ell}$ are relevant to the design of the ℓth loop. A relationship similar to (10.53) for $Q^{\ell}(s)$ is useful in the choice of $K_b(s)$:

$$Q^{\ell}(s) \triangleq [F^{\ell}(s)]^{-1}GK_a[K_{b1}, \ldots, K_{b\ell}] = [F^{\ell}(s)]^{-1}GK_aK_b^{\ell}(s) \tag{10.58}$$

$$\bar{Q}^{\ell}(s) = [F^{\ell}(s)]^{-1}GK_a[K_{b1}, \ldots, K_{b\ell}K_{b(\ell+1)}]$$

$$= [F^{\ell}(s)]^{-1}GK_aK_b^{\ell}K_{b(\ell+1)} = Q^{\ell}(s)K_{b(\ell+1)} \tag{10.59}$$

$$\bar{f}_{\ell(s)} = 1 + h_\ell(s)\bar{q}_{\ell\ell}^{\ell-1}(s) \tag{10.60}$$

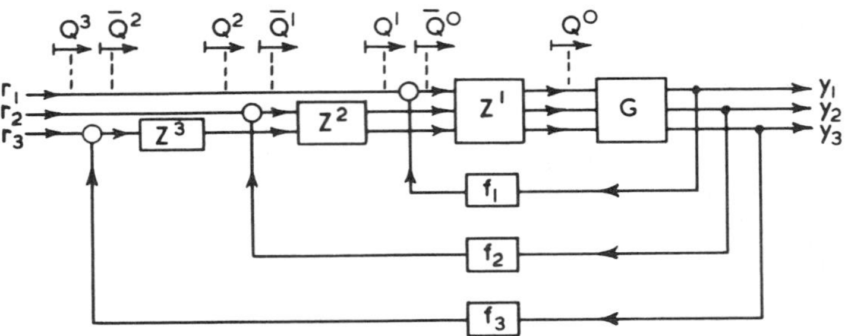

Fig. 10.13 Interpretation of (10.58) and (10.59).

Defining the ℓth stage of design as that stage when $h_1(s), \ldots, h_{\ell-1}(s)$ and $k_{b1}, \ldots, k_{b(\ell-1)}$ have been found so that $Q^{\ell-1}(s)$ is known $[\bar{Q}^0(s) = G(s)K_a]$. The design of the $(\ell + 1)$th stage is based first on the choice of $k_{b\ell}$, so that $\bar{Q}^{\ell-1} = Q^{\ell-1}K_{b\ell}$ yields a $\bar{q}_{\ell\ell}^{\ell-1}$ satisfactory from stability viewpoint (i.e., has its zeros on the left half of the s plane—a necessary condition). This involves only the first column of Z_ℓ in $K_{b\ell}$ (see Fig. 10.13). This is followed by the computation of $h_\ell(s)$ so that $f_\ell(s)$ has acceptable stability properties by following the design stages of $H^\ell(s)$ described earlier. This completes the $(\ell + 1)$ stage of the design procedure. These steps are repeated for $\ell \in 1, \ldots, m$. An illustrative design example is outlined below.

Design example 10.3 [11]: SRD method We have a system with transfer function matrix G(s) [a slightly modified form of (10.14)] given by

$$G(s) = \begin{bmatrix} \frac{1-s}{(1+s)^2} & \frac{1/3-s}{(1+s)^2} \\ \frac{2-s}{(1+s)^2} & \frac{1-s}{(1+s)^2} \end{bmatrix} \tag{10.61}$$

For the system, $\det\{G(s)\} = (s+1)/3(s+1)^4$, which is acceptable; but neither $g_{11}(s)$ nor $g_{22}(s)$ is satisfactory; hence precompensation is necessary.

To compute Q(s), let K(s) be chosen as below. The factors influencing the choice of K(s) are considered later. The design stages indicated earlier are followed:

(1) $$K(s) = K_a K_b = \begin{bmatrix} 1 & 0 \\ -2 & 1 \end{bmatrix} \quad Q^0(s) = G(s)K(s) = \frac{1}{(1+s)^2}\begin{bmatrix} \frac{1}{3}+s & \frac{1}{3}-s \\ s & 1-s \end{bmatrix} \tag{10.62}$$

(2) For $\ell = 1$; $f_\ell(s) = f_1(s) = 1 + h_1(s)q_{11}^0(s) = 1 + \dfrac{h_1(s)(\frac{1}{3} + s)}{(1+s)^2}$

$$= \frac{(1+s)^2 + (\frac{1}{3} + s)h_1(s)}{(1+s)^2} = \frac{\alpha_1(s)}{(1+s)^2} \qquad (10.63)$$

At this stage the choice of $H^1(s)$ may be made such that the zeros of $f_1(s)$ lie on the left half of the s-plane.

(3) $\ell = 1 \neq m = 2$.

(4) $\tilde{f}_1(s) = h_1(s)/f_1(s)$ with $f_1(s)$ as in (10.63).

(5) $Q^1(s) = Q^0(s) - \dfrac{h_1(s)}{f_1(s)} q_{c1}^0(s) q_{1r}^0(s)$

$$= \frac{1}{(1+s)^2}\begin{bmatrix} \frac{1}{3} + s & \frac{1}{3} - s \\ s & 1 - s \end{bmatrix} - \frac{h_1(s)}{\alpha_1(s)(1+s)^2}\begin{bmatrix} \frac{1}{3} + s \\ s \end{bmatrix}\left[(\frac{1}{3} + s)(\frac{1}{3} - s)\right]$$

$$= \frac{1}{(1+s)^2}\begin{bmatrix} \frac{1}{3} + s & \frac{1}{3} - s \\ s & 1 - s \end{bmatrix} - \frac{h_1(s)}{(1+s)^2\alpha_1(s)}\begin{bmatrix} (\frac{1}{3} + s)^2 & (\frac{1}{3} - s)(\frac{1}{3} + s) \\ s(\frac{1}{3} + s) & s(\frac{1}{3} - s) \end{bmatrix}$$

$$= \frac{1}{\alpha_1(s)}\begin{bmatrix} \frac{1}{3} + s & \frac{1}{3} - s \\ s & (1+s)^2(1-s) + h_1(s)(1+s)/3 \end{bmatrix} \qquad (10.64)$$

Comment:

$$q_{22}^1(s) = \frac{(1+s)^2(1-s) + h_1(s)(1+s)/3}{\alpha_1(s)} = \frac{(1+s)\{1 + h_1(s)/3 - s^2\}}{(1+s)^2 + (\frac{1}{3} + s)h_1} \qquad (10.65)$$

Thus if $h_1(s) = h_1$, where h_1 is a finite constant, (10.65) shows that $q_{22}^1(s)$ will have a zero at $\sqrt{(1 + h_1/3)}$ on the right half of the s-plane. Thus the return ratio in the first loop is unstable for finite h_1. If h_1 is chosen such that $h_1 \to \infty$, then

$$f_1(s) \simeq \frac{h_1(\frac{1}{3}+s)}{(1+s)^2}; \quad \tilde{f_1}(s) = \frac{h_1(s)}{f_1(s)} \simeq \frac{(1+s)^2}{\frac{1}{3}+s}; \quad q_{11}^0(s) = \frac{1}{(\frac{1}{3}+s)(1+s)^2} \tag{10.66}$$

which is stable by Theorem 10.5. With high gain in the first loop, the return difference in the second loop is

$$q_{22}^1(s) = \frac{1}{3(1+s)(\frac{1}{3}+s)} \tag{10.67}$$

which is also stable and can be easily controlled.

Choice of K(s): As seen above, the choice of K(s) in (10.62) provides a zero at $+\sqrt{1+h_1/3}$. A different choice of K may be made to avoid this zero on the right half of the s-plane. Additional compensation terms in K(s) may be included to improve dynamic performance.

10.4 DESIGN THROUGH DYADIC TRANSFORMATION [3, 12, 13, 14]*

The introduction of dyadic transfer function matrices in the analysis of multivariable systems was motivated by the physical insight provided by the concepts of dyad and dyadic expansion in the design of nuclear vector spatial control problems [12]. This resulted in the development of a generalized method of designing linear multivariable system. The relevant definitions and properties of dyadic are first developed; these are then followed by their use in system design.

<u>Definition 10.1: Dyad</u> A dyad D of order m × n is a matrix of this order which can be expressed as a (column) · (row) product:

$$D_{m\times n} = \alpha_{m\times 1}\beta_{1\times n} \rightarrow \rho(D) < 1 \tag{10.68}$$

where $\rho(D) = 0$ if $\alpha = \beta \equiv 0$, in which case $D \equiv 0$. It is thus possible to extract nonzero scalar factors k_1 and k_2 from α and β, respectively, such that $\alpha = k_1\bar{\alpha}$ and $\beta = k_2\bar{\beta}$. Then from (10.68),

$$D = k_1k_2\alpha\beta = k\bar{\alpha}\bar{\beta} \qquad (k = k_1k_2) \tag{10.69}$$

Thus, in general, a dyad may be defined as the product of a scalar, a column vector, and a row vector. This observation forms the basis of the analytical developments in this section.

*Derivations within sections are based on the original work of Owens and closely follows Refs. 3 and 12.

Theorem 10.7 If M is a p-square diagonal matrix, α_{ic} is the ith column of the matrix L of order m × p, and β_{ir} is the ith row of the matrix N of order p × n, then the triple product matrix LMN of order m × n is given by*

$$LMN = \{\alpha_{ic}\} \operatorname{diag}\{m_i\}\{\beta_{ir}\} = \sum_{i=1}^{p} m_i \alpha_{ic}\beta_{ir} \tag{10.70}$$

The proof of the theorem is straightforward. It is important to observe from the above that if the diagonal matrix M is chosen to be an identity matrix, then

$$LN = \sum_{i=1}^{p} \alpha_{ic}\beta_{ir} \tag{10.71}$$

which expresses the product of two matrices in terms of the columns of the first matrix (L) and the rows of the second (N) (instead of the usual definition of row of the first and the column of the second). The case that is of interest in the context of design of m-input and m-output multivariable systems is the case where m = n = p and the dyads in the summation are all square. This leads to the formulation of the more generalized relation through the following theorem [3].

Theorem 10.8 Any p-square matrix M with rank r can be expanded in the following dyadic form:

$$M = LI_{p,r}N = \sum_{i=1}^{r} \alpha_{ic}\beta_{ir} \tag{10.72}$$

where L and N are p-square matrices, and

$$I_{p,r} = \begin{bmatrix} I_{r\times r} \\ \\ 0_{p-r,r} \end{bmatrix}, \quad \begin{array}{l} \rightarrow I_{rr} \text{ being a r-square identity matrix} \\ \\ \rightarrow 0_{p-r,r} \text{ being a null matrix of order } (p-r)\times r \end{array}$$

Proof of the theorem is given through the following derivations. Let there be a nonsingular p-square matrix A, then the following identity holds good for any arbitrary p-square matrix M with rank r < p.

$$M = AA^{-1}M = AM_1, \quad \text{where } M_1 = A^{-1}M \tag{10.73}$$

Thus $\rho(M_1) = r$ that is, (p − r) rows of M_1 are linearly dependent. Let K_1 be a permutation matrix [det $\{K_1\} = \pm 1$] which makes the first r rows of M_2 linearly independent, where

$$M_2 = K_1M_1 \tag{10.74}$$

*Subscripts ic and ir are used to denote column and row vectors.

Thus the last (p - r) rows of M_2 are linearly dependent. Let K_2 be another matrix, representing a sequence of elementary row operations [det $\{K_2\} = 1$], which makes these last (p - r) rows of M_3 null, while the first r rows are left unchanged, where

$$M_3 = K_2 M_2 \tag{10.75}$$

Finally, let M_4 be another matrix such that the first r rows of M_4 are identical to the first r rows of M_3, and the last (p - r) rows of M_4 are arbitrary, such that $\rho(M) = p$, then by using the definition of $I_{p,r}$ above,

$$M_4 = I_{p,r} M_3 \tag{10.76}$$

Thus combining (10.73) to (10.76),

$$M = AK_1^{-1} K_2^{-1} I_{p,r} M_4 = LI_{p,r} N \tag{10.77}$$

where $L = AK_1^{-1}K_2^{-1}$ and $N = M_4$: both nonsingular matrices, which completes the proof. A few comments are in order.

Since the choice of the last (p - r) rows of M_4 is completely arbitrary apart from the constraint that $\rho(M_4) = p$, and there are an infinite number of ways in which A may be chosen, the expansion of M according to (10.77) is not unique.

Combination of (10.71) and (10.73) to (10.77) gives

$$M = LI_{p,r} N = \sum_{i=1}^{r} \alpha_{ic} \beta_{ir} \tag{10.78}$$

which suggests that any p-square matrix of rank $r < p$ may be expanded as the sum of r dyads of the same order, in which the set of columns α_{ic} and the set of rows β_{ir}, $i \in 1, \ldots, r$, are both linearly independent sets. Since the number of dyads can never be less than r whatever values of L and N are taken, α_{ic} and β_{ir} are therefore always nonnull vectors. This leads to the following definition [3].

Definition 10.2: Minimal dyadic expansion A minimal dyadic expansion of a p-square matrix M of rank r ($<p$) is any sum of r dyads such that their column factors and their row factors are both linearly independent sets.

Finally, as a generalized form of dyadic expansion, (10.78) can be expanded as

$$M = \sum_{i=1}^{r} m_i \alpha_{ic} \beta_{ir} = L \operatorname{diag}\{m_i\} N \tag{10.79}$$

where L and N are nonsingular matrices and diag $\{m_i\}$ is a diagonal matrix of order p in which the last (p - r) elements are zeros and $\Pi_{i=1}^{r} m_i \neq 0$, α_{ic} and β_{ir} being nonzero scalar multipliers.

Definition 10.3: Dyadic transfer-function matrix An m-square nonsingular matrix G(s) over a field of real rational functions of s (and hence over part of the broader field of complex rational functions, F) is said to be a dyadic transfer function matrix if G(s) can be expanded as

$$G(s) = P_1\Gamma(s)P_2 \tag{10.80}$$

where $\Gamma(s) = \text{diag}\{g_{ii}(s)\}$, $\Pi_{i=1}^{m} g_{ii}(s) \neq 0$, and P_1 and P_2 are numeric (i.e., defined over F) and nonsingular.

In Definition 10.3, which is restricted to the real transfer-function matrix G(s), the three factors on the right of (10.80) may assume complex form. The practical restriction of dyadic transfer matrices is that P_1 and P_2 must be numeric, a constraint that may not be satisfied by transfer functions of many physical systems. In these cases the transfer function matrix may be approximated to a form that satisfies (10.80); this is considered later.

Synthesis [12]

Using Definition 10.3 and the property in (10.79), the following expression may be developed:

$$G(s) = \sum_{i=1}^{m} g_i(s)\alpha_{ic}\beta_{ir} \tag{10.81}$$

where $\{g_i(s)\}$ are rational scalar transfer functions and α_{ic} and β_{ir} are linearly independent frequency independent real vectors satisfying the orthogonality relationship

$$\beta_{ir}\beta_{jc} = \delta_{ij}, \text{ where } \delta_{ij} = \begin{cases} 1 & \text{for } i = j \\ 0 & \text{for } i \neq j \end{cases} \tag{10.82}$$

Combining (10.81) and (10.82) gives us

$$G(s)\beta_{jc} = \sum_{i=1}^{m} g_i(s)\alpha_{ic}(\beta_{ir}\beta_{jc}) = \sum_{i=1}^{m} g_i(s)\alpha_{ic}\delta_{ij} = g_j(s)\alpha_{jr}, \quad j \in 1, \ldots, m \tag{10.83}$$

The vectors $\{\alpha_{ic}\}$ therefore are the eigenvectors of G(s) with corresponding eigenvalues $\{g_i(s)\}$, which implies that an input to the system of shape α_i produces an output of the same form. In general, $\beta_{ir}\alpha_{jc} \neq 0$, and by following the properties of a dyad, it can be shown that the vectors α_{jc}, $j \in 1, \ldots, m$, in (10.83) are linearly independent if G(0) is nonsingular. The concept of design synthesis basis on dyadic transformation may now be developed.

Let $g_i(0) = 1$ for any $i \in 1, \ldots, m$. Then

$$\beta_{ir}[G(0)]^{-1}\alpha_{jc} = \delta_{ij} \quad \text{for } i, j \in 1, \ldots, m \tag{10.84}$$

Let a feedforward controller of the form $K(s) = [G(0)]^{-1}$ be introduced in the system with dyadic transfer-function matrix G(s) (which will make the augmented system noninteracting in the steady state). The forward-path transfer function of the system is given by

$$Q_1(s) = G(s)[G(0)]^{-1} = \sum_{i=1}^{m} g_i(s)\alpha_{ic}\gamma_{ir} \tag{10.85a}$$

where

$$\gamma_{ir} = \beta_{ir}[G(0)]^{-1} \tag{10.85b}$$

and the various channels in $Q_1(s)$ are noninteracting, since

$$Q_1(s)\alpha_{ic} = g_i(s)\alpha_{ic} \tag{10.86}$$

Thus the vectors $\{\alpha_{ic}\}$ and the transfer functions $\{g_i(s)\}$ are the eigenvectors and eigenvalues of $Q_1(s) = G(s)[G(0)]^{-1}$, respectively. The result above may be used to design a generalized feedforward controller K(s) by structuring it in the form

$$K(s) = [G(0)]^{-1} \sum_{i=1}^{m} k_i(s)\alpha_{ic}\gamma_{ir} \tag{10.87}$$

where $\{k_i(s)\}$ are scalar transfer functions. From (10.83) to (10.86), G(s) K(s) may be expressed as

$$Q_2(s) = G(s)K(s) = \sum_{i=1}^{m} g_i(s)k_i(s)\alpha_{ic}\gamma_{ir} \tag{10.88a}$$

which suggests that G(s)K(s) is also dyadic; combining (10.84) and (10.87), we have

$$G(s)K(s)\alpha_{ic} = g_i(s)k_i(s)\alpha_{ic}, \quad i \in 1, \ldots, m \tag{10.88b}$$

Thus α_{ic} are also the eigenvectors of the system with transfer-function matrix G(s)K(s). The closed-loop transfer-function matrix with unity feedback and dyadic K(s) in (10.87) is thus given by

$$R(s) = [I + Q_2(s)]^{-1}Q_2(s) = \sum_{i=1}^{m} \frac{g_i(s)k_i(s)}{1 + g_i(s)k_i(s)} \alpha_{ic}\gamma_{ir} \tag{10.89}$$

which is dyadic. Stability characteristics of the closed-loop system therefore may be analyzed by the m independent channels in (10.89) with transfer functions

$$r_i(s) = \frac{g_i(s)k_i(s)}{1 + g_i(s)k_i(s)} \qquad \text{for any } i \in 1, \ldots, m \tag{10.90}$$

Controller structure

For most open-loop transfer-function matrices, the dyadic forms of (10.81) will have to be evaluated, thus requiring the need to compute the vectors α_{ic} and β_{ir}, $i \in 1, \ldots, m$. Relationships between α_{ic}, β_{ir}, and $g_i(s)$, and P_1, P_2, and $\Gamma(s)$ in (10.80), will therefore be required. This is derived below.

Let the compensation transfer function for the system (Fig. 10.14) be chosen as

$$K(s) = P_2^{-1} \operatorname{diag} \{k_i(s)\} P_1^{-1} \tag{10.91}$$

The closed-loop transfer function then becomes

$$\begin{aligned}
R(s) &= [I + G(s)K(s)]^{-1}G(s)K(s) \\
&= [I + P_1 \operatorname{diag} \{g_i(s)\} P_2]^{-1}P_2^{-1} \operatorname{diag} \{k_i(s)\} P_1^{-1} \\
&= P_1[I + \operatorname{diag} \{g_i(s)\}]^{-1} \operatorname{diag} \{k_i(s)\}P_1^{-1} \\
&= P_1 \operatorname{diag} \left\{\frac{g_i(s)}{1 + g_i(s)k_i(s)}\right\} P_1^{-1} \\
&= \sum_{i=1}^{m} \alpha_{ic} \frac{g_i(s)}{+ g_i(s)k_i(s)} \gamma_{ir} \quad \text{from (10.89)}
\end{aligned}$$

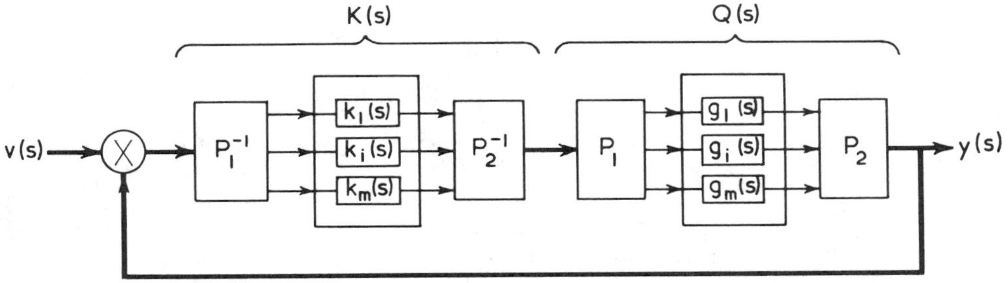

Fig. 10.14 Schematic of the closed-loop system with dyadic Q(s) and K(s).

The following relationship between (10.80) and (10.81) may now be established: For dyadic expansion of G(s),

$$P_1 = \{\alpha_{ic}\}; \quad P_2 = \{\beta_{ir}\}; \quad P_1^{-1} = \{\gamma_{ir}\} \tag{10.92a}$$

If dyadic expansion of $Q_1(s) = G(s)[G(0)]^{-1}$ is required, then

$$P_1 = \{\alpha_{ic}\}; \quad P_2 = \{\beta_{ir}\}[G(0)]^{-1}; \quad P_1^{-1} = \{\gamma_{ir}\} \tag{10.92b}$$

which are compatible with (10.81), (10.84), and (10.85). A numerical method of computing P_1 and P_2 for a given G(s) is considered below.

Design example 10.4 [12]: Dyadic expansion method For most open-loop transfer-function matrices, the dyadic form of (10.80) or (10.81) is unknown. Numerical methods of deriving the vectors α_{ic} and β_{ir} and the scalar transfer functions $g_i(s)$ are outlined here through the design examples considered in Sec. 10.1 (10.14), where

$$G(s) = \frac{1}{(1+s)^2}\begin{bmatrix} 1-s & 2-s \\ \frac{1}{3}-s & 1-s \end{bmatrix} \tag{10.93}$$

The first stage of the design procedure is to compute the eigenvectors and the eigenvalues of $Q_1(s)$ in (10.85):

$$Q_1(s) = G(s)[G(0)]^{-1} = \frac{1}{(1+s)^2}\begin{bmatrix} 1-2s & 3s \\ -2s & 1+3s \end{bmatrix} \tag{10.94}$$

The eigenvalues of $Q_1(s)$ are given by

$$\det\{g(s)I - Q_1(s)\} = 0 \rightarrow \begin{vmatrix} g(s) - \dfrac{1-2s}{(1+s)^2} & \dfrac{3s}{(1+s)^2} \\ \dfrac{2s}{(1+s)^2} & g(s) - \dfrac{1+3s}{(1+s)^2} \end{vmatrix} = 0$$

which gives

$$g_1(s) = \frac{1}{1+s}; \quad g_2(s) = \frac{1}{(1+s)^2} \tag{10.95}$$

The eigenvectors α_{1c} and α_{2c} associated with $g_1(s)$ and $g_2(s)$ above may now be computed by using (10.86), which for the system above becomes

$$\begin{bmatrix} g_i(s) - \frac{1-2s}{(1+s)^2} & -\frac{3s}{(1+s)^2} \\ \frac{2s}{(1+s)^2} & g_i(s) - \frac{1+3s}{(1+s)^2} \end{bmatrix} \begin{bmatrix} \alpha_{1i} \\ \alpha_{2i} \end{bmatrix} = 0, \quad i \in 1, 2 \tag{10.96}$$

Solution of (10.96) gives the frequency-independent vectors

$$\alpha_{1c} = \begin{bmatrix} \alpha_{11} \\ \alpha_{22} \end{bmatrix} = \begin{bmatrix} 1 \\ 1 \end{bmatrix} \quad \text{and} \quad \alpha_{2c} = \begin{bmatrix} \alpha_{12} \\ \alpha_{22} \end{bmatrix} = \begin{bmatrix} 3 \\ 2 \end{bmatrix} \tag{10.97a}$$

The matrix P_1 in (10.80) is then given by

$$P_1 = \{\alpha_{1c} \ \ \alpha_{2c}\} = \begin{bmatrix} 1 & 3 \\ 1 & 2 \end{bmatrix} \tag{10.97b}$$

The vectors $\{\beta_i^T\}$ may now be computed using (10.84):

$$\beta_{ir} = [-1 \quad -1], \quad \beta_{2r} = \begin{bmatrix} \frac{2}{3} & 1 \end{bmatrix} \tag{10.98a}$$

The matrix P_2 in (10.80) is thus given by

$$P_2 = \begin{bmatrix} \beta_{1r} \\ \beta_{2r} \end{bmatrix} = \begin{bmatrix} -1 & -1 \\ \frac{2}{3} & 1 \end{bmatrix} \tag{10.98b}$$

Comment: In vector form, the dyadic expansion of G(s), from (10.81), is given by

$$G(s) = \sum_{i=1}^{m} g_i(s)\alpha_{ic}\beta_{ir} = \sum_{i=1}^{2} \alpha_{ic} g_i(s)\beta_{ir}$$

$$\equiv \frac{1}{1+s}\begin{bmatrix} 1 \\ 1 \end{bmatrix} [-1 \quad -1] + \frac{1}{(1+s)^2}\begin{bmatrix} 3 \\ 2 \end{bmatrix} \begin{bmatrix} \frac{2}{3} & 1 \end{bmatrix}$$

$$\equiv \underbrace{\begin{bmatrix} 1 & 3 \\ 1 & 2 \end{bmatrix}}_{P_1} \underbrace{\begin{bmatrix} \frac{1}{1+s} & 0 \\ 0 & \frac{1}{(1+s)^2} \end{bmatrix}}_{\Gamma(s)} \underbrace{\begin{bmatrix} -1 & -1 \\ \frac{2}{3} & 1 \end{bmatrix}}_{P_2}$$

The vector $\{\gamma_i\}$ may now be computed to obtain the structure of the compensator K(s) by using (10.85b) and (10.92b):

$$\gamma_{1r} = [-2 \quad 3] \quad \text{and} \quad \gamma_{2r} = [1 \quad -1] \rightarrow \begin{bmatrix} \gamma_{1r} \\ \gamma_{2r} \end{bmatrix} = \begin{bmatrix} -2 & 3 \\ 1 & -1 \end{bmatrix} = P_1^{-1}$$

The controller transfer functions $k_1(s)$ and $k_2(s)$ are now chosen for each of the two forward paths; then from (10.87), the controller matrix is given by

$$K(s) = \underbrace{\begin{bmatrix} 3 & -6 \\ -1 & 3 \end{bmatrix}}_{[G(0)]^{-1}} \left\{ k_1(s) \underbrace{\begin{bmatrix} 1 \\ 1 \end{bmatrix}}_{\alpha_{1c}} \underbrace{[-2 \quad 3]}_{\gamma_{1r}} + k_2(s) \underbrace{\begin{bmatrix} 3 \\ 2 \end{bmatrix}}_{\alpha_{2c}} \underbrace{[1 \quad -1]}_{\gamma_{2r}} \right\}$$

$$\equiv \underbrace{\begin{bmatrix} -3 & -3 \\ 2 & 3 \end{bmatrix}}_{P_2^{-1}} \begin{bmatrix} k_1(s) & 0 \\ 0 & k_2(s) \end{bmatrix} \underbrace{\begin{bmatrix} 1 & 3 \\ 1 & 2 \end{bmatrix}}_{P_1} \tag{10.100}$$

The closed-loop transfer function matrix from (10.89) is given by

$$R(s) = \underbrace{\frac{g_1(s)k_1(s)}{[1+g_1(s)k_1(s)]}}_{r_1(s)} \begin{bmatrix} 1 \\ 1 \end{bmatrix} [-2 \quad 3] + \underbrace{\frac{g_2(s)k_2(s)}{[1+g_2(s)k_2(s)]}}_{r_2(s)} \begin{bmatrix} 3 \\ 2 \end{bmatrix} [1 \quad -1]$$

$$\equiv \underbrace{\begin{bmatrix} 1 & 3 \\ 1 & 2 \end{bmatrix}}_{P_1} \begin{bmatrix} r_1(s) & 0 \\ 0 & r_2(s) \end{bmatrix} \underbrace{\begin{bmatrix} -2 & 3 \\ 1 & -1 \end{bmatrix}}_{P_1^{-1}} \tag{10.101}$$

Although this design method has one significant advantage in terms of physical realizability in that P_1 and P_2 are real constant matrices, it is perhaps appropriate to indicate the following aspects of this design procedure.

1. The inclusion of $[G(0)]^{-1}$ in the forward path may not always be necessary. Although such a controller does not affect the eigenvalues or the eigenvectors of G(s), its use may, in some cases, make the numerical computation easier by providing diagonal dominance in $G(s)[G(0)]^{-1}$.

2. For some physical systems, the open-loop transfer function matrix may not have a dyadic expansion with P_1 and P_2 independent of frequency. In

such a case, approximate dyadic expansion may be used. The expression for $G_a(s)$, the approximate dyadic expansion of G(s) (which is assumed not to be dyadic), is given below. Let $S = \{\alpha_i\}$ be the set of eigenvalues of $G_2(s)$ defined as

$$G_2(s) = \lim_{s \to 0} \frac{1}{s}\{[G_1(s)]^{-1} - I_m\}; \; G_1(s) = G(s)[G(0)]^{-1} \tag{10.102}$$

The dyadic approximation of G(s) is then given by [12]

$$G_a(s) = \sum_{i=1}^{m} \{\gamma_{ir} G_1(s)\alpha_{ic}\}\alpha_{ic}\gamma_{ir} G(0) \equiv G(s) \quad \text{if } G(s) \text{ is dyadic, where } \{\gamma_{ir}\} = S^{-1} \tag{10.103}$$

The stability of any system, designed on the basis of the foregoing approximation of the system transfer-function matrix, will have to be checked since the approximation above is valid only for low frequencies. Thus a system with $G_a(s)$ and the resulting K(s) in the feedforward path and unity feedback may be stable, while the physical system [with transfer-function matrix G(s)] with K(s) computed on the basis of $G_a(s)$ is actually unstable [12].

The analysis and design of multivariable systems using generalized frequency-domain concepts attained a very high state of development during the past decade. This is due primarily to the widespread availability of interactive facilities and the software capability of modern digital computers. As a consequence, a variety of computer-aided-design methods for linear time-invariant systems exists. In general, frequency-domain design methods provide a good insight into system structure and performance and an opportunity to make many engineering judgments and practical trade-offs such as reduction of interaction versus controller complexity, and a compromise between stability and integrity. Despite these practical advantages (and the associated demand on software support), it is not possible to identify one (of the many) frequency-domain design techniques as appropriate for most design problems.* Despite the existence of an extensive literature [3,4,16-22]† only four synthesis methods have been included in this chapter, with a vew primarily to introducing the flavor of some of the more established methods which fit in best with the analytical aspects covered in earlier chapters. For an introduction to many other aspects of analysis/design of multivariable systems, the reader is advised to consult Ref. 7.

*See Ref. 15 for a comparison of some frequency-domain methods through experimental results.

†A survey of some recent results on linear multivariable control appears in Ref. 22.

REFERENCES

1. Rosenbrock, H. H., "Design of multivariable control systems using the inverse Nyquist array," Proc. IEE, 116: 1929-1936 (1969).
2. McMorran, P. D., "Design of gas-turbine controller using inverse Nyquist method," Proc. IEE, 117: 2050-2056 (1970).
3. Layton, J. M., Multivariable Control Theory, Peter Peregrinus, London, 1976.
4. Rosenbrock, H. H., Computer-Aided Control System Design, Academic Press, New York, 1974.
5. Rosenbrock, H. H., "Inverse Nyquist array method," in Modern Approaches to Control System Design, N. Munro (Ed.), Peter Peregrinus, London, 1979, Chap. 5.
6. MacFarlane, A. G. J., and Karcanias, N., "Poles and zeros of multivariable systems: a survey of the algebraic, geometric and complex-variable theory," Int. J. Control, 24: 33-79 (1976).
7. MacFarlane, A. G. J., and Postlethwaite, I., "The generalised Nyquist stability criterion and multivariable root-loci," Int. J. Control, 25: 81-127 (1977).
8. Belletrutti, J. J., and MacFarlane, A. G. J., "Characteristic loci techniques in multivariable-control-system design," Proc. IEE, 118: 1291-1297 (1971).
9. Mayne, D. Q., "The design of linear multivariable systems," Automatica, 9: 201-207 (1973)
10. Mayne, D. Q., "The effect of feedback on linear multivariable systems," Automatica, 10: 405-412 (1974).
11. Mayne, D. Q., "Sequential return-difference-design," in Modern Approaches to Control System Design, N. Munro (Ed.), Peter Peregrinus, London, 1979, Chap. 8.
12. Owens, D. H., "Dyadic approximation method for multivariable control-system analysis with a nuclear-reactor application," Proc. IEE, 120: 801-809 (1973).
13. Owens, D. H., Feedback and Multivariable Systems, Peter Peregrinus, London, 1978.
14. MacFarlane, A. G. J., "Relationships between recent developments in linear control theory and classical design techniques," Parts 4 and 5, Meas. Control, 8: 319-323, 371-375 (1975).
15. Fisher, D. G., and Kuon, J. F., "Comparison and experimental evaluation of multivariable frequency-domain design techniques," Preprints 4th IFAC Symp. Multivariable Technological Systems, New Brunswick, Pergamon Press, Oxford, 1977, pp. 453-462.
16. Wolovich, W. A., Linear Multivariable Systems, Springer-Verlag, New York, 1974.
17. MacFarlane, A. G. J., Frequency-Response Methods in Control Systems, IEEE Press, New York, 1979.

18. MacFarlane, A. G. J., "The development of frequency-response methods in automatic control," Trans. IEEE, AC-24: 250-265 (1979).
19. MacFarlane, A. G. J., and Postlethwaite, I., A Complex Variable Approach to the Analysis of Linear Multivariable Feedback Systems, Springer-Verlag, West Berlin (1979).
20. Munro, N. (Ed.), Modern Approaches to Control Systems Design, Peter Peregrinus, London, 1979.
21. Munro, N., and Patel, R. V., Multivariable System Theory and Design, Pergamon, Oxford, 1981.
22. Trans. IEEE, AC-26, No. 1: Special Issue on Linear Multivariable Control Systems, February 1981.

Bibliography for Part II

Computation Aspects

Akashi, H., Proc. 8th IFAC World Congr., Kyoto, Vol. 3: Computer Aided Design, Sessions 51 and 52, 1981, pp. 1587-1629.

Cuenod, M. A. (Ed.), Proc. IFAC Symp. Computer Aided Design of Control Systems, Zurich, 1979.

Davison, E. J., and Gesing, W., "A systematic design procedure for the multivariable servo-mechanism problem," Natl. Eng. Conf. Alternative Methods for Multivariable Control System Design, Chicago, 1977.

de-Jong, L. S., "Numerical aspects of realization algorithms in linear systems theory," Ph.D. thesis, University of Eindhoven, The Netherlands, 1975.

Dyer, P., and McRaynolds, S. R., The Computation and Theory of Optimal Control, Academic Press, New York, 1970.

Horrisberger, H. P., and Belanger, P., "Solution of the optimal constant output feedback problem by conjugate gradients," Trans. IEEE, AC-19: 434-435 (1975).

IEE Conf. Computer-Aided Control System Design, IEE Publ. 96, 1973.

Landau, I. D., and Bethoux, G., "Algorithms for discrete time model adaptive system," Proc. 6th IFAC World Congr., Boston, Vol. 1, 1975, Paper 58.4.

Laub, A. J., "Efficient multivariable frequency response calculations," Trans. IEEE, AC-26: 47-65 (1981).

Kauffman, H., "Computation of output feedback gains for linear stochastic systems using Zangwill-Powell method," Proc. JACC, San Francisco, 1977, pp. 1576-1581.

Kung, S., Kailath, T., and Morf, M., "Fast and stable algorithms for minimum design problem," Preprints 4th IFAC Symp. Multivariable Technol. Syst., New Brunswick, Pergamon Press, Oxford, 1977, pp. 97-104.

Longbottom, G., and Gill, K. F., "Dynamic modelling and multi-axis attitude control of a highly flexible satellite," Preprints 7th IFAC World Congr., Helsinki, Vol. 4, 1978, pp. 2435-2439.

Polak, E., "On a class of computer-aided design problems," Preprints, 7th IFAC World Congr., Helsinki, Vol. 4, 1978, pp. 2443-2449.

Polak, E., and Mayne, D. Q., "An algorithm for optimization problems with functional inequality constraints," IEEE Trans., AC-21: 184-193 (1976).

Polak, E., and Trahan, R., "An algorithm for computer-aided design problems," Proc. IEEE Control Decis. Conf., 1976, pp. 537-542.

Rosenbrock. H. H., Computer-Aided Control System Design, Academic Press, New York, 1974.

Yuan, J. S. C., and Wonham, W. M., "An approach to on-line adaptive decoupling," Proc. IEEE Control Decis. Conf., 1977, pp. 853-857.

Zahir, K. M., and Slivinsky, C., "State variable feedback in computer-controlled multivariable systems," Trans. IEEE, AC-19: 404-407 (1974).

General Aspects

Ackerman, J., "On the synthesis of linear control systems with specified characteristics," Proc. 6th IFAC World Congr., Boston, Vol. 1, 1975, Paper 43.1.

Bohn, E. V., "Use of matrix transformations and system eigenvalues in the design of linear multivariable control systems," Proc. IEE, 110: 989-997 (1963).

Ferreira, P. M. G., "Invariance of infinite zeros under feedback," Int. J. Control, 35: 535-543 (1982).

Kreisselmeir, G., "Stabilization of linear systems by constant output feedback using the Riccati equation," Trans. IEEE, AC-20: 556-557 (1975).

Krough, B., and Cruz, J. B., "Design of sensitivity-reducing compensator design using observers," Trans. IEEE, AC-23: 1058-1062 (1978).

Loscutoff, W. V., "Arbitrary pole placement with limited number of inputs and outputs," ASME Trans. J. Dyn. Syst. Meas. Control, 96: 322-326 (1974).

Miller, R. A., "On the design of dynamic output feedback regulators for linear systems," Int. J. Control, 21: 545-559 (1975).

Miller, R. J., and Mukundan, R., "On designing reduced-order observers for linear time-invariant systems subject to unknown inputs," Int. J. Control, 35: 183-188 (1982).

Morgan, B. S., "Computational procedure for the sensitivity of an eigenvalue," Electron. Lett., 2: 197-198 (1966).

Mueller, G. S., and Nghiem, L. X., "Design of a dynamic output feedback compensator," Proc. IEE, 125: 343-345 (1978).

Munro, N., "Pole assignment," Proc. IEE, 126: 549-554 (1979).

Patel, R. V., "On the computation of numerators of transfer functions of linear systems," Trans. IEEE, AC-18: 400-401 (1973).

Patel, R. V., "On computing the invariant zeros of multivariable systems," Int. J. Control, 24: 145-146 (1976).
Proc. IEE, Special Issue on Multivariable Theory, 126(6): (1979).
Rosenbrock, H. H., "Sensitivity and an eigenvalue to changes in the matrix," Electron. Lett., 1: 278-279 (1965).
Wright, W. C., "An efficient, computer-oriented method for the stability analysis of large multivariable systems," ASME J. Basic Eng., 92: 279-286 (1970).

State Feedback

Anderson, B. D. O., and Luenberger, D. G., "Design of multivariable feedback systems," Proc. IEE, 114: 395-399 (1967).
Aström, K. J., "Why use adaptive techniques for storing large tankers?" Int. J. Control, 32: 689-708 (1980).
Crossley, T. R., and Porter, B., "Modal theory of state observers," Proc. IEE, 118: 1835-1835 (1971).
Davison, E. J., "On pole assignment in multivariable linear systems," Trans. IEEE, AC-13: 747-748 (1968).
Davison, E. J., "The feedforward control of linear multivariable time-invariant systems," Automatica, 9: 561-573 (1973).
Davison, E. J., "The systematic design of control systems for large multivariable linear time-invariant systems, Automatica, 9: 441-452 (1973).
Davison, E. J., and Smith, H. W., "Pole assignment in linear time-invariant multivariable systems with constant disturbances," Automatica, 7: 489-498 (1971).
Gopinath, B., "On the control of linear multiple input-output systems," Bull. Syst. Tech. J., 50: 1063-1081 (1971).
Ichikawa, K., "An approach to the synthesis model-reference adaptive control systems," Int. J. Control, 32: 175-190 (1980).
Kreisselmeier, G., "Adaptive control via adaptive observation and asymptotic feedback matrix synthesis," Trans. IEEE, AC-25: 717-722 (1980).
Kritelis, N. J., "State feedback integral control with 'intelligent' integrators," Int. J. Control, 32: 465-473 (1980).
Landau, I. D., "A survey of model-reference adaptive techniques—theory and applications," Automatica, 10: 353-379 (1974).
Landau, I. D., and Courtiol, B., "Design of multivariable adaptive model following control systems," Automatica, 10: 483-494 (1974)
Lee, G., Jordan, D., and Sohrwardy, M., "A pole assignment algorithm for multi-variable control systems," Trans. IEEE, AC-24: 357-362 (1979).

Moore, B. C., "On the flexibility offered by state feedback in multivariable systems beyond closed-loop eigenvalue assignment," Trans. IEEE, AC-21: 689-692 (1976).

Morse, A., "Global stability of parameter-adaptive control systems," Trans. IEEE, AC-25: 433-439 (1980).

Park, H., and Seborg, D. E., "Eigenvalue assignment using proportional-integral feedback control," Int. J. Control, 20: 517-523 (1974).

Porter, B., "Design of multi-loop modal control systems for plants having complex eigenvalues," Meas. Control, 1: T61-T68 (1968).

Porter, B., and Crossley, T. R., Modal Control, Taylor & Francis, London, 1972.

Ramar, K., and Gourishankar, V., "Utilization of the design freedom of pole assignment feedback controllers of unrestricted rank," Int. J. Control, 24: 423-430 (1976).

Shaked, U., "Design in general model following control systems," Int. J. Control, 25: 57-79 (1977).

Simon, J. D., and Mitter, S. K., "A theory of modal control," Inf. Control, 13: 316-353 (1968).

Wang, S. H., and Desoer, C. A., "The exact model matching of linear multivariable systems," Trans. IEEE, AC-17: 347-348 (1972).

Output Feedback

Ahamari, R., "Pole assignment in linear systems with structurally constrained dynamic compensators," Int. J. Control, 24: 843-851 (1976).

Ahamari, R., and Vacroux, A. G., "Approximate pole placement in linear multi-variable systems using dynamic compensators," Int. J. Control, 18: 1329-1336 (1973).

Ahamari, R., and Vacroux, A. G., "On the pole assignment in linear systems with fixed order compensators," Int. J. Control, 17: 397-404 (1973).

Brasch, F. M., and Pearson, J. B., "Pole-placement using dynamic compensators," Trans. IEEE, AC-15: 34-43 (1970).

Davison, E. J., "On pole assignment in linear systems with incomplete state feedback," Trans. IEEE, AC-15: 348-351 (1970).

Davison, E. J., and Chatterjee, R., "A note on pole assignment in linear systems with incomplete state feedback," Trans. IEEE, AC-16: 98-99 (1971).

Davison, E. J., and Wang, S. H., "On pole assignment in linear multivariable systems using output feedback," Trans. IEEE, AC-21: 516-518 (1975).

Godbout, L. F., and Jordan, D., "Gradient matrices for output feedback systems," Int. J. Control, 32: 411-433 (1980).

Ichikawa, K., "Output feedback stabilization," Int. J. Control, 16: 513-522 (1972).

Kimura, H., "Pole assignment by gain output feedback," Trans. IEEE, AC-20: 509-516 (1975).

Kimura, H., "On pole-assignment by output feedback," Int. J. Control, 28: 11-22 (1978).

Koschmann, J. E., "Compensator design for linear multivariable systems," Trans. IEEE, AC-20: 113-116 (1975).

Kreisselmeier, G., "Stabilization of linear systems by constant output feedback using the Riccati equation," Trans. IEEE, AC-20: 556-557 (1975).

Lee, G., Jordan, D., and Sohrwardy, M., "A pole assignment algorithm for multi-variable control systems," Trans. IEEE, AC-24: 357-362 (1979).

Miller, L. F., Cochran, R. G., and Howze, J. W., "Output feedback stabilization by minimization of a spectral radius function," Int. J. Control, 27: 455-462 (1978).

Miller, R. A., "On the design of dynamic output feedback regulators for linear systems," Int. J. Control, 21: 545-559 (1975).

Munro, N., "Further results on pole-shifting using output feedback," Int. J. Control, 20: 775-786 (1974).

O'Reilley, J., "Minimal-order observers for linear multivariable systems with unmeasurable disturbances," Int. J. Control, 28: 743-751 (1978).

Paraskevopoulous, P. N., "Modal control by output feedback," Int. J. Control, 24: 209-216 (1976).

Paraskevopoulous, P. N., "A general solution to the output feedback eigenvalues-assignment problem," Int. J. Control, 24: 509-528 (1976).

Seraji, H., "Pole assignment using dynamic compensators with prespecified poles," Int. J. Control, 22: 271-279 (1975).

Seraji, H., "Design of multivariable PID controllers for pole placement," Int. J. Control, 32: 661-668 (1980).

Sirisena, H. R., and Choi, S. S., "Optimal pole placement in linear multivariable systems using dynamic output feedback," Int. J. Control, 21: 661-671 (1975).

Sirisena, H. R., and Choi, S. S., "Pole placement in prescribed regions of the complex plane using output feedback," Trans. IEEE, AC-20: 810-812 (1975).

Sirisena, H. R., and Choi, S. S., "Minimal order compensators for decoupling and arbitrary pole placement in linear multivariable systems," Int. J. Control, 25: 755-767 (1977).

Sridhar, B., and Lindorff, D. P., "Pole placement with constant gain output feedback," Int. J. Control, 18: 993-1003 (1973).

Tarokh, M., "Output-feedback stabilization and pole assignment," Int. J. Control, 31: 399-408 (1980).

Vardulakis, A. I., "The structure of multivariable systems under the action of constant output feedback control laws," Int. J. Control, 27: 261-269 (1978).

Youla, D. C., Bongiorno, J. J., and Lu, C. N., "Single-loop feedback-stabilization of linear multivariable dynamical plants," Automatica, 10: 159-173 (1974).

Observers

Balestrino, A., and Celentano, G., "Pole assignment in linear multivariable systems using observers of reduced order," Trans. IEEE, AC-24: 144-146 (1979).

Cavin, R. K., Thisayakorn, C., and Howze, J. W., "On the design of observers with specified eigenvalues," Trans. IEEE, AC-20: 568-570 (1975).

Fortmann, T. F., and Williamson, D., "Design of low-order observers for linear feedback control laws," Trans. IEEE, AC-17: 301-308 (1972).

Ho, B. L., and Kalman, R. E., "Effective construction of linear state variable models from input-output functions," Proc. 3rd Annu. Allerton Conf., 1965, pp. 449-459.

Ikeda, M., Maeda, H., and Kodama, S., "Estimation and feedback in linear time-varying systems: a deterministic theory," SIAM J. Control, 13: 304-326 (1975).

Iwai, Z., Mano, K., and Inada, K., "An adaptive observer for single-input single-output linear systems with inaccessible input," Int. J. Control, 32: 159-174 (1980).

Kawaji, S., "Design procedure of observer for linear functions of the state," Int. J. Control, 32: 381-395 (1980).

Kraft, L. G., "A control structure using adaptive observer," Trans. IEEE, AC-24: 804-806 (1979).

Kreisselmeier, G., "Algebraic separation in realizing a linear state feedback control law by means of an adaptive observer," Trans. IEEE, AC-25: 238-243 (1980).

Kreisselmeier, G., "Adaptive control via adaptive observation and asymptotic feedback matrix synthesis," Trans. IEEE, AC-25: 717-722 (1980).

Kudva, P., and Gourishankar, V., "Optimal observers for the state regulation of linear continuous-time plants," Int. J. Control, 26: 115-120 (1977).

Kudva, P., Viswanadham, N., and Ramakrishna, A., "Observers for linear systems with unknown inputs," Trans. IEEE, AC-25: 113-115 (1980).

Lüders, G., and Narendra, K. S., "Stable adaptive schemes for state estimation and identification of linear systems," Trans. IEEE, AC-19: 841-847 (1974).

Munro, N., "Computer-aided-design procedure for reduced order observers," Proc. IEE, 120: 319-324 (1973).

Nagata, A., Nishimura, T., and Ikeda, M., "Linear function observer for linear discrete-time systems," Trans. IEEE, AC-20: 401-407 (1975).

Newmann, N. M., "Design algorithm for minimal order Luenberger observers," Electron. Lett., 5: 390-392 (1969).

Roman, J. R., and Bullock, T. E., "Design of minimal order stable observers for linear functions of the state via realization theory," Trans. IEEE, AC-20: 613-622 (1975).

Schumachar, J. M., "On the minimal stable observer problem," Int. J. Control, 32: 17-30 (1980).

Wonham, W. M., "Dynamic observers—geometric theory," Trans. IEEE, AC-15: 258-259 (1970).

Decoupling

Alonso-Concheiro, A., "On the definitions of non-interaction for linear time-varying systems," Int. J. Control, 28: 917-925 (1978).

Bayomi, M. M., and Duffield, T. L., "Output feedback decoupling and pole-placement in linear time-invariant systems," Trans. IEEE, AC-23: 142-143 (1977).

Boksenbom, A. S., and Hood, R., "Generalized algebraic method applied to control analysis of complex engine types," NCA-TR-980, NACA, Washington, D.C., 1949.

Cook, P. A., "Approximate decoupling by dynamic compensation," Int. J. Control, 28: 847-852 (1978).

Descusse, J., "State feedback decoupling with stability of linear constant (A, B, C, D) quadrupies," Trans. IEEE, AC-25: 739-743 (1980).

El-Bagoury, M. A., and Bayomi, M. M., "Decoupling of multivariable systems using output feedback," Trans. IEEE, AC-23: 146-149 (1977).

Gilbert, E. G., and Pivnichny, J. R., "A computer program for the synthesis of decoupling multivariable feedback systems," Trans. IEEE, AC-14: 652-659 (1969).

Hirzinger, G., "Decoupling multivariable systems by optimal control technique," Int. J. Control, 22: 157-168 (1975).

Howze, J. W., and Pearson, J. B., "Decoupling and arbitrary pole placement in linear systems using output feedback," Trans. IEEE, AC-15: 660-663 (1970).

Kavanagh, R. J., "Noninteraction in linear multivariable systems," Trans. AIEE, 76: 95-100 (1957).

Loussiouris, T. G., "A frequency domain approach to the block decoupling problem," Int. J. Control, 32: 443-464 (1980).

Mazumdar, A. K., and Choudhury, A. K., "On the decoupling of linear multivariable systems," Int. J. Control, 17: 225-256 (1973).

Morse, A. S., and Wonham, W. M., "Decoupling and pole assignment by dynamic compensation," SIAM J. Control, 8: 317-337 (1970).

Morse, A. S., and Wonham, W. M., "Triangular decoupling of linear multivariable systems," Trans. IEEE, AC-15: 447-449 (1970).

Power, H. M., "Simplification and extension of the Falb-Wolovich decoupling theory," Int. J. Control, 25: 805-818 (1977).

Sato, S. M., and Lopresti, P. V., "New results in multivariable decoupling theory," Automatica, 7: 499-508 (1971).

Silverman, L. M., "Decoupling with state feedback and compensation," Trans. IEEE, AC-15: 487-489 (1970).

Sirisena, H. R., and Choi, S. S., "Minimal order compensators for decoupling and arbitrary pole placement in linear multivariable systems," Int. J. Control, 25: 755-767 (1977).

Stoyle, P. N. R., and Vardulakis, A. I. G., "The mechanism of decoupling," Int. J. Control, 29: 589-605 (1979).

Tripathi, A. N., "Decoupling multivariable systems by state feedback: sensitivity considerations," Proc. IEE, 119: 737-742 (1972).

Vardulakis, A. I. G., and Stoyle, P. N. R., "An algorithm for decoupling and maximal pole assignment in multivariable systems by the use of state feedback," Trans. IEEE, AC-24: 362-365 (1979). Corrections: ibid., p. 809.

Viswanadhan, N., "Uniform decoupling and stabilization by output feedback," Int. J. Control, 21: 451-463 (1975).

Wang, S. H., "Design of precompensator for decoupling problem," Electron. Lett., 6: 739-741 (1970).

Wang, S. H., and Davison, E. J., "Design of decoupled control systems: a frequency-domain approach," Int. J. Control, 21: 529-536 (1975).

Wolivich, W. A., "Output feedback decoupling," Trans. IEEE, AC-21: 148-149 (1975).

Wonham, W. M., and Morse, A. S., "Decoupling and pole assignment in linear multi-variable systems: a geometric approach," SIAM J. Control, 8: 1-19 (1970).

Frequency Domain Design/Synthesis

Cheng, L., and Pearson, J. B., "Frequency domain synthesis of multivariable linear regulators," Trans. IEEE, AC-23: 3-15 (1978).

Hawkins, D. J., "Pseudo-diagonalization and the inverse-Nyquist array method," Proc. IEE, 119: 337-342 (1972).

Hodge, S. S., "Design procedure relating open- and closed-loop diagonal-dominance," Proc. IEE, 118: 927-930 (1971).

Kinnen, E., and Liu, D. S., "Linear multivariable control system design with root-loci," Trans. AIEE, 81: 41-44 (1962).

Kouvaritakis, B. , and MacFarlane, A. G. J., "Geometric approach to analysis and synthesis of system zeros," Part I: Square systems, and Part II: Non-square systems, Int. J. Control, 23: 149-166, 166-181 (1976).

Kouvaritakis, B. A., "Theory and practice of the characteristic locus design method," Proc. IEE, 126: 542-548 (1979).

Kouvaritakis, B., and Shaked, U., "Asymptotic behaviour of root-loci of linear multivariable systems," Int. J. Control, 23: 297-340 (1976).

Mayne, D. Q., "The design of linear multivariable systems," Automatica, 9: 201-207 (1973).

Mayne, D. Q., "Sequential design of linear multivariable systems," Proc. IEE, 126: 568-572 (1979).

MacFarlane, A. G. J., and Kouvaritakis, B. A., "A design technique for linear multivariable feedback systems," Int. J. Control, 25: 837-874 (1977).

MacFarlane, A. G. J., and Postlethwaite, I., "The generalized Nyquist stability criterion and multivariable root-loci," Int. J. Control, 25: 81-127 (1977).

McMorran, P. D., "Parameter sensitivity and inverse Nyquist method," Proc. IEE, 118: 802-804 (1971).

Nicholson, H., "Sequential formulation of the multimachine-power-system and multivariable-control-system problems," Proc. IEE, 118: 1642-1647 (1971).

Owens, D. H., "Multivariable-control-system design concepts in failure analysis of a class of nuclear-reactor spatial-control systems," Proc. IEE, 120: 119-125 (1973).

Owens, D. H., "Dyadic expansion for the analysis of linear multivariable systems," Proc. IEE, 121: 713-716 (1974).

Owens, D. H., "Dyadic expansion, characteristic loci and multivariable-control systems design," Proc. IEE, 122: 315-320 (1975).

Owens, D. H., "Root-loci concepts for Kth-order-type multivariable structures," Proc. IEE, 123: 933-940 (1976).

Owens, D. H., "Dyadic expansions and their applications," Proc. IEE, 126: 563-567 (1979).

Owens, D. H., "Compensation theory for multivariable root-loci," Proc. IEE, 126: 538-541 (1979).

Postlethwaite, I., "A generalized inverse Nyquist stability criterion," Int. J. Control, 26: 325-340 (1977).

Postlethwaite, I., "The asymptotic behaviour, the angles of departure, and the angles of approach, of the characteristic frequency loci," Int. J. Control, 25: 677-695 (1977).

Retallack, D. G., "Extended root-locus technique for design of linear multivariable feedback systems," Proc. IEE, 117: 618-622 (1970).

Rosenbrock, H. H., "Progress in the design of multivariable control systems," Meas. Contr., 4: 9-11 (1971).

Rosenbrock, H. H., "The stability of multivariable systems," Trans. IEEE, AC-17: 105-107 (1973).

Rosenbrock, H. H., Computer-Aided Control System Design, Academic Press, London, 1974.

Rosenbrock, H. H., and McMorran, P. D., "Good, bad or optimal?" Trans. IEEE, AC-16: 552-554 (1971).

Shaked, U., "The angles of departure and approach of the root-loci in linear multivariable systems," Int. J. Control, 23: 445-457 (1976).

Shaked, U., "The intersection of the root-loci of multivariable systems with the imaginary axis," Int. J. Control, 25: 603-607 (1977).

Vardulakis, A. I., "Generalized root locus assignment of all the poles of a multivariable system by output feedback," Int. J. Control, 23: 39-47 (1976).

Problems for Part II

1. Derive suitable full-rank state feedback matrices which would yield the desired closed-loop poles for the following systems:

(a) $$A = \begin{bmatrix} 1 & 0 & 0 \\ 0 & 1 & 1 \\ 0 & 0 & 2 \end{bmatrix}; \quad B = \begin{bmatrix} 1 & 0 \\ 0 & 0 \\ 0 & 1 \end{bmatrix}$$

desired closed-loop characteristic polynomial: $s^3 + s^2 + 90s + 8$

(b) $$A = \begin{bmatrix} 1 & 1 & -2 \\ 2 & 0 & -2 \\ 4 & 2 & -5 \end{bmatrix}; \quad B = \begin{bmatrix} 0 & 0 \\ 1 & 0 \\ 0 & 1 \end{bmatrix}$$

desired closed-loop poles: -3, -3, and -3.

2. Derive a suitable unity-rank state feedback matrix ($F = pq$) to move the closed-loop poles of the system

$$A = \begin{bmatrix} 1 & 1 & -2 \\ 2 & 0 & -2 \\ 1 & 2 & 1 \end{bmatrix}; \quad B = \begin{bmatrix} 1 & 0 & 0 \\ 0 & 0 & 1 \\ 1 & -2 & 0 \end{bmatrix}$$

to -1, -1, -2 with $p = [5 \quad 1 \quad 0]^T$.

3. The desired closed-loop performance of the system S(A, B, C) with

$$A = \begin{bmatrix} 1 & -5 & 21 & -55 \\ 8 & -16 & 36 & -89 \\ 16 & -29 & 53 & -126 \\ 6 & -10 & 15 & -34 \end{bmatrix}; \quad B = \begin{bmatrix} 2 & 1 \\ 3 & 2 \\ 4 & 3 \\ 1 & 1 \end{bmatrix}; \quad C = \begin{bmatrix} 6 & -14 & 9 & -5 \\ 0 & -1 & 1 & -1 \end{bmatrix}$$

is achieved by using the state feedback control law

$$u(t) = \begin{bmatrix} -168 & 339 & -199 & 87 \\ 6 & -19 & 14 & -10 \end{bmatrix} x(t) + v(t)$$

Derive the output feedback matrix which will produce the same closed-loop poles.

4. Derive suitable unity-rank output feedback matrices for the following systems needed to obtain the specified closed-loop poles:

(a) $$A = \begin{bmatrix} -3 & 2 & 0 \\ 4 & -5 & 1 \\ 0 & 0 & -3 \end{bmatrix}; \quad B = \begin{bmatrix} 0 & 1 \\ 1 & 0 \\ 0 & 1 \end{bmatrix}; \quad C = \begin{bmatrix} 1 & 0 & 0 \\ 0 & 1 & 0 \end{bmatrix}$$

desired closed-loop poles: -8, -9, -10.

(b) $$A = \begin{bmatrix} 1 & 1 & -2 \\ 2 & -3 & -2 \\ 8 & 2 & -7 \end{bmatrix}; \quad B = \begin{bmatrix} 0 & 0 \\ 1 & 0 \\ 0 & 1 \end{bmatrix}; \quad C = \begin{bmatrix} 0 & 1 & 0 \\ 0 & 0 & 1 \end{bmatrix}$$

desired closed-loop poles: -1, -2, -3.

5. Derive a suitable output feedback control law for the system

$$A = \begin{bmatrix} 0 & 1 & 0 \\ 4 & 2 & 3 \\ 1 & 0 & 1 \end{bmatrix}; \quad B = \begin{bmatrix} 0 & 0 \\ 1 & 0 \\ 0 & 1 \end{bmatrix}; \quad C = \begin{bmatrix} 0 & 1 & 0 \\ 1 & 0 & 1 \end{bmatrix}$$

needed to obtain $s^3 + s^2 + s + 1$ as its closed-loop characteristic polynomial.

6. Derive the necessary P + D controller such that the closed-loop poles of the system

$$A = \begin{bmatrix} 0 & 1 & 0 & 0 & 0 \\ 0 & -1 & 1 & 0 & 0 \\ 0 & 0 & 0 & 1 & 0 \\ 0 & 0 & 0 & 0 & 1 \\ 0 & 0 & 0 & 0 & 0 \end{bmatrix}; \quad B = \begin{bmatrix} 0 & 0 \\ 1 & 0 \\ 0 & 0 \\ 0 & 0 \\ 0 & 1 \end{bmatrix}; \quad C = \begin{bmatrix} 1 & 0 & 0 & 0 & 0 \\ 0 & 0 & 1 & 0 & 0 \end{bmatrix}$$

are at -1, -1, -2, -2, -3.

7. Derive an appropriate identity observer for the system

$$\dot{x}(t) = \begin{bmatrix} -5 & 3 & 4 \\ -4 & 2 & 4 \\ -3 & 3 & 2 \end{bmatrix} x(t) + \begin{bmatrix} 1 & 0 \\ 0 & -1 \\ 2 & 0 \end{bmatrix} u(t); \quad y(t) = \begin{bmatrix} -2 & 0 & 1 \\ 0 & 1 & 1 \end{bmatrix} x(t)$$

which would yield $s^3 + 3s^2 + 7s - 2$ as its closed-loop characteristic polynomial through $u(t) = k\hat{x}(t) + v(t)$.

8. Derive a suitable reduced-observer feedback control law for

$$\dot{x}(t) = \begin{bmatrix} -2 & -1 & 1 \\ 1 & 0 & 1 \\ -1 & 0 & 1 \end{bmatrix} x(t) + \begin{bmatrix} 1 \\ 1 \\ 1 \end{bmatrix} u(t); \quad y(t) = \begin{bmatrix} 1 & 0 & 0 \\ 0 & 1 & 0 \end{bmatrix} x(t)$$

to move the closed-loop poles to: $-1 \pm j1$, -2, -3. Use the normalization method in Sec. 8.3d.

9. Derive suitable state and output feedback decoupling matrices for the system

$$\dot{x}(t) = \begin{bmatrix} 0 & 1 & 0 \\ 2 & 3 & 0 \\ 1 & 1 & 1 \end{bmatrix} x(t) + \begin{bmatrix} 0 & 0 \\ 1 & 0 \\ 0 & 1 \end{bmatrix} u(t); \quad y(t) = \begin{bmatrix} 1 & 1 & 0 \\ 0 & 0 & 1 \end{bmatrix} x(t)$$

10. Show that the system described by the transfer-function matrix

$$G(s) = \begin{bmatrix} \dfrac{1}{s^3} & \dfrac{s+2}{s^3} \\ \dfrac{1}{s+1} & \dfrac{1}{s+1} \end{bmatrix}$$

can be decoupled by using an output feedback matrix of the form

$$H(s) = \begin{bmatrix} \dfrac{z_{11}(s)}{p_{11}(s)} & -1 \\ 0 & \dfrac{z_{22}(s)}{p_{22}(s)} \end{bmatrix}$$

11. Derive an appropriate feedforward compensator to improve the diagonal dominance of

$$G(s) = \begin{bmatrix} \frac{1}{s+1} & \frac{3}{8s+4} \\ \frac{3}{8s+4} & \frac{2}{s+2} \end{bmatrix}$$

12. Plot the inverse characteristic loci of the following transfer functions:

(a) $$G(s) = \frac{1}{s^2+0.6s+1}\begin{bmatrix} 0.5(s+2) & 4(s+0.1) \\ 0.3(s+8.3) & 3.2(s+1.25) \end{bmatrix}$$

(b) $$G(s) = \frac{1}{1.25(s+1)(s+2)}\begin{bmatrix} s-1 & s \\ -6 & s-2 \end{bmatrix}$$

13. Plot the Gershgorin bands of $\hat{G}(s)$ for

$$G(s) = \frac{1}{(1+s)(2+s)}\begin{bmatrix} 1 & -1 \\ s^2+s-4 & 2s^2-s-8 \end{bmatrix}$$

14. Plot the inverse Nyquist arrays and Gershgorin bands with $K(s) = \hat{G}(0)$ for the following transfer function matrix:

$$G(s) = \frac{1}{(1+2s)(3+2s)}\begin{bmatrix} 4(1+s) & -2 \\ -2 & 4(1+s) \end{bmatrix}$$

APPENDIXES

Appendix 1
Stability Concepts

Stability, like controllability and observability, discussed in Chapter 3, is a qualitative property of an engineering system which is not influenced by the states of the system or (bounded) input signals. The importance of the concept of stability is emphasized by the fact that almost all workable systems are designed to be "stable." If a system is not stable, it is usually of no use in practice.

The degree of complexity of stability analysis of dynamic systems increases rapidly as the mathematical models change from linear time-invariant to linear time-varying and nonlinear. A variety of stability criteria are applicable to linear time-invariant systems, whereas only a limited number of methods are applicable to linear time-varying and nonlinear systems. Although stability theory is a vast subject, most of the analytical criteria available in the literature are quite limited in their applications. In view of this, this short chapter contains only a few stability definitions, concepts, and criteria associated with linear systems to (1) illustrate how they may be related to the behavior and properties of physical systems, and (2) form a basis for multivariable stability analysis, presented in Chap. 5.

A1.1 STABILITY DEFINITIONS [1–4]

A few definitions that are used frequently in stability analysis are given here.

Definition A1: System The system considered here is defined by

$$x = f(x, t) \tag{A1}$$

where $x = \{x_i\}$, $i \in 1, \ldots, n$, is a state vector and $f(x, t)$ is an n-vector whose elements are functions of $x_1, x_2, \ldots, x_n$ and t. Uniqueness of solutions and a continuous dependence of solutions of (A1) on initial conditions are assumed. Denoting the solution of (A1) as $\phi(t; x_0, t_0)$, where $x = x_0$ at $t = t_0$ and t is the observed time,

$$\phi(t_0; x_0, t_0) = x_0 \tag{A2}$$

Definition A2: Motion In defining stability, asymptotic stability, and instability, the word "motion" is frequently used. A motion is defined as a trajectory starting from any state or any point in the n-dimensional state space.

Definition A3: Equilibrium States In the system defined by (A1), a state x_e, where

$$f(x_e, t) = 0, \quad \forall t \tag{A3}$$

is called an equilibrium state of the system. If the system is linear time-invariant, and if $f(x, t) = Ax(t)$, then there exists only one equilibrium state if A is nonsingular, and there exist infinitely many equilibrium states if A is singular. For nonlinear systems there may be one or more equilibrium states. The equilibrium states correspond to the constant solutions of the system ($x = x_e$, $\forall$ t). The determination of the equilibrium state does not involve the solution of the system's differential equation (A1), but only the solution of (A3). If equilibrium states are isolated from each other, they are called isolated equilibrium states. Any isolated equilibrium state can be shifted to the origin of the coordinates, or $f(0, t) = 0$, by a translation of coordinates. Most stability theorems are concerned with isolated equilibrium states.

Definition A4: Stability in the Sense of Lyapunov (i.s.L.) An equilibrium state x_e of (A1) is said to be stable i.s.L. if for each real number $\epsilon > 0$ there is a real number $\delta(\epsilon, t_0) > 0$ such that the inequality

$$\|x_0 - x_e\| < \delta \quad \text{implies} \quad \|\phi(t, x_0, t_0) - x_e\| < \epsilon, \quad \forall t > t_0 \tag{A4}$$

The real number δ depends on ϵ and, in general, t_0. If δ does not depend on t_0, the equilibrium state is said to be uniformly stable. The solution of (A1) is said to be bounded if for a given $\delta > 0$, there exists a constant $\epsilon(\delta, t_0)$ such that

$$\|x_0 - x_e\| < \delta \quad \text{implies} \quad \|\phi(t, x_0, t_0) - x_e\| < \epsilon(\delta, t_0), \quad \forall t > t_0 \tag{A5}$$

If ϵ does not depend on t_0, the solution is said to be uniformly stable. The physical significance of stability i.s.L. is that given a distance ϵ from the equilibrium point, the response must remain within this distance for all $t > 0$. One can then find a second distance δ so that when the initial conditions are anywhere within this distance δ of x_e, the response stays within ϵ of x_e, as graphically shown, for a second-order system, in Fig. A1. Thus given the larger circle E of radius ϵ, there exists a small circle D of radius δ, such that any initial condition within E produces a response that stays within D.

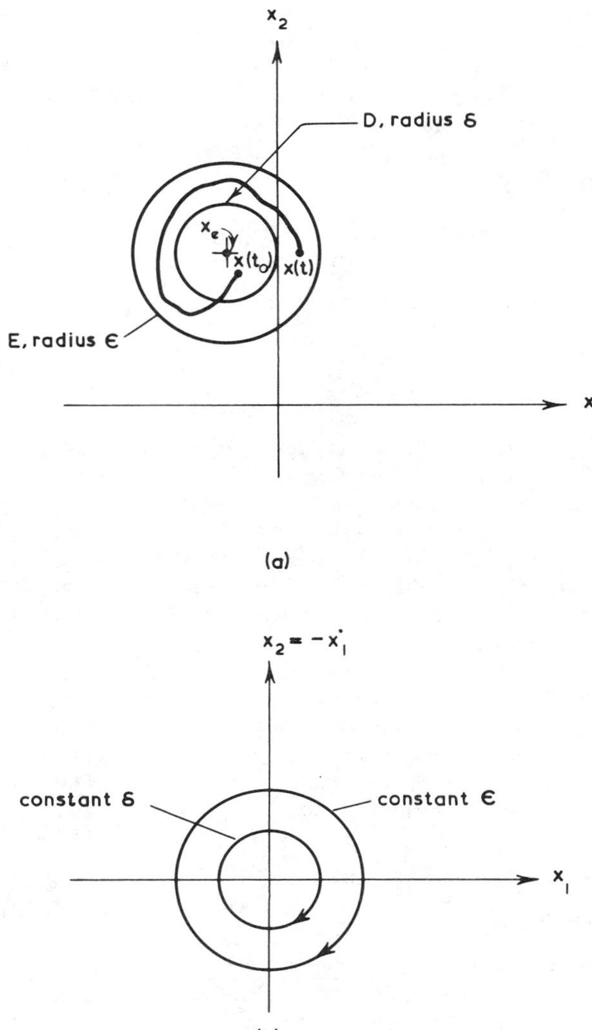

Fig. A1 Graphical interpretation of stability i.s.L.: (a) uniformly stable; (b) harmonic oscillator. $\ddot{x}_1 + x_1 = 0$.

The practical significance of this definition is that it prohibits the response after a disturbance from moving toward infinity. In terms of transfer function, it signifies that all poles of the system are inside the left-half s-plane or at worst on the imaginary axis (Appendix 2).

Definition A5: Asymptotic Stability An equilibrium state x_e of the system in (A1) is asymptotically stable if it is stable and if every solution starting at a state x_0 (which is sufficiently near x_e) converges to x_e as t increases indefinitely. That is, given two real numbers $\delta > 0$ and $\mu > 0$, there are real numbers $\epsilon > 0$ and $T(\mu, \delta, t_0)$ such that

$$\lim_{t \to \infty} x(t) = x_e \quad \text{with} \quad \|x_0 - x_e\| < \delta$$

implies

$$\| \phi(t; x_0, t_0) - x_e \| < \epsilon, \quad \forall\, t > t_0 \tag{A6}$$

and

$$\| \phi(t; x_0, t_0) - x_e \| < \mu, \quad \forall\, t > t_0 + T(\mu, \delta, t_0)$$

The definition of asymptotic stability stems from the fact that for practical purposes, it is useful to identify the "tendency" of the system to return to its equilibrium point after a disturbance. Asymptotic stability adds to the idea of stability the further requirement that every trajectory starting within the inner circle (D) in Fig. A1 must eventually approach the equilibrium point. Any disturbance within E from the equilibrium point results in a system response which not only stays within E (since it is stable) but also finally settles back to the original equilibrium.

Asymptotic stability, in practice, is more important than stability i.s.L. since asymptotic stability is a local concept. Asymptotic stability may not imply stability in general, and some knowledge of the size of the largest region of asymptotic stability is usually necessary. The largest region of asymptotic stability is called the domain of attraction. It is that part of the state space in which asymptotically stable motions originate. In other words, every motion originating in the domain of attraction is asymptotically stable.

Definition A6: Asymptotic Stability in the Large If asymptotic stability holds for all points in the state space from which motions originate, the equilibrium state is said to be asymptotically stable in the large. That is, the equilibrium state x_e of (A1) is asymptotically stable in the large if it is stable and if every solution of (A1) converges to x_e as $t \to \infty$. Consequently, a necessary condition for asymptotic stability in the large is that there be only one equilibrium state in the whole state space. In engineering design, the desirable feature is asymptotic stability in the large. If this kind of stability does not exist, the problem becomes that of determining the largest region of asymptotic stability. However, this is difficult to achieve; for practical purposes, it is sufficient to determine a region of asymptotic stability large enough so that no disturbance will exceed it.

Definition A7: Instability An equilibrium state is said to be unstable if it is neither stable nor asymptotically stable.

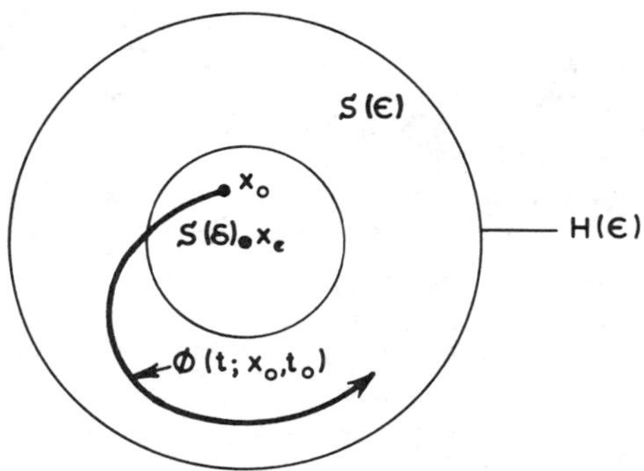

Fig. A2 Equilibrium state x_e.

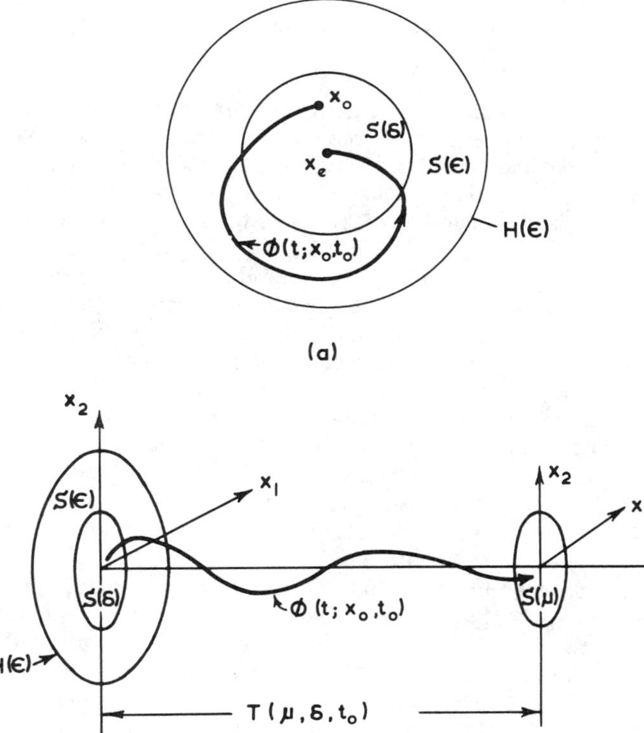

Fig. A3 Asymptotically stable system: (a) equilibrium state trajectory; (b) trajectory of x_e with time.

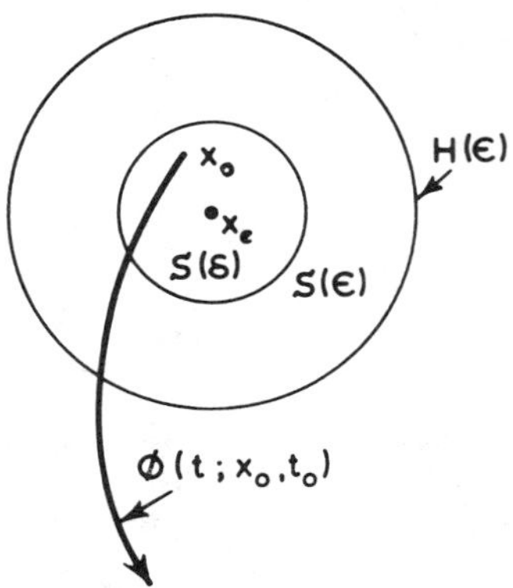

Fig. A4 Trajectory of an unstable equilibrium state.

The concepts of stability above are illustrated below with reference to a second-order system; $S(\epsilon)$ and $S(\delta)$ are, respectively, circular regions of radius $\epsilon > 0$ and $\delta > 0$ about the equilibrium state x_e. In the case of an n-dimensional state space, any closed bounded surface can be used in place of circles to define regions $S(\epsilon)$ and $S(\delta)$. $S(\epsilon)$ consists of the states x satisfying $\|x - x_e\| < \epsilon$.

Figure A2, representing a stable equilibrium state x_e, shows that corresponding to each $S(\epsilon)$ there is an $S(\delta)$ such that a motion starting at state x_0 in $S(\delta)$ does not leave $S(\epsilon)$, or x_0 in $S(\delta)$ implies that $\phi(t;x_0, t_0)$ is in $S(\epsilon)$ for all $t > t_0$. That is, a motion starting in $S(\delta)$ never reaches the boundary circle $H(\epsilon)$ of $S(\epsilon)$.

Figure A3a shows an asymptotically stable equilibrium state x_e with a representative trajectory starting from x_0, where Fig. A3b is a plot of the representative trajectory as time elapses; $S(\mu)$, like $S(\epsilon)$ and $S(\delta)$, is a circular region of radius $\mu > 0$ about x_e. If x_e is asymptotically stable, every motion starting at a state x_0 in $S(\delta)$ converges [without leaving $S(\epsilon)$] to the origin as time increases indefinitely.

Figure A4 shows an unstable equilibrium state x_e and a representative trajectory starting from x_0. In this case, for some real numbers $\epsilon > 0$ and $\delta > 0$, however small, there is always in the circular region $S(\delta)$ of a state x_0 such that the motion starting from this state reaches the boundary circle $H(\epsilon)$

Finally, a few general stability definitions relating the system input and output are given below.

By following the definitions in Sec. 2.3 for a homogeneous relaxed system, the input/output relationship of a linear time-varying system can be expressed as

$$y(t) = \int_{-\infty}^{t} g(t, \tau)u(\tau)\, d\tau, \quad \forall\, t \text{ in } (-\infty, \infty) \tag{A7}$$

where $g(t, \tau)$ is the impulse response of the system and, by definition, is the output measured at time t due to an impulse function input applied at time τ.

In order to formulate the stability criteria of a system, it is relevant to specify under what conditions the output variable y(t) may have the same properties as input control u(t). For example, if the input is bounded, that is,

$$|u(t)| < k_1 < \infty, \quad \forall\, t \tag{A8}$$

it may be sufficient in the stability analysis to ascertain the condition(s) under which there will exist a constant k_2 such that

$$|y(t)| < k_2 < \infty, \quad \forall\, t \tag{A9}$$

Some of the specific constraints that may be put on the input control are

1. Finite input energy; that is, $\left(\int_{-\infty}^{\infty} u(t)^2\, dt\right)^{\frac{1}{2}} < k < \infty$
2. Periodic input function; that is, $u(t + t_1) = u(t) = u(t - t_1)$ for finite values of t_1
3. Constant input control; that is, $u(t) = u_0, \forall\, t$.

Although various different stability conditions can be introduced according to different input/output properties, the one most commonly used in linear systems, bounded-input/bounded-output stability (BIBO), will be used here. The definition of such stability is given below.

Definition A8: BIBO Stability A relaxed time-varying system is said to be bounded-input/bounded-output stable (BIBO stable) only if for any bounded input the output is bounded. [It is important to stress that the condition of relaxedness (Sec. 2.3) is essential in BIBO stability.)

Theorem A1 A relaxed single-input/single-output system defined by $y(t) = \int_{-\infty}^{\infty} g(t, \tau)u(\tau)\, d\tau$ is bounded-input/bounded output stable if there exists a finite number k such that

$$\int_{-\infty}^{\infty} g(t, \tau)\, d\tau < k < \infty \tag{A10}$$

for all values of t in $(-\infty, \infty)$.

Theorem A2 For a relaxed system defined by (A1), if $\int_{-\infty}^{\infty} g(t)\, dt < k < \infty$, for some k, then the following applies:

1. If the input is a periodic function with period T [i.e., $u(t) = u(t+T)$ for all $t > 0$], the output tends to a periodic function with the same period T (not necessarily of the same waveform).
2. If the input is bounded and tends to a constant, the output will tend to a constant.

By using the definition in (A10), if a time-invariant system that is described by

$$y(t) = \int_0^t g(t - \tau)u(\tau)\, d\tau$$

is bounded-input/bounded-output stable, then if the input has certain property, the output will have the same property. For the time-varying system, however, this is not necessarily true. This leads to the following theorem.

Theorem A3 For the relaxed time-varying system with input/output relation

$$y(t) = \int_0^t g(t - \tau)u(\tau)\, d\tau$$

where

$$\int_0^{\infty} g(t)\, dt < k < \infty \tag{A11}$$

if the input control is $u(t) = \sin \omega t$ for $t > 0$, then the output response is described by $y(t) \rightarrow |g(j\omega)| \sin(\omega t + \theta)$ as $t \rightarrow \infty$, where $\theta = \tan^{-1}[\text{Im } g(j\omega)/\text{Re } g(j\omega)]$, and g(s) in the Laplace transform of $g(\cdot)$; $\text{Re}(\cdot)$ and $\text{Im}(\cdot)$ denote the real and the imaginary part of $g(\cdot)$ given by

$$\text{Re } g(j\omega) = \int_0^{\infty} g(t) \cos \omega t\, dt; \quad \text{Im } g(j\omega) = -\int_0^{\infty} g(t) \sin \omega t\, dt \tag{A12}$$

This theorem (proof not considered here) shows that if the input signal to a bounded-input/bounded-output stable time-invariant relaxed system is a sinusoidal function, the output response after the transient dies out is also a sinusoidal function. Furthermore, from this sinusoidal output, the magnitude and phase of the transfer function at that frequency can be read out directly. This fact is often used in practice to measure the transfer function of a linear time-invariant relaxed system.

Since linear time-invariant systems are often described by transfer functions, it is useful to study the stability conditions in terms of transfer functions. [If g(s) is not a rational function of s, the stability condition cannot be easily stated in terms of g(s).]

Theorem A4 A relaxed single-input/single-output system that is described by a proper rational function g(s) is bounded-input/bounded-output only if all poles of g(s) are in the open left half of the s-plane or, equivalently, all poles of g(s) have negative real parts. (The open left-half s-plane means the left half of the s-plane excluding the imaginary axis. On the other hand, the closed left-half s-plane is the left half of the s-plane including the imaginary axis.)

The proof of this theorem follows from the assumption that g(s) is a proper rational function, as it can then be expanded by partial fraction expansion into a sum of finite number of terms of the form $\beta/(s - \lambda_i)^k$ and, possibly a constant, where λ_i is a pole of g(s). Consequently, g(t) is a sum of a finite number of the term $t^{k-1}e^{\lambda_i t}$ and possibly a δ-function. It is easy to show that $t^{k-1}e^{\lambda_i t}$ is absolutely integrable only if λ_i has a negative real part (see also Theorem A8). Hence the proof. It should be noted that stability of a single-input/single-output system is not directly dependent on the zeros of g(s).

A1.2 STABILITY THROUGH $\phi(\cdot)$ [4-6]

By using the definitions above, a notion of stability may be established in terms of the state variables through the state transition matrix $\phi(\cdot)$ (Sec. 2.2), giving a stability condition in the time domain as opposed to the frequency domain as given by Theorem A4. The basic criterion is stated in the following theorem.

Theorem A5 For any finite initial state $x(t_0)$ of a stable system, there exists a positive number M [which depends on $x(t_0)$] such that

$$(1) \quad \|x(t)\| < M, \ \forall\, t > t_0 \tag{A13a}$$

and

$$(2) \quad \lim_{t \to \infty} \|x(t)\| = 0 \tag{A13b}$$

where $\|x(t)\|$ represents the norm of the state vector x(t) and is given by

$$\|x(t)\| = \left\{ \sum_{i=1}^{n} [x_i(t)]^2 \right\}^{\frac{1}{2}} \tag{A14}$$

The physical interpretation of this theorem is that the transition of the state (vector) for any $t > t_0$ as represented by the norm of the vector x(t) must be bounded (A13a), and that the system must reach its equilibrium point as t

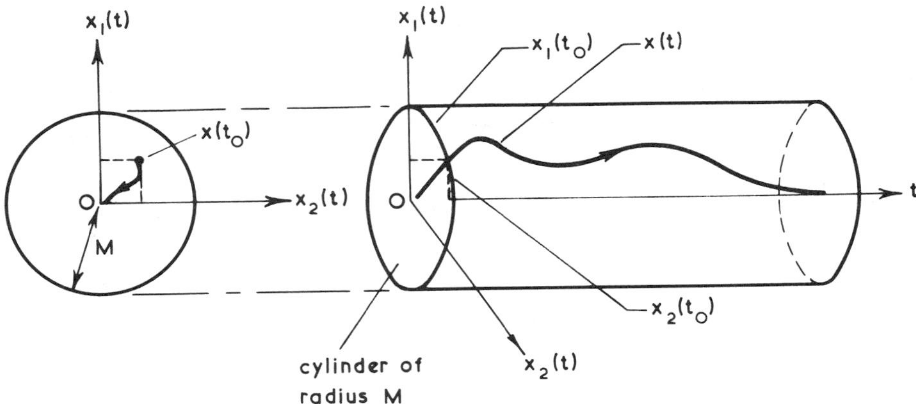

Fig. A5 Illustration of (A13a) and (A13b) for a second-order system.

approaches infinity (A13b). For a second-order stable system, Fig. A5 shows that trajectory of the transition of x(t) (for $t > t_0$) and a cylinder of radius M forms an upper bound for the trajectory point for all $t > t_0$. As $t \to \infty$, the equilibrium point of the system $x(t) \to 0$.

To relate this stability criterion to $\phi(\cdot)$, the following relationship (2.28) for zero input response (Sec. 2.2) is used.

$$x(t) = \phi(t, t_0)x(t_0) \tag{A15}$$

Therefore,

$$\begin{aligned} \|x(t)\| &= \|\phi(t, t_0)x(t_0)\| \\ &\leq \|\phi(t, t_0)\| \, \|x(t_0)\| \end{aligned}$$

and

$$\lim_{t\to\infty} \|x(t)\| \leq \lim_{t\to\infty} \|\phi(t, t_0)\| \, \|x(t_0)\| \tag{A16}$$

Since $x(t_0)$ is finite, from (A13) and (A15) for zero input,

$$\lim_{t\to\infty} \|\phi(t, t_0)\| = 0 \tag{A17}$$

Also

$$\phi(t) \equiv \mathscr{L}^{-1}[sI - A]^{-1} \equiv \mathscr{L}^{-1} \frac{\text{adj }(sI - A)}{\det\{sI - A\}}, \quad \forall\, t > 0$$

det $\{sI - A\} = 0$ represents the characteristic equation, and the time response of $\phi(\cdot)$ is controlled by the roots of the characteristic equation. Combining this with Theorem A4, for a system described by $\dot{x}(t) = Ax(t)$, the condition in (A17) requires that the roots of

$$p(s) = \det\{sI - A\} = 0 \tag{A18}$$

must all have negative real parts. Thus the regions of stability and instability on the s-plane may be clearly identified as illustrated in Fig. A6, where the imaginary axis, excluding the origin, belongs to the unstable region.

It is perhaps worth noting here that this interpretation of Theorem A5 is strictly valid for linear systems. The stability of a nonlinear system, affected by the input condition and the initial state, needs a variety of other definitions.

In view of the definitions of stability given in Sec. A1.1, the following theorems, derived as a direct extension of the development above, are stated without any proofs.

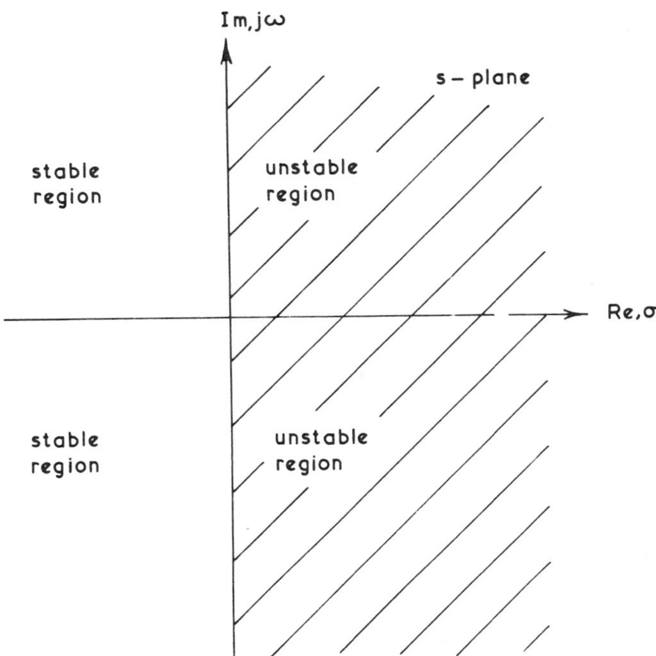

Fig. A6 s-Plane with stable and unstable regions.

Theorem A6 If A, B, and C are bounded on $(-\infty, \infty)$, the linear time-varying system defined by

$$\begin{bmatrix} \dot{x}(t) \\ y(t) \end{bmatrix} = \begin{bmatrix} A(t) & B(t) \\ C(t) & 0 \end{bmatrix} \begin{bmatrix} x(t) \\ u(t) \end{bmatrix} \tag{A19}$$

is bounded-input/bounded-output stable only if the unforced system $\dot{x}(t) = A(t)x(t)$ is asymptotically stable.

Proof of this theorem is excluded here.

Combining the theorems above and controllability and observability properties of linear systems, the following theorem may be stated.

Theorem A7 A time-varying linear system S(A, B, C, t) is stable if it is bounded-input/bounded-output stable. If S(A, B, C, t) is BIBO stable and both controllable and observable, it is also asymptotically stable.

A1.3 STABILITY THROUGH H(·) [5, 6]

The BIBO definition of stability given above leads to an important constraint on the impulse response of any stable system. For a single-input/single-output linear system, the time response of the output variable for any arbitrary input control u(t) may be expressed as

$$y(t) = \int_0^\infty h(\tau)u(t - \tau)\, d\tau \tag{A20}$$

where h(·) is the impulse response of the system. Thus

$$\|y(t)\| = \left\| \int_0^\infty h(\tau)u(t - \tau)\, d\tau \right\| \leq \int_0^\infty \|h(\tau)\|\, \|u(t - \tau)\|\, d\tau \tag{A21}$$

If the input signal u(t) is constrained to be bounded as

$$u(t) \leq k < \infty, \quad \forall t > t_0$$

then

$$\|y(t)\| \leq k \int_0^\infty \|h(\tau)\|\, d\tau \tag{A22}$$

Thus for BIBO stability, the output satisfies the inequality

$$y(t) \leq k_2 < \infty \tag{A23}$$

Hence

$$k_1 \int_0^\infty \|h(\tau)\| \, d\tau \leq k_2 < \infty$$

that is,

$$\int_0^\infty \|h(\tau)\| \, d\tau \leq \frac{k_2}{k_1} = k < \infty \tag{A24}$$

Since integration represents summation, (A24) signifies that the area under the absolute-value curve of the impulse response h(·), evaluated from t = 0 to t = ∞, must be finite if the system were to be BIBO stable.

The condition for stability above may be translated directly into the transfer function g(s), since

$$g(s) = \int_0^\infty h(t)e^{-st} \, dt$$

Therefore,

$$\|g(s)\| \leq \int_0^\infty \|h(t)\| \, \|e^{-st}\| \, dt \tag{A25}$$

Assuming that g(s) is a proper rational function [i.e., the degree of numerator polynomial (≡ number of zeros) is less than or equal to the degree of denominator polynomial (≡ number of poles)], it can be expanded by partial fraction expansion of the form

$$g(s) = \sum_i g_i(s) + \beta_0 = \sum_i \frac{\beta_i}{(s - \lambda_i)^{ki}} + \beta_0 \tag{A26}$$

where λ_i is a pole of g(s). Thus when $s = \lambda_i$, for any i, $\|g(s)\| = \infty$. Also, $\lambda_i = \sigma_i + j\omega_i$ and the absolute value of $e^{-\lambda_i t} = \|e^{-\sigma_i t}\|$. Equation (A26) implies that

$$\infty \leq \int_0^\infty \|h(t)\| \, \|e^{-\sigma_i t}\| \, dt \qquad \text{for } s = \lambda_i$$

Thus if one or more roots have positive real parts (i.e., σ_i is positive) or are on the imaginary axis ($\sigma_i \equiv 0$), the equation above may be written as

$$\infty \leq \int_0^\infty k_3 \|h(t)\| \, dt \tag{A27}$$

where

$$\|e^{-\sigma t}\| \begin{cases} < k_3 & \text{for Re (s)} = \sigma > 0 \\ = 1 & \text{for Re (s)} = \sigma = 0 \end{cases}$$

It is seen that (A27) suggests that the output response of the system monotonically increased for a bounded input—which contradicts the definition of stability given earlier. Thus for a system to be stable, the poles λ_i must not have any positive real parts. This is stated through the following theorem.

Theorem A8 A single-input/single-output system with proper rational transfer function g(s) is BIBO stable if all the poles of g(s) are in the left half of the s-plane or, equivalently, all poles of g(s) have negative real parts. Since

$$\mathscr{L}^{-1}[g_i(s)] = \mathscr{L}^{-1}\left[\frac{\alpha_i}{(s - \lambda_i)^{ki}}\right] = t^{k_i - 1} e^{\lambda_i t} \tag{A28}$$

it is easy to show that $g_i(s)$ is absolutely integrable if λ_i has a negative real part. It is to be noted here that the stability criterion given above does not depend on the zeros of g(s).

A1.4 CONCEPT OF FEEDBACK [5-7]

If a control system has an input signal (scalar or vector) that is generated by some means external to the system (i.e., not influenced by the behavior of the system), the system is said to be an open-loop control system. The input signal may be generated by several means; for example, for the system in Fig. A7, various input controls may have been based on long-term forecasts on power demand, long-term effects of ash disposals and the influence of statutory rules, and union cooperation; but the input control u(t) does not change no matter what the immediate output of the power plant is. This system thus is an open-loop system (but see later). Another example of an open-loop system is a footballer striking a ball (Fig. A8). When hitting the ball, he has some idea (perhaps a very good idea if he is very experienced) of the dynamics of a football. By using his judgment, he kicks the ball at the "correct" position to force it to follow a "best possible" trajectory for the desired final location. Once the ball has been hit, he has no control over its movement, and there is no guarantee that the ball will reach its desired final location; for example, a sudden change of direction of wind will move the ball away from its initial (projected) trajectory. This is another example of an open-loop system.

It may be clear that open-loop systems may not be suitable for applications where a "tight association" between "demanded" input and actual output is required. In the football example, this is apparent as there is always

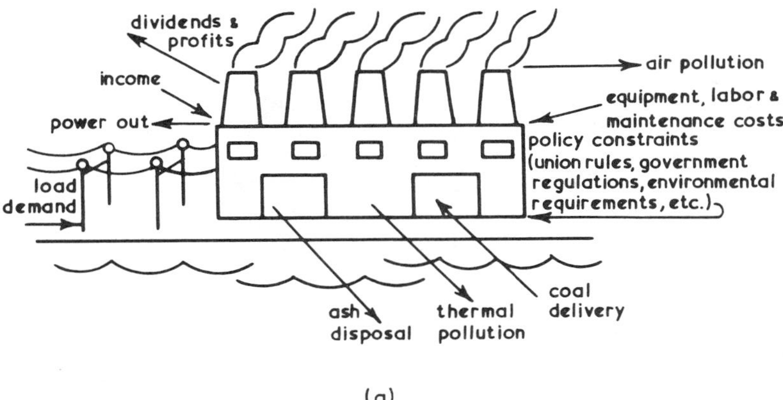

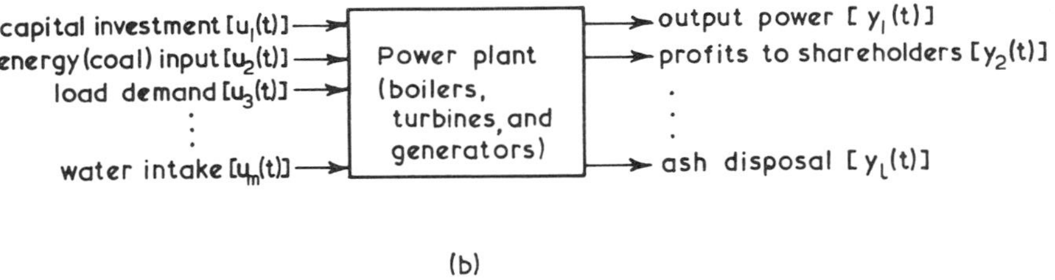

Fig. A7 Representation of a power plant: (a) input/output variables; (b) simplified block diagram.

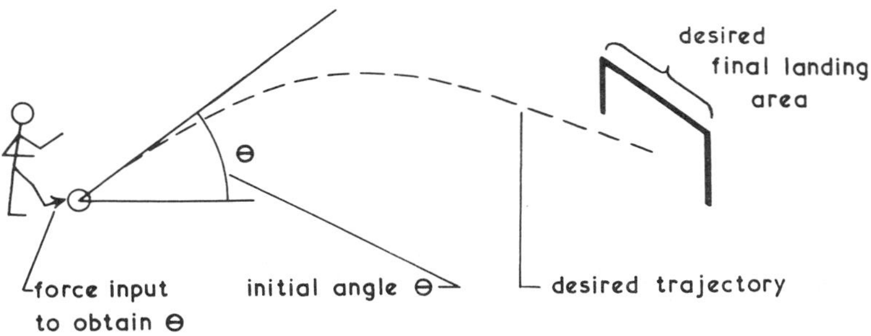

Fig. A8 Open-loop system.

likely to be some discrepancy between the desired and (actual) final landing positions. This leads to the notion of closed-loop control. In closed-loop control, the input (scalar or vector) is evaluated on a continuing basis by the output (≡ performance or behavior) of the system itself. Such a system is also called a feedback control system, since the output is fed back into the input terminal. An example of a closed-loop system is the steering of a motor car (Fig. A9), where the driver continually adjusts the angle of the steering wheel (≡ input) of the car to keep its position with respect to a datum line. If the vehicle's position is as desired, no change in the input signal is necessary.

Having introduced the notion of feedback control, it is perhaps useful to note that feedback may be a "viewpoint." That is, in some systems there may not be a direct way to ascertain the existence of feedback—and the answer may depend on one's "reference." An example of such a case is the dc circuit in Fig. A10. The signal flow graph of the system clearly shows that the current flow i is influenced by R_0. But if one were looking at e_0, the output variable is fixed for a given e_i. Thus, any system may be considered as being with or without feedback, depending on the precise definition of the output variable and input control. For example, the power plant considered earlier is, at least partially, a closed-loop system, as it is most

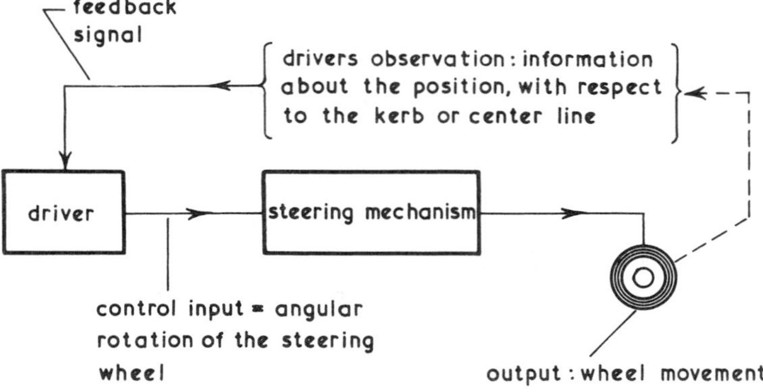

Fig. A9 Closed-loop system: steering of a motor car.

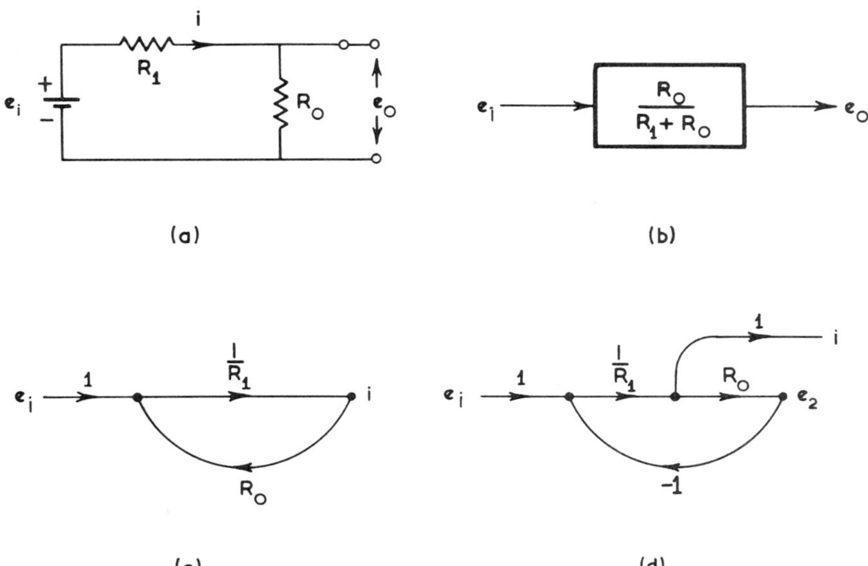

Fig. A10 Dc network and its representations.

unlikely that the plant will be operating at its full capacity* at night when the demanded power is less than the peak (day) demand. With this notion, it may well be argued that most natural dynamic systems have some kind of feedback.

It is quite apparent that a system with feedback is likely to be more complex, at least in a mathematical sense, than a system without it. Despite this added complexity, most engineering dynamic systems† have feedback. The need for feedback stems from the requirement that any physical system (1) must be stable, and (2) must display a clear relationship between output variables and input controls. Although there is a specific relationship between the output and the input in the automotive system in Fig. A9, the system is unstable in the absence of any feedback (≡ blindfolded driver). This is because of the absence of any information about the current position of the car. The football system, on the other hand, is stable, but there is no a priori relationship (in the strictest sense) between the input force/angle and the output position. Thus if the final location of the ball is to be guaran-

*Assuming, of course, that it is an isolated plant (i.e., not connected to any grid system).

†Anything human-made, broadly speaking.

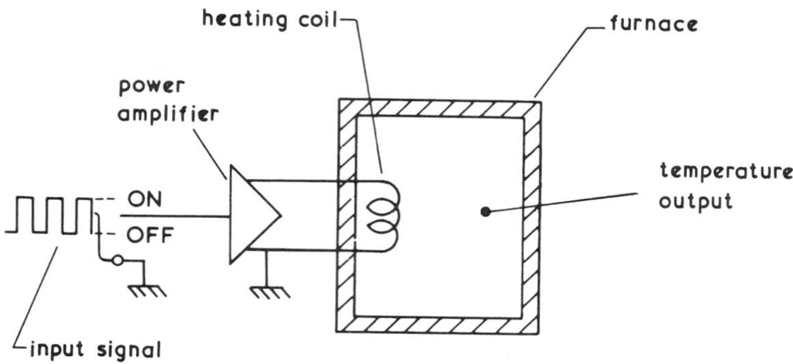

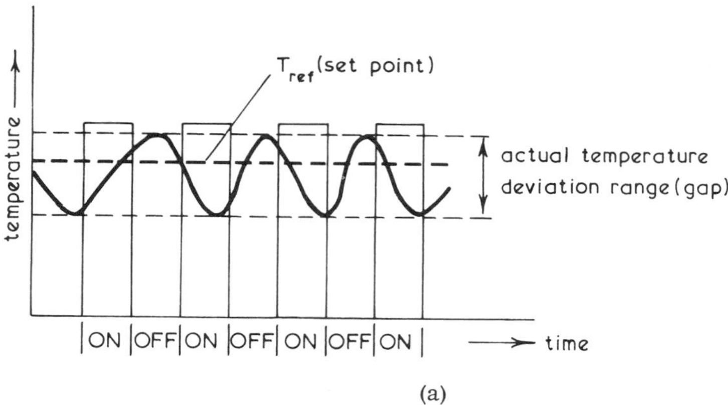

(a)

Fig. A11 (a) Open-loop furnace; (b) closed-loop system—the ratio of on/off time is controlled by the feedback loop (reduced gap).

teed, some kind of feedback must be incorporated, as is done in guided missiles.

In most engineering systems, however, the critical design requirement is to operate the system satisfactorily even when certain parameters or some parts of it are changing significantly. Thus the need for a feedback loop is always assessed on the basis of the extent of control to be exerted to achieve the acceptable relationship between the output variable and input control. For a class of systems, it may be quite possible to satisfy this requirement in open-loop through a suitable control input calculated from

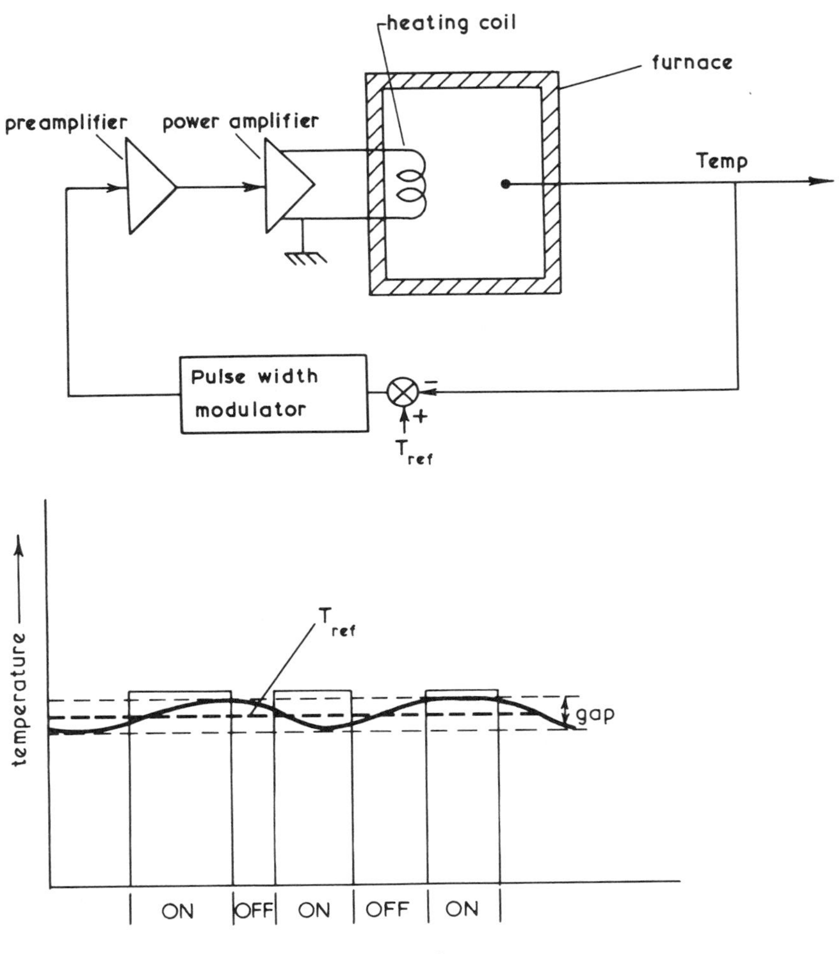

(b)

the known behavior of the system, as in the temperature control system in Fig. A11a. If, however, the resulting oscillation in the temperature output is unacceptable (or unacceptably high), closed-loop control, as in Fig. A11b, may be necessary.

Since qualitative behavior of the output is generally obtainable from stability analysis, a few classical methods of determining the stability of linear systems, based on the criteria in Secs. A1.1 to A1.3, are outlined in Appendix 2.

Appendix 2
Tests for Stability

The stability criteria developed in Sec. A1 in terms of impulse response $h(\cdot)$, the state transition matrix $\phi(\cdot)$, or the roots of the characteristic equation $\det\{\lambda I - A\} = 0$ require a significant amount of numerical computation to implement. As a result, a number of "applicable" methods have been developed; these are briefly outlined here.

A2.1 ROUTH-HURWITZ CRITERION [5-7]

This algebraic method tests the existence of any roots of the characteristic equation in the right-half of the s-plane. The method thus is useful to ascertain the absolute stability of the system.

Let the characteristic equation of the linear system under consideration be

$$p(\lambda) = p_0\lambda^n + p_1\lambda^{n-1} + \cdots + p_{n-1}\lambda + p_n = 0 \tag{A29}$$

where p_i, $i \in 1, \ldots, n$, are real numbers and are related to the roots of $p(\lambda)$ as

$$-\sum_{i=1}^{n} \lambda_i = \frac{p_1}{p_0}; \qquad \sum_{i,j=1}^{n} \lambda_i\lambda_j = \frac{p_2}{p_0}; \qquad -\sum_{\substack{i,j,k=1 \\ i\neq j\neq k}}^{n} \lambda_i\lambda_j\lambda_k = -\frac{p_3}{p_0}; \qquad \cdots;$$

$$(-1)^n\lambda_1\lambda_2\cdots\lambda_n = \frac{p_n}{p_0} \tag{A30}$$

Thus for (A29) to have no roots with positive real part, it is necessary, but not sufficient, that

1. All the coefficients of (A29) have the same sign.
2. None of the coefficients of (A29) vanish.

The necessary and sufficient condition that all the roots of the nth-order characteristic polynomial in (A29) lie on the left half of the s-plane is that the Hurwitz determinants H_k of (A29), $k = 1, \ldots, n$, be positive, where

$$H_k = \underbrace{\begin{vmatrix} a_1 & a_3 & \cdots & a_{2k-1} \\ a_0 & a_2 & \cdots & a_{2k-2} \\ \vdots & & & \\ 0 & & & a_k \end{vmatrix}}_{k \text{ columns}} \left.\vphantom{\begin{vmatrix} a \\ a \\ a \\ a \end{vmatrix}}\right\} k \text{ rows} \tag{A31}$$

where the coefficients corresponding to $k > n$ or $k < 0$ are replaced by zeros. Thus for absolute stability of the system with $p(\lambda) = 0$ as a characteristic polynomial, $a_0 > 0$, $H_k > 0$, for all $k = 1, 2, \ldots, n$.

The method above provides an analytical procedure for identifying the location of the poles of the system with respect to the imaginary axis rather than their exact location, which may be obtained only by solving (A29). Thus the Routh-Hurwitz criterion gives a qualitative answer to stability (i.e., whether a system is stable or not), which in some cases may not be adequate. This leads to the formulation of the following graphical criterion.

A2.2 NYQUIST CRITERION [5-7]

By observing the pattern of the Nyquist plot of the open-loop transfer function (Fig. A12), this method gives information on the difference between the number of poles and zeros of the closed-loop transfer function (Fig. A13) that are on the right half of the s-plane. Salient features of this method are:

1. It provides information on absolute stability as in the Routh-Hurwitz criterion.
2. It provides some indication on the degree of stability of a stable system and indicates how the system stability can be improved.
3. It provides information regarding the frequency response of the system.

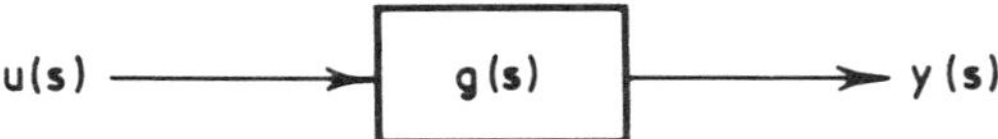

Fig. A12 Open-loop system.

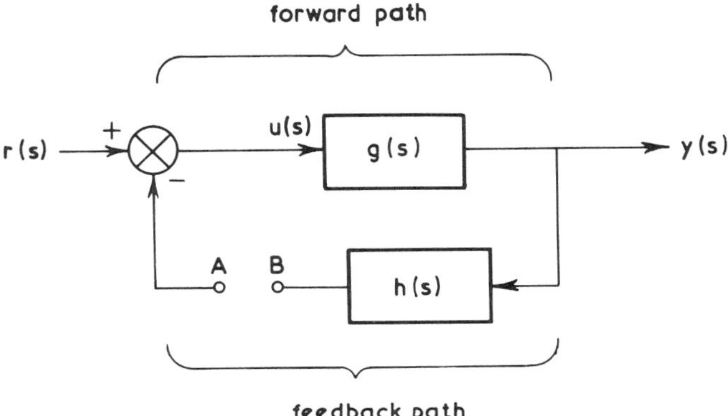

Fig. A13 Closed-loop system when A and B are connected.

If the loop in Fig. A13 is closed by connecting A to B, stability depends on the zeros of $[1 + g(s)h(s)]$, or the points at which

$$g(s)h(s) = -1 \tag{A32}$$

The system is stable if all roots of (A32) lie within the left half of the s-plane. The Nyquist criterion may now be stated.

Theorem A9: Nyquist Criterion The closed-loop system whose open-loop transfer function is $g(s)h(s)$ is stable if and only if

$$En = P_1 - P_2$$

where En is the number of clockwise encirclements of the point $(-1 + j0)$ in the $g(s)h(s)$ plane; P_1 the number of poles of the transfer function $g(s)/[1 + g(s)h(s)]$ inside the right-half s-plane, and P_2 the number of poles of $[g(s)h(s)]$ inside the right half of the s-plane.

The following points may be made from Theorem A.9.

1. Poles and zeros
   Loop-gain zeros = zeros of $g(s)h(s)$
   Loop-gain poles = poles of $g(s)h(s)$
   Closed-loop poles = poles of $g(s)/[1 + g(s)h(s)]$ = zeros of $[1 + g(s)h(s)]$ = roots of the characteristic equation of the closed-loop system
2. For a stable feedback system, there is no restriction on the location of the poles and zeros of the loop gain $g(s)h(s)$, but the closed-loop poles must all be located in the left half of the s-plane.

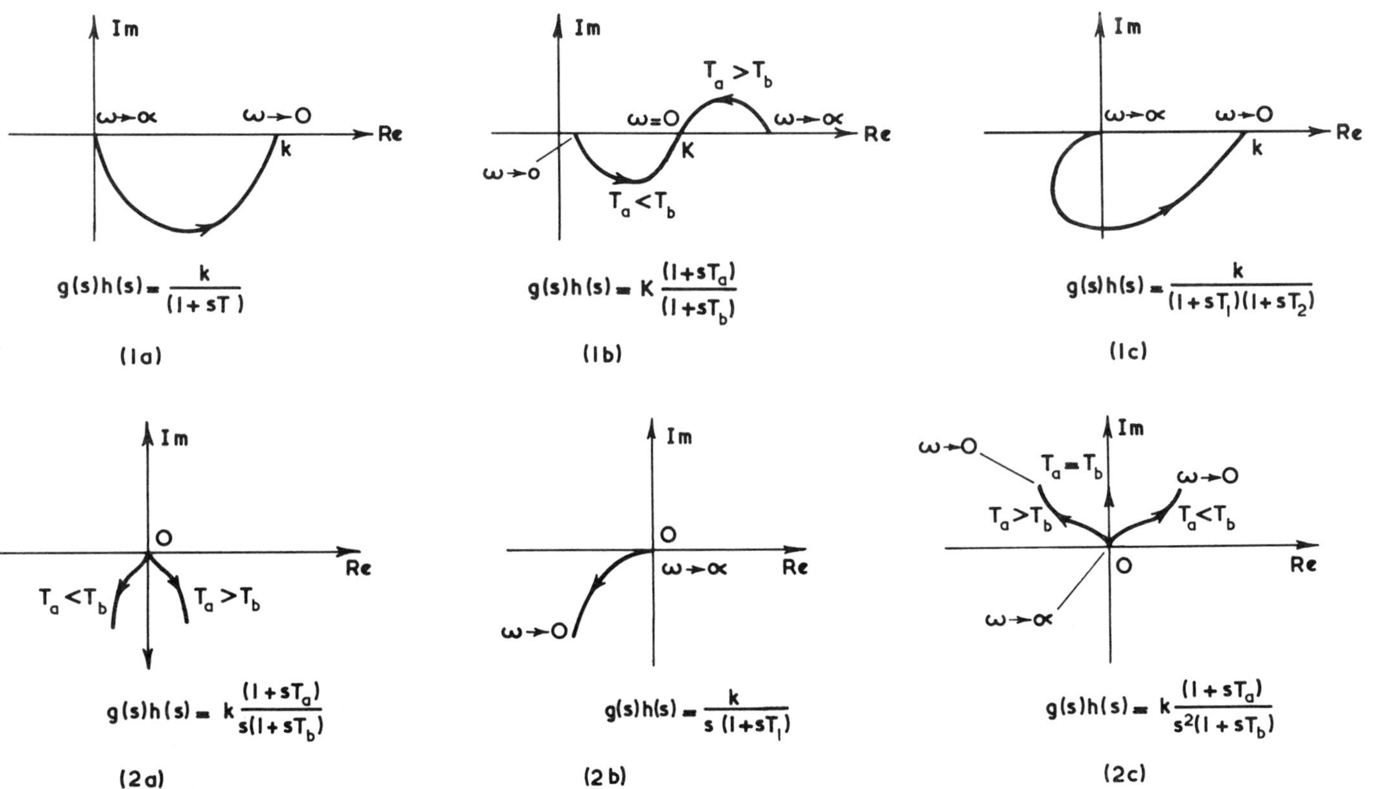

$g(s)h(s) = \frac{k}{(1+sT)}$

(1a)

$g(s)h(s) = K\frac{(1+sT_a)}{(1+sT_b)}$

(1b)

$g(s)h(s) = \frac{k}{(1+sT_1)(1+sT_2)}$

(1c)

$g(s)h(s) = k\frac{(1+sT_a)}{s(1+sT_b)}$

(2a)

$g(s)h(s) = \frac{k}{s(1+sT_1)}$

(2b)

$g(s)h(s) = k\frac{(1+sT_a)}{s^2(1+sT_b)}$

(2c)

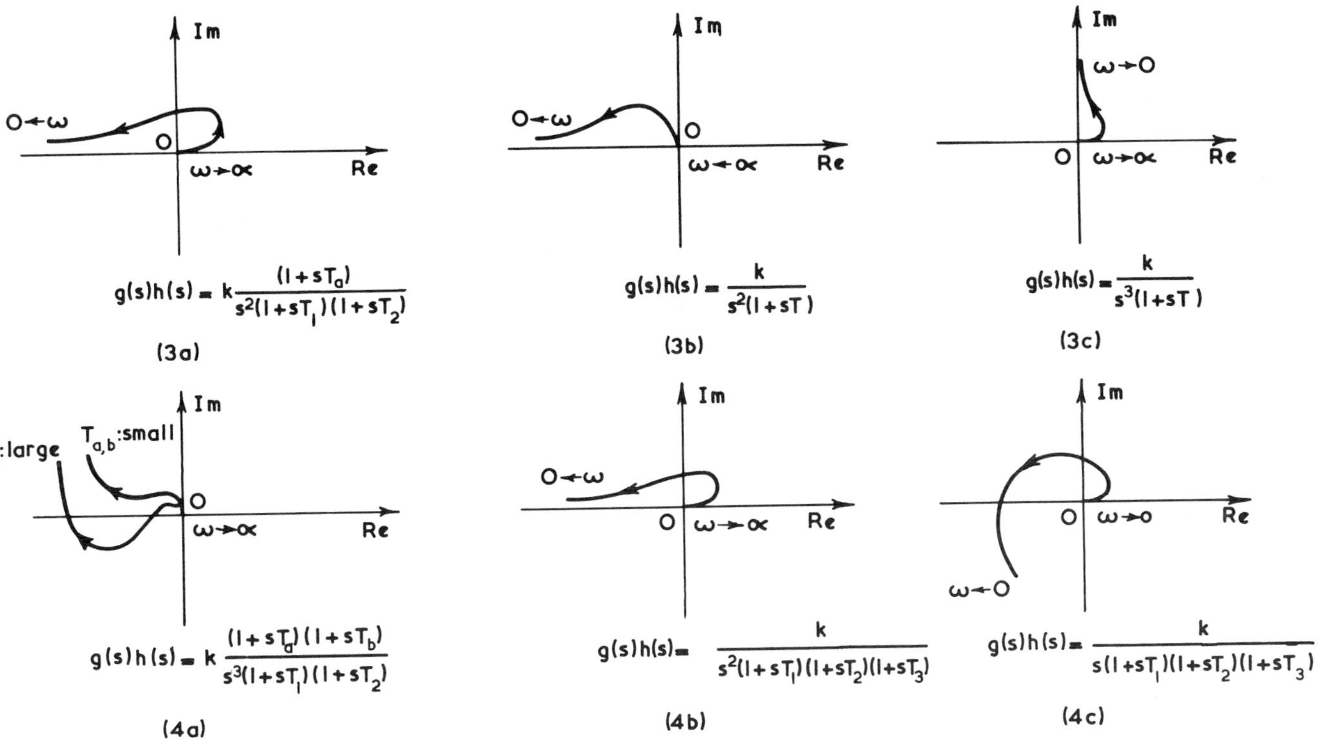

Fig. A14 Several typical Nyquist plots.

Some typical Nyquist plots are shown in Fig. A14. The Nyquist criterion may be generalized to a wider class of criteria, two of which are stated below [8].

Theorem A10: Popov Stability Criterion For any given initial value, the zero input solution of (A1) is bounded and tends to zero as time increased if there exists a real number α and an arbitrary small positive constant ϵ such that

$$\operatorname{Re}\left\{(1 + j\omega\alpha)g(j\omega)\right\} + \frac{1}{\beta} > \epsilon > 0, \quad \forall \omega > 0 \tag{A33}$$

where g(s) is the transfer function of the system.

A consequence of this constraint on the Nyquist plot of $g(j\omega)$ is that it lies to the right of the straight line with nonzero slope passing through the point $(-(1/\beta) + j0)$ provided that $\lim_{\omega \to \infty} g(j\omega) > 1/\beta$, as illustrated in Fig. A15. The Nyquist criterion may be generalized to many other stability criteria [7-10]. One graphically simple criterion is stated below without detailed proof [8,9].

Let the transfer function matrix g(s) be represented by

$$g(s) = \frac{q(s)}{p(s)} \tag{A34}$$

where $\delta\{p(s)\} > \delta\{q(s)\}$ and $\{p(s), q(s)\}$ are relatively prime. Then for zero

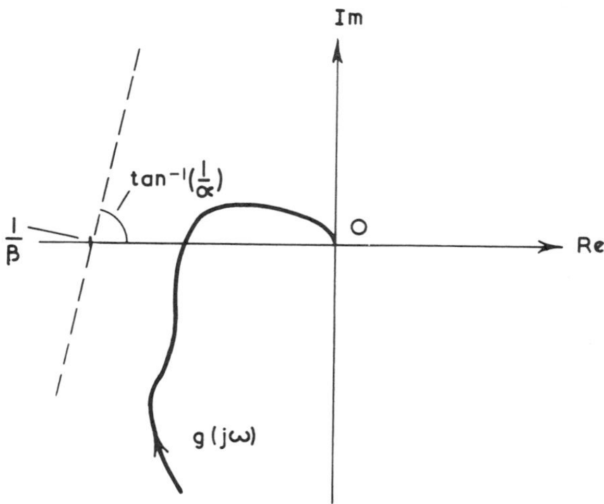

Fig. A15 Popov criterion stable system.

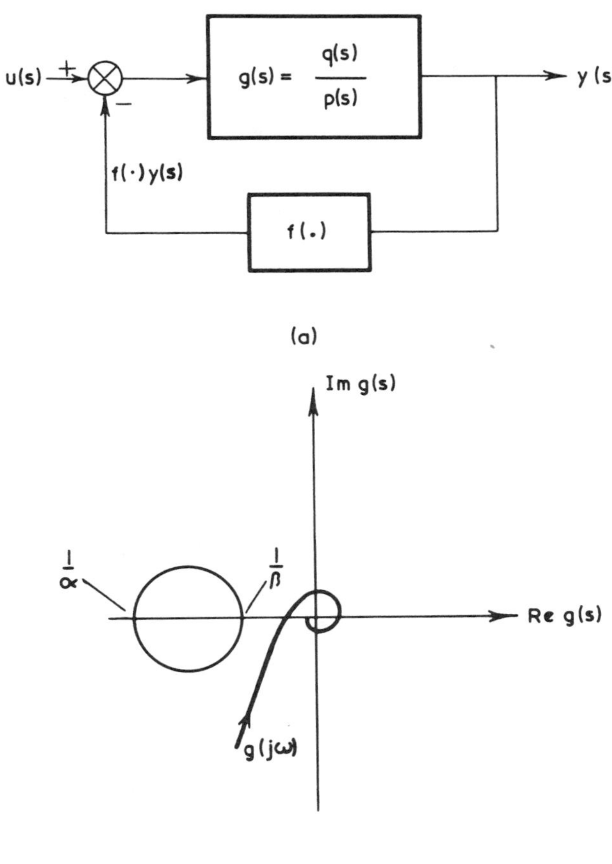

Fig. A16 (a) Block diagram of the general configuration; (b) circle criterion illustration for a stable system.

input $[u(t) \equiv 0]$, the equation of motion of the system described by (A34) may be written in any one of the following forms (Fig. A16a):

$$p(D)x + f(t)q(D)x(t) = 0; \quad y(t) = q(D)x(t)D = \frac{d}{dt}(\cdot)$$

where $c(sI - A)^{-1}b = g(s)$ and

$$y(t) = ce^{AT}x(t_0) - \int_{t_0}^{T} ce^{A(t-\tau)}bf(\tau)y(\tau)\,d\tau \tag{A35}$$

The following theorem may now be stated [9].

Theorem A11: Circle Criterion Let $\dot{x}(t) + Ax(t) + bu(t)$, $y = cx(t)$ be an irreducible realization of $g(s) = q(s)/p(s)$, where $\{p(s), q(s)\}$ are relatively prime. If p(s) has no roots with positive real parts, then

1. All solutions of (A35) are bounded if $0 \leq \beta \leq f(t) < \alpha$ and the Nyquist plot of g(s) does not encircle or intersect the open disk, which is centered on the negative real axis of the g(s) plane and has as diameter the segment of the negative real axis $(-1/\alpha, -1/\beta)$, as shown in Fig. A16b.
2. All solutions are bounded and go to zero at an exponential rate if there is some $\epsilon > 0$ such that

$$0 \leq \beta + \epsilon \leq f(t) < \alpha - \epsilon$$

and the Nyquist plot behaves as in (1) above.

It is to be noted that while generalizing the Nyquist criterion, the circle criterion provides a sufficient, but not necessary condition for stability.

Finally, some of the basic concepts associated with the Nyquist criterion are outlined here, with a view to establishing a generalized Nyquist

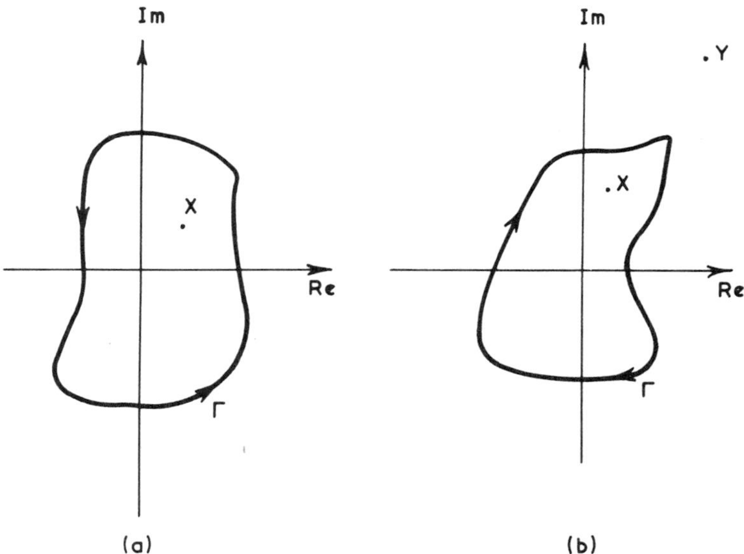

Fig. A17 Encircled points and regions.

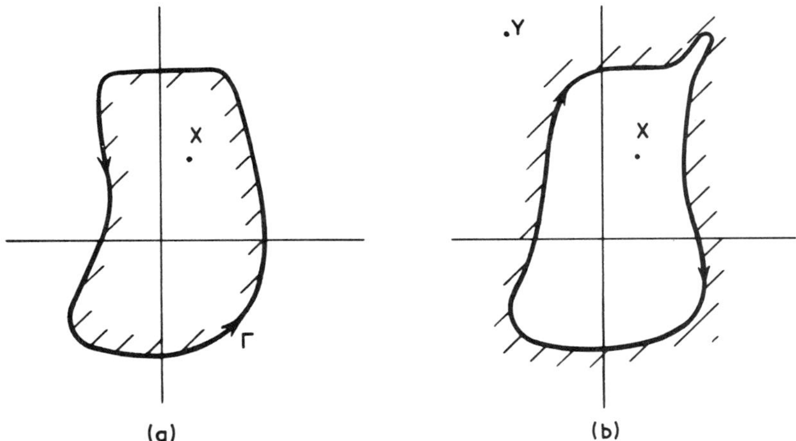

Fig. A18 Areas enclosed by Γ shown with shaded lines.

criterion for single-input/single-output systems [5-7]. These concepts and the resulting generalized stability criterion are extended to multivariable systems in Sec. 5.2.

Definition A9: Encircled A point X in the complex frequency plane is said to be encircled by a closed path Γ if it is found inside Γ. The direction of encirclement is often indicated by an arrow in the path and may be either clockwise or counterclockwise. Thus the point X is encircled in Fig. A17, while the point Y is not.

When a closed path encircles a set of points, it is said that the region inside the closed paths, Γ, is encircled by the path in the indicated direction.

Definition A10: Enclosed A point, or a region, is said to be enclosed by a closed path if it lies to the left of the path when the path is traversed in a prescribed direction. With this definition, the point X in Fig. A18a is enclosed by Γ, whereas it is not enclosed in Fig. A18b. In the latter figure, the point Y is enclosed by Γ.

Principle of the argument

If f(s) is a single-valued rational function that is analytic everywhere in the complex s-plane, then by a one-to-one correspondence, for each point of analyticity in the specified region on the s-plane, one can identify a corresponding point on the f(s)-plane. Thus for a continuous closed path Γ_s on s-plane, where each point on Γ_s is in the specified region in which f(s) is analytic, there exists a closed curve Γ_f mapped by the function f(s) onto the f(s)-plane (Fig. A19). The f(s) locus mapped in the f(s)-plane will encircle

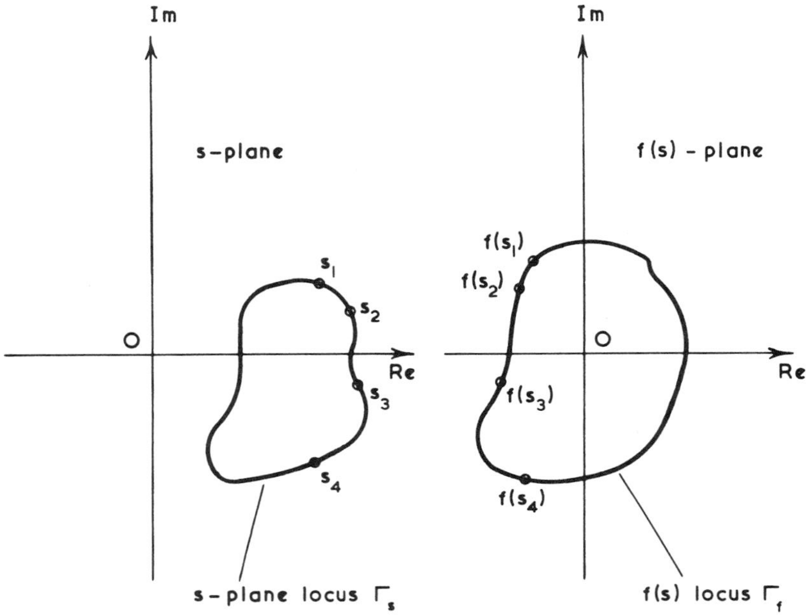

Fig. A19 Correspondence between s-plane locus and f(s) plane locus.

the origin as many times as the difference between the number of zeros and the number of poles of f(s) that are encircled by the s-plane locus Γ_s. Thus, if

En = number of encirclements of the origin made by f(s) locus Γ_f in the f(s)-plane

Np and Nz = number of poles and zeros of f(s) circled by the s-plane locus Γ_s in the s-plane

then

$$\text{En} = \text{Nz} - \text{Np} \tag{A36}$$

A summary of the various cases is given below [5] where CW = clockwise, CCW = counterclockwise, and NE = no encirclement.

N = Nz - Np	Direction of the s-plane locus	f(s)-plane locus: encirclement of the origin	
		Number	Direction
N = 0	CW	0	NE
	CCW		NE
N < 0	CW	N	CCW
	CCW		CW
N > 0	CW	N	CW
	CCW		CCW

The Nyquist criterion in Theorem A9 is a direct application of the above with Γ_s as the Nyquist path shown in Fig. A20. The small semicircles around the poles on the imaginary axis represent those poles that provide a persistent but bounded oscillatory motion, and hence correspond to BIBO

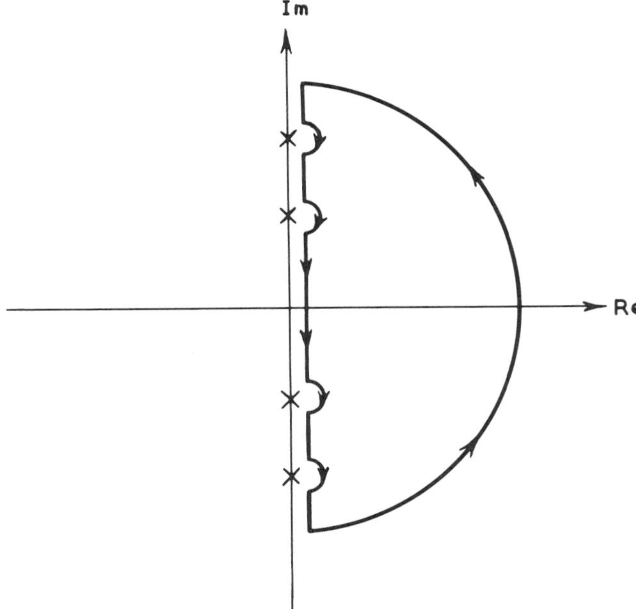

Fig. A20 Nyquist path.

modes. Once the Nyquist path is specified, the stability of the closed-loop system in Fig. A13 can be determined by plotting the

$$f(s) = 1 + g(s)h(s) \tag{A37}$$

locus with the values of s along the Nyquist path, and investigating the behavior of the f(s) plot with respect to the origin.

Since in general, g(s) and h(s) are known, it is more convenient to sketch the Nyquist plot of g(s)h(s) rather than f(s), the difference between these two cases being a shift of the imaginary axis, where the (-1 + j0) point of the g(s)h(s)-plane corresponds to the origin of the f(s)-plane. This leads to the following.

Theorem A12: Generalized Nyquist Stability Criterion For a stable closed-loop system, the Nyquist plot of g(s)h(s) should encircle the (-1+j0) point (≡ critical point) as many times as there are poles of g(s)h(s) in the right-half s-plane. The encirclements, if any, must be made in the clockwise direction.

By using the notations as in (A36) for g(s)h(s), since the number of right-half zeros (Nz) for a stable system is zero, then, by the criterion above,

$$En = -Np \tag{A38}$$

Furthermore, in the majority of physical systems, g(s) and h(s) are stable transfer functions (i.e., Np = 0); hence (A38) becomes

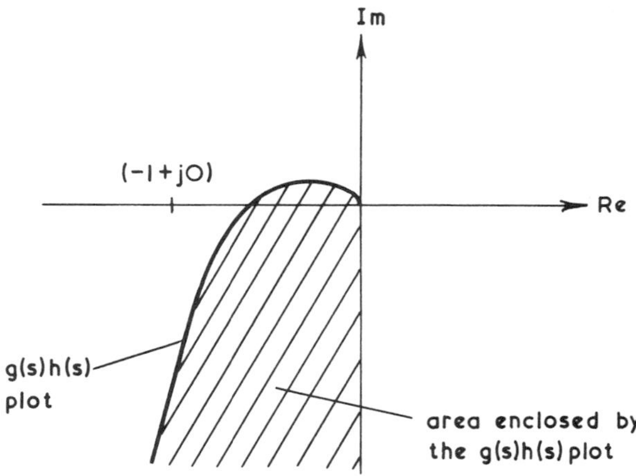

Fig. A21 Nyquist plot g(s)h(s) for a stable closed-loop system.

$En = 0$

In cases when $Np = 0$ but $En = Nz \neq 0$, the criterion above may be modified to read as follows.

Theorem A13 If the loop-gain function g(s)h(s) is a stable function, (i.e., $Np = 0$) for a stable closed-loop system, the Nyquist plot of g(s)h(s) must not enclose the critical point $(-1 + j0)$. (Figure A21 shows the Nyquist plot of a stable closed-loop system.)

A2.3 ROOT-LOCUS PLOT

The root-locus diagram indicates graphically the movement of the roots of the characteristic equation of a system when its open-loop gain is varied. This method is especially useful when a system has been designed with a constant feedforward gain k with a unity-gain feedback loop, as shown in Fig. A22, for which the closed-loop poles are the roots of

$$d(s) + kn(s) = 0 \qquad \text{(A39)}$$

Thus the final choice k may be made after the system has been designed to meet certain requirements specific to a particular application (e.g., damping or natural frequency). The gain margin is defined as the ratio of the value of k at an imaginary axis crossover to the design value of k. If the root locus does not cross over the $j\omega$ axis, the gain margin is infinity. Thus root-locus plots give quantitative measures of gain margin (≡ how much the gain—expressed in decibels—must be increased to cause instability) and phase margin (≡ how much the phase lag—expressed in degrees—must be increased to cause instability); these are considered further in the following section. A few typical root loci for a selection of open-loop pole-zero configurations are shown in Fig. A23.

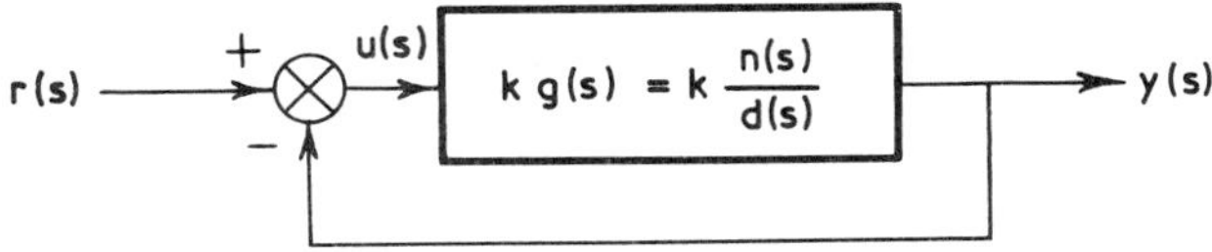

Fig. A22 Unity feedback system with constant open-loop gain k.

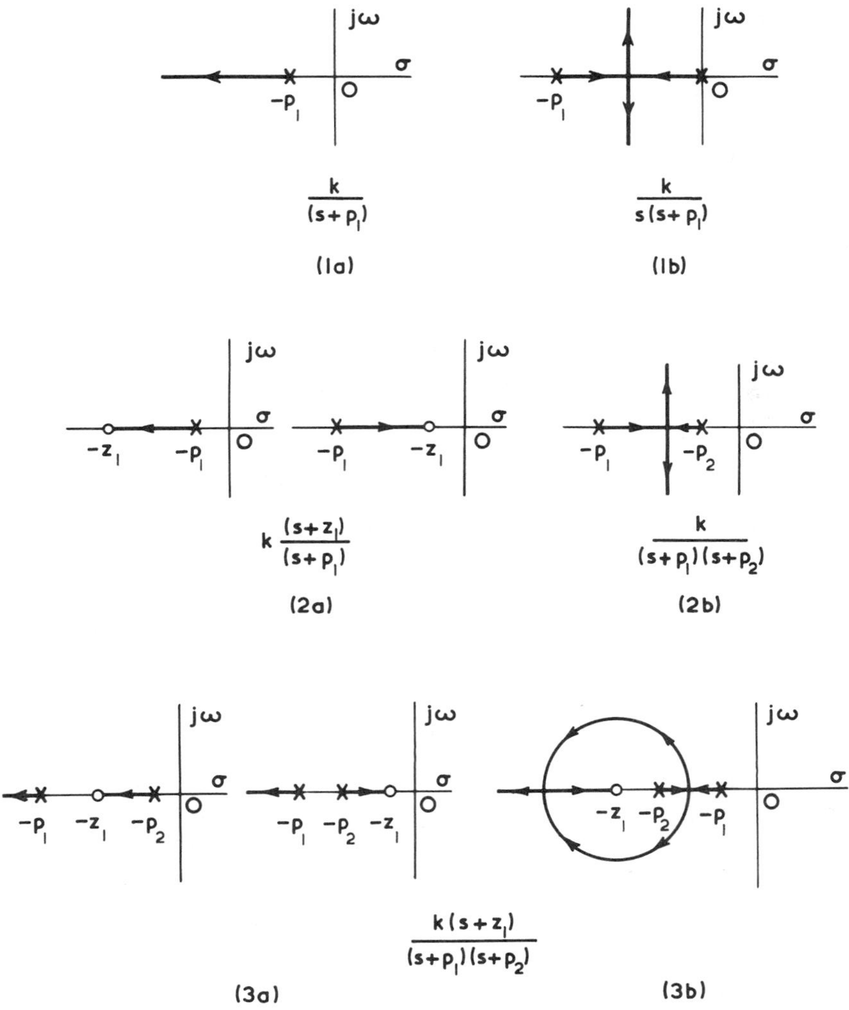

Fig. A23 Sketches of a few typical root loci.

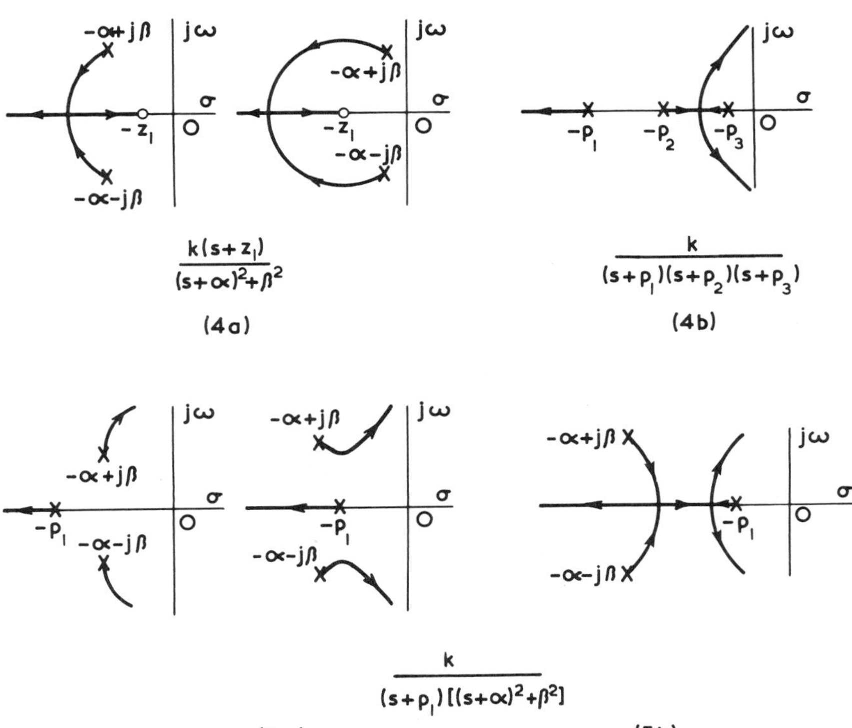
-α+jβ
jω
σ
-z1
O
-α-jβ
-α+jβ
jω
σ
-z1
O
-α-jβ
$\frac{k(s+z_1)}{(s+\alpha)^2+\beta^2}$
(4a)
jω
σ
-p1
-p2
-p3
O
$\frac{k}{(s+p_1)(s+p_2)(s+p_3)}$
(4b)
jω
-α+jβ
σ
-p1
-α-jβ
O
-α+jβ
jω
σ
-p1
O
-α-jβ
-α+jβ
jω
σ
-p1
O
-α-jβ
$\frac{k}{(s+p_1)[(s+\alpha)^2+\beta^2]}$
(5a)
(5b)

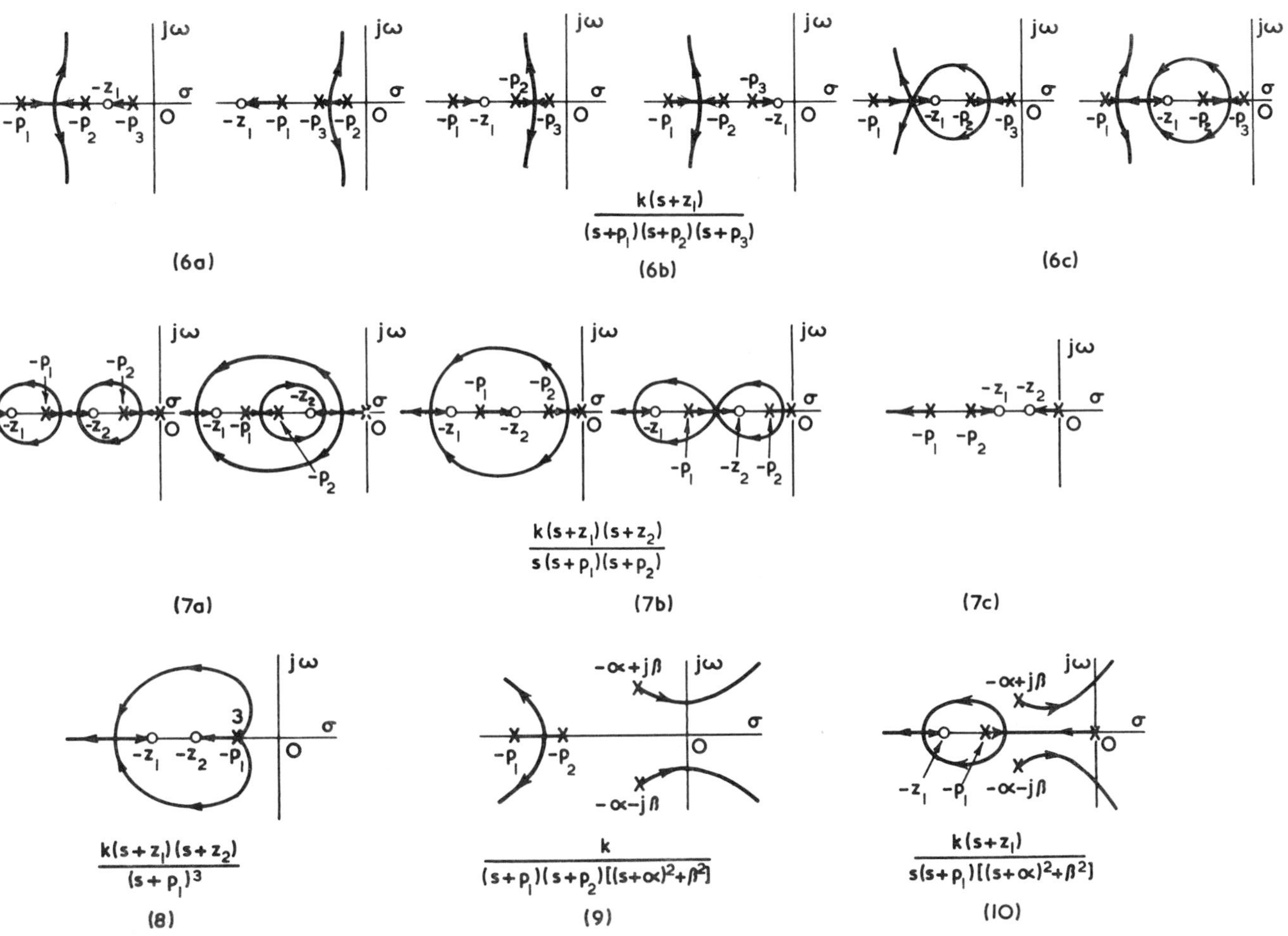

k(s+z1) / (s+p1)(s+p2)(s+p3)
(6a)
(6b)
(6c)
k(s+z1)(s+z2) / s(s+p1)(s+p2)
(7a)
(7b)
(7c)
k(s+z1)(s+z2) / (s+p1)^3
(8)
k / (s+p1)(s+p2)[(s+α)^2+β^2]
(9)
k(s+z1) / s(s+p1)[(s+α)^2+β^2]
(10)

A2.4 BODE DIAGRAM

The Bode plot of the loop transfer function g(s)h(s) in Fig. A13 may be used to determine the stability of the closed-loop system in the same sense as the Nyquist criterion. Bode plots consist of two graphs: the magnitude of $g(j\omega)h(j\omega)$ and the phase angle of $g(j\omega)h(j\omega)$, both plotted as a function of frequency ω, usually on logarithmic scales. These plots illustrate the relative stability of a system with minimum of computational effort, especially when experimental frequency response is available.

The relative stability indicates "gain margin" and "phase margin" on Bode plots as defined earlier and illustrated in Fig. A24. Since 0 dB corresponds to a magnitude of 1, the gain margin is the number of decibels that

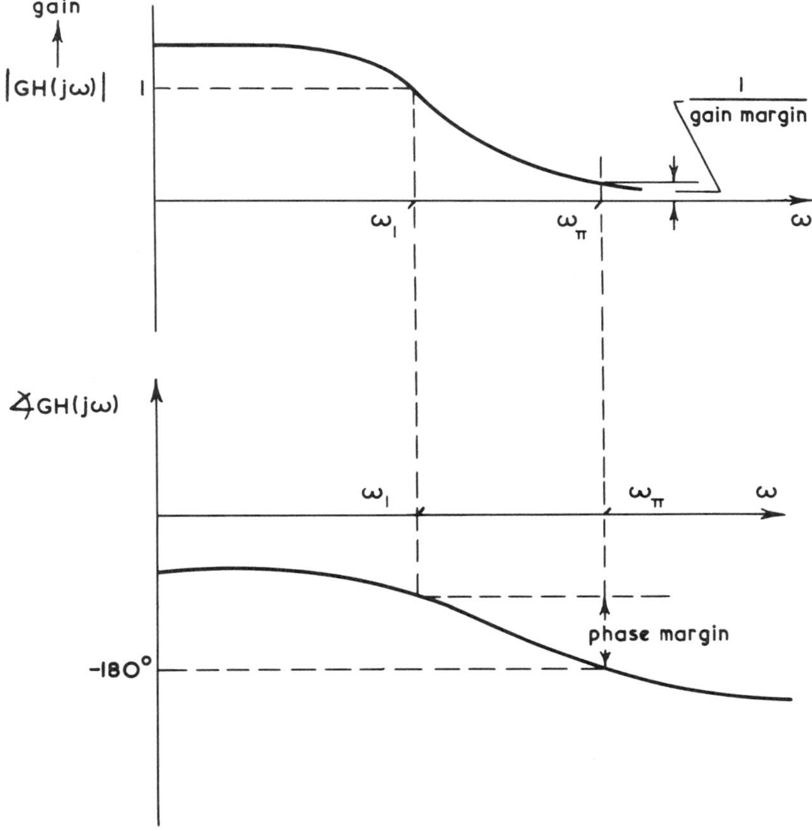

Fig. A24 Graphical representation of gain and phase margins on a Bode plot.

$|g(j\omega)h(j\omega)|$ is below 0 dB at the phase crossover frequency ω_π. The phase margin is the number of degrees $g(j\omega)h(j\omega)$ is above -180° at the gain crossover frequency $\omega_1[|g(j\omega)h(j\omega)| = 1]$.

In most cases, positive gain and phase margins will ensure the stability of the closed-loop system. However, a Nyquist plot may be sketched to verify the absolute stability of the system. The open-loop frequency response of most systems is characterized by decreasing gain with increasing frequency, due to the usual excess of poles over zeros $[\delta\{d(s)\} > \delta\{n(s)\}]$. Thus the closed-loop frequency response of a unity feedback system $[h(s) = 1]$ may be approximated by a magnitude of 1 (0 dB) and a phase angle of 0° for frequencies below the gain crossover frequency ω_1 (Fig. A24). For $\omega > \omega_1$, the closed-loop response may be approximated by the magnitude and phase angle of $g(j\omega)$. Thus an approximate closed-loop bandwidth of many systems is the gain crossover frequency ω_1.

A2.5 LYAPUNOV STABILITY ANALYSIS*

Contrasted with the classical stability criteria of Hurwitz, Nyquist, and Bode, the first and the second (or direct) methods of Lyapunov can be used to determine the stability behavior of free or unforced linear or nonlinear, stationary, or time-varying as well as multivariable systems within the same framework. The second method provides sufficient conditions for asymptotic stability of the equilibrium state without solving the state equation. Various properties and aids to finding Lyapunov functions are discussed briefly here.

A2.5.1 First Method of Lyapunov [1, 2]

This method consists of a procedure using the explicit form of the solution for stability analysis, each state being investigated separately, for the autonomous nonlinear system with vector state equation, $x \in X^n$,

$$\dot{x}(t) = f(x) \tag{A40}$$

where $f(x)$ is continuously differentiable in x_i, $i \in 1, \ldots, n$. By the introduction of a new vector,

$$z = x - x_e \tag{A41}$$

the equilibrium state x_e can be shifted to the origin. This, followed by Taylor series expansion of (A40) about $x = x_e$, gives

$$\dot{z} = Az + G(z)z \tag{A42}$$

*This section follows closely Refs. 1 to 3.

where A is the n × n Jacobian matrix given by

$$A = \begin{bmatrix} \frac{\delta f_1}{\delta x_1} & \frac{\delta f_1}{\delta x_2} & \cdots & \frac{\delta f_1}{\delta x_n} \\ \frac{\delta f_2}{\delta x_1} & \frac{\delta f_2}{\delta x_2} & \cdots & \frac{\delta f_2}{\delta x_n} \\ \vdots & \vdots & & \\ \frac{\delta f_n}{\delta x_1} & \frac{\delta f_n}{\delta x_n} & \cdots & \frac{\delta f_n}{\delta x_n} \end{bmatrix} \tag{A43}$$

and f_i, $i \in 1, \ldots, n$, are the n components of f(x). All partial derivatives in the Jacobian matrix A are evaluated at the equilibrium state in (A41), the n × n matrix G(z) containing terms from the higher-order derivatives in the Taylor series expansion, elements of which vanish at the equilibrium state. Thus (A42) can be linearized near the origin as

$$\dot{z} = Az \tag{A44}$$

which is the first approximation to the nonlinear equation (A41) [i.e., linearized form of (A41)].

It was first shown by Lyapunov that if all the eigenvalues of the constant matrix A in (A44) have nonzero real parts, the stability of the equilibrium state x_e of the original nonlinear equation (A40) is the same as that of the equilibrium state z = 0 of the linearized equation (A44). Hence if the eigenvalues of A have negative real parts, the equilibrium state x_e is asymptotically stable and all solutions of the system given by (A40), with initial state x(0) sufficiently close to x_e, approaches x_e as $t \to \infty$, and if at least one of the eigenvalues of A has a positive real part, then the equilibrium state x_e is unstable. If, however, at least one of the eigenvalues of A has a zero real part, the local stability behavior of the equilibrium state $x = x_e$ of the system in (A40) cannot be determined by (A44), a case called the critical case. In the critical case the local stability behavior of the equilibrium state of the system depends on higher-order terms of the Taylor series expansion of f(x). For example, if one eigenvalue of the Jacobian matrix A is zero at the equilibrium state, this state is asymptotically stable if in its neighborhood A has only eigenvalues with negative real parts. The equilibrium state is unstable, however, if in its neighborhood there is at least one eigenvalue with a positive real part. The first method of Lyapunov is thus concerned with stability in the small, that is, whether a motion originating from a neighborhood of an equilibrium state x_e will approach x_e as $t \to \infty$. This method does not give any indication of how small the neighborhood must be in order that the motion approach x_e as $t \to \infty$. Stability analysis in the large is generally more difficult than stability analysis in the small. The second method of Lyapunov, discussed in the following section, analyses stability in the large as well as local stability of linear and nonlinear systems.

A2.5.2 Second Method of Lyapunov [1-3]

This method provides some information on the stability of the equilibrium states of linear and nonlinear systems without obtaining the solution x(t). The information obtained by this method is precise and involves no approximation.

This second (or direct) method is based on a generalization of the idea that if the system has an asymptotically stable equilibrium state, the stored energy of the system displaced within the domain of attraction decays with increasing time until it finally assumes its minimum value at the equilibrium

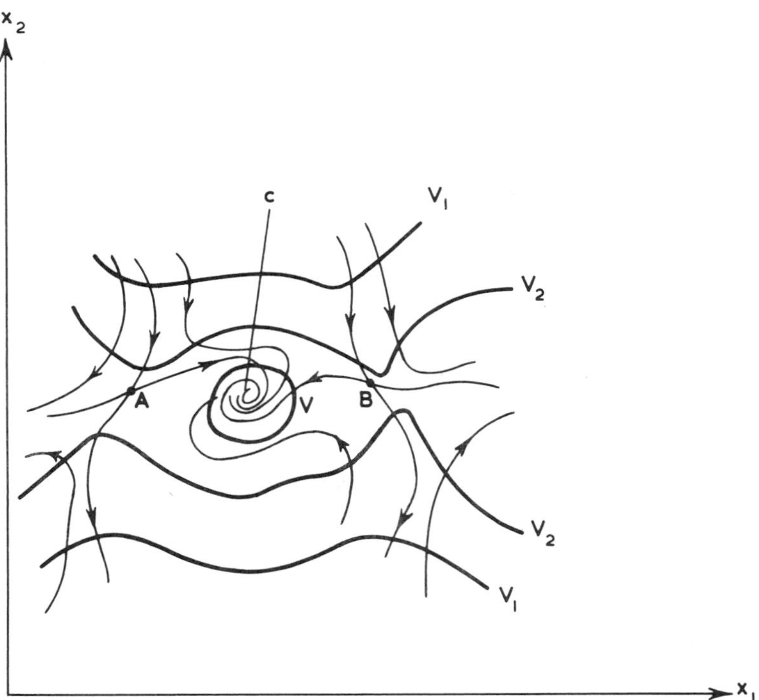

Fig. A25 Trajectories of V(·) for a second-order system (all trajectories except the singular points move "downhill" along the V-surface). (From Ref. 11.)

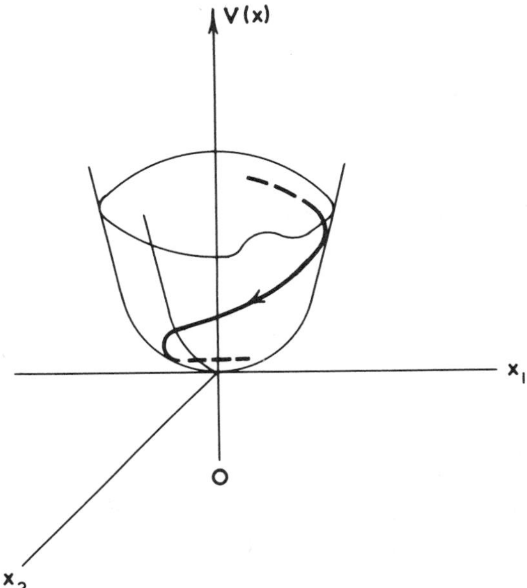

Fig. A26 Lyapunov function V(x).

state. This method involves the determination of a fictitious "energy" function called a Lyapunov function, which has a more general meaning than that of energy and is more widely applicable [11] (Fig. A25).

Lyapunov functions are functions of $x_1, x_2, \ldots, x_n$ and t and denoted by $V(x_1, x_2, \ldots, x_n, t)$ or $V(x, t)$ or $V(x, \bar{x}, t)$, where the bar denotes the complex conjugate. The sign behavior of $V(\cdot)$ and its time behavior $dV(\cdot)/dt$ (Fig. A26) give information on stability, asymptotic stability, or instability of the equilibrium state under consideration without directly solving (A40), as given in the following theorem.

Theorem A14: Lyapunov Stability Theorem Suppose that for a system defined by

$$\dot{x}(t) = f(x, t) \quad \text{with } f(0, t) = 0, \quad \forall\, t \tag{A45}$$

there exists a scalar function V(x, t) which has continuous first partial derivatives. If V(x, t) satisfies the following conditions:

1. V(x, t) is positive definite [that is, $V(0, t) = 0$] and $V(x, t) \geq \alpha(\|x\|)$ $\forall\, x \neq 0$ and $\forall\, t$, where α is a continuous nondecreasing scalar function such that $\alpha(0) = 0$;

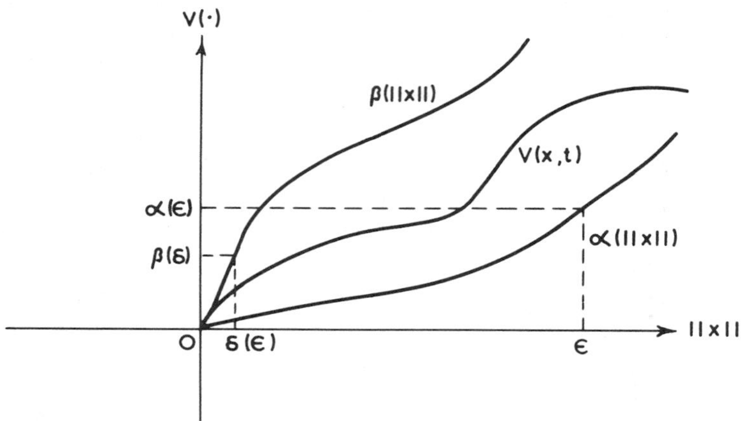

Fig. A27 Curves of $\alpha(\|x\|)$, $\beta(\|x\|)$, and V(x, t).

2. the total derivative V(x, t) is negative $\forall\, x \neq 0$ and $\forall\, t$, or $\dot{V}(x,t) \leq -\gamma(\|x\|) < 0 \;\forall\, x \neq 0$ and $\forall\, t$, where γ is a continuous nondecreasing scalar function such that $\gamma(0) = 0$;
3. there exists a continuous, nondecreasing scalar function such that $\beta(0) = 0$ and $V(x,t) \leq \beta(\|x\|) \;\forall\, t$;
4. $\alpha(\|x\|) \to \infty$ as $\|x\| \to \infty$;

then the origin of the system, x = 0, is uniformly asymptotically stable in the large.

Proof of the theorem is not considered here, but the definitions of V(·), α, and β are shown in Fig. A27.

A2.5.3 Lyapunov Function [2]

The Lyapunov function V(x, t) is a scalar positive-definite function, and it is continuous together with its first partial derivatives (with respect to its arguments) in the region Ω about the origin and has a time derivative which, when taken along the trajectory, is negative definite (or semidefinite). Since $\dot{V}(x,t)$ is the total derivative of V(x, t) with respect to t along a solution of the system, $\dot{V}(x,t) < 0$ implies that V(x, t) is a decreasing function t. A Lyapunov function is not unique in a given system. The simplest positive-definite function is a quadratic form

$$V(x) = \langle x^T, Qx \rangle = \sum_{i=1}^{n} \sum_{j=1}^{n} q_{ij} x_i x_j, \quad i, j \in 1, \ldots, n \tag{A46}$$

In general, Lyapunov function may not be of a simple quadratic form. For any Lyapunov function, the lowest-degree terms in V must be even. This can be seen as follows: If

$$\frac{x_i}{x_n} = y_i, \quad i \in 1, \ldots, (n - 1) \tag{A47}$$

then in the neighborhood of the origin, only the lowest-degree terms become dominant and thus V(x) can be written as $V(x) = x_n^p V(y_1, y_2, \ldots, y_{n-1}, 1)$. If y_i is fixed, then $V(y_1, y_2, \ldots, y_{n-1}, 1)$ is a fixed quantity. For p odd, x_n^p can assume both positive and negative values near the origin, which means that V(x) is not positive definite. Hence p must be even.

The second method of Lyapunov is simple to apply in principle, but in practice it is not necessarily easy for many cases, because no general method for constructing Lyapunov functions is available at present, although there is a great flexibility in choosing such a function. Failure to find a suitable V(·) function does not imply instability, but merely means that the stability condition cannot be guaranteed.

Several stability theorems based on the criteria above may be formulated; one such theorem, especially appropriate for direct stability tests for a linear time-invariant unforced (autonomous) system, is given below [3,12-14]

Theorem A15 The equilibrium (≡ zero) state of the linear time-invariant system in (A45) is asymptotically stable only if given any positive-definite Hermitian matrix Q (or positive-definite real symmetric matrix Q), there exists a positive-definite Hermitian matrix P (or positive-definite real symmetric matrix P) such that

$$A^*P + PA = -Q \tag{A48}$$

Proof of the theorem follows by choosing the scalar function $[x(t)]^T P[x(t)]$ as a Lyapunov function.

Example A1 [4]: Stability Test of a System with Dynamic Equations

$$\begin{aligned}
-k_x \omega_x &= I_x \dot{\omega}_x - (I_y - I_z)\omega_y \omega_z \\
-k_y \omega_y &= I_y \dot{\omega}_y - (I_z - I_x)\omega_z \omega_x \\
-k_z \omega_z &= I_z \dot{\omega}_z - (I_x - I_y)\omega_x \omega_y
\end{aligned} \tag{A49}$$

These equations represent the dynamics of a rigid body about the principal axes of inertia. The Euler equations above may be used to describe the motion of an inherently stable earth satellite with a suitable choice of state variables. Equation (A49) may be written in state-space form as

$$
\underbrace{\begin{bmatrix} \dot{x}_1(t) \\ \\ \dot{x}_2(t) \\ \\ \dot{x}_3(t) \end{bmatrix}}_{\dot{x}(t)} = \underbrace{\begin{bmatrix} -\frac{k_x}{I_x} & \frac{I_y}{I_x} x_3(t) & -\frac{I_z}{I_x} x_2(t) \\ \\ -\frac{I_x}{I_y} x_3(t) & -\frac{k_y}{I_y} & \frac{I_z}{I_y} x_1(t) \\ \\ \frac{I_x}{I_z} x_2(t) & -\frac{I_y}{I_z} x_1(t) & -\frac{k_z}{I_z} \end{bmatrix}}_{A(x)} \underbrace{\begin{bmatrix} x_1(t) \\ \\ x_2(t) \\ \\ x_3(t) \end{bmatrix}}_{x(t)} \tag{A50}
$$

For $x(t) \equiv 0$,

$$
A(0) = \begin{bmatrix} -\frac{k_x}{I_x} & 0 & 0 \\ \\ 0 & -\frac{k_y}{I_y} & 0 \\ \\ 0 & 0 & -\frac{k_z}{I_z} \end{bmatrix} < \infty
$$

Thus the system is stable at the equilibrium point (origin of the state space). $V(x)$ may be chosen by using Theorem A15:

$$
V(x) = \langle x^T P x \rangle
$$

where P is a positive-definite real symmetric and may be chosen as

$$
P = \operatorname{diag}\{I_x^2, I_y^2, I_z^2\}
$$

giving

$$
V(x) = [x_1(t), x_2(t), x_3(t)] \begin{bmatrix} I_x^2 & & 0 \\ & I_y^2 & \\ 0 & & I_z^2 \end{bmatrix} \begin{bmatrix} x_1(t) \\ x_2(t) \\ x_3(t) \end{bmatrix}
$$

$$
= I_x^2 x_1^2 + I_y^2 x_2^2 + I_2^2 x_3^2, \quad \text{dropping (t) for convenience} \tag{A51}
$$

where

$\sqrt{V(x)} \equiv$ norm of total angular momentum $\rightarrow$ a measure of stored energy

Combining (A51) and (A48), we have

$$-Q = [A(x)]^T P + P[A(x)]$$

or

$$Q = \text{diag}\{2k_x I_x, 2k_y I_y, 2k_z I_z\} \tag{A52}$$

which is positive definite and real since the stiffness parameters k_x, k_y, k_z are positive, and consequently the equilibrium is asymptotically stable in the large.

In applying Theorem A15 for stability analysis of linear time-invariant systems, for a given Q a Hermitian matrix P (or real symmetric matrix P) may be derived by equating the matrices

$$A^*P + PA \quad \text{and} \quad -Q$$

element by element. This results in $n(n + 1)/2$ linear equations for determination of the elements $p_{ij} = p_{ij}^*$ of P.† These simultaneous equations are solvable unless an eigenvalue of A or the sum of a pair of eigenvalues of A is zero. In other words, if the eigenvalues of A are defined as $\lambda_1, \lambda_2, \cdots,$ each repeated as often as its multiplicity as a root of the characteristic equation, and if for every sum of the two roots, $\lambda_j + \lambda_k \neq 0$, then the matrix P is uniquely determined by the matrix Q. By applying Sylvester's criterion for $x^*(t)Px(t)$ to be positive definite, that is,

$$p_{11} > 0; \quad \begin{vmatrix} p_{11} & p_{12} \\ p_{21} & p_{22} \end{vmatrix} > 0; \quad \begin{vmatrix} p_{11} & p_{12} & p_{13} \\ p_{21} & p_{22} & p_{23} \\ p_{31} & p_{32} & p_{33} \end{vmatrix} > 0, \ \ldots$$

(where $p_{ij} = p_{ji}^*$), asymptotic stability of the origin of the system can be ascertained.

An alternative form of Theorem A15, which is useful in terms of numerical computation, is given below.

<u>Theorem A16</u> A necessary and sufficient condition for $x(t) = 0$ to be an asymptotically stable solution of $\dot{x}(t) = A\dot{x}(t)$ is that there exists a positive-definite Hermitian (or positive-definite real symmetric) matrix P satisfying the equation

†p^* is the conjugate of p.

$$A^*P + PA = -I \tag{A53}$$

where I is an $(n \times n)$ identity matrix.

Finally, since the Lyapunov function of a given system is not unique, the following comments are relevant.

1. Failure in finding a $V(\cdot)$ function to show stability, asymptotic stability, or instability of the equilibrium state under consideration can give no information on stability.
2. Although a particular V function may prove that the equilibrium state under consideration is stable or asymptotically stable in the region Ω which includes this equilibrium state, it does not necessarily imply that the motions are unstable outside the region Ω.
3. For a stable of asymptotically stable equilibrium state, a V function with the required properties always exists.

Example A2 [2]: Derivation of Lyapunov function through Theorem A16 with

$$A = \begin{bmatrix} -1 & -2 \\ 1 & -4 \end{bmatrix} \tag{A54}$$

Substitution of the values of A in (A53) gives

$$\begin{bmatrix} -1 & 1 \\ -2 & -4 \end{bmatrix}\begin{bmatrix} p_{11} & p_{12} \\ p_{21} & p_{22} \end{bmatrix} + \begin{bmatrix} p_{11} & p_{12} \\ p_{21} & p_{22} \end{bmatrix}\begin{bmatrix} -1 & -2 \\ 1 & -4 \end{bmatrix} = \begin{bmatrix} -1 & 0 \\ 0 & -1 \end{bmatrix}$$

or

$$P = \frac{1}{60}\begin{bmatrix} 23 & -7 \\ -7 & 11 \end{bmatrix} \rightarrow \text{positive-definite symmetric}$$

Thus

$$V(x) = \langle x^T P x \rangle = \frac{1}{60}[x_1 \ \ x_2]\begin{bmatrix} 23 & -7 \\ -7 & 11 \end{bmatrix}\begin{bmatrix} x_1 \\ x_2 \end{bmatrix}$$

$$= \frac{1}{60}(23x^2 - 14x_1x_2 + 11x_2^2)$$

and

$$\dot{V}(x) = \frac{1}{60}(46x_1\dot{x}_1 - 14\dot{x}_1x_2 - 14x_1\dot{x}_2 + 22x_2\dot{x}_2) = -(x_1^2 + x_2^2) \leq 0$$

Hence the system is asymptotically stable.

Example A3 [8]: Stability test for the system described by

$$\ddot{y}(t) + 3\dot{y}(t) + 2y(t) = u(t) \tag{A55}$$

The state-space representation of the unforced system is

$$\begin{bmatrix} \dot{x}_1(t) \\ \dot{x}_2(t) \end{bmatrix} = \begin{bmatrix} 0 & 1 \\ -2 & -3 \end{bmatrix} \begin{bmatrix} x_1(t) \\ x_2(t) \end{bmatrix} \tag{A56}$$

and by using (A53),

$$\begin{bmatrix} 0 & -2 \\ -1 & -3 \end{bmatrix} \begin{bmatrix} p_{11} & p_{12} \\ p_{12} & p_{22} \end{bmatrix} + \begin{bmatrix} p_{11} & p_{12} \\ p_{12} & p_{22} \end{bmatrix} \begin{bmatrix} 0 & 1 \\ -2 & -3 \end{bmatrix} = \begin{bmatrix} -1 & 0 \\ 0 & -1 \end{bmatrix}$$

from which

$$P = \frac{1}{4} \begin{bmatrix} 5 & 1 \\ 1 & 1 \end{bmatrix} \rightarrow \text{positive definite}$$

The Lyapunov function is therefore given by

$$\begin{aligned} V(x) &= [x_1 \quad x_2]P \begin{bmatrix} x_1 \\ x_2 \end{bmatrix} \\ &= \frac{1}{4}(5x_1^2 + 2x_1x_2 + x_2^2) \\ \dot{V}(x) &= \frac{1}{4}(5x_1\dot{x}_1 + 2x_1\dot{x}_2 + 2\dot{x}_1x_2 + 2x_2\dot{x}_2) \\ &= -(x_1^2 + x_2^2 + \frac{1}{4}x_1x_2) \end{aligned} \tag{A57}$$

from which no conclusion may be made about the negative definiteness of $\dot{V}(x)$. Hence the Lyapunov function in (A57) does not yield any conclusion about the stability of the system in (A56).

If, however, the Lyapunov function were chosen to be a positive-definite function of the form

$$V(x) = x_1^2 + \frac{1}{2}x_2^2 \tag{A58}$$

then

$$\dot{V}(x) = 2x_1\dot{x}_1 + x_2\dot{x}_2 = -3x_2^2 \leq 0$$

Hence asymptotic stability of the system is predicted. Another choice of V(x) which also predicts asymptotic stability is

$$\begin{aligned} V(x) &= x_1^4 + \frac{1}{2}x_2^4 \\ \dot{V}(x) &= -6x_2^2 \leq 0 \end{aligned} \qquad \text{(A59)}$$

It is to be noted here that the stability region defined by (A57) to (A59) are all different. The existence of different stability regions is graphically illustrated through the following example [13].

Example A4: Stability regions of the system described by

$$\begin{aligned} \dot{x}_1(t) &= 2\{x_1(t)\}^2 x_2(t) - x_1(t) \\ \dot{x}_2(t) &= x_2(t) \end{aligned} \qquad \text{(A60)}$$

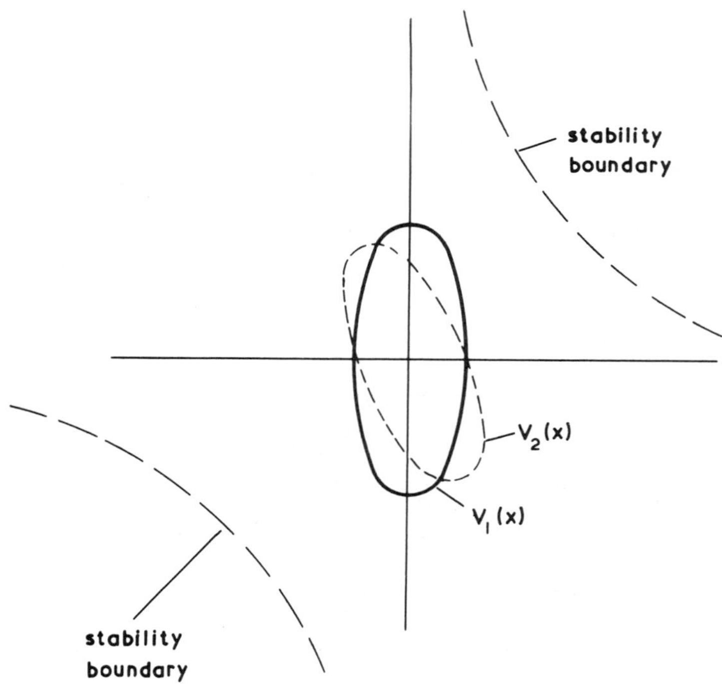

Fig. A28 Stability regions defined by $V_1(x)$ and $V_2(x)$ in (A61).

The stability boundary of this system is defined by $x_1(t)x_2(t) = 1$, with an equilibrium stability at the origin. The stability regions given by

$$V_1(x) = x_1^2 + x_2^2$$

and (A61)

$$V_2(x) = x_1^4 + x_2^4$$

are qualitatively shown in Fig. A28.

Appendix 3
Controllability, Observability, and Decoupling

This appendix presents an outline of the decoupling synthesis procedure utilizing the concepts of controllability and observability (Sec. 3.1). The first section extends the notion of output controllability to strong controllability followed by a definition of decoupling. The problem of state feedback decoupling is then formulated. This is followed by an alternative definition of closed-loop decoupling and a synthesis procedure based on observability index.

A3.1 OUTPUT CONTROLLABILITY AND DECOUPLING [15-19]

From the derivations of input/output equations in Sec. 3.1, the ith output response $y_i(t)$ of the open-loop system $S(\cdot)$ may be expressed as

$$y_i(t) = Q_i \int_{t_0}^{t} \alpha_k(\tau)u(t - \tau)\, d\tau \tag{A62}$$

where

$$e^{At} = \sum_{k=0}^{n-1} \alpha_k(t)A^k \quad \text{and} \quad Q = \{Q_i\} = [CB \quad CAB \cdots CA^{n-1}B]$$

is the output controllability matrix. The relationship in (A62) may thus be expressed as

$$y_i(t) = \sum_{k=0}^{n-1} [C_iA^kB \int \alpha_k(\tau)u(t - \tau)\, d\tau] \quad i \in 1, \ldots, m \tag{A63}$$

The significance of output controllability in the context of decoupling may now be studied from (A63). For complete output controllability, one should be able to choose m output variables and evaluate the corresponding elements of the m-input control u(t). Such a solution exists only when Q has rank m, that is, all rows of Q are linearly independent. The following definitions are relevant in the subsequent development.

Definition A11 A system is said to be strongly controllable (state or output) if it is controllable by each control input separately, while all others are zero; otherwise, it is weakly controllable.

Definition A12 A system is said to have complete input/output noninteraction if there is a one-to-one correspondence between each output and each controlling input.

The structure of (A63) shows that, in general, the output $y_i(t)$ is affected by all elements of the input control vector, as such changes in any one element of u(t) causes change in all output variables. If, however, the rows of $C_iB, \ldots, C_iA^kB, \ldots, C_iA^{n-1}B$ were such that

$$C_iA^kB = \lambda_{ik}e_i, \quad k \in 0, \ldots, (n-1) \text{ and } i \in 1, \ldots, m \tag{A64}$$

where λ_{ik} is a nonzero scalar and $I = \{e_i\}$, each output variable can be independently controlled by only one corresponding input. This leads to the following theorem.

Theorem A17 A system is said to be completely decoupled if it is strongly and completely controllable.

Proof of the theorem follows from (A62) and (A63) and the definitions above.

Complete input/output noninteraction is thus synonymous with strong and complete output controllability, and this leads to the following theorem.

Theorem A18 A system can be completely decoupled by proportional state feedback only if it is completely output controllable.

Theorem A19 If there exists a nonsingular $(m \times m)$ matrix G such that the product of each of $(m \times m)$ matrices $Q_0, Q_1, \ldots, Q_{n-1}$ and G is diagonal, then any ith $(i \in 1, \ldots, m)$ row of Q_k, $k \in 0, \ldots, (n-1)$, is a scalar multiple of the ith row of Q_{k-1}.

Proof: Let G exist; then by hypothesis,

$$Q_kG = \Lambda_k \tag{A65}$$

where $\Lambda_k = \text{diag}(\lambda_{ik})$, $k \in 0, \ldots, (n-1)$, $\lambda_{ik} \neq 0 \ \forall i, k$. Let $\Lambda = \text{diag}\{\lambda_i\}$, $\lambda_i \neq 0, \forall i$, so that

$$\Lambda\Lambda_{k-1} = \Lambda_k = Q_kG$$

that is,

$$Q_k = \Lambda Q_{k-1} \qquad (\text{since } \Lambda_{k-1} = Q_{k-1}G)$$

or

$$q_{i,k} = \lambda_i q_{i,k-1} \tag{A66}$$

This completes the proof.

Theorem A20 For a completely output controllable system S(A, B, C) there exists a state feedback decoupling pair (F, G).

Proof: From (A63) the ith output of the closed-loop system is given by

$$y_i(t) = \sum_{k=0}^{n-1} \left[C_i(A + BF)^k BG \int_{t_0}^{t} \gamma_k(\tau)u(t - \tau)\, d\tau \right] \tag{A67}$$

where

$$e^{(A+BF)t} = \sum_{k=0}^{n-1} \gamma_k(t)(A + BF)^k$$

The closed-loop system is thus decoupled if

$$C_i(A + BF)^k BG = \lambda_{ik} e_i \neq 0 \qquad \text{for } i \in 1, \ldots, m;\ k \in 0, \ldots, n-1$$

or

$$C(A + BF)^k BG = \Gamma_k \tag{A68}$$

where Γ_k is an m-square diagonal matrix for any k = 0, ..., (n - 1), not all Γ_k's are singular. If $\bar{Q}$ is the output controllability matrix of the closed-loop system, then

$$\bar{Q} = [\bar{Q}_0 \cdots \bar{Q}_k \cdots \bar{Q}_{n-1}]G$$

where

$$\bar{Q}_k = C(A + BF)^k B$$

By the definition and theorem above, the closed-loop system is decoupled if

$$\bar{Q}_k G = \Gamma_k, \quad \forall k \tag{A69}$$

A3.2 EXISTENCE OF A DECOUPLING PAIR (F, G) [17, 19]

Assuming that a G exists, (A69) is satisfied if the rows of $\bar{Q}_k$ for any k are scalar multiples of the rows of $\bar{Q}_{k-1}$, that is, if

$$\rho[\bar{Q}^i(F)] = 1, \quad \forall\, i \in 1, \ldots, m \tag{A70}$$

where

$$\bar{Q}^i(F) = \begin{bmatrix} C_i(A+BF)^{n-1}B \\ \vdots \\ C_i(A+BF)^k B \\ \vdots \\ C_i B \end{bmatrix}$$

Equation (A70) provides the criterion for evaluating the class of decoupling state feedback matrix F. Combining (A65), (A66), (A68), and (A69), the output controllability matrix with decoupling F becomes

$$\bar{Q}(F) = \bar{Q}_i(F) = \{q_{i,0} : \cdots\ q_{i,k} : q_{i,n-1}\}G = [\Gamma_0 \cdots \Gamma_k \cdots \Gamma_{n-1}] \tag{A71}$$

where

$$\bar{Q}^i(F) \equiv \begin{bmatrix} q_{i,n-1} \\ \vdots \\ q_{i,k} \\ \vdots \\ q_{i,0} \end{bmatrix}$$

If each of the output variables of the closed-loop system S(A, B, C, F, G) is to be connected to an input control, not all rows of Γ can be null, so there is at least one k for which Γ_k is nonsingular such that

$$\begin{bmatrix} q_{1,k} \\ \vdots \\ q_{i,k} \\ \vdots \\ q_{m,k} \end{bmatrix} G = \Gamma_k \rightarrow \text{nonsingular} \tag{A72}$$

Since G is nonsingular, a precondition for the existence of F, (A72) shows that the row vectors $q_{1,k}$, $q_{i,k}$, ..., $q_{m,k}$ are linearly independent, which when combined with (A72) shows that $\bar{Q}$ has rank m. This completes the proof.

A3.3 COMPUTATION OF G

Using the definitions of d_i and B* in Sec. 9.1 and the derivation above, for closed-loop decoupling,

$$C_i(A + BF)^{di}BG = \lambda_{i,di}e_i, \quad \forall i \in 1, \ldots, m$$

that is,

$$C_iA^{di}BG = \lambda_{i,di}e_i$$

that is,

$$B^*G = \Gamma$$

or

$$G = (B^*)^{-1}\Gamma \tag{A73}$$

where

$$B^* = \{C_iA^{di}B\}, \qquad \Gamma = \operatorname{diag}\{\lambda_{1,d1} \cdots \lambda_{idi} \cdots \lambda_{mdm}\}$$

<u>Theorem A21</u> For any state feedback decoupleable system, the rank of T^i is (di + 1), where

$$T^i = \begin{bmatrix} C_iA^{di} \\ C_iA^{di-1} \\ \vdots \\ C_i \end{bmatrix} \tag{A74}$$

This follows from the definition of di.

Since the criterion for closed-loop decoupling is that the rank of $\bar{Q}^i(F)$ is 1 for all i, it follows from (A70) that closed-loop decoupling also implies

$$\rho[N^i(F)] = 1, \quad \forall i \tag{A75}$$

where

$$\bar{Q}^i(F) = N^i(F)B$$

Combination of (A74) and (A75) leads to the following theorem.

Theorem A22 For any system with nonsingular B*, there exists a class of F such that

$$\rho[R^i(F)] = di + 1, \quad i \in 1, \ldots, m \tag{A76}$$

where

$$R^i(F) = \begin{bmatrix} N^i(F) \\ T^i(F) \end{bmatrix}$$

The observability matrix of the closed-loop system S(A, B, C, F, G) is

$$\bar{R}(F) = \begin{bmatrix} C \\ C(A+BF) \\ \vdots \\ C(A+BF)^k \\ \vdots \\ C(A+BF)^{n-1} \end{bmatrix} = \begin{bmatrix} R^1(F) \\ \vdots \\ R^i(F) \\ \vdots \\ R^{n-1}(F) \end{bmatrix}, \quad \text{where } R^i(F) \text{ is as given by (A76)}$$

Therefore,

$$\rho[\bar{R}(F)] = \sum_{i=1}^{m} (di + 1) = m + \sum_{i=1}^{m} di \tag{A77}$$

Theorem A23 The number of modes observable in any decoupled system is $(m + \Sigma_{i=1}^{m} di)$.

Since the number of observable modes is equal to the rank of the observability matrix, proof of the theorem follows from (A77). Furthermore, for a completely state controllable system, the number of assignable poles with state feedback which affect the input/output response is equal to the observability index. The invariance of state controllability under state feedback therefore implies that the number of poles that can be specified while simultaneously decoupling a completely state controllable system by state feedback is $(m + \Sigma_{i=1}^{m} di)$.

A3.4 COMPUTATION OF F

By the property of di and from (A76),

$$C_i A^{di} BF = C_i (A + BF)^{di+1} - C_i A^{di+1}$$

$$C_i (A + BF)^m = \text{linear combination of the rows of } T^i \text{ for } j \in (di + 1), \ldots, n - 1$$

that is,

$$C_i (A + BF)^{di+1} = \sum_{k=0}^{di} m_{ik} C_i A^k, \quad \text{where } m_{ik} = \text{scalar} \neq 0, \ \forall \, i \text{ and } k$$

Thus

$$B_i^* F = \sum_{k=0}^{di} m_{ik} C_i A^k - A_i^*, \quad i \in 1, \ldots, m$$

or

$$B^* F = \sum_{k=0}^{\delta} M_k C A^k - A^*$$

$$F = (B^*)^{-1} \left(\sum_{k=0}^{\delta} M_k C A^k - A^* \right) \tag{A78}$$

where

$$M_k = \text{diag}\,\{m_{1k}, \ldots, m_{ik}, \ldots, m_{mk}\}, \text{ and } \delta = \max di$$

Example A5 [17] The following system is considered here to illustrate the relevance of output controllability in state feedback decoupling.

$$(S, A, B, C)\text{:} \quad A = \text{diag}\,\{-1, -2, -3, -4\}; \quad B = \begin{bmatrix} 1 & 1 & 0 \\ 0 & 1 & 0 \\ 0 & 1 & 0 \\ 0 & 0 & 1 \end{bmatrix}; \quad C = \begin{bmatrix} 1 & 0 & 1 & 1 \\ 0 & 1 & 0 & 0 \\ 0 & -1 & 1 & 0 \end{bmatrix} \tag{A79}$$

The output controllability matrix for the system is

$$Q = [CB \mid CAB \mid CA^2B \mid CA^3B]$$

$$= \left[\begin{array}{ccc|ccc|ccc|ccc} 1 & 2 & 1 & -1 & -4 & -4 & 1 & 10 & 16 & -1 & -28 & -64 \\ 0 & 1 & 1 & 2 & 0 & -4 & -6 & -2 & 0 & 14 & 6 & -64 \\ 1 & 1 & 0 & -3 & -4 & 0 & 7 & 12 & 16 & -15 & -34 & 0 \end{array}\right]$$

$\rho(Q) = 2 < m = 3$. The system is not output controllable. Also, the system is not state feedback decouplable, since the rows of B^*, taken as the first nonzero $(1 \times m)$ row vectors from each row of Q above (Theorem A20), are linearly dependent, that is,

$$\rho(B^*) = \rho\left\{\begin{bmatrix} 1 & 2 & 1 \\ 0 & 1 & 1 \\ 1 & 1 & 0 \end{bmatrix}\right\} = 2 < m$$

The conclusion above may also be derived by using the open-loop transfer function matrix of the system.

$$B^* = \lim_{s \to \infty} s^{di} G(s) = \lim_{s \to \infty} \begin{bmatrix} \frac{s}{s+1} & \frac{2(s+2)s}{(s+1)(s+2)} & \frac{s}{s+4} \\ 0 & \frac{s}{s+2} & \frac{s}{s+4} \\ \frac{s}{s+1} & \frac{2(s+2) - (s+1)(s+3)]s}{(s+1)(s+2)(s+3)} & 0 \end{bmatrix} \tag{A80}$$

Because of the normalized form of (A79), it is easy to see that the system is not output controllable, since $\rho(C) = \rho(C_n) = 2 < m = 3$.

<u>Example A6: Consideration of an output controllable system</u>

$$S(A, B, C): \; A = \text{diag}\{-1, -3, -4\}; \quad B = \begin{bmatrix} 0 & 1 \\ 1 & 0 \\ 1 & 0 \end{bmatrix}; \quad C = \begin{bmatrix} 1 & 1 & -1 \\ 0 & 1 & 0 \end{bmatrix} \tag{A81}$$

The output controllability matrix is

$$Q = \begin{bmatrix} 0 & 1 & 1 & -1 & -7 & 1 \\ 1 & 0 & 3 & 0 & 9 & 0 \end{bmatrix} \tag{A82}$$

The transfer-function matrix is

$$G(s) = \begin{bmatrix} \dfrac{1}{(s+3)(s+4)} & \dfrac{1}{s+1} \\ \dfrac{1}{s+3} & 0 \end{bmatrix}$$

From the above,

$$B^* = \begin{bmatrix} 0 & 1 \\ 1 & 0 \end{bmatrix} \quad \text{and} \quad d_1 = d_2 = 0$$

The number of modes that can be controlled freely while decoupling is $\Sigma\,(di + 1) = 2$. The state feedback decoupling pair is

$$F = \begin{bmatrix} 0 & m_{02}+3 & 0 \\ m_{01}+1 & m_{01}+3 & -m_{01}-4 \end{bmatrix}; \quad G = \begin{bmatrix} 0 & 1 \\ 1 & 0 \end{bmatrix} \tag{A83}$$

where $M_0 = \text{diag}\,\{m_{01}, m_{02}\}$ and $\Gamma = I$. The closed-loop transfer function matrix is given by

$$R(s) = \begin{bmatrix} \dfrac{(s+4)(s-m_{02})}{(s+4)(s-m_{01})(s-m_{02})} & 0 \\ 0 & \dfrac{(s+4)(s-m_{01})}{(s+4)(s-m_{01})(s-m_{02})} \end{bmatrix} \tag{A84}$$

showing one $[= n - \Sigma\,(di + 1)]$ pole-zero cancellation.

References for the Appendixes

1. Kalman, R. E., and Bertram, J. E., "Control system analysis and design via the second method of Lyapunov: I. Continuous systems," ASME J. Basic Eng., 82: 371-393 (1960).
2. Ogata, K., State Space Analysis of Control Systems, Prentice-Hall, Englewood Cliffs, N.J., 1967.
3. Chen, C. T., Introduction to Linear System Theory, Holt, Rinehart and Winston, New York, 1970.
4. DeRusso, P. M., Roy, R. J., and Close, C. M., State Variables for Engineers, Wiley, New York, 1965.
5. Kuo, B. C., Automatic Control Systems, Prentice-Hall, Englewood Cliffs, N.J., 1967.
6. Dorf, R. C., Modern Control Systems, Addison-Wesley, Reading, Mass., 1976.
7. Truxal, J. G., Introductory System Engineering, McGraw-Hill, New York, 1972.
8. Atherton, D. P., Stability of Nonlinear Systems, Research Studies Press, Chichester, West Sussex, England, 1981.
9. Brockett, R. W., "The status of stability theory for deterministic systems," Trans. IEEE, AC-12: 596-606 (1967).
10. DiStefano, J. J., Stubberud, A. R., and Williams, I. J., Feedback and Control Systems, Schaum's Outline Series, McGraw-Hill, New York, 1967.
11. Stern, T. E., Theory of Nonlinear Networks and Systems, Addison-Wesley, Reading, Mass., 1965.
12. Brockett, R., Finite Dimensional Linear Systems, Wiley, New York, 1970.
13. Hewitt, J. R., and Storey, C., "Comparison of numerical methods in stability analysis," Int. J. Control, 10: 687-701 (1969).
14. Fortmann, T. E., and Hitz, K. L., An Introduction to Linear Control Systems, Marcel Dekker, New York, 1977.
15. Kalman, R. E., "Mathematical description of linear dynamical systems," SIAM J. Control, 1: 152-192 (1963).

16. Kreindler, E., and Sarachik, P. E., "On the concept of controllability and observability of linear dynamical systems," Trans. IEEE, AC-9: 129-136 (1964).
17. Sinha, P. K., "Controllability, observability and decoupling of multivariable systems," Int. J. Control, 26: 603-620 (1977).
18. Mufti, I. H., "On the observability of decoupled systems," Trans. IEEE, AC-14: 415-416 (1969).
19. Falb, P. L., and Wolovich, W. A., "Decoupling in the design and synthesis of multivariable control systems," Trans. IEEE, AC-12: 651-659 (1967).

Index

Adaptive control, 432
  model reference (following), 432, 470
Algebraic function, 336-338
  branch point, 336
  ordinary point, 336
  poles, 337
  pole polynomial, 337, 344, 345
  zero, 337
  zero polynomial, 337, 344, 345

Basis, 17
Block diagrams, 80

Canonical decomposition, 188
Canonical form, 187
Cayley-Hamilton technique, 53
Characteristic frequency, 121
Characteristic function, 336
  inverse, 352
Characteristic loci, 334-340, 574, 576
  design method, 576
Characteristic polynomial, 32, 121
  loop-gain, 333
  open-loop, 333
  stability, basic conditions (see also Stability), 311-315
Compensation, 476
  decoupling for, 545, 551
  feedback, 310, 331
  feedforward (cascade), 331
  output feedback, 476
Connectivity, 187
Control, 327 (see also Compensation)
  adaptive, 432
  decoupling, 525
  observer, through, 517
  observer, through state feedback, 520
  optimal, 329, 426
  output feedback, 441
  state feedback, 407
Controllability, 160, 168, 192
  index, 164
  input decoupling zeros and, 195
  matrix, 164, 169
  transformation under, 172
Controllability and observability, 159
  cost function and, 330
  input-output decoupling zeros and, 194
Controllable form, 179
Controllable subspace, 180, 192
Cost function, 326
Cyclicity, 454

Decoupling, 525
  controllability, 536
  matrix fraction, through, 556
  output feedback, 536, 540
  pair, 536, 539
  precompensation for, 545, 551
  stability, 560
  state feedback, 525-529, 534

Decoupling zeros, 149, 150
  controllability and observability and, 194-198
Determinant, 20
  cofactors, 21
  Laplace expansion, 21
  minors, 21
  partitioned, 24
Diagonal dominance, 359
Disturbance rejection, 373
Duality theorem, 166
Dyad, 592
  minimal expansion, 594
Dyadic design, 592, 595
Dyadic transfer function matrix, 595
Dynamic equations, 86
  irreducible, 190

Eigenvalues, 32 (see also Poles)
  shift, 410 (see also Pole placement)
Eigenvectors, 33
Extenden principle of the argument, 342
External disturbance, 373

Feedback (see Appendix 1):
  nonunity, 310
  output (see Output feedback)
  state (see State, feedback)
  unity, 310
Full-rank state feedback, 408
  multivariable system, 413
  single-input single-output system, 408
Functional controllability, 171

Generalized inverse Nyquist stability criterion, 351-359
Generalized Nyquist stability criterion, 341-351
Gershgorin bands (disk), 362
Gershgorin circles, 361
Gershgorin theorem, 361

Input decoupling zeros, 149
  controllability and, 195
Input-output decoupling zeros, 209-218
  controllability and observability and, 209
Instability, 310
Integrator decoupled system, 310
Integrity, 380-382
Interaction, 375
  index, 377, 379
  minimization of, 376
Inverse characteristic function, 352
Inverse Nyquist array, 564
  principal loops, 568
  stability tests, 566
Inverse transfer function matrix, 565
Irreducible dynamic equations, 190
Irreducible system, 190

Jacobi matrix, 10
Jordan canonical (normal) form, 8, 47

Kalman's duality theorem, 166

Linear output feedback control law, 441
Linear spaces, 13
Linear state feedback control law, 408
Loop-gain characteristic polynomial, 333

Mathematical model, 77
Matrices (see also Matrix):
  basic operations of, 11
  diagonalization of, 41
  elementary transformation of, 39
  equivalent, 39
  modal transformation of, 46
  normalization of, 46
  operations of by partitioning, 13
  powers of, 49
  row/column equivalence, 62

[Matrices]
rules of operations of, 11
similarity transformation of, 45
special forms of, 18-19
system, 94
types of, 6-10
Matrix, 3 (see also Matrices)
adjoint, 24
characteristic equation of a, 32
characteristic polynomial of a, 32, 121
column, 5
controllability, 164, 169
diagonal, 5
dynamics, 94
eigenvalues of a, 32
eigenvectors of a, 33
canonical form, 9
generalized inverse, 26, 442-445
Hankel, 301
impulse response, 98
input, 94
inverse of a, 25-31
Jacobi, 10
Jordan canonical (normal) form, 8, 47
minimal polynomial of a, 51
modal, 33
nonsingular, 12, 60
observability, 165
output, 94
output controllability, 169
permutation, 565, 589
polynomial, 49, 60
proper and improper transfer function, 115
rank, 12
rational, 69
residue, 120
return-difference, 331
return-ratio, 331
row, 5
Schwarz form, 10
singular, 24, 60
square, 5

[Matrix]
state controllability, 164
system, 38
Toeplitz, 10
trace of a, 32
transfer function, 113
transmission, 94
unimodular, 60
Matrix fraction description, 72
Minimal dyadic expansion, 594
Model, 77
Model following control, 432, 470
Modulo, 269
Multivariable systems, 83
identity observer, 488
reduced order observer, 496

Noninteraction, 525 (see also Decoupling)

Observability, 164, 165, 167, 204
index, 166
matrix, 165
output decoupling zeros, 203-209
transformation, under, 172
Observable form, 185
Observer, 481
control through, 517
full-order, 482
identity, 482, 487, 488
normalization, through, 505
observable form, from, 499
reduced-order, 482, 491, 496
Open-loop (see Appendix 1):
characteristic polynomial, 333
Optimal control, 329, 426
Order, 85, 137
decoupling zeros and, 225
least, 220
reduction of, 219
Output controllability, 168
decoupling and, 536
matrix, 169
Output decoupling zeros, 149, 203
observability and, 203-209

Output (functional) controllability, 171
Output equation, 87
Output feedback, 323
  decoupling, 536, 540
  extension from state feedback, 441
  linear (proportional), 441
  observability, and, 455
  proportional-plus-differential control, 464
  proportional-plus-error integral control, 454
  unity rank, 448

Parameter variation, 336
Permutation matrix, 565, 589
Pole placement, 317, 407
Poles, 120, 123, 132, 154
  decoupling zeros and, 226
Poles and zeros, 119 (see also System, zeros)
  system matrix, 226
  transfer function matrix, 226
Pole shifting (see Pole placement)
Pole-zero cancellation, 169
  controllability and, 170
  input decoupling zeros and, 199
  observability and, 170
Polynomial, 58
  pole, 121, 124, 344, 351
  proper and improper, 60
  relatively prime, 60, 62
  zero, 124, 337, 344, 351
  zeros of a, 59
Polynomial function description, 133
  state space and, 135
  system matrix in, 135
  transfer function and, 134
Polynomial matrices, 60
  congruent modulo, 269
  greatest common divisor (left/right), 62
  modulo, 269
  relatively prime (co-prime), 62
  row/column proper, 64
[Polynomial matrices]
  row/column reduced, 64
  single (normal) form of, 66
Polynomials, 58
  co-prime, 60
  greatest common divisor of, 60
  invariant, 67
  monic, 59
Principal loops, 568
Proportional-plus-derivative control, 464
Proportional-plus-error integral control, 454
Proportional output feedback control, 441

Quadratic cost function, 326
Quadratic performance criterion, 326

Rational function, 59
Rational matrices, 69
  equivalence of, 269
  Smith-McMillan form of, 70, 122
Realization, 191, 231
  irreducible, 191
  irreducible, companion form, 269,
  irreducible, direct sum, 256
  irreducible, Jordan form, 260
  matrix-fraction form, 276, 287, 295
  minimal, 232, 301
  nonminimal, block form, 240
    controllable form, 244
    observable form, 248
  normal form, 252
  order of minimal, 301
  residue matrix, through, 253
  single-input single-output systems, of, 234
Reconstructability, 489
Remainder theorem, 61
Return-difference matrix (operator), 332
Return-ratio matrix (operator), 331

Scalar, 14
Sensitivity, 366-373
Separation property, 519
Sequential-return difference method, 584, 588
Signal flow graph, 81
Stability (see also Appendix 1 and 2):
  asymptotic, 308, 310
  bounded-input bounded-output (BIBO), 306
  characteristic loci and, 334-340
  characteristic polynomial and, 311
  decoupling and, 560
  diagonal dominance and, 359-366, 566
  generalized conditions, 331-366
  generalized Nyquist criterion, 341
  inverse characteristic loci and, 356
  preliminary conditions, 312-315
  sequential-return difference and, 587
  total, 307-309
Stabilizability by output feedback, 324
Stabilizability by state feedback, 323
State, 84
  controllability, 160, 192
    condition of, 164, 167, 198
    matrix, 164
  equation, 87
    normalized, 101
  estimator (see Observer)
  feedback, 315
    adaptive, 432
    controllability and, 317, 320, 330
    control laws, 407
    decoupling, 525
    extension to output feedback, 34
    full rank, 408
    linear optimal control, 426
    matrix, 316
    multivariable systems of, 320, 413, 416
    observer (see Observer)
[State; feedback]
    optimal, 329, 427
    pole placement by, 317, 407
    single-input single-output systems of, 317, 408
    unity rank, 416
    zero assignment for, 421
  solutions of, 98
  space, 87
    representation, 94
  transition matrix, 96
  variables, 77, 84
Structure theorem, 290
Subspace and dual space, 17
Sylvester's theorem, 52
System, 78
  matrix, 94, 135, 137, 138
  similarity (se), 144
  standard forms, 134
  state space representation, 94
  strictly equivalent (sse), 137
  transmission zeros of a, 152
  zeros, 119, 123, 154
Systems, 78
  canonical forms, 179-188
  classification of, 78-80
  decoupling zeros of, 149-156
  equivalent (se), 143
  forced response of, 96
  free response of, 95
  input-output representation of, 111
  invariant zeros of, 152
  least order of, 220
  linear, 79
  multi-input multi-output, 83
  multivariable, 83
  nonlinear, 79
  order of, 85, 137
  poles of, 120, 123
  polynomial function representation of, 133
  reduction of, order of, 219-228
  responses of, 77, 95
  single-input single-output, 83
  time invariant, 94

[Systems]
time-varying, 94
zero-input response of, 95
zero-state response of, 95

Transfer function, 113
Transfer function matrix, 114
characteristic polynomial of a, 121
closed-loop, 331
dyadic, 595
feedback, 331
feedforward, 331
inverse, 74, 565
Laplace expansion, 126
least order, 121
matrix fraction, 129, 277
McMillan degree, 124
open-loop, 331
pole-zero computation, 124
realization, 231
Smith-McMillan form, 122

[Transfer function matrix]
system matrix and, 134
poles of a, 120, 123, 132, 154
zeros of a, 119, 123, 132, 148
Transformation, 38
controllability and observability, 172
Transmission blocking, 145

Vectors, 13
column, 5
dimension of, 16
dyadic product of, 15
linear independence of, 16
linear spaces and, 13
orthogonal, 15
row, 5
scalar product of, 15

Zero, 119, 123, 132, 148, 154
assignment, 421
polynomial, 124, 337, 344, 351